Johnbosco C. Egbueri · Joshua O. Ighalo ·
Chaitanya B. Pande
Editors

Climate Change Impacts on Nigeria

Environment and Sustainable Development

Springer

Editors
Johnbosco C. Egbueri
Chukwuemeka Odumegwu Ojukwu
University
Uli, Nigeria

Joshua O. Ighalo
Department of Chemical Engineering
Nnamdi Azikiwe University
Awka, Nigeria

Chaitanya B. Pande
Sant Gadge Baba Amravati University
Amravati, Maharashtra, India

Indian Institute of Tropical Meteorology
Pune, India

ISSN 2352-0698 ISSN 2352-0701 (electronic)
Springer Climate
ISBN 978-3-031-21006-8 ISBN 978-3-031-21007-5 (eBook)
https://doi.org/10.1007/978-3-031-21007-5

This Springer imprint is published by the registered company Springer Nature Switzerland AG
The registered company address is: Gewerbestrasse 11, 6330 Cham, Switzerland

Contents

Chapter 1
Current Effect and Projected Implications of Climate Change on Nigeria's Sustainable Development Plan

Steve Oshiokhai Eshiemogie, Joshua O. Ighalo, Michael Adekanbi, Titilope Banji, Stanley Aimhanesi Eshiemogie, Raymond Okoh, Chinenye Adaobi Igwegbe, Adewale George Adeniyi, Adedapo O. Adeola, and Kanika Dulta

Abstract The danger of climate change on the planet is now more glaring than it has been in the past. This emphatic evidence of the potential threat of climate change on human security has provoked serious and immediate action from relevant stakeholders—including governments of nations and concerned global organizations, to work collaboratively and see to an ultimate reduction in activities that promote global temperature rise. However, even though the rise in the earth's temperature would be a worldwide occurrence which would be felt globally, the impact of this increase would be unevenly distributed among global geographical locations. Specifically, the impact of climate change would be largely disastrous in developing areas of the world, particularly areas of sub-Saharan Africa which have limited available facilities and infrastructures needed to combat the negative effects of climate change. In developing countries like Nigeria, the effects of climate change would lead to a disruption in environmental sustainability, economic growth, social development, as well as energy availability—all of which are collectively tagged the nexus of sustainable development. This chapter reviews the current effect and projected impact of

S. O. Eshiemogie (✉) · S. A. Eshiemogie · R. Okoh
Department of Chemical Engineering, University of Benin, P. M. B. 1154, Benin, Nigeria
e-mail: steve.eshiemogie@eng.uniben.edu; steveeshiemogie@gmail.com

J. O. Ighalo · C. A. Igwegbe
Department of Chemical Engineering, Nnamdi Azikiwe University, P. M. B. 5025, Awka, Nigeria

J. O. Ighalo · A. G. Adeniyi
Department of Chemical Engineering, University of Ilorin, P. M. B. 1515, Ilorin, Nigeria

M. Adekanbi · T. Banji
Department of Mechanical Engineering, University of Ibadan, Ibadan, Nigeria

A. O. Adeola
Department of Chemical Sciences, Adekunle Ajasin University, Akungba-Akoko, Ondo State, Nigeria

K. Dulta
Faculty of Applied Sciences and Biotechnology, Shoolini University, Solan 173229, India

J. C. Egbueri et al. (eds.), *Climate Change Impacts on Nigeria*, Springer Climate,
https://doi.org/10.1007/978-3-031-21007-5_1

climate change on Nigeria's sustainable development plan. It discusses the impact of climate change on Nigeria's environment, economy, social development, and energy sector, and also provides recommendations for mitigating these effects.

Keywords Nigeria · Climate change · Sustainable development · Environmental development · Economic development

Introduction

The altering of the earth's climate has become a new phantom troubling present societies. Although the conceptual premise of a global temperature increase has been well-known for about a century ago, researchers did not believe it was possible until more recently (Urry 2015). Climate change is majorly caused by human activities which primarily involve the release of greenhouse gases to the atmosphere (Eshiemogie et al. 2022). Typically, these gases accumulate in the atmosphere and build up to unacceptable levels capable of threatening human security around the world. Karl and Trenberth (2003), report that since 1965, carbon dioxide emission into the atmosphere has increased by over 31% and would continue to do so until proper measures are taken to curb the menace. Other human activities such as urbanization and deforestation also promote climate change (Tabor et al. 2018). Climate change has the potential to negatively impact the earth, leading to problems such as global warming, melting of the glaciers, rising sea levels, flooding, famine, and other disasters. Though climate change is a global problem, its impact is likely to be largely felt by countries in the underdeveloped world (Koubi 2019). While most developed countries have put in adequate machineries to cushion the effects of climate change, underdeveloped countries struggle to use the meager resources at their disposal to meet other pressing necessities needed for survival.

In recent years, there has been a growing usage of the term, "sustainable development." However, the phrase remains ambiguous, without a singular worldwide acceptable definition (Jabareen and sustainability 2008). Broadly, the term sustainable development is used to discuss a variety of social, economic, and environmental issues (Hopwood et al. 2005), all of which can be categorized as the nexus of sustainable development (Fig. 1.1). To address these issues, concerned bodies have come up with numerous strategic plans, typically called the sustainable development plans.

Over the past few decades, the Nigerian climate has witnessed constant changes due to activities that degrade the atmosphere, such as carbon emissions from vehicles, bush burning, deforestation, urbanization, among others (Ogbo et al. 2013). The release of greenhouse gases (GHGs) such as carbon dioxide (CO_2), nitrogen oxide (N_2O), and methane (CH_4) in Nigeria, has gone beyond acceptable limit, and has made a significant contribution to global warming (Idowu et al. 2011; Eshiemogie et al. 2022). The visible impacts of climate change in Nigeria has resulted in situations that threaten the attainment of sustainable development goals. Some of these

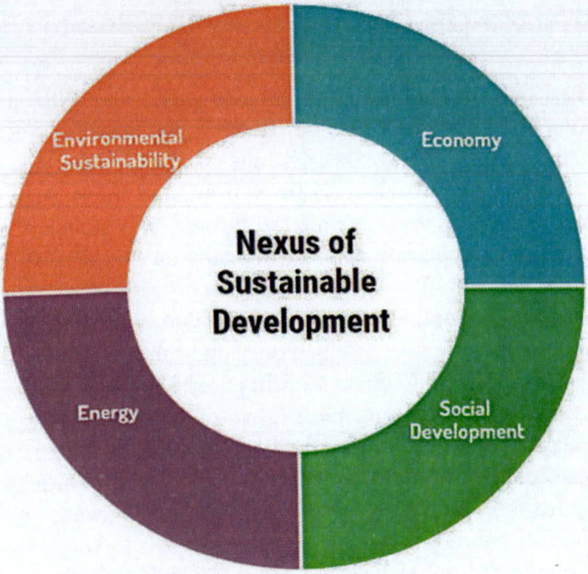

Fig. 1.1 Nexus of Sustainable Development

adverse effects include variable rainfall, land degradation, irregular weather conditions, drought and desertification, increased sea levels, flooding, heat waves, and others (Haider 2019). Nevertheless, sustainable development plans have been put in place both globally and in Nigeria to provide solutions to the menace of climate change. The United Nations General Assembly (with its 193 member states) set up 17 global goals that are to be achieved by 2030 to ensure a better future for humanity (Raimi et al. 2020). Efforts have also been made to meet up with the goals applicable to Nigeria's development some of which includes the establishment of SDG inclined policies such as the National Renewable Energy and Energy Efficiency Policy (NREEEP) of 2015, and the Sustainable Energy for all Action Agenda, which was introduced to reduce fossil fuel energy reliance in Nigeria by 30% before the year 2030 (Eshiemogie et al. 2022). However, Nigeria has struggled in meeting these goals despite being Africa's largest economy (Yetano Roche et al. 2020). Some of these struggles can be traced to prevailing factors in different spheres, and climate change is one of these things that hinder Nigeria from meeting her goals. This chapter studies the impact of climate change in Nigeria's sustainable development plan, mainly as it relates to the environment, the economy, social development, as well as the energy sector of the country.

Why Should Nigeria be Concerned About Climate Change?

Nigeria is affected by climate change in various ways, and this calls for concern. Beyond environmental impacts, other detrimental impacts on demography, security, social system, and the likes exist. Groups of people like rural dwellers who gather wood for cooking, cultivate food for selling, and fetch drinking water from nearby streams for daily survival, would be thrown into poverty when the availability of these resources become affected by climate change (Haider 2019). For example, after a survey on 1200 farmers in 20 rice producing states in Nigeria, it was observed that the production of rice is at risk due to climate change effects on rainfall, leaving the available rice produce only enough to meet half of the nation's need (Ajetomobi et al. 2010). Climate Conflicts over resources in areas threatened with desertification, such as the northern areas of Nigeria, also result in security challenges like clashes between farmers and herdsmen (Haider 2019). More so, the impacts of climate change on agriculture affects the economy, leads to food insecurity, and unsettles the various sectors of the country (Ogbo et al. 2013). Nigeria has over 200 million people, of which about 70% live in rural areas where they directly depend on natural resources for survival (Amobi and Onyishi 2015; Ighalo and Adeniyi 2020). Climate change also has direct impact on human health in form of heat waves, increased spread in airborne diseases, air pollution, and others (Haider 2019). Nigeria already has problems involving these, and it should not be allowed to proceed further. The government and the populace should take steps on curtailing the harmful effects of climate change. Energy generation and supply in Nigeria is also threatened by climate change in form of damage to transmission lines and reduced rainfall that affects hydroelectric sectors (Haider 2019). With the goals in place for a sustainable future and the increasing population and energy use, a solution must be found for climate change effects in Nigeria. In the subsequent sections, the impact of climate change on Nigeria's environment, social development, the economy, and the energy sector, are extensively discussed.

Climate Change and Its Impact on Nigeria's Environment

Different effects have stemmed from climate change in Nigeria, especially on the environment. As a result of this, humans, plants, and animals have had a fair share of the disadvantageous consequences. As climate change continues to occur, global warming causes increase in global temperature. It is evident as between 1901 and 2005 (for 105 years), a study revealed that global temperature increased by 1.1 °C (Akpodiogaga-a and Odjugo 2010). In Nigeria, between 2005 and 2019, the atmospheric temperature has been reported to further increase by 0.3 °C, and the discomfort has been close to getting unbearable as observed in various ecological zones of the country (Haider 2019). If left to increase this way, it is projected that by 2100, Nigeria would experience the middle and high risk temperature increase of 2.5 °C and 4.5 °C respectively (Akpodiogaga-a and Odjugo 2010). This could lead to severe heat waves capable of unleashing sicknesses and discomfort on the Nigerian people.

Another characteristic effect of climate change on the environment is increasing rainfall in coastal areas and decreasing rainfall in continental interiors. This has been observed in Nigeria, as rainfall has decreased steadily in most parts aside from the coastal areas like Calabar and Warri where it has increased (Akpodiogaga-a and Odjugo 2010). In a research that involved collection of data on the rainfall in the south-south of Nigeria for over forty (40) years, it was revealed that variation in rainfall in Nigeria is very evident and can impact agricultural activities negatively (Ukhurebor and Abiodun 2018). This could lead to severe food crisis in the country.

Some other effects of climate change on the environment include land degradation and desertification. This occurs mostly as a result of increased flooding coupled with bad human practices like indiscriminate felling of trees. Ultimately, this leaves initially good land in bad state, making it impossible to utilize the nation's land for profitable ventures. In Nigeria, the reduction in total rainfall, increased rate of evaporation, and intense storms, have been reported to enhance desert encroachment from the northern region to the southern parts of the country (Gisladottir and Stocking 2005). This irregular precipitation and increasing temperature in Nigeria, has led to drying and degradation of the soil. Eutrophication, which is the loss of essential nutrients in soil (majorly nitrogen and phosphorus), is also a problem caused by climate change, and has the potential to reduce the fertility of Nigeria's soil. To ensure agricultural and environmental sustainability in Nigeria, the issue of climate change must be swiftly and adequately addressed.

Climate Change and Its Impact on Nigeria's Economy

Beyond the environmental effects of climate change in Nigeria, the economy of the country also suffers setbacks as well. It directly affects sectors that serve as the bedrock for the economy of most developing countries in Africa, with Nigeria inclusive. These include agriculture, tourism, oil and gas, as well as other relevant sectors that serve as a nation's main power in international trades. Fundamental human needs like food security, water availability, health and shelter can also be in jeopardy as a result of climate change (Matemilola et al. 2019). Issues like this and other problems related to the scarcity of fundamental resources needed for living often lead to a hike in price, which directly worsens the economic situation cum poverty level in a country and also makes climate change a threat to sustainable living.

Asset accrual is a major driver of poverty reduction and the disasters caused by climate change destroy properties and cause deterioration of assets, thereby slowing down the rate at which people break free from poverty (Hallegatte et al. 2018). In conjunction with this, these natural disasters are also markers that affect the rate at which investors set up businesses in a country. For example, foreign investors may decide to invest less or pull their resources to other favorable locations if the possibility of drought or famine in a country increases as a result of changes in climate (Hallegatte et al. 2018). Human capital, which is a critical tool needed for national economic growth, is also hugely scourged by climate change (Hallegatte et al. 2018). All of these could make a developing country become poorer and grossly affect their economy. Majorly, the poorer countries of sub-Saharan Africa, which often lack the means to deal with the impacts of climate change, bear huge losses (Matemilola et al. 2019). Most developing countries' government budgets are frequently strained by the costs of establishing climate change coping systems, and Nigeria also experiences this problem as well (Abidoye and Odusola 2015). Due to the spatial location of countries in Africa on the lower latitude, the region is already at a disadvantage, as about 80% of the damages from climate change are concentrated in this location (Abidoye and Odusola 2015).

Climate change often has a huge negative impact on economies that rely heavily on natural resources (Ogbuabor et al. 2017). This situation applies to Nigeria. In fact, the economic space of most African countries is primarily driven by climate change dynamics (Ogbuabor et al. 2017). The total economy, as well as critical sectors that underlie economic performance, are extremely vulnerable (Abidoye and Odusola 2015). For example, Lagos Nigeria, which is a major commercial center in the country, also has a fair share of disasters caused by climate-based activities, and this forces investors and even citizens to change the location of their businesses. The state is Nigeria's industrial, commercial, and business powerhouse, contributing 35.6% of the country's GDP in 2010, estimated at $80.61 billion, and accounting for 62.3% of non-oil GDP (Elias and Omojola 2015). There is more than 10,000 industrial and commercial activities in Lagos Metropolitan Area (Elias and Omojola 2015). The Internally Generated Revenue (IGR) of Lagos State was $30.106 billion, $32.827 billion, and $35.467 billion in 2010, 2011, and 2012 respectively, according to the National Bureau of Statistics (Elias and Omojola 2015). Although the IGR shows a progressive trend, a huge portion of this revenue, which could be used to execute profitable projects, is rather spent on fighting climate change or battling the aftermath of climate activities such as flooding, diseases, food scarcity, and power outages among others. In the year 2011, for example, the Lagos state Government spent over 200 million USD to fight flooding which was ravaging both the Lagos island and mainland (Adelekan 2016). Lagos has the potential to generate more revenue through activities like farming, but the aftereffects accompanying disasters caused by climate change, have crippled some of the sectors that ought to contribute to increasing the state's IGR. For Lagos, increasing climate change risk has the tendency to worsen with time if adequate counter-measures are not taken. Flooding, sea level rise, storm surge, high temperatures, and high rainfall intensity are among the prominent hazards. It is important to also state that the susceptibility level of Lagos toward climate change disasters, is heightened by the dense Lagos population, low-lying geography, and coastal location (Elias and Omojola 2015). The aftermath of climate change activities in Lagos can lead to a downturn in the economy of the country at large, as the state contributes a lot to Nigeria's income and would definitely affect a lot of people as a lot of Nigerians are residing in Lagos. The remaining part of this section discusses the effect of climate change on the key sectors that serve as a backbone to Nigeria's economy.

Agriculture is one of the most important sectors of the Nigerian economy, accounting for roughly 20% of the country's GDP, and more than 80% of the rural population relies on it for their survival and livelihood (Shiru et al. 2018). In the third quarter of 2017, the agricultural industry contributed 46.71% to the national gross domestic product (GDP) of Nigeria (Olayide and Alabi 2018). Climate change, on the other hand, has a considerable impact on this sector, and the consequences may be more severe in the future than currently anticipated (Shiru et al. 2018). Agriculture, food insecurity, and climate risk are all closely interwoven, as a result, any gains in agricultural growth made over a given era can simply be eroded as a result of climate change (Olayide and Alabi 2018). The floods of 2012 alone cost the country an estimated US$16.9 billion (Olayide and Alabi 2018). Climate forecasts for Nigeria indicate a rise in temperature of 1.1 to 2.5 °C by 2060, as well as a 260-day increase in extreme heat days by 2100, compared to 10 days in 1990 (Liverpool-Tasie et al. 2019). Greater unpredictability in rainfall, dry spells during rainy seasons, and increased irregularity in the amount of precipitation once the rainy season begins have all been recorded in past studies (Liverpool-Tasie et al. 2019). All of these have an impact on crop yields coupled with other agricultural activities in Nigeria as the agricultural production systems in Nigeria are paradoxically more reliant on natural resources, most especially rainfall (Olayide and Alabi 2018). Climate change does not only affect crops, but also has huge effect on animal farming. Changes in temperature and rainfall patterns, for example, have been linked to reduction in feed sources, increased animal mortality, smaller herd sizes, reduced water sources during the dry season, decreased animal productivity, and the emergence of new animal diseases by pastoralists and livestock farmers across Africa (Liverpool-Tasie et al. 2019). Reduced available pasture land, reduced surface water resources for animals, increased salinity of water resources for animals, increased salinity at watering points due to increased temperature and evaporation in the face of reduced rainfall contribute to a decrease in livestock production and all these occurrences are consequences of change in climate (Ogbuabor et al. 2017). This suggests a decrease in livestock output, leading to a decrease in the supply and availability of animal protein, such as meat, eggs, milk, and other animal products like hides and skins (Ogbuabor et al. 2017). The availability of water supplies and arable land for agriculture is impaired by irregular rainfall and this adversely affects agricultural yields (Ogbuabor et al. 2017). None of the ends of extreme rainfall conditions is favorable to agriculture. Excessive rainfall damages arable land, degrades cultivated crops, promotes weed growth, and increases post-harvest losses, whereas a significant reduction in rainfall results in drier land, lower water levels in streams and rivers, and an increase in farmers' search for irrigation water, resulting in valuable man-hour losses and crop yield reductions (Ogbuabor et al. 2017). Climate change has a negative impact on forestry in Nigeria due to erosion and high wind, resulting in a decrease in forest products such as wood and cane (Ogbuabor et al. 2017). As a result, there is a decrease in forestry output and poor income, as well as an increase in the cost of construction and furnishing materials (Ogbuabor et al. 2017). Between 2004 and 2009, the cost of deforestation and losses in non-timber forest products in Nigeria was estimated to be N120 billion per year, or 1.7% of GDP in 2003 (Ogbuabor et al. 2017). Hence, climate change is declining the agricultural production capacity of Nigeria and this could become more intense in a few years' time if proper measures are not taken. This in turn downsizes the country's economy and then worsens the finances of the country.

Nigeria ranks high in terms of global oil production and the country's economy is hugely dependent on the proceeds of oil and gas trades. The oil sector is an integral part of Nigeria's economy and any issue that affects this sector can have a consequential effect on the economic stability of the country. However, the world's attention has shifted to greening global production processes, transitioning from non-renewable energy sources to clean and renewable sources, researching alternative fuels as well as encouraging people to patronize clean technologies (Murshed et al. 2021). This is particularly to reduce greenhouse gas emissions caused by the widespread use of fossil fuels and also eliminate activities that have gross effects on the climate (Murshed et al. 2021). CO_2 emissions from fossil fuel combustion account for almost three-quarters of annual anthropogenic greenhouse gas emissions globally (Erickson et al. 2018). In most scenarios, CO_2 emissions from fossil fuels must decline rapidly to near zero in the second half of this century to keep global temperature rise well below 2 °C and avoid disastrous climate change (Erickson et al. 2018). The United Nations' Sustainable Development Goals (SDG) agenda has also called for global pledges to incorporate renewable energy resources into global energy mixes in a bid to tackle the climate change problems caused by the use of fossil fuels (Murshed et al. 2021). Some governments have recently begun to diversify their portfolios to include measures aimed at reducing fossil fuel extraction and supply, and a few have chosen to forego major new fossil fuel supply infrastructure for climate reasons (Erickson et al. 2018). The continuation of this translates to the fact that there would be a decline in the demand for Nigeria's oil, as more countries keep prioritizing non-fossil-based resources. All of these could culminate in a fatal blow to the economic situation of Nigeria in the future. Apart from the environmental challenges associated with the combustion of oil and gas, there are other non-climate friendly activities affiliated with the exploration of the resources (Ighalo et al. 2020a, b). Oil exploration and production have a severe influence on neighboring local populations because of the environmental and socioeconomic issues it poses over time, threatening subsistence or peasant farming and biodiversity, as well as their entire social existence and survival (Ighalo et al. 2020a, b; Alemzero et al. 2021). Oil spills have a variety of repercussions on local populations, including farming land, fishing, and water bodies for bathing at the community level, as polluted water bodies have an impact on human health (Nwabueze and Ighalo 2020; Alemzero et al. 2021).

Tourists are climate-sensitive, and the need to explore the earth, the seas, and the skies, is a core factor that continues to drive tourism (Bigano et al. 2006). As a result, tourism is vulnerable to climate change. Nigeria offers a diverse range of natural tourism attractions, including beaches and abundant ecological attractions such as mountains, waterfalls, lakes, hills, and caverns, as well as a rich heritage of traditional arts and crafts (Dillimono and Dickinson 2015). Tourism has been suggested as a possible source of economic development for Nigeria if the country's political situation improves (Dillimono and Dickinson 2015). The tourism industry is thought to be one of the most vulnerable sectors of the economy to climate change, due to its exposure to many direct and indirect impacts of climate change on tourism destinations, tourist activities, and behaviors (Ma and Kirilenko 2020). The acute reliance of destinations on climate-related parameters such as temperature, sun, and snow cover is a major direct impact contributing to the tourism sector's high vulnerability to climate change (Ma and Kirilenko 2020). The seasonality of tourism, tourists' choice of destinations, accessible tourist activities and attractions, and overall satisfaction with a holiday are all influenced by the weather (Hoogendoorn and Fitchett 2018). Climate change has the potential to influence the appeal of tourism locales and regions, as climate plays a crucial part in the comparative selection of a tourist destination (Hoogendoorn and Fitchett 2018). Progressive changes in a location's climate, as well as increased dangers from natural catastrophes such as storms, flooding, and sea level rise, can cause areas to become increasingly unsuitable for tourism (Hoogendoorn and Fitchett 2018). Tourism's contribution to GDP and employment indicates the industry's economic importance. Between 1995 and 1998, the tourist sector contributed $166 billion to the continent's GDP, which increased to $336 billion between 2011 and 2014 (Uduji et al. 2019). Tourism generated 6.8% of Africa's GDP in 1995–1998, 9.6% in 2005–2008, and 8.5% in 2011–2014, as a percentage of GDP (Uduji et al. 2019). Tourism could potentially be a lifeline for Nigeria when fossil fuel exchanges lose their value on the international market (Uduji et al. 2019). Climate change could make the country not attractive as a destination for tourists and this downsizes the financial gains from tourism. The competitive advantage of particular sites will be altered due to changes in the duration and quality of climate-dependent tourism seasons (Hoogendoorn and Fitchett 2018). Businesses that thrive on tourist activities would also be largely affected. Oil bunkering activities in the Niger Delta have had a significant impact on the region's ecosystem and air quality (Bodo et al. 2020). This also impacts the rate at which tourists would select tourism centers in the region.

One of the possible adaptive reactions to climate change is migration (Chen and Caldeira 2020). Nigeria, after India, may be the second country in this regard, with the biggest number of individuals for whom climate change could provide an additional incentive to move (Chen and Caldeira 2020). People frequently consider relocating from hotter climates to cooler or more temperate climates (Li and Samimi 2022). In general, sub-Saharan Africa has a growing population with significant displacement in Africa, while also accounting for a significant percentage of migration flows to Europe, North America, and other parts of the world (Li and Samimi 2022). Nigeria holds a huge share in the number of people emigrating from Africa. It is also common for Nigerians to change locations within the country, especially cases of Northerners relocating to Southern Nigeria. In a study conducted by Abbas (2017), it was discovered that a large number of people have been compelled to migrate from North Eastern Borno because of the state's sensitivity to climate change. Migrants' destinations are spread across Nigeria, Cameroon, and Chad (Abbas 2017). All these have effects on the economic situation of the country.

The future of the economy of Nigeria is not safe as the sectors that are meant to contribute to its stability and improvement are in jeopardy because of climate-based activities. It has been established earlier that climate change has an impact on agriculture, oil and gas, tourism and migration and these form a tangible portion of Nigeria's GDP. Breakdown in these sectors would cripple the country's economy and worsen the living conditions of individuals residing in Nigeria.

Effect of Climate Change on Nigeria's Social Development

The indirect impact of climate change is grossly felt in most sectors that are instrumental to social development in Nigeria. Climate change affects public health and transportation, particularly in slum communities in developing countries and low-lying coastal towns conjoined with limiting households' ability to raise living standards in a variety of ways, including through its effects on income, food security, and health (Collier et al. 2008; Randell and Gray 2019). Quality education, good health sector, improved standard of living, coupled with stable economic conditions are pivotal in any country's growth plan and they do serve as part of the metric used in qualifying the development level of a country. However, the natural disasters erupting from any climate change event have the potential to cause harm or slow down national development. For example, extreme temperature, flood, severe drought, and other environmental consequences of climate change can affect schooling and cause a long-term impact on a country's educational sector (Randell and Gray 2019). Children's education is compromised when schools are destroyed by extreme weather events or when families can no longer afford to take their children to school due to the loss of businesses and income to climate-based activities (Sanson et al. 2019). Some children are even forced to engage in income-generating activities while still in school to make ends meet (Randell and Gray 2019). This in turn leads to a decline in school attendance and can further widen the number of out of school children in Nigeria. Owing to the fact that females' education is less valued in many regions of the world, this situation impacts them more than boys (Sanson et al. 2019). Climate change causes drought, which may have an indirect impact on the education of the girl child. In many rural parts of Nigeria, females are often responsible for house chores in the house, which often include making water available for the whole household and cases of water scarcity may cause them to go long distances in search of water. All of these could result in stress before the resumption of daily school activities, low school attendance, and being unable to concentrate during classes. By cultivating skills, intellectual ability, and career prospects, formal education contributes to poverty alleviation and economic development (Randell and Gray 2019). Hence, a reduction in educational attainment due to climate change may have a compounding effect on national development, amplifying the direct effects of climate change over time (Randell and Gray 2019).

Higher average temperatures, as well as higher peak temperatures, will have an impact on health. High peak temperatures (over 30 °C) will raise mortality, especially in large cities (Collier et al. 2008). This temperature range is the upper limit temperature often recorded in Northern Nigeria and 27–29 °C, which is the general temperature range of many regions in Nigeria, is close to the ideal temperature for the development of sporozoites in mosquitoes (Okunlola and Oyeyemi 2019). The typical rainfall patterns in Southern Nigeria create shallow water pockets that are ideal for the reproduction of *Anopheles gambiae*, the malaria parasite's major mosquito vector (Okunlola and Oyeyemi 2019). Temperature, rainfall, and humidity, for example, all have a significant impact on mosquito distribution and development rates (Collier et al. 2008). The longevity of mosquitos, their rate and frequency of reproduction, mosquito blood-feeding patterns, and extrinsic incubation times are all influenced by climate trends (Giesen et al. 2020). Rainfall, particularly when it is heavy, washes away many of the mosquito vectors of malaria parasites' breeding sites, but temperature controls the duration of mosquito larvae development in the environment and parasite development within the vector (Okunlola and Oyeyemi 2019). It is widely known that the plasmodium transmitted by mosquitoes is the major cause of Malaria. According to WHO estimates, climate change is responsible for 3% of all malaria deaths globally, particularly among children under the age of five (Giesen et al. 2020). Weather conditions that are warm and humid, such as those seen in tropical settings, are perfect for mosquito breeding (Giesen et al. 2020). Apart from this, excessive rainfall or floods can damage health infrastructures and buildings in the country, thereby causing the government to redirect funds to mitigate the issues. Efforts to get vaccines to reach some rural communities, in case of an outbreak of diseases, can be forestalled by climate-related activities. Getting funds and setting up coping mechanisms to cater for the damages done to the health sector of Nigeria by climate change activities could frustrate development.

The loss of tree and vegetation cover exposes the land to environmental dangers and degradation, reducing food production and land productivity inexorably (Ikhuoso et al. 2020). Since the end of the 1990s, deforestation and the decline of vegetation cover have grown by about 79.1%. Nigeria lost 39.2% of its vegetative cover between 1990 and 2005 and the country loses an average of 409,700 hectares of forest each year (Ikhuoso et al. 2020). This can lead to desertification and drought, which creates difficulty in cattle grazing. Cattle grazing have sparked a long-running public debate, with conflicts raging between landowners, pastoralists, and environmentalists striving to eliminate damaging land use regimes (Ikhuoso et al. 2020). Climate change contributed a new dimension in the land use and management debate. Conflicts may arise as a result of rivalry for restricted and scarce resources (Ani et al. 2021). The crisis between herders and farmers in Nigeria has been connected to the depletion of natural resources brought on by climate change (Lenshie et al. 2021). Water scarcity, illegal land encroachment, and cattle feeding are all linked to climate change and the farmer-herdsmen dilemma (Ikhuoso et al. 2020). Cattle are mostly reared in Northern Nigeria and the region is grossly affected by the scarcity of resources caused by climate change, which propels most of the farmers to migrate toward Southern

Nigeria (Ani et al. 2021). This southward influx of herders and their cattle has exacerbated herder-farmer crisis, particularly in Central Nigeria, over the previous two decades, with an unprecedented number of casualties (Lenshie et al. 2021). As of 2018, herder-farmer clashes in North Central Nigeria had claimed the lives of around 4000 individuals. Farms and other material resources have been destroyed in unquantifiable numbers (Lenshie et al. 2021). While farmers have blamed the conflict on animals damaging their crops and polluting water sources, herders have said that rustling, cattle theft, and the prohibition of traditional grazing are the main causes of conflict (Lenshie et al. 2021). This crisis worsen the security situation of the country, as it has led to an increase in arms trafficking and some herders exploited this to perpetrate other vicious acts like rape, robbery, and kidnapping (Ani et al. 2021). Climate change, with its attendant drought and desertification, is also linked to the increase of insurgency in Nigeria's north-eastern region (Ani et al. 2021). Cattle rearing is a lucrative investment and it has been established that this business may become more difficult for farmers due to climate issues and it will hugely affect the country. The secondary effect of the herder-farmer crisis on the security situation of Nigeria makes climate change a threat to the development of Nigeria. It has been established that climate change can cripple security and education, affect the health sector of Nigeria, and also pose a serious threat to the living conditions of individuals residing in the country. Any developing country trying to overcome this unfriendly situation will need to deploy extra measures to tackle the issues of climate change as there is a tendency for the country to experience a slow pace of development coupled with damage to existing national assets and resources.

Climate Change and Its Impact on Nigeria's Energy Sector

Like other pillars of sustainable development, climate change has its effect on the energy generation in Nigeria. With over 200 million people residing in Africa's most populous nation, the need for sustainable energy in Nigeria cannot be overemphasized. Nigeria's power supply is largely met by thermal power plants and hydropower stations (Eshiemogie et al. 2022). Hydropower generation accounts for approximately 20% of Nigeria's total generated power (Eshiemogie et al. 2022) and is the major source of renewable energy in Nigeria, contributing power to the electrical grid (Machina and Sharma 2017). Nigeria's hydropower is primarily supplied by the Kainji dam and the Jebba dam, both located in Niger state. These dams were respectively designed to generate 760 Mega Watts and 540 Mega Watts of electricity. However, due to insufficient inflow of water, both dams have been unable to meet intended design targets (Emoabino et al. 2007). Climate change which is responsible for the encroaching desertification from the north to the south of Nigeria, has led to reduced rainfall in Nigeria and a decrease in hydropower. In a study by Machina and Sharma (2017), it was observed that the trend of rainfall in Minna, Nigeria, is on a steady decline while the temperature in the region increases steadily, leading to increase in evaporation rate. Ultimately, this causes a reduction in water level of reservoirs needed to sustain hydropower generation, and any further unfavorable effects of climate change on water supply to the dams, could pose a significant threat to Nigeria's already fragile hydroelectric power generation (Sola and Adeniran 2009). For thermal power generation involving the combustion of natural gas, there has been persistent calls for nations around the world to decarbonize (Eshiemogie et al. 2022), this however has not had any effect on the use of natural gas for energy generation in Nigeria, as both the gas produced and utilized in Nigeria has shown an increasing trend since 2002 (Diugwu et al. 2013).

Conclusion and Recommendation

This chapter studies the impact of climate change on Nigeria's sustainable development plan. It can be seen that climate change has affected various sectors which are crucial for sustainable development in Nigeria. Nigeria's environment for instance, has been seriously affected by climate change. Disasters such as flooding, famine, drought, are becoming common occurrence in the Nigerian ecosystem, and these issues have been found to have direct consequences on the Nigerian economy. Also, climate change has been found to have negative effect on Nigeria's social development, leading to issues such as herders-farmers' crisis. The power sector has also experienced its own setbacks, with decreasing water levels in the nation's hydropower reservoirs leading to drop in power supply and frequent blackouts.

To bring an end to these issues and ensure sustainable development, the Nigerian government and citizens, need to collectively put resources together and prioritize the fight against climate change. The following paragraphs present some recommendations.

For environmental sustainability, the federal and state governments of the Nigerian nation, should enact and enforce strict laws against the prevalent indiscriminate felling of trees for furniture making, pulp production, and energy generation. Nigerian forests should be conserved and protected by standby forest guards. Aside enforcing laws against deforestation, the government and other concerned bodies need to also carry out climate awareness exercises that will be focused on educating the masses on the dangers of deforestation on the environment. When citizens receive basic education on these issues, they would be more environmentally aware, and ultimately become cautious of their activities on the environment. Also, since most rural dwellers engage in tree-felling activities largely for the purpose of energy generation for domestic cooking (firewood), government and private investors could intervene by providing affordable clean cooking alternatives for domestic cooking in rural communities, like clean energy stoves. While for those involved in felling of trees for furniture and paper making, government should ensure that such entities cultivate their own trees for wood production. This would ensure that natural forests are preserved and prevented from being destroyed.

For the Nigerian economy which is largely dependent on crude oil trade, the effect of climate change on global crude prices should spur the government to reduce reliance on crude oil trade, and diversify export. More focus should be given to strengthening the agricultural sector, the tourism industry, as well as other sectors that have faced some setbacks as a result of climate activities.

For social development, the government and other relevant stakeholders need to invest massively in the education sector. Educating the country's citizens is a crucial step in the fight against climate change and its effect on Nigeria's sustainable development plan. When the citizens are well-educated, they become enlightened and abstain from activities that promote climate change, such as indiscriminate disposal of organic wastes, burning of bushes, deforestation, etc. Educated citizens would also contribute to the development of other sectors needed for sustainable growth, such as the transport sector, health care, and housing, among others.

For energy sustainability, Government should begin exploring renewable energy options for energy generation. This would not only be a plus for energy generation, but would also create jobs for the Nigerian workforce who would be employed to work in renewable energy industries. Exploring other forms of energy generation could also usher in the breakthrough of stable electricity, which is needed to sustain the nation's small- and large-scale businesses, ultimately improving the economy of the country.

A major limitation during this study was obtaining materials focused on policy recommendations toward mitigating the effect of climate change in Nigeria. Future study should focus on exploring feasible pathways that could be explored in order to mitigate climate change and its effects in Nigeria's sustainable development plan.

Funding There was no external funding for the study.

Conflict of Interest The authors declare that there are no conflicts of interest.

Compliance with Ethical Standards The article does not contain any studies involving human or animal subjects.

CReDiT authorship contribution statement

Steve Oshiokhai Eshiemogie; Conceptualisation, Methodology, Writing—original draft, Writing—review & editing, Visualisation, Validation, Project administration.

Joshua O. Ighalo; Conceptualisation, Methodology, Writing—original draft, Writing—review & editing, Visualisation, Validation, Project supervision.

Michael Adekanbi; Writing—original draft, Writing—review & editing, Visualisation, Validation.

Titilope Banji; Writing—original draft, Writing—review & editing, Visualisation, Validation.

Stanley Aimhanesi Eshiemogie; Writing—original draft, Writing—review & editing, Visualisation, Validation.

Raymond Okoh; Writing—original draft, Writing—review & editing, Validation.

Chinenye Adaobi Igwegbe; Writing—review & editing, Validation.

Adewale George Adeniyi; Writing—review & editing, Validation.

Adedapo O Adeola; Writing—review & editing, Validation.

Kanika Dulta; Writing—review & editing, Validation.

References

Abbas AM (2017) Climate change and forced migration from ngala and kala-balge LGAs, NE borno state, Nigeria. Global changes and natural disaster management: geo-information technologies, 141–151.https://doi.org/10.1007/978-3-319-51844-2_12

Abidoye BO, Odusola AF (2015) Climate change and economic growth in Africa: an econometric analysis. J African Econ 24(2):277–301. https://doi.org/10.1093/jae/eju033

Adelekan IO (2016) Flood risk management in the coastal city of Lagos, Nigeria. J Flood Risk Manag 9(3):255–264. https://doi.org/10.1111/jfr3.12179

Ajetomobi JO, Abidun A, Hassan RM (2010) Economic impact of climate change on irrigated rice agriculture in Nigeria. https://doi.org/10.22004/ag.econ.95778

Akpodiogaga-a P, Odjugo O (2010) General overview of climate change impacts in Nigeria. J Hum Ecol 29(1):47–55. https://doi.org/10.1080/09709274.2010.11906248

Alemzero DA, Iqbal N, Iqbal S, Mohsin M, Chukwuma NJ, Shah BA, Research P (2021) Assessing the perceived impact of exploration and production of hydrocarbons on households perspective of environmental regulation in Ghana. Environ Sci Pollut Res 28(5):5359–5371.https://doi.org/10.1007/s11356-020-10880-3

Amobi D, Onyishi T (2015) Governance and climate change in Nigeria: a public policy perspective. J Policy Develop Studies 9(2):199–209

Ani KJ, Anyika VO, Mutambara E (2021) The impact of climate change on food and human security in Nigeria. Int J Climate Change Strateg Manag, 148–167.https://doi.org/10.1108/IJCCSM-11-2020-0119

Bigano A, Hamilton JM, Tol RS (2006) The impact of climate on holiday destination choice. Climate Change 76(3):389–406.https://doi.org/10.1007/s10584-005-9015-0

Bodo T, Gimah BJ, Seomoni KJ (2020) Illegal oil bunkering in the niger delta region of Nigeria: a challenge to nigeria's development. Europ Scientific J 16(29):134–154. https://doi.org/10.19044/esj.2020.v16n29p134

Chen M, Caldeira K (2020) Climate change as an incentive for future human migration. Earth Syst Dyn 11(4):875–883. https://doi.org/10.5194/esd-11-875-2020

Collier P, Conway G, Venables TJOROEP (2008) Climate change and Africa. Oxford Rev Econ Policy 24(2):337–353.https://doi.org/10.1093/oxrep/grn019

Dillimono HD, Dickinson JE (2015) Travel, tourism, climate change, and behavioral change: travelers' perspectives from a developing country, Nigeria. J Sustaiable Tourism 23(3):437–454. https://doi.org/10.1080/09669582.2014.957212

Diugwu IA, Ijaiya MA, Mohammed M, Egila AE (2013) The effect of gas production, utilization, and flaring on the economic growth of Nigeria. Nat Resour 4:341–348https://doi.org/10.4236/nr.2013.44041

Elias P, Omojola AJ (2015) Case study: The challenges of climate change for Lagos, Nigeria. Current Opinion Environ Sustain 13:74–78. https://doi.org/10.1016/j.cosust.2015.02.008

Emoabino IU, Waheed AA, Bamgboye O (2007) Tail water recycling for higher efficiency in hydropower–case study of Kainji–Jebba hydropower dams in Nigeria. International conference on small hydropower-hydro Sri Lanka, Citeseer

Erickson P, Lazarus M, Piggot GN (2018) Limiting fossil fuel production as the next big step in climate policy. Nat Clim Chang 8(12):1037–1043. https://doi.org/10.1038/s41558-018-0337-0

Eshiemogie SO, Ighalo JO, Banji TI (2022) Knowledge, perception and awareness of renewable energy by engineering students in Nigeria: a need for the undergraduate engineering program adjustment. Cleaner Eng Technol. https://doi.org/10.1016/j.clet.2021.100388

Giesen C, Roche J, Redondo-Bravo L, Ruiz-Huerta C, Gomez-Barroso D, Benito A, Herrador Z (2020) The impact of climate change on mosquito-borne diseases in Africa. Pathogens Global Health 114(6):287–301. https://doi.org/10.1080/20477724.2020.1783865

Gisladottir G, Stocking M (2005) Land degradation control and its global environmental benefits. Land Degrad Dev 16(2):99–112. https://doi.org/10.1002/ldr.687

Haider H (2019) Climate change in Nigeria: impacts and responses. K4D Helpdesk Report 675. Brighton, UK: Institute of Development Studies http://opendocs.ids.ac.uk/opendocs/handle/20.500.12413/14761

Hallegatte S, Fay M, Barbier E (2018) Poverty and climate change: introduction. Environ Dev Econ 23(3):217–233. https://doi.org/10.1017/S1355770X18000141

Hoogendoorn G, Fitchett J (2018) Tourism and climate change: a review of threats and adaptation strategies for Africa. Curr Issue Tour 21(7):742–759. https://doi.org/10.1080/13683500.2016.1188893

Hopwood B, Mellor M, O'Brien G (2005) Sustainable development: mapping different approaches. Sustain Dev 13(1):38–52. https://doi.org/10.1002/sd.244

Idowu AA, Ayoola SO, Opele A, Ikenweiwe N (2011) Impact of climate change in Nigeria. Marine Sci 2(2):145–152. https://ir.unilag.edu.ng/handle/123456789/6309

Ighalo JO, Adeniyi AG (2020) A comprehensive review of water quality monitoring and assessment in Nigeria. Chemosphere 260:127569. https://doi.org/10.1016/j.chemosphere.2020.127569

Ighalo JO, Adeniyi AG, Otoikhian KS (2020a) Recent advances in environmental protection of oil polluted surface and groundwater in the nigerian context. J Eng Exact Sci 6(3):416–420. https://doi.org/10.18540/jcecvl6iss3pp0416-0420

Ighalo JO, Enang WP, Nwabueze QA (2020b) "Re-evaluating the problems of gas flaring in the nigerian petroleum industry. World Scientific News 147:76–87. https://bibliotekanauki.pl/articles/1031825

Ikhuoso OA, Adegbeye M, Elghandour M, Mellado M, Al-Dobaib S, Salem AJ (2020) Climate change and agriculture: the competition for limited resources amidst crop farmers-livestock herding conflict in Nigeria-A review. J Cleaner Prod 272:123104. https://doi.org/10.1016/j.jclepro.2020.123104

Jabareen Y (2008) A new conceptual framework for sustainable development. Environ Develop Susyainability 10(2):179–192. https://doi.org/10.1007/s10668-006-9058-z

Karl TR, Trenberth KE (2003) Modern global climate change. Science 302(5651):1719–1723. https://doi.org/10.1126/science.1090228

Koubi V (2019) Sustainable development impacts of climate change and natural disaster. Background paper prepared for Sustainable Development Outlook 2019 of the United Nations

Lenshie NE, Okengwu K, Ogbonna CN, Ezeibe C (2021) Desertification, migration, and herder-farmer conflicts in Nigeria: rethinking the ungoverned spaces thesis. Small Wars Insurgencies 32(8):1221–1251. https://doi.org/10.1080/09592318.2020.1811602

Li Q, Samimi C (2022) Sub-Saharan Africa's international migration constrains its sustainable development under climate change. Sustain Sci, 1–25.https://doi.org/10.1007/s11625-022-01116-z

Liverpool-Tasie LSO, Sanou A, Tambo J (2019) Climate change adaptation among poultry farmers: evidence from Nigeria. Clim Change 157(3):527–544. https://doi.org/10.1007/s10584-019-02574-8

Ma S, Kirilenko AP (2020) Climate change and tourism in English-language newspaper publications. J Travel Res 59(2):352–366. https://doi.org/10.1177/0047287519839157

Machina MB, Sharma SJ (2017) Assessment of climate change impact on hydropower generation: a case study of Nigeria. Int J Eng Technol Sci Res 4(8)

Matemilola S, Adedeji OH, Elegbede I, Kies F (2019) Mainstreaming climate change into the EIA process in Nigeria: perspectives from projects in the Niger Delta Region. Climate 7(2):29.https://doi.org/10.3390/cli7020029

Murshed M, Tanha MM (2021) Oil price shocks and renewable energy transition: empirical evidence from net oil-importing South Asian economies. Energy Ecol Environ 6(3):183–203.https://doi.org/10.1007/s40974-020-00168-0

Nwabueze QA, Ighalo JO (2020) Production and economic analysis of enhanced oil recovery (EOR) by water flooding: a case study of Reservoir OB-48 In the Niger Delta. Pet Coal 62(1):103–113

Ogbo A, Lauretta NE, Ukpere W (2013) Risk management and challenges of climate change in Nigeria. J Hum Ecol 41(3):221–235. https://doi.org/10.1080/09709274.2013.11906570

Ogbuabor JE, Egwuchukwu EI (2017) The impact of climate change on the Nigerian economy. Int J Energy Econ Policy 7(2):217–223

Okunlola OA, Oyeyemi OT (2019) Spatio-temporal analysis of association between incidence of malaria and environmental predictors of malaria transmission in Nigeria. Sci Rep 9(1):1–11. https://doi.org/10.1038/s41598-019-53814-x

Olayide OE, Alabi T (2018) Between rainfall and food poverty: assessing vulnerability to climate change in an agricultural economy. J Clean Prod 198:1–10. https://doi.org/10.1016/j.jclepro.2018.06.221

Raimi MO, Adindu IB, Udensi EO, Funmilayo AA, Opufou T, Deinkuro NS, Adekunle AP, Adeniji OA, Care C (2020) Health impact assessment: expanding public policy tools for promoting sustainable development goals (SDGs) in Nigeria. EC Emerg Merd Critical Care 4(9):95–107

Randell H, Gray C (2019) Climate change and educational attainment in the global tropics. Proc Natl Acad Sci 116(18):8840–8845. https://doi.org/10.1073/pnas.181748011

Sanson AV, Van Hoorn J, Burke SE (2019) Responding to the impacts of the climate crisis on children and youth. Child Develop Perspect 13(4):201–207. https://doi.org/10.1111/cdep.12342

Shiru MS, Shahid S, Alias N, Chung E (2018) Trend analysis of droughts during crop growing seasons of Nigeria. Sustainability 10(3):871. https://doi.org/10.3390/su10030871

Sola AP, Adeniran J (2009) Global climate change and sustainable water management for energy production in the Niger Basin of Nigeria. IOP conference series. Earth and Environmental Science, IOP Publishing

Tabor K, Hewson J, Tien H, González-Roglich M, Hole D, Williams J (2018) Tropical protected areas under increasing threats from climate change and deforestation. Land 7(3):90. https://doi.org/10.3390/land7030090

Uduji JI, Okolo-Obasi EN, Asongu SA (2019) Does CSR contribute to the development of rural young people in cultural tourism of sub-Saharan Africa? Evidence from the Niger Delta in Nigeria. J Tour Cult Chang 17(6):725–757. https://doi.org/10.1080/14766825.2018.1502777

Ukhurebor K, Abiodun I (2018) Variation in annual rainfall data of forty years (1978–2017) for South-South, Nigeria. J Appl Sci Environ Manag 22(4):511–518. https://doi.org/10.4314/jasem.v22i4.13

Urry J (2015) Climate change and society. Why the social sciences matter. Springer, pp 45–59.https://doi.org/10.1057/9781137269928_4

Yetano Roche M, Verolme H, Agbaegbu C, Binnington T, Fischedick M, Oladipo EOJCP (2020) Achieving sustainable development Goals in Nigeria's power sector: assessment of transition pathways. Climate Policy 20(7):846–865. https://doi.org/10.1080/14693062.2019.1661818

Chapter 2
Effect of Climate Change on Air Quality: A Nigerian Perspective

Mohammad Yusuf, Huma Warsi Khan, Mukarram Beg, Bernard Chukwuemeka Ekeoma, Afreen Nishat, and Amani Lutfi Al-Othman

Abstract The entirety of the ecosystem today is subject to incessant global warming and climate changes with the aftermaths and risks being a great cause for concern for both developed and developing nations. Factors leading to the exposure of the natural environment to these risks are mainly anthropogenic with the most significant factor being the high dependence on fossil fuels which increase the concentration of two major greenhouse gases (GHGs), i.e., CO_2 and CH_4. Consequently, the world is faced with not just a rapid rise in global temperatures, but on a wider spectrum, the bizarre behaviour of different ecosystems of the world with the quality of air not left out. This is evidenced by the several health issues emanating from the intake of polluted air by human beings. This study highlights how the air quality has impacted on the climate changes using Nigeria as a case study. Climatic changes such as prolonged dry and sunny season results in wildfires which obviously degrade the quality of air by reducing the oxygen content and introducing smoke and particulate. The population explosion and hence, the need for more fossil fuels and limited oxygen in the metropolitan cities of Nigeria such as Lagos is the major reason for the poor air quality index (AQI) relative to other suburban and rural areas. The

---

M. Yusuf (✉)
Department of Petroleum Engineering, Universiti Teknologi PETRONAS, 32610 Bandar Seri Iskandar, Perak, Malaysia
e-mail: yusufshaikh.amu@gmail.com

Institute of Hydrocarbon Recovery, Universiti Teknologi PETRONAS, 32610 Bandar Seri Iskandar, Perak, Malaysia

H. W. Khan · B. C. Ekeoma
Department of Chemical Engineering, Universiti Teknologi PETRONAS, 32610 Bandar Seri Iskandar, Perak, Malaysia

M. Beg
Department of Petroleum Engineering, Chandigarh University, Mohali 160055, Punjab, India

A. Nishat
Department of Petroleum Studies, Aligarh Muslim University, Aligarh 202002, India

A. L. Al-Othman
Department of Chemical Engineering, American University of Sharjah, PO. Box 26666, Sharjah, United Arab Emirates

J. C. Egbueri et al. (eds.), *Climate Change Impacts on Nigeria*, Springer Climate,
https://doi.org/10.1007/978-3-031-21007-5_2

continuous rise in global warming is characterized by increments in the presence of toxic components like sulphur and nitrogen oxides, particulate matter ($PM_{2.5}$) and ozone (O_3) levels in the atmosphere. In this study, the causes of the deterioration in air quality, current trends/scenario in Nigeria and effective solutions to remedy the situation are discussed in detail.

Keywords Global warming · Greenhouse gases · Ecosystem · Air quality index · Pollutants

Abbreviations

AQI	Air Quality Index
VOCs	Volatile Organic Compounds
POPs	Persistent Organic Pollutants
GHGs	Greenhouse gases
PM	Particulate Matter
MMT	Million Metric Tonnes
CO	Carbon Monoxide
LPG	Liquefied Petroleum Gas
CEC	Contaminant of Concern
PBPs	Petroleum-based Products

Introduction

With the progress of technology, living standard of humans increases and we run more towards urbanization at a faster pace due to the explosion in population growth (Qureshi et al. 2022). It is projected that most low- and middle-income countries will experience expeditious urbanization with the net effect being a larger half of the global population residing in urban areas by 2030. Urbanization gives rise to industrialization and infrastructural development and various other factors which contribute to environmental pollution. Increase in anthropogenic activities and more consumption of natural resources, are major consequences of urbanization and industrialization which lead to deteriorated condition of the earth (Liang et al. 2019). Pollution prevalent around the globe is a major concern for developed as well as for developing countries as it leads to climate change. Climate change refers to a deviation in the climatic condition which can be statistically identified by studying changes in the mean or the variability of its features that persists for a long duration (Yusuf et al. 2022).

Air pollution which is an important environmental issue considering its severe consequences involves a reduction in the quality of air by the introduction of pollutants including but not limited to chemicals, biological materials and particulate matter thereby causing harm to living beings and the natural environment. Air pollution is predominantly due to the production and use of energy (Heald and Kroll 2021). Chemicals and gases are released into the air by the combustion of fossil fuels which in a destructive feedback loop, does not only result in climatic changes but is also exacerbated by it. Air pollution in the form of CO_2, CH_4 and smog elevates global temperatures and introduces ultraviolet radiations (Heald and Kroll 2021). Climate change also promotes the formation of allergenic air pollutants, including mould (owing to damp conditions resulting from intense weather and escalated flooding) and pollen (from longer spells of pollen season) (Deng et al. 2020). Climate change alters the composition of the global atmosphere over comparable time periods. Wildfires are common in wooded areas during long durations of dry season, generally due to a lack of precipitation (Yusuf et al. 2021). The smoke and carbon monoxide formed increases the concentration of atmospheric carbon and creates warmer conditions via greenhouse effect. Climate changes have altered the composition of the global atmosphere over the past few decades. In the past decade, the climate have undergone several changes of which CO_2 concentration in the atmosphere exceeding 400 ppm, increase in global mean temperature, Sea rise by over 1.6 inches, ice level reduction (Antarctic ice mass loss) and increased frequency of extreme weather scenarios (storms, droughts, floods, heatwaves) are notable (Yusuf et al. 2020). The interactions between climate change and air quality with the accompanying outcomes is shown in Fig. 2.1.

The effects of climate change are more prominent in developing nations and countries located in Africa has been considered to be exposed to the highest risk (Fisher et al. 2021). They attributed the difficulty in achieving cooperation for air pollution control in developing countries to their primary concern and focus on attaining the basic needs of food, shelter and employment for the populace. Air pollution now ranks second to AIDS in the highest causes of death in Africa and accounts for more deaths than road accidents and drug-related mortality (Fisher et al. 2021). Deaths related to air pollution result from lower respiratory infections, ischaemic heart disease, neonatal disorders, chronic obstructive pulmonary disease and stroke (Fisher et al. 2021). Patterns of deaths and diseases attributable to air pollution vary across Africa with the peak rates predominant in countries with very low social development indices (Abbafati et al. 2020). In contrast to the declining household air pollution-related deaths, deaths related to ambient air pollution are on the increase from 26 deaths per 100,000 population in 1990 to 29 per 100,000 in 2019 (Abbafati et al. 2020). The effects of air pollution can be categorized under health and environmental effects. The aforementioned diseases are under the health effects while occurrences such as global climate change, acid rain, eutrophication, haze, ozone depletion, crop and forest damage are environmental effects. Nigeria accounts for 2.64% of the world's population and is the most populated African country (Reed and Mberu 2014). This makes Nigeria a good yardstick to adequately represent and characterize the general situation in Africa with respect to studies on

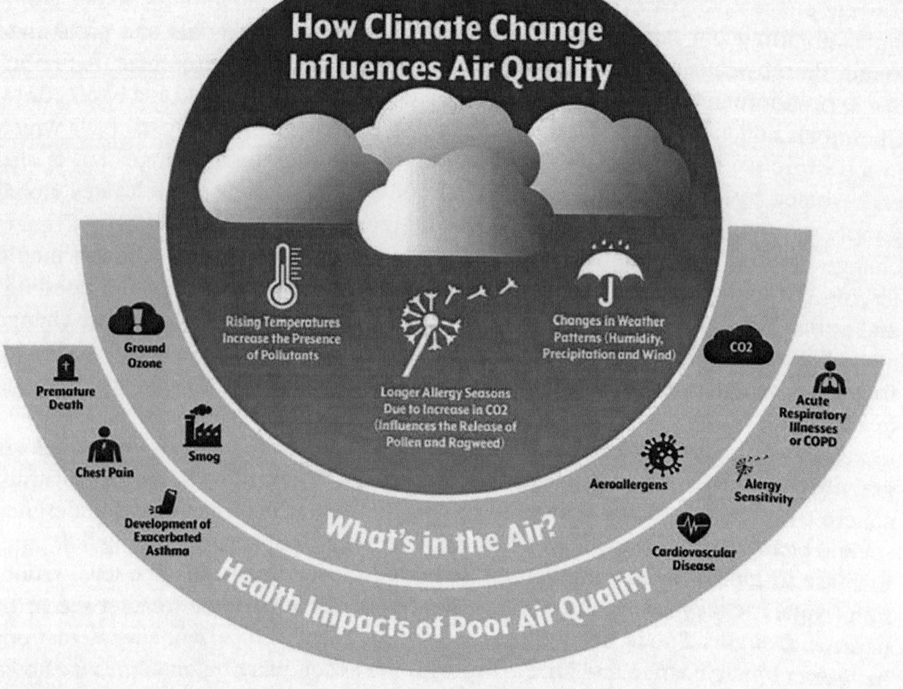

Fig. 2.1 Climate change, air quality and associated impacts (AQI India 2019)

the level and impacts of climatic changes. Lagos is Nigeria's most populated state and boasts of residents from multifarious regions of the country. In this regard, the state is the commercial hub of Nigeria with over 70% of the nation's industrial, trading and other economic activities based in it. The increasing global warming is leading to constant rising oxides of sulphur and nitrogen along with elevated particulate matter (PM2.5) and ozone (O_3) levels in the city's air. The contributing factors to pollution in Nigeria are dependence on carbon generating sources for cooking, waste combustion and traffic pollution. At home, due to the constant power interruption, many Nigerians use generators, which release toxic fume often in unventilated areas. The unregulated emission from cars on the street make the problem worse. It can be inferred that the rise in pollution levels in Africa can be traced to the fact that there is fast economic and industrial development without the adequate technology. In Nigeria, Oil, Agriculture, telecom industries are driving the industrial growth at certain environmental cost.

Nigeria has been subject to the negative realities of environmental degradation associated with oil development since oil was first discovered in the 1950s. The advancement of the country's oil industry, combined with rapid population growth and a lack of environmental management, led to significant damage to Nigeria's environment particularly in the Niger Delta Region. Niger Delta is the petroleum-rich region and hence is comprised of the oil producing states namely Abia, Akwa-Ibom, Bayelsa, Cross River, Delta, Edo, Imo, Ondo and Rivers. The sources of air pollution in the Niger Delta are industrial emissions, gas flaring and pipeline explosions (Williams et al. 2019). The region is home to several oil and non-oil related industries that emit various kinds of air pollutants thereby compounding the problem of gaseous and particulate pollutants in the air. (Elum et al. 2016) reported that the primary air pollutants in Niger Delta Area gases are CO, CH_4, SO_2, N_2O, NO_2, NH_3, VOCs and Particles: PM10, PM2.5, PAHs, Dioxins PAHs, Heavy metals. Nigeria held the highest record (19.79%) of global natural gas flaring back in the year 2000 and contributed 46% of Africa's total gas flared per tonne of oil produced (Elum et al. 2016). Successive governments have made progress by reducing gas flaring by 70% between 2000 and 2020, according to the International Energy Agency, due to more stringent penalties and incentives to capture and sell the gas (Olujobi 2020).

This study aims to elucidate the interaction between climate change and air quality in Nigeria, by taking an in-depth look on the several factors prevalent in Nigeria, and working in synergy to disrupt the natural climatic conditions and promote air pollution. In this study, air pollution impacts on health and economy will be quantified, and opportunities for climate stabilization and air pollution control identified. The severity of the incessant climatic changes and uncontrolled rise in air pollution will be highlighted and control measures suggested with regard to preventive and curative dimensions. To better understand the role played by the primary pollutants in altering the natural environment and reducing air quality, each key pollutant will be discussed in detail to highlight its intrinsic properties and mechanism of causing air pollution.

Major Air Pollutants

Air pollutants are biological, chemical, physical or radioactive substance in the atmosphere in a concentration more than their permissible limits and therefore pose harmful effects to humans, animals and vegetation. There is a variety of these pollutants which have been identified to be harmful to the environment and they have broadly been classified as anthropogenic (a consequence of human activities) and natural air pollutants. Air pollutants are also classified as primary when directly emitted into the atmosphere and as secondary when they originate from various physiochemical processes in air due to climatic conditions. The effects of severe climatic conditions can cause the increase in secondary air pollutants and these air pollutants can severely damage the air quality. Major air pollutants are discussed below.

Nitrogen Oxides (NO_x)

The three oxides of nitrogen which are significant in polluting the atmosphere are: nitrous oxide (N_2O), nitric oxide (NO) and nitrogen dioxide (NO_2). These oxides are produced in substantial amounts by fossil fuels combustion and by the action of microbes on nitrogenous organic matter and chemical fertilizers used in agricultural fields. About 70–80% of the total N_2O input originates from natural sources and the disintegration of nitrogenous fertilizers easily further reacts to give rise to NO and NO_2 which are much more toxic. NO_2 itself readily forms nitric acid (HNO_3) on reacting with water vapour (Islam et al. 2022). Nitric oxide accounts for the ozone layer depletion while NO_2 is a primary constituent of photochemical smog or oxidizing type of air pollution and contributes to the occurrence of acid rain. In a natural scenario, in the event of lightning, the vicinity of lightning bolts in air is heated to a very high temperature and N_2 and O_2 gases present in the air react to form the nitric oxide. The nitric oxide is further converted into nitrogen dioxide in presence of oxygen. These two natural events are represented by Eqs. 2.1 and 2.2 respectively.

$$N_2 + O_2 \rightarrow NO \tag{2.1}$$

$$NO + O_2 \rightarrow NO_2 \tag{2.2}$$

Considering NO_x emissions as a result of human activities, the formation of NO_x from nitrogen and oxygen gases is usually at elevated temperatures and these conditions are abundant in the human activities of the burning fossil fuels for energy usage in transportation and industrial applications. In rural areas of Nigeria with high rate of poverty, charcoal, wood and coal are preferred domestic fuels which cause serious pollution compared to modern energy sources such as electricity and LPG which have lower NO_x emissions (Urhie et al. 2020). On the contrary, the combustion of hydrocarbons in vehicles is the major factor contributing to the NO_x emissions in the urban areas characterized by automobiles in abundance (Urhie et al. 2020).

Sulphur Oxides (SO_x)

Sulphur dioxide (SO_2) which is a colourless gas characterized by its pungent and suffocating smell is the most common sulphur pollutant. It occurs in fossil fuels as a natural pollutant and is emitted when these fossil fuels are burnt with the possibility of further oxidation to form Sulphur trioxide (SO_3). Acid rain results from the formation of sulphuric acid (H_2SO_4) and/or weak sulphurous acid (H_2SO_3) when SO_2 get dissolved in water vapour (He et al. 2016). In addition to these acids, the formation of iron sulphates also occurs in a variety of solid and liquid particulates. These acids and sulphates are the most toxic forms of sulphur when health impacts are considered

(Fisher et al. 2021). Oil and gas exploration and production activities is the key SO_x emission source in Nigeria (Okedere et al. 2021b).

Carbon Oxides (COx)

Complete and incomplete combustion (under oxygen deficient condition) of fossil fuels and organic matter in the presence of oxygen results in CO_2 and CO formation, respectively. About 52% of the total atmospheric pollution is attributed to the colourless, odourless, non-irritating but very poisonous CO gas, thereby making it the most important atmospheric pollutant (Rose et al. 2017). Automobile exhaust systems stand out as the most prominent source of CO pollution with other sources being the various metallurgical operations, domestic combustion of coal, dried organic matter, fuel gases, etc. (Okedere et al. 2021b). Considering its potency even at very minute concentrations and ability to hinder the delivery of oxygen through the vital organs of the body, CO is deemed to be very dangerous to human health and can lead to death (Rose et al. 2017). CO_2 is a colourless, odourless, non-toxic greenhouse gas (GHG) which is produced in large amounts from the respiration of living things. Incineration of limestone during cement production, metallurgical operations and several industrial practices also yield large amounts of this gas (Okedere et al. 2021b).

Ozone (O_3)

The breaking up of molecular oxygen (O_2) to produce oxygen atoms (O) at UV wavelengths lower than 242 nm (photolysis) and the subsequent reaction of the oxygen atoms with O_2 yields ozone (O_3). The breakup of O_2 is caused by the solar irradiation of the upper atmospheric layers at altitudes above 30 km. Due to its highly reactive nature, Ozone facilitates the oxidation of various macromolecules within a biological system and produces free radicals responsible for several induced pulmonary or extra pulmonary damages including damages to the DNA and carcinogenesis (Kim et al. 2019).

Particulate Matter (PM)

Particulate matter (PM) is a complex mixture of solids and liquids including carbon, complex organic chemicals, mineral dust, nitrates and water suspended in the air (Mukherjee and Agrawal 2018). Industry, agriculture and transport generates PM, which can be carried in air currents across different regions owing to their light weight. PM particle size determines their behaviour in air, the height to which they can be carried up by air currents and their influence on organisms inhaling them or on the exchange of gases by plants through their stomata (Mukherjee and Agrawal 2018). Total suspended particulates (TSP) refer to particles of 10 microns (μm) diameter and above, such as dust, dirt, soot or smoke that are large or dark enough to be visible to the naked eyes. However, the particles that pose a more serious threat are the smaller particles (fine particles) with diameter less than 10 μm (PM10) and 2.5 μm (PM2.5). The smallest fine particles, less than 0.1 μm in diameter, are called ultrafine particles (Xing et al. 2016).

Hydrocarbons and Volatile Organic Compounds (VOCs)

These are emitted during the processes of refining, coal production, transportation, biomass decomposition, landfilling of municipal wastes and utilization of petroleum-based products (Sonibare 2010). They could be aliphatic (straight and branch structure) or aromatic (ring structure). Methane and benzene are typical examples of the aliphatic and aromatic structures respectively. They pollute the atmosphere and are toxic to human health. Also, they play precursor role in the formation of photochemical smog and ozone (Ahmed Laskar and Younus 2019).

Sources of Air Pollution Affecting Nigeria's Air Quality

There are two main sources of air pollution that influence air quality in Nigeria, and they are natural and anthropogenic sources. The anthropogenic sources are predominant especially in the urban areas.

Natural Sources of Air Pollution in Nigeria

Natural sources of air pollution in Nigeria are as follows:

Sand dust

Conventionally emanates from storms in the arid and semi-arid regions of Northern Nigeria, and travels over long distances. An ideal example is the harmattan dust which originates from the Faya Largeau area of the Chad basin and is typically felt during the months of December and January in Nigeria (Asubiojo 2017; Tawari and Abowei 2012). The ferocity and duration of the harmattan dust is greatest in the northern part of Nigeria (arid and semi-arid region of Nigeria) compared to the southern region of the country. For example, in contrast to the 329 g/m^3 of dust recorded in Ile-Ife (Southern Nigeria), a substantially higher value of 1033 g/m^3 was recorded in Kano (Northern Nigeria) during the 1993 harmattan period (Asubiojo 2017).

Methane

This results in the atmosphere from the metabolism of bacteria on wetland decaying plant materials and on cellulose in the stomach of termites and ruminant animals (Asubiojo 2017). Large amounts of methane are generated as a by-product from the digestion process of these animals which involves enteric fermentation stimulated by the microorganisms in the guts of these animals (Asubiojo 2017). The methane is either exhaled by the animals or via flatulence. Rearing of ruminant animals like cows, rams and goats is a major business in Nigeria especially in the northern part of the country.

Radon gas

This is a colourless, odourless and hazardous radioactive gas that results from the radioactive decay of radium within the Earth's crust. It is known to cause lung cancer and easily accumulates in confined areas (Tawari and Abowei 2012).

Anthropogenic Air Pollution Sources in Nigeria

Sources in this class and prevalent in Nigeria include:

Automobile emissions

This is most obvious in the urban areas and worsens with the per capita increase in vehicle ownership. Report from Nigeria's National Bureau of statics showed that there was a huge increase in the number of registered cars from 11.6 million in the fourth quarter of 2017 to 11.8 million in that of 2018 (NBS 2019). Although there is no documentation of the annual contribution of the transportation system in Nigeria to air pollution, several researches have been carried out and the inferences made confirm that transportation in Nigeria is a major driver of anthropogenic emissions (Fakinle et al. 2020). A good number of the vehicles plying the roads in Nigeria are old and have worn-out internal combustion engines, with the end result being massive emission of CO due to incomplete combustion of fuel (Atubi 2015). Atubi (2015) reported that vehicular atmospheric pollutants can either be exhaust or non-exhaust emissions. The exhaust emission includes products from the combustion of fuel such as CO, CO_2, NO, volatile organic compounds (VOCs) and toxic heavy metals containing particulate matter. Particles from tyre wear, brake wear, road surface abrasion and resuspension by road traffic are under the category of non-exhaust emissions (Atubi 2015).

Petroleum Industries

Nigeria being an oil producing country has a petroleum-based economy and is subject to the environmental issues associated with oil and gas activities. The Niger Delta region which is the hub for oil and gas industries in Nigeria due to the abundance of crude is most exposed to the pollution and health risks (Okoye et al. 2022). The primary units involved in refineries include atmospheric distillation column, vacuum

Table 2.1 Emissions from Nigerian refineries (Oladimeji et al. 2015a, b)

Pollutant	Amount (tonnes/per annum)
VOC	147,212
PM	1,217
SO_2	45,124
NO_x	167,570
CO	242,469

distillation columns, alkylation unit, fluid cracking unit, etc. Separation of CO_2, NO_2, H_2S, SO_2, VOCs, particulate matter, etc., are achieved in these units and emitted to the atmosphere (Oladimeji et al. 2015a, b). These emissions can occur at any stage or unit. Table 2.1 presents the emissions from Nigerian refineries. Aside refineries, the upstream sector of Nigerian petroleum industry contributes significantly to air pollution with gas flaring being the most notable activity increasing emission levels (Okoye et al. 2022). Gas flaring is the term associated with the disposal of flammable waste gases for oil exploration activity and even with the deliberate efforts of the government to cut this detrimental process down, Nigeria still ranks seventh as of 2022 in the global ranking of gas flaring nations (Okoye et al. 2022).

Gasoline Generators

The current status quo of Nigeria as a developing country is also reflected in her energy sector as the country is yet to achieve uninterrupted power supply. This explains the presence of numerous gasoline generators in households, businesses and industries thereby contributing heavily to the aforementioned exhaust emissions especially CO. The small gasoline generator market in Nigeria was estimated to be 22 million in 2018 and projected to reach 30 million by 2030 (Adeniran et al. 2017). Recently, (Giwa et al. 2019) reported that the average emission levels of CO_2, CO and PM2.5 from gasoline generators is as high as 710 ± 19.1 ppm, 83 ± 4.0 ppm and 83 ± 4.1 $\mu g/m^3$, respectively.

Mining

Mining of solid minerals from the earth's surface involves the processes of land site clearing, drilling, blasting, hauling, collection and transportation and these contribute heavily to the release of harmful gases like CO and CH_4 and particulate matter into the atmosphere (Omotehinse and Ako 2019). Mining activities are predominant and actively taking place in Amansiodo, Inyi and Ezinmo coal mine in Enugu State, Ogboyoga and Okaba coal mine in Kogi State, Okpara mine in Anambra State, Owupka mine in Benue State, Maiganga coal mine in Gombe State and Tin and Columbine mine in Jos (Omotehinse and Ako 2019).

Domestic Cooking

Domestic cooking in Nigeria involves the combustion of fuel-woods, sawdust, charcoal, kerosene and liquefied petroleum gas (LPG) (Fakinle et al. 2020). This is another important contributor to atmospheric air pollution in Nigeria. In 2017, Lagos state alone had an estimated fuel-wood usage of about 1,707,634,110 kg (Fakinle et al. 2020). Fuel-woods, sawdust and charcoal are the major domestic cooking fuel in the rural areas in Nigeria and biomass combustion has been reported to release pollutants mostly PM, CO, NO_x and PAHs with adverse human health implications and mortality (Akinrinola et al. 2014). Research on indoor air emissions performed by (Ubuoh and Nwajiobi 2018) showed that fuel-woods emitted more CO, $PM_{2.5}$, NO_2 and SO_2 than most other fuels. Aside fuel-woods, kerosene is another major domestic cooking fuel used in Nigeria. (Fakinle et al. 2020) reported that PM_{10}, SO_2, CO and NO_x emissions from the use of kerosene for domestic cooking in Lagos alone increased by 12.8% each between 2009 and 2014. Currently, the contribution of domestic cooking to air pollution in the urban areas is gradually reducing, due to the adoption of LPG which is a cleaner and more environmentally friendly as the major cooking fuel.

Municipal Waste Disposal

This is a major environmental issue in Nigeria, especially. The disposal of solid waste from multiple sources including domestic, municipal, agricultural and industrial sources is a major environmental issue in Nigeria. Burning and open dumping of refuse are still very common practices in the very big cities in Nigeria (Fakinle et al. 2020). Open dumping encourages the biodegradation of organic waste to produce methane and other gaseous toxins while burning of solid wastes generates VOCs, CO_2, CO, SO, NO, Total Hydrocarbons (THCs), Polycyclic Aromatic Hydrocarbons (PAHs), dioxins, Polychloro Biphenyls (PCBs) into the atmosphere (Okedere et al. 2021b).

Bush burning, land use and agriculture

In Nigeria, the indiscriminate burning of grasslands, vegetation and forest is very common in the dry season for the reasons of farmland clearing for planting, animal hunting and soil fertility improvement (Izah et al. 2017). Nigeria is ranked to be among the top 13 countries where bush burning is widespread (Izah et al. 2017). Air pollutants including CO, CO_2, PM, NO_x, SO_2, VOCs, PAHs are emitted by bush burning.

Climate Change and Disturbed Ecosytem in Nigeria: Aftermath of Air Pollution

Climate change is being exacerbated in Nigeria due to the country's high emissions of GHGs, PMs and VOCs. Overheating, acid rains, dwindling water supplies, drought, deforestation and low crop yield are just a few of the impacts of climate change in Nigeria (Pona et al. 2021). Infra-red radiations from gas flaring also warms the atmosphere and has detrimental effects on the ecosystem among which climate change is notable (Okedere et al. 2021a). Environmental scientists associate climate change effects with the depletion of the atmosphere's ozone layer. The ozone layer which shades the earth from the high heat intensity of the sun is depleted when the pollutant gases highlighted in this study are released into the atmosphere from human and natural factors. Climate change reflects the changes in the mean daily weather conditions such as rainfall, temperature, humidity and sunshine of a region over an extended period. Climate change in Nigeria is a menace to the economic growth of sectors dependent on climatic conditions such as agriculture, construction, tourism, energy, etc.

Climate change influences the natural wind patterns and is the source of the storms that affect Nigeria's farming sector, housing units and also give rise to bush fires, forest exploitation, decreased water quality and coastal erosion in Nigeria (Malley et al. 2021). The vegetation in Nigeria's southern regions (which are mostly recognized for receiving a lot of rain) is currently struggling to cope with fluctuations in the sequence of rainfall caused by air pollution-related climate changes (Alaanuloluwa Ikhuoso et al. 2020). Freshwater supply (surface and ground water), hydrological system stability and water recycling in Nigeria are all threatened by deforestation, according to the study by (Alaanuloluwa Ikhuoso et al. 2020). Due to decreasing rainfall and increased evapo-transpiration, many big water bodies are suffering significant reductions in flow rate and network length (Durodola and Durodola 2019). The impact of climate changes not only leads to shortages in rainfall, but can also cause excessive rainfall which is the origin of flooding especially in the riverine area of Nigeria. Floods displaced about 1.9 million Nigerians in 2019 (Echendu 2020). Nigeria is among the top five countries losing the most wood-landed areas per year (−3.7%) (Okon et al. 2021). As a result, ecosystem services, such as habitat protection for wildlife and plants, mangroves, mushrooms, cane, etc., wetlands and rivers are affected. The urgent needs of providing habitats for wildlife and flora, regulating the hydrologic cycle, and protecting the environment from catastrophic weather events were all stimulated by abrupt climatic changes related to air pollution (Onah et al. 2016).

In Nigeria, the effects of climate change are evident in rising sea levels and coastal erosion, as well as in the country's more unpredictable weather patterns, which have led to flooding and food shortages, most recently in Jigawa State (Echendu 2020). Gully erosion has ravaged numerous villages, particularly in Edo and Anambra States; the Lake Chad has nearly dried up due to chronic drought; and in the north, desert encroachment has persisted for long periods of time (Asubiojo 2017). Kanji, Jebba and Shiroro's hydropower facilities have been badly affected by conflicts over the use of water resources among diverse economic sectors, which is critical to the country's electricity supply and accounts for one-third of its installed power generation capacity (Olayini 2021). Environmental data collection in a standardized manner is lacking, making it nearly impossible to estimate the severity of the impact of climate change on: agricultural and food supply, flooding and erosion, health risks, water resources, wildlife, CO_2 emission rates and trends in temperature increase. The destruction of the natural ecosystem in different states of Nigeria by occurrences emanating from the existence of air pollution and climate change as mutually inclusive events is shown in Fig. 2.2.

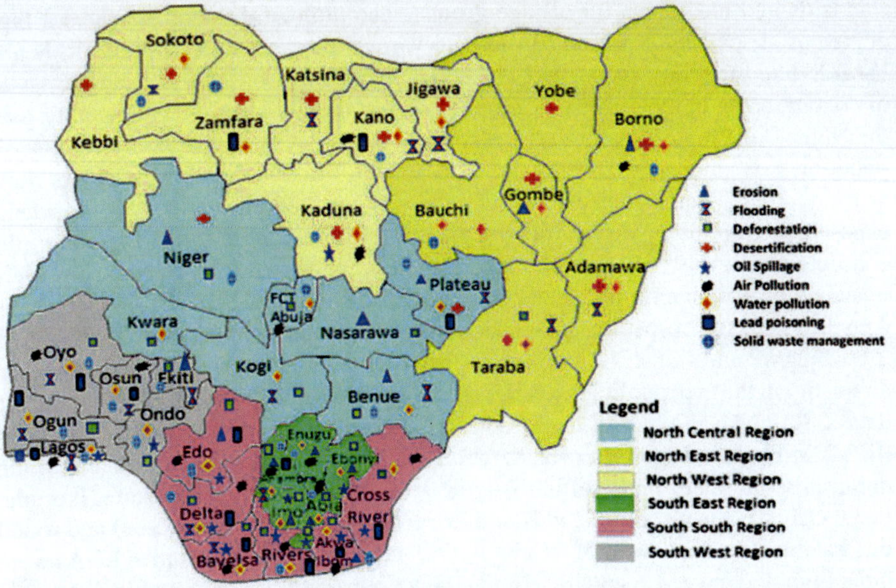

Fig. 2.2 Environmental problems resulting from air pollution and climate change in different states of Nigeria (Pona et al. 2021)

Impact of Climate Change on Nigeria's Air Quality

Climate change negatively affects the quality of air we breathe. The air quality of an environment is contingent on the atmospheric processes of the location and is said to be poor when the identified air pollutants exist at level higher than accepted standards (Pona et al. 2021). Air quality strongly depends on weather and thus is very sensitive to climate changes (Pona et al. 2021). Higher temperatures which is a consequence of climate changes related to air pollution leads to an increase in allergens and harmful air pollutants (Urhie et al. 2020). Longer warm seasons imply longer pollen seasons, which can escalate allergic sensitizations and asthma episodes and plummet productive work (Mannucci and Franchini 2017). Higher temperatures associated with climate change can also lead to an increase in ozone concentration, a harmful air pollutant thereby reducing air quality (Fisher et al. 2021). The impact of air pollution-derived climate changes is very obvious in Nigeria. Research on air quality carried out in Ilorin metropolis of Kwara state, Nigeria showed that the air quality index (AQI) in this region relative to international and national air quality standards is very poor and a risk to the inhabitants' health (Olalekan et al. 2018). Figure 2.3 shows the study area and in comparison to standards set by Nigeria's Federal Ministry of Environment (FMEV), results of this research revealed moderate mean levels of oxygen (97.65 ppm), PM2.5 (56.15 μg/m^3) and PM10 (28.81 μg/m^3), whereas VOC (240.8 ppm), formaldehyde (2570.65 mg/m^3) and combustible (107.23 LEL %) had high AQI because they were present at levels far higher than accepted standards and hazardous for all age groups. The high AQI score indicates a high degree of air pollution, which can lead to various diseases and, as a result, health hazards like breathing issues and leukaemia. This was ascribed to the high levels of contaminants in the air due to car and industrial emissions. The air quality over the Niger Delta region associated with abrupt climatic changes due to incessant air pollution was studied by (Weli et al. 2018) using Porthacourt city in Rivers State as a case study. The study revealed that the city had bad air quality with residents exposed to respiratory and cardiovascular problems. According to the study, the concentration levels of black carbon across different areas of Porthacourt city were relatively high with minimum and maximum mean values of $2.33 \pm 1.66\ \mu g/m^3$ and $4.66 \pm 5.43\ \mu g/m^3$ respectively in the morning. Mean values in the evening were $1.25 \pm 0.62\ \mu g/m^3$ and $4.75 \pm 3.13\ \mu g/m^3$. (Njoku et al. 2016) in their study of the ambient air quality in the metropolitan city of Lagos, aimed to show the air quality monitoring data of SO_2, NO, NO_x, CO, CO_2 gaseous pollutants and suspended particulate matter (SPM) in Lagos taking into consideration the industrial, residential, open burning dumpsites and major high traffic road intersection. They sampled Kosofe, Ikorodu, Ikeja, Oshodi, Apapa and Lagos Island Local Government areas (LGAs) and found out that the measured levels of NO, NO_2, CO and SPM in all the sampled LGAs were quite high and exceed regulatory limits. However, there was no notable difference except in SPM (at all the sampling points), and NO_2 and NO (only in major traffic intersection). AQI computed by them indicated that the ambient air can be described as poor for SPM, varied from fair to very poor in CO, NO and NO_2.

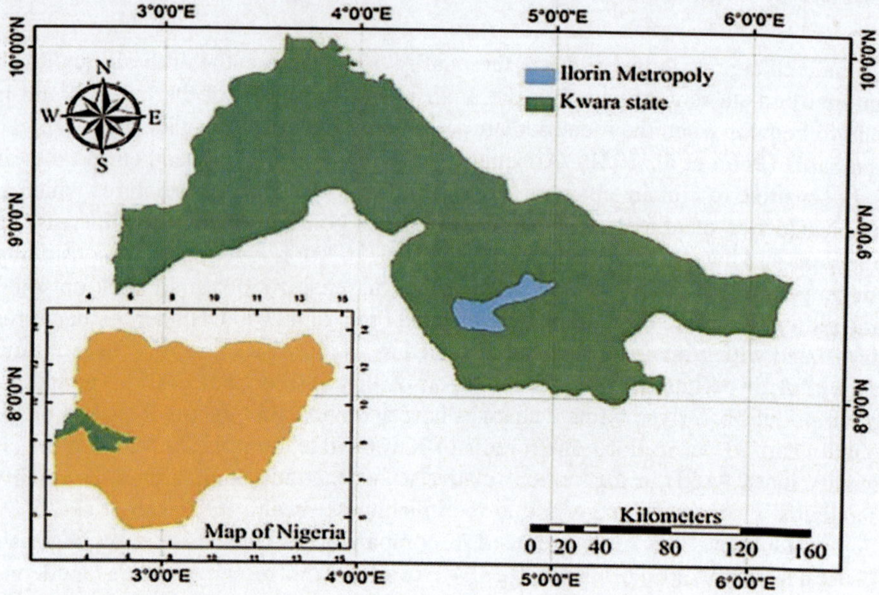

Fig. 2.3 Map of Kwara State showing the study area (Olalekan et al. 2018)

Effects of Air Pollution on Human Health

Several researches have been carried out to determine the impacts of air pollution from the different sources outlined, on the health of Nigerian citizens. (Olowoporoku et al. 2012) after their study, concluded that people living in Nigerian cities with higher levels of air pollution were more susceptible to diseases. (Nwachukwu et al. 2012) investigated the relationship between air pollutants and diseases, and discovered that pneumonia, tuberculosis and upper respiratory diseases experienced by a test group in Rivers State had their roots from exposure to air pollution. They also found out that members of the test groups residing in areas in close proximity to gas flaring sites had higher risks of elevated blood pressure, hypertension and circulatory diseases. Exposure to emissions from thermal plants and gasoline generators was also reported to have detrimental effects according to the study carried out (Giwa et al. 2019). In this study, it was proved that a good number of lung cancer and asthma cases were traceable to fumes from these energy generating utilities and Nigerians exposed to these fumes stood a higher chance of getting these ailments. Several hospital admission cases as a result of exposure to high levels of CO emission from generators and leading to restlessness, convulsion, breathlessness, sphincteric incontinence, foaming in the mouth and death have occurred in Nigeria (Roth et al. 2011).

Traffic pollution have also had its fair share of negative health impact in Nigeria by posing adverse effects on ambient and outdoor air quality (Aliyu and Botai 2018). A study carried out by (Aliyu and Botai 2018) showed that about 6.7% of Zaria residents exposed to outdoor concentration of air pollutants beyond the safe threshold had respiratory problems. The report given by (Croitoru et al. 2020) revealed that Lagos residents are highly exposed to air pollution-related morbidity and mortality relative to most other African cities. Open dumpsites which are common in Nigeria is responsible for a wide range of health problems including irritation of the skin, nose and eyes, gastrointestinal problems, psychological disorders and allergies (Omang et al. 2021). This was evident in a study conducted in Bekwarra Local Government Area of Cross River State whereby the residents living in areas where open refuse dumps were poorly managed had more experience of the associated health problems (Omang et al. 2021). Irrespective of the source of pollution, a common factor is that closer proximity to the pollution source connotes higher exposure to the associated health risks. This explains the statistics of Lagos and Niger Delta having the highest number of health cases related to air pollutions as they are the commercial and petroleum production hub of Nigeria.

Air Pollution Control

Taking cognizance of the discussed sources of air pollution in Nigeria, control of air pollution in Nigeria can be achieved through the following:

- Source reduction of emissions by adopting the use of clean energy sources, industries, home transportation systems. The use of electricity, green renewable energy sources like solar and wind energy, cooking gas rather than fossil fuels should be encouraged. Industries should be proactive in monitoring the concentration of pollutants in their emissions and adopt the use of scrubbers that remove or significantly reduce the concentration of the toxic components.
- Enforcement of legislations in favour of air pollution control across the country in order to regulate air quality. According to (Ayoade 2012), this can be done through setting benchmarks for pollution levels and/or limiting emissions for various pollutant types, examining new substances in order to ascertain their safety before use, requiring a detailed evaluation of the environmental implications of an activity or project before it is carried out and taxing of industries according to the amount of pollutants they emit (Polluter Pays Principle).
- Regular servicing of vehicles and use of catalytic converters to reduce the potency of exhaust emissions should be imbibed.
- Efficient waste management practices should be adopted. If open dumping cannot be totally stopped in all areas due to the costs associated with using modern waste management technologies, regular waste collection and incineration in controlled areas should then be practised.
- Planting of trees to increase the release of oxygen should be encouraged. Also, the planting of vegetation around mining and quarrying sites to avert the escape of dust to the environment should be adopted.
- Air monitoring stations should be set up by the government especially at the location known for high emission levels so as to provide continuous and accurate information on air quality.
- Enlightenment programmes should be organized to make the impacts of air pollution and the control measures known to the general public. This information can be passed through community seminars, television and radio programmes, social media circulation and free newsletters.

Conclusion

The impacts of climate change emanating from air pollution in Nigeria has been elucidated in this study. Major air pollutants which accumulate over time leading to the net effect of abrupt climate changes were discussed and their sources identified. A special focus of the study was the interaction between climate change, air quality and human health. The quality of life in any location is contingent on the level of pollutant emissions hence, the need for conscious efforts to be directed towards achieving a good air quality. Based on the mitigation approaches suggested, it is evident that air pollution control can only be achieved by a collective effort of the individuals, organizations and the government. Health issues related to air pollution were shown to be the subject to the environmental implications of air pollution. This is so because the type of disease predominant in an area is dependent on the particular air pollutant prevalent in the given environment. These diseases can be Chronic Obstructive Pulmonary Diseases (COPDs), such as cough, shortness of breath, wheezing, asthma, pulmonary insufficiency, cardiovascular diseases and in extreme cases death. Considering the fact that the future health banes of air pollutants will depend on actions taken to mitigate the responsible emissions, comprehensive knowledge of pollutant emissions under multiple development pathways is needed. Stringent regulatory control technologies are needed now more than ever in Nigeria bearing in mind that emission levels have already exceeded the forecasted trends.

Acknowledgements The authors appreciate Universiti Teknologi PETRONAS, Malaysia for the provision of resources in carrying out this study.

References

Abbafati C, Abbas KM, Abbasi-Kangevari M, Abd-Allah F, Abdelalim A, Abdollahi M, Abdollahpour I, Abegaz KH, Abolhassani H, Aboyans V, Abreu LG, Abrigo MRM, Abualhasan A, Abu-Raddad LJ, Abushouk AI, Adabi M, Adekanmbi V, Adeoye AM, Adetokunboh OO, … Murray CJL (2020) Global burden of 87 risk factors in 204 countries and territories, 1990–2019: a systematic analysis for the Global Burden of Disease Study 2019. The Lancet396(10258):1223–1249.https://doi.org/10.1016/S0140-6736(20)30752-2/ATTACHMENT/A38D7E68-5900-4997-AA22-E0B9885F4B45/MMC2B.PDF

Adeniran JA, Yusuf RO, Amole MO, Jimoda LA, Sonibare JA (2017) Air quality impact of diesel back-up generators (BUGs) in Nigeria's mobile telecommunication base transceiver stations (BTS). Manag Environ Qual Int J 28(5):723–744. https://doi.org/10.1108/MEQ-09-2015-0168/FULL/PDF

Ahmed Laskar A, Younus H (2019) Aldehyde toxicity and metabolism: the role of aldehyde dehydrogenases in detoxification, drug resistance and carcinogenesis. Drug Metab Rev 51(1):42–64. https://doi.org/10.1080/03602532.2018.1555587

Akinrinola FS, Darvell LI, Jones JM, Williams A, Fuwape JA (2014) Characterization of selected Nigerian biomass for combustion and pyrolysis applications. Energy Fuels 28(6):3821–3832. https://doi.org/10.1021/EF500278E

Alaanuloluwa Ikhuoso O, Adegbeye MJ, Elghandour MMY, Mellado M, Al-Dobaib SN, Salem AZM (2020) Climate change and agriculture: the competition for limited resources amidst crop farmers-livestock herding conflict in Nigeria—a review. J Clean Prod 272:123104. https://doi.org/10.1016/J.JCLEPRO.2020.123104

Aliyu YA, Botai JO (2018) An exposure appraisal of outdoor air pollution on the respiratory well-being of a developing city population. J Epidemiol Global Health 8(1–2):91. https://doi.org/10.2991/J.JEGH.2018.04.002

AQI India (2019) Air pollution and climate change can make your allergies worse. https://www.Aqi.in/Blog/Air-Pollution-Climate-Change-Can-Make-Your-Allergies-Worse/

Asubiojo OI (2017) Pollution sources in the Nigerian environment and their health implications. IFE J Sci 18(4):973–980. https://doi.org/10.4314/ijs.v18i4

Atubi AO (2015) Transport and the environment: towards reducing road traffic emissions in Nigeria. AFRREV STECH Int J Sci Technol 4(1):58–78. https://doi.org/10.4314/stech.v4i1.6

Ayoade J (2012) Introduction to building and urban climatology (1st Edition). Agbo Areo Publishers

Croitoru L, Akpokodje J, Chang J, Kelly A (2020) The cost of air pollution in Lagos. World Bank Group

Deng SZ, Jalaludin BB, Antó JM, Hess JJ, Huang CR (2020) Climate change, air pollution, and allergic respiratory diseases: a call to action for health professionals. Chin Med J 133(13):1552. https://doi.org/10.1097/CM9.0000000000000861

Durodola OS, Durodola OS (2019) The impact of climate change induced extreme events on agriculture and food security: a review on Nigeria. Agric Sci 10(4):487–498. https://doi.org/10.4236/AS.2019.104038

Echendu AJ (2020) The impact of flooding on Nigeria's sustainable development goals (SDGs) 6(1). https://doi.org/10.1080/20964129.2020.1791735

Elum ZA, Mopipi K, Henri-Ukoha A (2016) Oil exploitation and its socioeconomic effects on the Niger Delta region of Nigeria. Environ Sci Pollut Res 23(13):12880–12889. https://doi.org/10.1007/S11356-016-6864-1

Fakinle BS, Odekanle EL, Olalekan AP, Ije HE, Oke DO, Sonibare JA (2020) Air pollutant emissions by anthropogenic combustion processes in Lagos, Nigeria. http://www.Editorialmanager.Com/Cogenteng, 7(1):1808285. https://doi.org/10.1080/23311916.2020.1808285

Fisher S, Bellinger DC, Cropper ML, Kumar P, Binagwaho A, Koudenoukpo JB, Park Y, Taghian G, Landrigan PJ (2021) Air pollution and development in Africa: impacts on health, the economy, and human capital. The Lancet Planetary Health 5(10):e681–e688. https://doi.org/10.1016/S2542-5196(21)00201-1

Giwa SO, Nwaokocha CN, Kuye SI, Adama KO (2019) Gas flaring attendant impacts of criteria and particulate pollutants: a case of Niger Delta region of Nigeria. J King Saud Univ Eng Sci 31(3):209–217. https://doi.org/10.1016/J.JKSUES.2017.04.003

He H, Vinnikov KY, Li C, Krotkov NA, Jongeward AR, Li Z, Stehr JW, Hains JC, Dickerson RR (2016) Response of SO_2 and particulate air pollution to local and regional emission controls: a case study in Maryland. Earth's Future 4(4):94–109. https://doi.org/10.1002/2015EF000330

Heald CL, Kroll JH (2021) A radical shift in air pollution. Science 374(6568):688–689. https://doi.org/10.1126/SCIENCE.ABL5978

Islam N, Warsi Khan H, Gari AA, et al (2022) Screening of ionic liquids as sustainable greener solvents for the capture of greenhouse gases using COSMO-RS approach: computational study. Fuel 330:125540. https://doi.org/10.1016/j.fuel.2022.125540

Izah SC, Angaye TCN, Aigberua AO, Nduka JO (2017) Uncontrolled bush burning in the Niger delta region of Nigeria: potential causes and impacts on biodiversity. Int J Molec Ecol Conser. https://doi.org/10.5376/IJMEC.2017.07.0001

Kim KJ, Shin J, Choi J (2019) Cancer risk from exposure to particulate matter and ozone according to obesity and health-related behaviors: a nationwide population-based cross-sectional study. Cancer epidemiology, biomarkers & prevention: a publication of the american association for cancer research. Cosponsored Amer Soc Prevent Oncol 28(2):357–362. https://doi.org/10.1158/1055-9965.EPI-18-0508

Liang L, Wang Z, Li J (2019) The effect of urbanization on environmental pollution in rapidly developing urban agglomerations. J Clean Prod 237:117649. https://doi.org/10.1016/J.JCLEPRO.2019.117649

Malley CS, Omotosho D, Bappa B, Jibril A, Tarfa P, Roman M, Hicks WK, Kuylenstierna JCI, de la Sota Sandez C, Lefèvre EN (2021) Integration of climate change mitigation and sustainable development planning: lessons from a national planning process in Nigeria. Environ Sci Policy 125:66–75. https://doi.org/10.1016/J.ENVSCI.2021.08.022

Mannucci PM, Franchini M (2017) Health effects of ambient air pollution in developing countries. Int J Environ Res Public Health 14(9):1048. https://doi.org/10.3390/IJERPH14091048

Mukherjee A, Agrawal M (2018) A global perspective of fine particulate matter pollution and its health effects. Rev Environ Contam Toxicol 244:5–51. https://doi.org/10.1007/398-2017-3/COVER/

NBS (2019) Road transport data

Njoku KL, Rumide TJ, Akinola MO, Adesuyi AA, Jolaoso AO (2016) Ambient air quality monitoring in metropolitan city of Lagos, Nigeria. J Appl Sci Environ Manag 20(1):178–185. https://doi.org/10.4314/jasem.v20i1

Nwachukwu A, Chukwuocha E, Igbudu O (2012) A survey on the effects of air pollution on diseases of the people of Rivers State, Nigeria. Afr J Environ Sci Technol 6(10):371–379

Okedere OB, Elehinafe FB, Oyelami S, Ayeni AO (2021a) Drivers of anthropogenic air emissions in Nigeria—a review. Heliyon 7(3):e06398. https://doi.org/10.1016/j.heliyon.2021.e06398

Okon EM, Falana BM, Solaja SO, Yakubu SO, Alabi OO, Okikiola BT, Awe TE, Adesina BT, Tokula BE, Kipchumba AK, Edeme AB (2021) Systematic review of climate change impact research in Nigeria: implication for sustainable development. Heliyon 7(9):e07941. https://doi.org/10.1016/j.heliyon.2021.e07941

Okoye LU, Adeleye BN, Okoro EE, Okoh JI, Ezu GK, Anyanwu FA (2022) Effect of gas flaring, oil rent and fossil fuel on economic performance: the case of Nigeria. Resour Policy 77:102677. https://doi.org/10.1016/J.RESOURPOL.2022.102677

Oladimeji T, Sonibare J, Odunfa M, Ayeni A (2015a) Modeling of criteria air pollutant emissions from selected Nigeria petroleum refineries. J Power Energy Eng 03(06):31–45. https://doi.org/10.4236/jpee.2015.36004

Oladimeji T, Sonibare J, Odunfa M, Ayeni A, Oladimeji T, Sonibare J, Odunfa M, Ayeni A (2015b) Modeling of criteria air pollutant emissions from selected Nigeria petroleum refineries. J Power Energy Eng 3(6):31–45. https://doi.org/10.4236/JPEE.2015.36004

Olalekan R, Timothy A, Enabulele C, Olalekan A (2018) Assessment of air quality indices and its health impacts in ilorin metropolis, kwara state. NIGERIA. Sci Park 4(4):60–74

Olayini OA, FOA (2021) Review of climate change and its effect on Nogeria eco system. Int J Environ Pollut Res 2(3):2013–2015

Olowoporoku AO, Longhurst JWS, Barnes JH (2012) Framing air pollution as a major health risk in Lagos, Nigeria. WIT Trans Ecol Environ 157:479–486. https://doi.org/10.2495/AIR120421

Olujobi OJ (2020) Analysis of the legal framework governing gas flaring in Nigeria's upstream petroleum sector and the need for overhauling. Soc Sci 9(8):132. https://doi.org/10.3390/SOCSCI9080132

Omang DI, John GE, Inah SA, Bisong JO (2021) Public health implication of solid waste generated by households in Bekwarra local government area. Afr Health Sci 21(3):1467. https://doi.org/10.4314/AHS.V21I3.58

Omotehinse AO, Ako BD (2019) The environmental implications of the exploration and exploitation of solid minerals in Nigeria with a special focus on Tin in Jos and Coal in Enugu. J Sustain Mining 18(1):18–24. https://doi.org/10.1016/J.JSM.2018.12.001

Onah NG, Alphonsus NA, Ekenedilichukwu E (2016) Mitigating climate change in Nigeria: African traditional religious values in focus. Mediterr J Soc Sci. https://doi.org/10.5901/MJSS.2016.V7N6P299

Pona HT, Xiaoli D, Ayantobo OO, Tetteh ND (2021) Environmental health situation in Nigeria: current status and future needs. Heliyon 7(3):e06330. https://doi.org/10.1016/J.HELIYON.2021.E06330

Qureshi F, Yusuf M, Kamyab H, et al (2022) Latest eco-friendly avenues on hydrogen production towards a circular bioeconomy: currents challenges, innovative insights, and future perspectives. Renew Sustain Energy Rev 168:112916. https://doi.org/10.1016/j.rser.2022.112916

Reed HE, Mberu BU (2014) Capitalizing on Nigeria's demographic dividend: reaping the benefits and diminishing the burdens. Etude de La Population Africaine = African Population Studies 27(2):319. https://doi.org/10.11564/27-2-477

Rose JJ, Wang L, Xu Q, McTiernan CF, Shiva S, Tejero J, Gladwin MT (2017) Carbon monoxide poisoning: Pathogenesis, management, and future directions of therapy. Am J Respir Crit Care Med 195(5):596–606. https://doi.org/10.1164/RCCM.201606-1275CI/SUPPL_FILE/DISCLOSURES.PDF

Roth D, Hubmann N, Havel C, Herkner H, Schreiber W, Laggner A (2011) Victim of carbon monoxide poisoning identified by carbon monoxide oximetry. J Emerg Med 40(6):640–642. https://doi.org/10.1016/j.jemermed.2009.05.017

Sonibare JA (2010) Air pollution implications of Nigeria's present strategy on improved electricity generation. Energy Policy 38(10):5783–5789. https://doi.org/10.1016/J.ENPOL.2010.05.029

Tawari C, Abowei J (2012) Air pollution in the Niger delta area of Nigeria. Int J Fisheries Aquatic Sci 1:94–117

Ubuoh E, Nwajiobi B (2018) Implications of different household cooking energy on indoor air quality in urban and semi-urban settlements in Imo, South Eastern Nigeria. J Appl Sci Environ Manag 22(5):725–729. https://doi.org/10.4314/jasem.v22i5.23

Urhie E, Afolabi A, Afolabi A, Matthew O, Osabohien R, Ewetan O (2020) Economic growth, air pollution and health outcomes in Nigeria: a moderated mediation model. http://www.Editorialmanager.Com/Cogentsocsci, 6(1). https://doi.org/10.1080/23311886.2020.1719570

Weli V, Adegoke J, Eyo B (2018) The incidence of soot and surface boundary layer meteorology in port Harcourt Metropolis, Nigeria. J OJ Climatol Weather Forecast 6(2)

Williams EA, Olalekan RM, Yarwamara EI, Modupe O (2019) Renewable energy sources for the present and future: an alternative power supply for Nigeria. Energy Earth Sci 2(2):18. https://doi.org/10.22158/EES.V2N2P18

Xing YF, Xu YH, Shi MH, Lian YX (2016) The impact of PM2.5 on the human respiratory system. J Thoracic Disease 8(1):E69. https://doi.org/10.3978/J.ISSN.2072-1439.2016.01.19

Yusuf M, Farooqi AS, Alam MA, et al (2020) Latest trends in Syngas production employing compound catalysts for methane dry reforming. IOP Conf Ser Mater Sci 991:12071. https://doi.org/10.1088/1757-899X/991/1/012071

Yusuf M, Farooqi AS, Keong LK, et al (2021) Contemporary trends in composite Ni-based catalysts for CO2 reforming of methane. Chem Eng Sci 229:116072. https://doi.org/10.1016/j.ces.2020.116072

Yusuf M, Bazli L, Abdullah B (2022) 12 - Challenges and remediation for global warming to achieve sustainable development. In: Dubey AK, Narang SK, Srivastav AL, et al. (eds) Woodhead Publishing Series in Energy. Woodhead Publishing, pp 243–257, https://doi.org/10.1016/B978-0-323-90396-7.00017-1

Chapter 3
Impacts of Climate Change on Sustainable Crop Production in Nigeria

E. E. Osuji, M. O. Olaolu, and A. C. Tim-Ashama

Abstract In recent times, climate change has exacerbated agricultural production in Nigeria causing serious economic downturns and low food production of households. The study evaluated the impacts of climate change on sustainable crop production in Nigeria. Multi-stage random sampling was used to select 201 crop farmers who cultivated different crop types ranging from cassava, rice, yam, maize, etc., using structured questionnaire. Data obtained were analyzed using descriptive statistics such as the mean, frequency, and percentage and inferential statistics such as ordinary least square multiple regression technique. Results showed that 73.6% of the crop farmers were married, 50.7% females, and over 42.0% were relatively educated and experienced. A mean generated income of ₦172, 000 was estimated from the total sales of varieties of crops produced by the farmers in the state. Crops like rice, 7850.23 kg, cassava, 6492.41 kg, and maize, 6903.19 kg dominated the crop cultivation with higher mean outputs. Age, gender, household size, education, farming experience, farm size, and extension contacts were positive and significantly influenced crop production. Climate variables such as temperature, rainfall, number of rainy days were negative while sunshine hours and relative humidity were positive and both significantly influenced crop production in the state. Inadequate capital, land fragmentations, and inadequate information concerning climate change were perceived as major production and climatic constraints. Climate change should be technically mitigated to avert future negative consequences. Intensive awareness on climate change and provision of climate information and/or early warning signals to the knowledge of the crop farmers were recommended.

Keywords Climate change · Sustainable output · Crop production · Farm income · Nigeria

E. E. Osuji (✉) · M. O. Olaolu
Faculty of Agriculture, Department of Agriculture, Alex Ekwueme Federal University Ndufu-Alike Ebonyi State, Abakaliki, Nigeria
e-mail: osuji.emmanuel@funai.edu.ng

A. C. Tim-Ashama
Department of Agricultural Science, Alvan Ikoku Federal College of Education Owerri, Owerri, Imo State, Nigeria

J. C. Egbueri et al. (eds.), *Climate Change Impacts on Nigeria*, Springer Climate,
https://doi.org/10.1007/978-3-031-21007-5_3

Introduction

Agriculture is important in the Nigerian economy, accounting for over 70% of jobs and 26% of gross domestic product, GDP (FAO 2020). Since the 2016 recession, the government has prioritized the sector, evolving diversification of the predominantly oil-based economy. Crop production in Nigeria is a dispersed activity, with more than 70% of rural smallholder farmers engaged in it (FAO 2021). Crop production refers to the process of growing crops for both domestic and commercial use. Nigeria's agricultural land area totals 70.8 million hectares (FAO 2021). This is divided into 34 million hectares of arable land, 6.5 million hectares for permanent crops, and 30.3 million hectares for pastures and meadows (FAO 2020). The principal crops grown by farming households in Nigeria are maize, rice, jute, yam, cassava, guinea corn, wheat, yam, etc. The agricultural industry is divided into four sub-sectors: livestock, fishery, fishing, and crop production (USDA 2021). Crop production is by far the most important sub-sector, accounting for about 90% of agriculture's nominal GDP, with livestock, forestry, and fishery accounting for the remaining 10% (Rahman and Chima 2018). Crop production accounted for 91.6% of the sector in the third quarter of 2019, with a 44.12% quarterly growth rate (FAO 2021). Low productivity in the industry is ascribed to a lack of use of improved seeds, poor input quality, limited or no access to loans, and an aging farming population. The agricultural industry is a major contributor to the greenhouse gas emissions and land use changes that are assumed to be causing climate change (FAO 2020). Agriculture, in addition to being a big land user and consumer of fossil fuels, contributes directly to greenhouse gas emissions through practices such as rice production and animal keeping (Enete and Onyenekwe 2021). Fossil fuels, land use, and agriculture have been the three main contributors of the observed increase in greenhouse gases during the last 250 years (Olayemi et al. 2021). Agriculture contributes to global warming by anthropogenic greenhouse gas emissions and the conversion of non-agricultural land, such as forests, to agricultural land (Olayemi et al. 2021). Agriculture, forestry, and land use change were predicted to account for 20–25% of worldwide annual emissions in 2010 (FAO 2019). The European Union's Scientific Advice Mechanism predicted in 2020 that crop production contributed 37% of overall greenhouse gas emissions, with this figure expected to rise by 30–40% by 2050 due to population expansion and dietary changes (Osuafor et al. 2021).

Climate change and agriculture are interconnected and occurs on a worldwide scale, with climate change's negative consequences affecting crop production both directly and indirectly (Anthony et al. 2021). Temperature, rainfall, frost, wind, relative humidity, etc., are all climatic factors that affect crop production. Climate change is already having an impact on agriculture, with the effects being unevenly spread around the world. Future climate change will almost certainly have a detrimental impact on crop production in low-latitude countries, while effects in higher latitudes may be favorable or negative (Mahmuda and Md-Mizanur 2021). Due to the sensitivity and vulnerability of temperature and rainfall changes, climate change has become one of the most pressing issues confronting Nigerian food crop production

(Lisettc et al. 2021). Changes in precipitation patterns, for example, increase the chance of short-term crop failure and long-term output losses, making climate variability a major concern for food and crop production (Oriangi et al. 2020). Climate change is a serious issue all across the world, having a significant impact on food and crop productivity (Saalu et al. 2020). Crop production, particularly rain-fed cultivation, is dependent on favorable climate conditions, which change all the time (Sophie 2019). Above all, climate change is known to have variant impacts on crop outputs, which could be positive or negative (Mendelsohn and Massetti 2017). Droughts, intense windstorms, evaporation, floods, unpredictable rainfall, rising temperatures, late and early showers, and other climatic changes are becoming more common in Nigeria and elsewhere (Mulenga et al. 2017). Temperature and rainfall patterns are changing as the world heats, and extreme events such as crop failures, droughts, famine, floods, forest fires, and low agricultural yields are becoming more common and severe, affecting food and crop production all across the world (Chia et al. 2016).

However, various studies had investigated the effect of climate change on agricultural production in Nigeria; Nzeh et al. (2016) looked at climate change and agricultural production in Nigeria; a review of status, causes, and consequences. Adams (2019) examined the impact of climate change on agricultural production in Nigeria. Apata (2011) X-rayed effects of global climate change on Nigerian agriculture. Dominic et al. (2017) analyzed of the effects of climate change on crop output in Nigeria, while Eregha et al. (2014) evaluated climate change and crop production in Nigeria. Despite the several studies carried out on climate change regarding crop production, none has X-rayed the literal impacts of climate change on sustainable crop production in Nigeria using several varieties of crop types and more climate variables; thus this study examined up to five climate variables influencing crop production unlike other previous studies that utilized only temperature and rainfall, hence the gap in knowledge. Furthermore, this study is an improvement in knowledge in terms of the year in study being 2020, while other studies were conducted in previous years. The specific objectives of this study were to identify the socio-economic characteristics of the crop farmers, estimate the revenue generated from the total sales of crops produced, analyze the different output of crops produced, analyze the socio-economic factors influencing crop production, isolate the impacts of climate change on sustainable crop production in Nigeria and identify the perceived constraints influencing sustainable crop production.

Materials and Methods

Study Area

The study was carried out in Ebonyi State, which is one of the states located in Southeast, Nigeria. The state was chosen due to its agrarian nature and intensive crop cultivation. Ebonyi State is made up of thirteen Local Government Areas—Abakaliki, Afikpo-North, Afikpo-South, Ebonyi-Central, Ezza-North, Ezza-South,

Ikwo, Ishielu, Ivo, Izzi, Ohaukwu, Ohaozara, and Onicha (EBSG 2021). According to the National Population Census Commission Abuja, the 2006 population estimate of the state is 2,176,947 persons (NPC 2007). The state is bordered to the north and northeast by Benue State, Enugu State to the west, Cross River State to the east and southeast, and Abia State to the southwest. The total land area is about 5,533 km^2 with *Latitude*: 6° 10′ 40.7028″ and *Longitude*: 7° 57′ 33.4296″ (EBSG 2021). The state has a tropical climate with two distinct seasons, namely the raining and dry seasons. The raining season runs from April to October, while the dry season runs from November to March. The temperature of the state is 28 °C with about 17% precipitation (EBSG 2021). The vegetation of the area is predominantly rainforest, which supports the cultivation of food and cash crops such as rice, maize, yam, cassava, oil palm, cocoa, cowpea, sweet potato, cocoyam, plantain, banana, melon, bambara nut, breadfruit, groundnut, and various vegetables and fruit trees. The topography of the state is made up of a table-land with some hilly terrain. The people are mostly crop farmers and engage in rain-fed farming, due to rainfall variations and unpredictability. Other sources of livelihood include trading activities as well as other occupations such as civil service, corporate businesses, etc. The state is plagued with food insecurity due to adverse consequences of climate change and over dependence on rain-fed agriculture practices which limits household income sources (Fig. 3.1).

Sample Selection

Multi-stage sampling technique was employed for sample selection. In the first stage, two local government areas (LGAs) were randomly drawn from each of the structural zones of the state namely, (Ebonyi-North, Ebonyi-South, and Ebonyi-Central) making six LGAs. In the second stage, two communities were randomly pulled from the above LGAs making a total of 12 communities. In the third stage, 4 villages were randomly picked from the above selected communities giving 48 villages. The fourth stage had five crop farmers randomly selected giving a total of 240 crop farmers. The list of the registered crop farmers which formed the sample frame were obtained from the state agricultural development agency. Drafted questionnaire on the specific objectives of the study were administered to the sample respondents using oral interviews and only 201 questionnaire were found useful for data analysis. However, before actual data collection, a pilot survey was conducted across the selected LGAs to standardize the questionnaire and ascertain its content validity and reliability.

Data Analysis

Data collected were analyzed after properly sorted out, coded, and entered into excel sheet. Descriptive statistics and ordinary least squares multiple regression technique were employed in analyzing the data. Descriptive statistics which comprises of

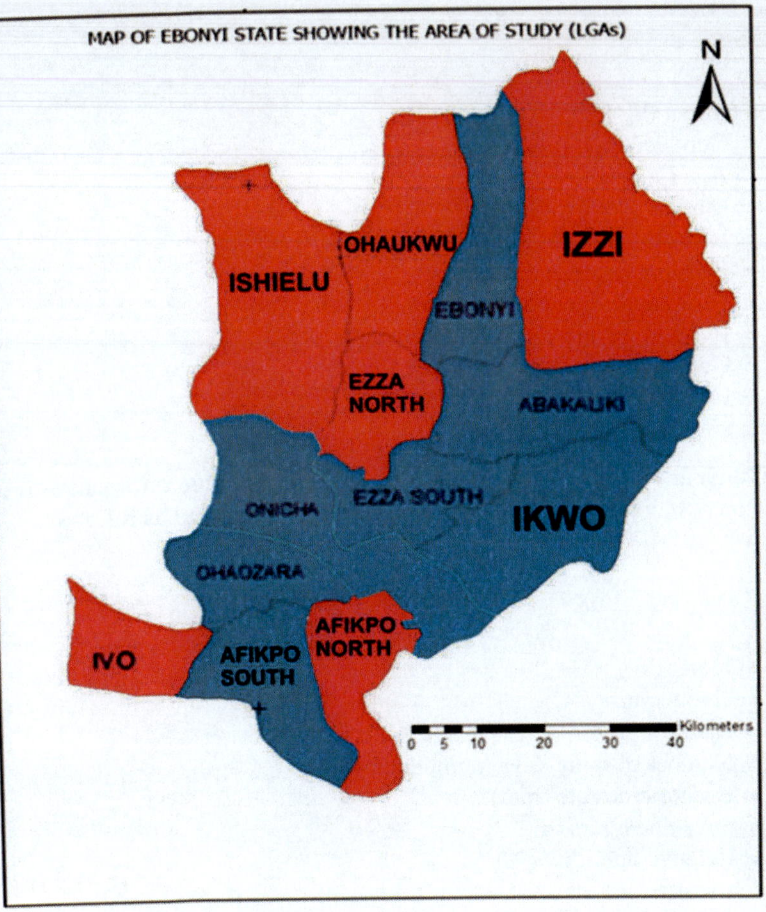

Fig. 3.1 Map of Ebonyi State, Nigeria. *Source* EBSG (2021)

the mean, frequency distribution, and percentages was used to analyze the socio-economic characteristics of the crop farmers, the revenue generated from the total sales of crops produced, the different output of crops produced, and the perceived constraints influencing sustainable crop production, while the ordinary least squares multiple regression technique was used to analyze the socio-economic factors influencing crop production and the impacts of climate change on sustainable crop production. These statistical tools were fully employed in data analysis due to its statistical relevance, appropriateness, and econometric fitness in handling such objectives in time past (John et al. 1998). The use of the statistical tools was also validated on econometric laws and principles as well as empirical theories (Samprit and Ali 2006). Ordinary least squares multiple regression technique on socio-economic factors influencing crop production was fully expressed as follows:

$$Y = f(X_1, X_2, X_3, X_4, X_5, X_6, X_7, X_8, X_9) + e \tag{3.1}$$

where
- Y = Output (kg)
- X_1 = Gender (male 1, female 0)
- X_2 = Age (years)
- X_3 = Marital status (married 1, single 0)
- X_4 = Education (years)
- X_5 = Farming experience (years)
- X_6 = Household size (No. of persons)
- X_7 = Extension contacts (No. of contacts)
- X_8 = Cooperative membership (yes 1, no 0)
- X_9 = Farm size (ha)
- e = error term.

Again, ordinary least squares multiple regression technique on impacts of climate change on sustainable crop production was further expressed as follows:

$$Y = f(X_1, X_2, X_3, X_4, X_5, X_6) + e \tag{3.2}$$

where
- Y = Output (kg)
- X_1 = Temperature (°C)
- X_2 = Rainfall (mm)
- X_3 = Number of rainy days (number of times)
- X_4 = Evaporation rate (mm)
- X_5 = Sunshine hours (h)
- X_6 = Relative humidity (%)
- e = error term.

Results and Discussion

Socio-Economic Characteristics of Crop Farmers

Table 3.1 shows the socio-economic characteristics of the crop farmers. Majority of the farmers, 50.7%, were females. This implies that the female farmers dominated crop production in the state more than their male counterparts. This seems to support the assertion that female crop farmers have dominated crop production in Nigeria in recent times (Adams 2019). The majority of crop farmers, 47.3%, were young and within their productive ages. This implies that the crop farmers were strong and energetic enough to carry out their crop production. Young farmers are known for greater farm productivity and outputs which transit to higher farm income (Eregha et al. 2014). Marital status showed that, married farmers, 73.6%, dominated the state crop production. This implies that the crop farmers were matured, responsible, and committed to their occupation. Marriage is reputed for the increase in labor availability and supply (family labor) which is being utilized for increased outputs (Dominic et al. 2017). The majority of crop farmers, 45.3%, attended secondary schools and this shows that they were relatively educated to enhance their crop production. Education is a mirror that reflects the potentials of the crop farmers in general (Apata 2011). It helps the farmers to gain requisite knowledge and understanding of farm production principles which ensures higher crop yields (Nzeh et al. 2016). Again, the mean farming experience was 16 years; this implies that the crop farmers were relatively experienced to face inherent crop production constraints. Farm experience helps the farmers to integrate farm production principles into practice and equips the farmers for greater crop production via efficient and effective resource and farm inputs utilizations (Godwin and Sydney 2021). The crop farmers had a mean household size of 7 persons. This is relatively large and could be utilized in crop cultivation. Large household sizes support farm labor availability and save time and financial resources that would have been expended to hire extra farm laborers (FAO 2020). Extension contacts had majority of the crop farmers, i.e., 56.2%. This implies that the crop farmers had contacts with extension agents and utilized the practical knowledge gained from such visits. Extension contacts are mediums of crop production revolutions both in the short- and long-run bases. They assist crop farmers with on-hand farm practical exposures and experiences which trigger increase in crop outputs and farm income (Richard and Fred 2020). Majority of the crop farmers, 61.7%, belongs to cooperative societies and stands the chance to receive first-hand information on agricultural innovation, inventions, and new practices. Also, it increases their chance to benefit and access agricultural farm inputs such as improved planting materials, seedlings, fertilizers, and other agro-chemicals and this has a way of enhancing crop production (Chisom and Katsushi 2021). The mean farm size was 1.5 hectares; this implies that the farmers cultivated less than 2 hectares of farmlands due to land scarcity and fragmentations. Land scarcity and fragmentation have been major crop production constraints that inhibit higher farm production; this limits extensive crop cultivation and promotes meager farm income of farming households (Martey et al. 2019).

Revenue Generated from Sales of Crops Produced

Table 3.2 shows the revenue raised from the sales of crops produced in the state. It indicated that the crop farmers generated different relative income from the marketing of their farm produce. About 2% of the crop farmers made below or ₦50, 000 from the sales of crops produced. This indicates the least amount of money realized from the marketing of produced crops in the state. 27.4% made between ₦51, 000–₦100, 000 and about 26% made between ₦101, 000–₦ 150, 000 indicate increase in amount generated from the total sales of produced crops in the state. The majority of the farmers, 35.8%, had income between ₦151, 000–₦200, 000. This implies that these

Table 3.1 Socio-economic characteristics of crop farmers

Socio-economic factors	Frequency	Percent
Gender		
Male	99	49.3
Female	102	50.7
Age		
21–30	10	5.0
31–40	77	38.3
41–50	95	47.3
51–60	19	9.5
Mean age	46	
Marital Status		
Single	21	10.4
Married	148	73.6
Separated/Divorced	10	5.0
Widow/Widower	22	10.9
Education		
Primary	67	33.3
Secondary	91	45.3
University	12	6.0
None	31	15.4
Farming Experience		
1–10	42	20.9
11–20	87	43.3
21–30	69	34.3
31–40	3	1.5
41–50	–	–
Mean farming experience	16	
Household size		
1–4	90	44.8
5–9	111	55.2
10–14	–	–
Mean household size	7	
Extension contacts		
Yes	113	56.2
No	88	43.9
Cooperative membership		
Yes	124	61.7
No	77	38.3

(continued)

Table 3.1 (continued)

Socio-economic factors	Frequency	Percent
Farm size		
0.1–1.0	66	32.8
1.1–2.0	125	62.2
2.1–3.0	10	5.0
Mean farm size	1.5	

Source Field survey data, 2020

Table 3.2 Revenue generated from sales of crops produced

Income	Frequency	Percent
≥₦ 50,000	4	2.0
₦ 51,000–₦ 100,000	55	27.4
₦ 101,000–₦ 150,000	52	25.9
₦ 151,000–₦ 200,000	72	35.8
₦ 201,000–₦ 250,000	10	5.0
₦ 251,000–₦ 300,000	8	4.0
Mean income	₦ 172,000	

Source Field survey data, 2020

groups of farmers realized a relatively meaningful income from their produce sales. Above 4% of the crop farmers had income between the range of ₦ 201, 000–₦ 250, 000 and ₦251, 000–₦300, 000, respectively. These indicate the highest profitable income raised from the total sales of the crops produced in the state. The mean income generated was ₦172, 000; this implies that the majority of the crop farmers made higher farm income above ₦170, 000 per annum. This further indicates the thriving nature of business sales and marketing activities and operations in the state (Kadurumba et al. 2019; Ejechi 2021; Esheya 2021).

Descriptive Statistics of Different Output of Crops Produced

The descriptive statistics of different output of crops produced is shown in Table 3.3. The table shows that cassava production dominated the root-tubers with a mean output of 6492.41 kg among yam, 4110.91 kg, cocoyam, 4079.10 kg, and sweet potato, 3341.95 kg. This indicates the true relevance and importance of cassava production which spans across the whole local government areas of the state (Ayinde et al. 2010). Cassava is known to be widely cultivated due to its food security and sustainability nature which distinguishes cassava among other tuber crops (Bosede et al. 2021). Across the staple crops, rice production dominated other staple crops such as maize, with a mean output of 7850.23 kg. No doubt, rice cultivation is the chief grown crop

Table 3.3 Descriptive statistics of different output of crops produced

Crops	No. of farmers		Unit	Mean output	Standard deviation	Minimum output	Maximum output
	F	%					
Fluted pumpkin	199	99.0	Kg	3094.20	80.3718	250.00	4000.00
Cassava	201	100	Kg	6492.41	57.3621	250.00	7900.00
Cucumber	111	55.2	Kg	3521.21	43.0630	150.00	3900.00
Tomato	99	49.3	Kg	2998.03	68.3011	300.00	3950.00
Cabbage	88	43.9	Kg	2202.41	55.1944	100.00	3800.00
Watermelon	106	52.7	Kg	3342.27	69.2374	250.00	4050.00
Yam	178	88.6	Kg	4110.91	49.1957	300.00	5400.00
Coco-yam	114	56.7	Kg	4079.10	63.2967	250.00	5500.00
Okra	119	59.2	Kg	3934.91	88.3724	200.00	4700.00
Garden egg	109	54.2	Kg	3492.16	66.5404	150.00	4070.00
Green pepper	107	53.2	Kg	3219.82	59.1170	100.00	4369.00
Cowpea	131	65.2	Kg	3930.44	24.3334	200.00	4400.00
Maize	201	100	Kg	6903.19	76.1387	300.00	7659.00
Plantain	121	60.2	Kg	3026.17	47.6620	100.00	4500.00
Banana	171	85.1	Kg	2322.07	69.1023	100.00	3900.00
Sweet Potato	160	79.6	Kg	3341.95	78.3402	150.00	4080.00
Rice	201	100	Kg	7850.23	61.3363	300.00	8090.00
Amaranthus	172	85.6	Kg	2336.29	46.2492	210.00	3082.00
Soya bean	128	63.7	Kg	3804.90	82.0331	200.00	4150.00
Groundnut	156	77.6	Kg	3090.14	46.2002	200.00	4500.97
Melon	188	93.5	Kg	5096.98	51.9394	250.00	6983.09
Pineapple	196	97.5	Kg	3048.09	63.9300	100.00	4800.00

Source Field survey data, 2020

in the state by virtue of the state being ranked top as rice producing state in the country (Onyeneke et al. 2021). In essence, almost every farming household grow and market rice in the state (Kadurumba et al. 2019). Among the vegetable/fruit production, okra dominated the group with a mean output of 3934.91 kg, followed by cucumber, 3521.21 kg, garden egg, 3492.16 kg, watermelon, 3342.27 kg, green paper, 3219.82 kg, fluted pumpkin, 3044.20 kg, tomato, 2998.03 kg, etc. The dominance of these vegetables could be as a result of their higher demands and nutritional compositions which trigger their cultivation in the state (Ejechi 2021). Across the legumes, melon dominated the production with a mean output of 5096.98 kg against cowpea, 3930.44 kg, soya bean, 3804.90 kg, and groundnut, 3090.14 kg. This indicates the superiority of melon crops in the state. The legumes in general enrich soil fertility and aid soil–water conservation at all times (Obidiegwu and Akpabio 2017). Other notable produced crops were plantain and banana with mean outputs of 3026.17 kg and 2322.07 kg. These two crops are seasonal grown crops which provide shade to newly planted crops in the nursery plantation (Iheke et al. 2021). However, the widely cultivated major crops in the state were rice, cassava, and maize, as depicted by their mean outputs. This further indicates high concentration of the crop farmers on these crops more than other crops in the state.

Socio-Economic Factors Influencing Crop Production

Table 3.4 shows the multiple regression results on influence of socio-economic factors on crop production in the state. Double-log functional form provided the best fit and was chosen as the lead equation because it had the highest value of the coefficient of multiple determinations (R^2), highest number of significant variables, and F-value, respectively. The R^2 of 0.870 indicated that 87% of the total variations in endogenous variable were explained by the socio-economic factors investigated. The coefficient of age was positive, and significant, implying that increase in age of the crop farmers increases crop outputs. Age is noted to be a strong determinant of crop production as productive age engenders high outputs and productivity (Ibukun et al. 2021). Gender was negative, and significant, implying that female farmers dominated the crop cultivation with more outputs. Recently, empirical studies have revealed the predominance of the female folks in crop production arising from gender equity and equality (Oluwatoba et al. 2021). This result further corresponds the result in Table 3.1. Household size was positive, and significant, implying that increase in household size of the crop farmers increases outputs. Large household size determines to a great extent, the family labor availability and supply which is a prerequisite for increased production (Ihemezie and Dallimer 2021). The coefficient of farm size was positive, and significant, implying that increase in farm size of the crop farmers will result in a corresponding increase in crop outputs. Farm sizes are very important in crop cultivation as it determines the extent of crop cultivation and adoption of new and improved agricultural practices (Paudel et al. 2019). Education was positive, and significant, implying that a 1% increase in education of crop farmers will result in 1% increase in crop outputs. Education enhances the knowledge and understanding of the crop farmers in intensifying their farming activities and incorporation of new cultivation practices and methods (Ibukun et al. 2021). Farming experience was positive, and significant, implying that any increase in farming experience of the crop farmers will result in a corresponding increase in crop outputs. Farming experience enhances both technical and economic efficiency of crop production resulting in increased outputs and income (Sara and Scandizzo 2017). Extension contacts were positive, and significant, implying that any increase in extension contacts of the crop farmers will increase crop production by 415%. Extension contacts is known to induce practical on-hand knowledge and new and/ or improved farming innovations, it is an agricultural transformation medium that enhances crop farmers' skills, knowledge, and expertise (Richard and Fred 2020).

Table 3.4 Socio-economic factors influencing crop production

Variable	Linear	Semi-log	Double-log	Exponential
Constant	101.162 (1.004)	2.1302 (3.101)***	31.0202 (2.402)**	4.1022 (1.320)
Age (X_1)	3.0350 (2.024)**	3.3221 (1.400)	5313.1 (2.331)**	3.3201 (0.149)
Gender (X_2)	−421.133 (−2.103)**	−5.1029 (−0.141)	−602.11 (−3.900)***	−3.8775 (−1.577)*
Household size (X_3)	443.012 (1.411)	3.2006 (1.111)	4165.0 (2.039)***	3.2351 (1.407)
Farm size (X_4)	5001.10 (1.039)	3.2045 (3.019)***	4204.32 (3.603)***	3.1025 (1.124)
Education (X_5)	440.156 (2.020)**	3.0320 (1.023)	3049.9 (3.701)***	4.4019 (3.014)***
Experience (X_6)	421.202 (1.332)	8.1515 (3.004)***	5342.1 (3.160)***	6.0502 (1.430)
Marital status (X_7)	702.06 (0.739)	4.4991 (1.015)	470.212 (1.332)	3.3405 (1.045)
Membership in cooperative (X_8)	6003.1 (1.401)	5.1146 (1.201)	5.0152 (1.013)	3.7118 (3.049)***
Extension contact (X_9)	8.7195 (1.721)*	4.9333 (1.473)	4.1542 (1.910)*	5.2401 (1.231)
R^2	0.789	0.708	0.870	0.601
F- ratio	14.79***	27.58***	44.18***	13.49***

***, **, * significant at 1%, 5%, and 10% Source: Field survey data, 2020

Impacts of Climate Change on Sustainable Crop Production

The Impacts of climate change on sustainable crop production was shown in Table 3.5 and was determined by the double-log functional form taking into cognizance its number of significant variables, highest F-value, and R^2. The R^2 value of 0.8401 indicated that 84% of the total variations in the endogenous variable were fully explained by the climate variables observed. Temperature was negative, and significant, implying that increase in temperature decreases sustainable crop production. Increased temperature builds up soil pathogen which breeds crop insect pests and diseases thus affecting crop outputs, income, and sustainable yields. Crop seedling reproductive and developmental stages are hampered by high temperatures, which reduce plant height and root extension leading to poor growth and yields (Oti et al. 2021). Rainfall was negative, and significant, implying that increase in rainfall decreases crop outputs. Increased rainfall serves as potential dangers to crop cultivation via soil erosion, water logs, and percolation, causing stunted and devastating root and shoot growth effects. Increased rainfall impairs planted field crops, reducing their yields and market values, and thus impeding general sustainability of crop production (Asrat and Simane 2018). Flooding can also occur as a result of increased rainfall intensity, thus reducing crop outputs dramatically and exacerbating the local food security situation in the state (Elia 2017). Number of rainy days was negative and significant, implying that increase in number of rainy days reduces yields and sustainable crop outputs via heavy flooding of farmlands and wearing away of top vegetative soils (Oti 2017). It equally decreases the nutrient contents and budget composition of the soil making it infertile and less effective (Ifeanyi-Obi and Asiabaka 2014). Increase in number of rainy days also favors the development of many plant and crop diseases, resulting in the increase in number of diverse disease issues in crop production following (Fierros-Gonzale and Lopez-Feldman 2021). Sunshine hours were positive, and significant, implying that a percent increase in sunshine hours increases crop outputs. Sunshine is required for crop growth and development and encourages photosynthetic activities of crop plants leading to sustainable yields and outputs (Igberi et al. 2022). It provides the energy plants' need to convert carbon dioxide and water into carbohydrates and oxygen. The carbohydrates produced by photosynthesis are used for crop vegetative and reproductive growth and to increase crop biomass (Oti 2017). The relevance of sunshine in crop production is very imperative especially during photosynthesis, the process used by plants and other autotrophic organisms to convert light energy, normally from the Sun, into chemical energy that can be used to synthesize carbohydrates and to fuel soil organisms' microbial activities (Onyeneke et al. 2021). Relative humidity was positive, and significant, implying that a 1% increase in relative humidity will result in a 1% increase in crop outputs. Relative humidity increases the moisture content of farmlands especially during dry seasons resulting in increased crop production and sustainable yield. Relative humidity promotes water relations of plants (transpiration), encourages leaf growth, and enhances photosynthesis, crop pollination, and economic yields. It favors seed growth and germination and enhances soil moisture supply at all times (Makuvaro et al. 2018).

Perceived Constraints Influencing Sustainable Crop Production

Perceived constraints influencing sustainable crop production was shown in Table 3.6. The table showed that all the crop farmers, 100% complained of inadequate capital, land fragmentation, and inadequate information concerning climate change. This implies that the crop farmers perceived these constraints as their major disturbing and challenging production constraints in the state. Capital has over the years being a source of concern and worry to crop farmers, as it controls and regulates other factors of production (Jianxu et al. 2020). The fragmentation of land makes it difficult for both intensive and extensive crop cultivation of farming households. Inadequate information concerning climate change causes serious farming disruptions and inadequacies, as farmers often fall victims of such unheard information thereby causing distortion to farming operations (Nzeh et al. 2016). Again, above 99.6% of the crop

Table 3.5 Impacts of climate change on sustainable crop production

Variable	Linear	Semi-log	Double-log	Exponential
Constant	6703.71 (1.404)	53.0215 (3.018)***	23.9434 (1.070)	12.9002 (1.922)**
Temperature (X_1)	−3.4030 (−4.031)***	−4.4019 (−3.041)***	−5101.02 (−2.571)**	−4.0514 (−1.092)
Rainfall (X_2)	−831.052 (−2.112)**	−4.0094 (−1.181)	−301.418 (−3.850)***	−50.687 (−1.350)
Number of rainy days (X_3)	−15.6120 (−1.313)	−4.1106 (−1.211)	−3103.05 (−4.901)***	−4.0113 (−0.304)
Evaporation rate (X_4)	−4012.92 (−1.352)	−4.51152 (−1.915)**	−6134.77 (−1.409)	−4.30459 (−3.325)***
Sunshine hours (X_5)	362.252 (1.350)	5.91213 (1.098)	33,957.49 (4.870)***	6.4018 (4.902)***
Relative humidity (X_6)	705.031 (1.013)	5.6015 (1.006)	6115.81 (3.901)***	6.9092 (1.009)
R^2	0.6031	0.7001	0.8401	0.6821
F- ratio	5.109***	8.901***	26.04***	11.22***

***, **, * significant at 1%, 5%, and 10% Source: Field survey data, 2020

farmers indicated problem of storage/processing facility, pests and disease attacks, and high cost of input materials as their major production constraints. Problem of storage/processing facilities impairs food storage and efficient processing in the long run leading to food shortage, availability, and supply (Oluwalana et al. 2019). Pests and disease attacks further worsen crop production resulting in food insecurity of households. High cost of input materials deters extensive crop cultivation as majority of the rural farmers lacks the financial capacity to acquire such costing inputs, thus this leads to low crop production (Oti et al. 2021). Other notable perceived production constraints were high rate of rural migration, 93.5%, inadequate farming lands, 70.6%, poor extension access and services, 57.2%, and high cost and low availability of labor supply, 52.7%. High rate of rural migration diminishes local content labor or farm work force that migrates to urban centers in search of greener pasture leaving farm work in the hands of old age laborers whose strength is inadequate or sufficient (Adekunle et al. 2020). Inadequate farming lands pose a serious production challenge as it reduces the effectiveness and efficiency of crop cultivation. Poor extension access and services makes it difficult for crop farmers to access new and improved cultivation methods and practices (Richard and Fred 2020). High cost and low availability of labor supply hinders farm production and productivity to a great extent; it decreases the efficiency of the crop farmers as they fail to engage the few available labor force due to high fares or charges (Sara and Scandizzo 2017). Put together, all these factors influenced sustainable crop production in the state.

Table 3.6 Perceived constraints influencing sustainable crop production

Perceived Constraints	Frequency	Percent
High rate of rural migration	188	93.5
High cost and low availability of labor supply	106	52.7
Inadequate farming lands	142	70.6
Poor extension access and services	115	57.2
Land fragmentation	201	100
Problem of storage/processing facility	200	99.5
Inadequate capital	201	100
Inadequate information concerning climate change	201	100
Pests and disease attacks	199	99.0
High cost of input materials	198	98.5

Source Field survey data, 2020

Conclusions

Climate change has become a common phenomenon altering planting and cropping seasons beyond the wish of the crop farmers. The findings of the study revealed that temperature, rainfall, number of rainy days, sunshine hours, and relative humidity were the climatic variables investigated in this study as they significantly influenced crop production both positively and negatively. The positive side of climate change enhanced crop production and increased crop farmer's outputs while the negative significances impaired crop outputs and yields. Again, age, household size, education, farming experience, farm size, and extension contacts were all positive and significantly increased crop production in the state. Gender appeared negative denoting the relevance of female farmers in crop production arising from gender sensitivity, equity, and equality. Results also showed that the crop farmers were young, married, educated, and experienced to carry out their crop production. An average income of ₦172, 000 was generated by the majority of the crop farmers from the sales and marketing of crops produced in the state. Rice, cassava, and maize dominated the entire crop production in the state as depicted by their higher mean outputs. Inadequate capital, land fragmentation, and inadequate information concerning climate change were perceived as major production constraints confronting the household crop farmers in the state.

Recommendations

The following recommendations were proposed based on the findings of the study.

i. Intensive and elaborate awareness on climate change and provision of climate information and/or early warning signals to the knowledge of the crop farmers to avert future possible negative impacts.
ii. Crop farmers should be encouraged by officials of metrological agencies to maximize the positive and/or favorable attributes of climate change to enhance sustainable crop production.
iii. Soft loans/grants should be granted to crop farmers by micro-finance banks with little or no interest rate to encourage the farmers and to enhance sustainable crop production in the state.
iv. Governments should formulate new agricultural policies or amend existing ones to accommodate crop farmers' plights and ensure sufficient food production in the state.

Acknowledgements The authors acknowledged contributions of Prof. N.C. Ehirim and Ministry of Agriculture and Natural Resources, Ebonyi State, Nigeria.

References

Adams OK (2019) Impact of climate change on agricultural production in Nigeria. Int J Sci Eng Res 10(3):257–263

Adekunle OA, Adefalu LL, Oladipo FO, Adisa RS, Fatoye AD (2020) Constraints to youth's involvement in agricultural production in Kwara State, Nigeria. J Agric Extension 13(1):102–108

Anthony L, Alabi OO, Ebukiba ES, Gamba V (2021) Factors influencing output of rice produced and choice of marketing outlets among smallholder farming households, Abuja. Nigeria. Sarhad J Agric 37(1):262–277

Apata TG (2011) Effects of global climate change on Nigerian agriculture: An empirical analysis. CBN J Appl Stat 2(1):31–46

Asrat P, Simane B (2018) Farmers' perception of climate change and adaptation strategies in Dabus watershed. North-West Ethiopia. Ecol Process 7(7):1–13

Ayinde OE, Ajewole OO, Ogunlade I, Adewumi MO (2010) Empirical analysis of agricultural production and climate change: a case study of Nigeria. J Sustain Dev Africa 12(6):275–283

Bosede NA, Praise DA, Ademola O, Romanus O (2021) Agro-productivity amidst environmental degradation and energy usage in Nigeria. Sci Rep 11:18940

Chia E, Tiani AM, Sonwa D, Perez-Teran A, Tchatchou B (2016) Securing well-being with the advent of climate hazards: case of forest-dependent communities in a landscape in the Congo basin. Int J Climate Change Strat Manag 8(2):175–193

Chisom U, Katsushi SI (2021) Are farmers "efficient but poor"? the impact of crop choices on agricultural productivity and poverty in Nigeria, Economics Discussion Paper Series EDP-2109. School of Social Sciences. The University of Manchester, Manchester M13 9PL

Dominic ZA, Sunday O, Adewara JI, Adama KT, Adzer GOA (2017) Analysis of the effects of climate change on crop output in Nigeria. Amer J Climate Change 6:554–571

EBSG, (2021). Ebonyi state government. Ebonyi state profile. Vegetation, natural and human resources. http://www.ebonyistate.gov.ng/profile.aspx#collathpse-4. Retrieved 24 August, 2021

Ejechi ME (2021) Analyses of production cost and marketing outlets for sweet potato farmers in Ebonyi State Nigeria. Nigerian Agric J 52(1):157–164

Elia E (2017) Farmers' awareness and understanding of climate change and variability in Central arid Tanzania. Univ Dares Salaam Library J 12(2):124–138

Enete AA, Onyenekwe CS (2021) Policy adjustments for enhanced agricultural production in Nigeria after covid-19 pandemic. Agro-Sci J Tropical Agric Food Environ Extension 20(4):28–35
Eregha PB, Babatolu JS, Akinnubi RT (2014) Climate change and crop production in Nigeria: an error correction modelling approach. Int J Energy Econ Policy 4(2):297–311
Esheya SE (2021) Economic analyses of gari processing in Ebonyi State, Nigeria. Nigeria Agric J 52(1):237–241
FAO (2019) Climate change and crop production. Food and Agriculture Organization of United Nations, Rome
FAO (2020) Statistical reports on crop production in Nigeria. Food and Agricultural Organization of the United Nations, Rome
FAO (2021) Food crop production in Nigeria. Food and Agricultural Organization of the United Nations, Rome
Fierros-Gonzalez I, Lopez-Feldman A (2021) Farmers' perception of climate change: A review of the literature for Latin America. Front Environ Sci 9:1–7
Godwin O, Sydney C (2021) Estimating the effect of terrorism on agricultural production in Nigeria. Afr Dev Rev 33(4):703–714. https://doi.org/10.1111/1467-8268.12607
Ibukun E, Senanu R, Okuboyejo OA, Odetunmibi BO (2021) An empirical investigation of acceptance, adoption and the use of E-agriculture in Nigeria. Heliyon 7(e07588):1–13
Ifeanyi-Obi CC, Asiabaka CC (2014) Impacts of climate change on sustainable livelihood of rural dwellers in southeast Nigeria. Nigerian J Agric Food Environ 10(3):71–77
Igberi CO, Osuji EE, Anuli RO, Paul CO, Ngozi OE (2022) Climate smart adaptive measure of yellow cassava, linkages and implications in Southeast Nigeria. Agrociencia 56(3):1–32. https://doi.org/10.47163/1646.Ag
Iheke OR, Onu DO, Egem MA (2021) Market orientation, innovation adoption and performance of food crops farmers in Abia State, Nigeria. Nigerian Agric J 52(1):188–220
Ihemezie EJ, Dallimer M (2021) Stakeholders' perceptions on agricultural land-use change, and associated factors. Nigeria. Environments 8(113):2–18. https://doi.org/10.3390/environments
Jianxu L, Mengjiao W, Li Y, Sanzidur R, Songsak S (2020) Agricultural productivity growth and its determinants in South and Southeast Asian Countries. Sustainability 12(4981):2–21. https://doi.org/10.3390/su12124981
John O, Rawlings SG, Pantula DAD (1998) Applied regression analysis: a research tool, second edition, springer-verlag New York, Inc., North Carolina State University Raleigh
Kadurumba C, Njoku IA, Achi O (2019) Multifactor productivity of smallholder yam farmers in Ebonyi State, Nigeria. Agricultura 16:(1–2):19–25 https://doi.org/10.18690/agricultura.15.1-2.19-34.2018
Lisette M, Denis JS, Germain B, Jérôme E, Jean-Marie K (2021) Building a framework towards climate-smart agriculture in the Yangambi landscape, Democratic Republic of Congo. Int J Climate Change Strategies Manag 12:44–49. https://doi.org/10.1108/IJCCSM-08-2020-0084
Mahmuda A, Md-Mizanur RS (2021) Impacts of climate factors influencing rice production in Bangladesh. Int J Environ Climate Change 11(1):43–52
Makuvaro V, Walker S, Masere TP, Dimes J (2018) Smallholder farmer perceived effects of climate change on agricultural productivity and adaptation strategies. J Arid Environ 152:75–82. https://doi.org/10.1016/j.jaridenv.2018.01.016
Martey E, Wiredu AN, Etwire PM, Kuwornu JK (2019) The impact of credit on the technical efficiency of maize-producing households in Northern Ghana. Agric Financ Rev 79(3):304–322
Mendelsohn RO, Massetti E (2017) The use of cross-sectional analysis to measure climate impacts on agriculture: theory and evidence. Rev Environ Econ Policy 11(2):280–298
Mulenga BP, Wineman A, Sitko NJ (2017) Climate trends and farmers' perceptions of climate change in Zambia. Environ Manage 59:291–306
NPC (2007) Reports of national population census commission Abuja, Nigeria
Nzeh EC, Uke PC, Attamah N, Nzeh DC, Agu O (2016) Climate change and agricultural production in Nigeria: a review of status, causes and consequences. Nigerian Agric Policy Res J 1(1):102–111

Obidiegwu J, Akpabio E (2017) The geography of yam cultivation in southern Nigeria: exploring its social meanings and cultural functions'. J Ethnic Foods 4(1):28–35

Olayemi OO, Abegunrin OO, Ogunwale OG, Ogunsola JO, Marizu JT, Adebayo AS, Olatunji BT (2021) Analyses of willingness to practice agriculture as enterprise among students of tertiary institutions in Ibadan. Oyo State. Nigerian Agric J 52(2):1–7

Oluwalana T, Okeleke SO, Akinbosoye TBS (2019) Economics analysis of small scale vegetable production in Odeda local government area of Ogun State. Direct Res J Soc Sci Educ Studies 6(9):127–132

Oluwatoba J, Omotilewa TS, Jayne MM, Adebayo B, Aromolaran LSO, Liverpool-Tasie TA (2021) A revisit of farm size and productivity: empirical evidence from a wide range of farm sizes in Nigeria. World Dev 146(105592):1–12

Onyeneke RU, Amadi MU, Njoku CL, Osuji EE (2021) Climate change perception and uptake of climate-smart agriculture in rice production in Ebonyi State, Nigeria. Atmosphere 12(1503):2–21. https://doi.org/10.3390/atmos121115

Oriangi G, Albrecht F, Di-Baldassarre G, Bamutaze Y, Mukwaya P, Ardö J, Pilesjö P (2020) Household resilience to climate change hazards in Uganda. Int J Climate Change Strategies Manag 12(1):59–73

Osuafor OO, Ude KD, Ositanwosu CO (2021) Effect of land degradation on maize yield in obudu local government area of cross river state. Nigeria. Nigerian Agric J 52(2):51–60

Oti OG (2017) Effects of livelihood factors on climate change adaptation and resource use efficiency of rural farm households in Southeast Nigeria. Unpublished PhD Thesis submitted to the Department of Agricultural Economics, University of Nigeria, Nsukka

Oti OG, Okoye CU, Obasi IO (2021) Awareness and perception of climate change among farmers in Nigeria: implications for food security. Nigerian Agric J 52(2):164–173

Paudel B, Zhang Y, Yan J, Rai R, Li L (2019) Farmers' perceptions of agricultural land use changes in Nepal and their major drivers. J Environ Manage 235:432–441

Rahman S, Chima CD (2018) Determinants of pesticide use in food crop production in Southeastern Nigeria. Agriculture 8(35):1–14. https://doi.org/10.3390/Agriculture8030035

Richard S, Fred M (2020) Effects of agricultural extension services on farm productivity in Uganda, AERC Research Paper 379. African Economic Research Consortium, Nairobi, Kenya

Saalu F, Oriaso S, Gyampoh B (2020) Effects of a changing climate on livelihoods of forest dependent communities: evidence from buyangu community proximal to kakamega tropical rain forest in Kenya. Int J Climate Change Strategies Manag 12(1):1–21

Samprit C, Ali SH (2006) Regression analysis by example, fourth edition, Wiley, Hobokon, New Jersey

Sara SP, Scandizzo L (2017) Farm size and productivity, a "direct-inverse-direct" relationship. Policy Research Working Paper 8127, World Bank Group, Development Economics Global Indicators

Sophie W (2019) Putting climate services in contexts: advancing multi-disciplinary understandings: introduction to the special issue. Clim Change 157:1–8. https://doi.org/10.1007/s10584-019-02600-9

USDA (2021) Crop production. National agricultural statistics service (NASS), Agricultural Statistics Board, United States Department of Agriculture, USDA, USA

Chapter 4
Implication of Land Use/Land Cover Dynamics on Accelerated Soil Erosion in Kereke Watershed of the Lower Benue Basin, North Central Nigeria

Godwin Akaayar Songu, Salau Wahab, Nenpanmun Temwa Molwus, Joshua Ade, Stephen Tersoo Yiye, and Bandi Garba Mohammed

Abstract Accelerated soil erosion especially in the form of gullies is one of the serious environmental problems enhanced by human activities that has threatened lives and properties in the Kereke watershed. Increase in temperature and late onset of rainfall due to climate change over the years has negatively influenced vegetal cover and pattern of land use/land cover changes in many parts of the world including the study area. Land use/land cover changes that result from anthropogenic activities such as farming practices and urban development as observed in the study area have serious influence on accelerated soil erosion in the watershed. This chapter assessed the implication of land use/land cover change on accelerated soil erosion in the Kereke watershed, with a view to ascertain the influence of human activities on soil erosion development in the area. Therefore, satellite data which includes Enhanced Thematic Mapper Plus (ETM+) 30 m resolution images of 2007, 2012 and Landsat 8 Operational Land Imager (OLI) 30 m resolution image of 2017 were used for land use/land cover analysis for a ten-year period (2007–2017) using ArcGIS 9.3 software. Nature of human activities in the study area was observed on the field, and well-known procedures and percentages were used to determine the rate of change in land use/land cover over the ten-year period. Findings of the study show that built-up area increased from 15.97 km^2 to 30.83 km^2 at the rate of 9.30% per annum over the ten-year period. While, cultivated land increased from 223.63 km^2 to 294.12 km^2 at annual rate of 6.30% per year. However, vegetal cover decreased from 299.28 km^2 to 78.13km^2 at an annual rate of −7.39% over the ten-year period. These observed land use/land cover changes in the study area influenced positive feedbacks on accelerated soil erosion in the area, by reinforcing the impact of raindrop, runoff, soil detachment

G. A. Songu (✉) · S. Wahab · N. T. Molwus · B. G. Mohammed
Department of Geography, Federal University of Kashere, Gombe, Gombe State, Nigeria
e-mail: godwinsongu22@gmail.com

S. T. Yiye
Department of Geography, Benue State University, Makurdi, Nigeria

J. Ade
Akawe Torkula Polytechnic, Makurdi, Benue, Nigeria

J. C. Egbueri et al. (eds.), *Climate Change Impacts on Nigeria*, Springer Climate,
https://doi.org/10.1007/978-3-031-21007-5_4

and entrainment, thereby influencing soil erosion development in the Kereke watershed. Findings of the study also suggest that, land use/land cover change resulting from climate change has negatively impacted soil erosion development in the area, especially during onset of rains in areas with decreased vegetation and bare soils. It is therefore recommended that the Federal Government of Nigeria should ensure strict compliance on deforestation laws in the Kereke watershed to reduce the rate of deforestation and controlled development in the area, for improved environmental management, and reduced climate change impacts as well as accelerated soil erosion due to human activities in the Kereke watershed.

Keywords Climate change · Land use/land cover · Accelerated soil erosion · Kereke watershed · Benue trough

Introduction

Land use/land cover change (LU/LCC) is a general term for human modification of the earth's terrestrial surface; and these changes encompass the greatest environmental concerns of human populations today, including climate change, biodiversity loss and the pollution of water, soils and air (Hassan et al. 2016). Igbokwe et al. (2008) emphasised that change in vegetation is probably the most important physical factor influencing soil erosion. A good cover of vegetation shields the soil from the impact of raindrops. It also binds the soil together, making it more resistant to runoff. A vegetative cover provides organic matter, slows runoff and filters sediment. On a graded slope, the condition of vegetal cover will determine whether erosion will be stopped or only slightly halted. A dense, robust cover of vegetation is one of the best protections against soil erosion (Igbokwe et al. 2008).

Natural vegetation breaks the fall of raindrops and helps to bind loose soil on slopes. It also reduces the frequency of overland flow by encouraging infiltration. Organic material, in the form of dead and decaying plant material, plays several important roles; it holds the topsoil together, increases permeability and provides a supply of nutrients (Songu 2019).

Accelerated soil erosion, therefore, is the type of erosion that results from the removal of natural vegetation by human activities such as farming, ranching, forestry and construction. The production of agricultural and forest products can be affected by excessive erosion (Hassan et al. 2016). The removal of natural vegetation greatly reduces soil protection, and the rates of soil erosion may accelerate in many folds. Land use pressure involving a modification of the land use/cover and the surface is affected through a general change in land use. For example, a change from forests and grass land with a dense matter or roots which inhibits surface runoff to ploughed land, which is susceptible to surface wash facilitates the formation of gully erosion, especially if ridges and heaps are aligned orthogonal to the slope contour (Faniran et al. 2006). Changes on the environment resulting from human activities have serious impact on soil erosion initiation especially accelerated gully erosion. Land use change such as deforestation, tillage, cropping and the conversion of land to pasture or range land, all increase the erodibility of the soil. Accelerated soil erosion occurs when rates of soil erosion increase as a direct or indirect result of human activities. This is a major environmental problem worldwide (Charlton 2008). The most important anthropological factor, according to chamberlain (1990), involves heavy population pressure on available land; and consequently over cultivation and overgrazing and climate change. In addition, the unusual poor cultivation methods involving heap making on steep slopes and alignment of ridges orthogonal to the slope contour, usually channel runoff water across such slopes to facilitate erosion. Also the frequent exposure of soils leads to desiccation and impoverishment of the top soil especially as the global temperatures are increasing due to climate change.

Climate change in many ways impacts land use/land cover change in a wide range of environment. The relationship between land use and the climate is complex. First, land cover is shaped by land use practices and in turn affects the global concentration of greenhouse gases. Second, while land use change is an important driver of climate change, a changing climate can lead to changes in land use and land cover. For example, farmers are shifting from their customary crops that have higher economic return from rain fed farming to irrigation farming under changing climatic conditions. Moreover, higher temperatures affect mountain snowpack and vegetation cover as well as water needed for irrigation, thereby creating need for usage of other water sources such as dams, boreholes and streams for irrigation farming. Moreso because forests and trees store carbon. When they are degraded or completely cleared, perhaps by fire or other forms of deforestation, the stored carbon has the potential to be released back into the atmosphere as carbon dioxide and contributes to global warming and climate change that affects the environment Hassan et al. (2016). Many studies revealed that accelerated soil erosion is consequent upon changes in land use/land cover as a result of human activities and climate change.

For instance, Ibitoye and Eludoyin (2010) identified human activities as some of the major determinants of accelerated soil erosion in a watershed. These effects are more pronounced where there are high intensities of human activities such as farming practices of slash and burn agriculture, cutting down of trees, making of heaps orthogonal to slope contours, abandoned excavated gutters, size of bare surface, nature of ground surface, development of pavements that enhances runoff generation on slopes; influences the initiation of soil erosion (Ibitoye and Eludoyin 2010). Where there is a negative modification of land use/cover of the surface and is affected through a general change in land use; accelerated erosion is likely to be a consequence. For example, a change from forests and grassland with a dense matter or roots to ploughed land, could make the area susceptible to surface erosion, and gully erosion is likely to be initiated over time. Abbas et al. (2018) in a study on land use/land cover change in Savannah area of Northern Nigeria reported that between 2001 and 2009 savanna area was reduced at the rate of 4% while the forested area was reduced at the rate of about 9% per annum. The implication of this finding is that, with the observed current negative trend on land cover change in the area, the soils will be exposed to raindrop impact and accelerated erosion is likely an environmental menace in such areas. Also, Abu et al. (2020) in their study on land use/land cover dynamics on vegetation cover in some parts of Biu plateau reported that changes in the extent of the area of land use/land cover from 2000–2015 resulted from an increase of about 2.69% in built-up area, 1.96% of water bodies, 13.26% in cultivated land and 7.78% of bare surface with a decrease of 13.09% in forest area and 12.6% in scrubland. The findings of this study suggest that increase in cultivated area, built-up area and decreased forested area and scrubland has implication on accelerated soil erosion in the area through positive feedbacks. Egbueri and Igwe (2022) in their study on erosion risk mapping of Anambra State in South-eastern Nigeria based on soil loss estimation using RUSLE model and geoinformatics. The RUSLE model was based on five erosion factors (rainfall erosivity (R), soil erodibility (K), topography (LS), vegetation cover (C) and anti-erosion practices (P). Findings of the study showed that the R-factor ranged from 460.51 to 582.08 MJ/mm/ha^{-1} h^{-1} year^{-1} whereas the K-factor ranged from 0.100 to 0.310 t/h/MJ^{-1}/mm^{-1} and the average annual soil loss ranged from 0 to over 6 t/ha^{-1} year^{-1}. The soil loss results showed that about 5% (242.2 km^2), 25% (1211 km^2), 30% (1453.2 km^2) and 40% (1937.6 km^2) of the total area have very low, low, medium and high erosion risks, respectively. The northern and western portions of the State were characterised by very low, low and mocharacterised by high soil loss rate. In a similar vein, Egbueri and Igwe (2021) conducted a study on the impact of hydrogeomorphological characteristics on gullying processes in erosion-prone geological units in parts of south-east Nigeria using integrated field survey, hydrological, geotechnical and geomorphological approaches. Findings of the study indicated that there are numerous surface water bodies and shallow groundwater systems. Both the surface waters and groundwater have a westward flow direction, from areas of high elevations on the Nanka Formation to areas of low elevations on the Ogwashi and Benin formations capable of influencing gully development in the area. Geotechnical results revealed that the soils are permeable, weak, easily dispersible and collapsible. Geomorphological analysis showed that the area is characterised by uneven badland topography, high gully slope gradients, concave slopes, poor land-use practices and low vegetation cover. Hence, the results of this study indicated that hydrogeomorphology and soil engineering properties substantially influence the gullying processes in the area.

In the Kereke watershed within the lower Benue trough, the area is typically agrarian, and there are places that were previously covered with grassland and trees; and presently those areas have been de-vegetated and converted to built-up areas with pavements. Also, engineering works involving the construction and excavation of artificial drainage channels such as the diversion channels excavated in North Bank town have been observed to influence accelerated erosion in the watershed. Despite the observed consequences of accelerated soil erosion in the study area, the literature review in this book chapter shows that there is dearth of data on the actual rates and extent of land use/land cover changes in the Kereke watershed and its influence on accelerated erosion that ought to be useful for soil erosion control and environmental management. It is in this regard, that this chapter determined the extent of change in land use/land cover and its influence on accelerated soil erosion in the Kereke watershed to bridge literature gaps, and enhance soil erosion control in the area. The specific objectives of this study are:

(i) To determine the extent of change in land use in the Kereke watershed from years 2007–2017.
(ii) To evaluate the extent of change in land cover of the study area.
(iii) To assess the influence of land use/land cover change on accelerated soil erosion in the Kereke watershed.

Materials and Methods

Description of the Study Area

The study was carried out in the Kereke watershed. Kereke watershed cover areas such as North Bank area, Agan, Asamakaha, Kyaan, Daudu, Yerewata, Iyorkya, Adugu, Agyaragu, Kadaroko, Agaza, Aloshi, Gaval among others in Obi, Keana, Guma and Makurdi Local Government Areas within the Benue Basin, North Central Nigeria. Kereke watershed lies between latitudes $7^\circ 36^{I}$ N and $8^\circ 20^{I}$ N and between longitudes $8^\circ 28^{I}$ E and $8^\circ 50^{I}$ E (Ministry of Lands and Survey Makurdi 2016). The Kereke watershed like other parts of the Benue Trough experiences wet and dry climate known as "Aw" by Koppen's climatic classification scheme. This type of climate is also called the tropical humid savanna climate (Tyubee 2005). The wet season which last for seven months starts from April and ends in October. Conversely, the dry season starts from November and ends in March. The climate is usually associated with torrential rainfall and high temperatures. The annual total rainfall ranges from 1000 to 2000 mm, with an average annual intensity of 44.85 mm/hr (NIMET 2018).

Temperatures are high throughout the year, averaging 27–31 °C. Though, it may occasionally rise to 37 °C in some days in March and April. The study area is confined in the Guinea Savanna region of the Benue valley. It is therefore characterised with shrubs, grasses and trees. Some of tree species in the study area include: Raphia palm (*Raphiasudanica*), sheabutter trees (*Butrospernumparadoxum)*, Malina (*Gmelinaarborea)*, Obeche (*Afzetia Africana*), locust bean trees, Mahogany, Iroko trees. Other economic trees found in the study area are Mango (*mangiferaindica)*, Cashew tree (*Anacardiunaccidentale*), Guava (*Psidiumguajava*), Pawpaw (*Caricapapaya*), Palm tree (*Elaeisguinniensis*), Banana (*Musaspp)* and Sweet orange (*citrussinensis*).

The geology is predominantly sandstone. The sedimentary basins found within the Benue trough aulacogen comprises of older and younger sedimentary rocks formed between Paleozoic and Cenozoic era of geologic time scale (Benkhelil 1989). The superficial geology of the Benue trough is largely meta-sediments, predominantly comprised of alluvial deposits made up of sand, clay, silt, gravel and pebbles and forms the dominant surface geology of the study area (Benkhelil 1989). The study area is endowed with erodible soils, which are predominantly deep, sandy and are derived from alluvial deposits. The soil is well drained and so the water holding capacity is limited; and more so, because the soil has moderate organic matter content which enhances permeability (Aneke 1991).

The Kereke watershed has an undulating terrain with slope angle exceeding 10° in some steep areas. Its general topography varies between 69 and 255 m as analysed from Satellite Imagery (SRTM) of the study area. The Kereke watershed is a 4th order drainage basin, with a total of fifty-nine 1st order streams, twenty-seven 2nd order streams, twelve 3rd order streams and one 4th order stream. The Kereke watershed has an average bifurcation ratio of 5.48. The moderate bifurcation ratio could be attributed to the sedimentary rocks which underlain the basin and moderate rainfall intensities. The mean bifurcation ratio of 5.48 implied that the dissection of the Kereke drainage basin is coarse and this could influence channelised flow especially at the lower segment of the basin and thereby influencing gully initiation and flooding in some parts of the basin. Figure 1a, b are location and relief maps of the study area.

(a)

Fig. 4.1 **a** Location of Kereke watershed. **b** DEM of Kereke watershed showing variation in relief. *Source* Authors' GIS Analysis from SRTM, 2019

(b)

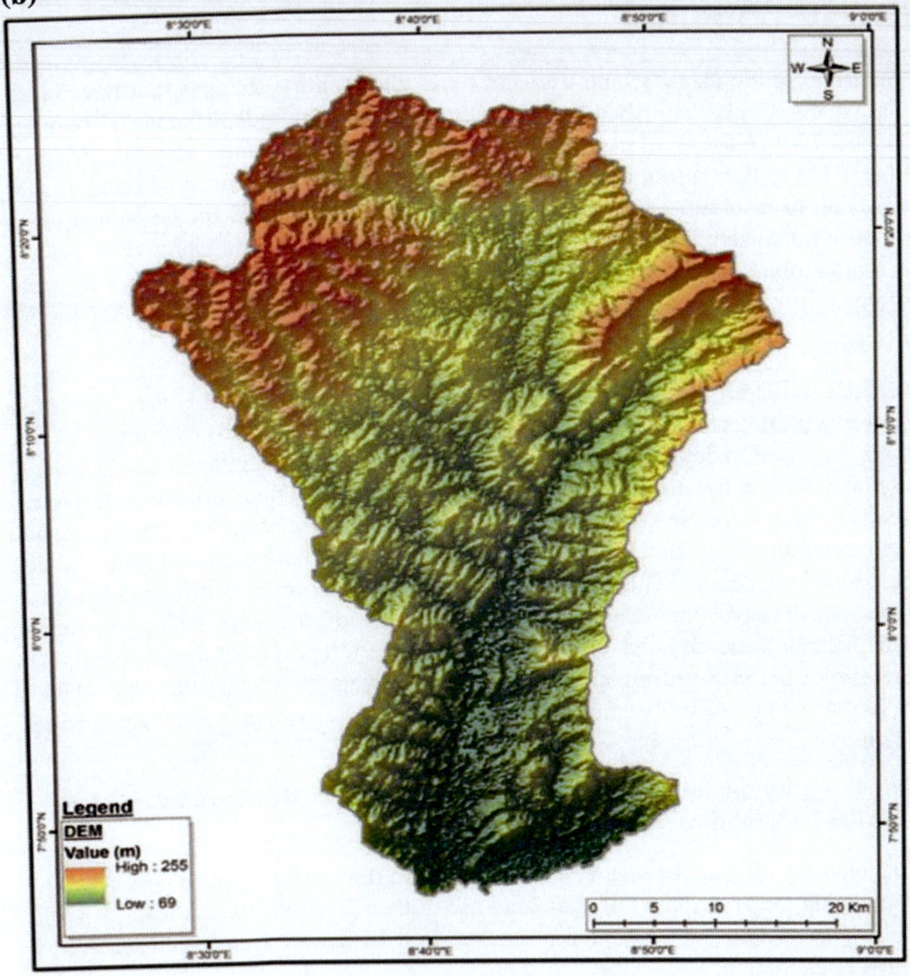

Fig. 4.1 (continued)

Methodology

Since the focus of this paper was on the implication of land use/land cover dynamics on accelerated soil erosion, data was collected on built-up area, cultivated/bare land, total vegetal cover and human activities in the study area. Data needed were derived from a combination of direct field observation and Geo-referenced data to assess the magnitude and percentage of change that has taken place over the ten years period. The methods of data collection are described below.

Techniques Used for Determination of Land Use/Land Cover Change in the Kereke Watershed

The data used to classify land use/land cover and identify changes that have taken place in the Kereke watershed were Satellite data which include: Enhanced Thematic Mapper Plus (ETM+) 30 m resolution image of 2007; Enhanced Thematic Mapper Plus (ETM+) 30 m resolution image of 2012 and Landsat 8 Operational Land Imager (OLI) 30 m resolution image of 2017. The images were used for classification to produce land use/land cover classes. This also provided the statistical data used to determine the change in land use/land cover extents over time. The SRTMs were sourced from United States Geological Survey (USGS) Earth Explorer online portal.

(i) **Image processing and classification**

In order to identify the changes in land use/land cover in the study area, satellites images were processed and classified to produce the land use/land cover maps, which were then used to determine the physical and quantitative extents of change of land use/land cover in the study area. The Satellite images of the study area were corrected geometrically to remove distortions and subsequently enhanced to improve visual interpretation. Image pixels were grouped into different land uses as outlined in the classification scheme. This was followed by classification into different land use types. Supervised classification system was employed because of its high accuracy. This helped to identify sets of pixels that accurately represented spectral variation present within each information region. The datasets were classified into classes of built-up area, natural vegetation, cultivated land, water bodies and bare surfaces (Table 4.1).

(ii) **Determination of the extent and rate of changes that have taken place in land use/land cover in the study area**

Calculations of changes that have taken place in the land uses were performed by calculating the area (Km^2) of each land use in their respective years using ArcGIS 10.5. These figures were copied and taken to a Microsoft Excel environment for further calculations.

Table 4.1 Classification Scheme adopted in the study

S/no	Land use/Land cover type	Interpretation
1	Built-up area	Villages, Urban Areas, Open Spaces/Recreational Areas, Commercial Areas, Institutional Areas
2	Natural vegetation	Shrub/Grassland, Forest cover, Riparian Vegetation
3	Cultivated land	Actively Cultivated Lands, Uncultivated Farm Lands, Rainfed Agricultural Lands and Irrigation Agricultural Lands
4	Water bodies	River, ponds and lakes
5	Bare surfaces	Ground surface without grasses, de-vegetated lands and bare rocky surfaces

Source Adopted from Njoku (1992)

Calculating the extent and rate of changes involved the three following steps:

i. Calculating thc magnitude of change.
ii. Percentage of change.
iii. Calculating the annual rate of change.

(a) **Magnitude of change**

Calculating the magnitude of change involved subtracting the area of the beginning year from that of the end year as shown in the equation presented in Eq. 4.1.

$$C = E - B \tag{4.1}$$

where C = Change
E = Area of the End year
B = Area of the Beginning year

(b) **Percentage change**

Percentage change was calculated by subtracting the beginning year from the end year divided by the beginning year and multiplied by 100 (Njoku 1992).

$$\%C = \left(\frac{E - B}{B}\right)100 \tag{4.2}$$

where
%C = Percentage Change,
E = Area of the End year,
B = Area of the Begin year

(c) **Annual rate of change**

In calculating the annual rate of change, the change (the difference between the two years) was divided by the number of study years (Njoku 1992).

$$AROC = \frac{E - B}{N} \tag{4.3}$$

where AROC = Annual Rate of Change
E = Area of the End year
B = Area of the Begin year
N = Number of the Study years.

Nature and Types of Human Activities

The nature of human activities in the study area were directly observed on the field. Observations were made on types of farming practices, land uses, nature of cultivation, types of crops cultivated, nature of vegetation and uses of cut down trees in the area.

Results and Discussion

The results of the study are presented below and discussed in relation to existing literature.

Land Use/Land Cover Change and Accelerated Soil Erosion

Land use/land cover changes are consequent upon human activities taking place in a watershed. Human activities are some of the external factors that influence accelerated soil erosion in a watershed; especially as regards agricultural activities and urban development. The Kereke watershed comprises of both urban and rural settlements. Watershed disturbances relating to agricultural practices and urban growth are oftentimes major drivers of soil erosion at a drainage basin scale. The expanse of farm lands, farming practices/systems, construction of houses and drains, removal of vegetal cover and road construction all impacts soil erosion initiation in diverse ways.

Data on cultivated land, total vegetal cover and built-up area were derived from STRM of the study area version, 2007, 2012 and 2017, to assess the magnitude and percentage of change that has taken place over the past ten years. This was meant to examine the influence of these environmental variables on accelerating soil erosion in the area. Land uses in Kereke watershed have undergone various changes in physical extent over the past ten years. The results of these changes are presented in tables below.

Extent and Rate of Change of Built-Up Area in Kereke Watershed

Built-up area in this context comprised areas made up of houses of different types, commercial, religious and institutional structures. Based on analyses of the satellite imageries, data on the extent and rate of change in built-up area in Kereke watershed between 2007 and 2017 is presented in Table 4.2 and Fig. 4.2.

It is observed in Table 4.2 and Fig. 4.2 that between 2007 and 2012 there was an increased change in built-up area from 15.97 km^2 to 24.21 km^2. Built-up area increased by 8.24 km^2 (51.59%) at a rate of 10.31% annually. Built-up area in the

Table 4.2 Extent and rate of change of built-up area in Kereke watershed (2007–2017)

Peroid	Built-up area		Extent of change		Annual rate of change	
	Year	(km^2)	(km^2)	%	km^2/Year	%/Year
2007–2012	2007	15.97	8.24	51.59	1.65	10.31
(5Years)	2012	24.21				
2012–2017	2012	24.21	6.62	27.34	1.32	5.47
(5Years)	2017	30.83				
2007–2017	2007	15.97	14.86	93.05	1.49	9.30
(10Years)	2017	30.83				

Source Authors' GIS Analysis from SRTM, 2019

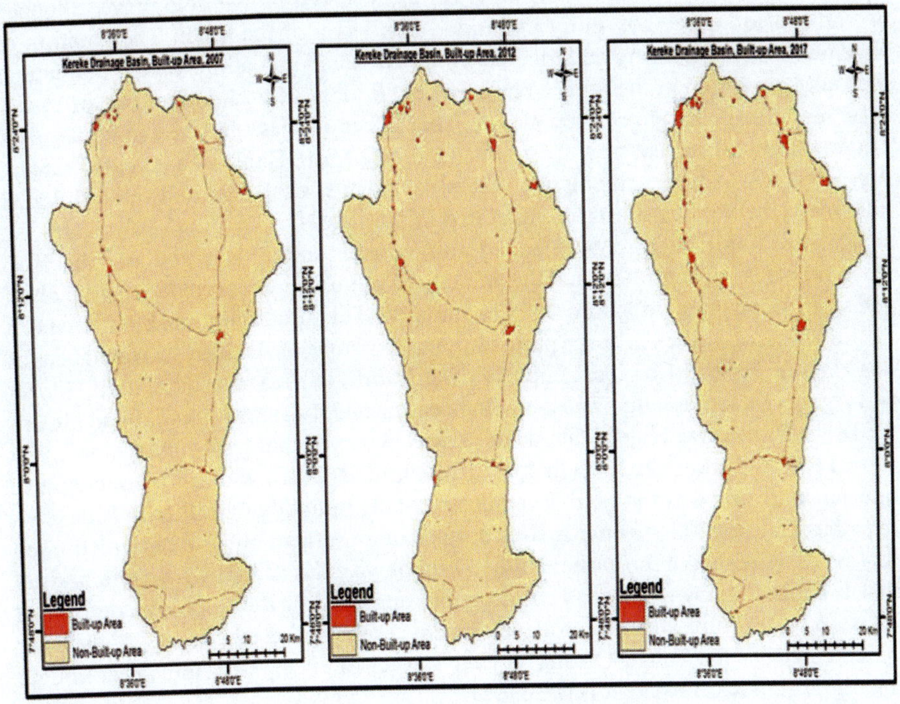

Fig. 4.2 Extent of change of built-up area in Kereke Watershed (2007–2017). *Source* Authors' GIS analysis, 2019

period 2012–2017 increased from 24.21 km^2 to 30.83 km^2, amounting to a change of 6.62 km^2 (27.34%) at an annual rate of 5.47% per year. In the overall period of 2007–2017, built-up area increased from 15.97 km^2 to 30.83 km^2. There was an observed increase in built-up area over the ten years by 14.86 km^2 (93.05%) at 9.30% rate per annum.

The extent of built-up area in the Kereke watershed increased from 15.97 km^2 in 2007 to 30.83 km^2 in 2017. This showed the extent of increase in the amount of pavements and roof tops that could contribute to surface runoff in the area. Where there is high percentage of pavement and roof tops that could impede infiltration, surface runoff may increase if there are no drains to contain it. However, built-up area represented only 8.28% of the Kereke watershed, with a basin area of 372.41 km^2. The implication of this change on hydrological processes especially the initiation of accelerated gully erosion is that pavements or concreted surfaces and roof tops are small in the study area, and therefore their impact on surface runoff along concentrated channels will be minimal.

In the urban areas of the Kereke watershed such as North Bank area, Agan, Daudu, Agyaragu among others housing development, construction of houses and drains among other watershed disturbances as observed on the field, have influence on soil erosion especially gully initiation in one way or the other. The creation of pavements due to housing and road development has implications on the magnitude of runoff generated in the area. In areas where drains are limited, volume of runoff is increased especially on steep slopes. This situation was observed in some parts of the watershed such at Low Cost housing estate North Bank area, Daudu, Aloshi, Agyaragu and Adugu settlements, where pavements, roof tops and limited drains influenced the volume of runoff generated in such areas.

This has implication on accelerated gully erosion initiation in the area, because when runoff is concentrated, and the velocity of runoff is increased due to high relief and steep slope gradient, gullies may be initiated. This situation probably influenced the initiation of gullies in some parts of the urban areas of the Kereke watershed, as observed in Federal Low cost housing estate North bank Makurdi, Daudu, Aloshi and Agyaragu settlements. These findings confirmed those of Mahmud and Umaru (2018) in Bida town, Niger State where it was observed that built-up area increased from 12,880.350 km^2 in 1995 to 15,971.638 km^2 in 2015, with a 31% increase in built-up land; and was found to have influenced the initiation of gullies in Bida town over the years resulting from increased surface runoff especially in areas with limited drains. Also, compare the earlier study by Igbokwe et al. (2008) whom emphasised that, urbanisation involving road constructions and building development contributed immensely to gully sites. Such developments were noticeable at the Ajali water scheme in Owa, the Enugu-Onisha highways, Nachi and Onyeama mine near Enugu; and the Agutaoweri Road at Ikwese town.

Extent and Rate of Change of Cultivated Land in the Kereke Watershed

Cultivated land includes all land meant for farming activities and open land preserved for cultivation via shifting cultivation system. Analyses of SRTM showed changes on the area of land under cultivation from 2007 to 2017. The extent and rate of change in cultivated land in the study area are shown in Table 4.3 and Fig. 4.3.

It is observed in Table 4.3 and Fig. 4.3 that the period 2007–2012 experienced increase in cultivated land from 223.63 km^2 to 250.44 km^2 representing a change of

Table 4.3 Extent and rate of change of cultivated land in Kereke watershed (2007–2017)

Cultivated land			Extent of change		Annual rate of change	
Period	Year	Extent (Km^2)	(km^2)	%	km^2/Year	%/Year
2007–2012	2007	223.65	26.79	11.97	5.36	2.39
(5Years)	2012	250.44				
2012–2017	2012	250.44	43.68	17.44	8.74	3.49
(5Years)	2017	294.12				
2007–2017	2007	223.65	70.47	31.51	14.09	6.30
(10Years)	2017	294.12				

Source Authors' GIS Analysis from SRTM, 2019

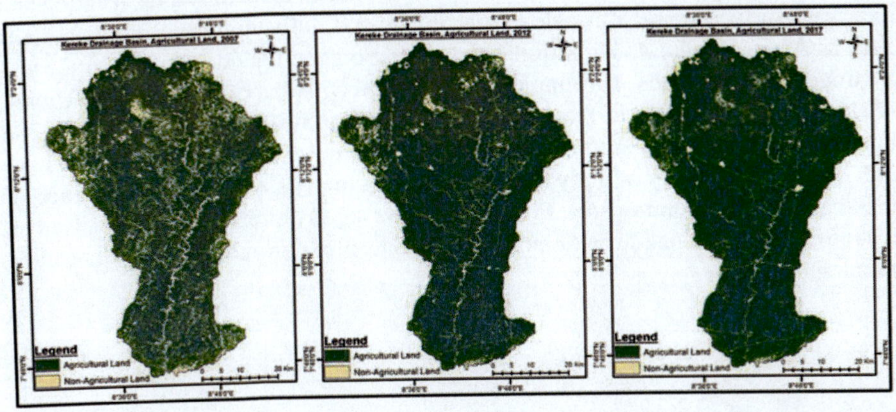

Fig. 4.3 Extent of change of cultivated land in Kereke Watershed (2007–2017). *Source* Authors' GIS analysis, 2019

26.79 km^2 (11.97%) with annual rate of 2.39% per year. A change from 250.44 km^2 to 294.12 km^2 in cultivated land was witnessed in the period 2012–2017 amounting to a change of 43.68 km^2 (17.44%) at annual rate of 3.49% per year. The overall period of 2007–2017 experienced an increase in cultivated land from 223.63 km^2 to 294.12 km^2 causing a change of 70.47 km^2 (31.51%) at an annual rate of 6.30% per year.

The change observed in cultivated land was positive and indicating that from the year 2007 to 2017 cultivated land increased from 223.63 km^2 to 294.12 km^2 causing a change at an annual rate of 6.30% per year (Table 4.3). This showed that cultivated land represented 78.97% of the Kereke watershed. Farming activities in the study area have been on the increase with impacts on hydrological processes in the study area. This is because cultivated lands expose the soils to desiccation and direct raindrop impact which could enhance soil detachment and entrainment.

The major agricultural practices observed on the field at the rural areas of the Kereke watershed during the course of this research were mainly cultivation of grains, root crops, lumbering for fire wood and construction. These agricultural activities especially tilling of the soil impact the environment in different ways. The traditional practice of slash and burn agriculture is still practiced in the study area. The clearing of land for farming exposes the land to direct raindrop impact especially during onset of rains. This enhances the dislodgement of loose parts of the earth which leads to the initiation of sheet, rills and subsequently gullies, especially areas with unconsolidated sandy soils as observed in the Kereke watershed.

Also, the traditional method of making heaps and ridges in the Benue valley plays significant role in enhancing surface runoff, when the heaps and ridges are made along slopes. The trenches in between the ridges could become the source points for the initiation of gully erosion. This scenario was also observed in the study area. With the large size of cultivated land in the study area (294.12 km^2), agricultural practices especially farming activities have been observed to contribute to soil desiccation and increased surface runoff downslope that influenced gully initiation in some areas such as Agan, Daudu, Akiki, Asamakaha among others; which in turn degrade and destroyed the farm lands. This finding corroborated with that of Mahmud and Umaru (2018) in Bida town, Niger State, where increased cultivated land from 10,304.280 km^2 in 2005 to 11,334.708 km^2 in 2015, with an increase of 22% was found to have influenced the intensity of gully erosion area. Figure 4.4 shows a typical method of making ridges in the study area. Figure 4.5 shows a gully erosion channel incised on a large rice farm at Daudu settlement where concentrated runoff was influenced by farming practices.

Fig. 4.4 Cleared and cultivated land for crops at Agan settlement. *Source* Authors' field work, 2019

Fig. 4.5 Incised Gully erosion channel on a large rice farm at Daudu settlement. *Source* Authors' field work, 2019

Extent and Rate of Change of Vegetation Cover in the Kereke Watershed

Results on total vegetal cover in the study area from the year 2007 to 2017 was analysed from SRTM of the study area and are contained in Table 4.4 and Fig. 4.6.

It is observed from Table 4.4 and Fig. 4.6 that vegetal cover decreased from year 2007 to 2012. The vegetation cover decreased from 299.28 km^2 to 114.37

Table 4.4 Extent and rate of change of vegetal cover in Kereke watershed (2007–2017)

Vegetal cover			Extent of change		Annual rate of change	
Period	Year	Extent (km^2)	(km^2)	%	km^2/Year	%/Year
2007–2012	2007	299.28	−184.91	−61.78	−36.98	−12.36
(5Years)	2012	114.37				
2012–2017	2012	114.37	−36.24	−31.68	−7.25	−6.34
(5Years)	2017	78.13				
2007–2017	2007	299.28	−221.15	−73.89	−22.11	−7.39
(5Years)	2017	78.13				

Source Authors' GIS Analysis, 2019

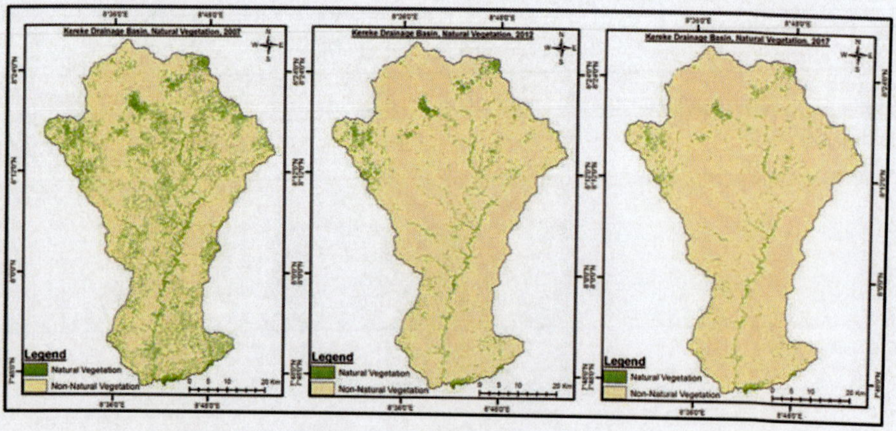

Fig. 4.6 Extent of change of vegetal cover in Kereke Watershed (2007–2017). *Source* Authors' GIS analysis, 2019

km^2 depicting a change of -184.91 km^2 (-61.78%) at an annual rate of -12.36%. The period 2012–2017 also experienced a negative change in vegetation cover from 114.37 km^2 to 78.13 km^2 representing a change of -36.24 km^2 (-31.68%) at an annual rate of -6.34% yearly. However, the overall period of 2007–2017 had vegetation reduced from 299.28 km^2 to 78.13 km^2 leading to a change of -221.15 km^2 (-73.89%) at an annual rate of -7.39%.

The above analysis on vegetation cover in the study area showed that there has been a high rate of loss in vegetal cover in the study area. This could be attributed to increased agricultural activities in the study area especially farming activities and lumbering. The present total vegetal cover represents only 20.97% of the Kereke watershed. The total vegetation coverage in the study area is considered low. The removal of vegetal cover for agricultural activities and urban development exposes the land surface to direct raindrop impact especially during high intensities and amount of rainfall. The study area experiences moderate amount and intensities of rainfall. The moderate amount and intensities of rainfall are capable of exerting intense energy on the earth surface, especially in areas where vegetal cover is limited and loose particles of soils will be detached.

Cutting of trees for firewood and building of houses is some of the major observable causes of loss in vegetal cover over the ten years in both urban and rural parts of the watershed. The inhabitants of the study area use firewood as a major means of cooking and timber for roofing of houses. The increased built-up area as observed in Table 4.2 implied that houses built especially in urban settlements of the study area were roofed using timber gotten from cut down trees; except in some rural parts of the basin that contains thatched houses that are not necessarily roofed with timber. However, such lands where those thatched houses are built, the land is cleared and trees removed for development. Vegetation loss in the study area, therefore, was

Fig. 4.7 Cut down tree at a house construction site in Agan settlement. *Source* Authors' field work, 2019

considered as one of the major drivers or factors of gully erosion initiation in the study area.

These findings confirmed those of Ibitoye and Eludoyin (2010) in Akure, where it was observed that clearance of vegetation at three proposed sites by the government of Ondo State at Ose-oba in Akoko Akure along Ilesa-Benin Road and along Oda road for starch plant, auto village and housing units, receptively, exposed the surfaces to rain wash and gully erosion with the tendency of causing channel widening and deepening. Also, Udosen (2013) in his study of gully erosion consequences in Uyo urban noted that, deforested areas were exposed to raindrop impact that led to soil entrainment and transportation, thereby initiating gully erosion. Figures 4.7 and 4.8 show cut down tree stem for house construction at Agan settlement and charcoal produced from dried and burnt wood at Kadaroko settlement.

Conclusion and Recommendations

Fundamental changes of any environment may be attributed to natural or anthropogenic activities. Due to the change in global climate, the land surface temperature has increased, which has affected land use/land cover change, vegetated areas, water resources among others, and such changes in land use/land cover due to climate change are responsible for various environmental problems such as soil erosion and flooding. Land use/land cover changes resulting from human activities and climate change have significant influence on the rate at which soil erosion is accelerated.

Fig. 4.8 Bags of charcoal made from burnt wood in Kadaroko settlement. *Source* Authors' field work, 2019

It is hereby, concluded that the influence of human activities remains one of the leading factors in activating soil erosion, especially the initiation and development of gullies in the Kereke watershed. This study revealed that there are positive changes in built-up area and cultivated land area in the Kereke watershed; thus, increasing the total land area under construction and crop cultivation. On the contrary, the total vegetal cover in the study area decreased tremendously over the ten-year period assessed, thereby influencing raindrop impact, runoff generation, soil detachment and gully erosion development in some parts of the watershed. Therefore, the following recommendations are made based on the findings of this study:

(i) The Federal government of Nigeria should ensure strict compliance on deforestation laws in the Kereke watershed, to help reduce the rate of deforestation in the area. This will enhance improved environmental management and reduced impact of raindrops on soil, thus limiting soil erosion initiation in the area.
(ii) Carbon sequestration should be encouraged by planting of more trees through afforestation programmes to help collect excess carbon in the atmosphere to reducing the impact of global warming and climate change.
(iii) Human activities such as making of ridges, engineering works and urban development among others, that influence accelerated soil erosion in the watershed

should be controlled especially along steep slopes, to minimise the development of erosional channels such as rills and gullies in the area.

(iv) More green area landscaping should be adopted by house owners rather than concreted surfaces, to enhance infiltration and reduce the impact of climate change viz-a-viz accelerated soil erosion in the study area through channelised flow.

Acknowledgements We sincerely acknowledge the management of Federal University of Kashere (FUK), Gombe State, especially the Head of Department of Geography, FUK for their support throughout the period of this study. Also, the sincere contributions of Professor Iorkua Salemkaan Aloysius and Professor Maxwell Ocheri of the Department of Geography, Benue State University, Makurdi to the success of this research are highly appreciated. Lastly, we acknowledge the untiring support of Mrs Esther Dooshima Songu towards the success of this study.

References

Abbas II, Bello OM, Abdullahi H (2018) Mapping and analyzing the land use-land cover of Nigeria between 2001 and 2009. MOJ Eco Environ Sci 3(3):197–205

Abu RD, Yakubu E, Songu GA, Thlakma RS, Abu GO, Gajere EN (2020) Impact of Landuse/Landcover dynamics on vegetation cover in part of Biu Plateau, Borno, Nigeria. Illorin J Bus Soc Sci 1(22):155–166, ISSN:1115-960X

Aneke DO (1991) Erodibility of Luvisol and Luvisil, Cambisol of Benue State. J Agric Sci Technol 1:50–53

Benkhelil J (1989) The origin and evolution of the cretaceous Benue trough. J Afr Earth Sc 8:257–282

Charlton RO (2008) Fundamentals of fluvial geomorphology. Routledge publishers, London

Chamberlain J (1990) Understanding soil erosion and its control. Volunteers in technical assistance. Report submitted for soil erosion control in London, UK

Egbueri JC, Igwe O (2022) Erosion risk mapping of Anambra State in Southeastern Nigeria: soil loss estimation by RUSLE model and Geoinformatics. Bulletin Eng Geol Environ 9(81). Retrieved from https://doi.org/10.1007/s10064-022-02589-z. 9th April, 2022

Egbueri JC, Igwe O (2021) The impact of Hydrogeomorphological characteristics on gullying processes in Erosion-prone geological units in parts of Southeast Nigeria. Geol Ecol Landscapes 3(5). Retrieved https://doi.org/10.1080/24749508.2020.1711637. 9th April, 2022

Faniran A, Jeje LK, Ebisemiju FS (2006) Essentials of geomorphology. Pent House Publications (NIG), Ibadan, pp 25–37

Hassan Z, Shabbir R, Ahmed SS, Malik AH, Aziz N, Butt A, Erum S (2016) Dynamics of Land use and Land cover Change (LULCC) using geospatial techniques: a case study of Islamabad Pakistan. Retrieved from https://doi.org/10.1186/s40064-016-2414-z

Ibitoye MO, Eludoyin A (2010) Urban growth and development of soil erosion inpart of the humid region of South West Nigeria. Paper presented at the FIG congress in Sydney, Australia

Igbokwe JI, Akinyede JO, Dang B, Alaga T, Ono MN, Nnodu VC, Anike LO (2008) Mapping and monitoring of the impact of gully erosion in South Eastern Nigeria with satellite, remote sensing and geographic information system. Int Arch Photogram Remote Sens Special Inf Sci 8(37):56–61

Mahmud HL, Umaru ET (2018) Impact of gully erosion and Landuse/Landcover in Bida Town, Niger State Nigeria. IIARD Int J Geogr Environ Manag 4(2):7–15

Ministry of Land and Survey, Makurdi (2016) Location of the Study Area

Njoku JD (1992) A multi—band assessment of SPOT imaging system for land and land cover mapping of Northwestern Nigeria. Unpublished MSc Thesis. Dept of Geography and Planning, University of Lagos, Akoka , Lagos

Nigerian Meteorological Weather Office, Makurdi (2018) Rainfall Data in Makurdi. Retrieved on 2nd February, 2018

Songu GA (2019) Thresholds of environmental factors of gully erosion in Kereke Drainage Basin of the Benue Trough. An Unpublished Ph.D thesis submitted to the Postgraduate School, Benue State University, Makurdi

Tyubee TB (2005) Spatial organisation of daily rainfall in the middle belt of Nigeria. The Benue Valley 41:69–71

Udosen CE (2013) Gully erosion and cities: an unwanted partnership. a paper presented at the faculty of Social sciences Seminar at CBN Hall, University of Uyo, pp 1–15

Chapter 5
Thermal Conditions in Artisanal Mine Sites: A Case Study of Ife Area, Southwest Nigeria

Iyanuoluwa E. Fatunmbi and Oluwagbenga O. I. Orimoogunje

Abstract In the past few decades, mining has recently become important in the southwestern Nigeria. Existing studies on the implications of mining activities in the region have mostly overlooked the capacity of activities around the sites to impact significant changes in the microclimate of neighbouring communities. The present study is thus an attempt to evaluate daytime changes in thermal conditions (in terms of air temperature and relative humidity, heat index and dew point temperature) and perceptions of 15 key respondents at the mining communities in Atakunmosa West Local Government in Osun State, Nigeria. Data were obtained at daytime (between 0900 and 1800 Nigerian Local Standard Time). Heat index and dew point temperature were computed for selected locations and the data were analyzed using geo-statistical and qualitative techniques. Results showed higher values of temperature and increased tendency for thermal stress at the mine sites than at locations far away from the site. Majority of residents (>60%) explained that heat-related illness (including dry skin, heat rash) has increased since the mining activities began in the area. The study concluded that temperature increase, and thermal and associated stress would be compounded in communities around the mine sites. Mine sites restoration, increased awareness on coping strategies to heat stress and inclusion of propensity for heat concerns in Environmental Impact Assessment for mining activities are therefore recommended.

Keywords Thermal conditions · Mine sites · Air temperature · Heat stress

I. E. Fatunmbi (✉) · O. O. I. Orimoogunje
Department of Geography, Obafemi Awolowo University, Ile-Ife, Nigeria
e-mail: iyanuoluwafatunmbi@gmail.com

O. O. I. Orimoogunje
e-mail: ooorimoogunje@oauife.edu.ng

J. C. Egbueri et al. (eds.), *Climate Change Impacts on Nigeria*, Springer Climate,
https://doi.org/10.1007/978-3-031-21007-5_5

Introduction

Mining as a primary activity has been in existence from ages. Many researches have been carried out to explain that almost, if not, all the countries in Africa are active in one or more possession and production of minerals (Aigbedon and Iyayi, 2007; Oladipo et al. 2014; Adesipo et al. 2020). Their economic importance on the well-being of the environment was also discussed (Ralph et al. 2018; Omotehinse and Ako 2019; Adesipo et al. 2020). However, over 95% of the mining activity is artisanal and another 95% of this portion is illegal (Mallo 2012). Studies (e.g. Eludoyin et al. 2017; Oyebamiji et al. 2018; Orimoogunje 2010) have been carried out on Southwestern Nigeria to assess the multifaceted impacts illegal and artisanal mining activities can have on a community. Their results reveal that illegal and small-scale mining activities often take place in areas that are rural and undeveloped. Though mining has its positive impacts on the economy (Soderholm and Svahn 2015; Ralph et al. 2018), most of its discussed effects in studies (e.g. Eludoyin et al. 2017; Oyebamiji et al. 2018; Adesipo et al. 2020) have been disastrous to the environment (Human health, Land, Water, Atmosphere and Vegetation).

Itagunmodi, a rural area and village (of the many others) in Atakunmosa West Local Government in Osun state, Nigeria, is notable for artisanal gold mining. Itagunmodi and some other close villages have undergone many researches in times past focused on the impact of the mining activity on soil, plants and water. However, most of these researches have overlooked the capacity of activities around the mining sites to impact significant changes in the microclimate of neighbouring communities. Martinez et al. (2021), amongst other authors, noted that the method of mining in the area is associated with various environmental challenges. For example, studies showed that mining activities in the area typically involves search and identification process that includes creation of slurry (process involving soil-water mix), and setting up of a sluice system with jute sack to collect the gold particles as described by Bibitayo (2018) and Tun et al. (2020). At the end, a rubber pan is typically used for the finishing process and with the gold particles poured into a bowl for further treatment with mercury. The method of gold exploitation has come with various challenges due to the inappropriate methods applied and its environmental consequences. Release of obnoxious gasses such as sulphur-dioxide from processing plants, runoffs from mine processes and leakages that are injurious to human and environmental health are attributes of many mine sites (Ralph et al. 2018; Martinez et al. 2021). Gold mining also typically produces silica-rich respirable dust particles that are capable of causing dangerous health issues and sometimes loss of lives (Andraos et al. 2018).

Consequently, as much as mining activity takes place, it continues to alter an area's environmental balance (Eludoyin et al. 2017). In developed countries, measures are being employed to mitigate the adverse effects of mining activities, though may be inadequate (Soderholm and Svahn 2015). However, in underdeveloped and developing countries as Nigeria, the effect of artisanal mining affects the environment strongly. Researchers have revealed the adverse effect of mining to the environment (Salami et al. 2003; Oyebamiji et al. 2018; Adesipo et al 2020), which includes its effect on soil, flora and fauna, water and chemical compositions (Orimoogunje 2010; Omotehinse and Ako 2019). This study aims to assess the impact of this primary and small-scale activity on the microclimate of the neighbouring communities.

Materials and Methods

Study Area

Itagunmodi is a region located within latitude 7°30′ N–7°36′ N and longitude 4°37′ E–4°42′ E, in Atakunmosa West Local Government, Osun state, Southwestern Nigeria (as seen in Fig. 5.1). Itagunmodi is close to towns such as Ile-ife, Ilesa in Osun state. The region covers an area of about 73.6 km^2 and labelled by tropical dry and wet climate pattern. It has mean annual temperature varying between 25.7 and 27.7 °C, while relative humidity over the area varies from 60 to 80%. The average rainfall of the study area is about 140 cm or approximately 1343 mm per year. The characteristics of the study area is a reflection of the parent materials and the climatic region within which it falls (Eludoyin et al. 2017). Itagunmodi's major land use is agriculture, in particular, the cultivation of cash crops such as plantain, oil palm, cocoa and kola. Also, it accommodates the plantation of food crops amongst which are yam, cassava and fruits. The soil is largely a mixture of sands and clays on the surface.

Data Collection and Methodology

Air temperature and relative humidity of different locations were obtained through north–south and west–east axes of selected region using portable weather station. The coordinates (x, y) of selected stations were also obtained with a handheld global positioning system (GPS, Magellan etrex model, ±10 m accuracy). Data were obtained at daytime (between 0900 and 1800 Nigerian Local Standard Time) at different locations within active and dormant mine sites' region and areas outside them within the study area. Heat index and dew point temperature were computed for selected locations and the data were analyzed using the National Weather Service Heat Index and Dew Point Temperature Calculator of the National Oceanic and Atmospheric Administration (NOAA). The heat index and dew point temperature are well known integrative thermal indices, while air temperature and relative humidity are considered as unitary indices. Both integrative and unitary indices have been found to be part of the many accepted indices for use in the regions (see Eludoyin and Oluwatunmise 2021).

Although studies have revealed that temperature variations are better noted at night, however, they could not be assessed for security reasons and lack of standard unmanned recording instruments for such measurements during the time of

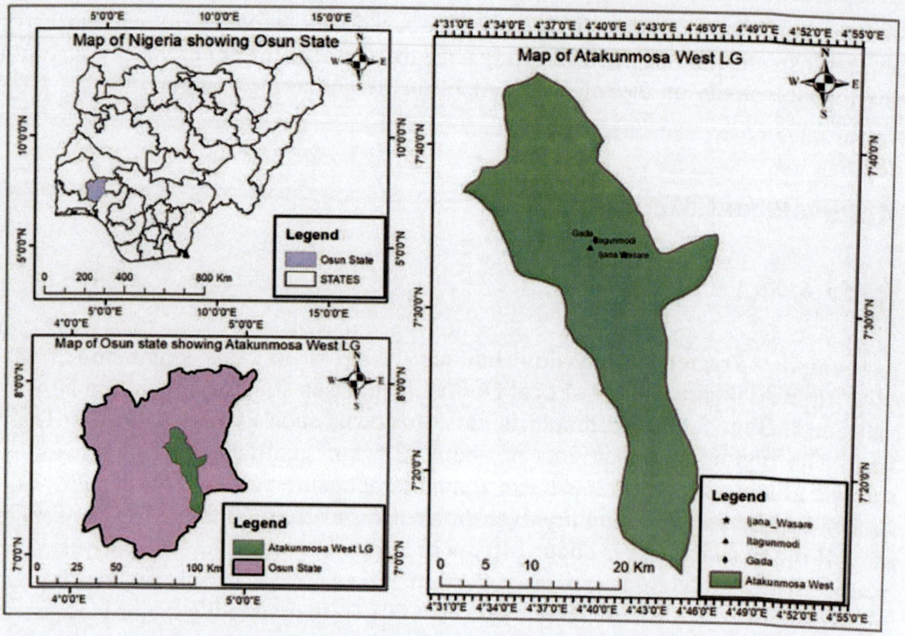

Fig. 5.1 Map of study area (*Source* Researcher)

data collection. The data were analyzed using geo-statistical and qualitative techniques. Readings were collected from portable weather station held about 1.5 m above the ground surface with hands stretched out for at least two minutes to minimize interference with land surface heat at each data point.

A set of structured questionnaire backed with in-depth interviews were administered to 15 purposively selected residents (surrounding village dwellers and miners) who were considered key informants for the study. The responses were analyzed using descriptive statistics and content analysis. The air temperature, relative humidity and thermal indices data were described using both descriptive and graphical analyses, including isotherms with PAleontological STatistical (PAST; Version 3). Trends were also examined using simple linear regression.

Results and Discussion

Statistical Analysis

Data were generally subjected to statistical analysis using Microsoft Excel (version 2013) and PAleontological STatistics Version 3.12 (Hammer et al. 2001), while descriptions of spatial variations in temperature values are drawn as Isolines. The relative humidity and temperature data, alongside the dew point temperature and heat index values, were further analyzed in Microsoft Excel to assess their respective mean and standard deviations. The relationship as well as the linear regression equation between the temperature and the relative humidity, heat index temperature and dew point temperature were assessed using Past 3.

Results

Relationship between air temperature and relative humidity, dew point temperature and heat index temperature.

Table 5.1 shows the mean, coefficient of variation and standard deviation of data collected from 88 points within and outside 2 active mine sites as well as 2 passive mine sites, in the morning and afternoon in Atakunmosa West Local Government.

Observably, data collected and analysis made (Table 5.1) show that the higher the temperature, the lower the relative humidity as seen in Fig. 5.2 and agreed by Ayanlade (2017). The mean temperature within the mine sites is seen to be greater than outside the mining sites while its relative humidity reduces. This is also observed in the mean heat index and mean dew point temperature of the mine sites (Table 5.1). However, the mean data obtained from within and outside the active mine sites are higher than the mean data obtained from passive mine sites. This depicts that the passive mine sites are beginning to get restored back to normal; temperatures, heat indexes and dew point temperatures are reducing while relative humidity is increasing. This is a pointer that the mining activity in the active mine sites has significant effect on the temperature, relative humidity, heat index and dew point temperature values of the area. By extension, this relationship could be similar for other mine sites.

On the other hand, heat index combines the air temperature and relative humidity to be explained. Heat index expresses in numerical values the relationship between comfort and discomfort. It explains how temperature (hotness) is felt on the body compared to the initial air temperature and humidity present in the atmosphere of a given location. Heat index and air temperature values collected and analyzed exhibit a direct relationship and increase as temperature increases as shown in Fig. 5.3. This is due to the intervening impact of relative humidity values.

The mean heat index (body heat) values are also seen to be higher within the mining sites than outside the mining areas (Table 5.1). Observably, the body-felt temperature, that is, heat index temperature, is greater than the air temperature, implying that the humidity level in the atmosphere does not permit for perspiration evaporation, thus resulting in excess body-felt temperature.

The coefficient of variance in the mine sites is observed to vary quite significantly from within to outside the mining sites (Table 5.1). On the average, the coefficient of variance within the mine sites is still seen to be greater than that of the outer areas of the mining sites. There is also a higher body-felt temperature within the mine sites

Table 5.1 Mine sites and their collected microclimatic parameter values

Mine sites	Areas	Temperature (°C)			Relative humidity			Heat index (°C)			Dew point temperature (°C)		
		Mean	C.O.V	S.D	Mean	C.O.V	S.D	Mean	C.O.V	S.D	Mean	C.O.V	S.D
Active	Within Mine Sites	38.87	8.18	3.18	46.38	12.14	5.63	48.86	11.73	5.73	25.36	7.37	1.87
	Outside Mine Sites	37.21	8.14	3.03	49.25	10.36	5.10	46.20	11.93	5.51	25.29	10.87	2.75
	Average	38.04	8.16	3.11	47.82	11.25	5.37	47.53	11.83	5.62	25.33	9.12	2.31
Passive	Within Mine Sites	32.03	17.64	5.65	55.07	24.50	13.49	35.63	21.67	7.72	21.25	10.07	2.14
	Outside Mine Sites	31.4	16.85	5.29	57.65	27.60	15.91	34.45	19.33	6.66	21.33	7.45	1.59
	Average	31.72	17.25	5.47	56.36	26.0	14.70	35.04	20.5	7.19	21.29	8.76	1.87

Source Field Survey Data Analysis, 2018

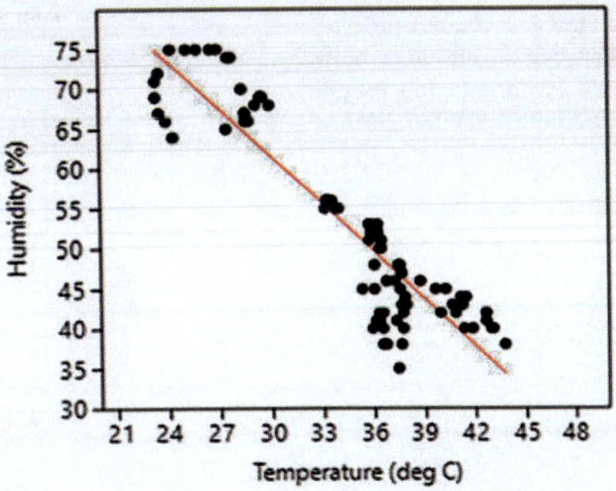

Fig. 5.2 Relationship between temperature and relative humidity values (morning and afternoon)

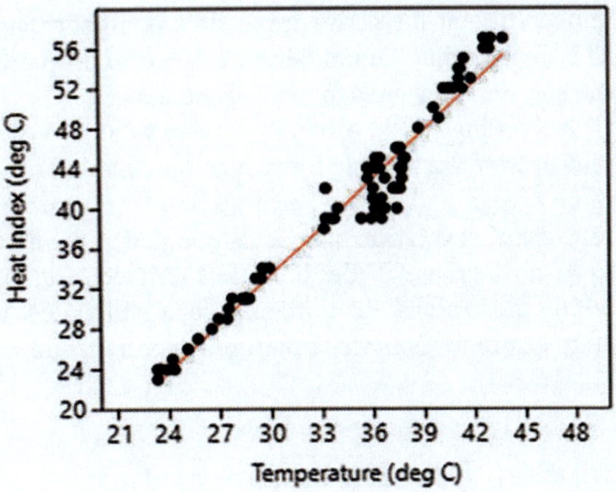

Fig. 5.3 Relationship between temperature and heat index temperature values (morning and afternoon)

than outside the mine sites. More to this, the average heat index values at the active mine sites are seen higher than those of the passive mine sites. This depicts that it is more likely for perspiration to evaporate faster at the passive mine sites than the active mine sites.

The aspect of finding the relationship between air temperature and dew point temperature may seem unimportant. However, to express how much the temperature of an environment must drop before dew can be seen also has an indirect relationship with temperature, as seen in Fig. 5.4. It states how much the environment is heated as well as the extent to which it must be cooled to experience any kind of precipitation.

With reference to Table 5.1, the mean dew point temperature of the active mine sites is seen to be slightly lesser within than outside the mine sites. This shows that the temperature outside the active mine sites must drop higher than within to experience dew. In the same wise, the mean dew point temperature outside the passive mine sites are seen slightly higher than the mean dew point temperature within the passive mine sites. More obviously on the contrary, the overall average mean dew point

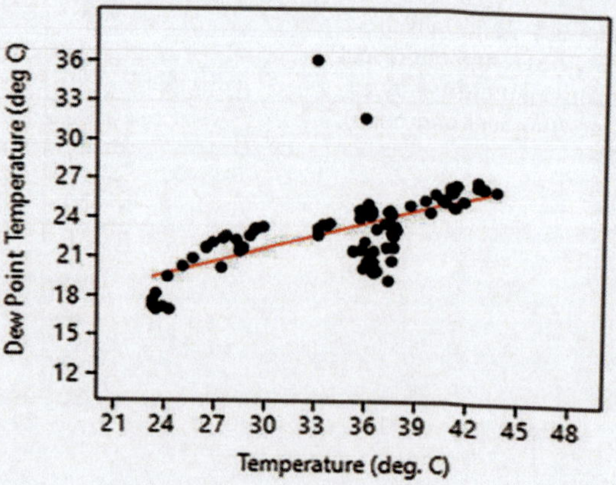

Fig. 5.4 Relationship between the air temperature and dew point temperature (morning and afternoon)

temperature at the active mine sites is higher than those of the passive mine sites. This depicts that the temperature needs to drop more at the active mine sites than at the passive mine sites to experience dew.

According to the urban heat island theory (Ayanlade 2017), places with higher temperature are notable for reduced population of vegetation that tend to make much more breeze and oxygen available to their surroundings. In the absence of or very low rate of vegetation in an area, coupled with an increasing temperature, there tends to be an increase in the body-felt temperature. Also, as air temperature increases within these areas, the temperature at which dew would be formed increases as the atmosphere loses more vapour to evapotranspiration.

Diurnal Variation in Temperature

Figure 5.5a and b show the diurnal variations in the mean temperature values in different land use in Atakunmosa West Local Government Local Government Area in the late parts of 2018. Figure 5a is an isotherm generated in the morning (09:00–10:00 LST) and Fig. 5.5b in the afternoon (12:00–13:00 LST). Figure 5a shows isotherm distribution lining in a north–south direction, except for hot spots with heat islands at Itagunmodi, Isaobi and an inverse heat island at Igun-Ibodi. On the other hand, isotherm lines southwest to East direction with no heat islands but higher temperature values at the rural parts of the map extract (Itagunmodi, Eriperi, Ijana Wasare, etc.). In Fig. 5.5a, highest temperature values were obtained in areas around mining sites 3 and 4, and towards areas southwards, as well as areas confluencing around Isaobi settlement (village). However, lower temperature values were seen at areas around Igun, Ibodi and along the main road on the map extract.

In Fig. 5.5b, highest temperature (44 °C) was obtained around the rural areas, most notably higher at areas southwest of mining sites 3 and 4 while the areas towards the Ilesha Township, north-eastward of mining sites 1 and 2, recorded the

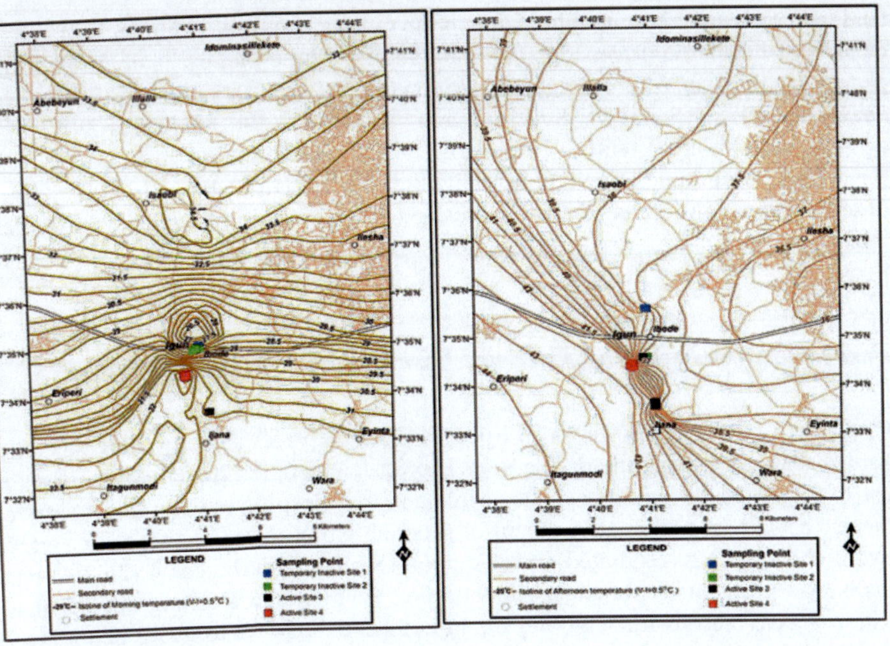

Fig. 5.5 **a** and **b**: Isoline maps showing morning and afternoon temperature variation around mine sites

low temperature values especially as seen on the nearest isotherm to Ilesha across the main road (36 °C). The outskirts of Ilesha town as shown in Fig. 5.5a and b appeared to contain lesser concretized, urbanized or cemented surfaces, and much more vegetation surfaces than the township of Ilesha. Similarly, the distribution pattern of Isotherms around the main road on Fig. 5.5a and b shows a closely lined pattern of Isotherm around the mining sites 1–4 (26.5–32.5 °C in the morning and 37–42 °C in the afternoon), depicting a close variation of temperature within the four mining sites, with mining sites 1 and 2 having lower temperature values and mining sites 3 and 4 having higher temperature values.

Summarily, Fig. 5.5a and b have temperature values closely packed around the mining sites 1–4 both in the morning and the afternoon. However, the mining sites 3 and 4 have higher temperature values than mining sites 1 and 2. Also, the rural areas and surrounding settlements (villages) exhibit higher temperature values than the CBD Ilesha Township. Incidentally, this forms the hotspots on the map, negating the theory of Urban Heat Island, even as much more vegetation is seen in both Fig. 5.5a and b, leaving the built-up areas of Ilesha north-eastward with much of tarred roads, higher population, denser housing structures and lesser vegetation than the other parts of the map.

These properties have been observed to be capable of absorbing and emitting heat (Samuels 2005; Ayanlade 2017; etc.), still, rural areas with more vegetation, scattered housing structures, lower human population and scattered untarred roads tend to absorb and emit more heat due to the mining activities rampant in the areas outskirt of the Ilesha Township (Fig. 5.5a and b). These pattern is contrary to the pattern obtained in Akure, Nigeria (Akinbode et al. 2008). Other studies carried out by scientists (Oke 1982; Svensson and Tarvainen 2004; McKenzie 2015 etc.) agree that urban areas tend to be warmer than the surrounding rural areas. From Fig. 5.5a and b, mining of gold plays a significant role in the 'abnormal' distribution of temperature. The areas exterior of Ilesha Township were generally warmer than the interior.

Perception on Effect of Thermal Discomfort in the Study Area

Of all the 15 people interviewed, 11 people (73.3%) prefer sleeping in their court-yards/backyards during the dry season. The other 4 people (26.7%) suggested the use of locally made hand fans. They explained that sleeping outside is not safe for them. They advised alternative means of generating breeze as they hope for a stable hydro-electronically generated power supply. Observably, all these 4 (26.7%) who dislike sleeping out are female while 9 of the 11 who prefer sleeping out are male. These 4 gave reasons for why they could not sleep outsides: to avoid being raped and to avoid excess mosquito bites.

As a result of excess heat, some of these population gave instances of the kind of sickness/diseases they have had in times past. This research reveals that 80% of the population experiences dry skin while 60% experiences heat rash. Some other suggested diseases/illnesses experienced more excessively since the activity had begun, malaria, dizziness, fainting and lots more, following the findings of Brousse et al. (2019) and Eludoyin and Oluwatunmise (2021).

This gave rise to the question of whether the season is as it used to be or changes abnormally. Of the total population, approximately 33% see the weather condition and seasonal change as normal while 67% see the changes as caused by some effects, thus being abnormal. The effect that was seen to be as a result of mining activity to have caused the abnormal weather condition was supported by 60% of the population who agreed that the weather is abnormal. The remaining 40% discussed that other factors gave rise to the weather change safe mining activity. The 40% of the population who discussed that the change in weather condition exist, but not due to mining activity, suggested other factors that could have given rise to weather change. These include deforestation, land degradation, atmospheric disturbances from anthropogenic activities, increasing population and nature (Spickett et al. 2011; Harden 2012; Parkes et al. 2019 etc.).

According to the statistics, of the population who discussed that the change in weather is not as a result of mining, 60% suggested that deforestation, cutting down vegetation and leaving land surface open are factors while all of them consented that land degradation causes excess heat as soil is being disturbed in the process of mining. About 40% of them agreed that the abnormal weather change is caused by anthropogenic activities such as constructions within the settlements. Around 40%

also discussed that the increasing population rate increases the temperature or heat in the environment. Lastly, 40% suggested that the change is a natural phenomenon.

Of the 67% of the total population who agreed that the season/weather is abnormal, all gave suggestions to means of solving the problems. Amongst these solutions are to send the gold miners packing, stop deforestation, charge land owners to court, etc. About 60% of these 67% argued that the mining activity be stopped and the miners be sent packing and mining activities be stopped or reduced largely. Around 10% of the 67% argued that the land owners who gave land out to the miners be charged to court.

Observably, many who suggested that the miners be sent packing away from the environment and mining be stopped or reduced, are indigenes and inhabitants of the settlements. On the other hand, the only person who suggested that land owners be charged to court is a chief in Ijana Wasare who claimed that he opposed giving out land to miners few years ago but was opposed by the council of chiefs. He further discussed that greed made them lease out land but they regret their actions as at present moment because their farmlands are being snatched totally away from them.

Generally, 5 people of the total respondents (15 people) discussed that the effect of mining on the environment has been positive while the remaining population (10 people) expressed their view as to the negative impact of mining on the environment. Observably, the 5 people who viewed the effects of mining to be positive are either Chiefs or Hausas (miners and miners' wives). The opinion to stop mining was supported by 11 people of 15 while 4 people explained why it should not be stopped. Some of the reasons given by these 11 people include the need to stop deforestation, reduce crop/food loss, stop soil degradation, ensure safe movements, stop selfish interest of land owners, end abnormal heat/storms, prevent sexual harassments and prevent inter-tribal fights/accidents. On the other hand, the 4 people who suggested that mining should not be stopped gave reasons such as making a living and encouraging population growth.

Discussion, Conclusion and Recommendation

The investigation aimed at assessing the impact of small-scale artisanal mining on the thermal weather of Itagunmodi and its environs, Atakunmosa West Local Government, Osun state, Nigeria. Mining activity within the mining area settlements has increased the population of the area as suggested by the inhabitants of the study area, and as well increasing the net income of the land owners in the areas. Yet, mining activity has been seen to have adverse effect on the weather as well as the biodiversity of the environment (Aigbedon and Iyayi 2007; Singh et al. 2021). Small-scale illegal mining has been seen to directly affect the environment, weather, man, animals, water and plants and threatens life existence in an area (Mallo 2012; Darko et al. 2021; Suglo et al. 2021 etc.). The air temperature has been seen to be higher within mine sites than surroundings and also exhibit an inverse relationship with relative humidity.

The body-felt temperature is also seen to be higher within mining environments and was also experienced largely by inhabitants of the surrounding settlements. The isotherms constructed display the variation in temperature from the closest urban settlement (Ilesha) to the mining areas, showing that the temperature is higher in mine sites than in urban areas, both in the morning and noon. There is also diurnal variation in the temperature as seen in the isotherms. This is due to the anthropogenic activities of mining that takes place around the sites. This is to say that it is possible for the air temperature within a rural area to be greater than that of an urban area, provided the small-scale artisanal mining comes in place (Owolabi et al. 2021).

Researches carried out have shown that temperature is evident in artisanal mine sites in developed countries, though measures to reduce them are being taken (Soderholm and Svahn 2015; Ralph et al. 2018). However, their effect on man's health and well-being as well as the weather has been seen irrelevant as many studies are only been restricted to areas such as its effect on water bodies, biodiversity, geomorphology and economy. This study has thus revealed that small-scale artisanal mining activity affects the weather of an environment greatly, and also has adverse effect on the well-being of man and the surrounding biodiversity.

In conclusion, temperature increase and associated stress would be compounded in communities around the mine sites. Increased awareness on coping strategies to heat stress and inclusion of propensity for heat concerns in Environmental Impact Assessment for mining activities are therefore recommended.

Acknowledgements Special gratitude to my co-author, Prof. O.O.I. Orimoogunje, for his advice and assistances to the completion of this beautiful research. I definitely will not forget the immense contribution of Dr. A.O. Eludoyin for his wonderful assistances too. I acknowledge the colleagues who assisted in data collection processes. I also recognize the incomparable cooperation of miners and in-dwellers of villages where data were collected—it was an awesome moment with you all. In the same vein, I acknowledge the comments of the reviewers in making this manuscript better. Thank you.

References

Adesipo AA, Akinbiola S, Awotoye OO, Salami AT, Freese D (2020) Impact of mining on the floristic association of gold mined sites in southwestern Nigeria. BMC Ecol 20(1):9

Aigbedon I, Iyayi SE (2007) Environmental effect of mineral exploration in Nigeria. Int J Phys Sci 2(2):33–38

Akinbode OM, Eludoyin AO, Fashae OA (2008) Temperature and relative humidity distributions in a medium-size administrative town in southwest Nigeria. J Environ Manag 87(1):95–105

Andraos C, Utembe W, Gulumian M (2018) Exceedance of environmental exposure limits to crystalline silica in communities surrounding gold mine tailings storage facilities in South Africa. Sci Total Environ 619–620(1):504–516

Ayanlade A (2017) Variations in Urban Surface Temperature: an assessment of land use change impacts over Lagos Metropolis. Clim Change 72(10):315–319

Bibitayo YA (2018) Socio-economic effect of illegal gold mining in Itagunmodi of Ijesha Region, Osun state, Nigeria. B.Sc. Dissertation in the Department of Geography, Obafemi Awolowo University, Ile-ife, Osun state, Nigeria
Brousse O, Georganos S, Demuzere M, Vanhuysse S, Wouters H, Wolff E, Linard C, Nicole PM, Dujardin S (2019) Using local climate zones in Sub-Saharan Africa to tackle urban health issues. Urban Clim 27:227–242
Darko HF, Karikari AY, Duah AA, Akurugu BA, Mante V, Teye FO (2021) Effects of small-scale illegal mining on surface water and sediment quality in Ghana. Int J River Basin Manag
Eludoyin AO, Ojo AT, Ojo TO, Awotoye OO (2017) Effects of artisanal gold mining activities on soil properties in a part of southwestern Nigeria. Cogent Environ Sci 3(1)
Eludoyin OM, Oluwatunmise O (2021) Thermal comfort and vulnerability of residents to heat stress in Ikare Akoko, Ondo State, Nigeria, Chapter 191. In: Filho WL, Luetz JM, Ayal DY (eds) Handbook of climate change management. Springer, Cham, New York, USA, pp 3273–3297
Hammer O, Harper DAT, Ryan PD (2001) PAST: paleontological statistics software package for education and data analysis. Palaeontol Electron 4(1):9Pp
Harden CP (2012) Framing and reframing questions of human-environment interactions. Ann Assoc Am Geogr 102(4):737–747
Mallo SJ (2012) Mitigating the activities of artisanal and small-scale miners in Africa: challenges for engineering and technological institutions. Int J Mod Eng Res 2(6):4714–4725
Martinez G, Restrepo-Baena OJ, Veiga MM (2021) The myth of gravity concentration to eliminate mercury use in artisanal gold mining. Extr Ind Soc 8(1):477–485
McKenzie L (2015) Climate change and planning for resilient, healthy urban environments. The Routledge handbook of planning for health and well-being: shaping a sustainable and healthy future, pp 254–270
Oladipo OG, Olayinka A, Awotoye OO (2014) Ecological impacts of mining on soils of southwestern Nigeria. Environ Explor Biol 12(1):179–186
Omotehinse AO, Ako BD (2019) The environmental implications of the exploration and exploitation of solid materials in Nigeria with a special focus on tin in Jos and coal in Enugu. J Sust Min 18(1):18–24
Oke TR (1982) The Energetic basis of the urban heat island. Q J R Meteorol Soc 108(455):1–24
Orimoogunje OO (2010) Land cover response to changes in forest resources utilization in southwestern Nigeria: GIS perspective. Ife Res Publ Geogr 9(1):83–97
Owolabi AO, Amujo K, Olorunfemi IE (2021) Spatiotemporal changes on land surface temperature, land and water resources of host communities due to artisanal mining. Environ Sci Pollut Res 28:36375–36398
Oyebamiji A, Amanambu A, Zafar T, Adewumi AJ, Akinyemi DS (2018) Expected impacts of active mining on the distribution of heavy metals in soils around Iludun-Oro and its environs, southwestern Nigeria. Cogent Environ Sci 4(1)
Parkes B, Cronin J, Dessens O, Sultan B (2019) Climate change in Africa: costs of mitigating heat stress. Clim Chang 154(3–4):461–476
Singh PK, Chualya SK, Singh VK (2021) Development of microwave radar based periphery surveillance system in a selected mining area. J Electromagn Waves Appl 35(4):529–548
Soderholm P, Svahn N (2015) Mining, regional development and benefit-sharing in developed countries. Resour Policy 45(c):78–91
Ralph O, Gilles N, Fon N, Luma H, Greg N (2018) Impact of artisanal gold mining on human health and the environment in the Batouri gold district, east Cameroon. Acad J Interdiscip Stud 7(1):25–44
Salami A, Jimoh MA, Muoghalu J (2003) Impact of gold mining on vegetation and soil in southwestern Nigeria. Int J Environ Stud 60(4):343–352
Samuels R (2005) Environmental accountability: users, buildings and energy. Glob Warm Built Environ 1(1):40–57
Spickett JT, Brown HL, Rumchev K (2011) Climate change and air quality: the potential impact on health. Asia Pac J Public Health 23(2):37–45

Suglo P, Effah P, Acheampong AA, Sunkari R, Yeboah A (2021) Effects of illegal mining on the environment, economy and agricultural productivity. Biochem Mol Biol 6(4):79–91
Svensson D, Tarvainen L (2004) The past and present urban heat Island of Beijing. Environ Sci 1(1)
Tun AZ, Wongsasuluk P, Siriwong W (2020) Heavy metals in the soils of placer small-scale gold mining sites in Myanmar. J Health Pollut 10(27):1–12

Chapter 6
Modelling and Prediction of Rainfall in the North-Central Region of Nigeria Using ARIMA and NNETAR Model

E. H. Chukwueloka and A. O. Nwosu

Abstract Modelling and predicting rainfall in research are essential because the inferences from the study will assist decision-makers, planners, and policymakers in mitigating the effects of drought or flooding in the environment. This chapter aims to fit time series models to rainfall data from seven states in the Nigerian north-central region. The data used for this research was obtained from NIMET (Jan 1989–Dec 2019). The rainfall data set was modelled and predicted using the conventional seasonal Autoregressive Integrated Moving Average (ARIMA) and Neural Network Times Series Autoregressive (NNETAR) models. The time plot sequence shows the time series data is stationary, and the Augmented Dick Fuller (ADF) test did not suggest otherwise. Furthermore, the Hegy and Canova-Hansen tests indicate seasonality in the data with order 1. When the ARIMA and NNETAR models were applied to the rainfall data set, the analysis revealed that the NNETAR model outperformed the ARIMA model in modelling and predicting the Ilorin, Jos, Lafia, Lokoja, and Minna rainfall data sets. In contrast, the ARIMA model outperformed the NNETAR model for predicting rainfall in Abuja and Makurdi. The fitted models were used to predict monthly rainfall in the north-central region for the next five years. The forecast suggests an expected increase in rainfall in Lafia, Abuja, and Minna. At the same time, an expected decrease in rainfall in Ilorin, Lokoja, Jos, and Makurdi states in the north-central region of Nigeria.

Keywords ARIMA · NNETAR · Stationarity · Seasonality · Rainfall

---

E. H. Chukwueloka (✉)
Mathematics Unit, School of Basic Sciences, Nigeria Maritime University, Okerenkoko, Delta State, Nigeria
e-mail: henry4real345@yahoo.com; henry.chukwueloka@nmu.edu.ng

A. O. Nwosu
Faculty of Marine Environmental Management, Department of Marine Environment and Pollution Control, Nigeria Maritime University, Okerenkok, Delta State, Nigeria
e-mail: agnes.nwosu@nmu.edu.ng

J. C. Egbueri et al. (eds.), *Climate Change Impacts on Nigeria*, Springer Climate,
https://doi.org/10.1007/978-3-031-21007-5_6

Introduction

Rainfall is necessary for all forms of life, all aspects of socioeconomic development, and the maintenance of healthy ecosystems (Ikot, The Environment and Socioeconomic Development 2021). The extent of rainfall distribution in a region generally determines a country's agricultural performance. As sustainable lifestyles have begun to gain ground in most countries, Africa's Gross Domestic Product (GDP) is solely dependent on agriculture and its resources, which are dependent on the region's climatic conditions (Tadesse 2010; Nyong et al. 2007; Bjornlund et al. 2020). Rainfall modelling and forecasting information are critical for planning and designing sustainable measures to carry out sustainable agricultural practices such as the best planting time, the best crop to strive for when and where, and a large harvest and it also determines all other human activities that improve food security, economic growth, and a healthy and safe environment for human development (Tuğrul 2019). Most time series methods, such as Moving Average (MA), Autoregressive Moving Average (ARMA), and others, have been utilised to evaluate rainfall data in diverse works of literature (Momani 2009; Kajuru et al. 2019).

According to Somvanshi et al. (2006), rainfall is a natural climatic event that remains difficult to predict due to climatic fluctuation and the precipitation forecasts are critical for agriculture, and the growth and development of plants, which contribute to Africa's economy. The author (Somvanshi et al. 2006) examined and projected mean annual rainfall data extending over 104 years in the Hyderabad region of India using ANN and ARIMA algorithms. The study demonstrated that the ANN model might be an effective forecasting tool for rainfall prediction, outperforming the ARIMA model. ARIMA is primarily a linear statistical technique for forecasting rainfall and time series to make future forecasts. While accurate rainfall estimation is critical for water resource planning, it can be tough to analyse places with a scarcity of rainfall stations. This makes rainfall estimation a legitimate worry when the appropriate method is used (Darji 2019).

In order to model and forecast monthly rainfall data from 1980 to 2011 collected by the Agrometeorological department of Junagadh Agricultural University, Dwivedi et al. (2019) employed a seasonal ARIMA and ANN model and the results of the study indicate that the seasonal ARIMA (1,0,0) (3,1,1)12 model was the most appropriate ARIMA model after assessing its performance and residuals. In contrast, the ANN back-propagation model (5-12-1) did well in calculating Junagadh, India's average annual rainfall. The study also shows that when comparing actual and predicted monthly rainfall, the ANN model performed better than the ARIMA model.

Akinbobola et al. (2018) used ARIMA and ARIMA to model and forecast rainfall in Nigeria's forest and Savannah eco-climatic areas. The seasonal Autoregressive Integrated Moving Average (ARIMA) model effectively modelled and predicted monthly rainfall in the study's selected stations. Pal and Mazumdar et al. (2019) used NNETAR and ARIMA to forecast the monthly rainfall volume during the monsoon season from 2019 to 2020 for the two districts in the Gangetic West Bengal meteorological sub-division of India. It was established that the NNETAR model outperformed the traditional ARIMA model for the district. ARIMA and its seasonal counterpart ARIMA and neural network had been used by many researchers to tackle problems on some weather variables prediction like air temperature, humidity, solar radiation, wind speed, and so on Wang et al. (2006), Jain and Kumar (2007), Di et al. (2014), Grigonyte and Butkeviciuye (2016), Lai and Dzombak (2020), Ray et al. (2021), Shad et al. (2022).

In this study, the NNETAR and ARIMA model techniques were used to model and forecast rainfall in Nigeria's north-central region.

Materials and Methods

Study Area

North-central region Nigeria lies between 8.57667 and 6.459444 latitude and 4.898056 and 6.66444 longitude. It consists of seven (7) states, namely, Niger State, Kwara State, Nasarawa State, Kogi State, Benue State, Plateau State, and the Federal Capital Territory (FCT). North-central Nigeria is the gateway between Northern Nigeria (North East and North West Nigeria) and Southern Nigeria (South East and South West Nigeria). North-central Nigeria's climatic characteristics are comprised of two distinct seasons; the rainy season starts from April through October; and the dry season usually starts from November to March. The rainy season peak at 1500 mm in August and the dry seasons peaked in February through March (see Fig. 6.1).

Data Collection

The information was obtained from the Nigerian Meteorological Agency (NiMet).

The Table 6.1 shows various meteorological stations in the north-central region of Nigeria.

Box-Jenkins Approach

As a result of the work of statistician George Box and Gwilym Jenkins in 1970, ARMA, ARIMA, and ARIMA models can now be predicted more accurately using

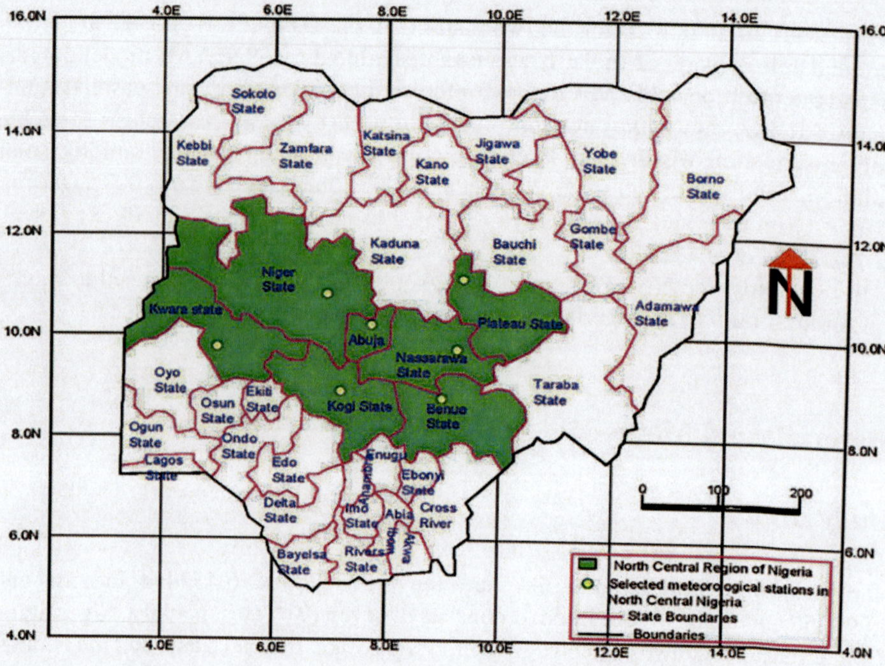

Fig. 6.1 Showing the map of north-central Nigeria and its selected metrological stations

Table 6.1 Met stations and rainfall durations

S/no.	Met stations	Durations
1	Abuja	1989–2019
2	Ilorin	1989–2019
3	Jos	1989–2019
4	Lafia	1989–2019
5	Lokoja	1989–2019
6	Makurdi	1989–2019
7	Minna	1989–2019

a procedure that uses the intuition of obtaining the most appropriate fit of a time series data to previous values of the same time series, and with the instinct of making better forecasting models. Iterative three-stage modelling is used in the model-building process outlined below.

Model-Identification and Selection

Box-Jenkins Model Identification.

(1) Stationarity and seasonality

The initial step to take when constructing a Box-Jenkins model is checking for the stationarity of the time series data and if there is significant seasonality to be fitted. Hence, the augmented Dickey-fuller test was used to determine stationarity and seasonality (ADF).

(2) Detecting stationarity

A time plot series may be used to determine stationarity. For a stationary series, the run sequence plot should depict a constant direction and scale. Additionally, it may be identified using autocorrelation plot. A non-stationarity is frequently shown by a slow decaying autocorrelation plot.

(3) Detecting seasonality

To detect a seasonality, we employ autocorrelation plot or a seasonal subseries.

(4) Differencing for stationarity

In order to attain stationarity, Box-Jenkins suggests a differencing strategy. However, Box-Jenkins models may be developed by curve fitting and minusing the fitted values from the initial data. For instance, ensuring that entirely the variables are "stationary," recognising seasonality for all the dependent series and utilising the plots of the autocorrelation and partial autocorrelation functions of the dependent time series to find out whether autoregressive or moving average component should be used in the model or otherwise.

Augmented Dickey-Fuller Test

An augmented Dickey-Fuller test (ADF) is used to determine if a time series sample has a unit root. It is an enhanced version of the Dickey-Fuller test that is applicable to a more complex collection of time series models. The test statistic, the augmented Dickey-Fuller (ADF) test uses, is a negative number. The more negative it is, the stronger the rejection of the hypothesis that there is still a unit root at some level of confidence (Mushtaq and Rizwan 2011).

Testing Procedure

The unit root test can then be performed with the null hypothesis $\gamma = 0$ and the alternative hypothesis of $\gamma < 0$. Once a value for the test statistic D $F_t = \frac{\widehat{\gamma}}{SE(\widehat{\gamma})}$ is obtained, it may be compared to the Dickey-Fuller test's critical value. Assume that the test statistic is less than (since this test is asymmetrical, we do not evaluate an absolute value) the crucial value (a more significant negative). In this scenario, the null hypothesis $\gamma = 0$ is rejected, and no unit root is present (Mushtaq and Rizwan 2011).

Model Parameter Estimation

When utilising computing algorithms to determine the optimal coefficients for the given ARIMA model, the maximum likelihood estimate or nonlinear least-squares estimation are two of the most often used approaches.

Model Diagnostic and Forecasting

Model-checking involves determining if the estimated model meets the requirements of a stationary univariate process. The residuals, in particular, should be independent of one another and have constant mean and variance throughout time. Misspecification may be detected by graphing the mean and variance of residuals with time and running a Ljung-Box test or plotting autocorrelation and partial autocorrelation of the residuals. If the estimate is inadequate, we must go back to step one and create a more accurate model. We utilise the proposed model to characterise and predict the time series observation for the next five years if the estimate is appropriate.

Probabilistic Models

2.4.1. Autoregressive (AR) models: In this study, the current value of the series of the rainfall data y_t may be described as a function of p past values (past rainfall data), $y_{t-1}, y_{t-2}, \ldots, y_{t-p}$, where p is the number of steps into the past that are needed to predict the current value.

The autoregressive model of order p, A.R. (p) is shown in Eq. 6.1:

$$y_t = \varphi_1 y_{t-1} + \varphi_2 y_{t-2} + \cdots + \varphi_p y_{t-p} + \varepsilon_t \tag{6.1}$$

y_t is a stationary series, $\varphi_1, \varphi_2, \ldots, \varphi_p$ are the parameters of A.R. ($\varphi_p \neq 0$), and ε_t is a Gaussian white noise series with mean zero and variance σ_w^2.

Moving Average (MA) Model

A series y_t is said to follow a moving average process of order q or M.A. (q) is shown in Eq. 6.2:

$$y_t = \varepsilon_t + \theta_1 \varepsilon_{t-1} + \theta_2 \varepsilon_{t-2} + \cdots + \theta_q \varepsilon_{t-q} \tag{6.2}$$

where $\theta_1, \theta_2, \ldots,$ are the M.A parameters.

Autoregressive Moving Average (ARMA)

ARMA models were developed for stationary time series using autoregressive, moving average, and mixed autoregressive moving average (ARMA) techniques. When developing a stochastic model to describe a stationary time series, it is usually better to start with a hybrid autoregressive moving average model. Both p and q were used to indicate the order of an ARMA model. The model parameters connect what occurs during period t to both the past values and the random errors that happened during past periods. A general ARMA model is shown in Eq. 6.3:

$$y_t = \varphi_1 y_{t-1} + \varphi_2 y_{t-2} + \cdots + \varphi_p y_{t-p} + \varepsilon_t + \theta_1 \varepsilon_{t-1} + \theta_2 \varepsilon_{t-2} + \cdots + \theta_q \varepsilon_{t-q} \tag{6.3}$$

Autoregressive Integrated Moving Average (ARIMA). Autoregressive integrated moving average models are abbreviated as ARIMA. The model performs well when the data display a steady or consistent trend across time with few outliers. They are sometimes referred to as Box-Jenkins. ARIMA is a subset of the autoregressive moving average (ARMA) model. This model is fitted to time series data in order to improve understanding of the data or to anticipate the next point in the series (forecasting). They are used in certain circumstances when data show non-stationary behaviour, and an initial differencing step may be used to eliminate the non-stationary behaviour. The model is often referred to as an ARIMA (p.d,q) model, where p, d, and q are non-negative integers indicating the order of the autoregressive, integrated, and moving average components of the model. ARIMA models are a critical component of the Box-Jenkins method for time series modelling. The models are specified in Eqs. 6.4 and 6.5:

$$\Phi(B)W_t = \theta(B)Z_t \tag{6.4}$$

$$(1 - B)^d X_t = \theta(B)Z_t \tag{6.5}$$

where d is the number of difference

W_t is an ARMA process and $W_t = (1 - B)^d X_t$

The model for X_t is non-stationary due to the fact that the A.R. operator just on the left has a d root on the unit circle, d is often 1, and the random walk is ARIMA (0,1,0).

Seasonal Autoregressive Integrated Moving Average (ARIMA)

When the data have a phenomenon that repeated the pattern in time series data, the series is denoted by Seasonal ARIMA (p, d, q) (P, D, Q) $_s$. The generalised form of ARIMA model is shown in Eqs. 6.6–6.9:

$$\Phi_p(B)\Phi_P\left(B^s\right)W_t = \theta_q(B)\theta_Q\left(B^s\right)Z_t \quad (6.6)$$

where

$$\Phi_P\left(B^s\right) = 1 - \Phi_1 B^s - \Phi_2 B^{2s} -,,, -\Phi_P B^{ps} \quad (6.7)$$

$$\theta_Q\left(B^s\right) = 1 - \theta_1 B^s - \theta_2 B^{2s} -,,, -\theta_Q B^{Qs} \quad (6.8)$$

$$\nabla_s^D = \left(1 - B^s\right)^D \quad (6.9)$$

where $\Phi_P(B^s)$ is a polynomial of seasonal autoregressive with order P and seasonal period m is denoted by SAR(P); $\theta_Q(B^s)$ is a polynomial of the seasonal moving average with Q, seasonal period S is denoted by SMA(Q), and ∇_s^D is the seasonal differencing operator.

Neural Network Autoregressive (NNETAR) Model

In this study, a feed-forward neural network is trained with lagged values of y as inputs and a single hidden layer with size nodes. The inputs correspond to lags 1–p and m–M.P., where m = frequency (y). If x_{reg} is given, its columns are utilised as inputs as well. If there are missing values in y or x_{reg}, the fit ignores the associated rows (and those that rely on them as lags). A total of repeated networks is fitted, each with a randomly chosen initial weight. These are then averaged when predictions are computed. The network has been taught to predict in a single step. Forecasts with several steps are generated recursively. The fitted model for non-seasonal data is referred to as an NNAR (p, k) model, where k is the number of hidden nodes. This is comparable to an A.R. (p) model, except that the functions are nonlinear. The fitted model for seasonal data is referred to as an NNAR (p, P, k) [m] model, which is comparable to an ARIMA (p,0,0)(P,0,0)[m] model but nonlinear. Time series prediction approaches such as Box-Jenkins, ARMA, or ARIMA all presume a linear relationship between the inputs and outputs. Neural networks have the advantage of being able to predict any nonlinear function without previous knowledge of the data series' features (Sunil et al. 2019).

Model Performance Measures

The present study used a comparative study on the two models' numerical performance (ARIMA & NNETAR) on the rainfall data for the north-central region of Nigeria. Forecast accuracy was evaluated using the Mean Absolute Error (MAE), the Root Mean Square Error (RMSE), and the Mean Absolute Scaled Error (MASE).

The Akaike information criterion (AIC) is a fine-grained approach for assessing a model's probability to predict/estimate future values based on in-sample fit. A good model is the one with the minimum AIC value among all other models.

$$\text{AIC} = 2K - 2ln\left(\hat{L}\right) \tag{6.10}$$

where L is the likelihood function, n represents the number of measured values, and k denotes the number of estimated parameters.

The root-mean-square deviation (RMSD) or root-mean-square error (RMSE) is a widely used statistic that expresses the difference between predicted values by a model (sample or population values) and actual values.

$$\text{RMSE} = \sqrt{\frac{1}{n}\sum_{j=1}^{n}(y_t - \hat{y}_t)^2} \tag{6.11}$$

where y_t denotes the observation of rainfall at the t-th time point and $\hat{y}_t$ denotes estimate of y_t then forecast error will be

$$\varepsilon = y_t - \hat{y}_t \tag{6.12}$$

Mean Absolute Error (MAE): The mean absolute error is a statistic that indicates the average degree of errors in a collection of predictions without regard for their direction. It is the average of the absolute differences between prediction and actual observation across the test sample, where all individual deviations are given equal weight.

$$\text{MAE} = \frac{1}{n}\sum_{t=1}^{n}|y_t - \hat{y}_t| \tag{6.13}$$

Mean Absolute Scaled Error (MASE): MASE is a metric for determining prediction accuracy. The prediction values' mean absolute error is divided by the in-sample one-step naïve forecast's mean absolute error.

$$\text{MASE} = \text{mean}\left(\frac{|e_j|}{\frac{1}{T-m}\sum_{t=m+1}^{T}|y_t - y_{t-m}|}\right) \tag{6.14}$$

Results and Discussion

Descriptive Statistics for the Seven Meteorological Stations in the North-Central Region of Nigeria

The descriptive statistics for the seven states in Nigeria's north-central area are displayed in Table 6.2. Lafia's meteorological station had the greatest cumulative rainfall of the seven states (74,722 mm), while Makurdi received the least (37,200 mm) from 1989 to 2020. In addition, Lafia receives the most rainfall per month (1151 mm), while Makurdi receives the least (397 mm). The average rainfall in these meteorological stations is also shown in Table 6.2.

Time Plots for the Seven Meteorological Stations in the North-Central Region of Nigeria

Figure 6.2 shows the time plot of the monthly series for Abuja, Ilorin, Jos, Lafia, Lokoja, Makurdi, and Minna, respectively, from 1989 to 2019. The figure revealed that there is an irregular pattern in the data series. Throughout the series, the mean was not constant, and the data series had a trend variation. The figure also shows that the data series are stationary; therefore, there is a need to carry out a unit root test (see Figs. 6.3, 6.4, 6.5 and 6.6).

Stationarity Test

To test for stationarity or unit root test, Augmented Dickey-Fuller test is used. Table 6.3 displays the stationarity results obtained when applied on the data set.

Table 6.3 suggests that the series for Abuja, Ilorin, Jos, Lafia, Lokoja, Makurdi, and Minna attain stationarity since the p-value (0.01) for the series is less than 5% level of significance. The Hegy and Canova-Hansen tests were conducted to determine the seasonality and order at which the seasonal variation is stationary in the series. Table 6.3 also shows seasonality in the series for all the meteorological stations in the north-central region of Nigeria with order 1.

Table 6.2 Descriptive statistics

Var	N	Range	Minimum	Maximum	Sum	Mean		Standard deviation	Variance
	Statistic	Statistic	Statistic	Statistic	Statistic	Statistic	Standard error	Statistic	Statistic
Abuja	372	720	0	720	45,044	121.08	6.810	131.354	17,253.807
Illori	372	615	0	615	39,424	105.98	5.847	112.775	12,718.199
Jos	372	867	0	867	40,020	107.58	6.395	123.340	15,212.861
Lafia	372	1151	0	1151	74,722	200.87	13.054	251.781	63,393.796
Lokoja	372	524	0	524	38,678	103.97	5.452	105.145	11,055.524
Makurdi	372	397	0	397	37,200	100.00	5.428	104.695	10,961.057
Minna	372	481	0	481	37,447	100.66	5.729	110.501	12,210.512

Source Authors calculation

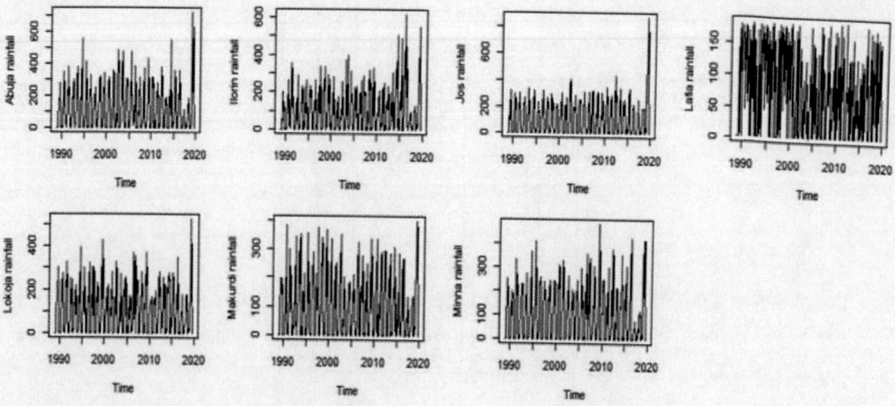

Fig. 6.2 Time plot for month rainfall series for the Metrological Stations in the north-central region of Nigeria

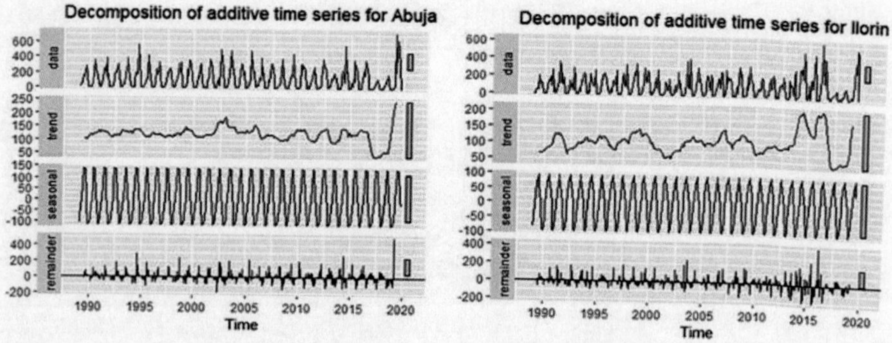

Fig. 6.3 Shows the additive time series decomposition into a random, seasonal, trend, and observed variation for Abuja and Ilorin rainfall data

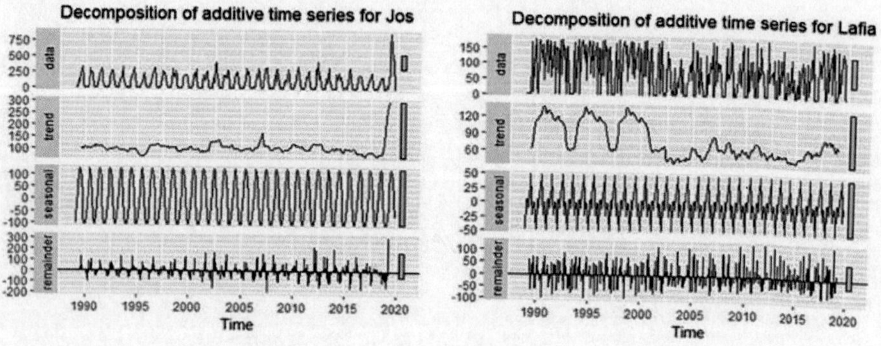

Fig. 6.4 Shows the additive time series decomposition into a random, seasonal, trend, and observed variation for Jos and Lafia rainfall data

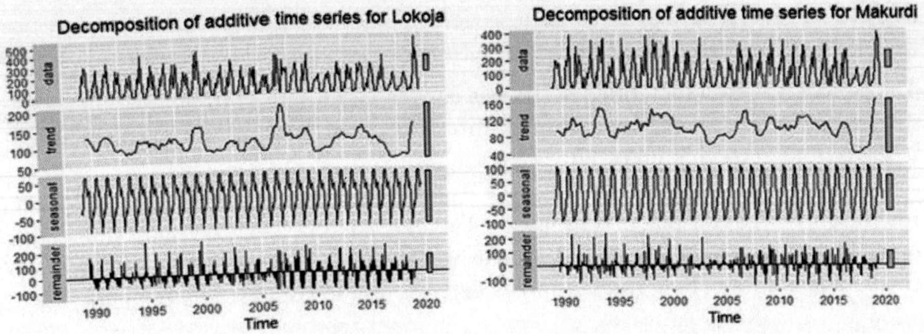

Fig. 6.5 Shows the additive time series decomposition into a random, seasonal, trend, and observed variation for Lokoja and Makurdi rainfall data

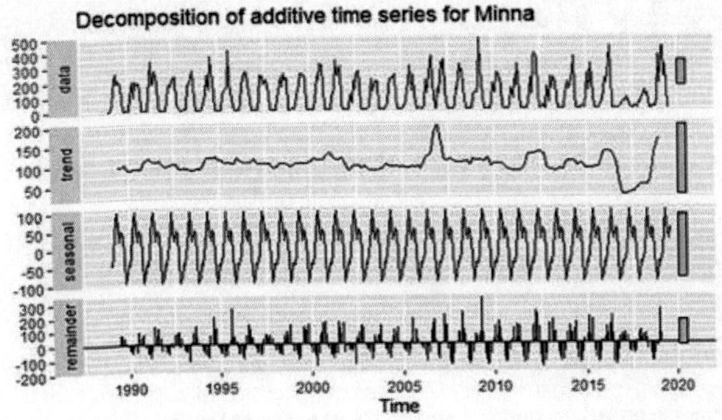

Fig. 6.6 Shows the additive time series decomposition into a random, seasonal, trend, and observed variation for Minna rainfall data

Table 6.3 Augmented Dickey-Fuller test for the series

Meteorological stations	df statistic	Lag-order	p-value	Seasonality	
				Decision	Order
Abuja	−10.854	7	0.01	Yes	1
Ilorin	−7.975	7	0.01	Yes	1
Jos	−8.6011	7	0.01	Yes	1
Lafia	−6.2428	7	0.01	Yes	1
Lokoja	−9.7871	7	0.01	Yes	1
Makurdi	−9.7561	7	0.01	Yes	1
Minna	−9.9813	7	0.01	Yes	1

Model Identification

The Autocorrelation Function (ACF) was used to identify the probability time series models that best describe the monthly rainfall series for the meteorological stations in the north-central region of Nigeria.

Figures 6.7 and 6.8 show the series ACF plot for monthly rainfall for Lokoja, Lafia, Jos, Ilorin, Abuja, Minna, and Makurdi. The plots suggest that the highest available order for the Autoregressive Model (AR) and Moving Average (MA) are between 0 and 9. The number of significant spikes determines the number of attainable orders.

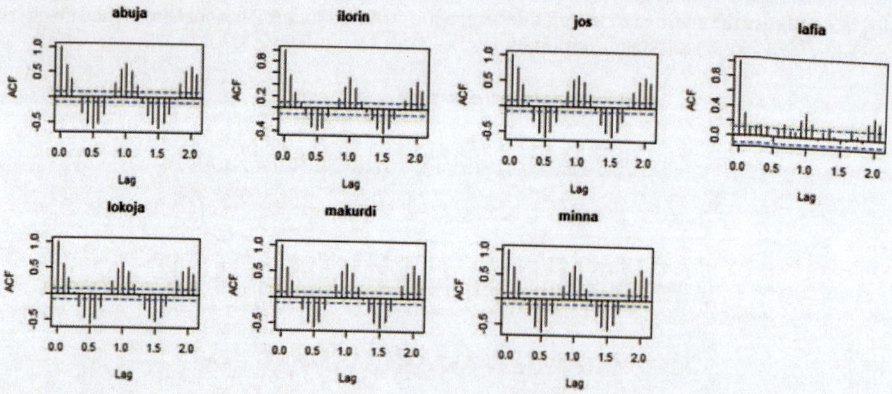

Fig. 6.7 ACF plots of monthly rainfall series for the Metrological Stations in the North-central region of Nigeria

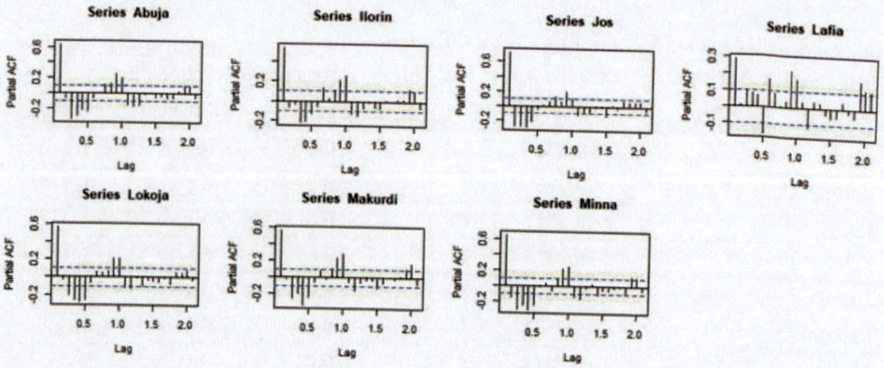

Fig. 6.8 PACF plots of monthly rainfall series for the Metrological Stations in the north-central region of Nigeria

Parameter Estimation

After the stationarity of the data has been attained and the ACF examined, we then concluded that the data series follows the seasonal Autoregressive Integrated Moving Average (ARIMA) model. Table 6.4 show the parameter estimation table for the model that best describes the monthly rainfall series for Abuja, Lafia, Lokoja, Makurdi, and Minna. Ilorin, Jos.

Model Diagnostic and Forecasting

The residual from each fitted ARIMA model was checked using the autocorrelation plots at different lags. The ACF plot does not show any autocorrelation among the residuals, which shows we can use the ARIMA model for forecasting. This section aims to check how well the selected models fit the actual data; this is an iterative procedure used to evaluate model aptness by examining whether the model assumptions are satisfied. Model adequacy is concerned with the residuals' analysis. To check the estimated models' adequacy, we first look at the standardised residual plot of the models. Figures 6.9, 6.10, 6.11, 6.12, 6.13, 6.14 and 6.15 shows the residual plots, histogram, and the ACF of the models fitted for the north-central rainfall data. The residuals' standardised time plots do not show any strange behaviour or specific direction; therefore, these plots do not suggest any major irregularities with the model. The histogram of the residuals has a bell shape, which is a good indication of the normality of the distribution (see Fig. 6.16).

Modelling of Rainfall Data Using NNETAR

A feed-forward neural network model was fitted to each of the metrological stations in north-central Nigeria using lagged values of the time series data as inputs and a single hidden layer with a number of nodes. On the data series, the Neural Network Autoregressive (NNETAR) model was used, and the best NNETAR model that suite the data series is as follows: Abuja has an NNAR(3, 1, 2) [12] model with 20 networks on the average, each of which is a 4-2-1 network with 13 weights; Ilorin has an NNAR (3,1,2) [12] model with an average of 20 networks, each of which is a 4-2-1 network with 13 weights; Jos has an NNAR(2,1,2) [12] model with an average of 20 networks, each of which is a 3-2-1 network with 11 weights; Lokoja has an NNAR(2,1,2) [12] model with an average of 20 networks, each of which is a 3-2-1 network with 11 weights; Makurdi has an NNAR(2, 1,2) [12] model with an average of 20 networks, each of which is a 3-2-1 network with 11 weights; and Minna has an NNAR(4,1,2) [12] model with an average of 20 networks, each with a 5-3-1 network with 22 weights (see Fig. 6.17 and Table 6.5).

Table 6.4 Model estimation of monthly rainfall series

Metrological station	Model	AR1	AR2	MA1	MA2	SMA1	SMA2	
Abuja ARIMA (2, 0, 2) (0, 1, 2)12	Estimate	1.4638	−0.7612	−1.2631	0.7204	−0.6363	−0.1235	
	S.E	0.1039	0.0859	0.0912	0.08	0.0711	0.0653	
	Performance criteria	R2	*AIC*	*Log-likelihood*	*ME*	*RMSE*	*MAE*	*MASE*
		5552	4119.71	−2052.86	1.474	72.674	46.636	0.864
Ilorin ARIMA (1, 0, 1) (0, 1, 1)12	*Model*	*AR1*	*MA1*	*SMA1*				
	Estimate	0.6872	−0.3877	−0.7151				
	S.E	0.0982	0.1252	0.0426				
	Performance criteria	R2	*AIC*	*Log-likelihood*	*ME*	*RMSE*	*MAE*	*MASE*
		6348	−2072.36	−2072.36	2.6351	78.0368	52.8908	0.8303
Jos ARIMA (1, 0, 0) (1, 1, 2)12	*Model*	*AR1*	*SAR1*	*SMA1*	*SMA2*			
	Estimate	0.5417	−0.5252	−0.2029	−0.4374			
	S.E	0.0446	0.479	0.4667	0.3471			
	Performance criteria	R2	*AIC*	*Log-likelihood*	*ME*	*RMSE*	*MAE*	*MASE*
		4389	4035.08	−2012.54	2.5446	64.8012	38.871	0.829
Lafia ARIMA (0, 1, 2) (2, 0, 0)12	*Model*	*MA1*	*MA2*	*SAR1*	*SAR2*			
	Estimate	−0.8333	−0.137	0.2195	0.1582			
	S.E	0.0512	0.0516	0.0535	0.0551			
	Performance criteria	R2	*AIC*	*Log-likelihood*	*ME*	*RMSE*	*MAE*	*MASE*
		3366	4053.61	−2021.81	0.7885	57.6242	48.7047	0.9561

(continued)

Table 6.4 (continued)

Metrological station	Model	AR1	AR2	MA1	MA2	SMA1	SMA2	
Lokoja ARIMA (2, 0, 2) (0, 1, 1)12	*Model*	*AR1*	*AR2*	*MA1*	*MA2*	*SMA1*		
	Estimate	1.6944	−0.962	−1.4885	0.7935	−0.8742		
	S.E	0.0331	0.0308	0.06	0.0459	0.0486		
	Performance criteria	*R2*	*AIC*	*Log-likelihood*	*ME*	*RMSE*	*MAE*	*MASE*
		4879	4055.68	−2021.84	−2.4156	68.2173	47.144	0.8721
Makurdi ARIMA (1, 0, 2) (1, 1, 2)12	*Model*	*AR1*	*MA1*	*MA2*	*SAR1*	*SMA1*	*SMA2*	
	Estimate	0.4936	−0.3356	0.1611	−0.6678	−0.0746	−0.449	
	S.E	0.1404	0.1435	0.0582	0.4875	0.501	0.3679	
	Performance criteria	*R2*	*AIC*	*Log-likelihood*	*ME*	*RMSE*	*MAE*	*MASE*
		4192	4018.88	−2002.44	0.2449	63.1537	44.6792	0.8238
Minna ARIMA (2, 0, 2) (0, 1, 1)12	*Model*	*AR1*	*AR2*	*MA1*	*MA2*	*SMA1*		
	Estimate	1.5854	−0.8424	−1.224	0.5559	−0.7502		
	S.E	0.0726	0.0677	0.1003	0.079	0.0583		
	Performance criteria	*R2*	*AIC*	*Log-likelihood*	*ME*	*RMSE*	*MAE*	*MASE*
		3993	3978.77	−1983.38	−0.3462	61.7099	41.8374	0.9069

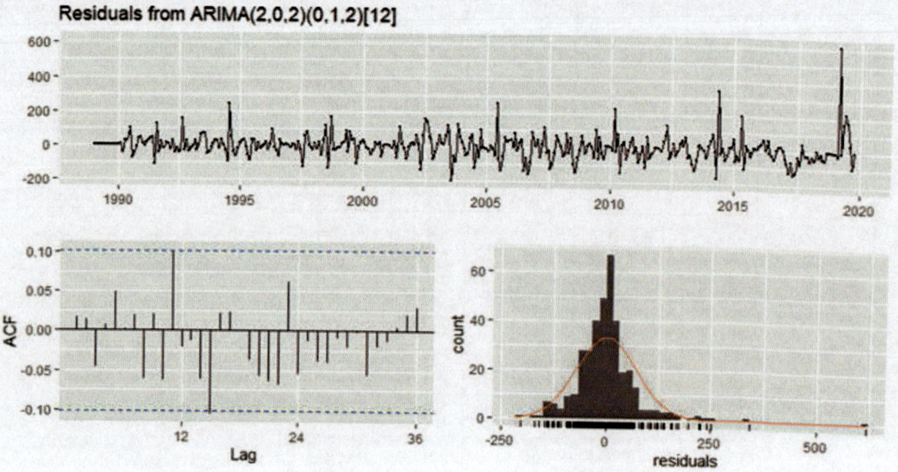

Fig. 6.9 Residual diagnostic for Abuja

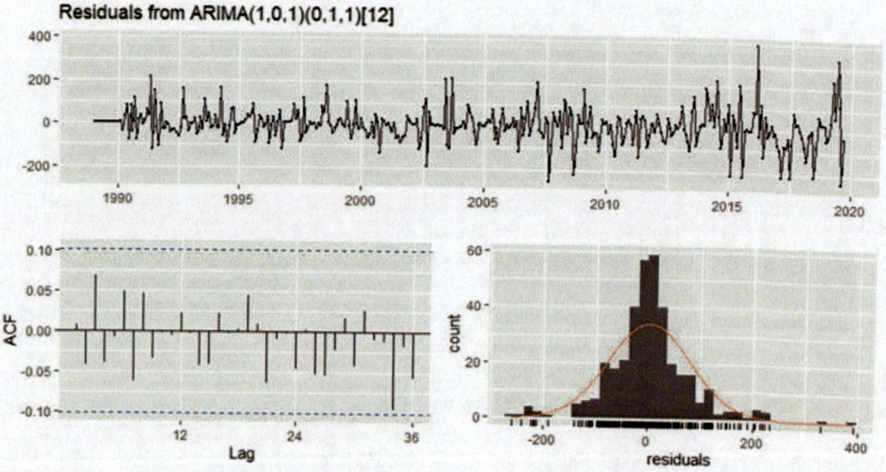

Fig. 6.10 Residual diagnostic for Ilorin

Predictive Performance Comparison Between ARIMA and NNETAR Model

The predictive performance is compared between ARIMA and NNETAR model for all the time series data under study. Predictive accuracy or performance is reported in Table 6.6.

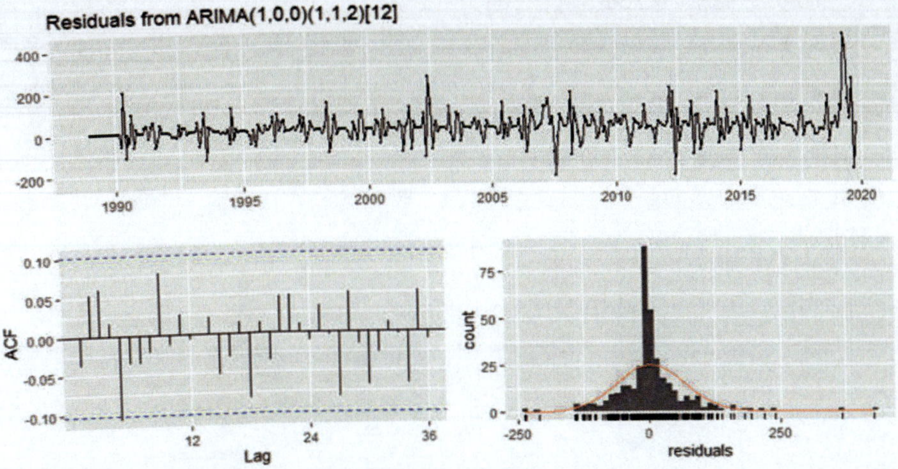

Fig. 6.11 Residual diagnostic for Jos

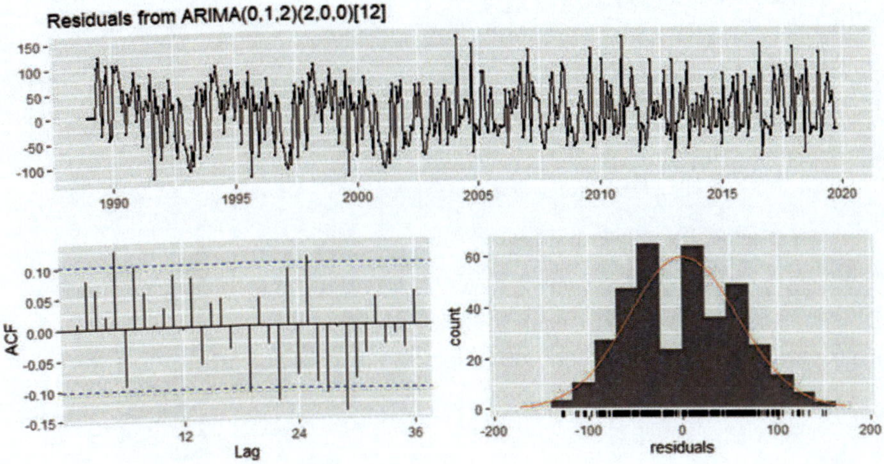

Fig. 6.12 Residual diagnostic for Lafia

Table 6.6 shows that the ARIMA model best predicted Abuja and Makurdi based on the values of diagnostic measures such as RMSE, MAE, and MASE. The NNETAR model, on the other hand, predicted Ilorin, Jos, Lafia, Lokoja, and Minna the best.

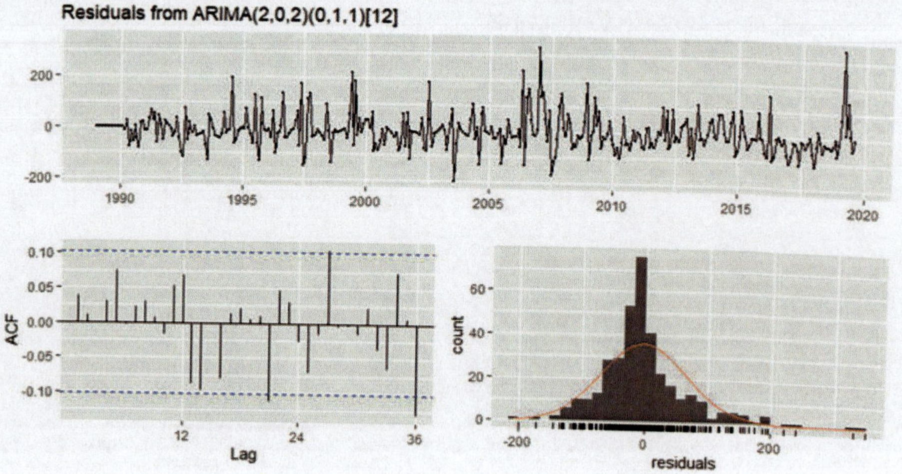

Fig. 6.13 Residual diagnostic for Lokoja

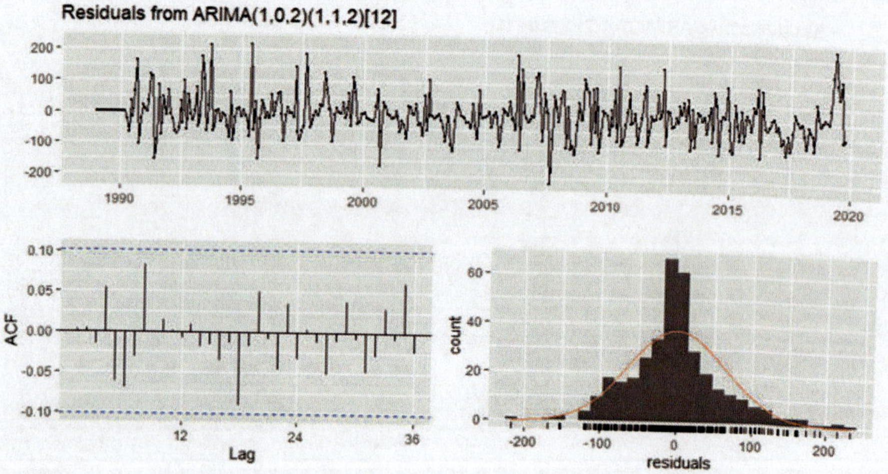

Fig. 6.14 Residual diagnostic for Makurdi

Conclusion

For the north-central area of Nigeria, where one of its obligations is to improve human existence and boost the nation's economy, rainfall is a major worry since it is necessary for all forms of life. Because the conclusions from this study will help decision-makers, planners, and policymakers lessen the consequences of drought or floods in the environment, research on modelling and predicting rainfall is crucial. The Nigerian Meteorological Agency provided the information for this study (NiMet). The meteorological station in Lafia recorded the most rainfall overall among the

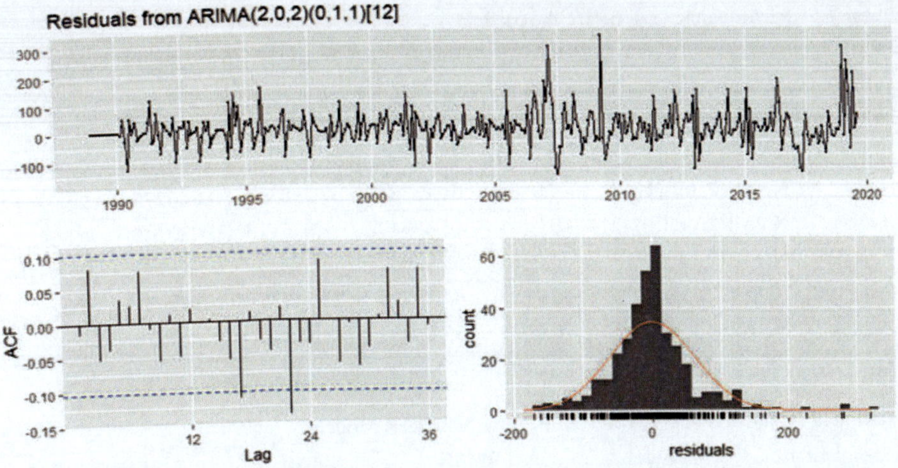

Fig. 6.15 Residual diagnostic for Minna

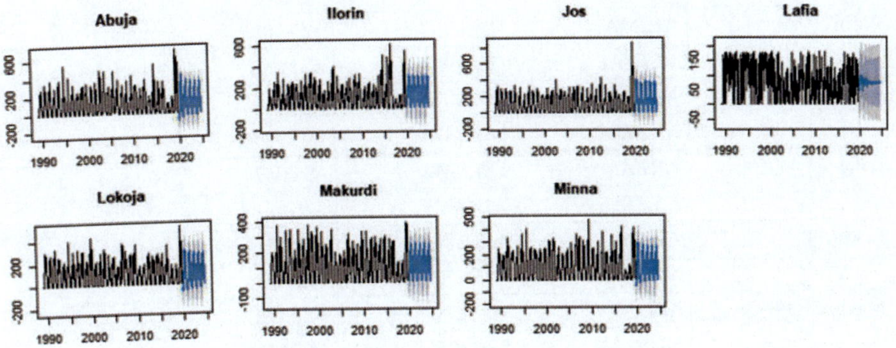

Fig. 6.16 ARIMA 5 years forecast plot for monthly rainfall for the Metrological Stations in the north-central region of Nigeria

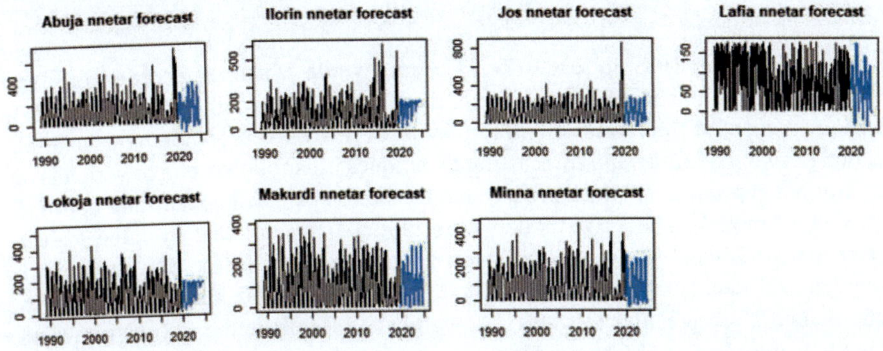

Fig. 6.17 NNETAR 5 years forecast plot for monthly rainfall for the Metrological Stations in the north-central region of Nigeria

Table 6.5 Performance Criteria for the models

Met stations	Model	ME	RMSE	MAE	MASE	ACF1
Abuja	NNAR(3,1,2)[12]	−0.0073	75.60725	46.04598	0.852792	−0.01375
Ilorin	NNAR(3,1,2)[12]	0.00802	74.79632	50.9982	0.80058	0.087496
Jos	NNAR(2,1,2)[12]	−0.0185	61.43784	39.39831	0.840275	0.002472
Lafia	NNAR(25,1,13)[12]	−0.0133	6.439192	1.056713	0.020745	0.009253
Lokoja	NNAR(2,1,2)[12]	−0.0322	65.97047	44.53025	0.823709	0.093785
Makurdi	NNAR(2,1,2)[12]	−0.0255	64.30594	45.74727	0.84353	−0.02433
Minna	NNAR(4,1,3)[12]	−0.0077	52.83834	33.49177	0.725982	−0.009

Table 6.6 Comparison between ARIMA and NNETAR Model

Met stations	Model	RMSE	MAE	MASE
Abuja	*ARIMA*	*72.674*	*46.636*	*0.864*
	NNETAR	75.6073	46.046	0.85279
Ilorin	ARIMA	78.0368	52.8908	0.8303
	NNETAR	*74.7963*	*50.9982*	*0.80058*
Jos	ARIMA	64.8012	38.871	0.829
	NNETAR	*61.4378*	*39.3983*	*0.84028*
Lafia	ARIMA	57.6242	48.7047	0.9561
	NNETAR	*6.43919*	*1.05671*	*0.02075*
Lokoja	ARIMA	68.2173	47.144	0.8721
	NNETAR	*65.9705*	*44.5303*	*0.82371*
Makurdi	*ARIMA*	*63.1537*	*44.6792*	*0.8238*
	NNETAR	64.3059	45.7473	0.84353
Minna	ARIMA	61.7099	41.8374	0.9069
	NNETAR	*52.8383*	*33.4918*	*0.72598*

seven states from 1989 to 2019 (74,722 mm), while Makurdi recorded the least (37,200 mm). Lafia likewise had the highest monthly rainfall of 720 mm, while Makurdi recorded the lowest monthly rainfall of 397 mm. A comprehensive application of two modelling and forecasting techniques is as follows: one of which is the traditional approach (ARIMA model) and the other is a machine learning approach (NNETAR model). These two models were applied to the data sets from the 7 seven-state meteorological stations in the north-central area of Nigeria. The time plot of the 7 states' data set was plotted to check stationarity, trend, and seasonality. The augmented Dickey-Fuller test was then conducted to affirm the stationarity of the data and the Hegy and Canova-Hansen tests indicate seasonality in the data with order 1. After all, Box-Jenkins methods were carried out. The data set was modelled using ARIMA first before the original data was applied to the NNETAR model, and the model performance metrics were used to determine which model better suited the data set (AIC, RMSE, MAE, and MASE). The model with thc lowest error was selected as the best model, and the prediction was carried out using that model. The ARIMA model and the NNETAR model were compared to determine which model better fits the data set in the seven states in Nigeria's north-central region. The ARIMA model best fitted the data sets for the meteorological stations in Abuja and Makurdi, while the NNETAR model best fitted the data sets in Ilorin, Jos, Lafia, Lokoja, and Minna. Based on these fitted models, we were able to forecast rainfall for the following five years.

Recommendation

I would recommend developing a hybrid model to incorporate the data set's long memory of volatility, as well as broader coverage of the research region.

References

Akinbobola A, Okogbue EC, Ayansola AK (2018) Statistical modeling of monthly rainfall in selected stations in forest and savannah eco-climatic regions of Nigeria. J Climatol Weather For 2018(6), S1. https://doi.org/10.4172/2332-2594.1000226

Bjornlund V, Bjornlund H, Van Rooyen AF (2020) Why agricultural production in sub-Saharan Africa remains low compared to the rest of the world–a historical perspective. Int J Water Resour Dev 36(sup1):S20–S53. https://doi.org/10.1080/07900627.2020.1739512

Box G, Jenkins G (1970) Time series analysis: forecasting and control

Darji M (2019) Rainfall forecasting using neural networks

Di C, Yang X, Wang X (2014) Hybrid neural network models for hydrologic time series forecasting. Plus One 9:e104663

Dwivedi DK, Kelaiya JH, Sharma GR (2019) Forecasting monthly rainfall using autoregressive integrated moving average model (ARIMA) and artificial neural network (ANN) model: a case study of Junagadh, Gujarat, India. J Appl Natl Sci 11(1), 35–41. https://doi.org/10.31018/jans.v11i1.1951

Grigonyte E, Butkeviciuye E (2016) Short-term wind speed foresting using ARIMA model. Energetika, 62

Ikot (2021) The environment and social and economic development nigerian aluminium smelter company. https://alscon.net/ikot-abasi-smelter/78-the-environment-and-socio-economic-development.html

Jain A, Kumar AM (2007) Hybrid neural network models for hydrologic time series forecasting. Appl Soft Comput 7:585–592

Kajuru J, Abdulkarim K, Muhammed M (2019) Forecasting performance of arima and arima models on monthly average temperature of zaria, Nigeria. ATBU J Sci Technol Educ 7(3):205–212. https://www.atbuftejoste.com/index.php/joste/article/view/823

Li Y, Dzombak DA (2020) Use of the autoregressive integrated moving average (ARIMA) model to forecast near-term regional temperature and precipitation. Weather For 35(3):959–976

Momani PE (2009) Time series analysis model for rainfall data in Jordan: case study for using time series analysis. Am J Environ Sci 5https://doi.org/10.3844/ajessp.2009.599.604

Mushtaq, Rizwan (August 17, 2011) Augmented Dickey-Fuller Test. SSRN: https://ssrn.com/abstract=1911068 or https://doi.org/10.2139/ssrn.1911068

Nyong A, Adesina F, Osman Elasha B (2007) The value of indigenous knowledge in climate change mitigation and adaptation strategies in the African Sahel. Mitig Adapt Strategy Glob Change 12:787 797

Somvanshi VK et al (2006) Modelling and prediction of rainfall using artificial neural network and ARIMA techniques. J Ind Geophys Union 10(2):141–151

Pal S, Mazumdar D (2019) Forecasting monthly rainfall using an artificial neural network 3:65–73

Ray S, Das SS, Mishra P, Al Khatib AMG (2021) Time series ARIMA modeling and forecasting of monthly rainfall and temperature in the south Asian countries. Earth Syst Environ 5(3):531–546.

Shad M, Sharma YD, Singh A (2022). Forecasting of monthly relative humidity in Delhi, Indian using ARIMA and ANN models. Modeling Earth Systems and Environment. https://doi.org/10.1007/s.40808-022-01385-8

Sunil S, Acharya S, Jogi AK (2019) Application of hybrid model for forecasting prices of jasmine flower in Bangalore, India. Int J Sci Technol Res 8(11)

Tadesse D (2010) The impact of climate change in Africa. Inst Secur Stud Ser 220(November):20. http://www.issafrica.org/uploads/Paper220.pdf %5Cn

Tuğrul KM (2019) Soil management in sustainable agriculture, sustainable crop production, Mirza Hasanuzzaman, Marcelo Carvalho Minhoto Teixeira Filho, Masayuki Fujita and Thiago Assis Rodrigues Nogueira, IntechOpen, https://doi.org/10.5772/intechopen.88319. https://www.intechopen.com/chapters/68683

Wang W, Van Gelder, Vrijling J, Ma J (2006) Forecasting daily sstreamflow using hybrid ANN models. J Hydrol 324:383–399

Chapter 7
Analysis of Rainfall Trend and Variability in Lagos, Southwestern Nigeria

Christian A. Njoku, Vincent N. Ojeh, Edward Naabil, Michael C. Ekwe, Maaku M. Joshua, Elizabeth Elijah, and Micheal W. Malgwi

Abstract The rainy season in the littoral city of Lagos has offered a lingering problem to water resource managers and the region's socioeconomic activities over the years. In a changing climate, it is critical to understand how seasonal wet regimes vary in order to predict floods and manage water resources. The aim of this research is to assess rainfall changes using the trend and variability of two independent meteorological stations located inside Lagos' coastal city. Monthly cumulative rainfall data was acquired and examined on monthly, seasonal, and annual scales for a 30-year period (1984–2013). Modified Mann–Kendall (MM-K) and Sen's slope estimator (Q) were used to estimate rainfall trends and statistical significance tests, while the Precipitation Concentration Index (PCI) and Rainfall Variability Index (RVI) were used to quantify rainfall variability (RVI). Rainfall increased from January to May and November to December, but declined from July to August at both stations. In June, August, and October, however, both stations showed a distinct mix of upward and downward trends: dry season rainfall increased across stations, whereas wet season and yearly rainfall increased in the Ikeja station and fell in the Lagos Roof station, respectively. For Ikeja and Lagos Roof stations, PCI wet and dry seasons range from 9 to 20 and 9 to 42, respectively, while annual PCI ranges from 11 to 28 for both

C. A. Njoku
Nigerian Meteorological Agency, Abuja, Nigeria

V. N. Ojeh (✉) · M. M. Joshua · E. Elijah
Department of Geography, Taraba State University, Jalingo, Nigeria
e-mail: vinceojehnetwork@gmail.com

E. Naabil
Department of Agricultural Engineering, Bolgatanga Technical University, Bolgatanga, Ghana
e-mail: naabil.edward@bolgatu.edu.gh

M. C. Ekwe
National Space Research and Development Agency, Lugbe, Nigeria

M. W. Malgwi
Department of Geography, Nigerian Army University, Bui, Borno State, Nigeria

J. C. Egbueri et al. (eds.), *Climate Change Impacts on Nigeria*, Springer Climate,
https://doi.org/10.1007/978-3-031-21007-5_7

stations. In the wet and dry seasons, the Lagos Roof station had higher positive RVI tendencies than the Ikeja station depicting a high variability of rainfall within Lagos.

Keywords Rainfall variability · Rainfall trend · Modified Mann–Kendall · Sen's slope · Seasonal variation · Climate change

Introduction

Climatic variables such as rainfall, temperature, and streamflow are often studied for the purpose of ascertaining the tendency for environmental change over time (Oguntunde et al. 2011a). Water resource planning and management are made easier through the analysis of long-term records of rainfall. However, the increasing rate of population and economic growth suggests that there will be an increasing demand for water resources (Oguntunde et al. 2011a). The climate projections for Nigeria have also indicated a continued rise in rainfall with severe impacts on human, agriculture, and water resources (e.g., Daramola et al. 2017; Abiodun et al. 2012). Therefore, it is imperative to continually study rainfall trends and variability, especially in a changing climate (e.g., Odjugo 2010; De Luis et al. 2011; Bello 1998).

Several studies on rainfall characteristics have been conducted at different scales in Nigeria (Oguntunde et al. 2011a; Ifabiyi and Ojoye 2013; Akinsanola and Ogunjobi 2014; Ogungbenro and Morakinyo 2014; Salami et al. 2016; Daramola et al. 2017). For instance, Oguntunde et al. (2011a) observed that spatial distribution of rainfall was highly dependent on latitude while no clear longitudinal relationship was seen. Rainfall trend over the Sahelian region of Nigeria also revealed that the drought episodes of 1972/73 and 1980 were a consequence of downward trend of rainfall amount in the 1970s and 1980s, while the upward trend in rainfall amount between 1990 and 2009 resulted in flooding over northern Nigeria within that period (Ifabiyi and Ojoye 2013). According to Ogungbenro and Morakinyo (2014), a significant transition of rainfall regimes from dry to wet, indicating an upward shift of rainfall distribution, were found in the climate zones of Nigeria. Similar findings have also been reported using observational (Akinsanola and Ogunjobi 2014) and model (Daramola et al. 2017) data across Nigeria. At the local scale, Oguntunde et al. (2011b) indicated a non-significant increasing trend in rainfall over Ibadan, Nigeria. At the basin scale, Animashaun et al. (2020) found varying results in rainfall changes, with some sub-basins showing negative rainfall trends, and the reverse is true for other sub-basins within the same climate zones. These confirm the previous findings of Oloruntade et al. (2017) over the Niger South basin. While the previous studies have provided useful information about rainfall changes and distribution across different parts of Nigeria and at different scales, the differing results of rainfall variability between stations within the same climatic zones or hydrological response units suggest a strong local influence on rainfall-induced mechanisms (Odunuga and

Oyebande 2007; Animashaun et al. 2020). The regional-scale assessment of rainfall may potentially lead to mischaracterizations of local rainfall distribution, with implications on agriculture and water resource management.

Therefore, the aim of the present study is to assess the trend and variability of rainfall in the coastal city of Lagos, southwestern Nigeria. The focus is on two independent weather stations, Ikeja and Lagos Roof, with different geographical characteristics namely Lagos mainland and Lagos Island, respectively. The difference in the geographical characteristics of the stations will help evaluate the argument about the potential local influence on rainfall variability (Animashaun et al. 2020). The findings will therefore enhance reliable seasonal rainfall predictions, flood forecasting, and early warning systems in the littoral mega city.

Materials and Methods

Study Area

The coastal mega city of Lagos in southwestern Nigeria is predominantly a commercial city and the hub of economic activities in the southern part of Nigeria, yet prone to consistent flooding activities. It lies between longitude 2°42′–3°42′ E and latitude 6°23′–6°41′ N (Fig. 7.1). It is one of the most populated cities in the entire globe, harboring over 24.6millions people (https://lagosstate.gov.ng/about-lagos/). The area is characterized by a tropical climate (Köppen-Geiger climate classification is Aw) with average annual rainfall estimated to be 1693 mm while the average temperature is 27 °C; the highest rainfall is observed in June while least rainfall occurs in December. The city experiences two distinct seasons, wet (April–October) and dry (November–March) (https://en.climate-data.org/africa/nigeria/lagos/lagos-552). The wet season is bimodal, with a little dry season in July/August (Omotosho 1988). Ikeja weather station lies on 39.4 m above the sea level and is sited inside an international airport located in the mainland area, while Lagos Roof weather station is on 14 m above the sea level located on the shoreline of the island area of Lagos, the stations are less than 20 km apart (Table 7.1). Hydrologically, the Owo River is one important surface water source in Lagos particularly in the western section of the city (Adegun, et al. 2015). It is the source of direct water intake for water supply by Lagos State Water Corporation to Amuwo-Odofin, Ojoand parts of Badagry Local Government Areas. Additionally, the area is drained by the lower Ogun River, which discharges into the Lagos Lagoon through a meandering system with several bifurcations due to erosional activities, sedimentation, and human activities. The sub-surface geology of Lagos indicates a complex lithology of alternating sequence of sand and clay deposits (Longe et al. 1987).

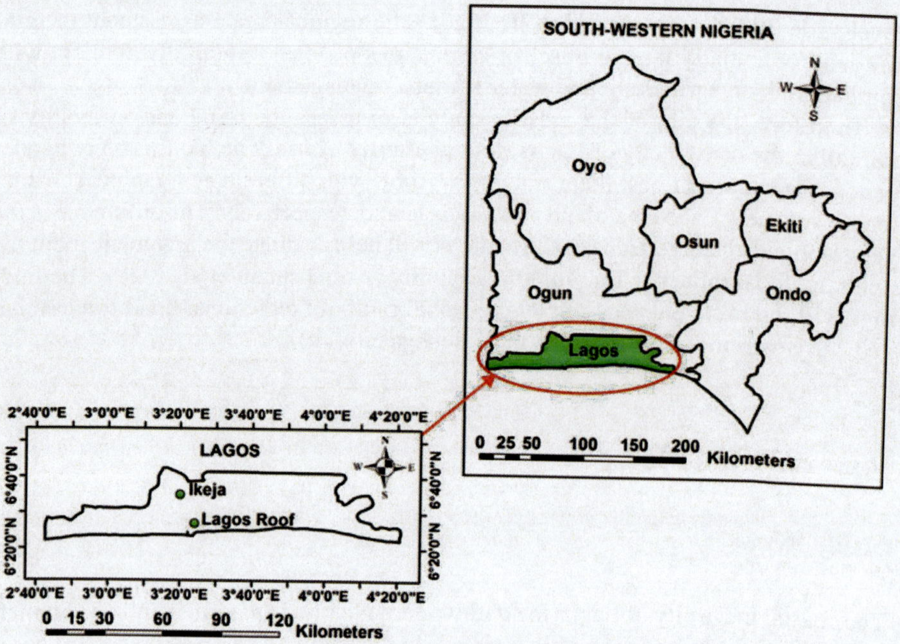

Fig. 7.1 Map of study area showing the selected weather stations

Table 7.1 Descriptions of stations location

Station name	Latitude (°N)	Longitude (°E)	Altitude (m)	Period of Data
Ikeja	06.35	03.20	39.4	1984–2013
Lagos Roof	06.27	03.24	14	1984–2013

Rainfall Data

Monthly accumulated rainfall measured by data spanning 1984–2013 (30 years) were utilized in this study. The data for the selected weather stations were daily rain gauge-observed rainfall amount aggregated to total monthly rainfall, and the data set was obtained from the data section of the Nigerian Meteorological Agency. The rainfall data covered all the months of the year and the choice of the period is as a result of the consistency in the data.

Trend Analysis

The statistical significance and the magnitude of the trend analysis in the time series data used in this study were determined using the modified Mann–Kendall (MMK) (Hamed and Rao 1998) and Sen's slope estimator (Sen 1968). Non-parametric methods such as Mann–Kendall (MK) (Mann 1945; Kendall 1975) and modified Mann–Kendall (MMK) have been used by researchers around the world to ascertain the trend in hydro-meteorological parameters (You and Wang 2004; Krishan et al. 2015; Hamed 2009), MK and MMK assume the null hypothesis (H_0) of no trend, and this is tested against the alternative hypothesis (H_A) of either a decreasing or increasing trend in the data series. Data are checked for randomness using the auto-correlation function; this is achieved by computing autocorrelation for the data set at different time lags. To check for significant autocorrelation in the data series for this study, the autocorrelation syntax acf () in R programming language was adopted, where significant autocorrelation coefficient existed in the series; the Bias-Corrected Pre-Whitening (BCPW) proposed by Hamed (2009) was performed before trend testing (Hamed 2008a, b; Hamed 2009). An important advantage of the MMK is that the effect of statistically significant autocorrelation coefficient is minimized or not expected in the resulting trend (Krishan et al. 2015; Abdul Aziz and Burn 2006; You and Wang 2004). Mann–Kendall statistics (*S*) is calculated as follows (Eq. 7.1):

$$S = \sum_{k=1}^{n-1} \sum_{j=k+1}^{n} \mathrm{sgn}\left(x_j - x_k\right), \tag{7.1}$$

where n is the length of data series, $k = 1,2,3,\ldots n-1, j = k+1, k+2, k+3, \ldots\ldots n$, x_j and x_k are values of sequential data, and x_j is a reference point of each of the data point to be compared with the rest of the data points x_k as shown in Eq. 7.2

$$\mathrm{sgn}\left(x_j - x_k\right) = \begin{bmatrix} 1 \text{ if } \left(x_j - x_k\right) > 0 \\ 0 \text{ if } \left(x_j - x_k\right) = 0 \\ -1 \text{ if } \left(x_j - x_k\right) < 0 \end{bmatrix}, \tag{7.2}$$

When length of data series is equal to 18, the statistics S is approximately Gaussian with mean $E(S)$ and variance Var(S) as follows (Eqs. 7.3 and 7.4):

$$E(S) = 0, \tag{7.3}$$

$$\mathrm{Var}(S) = \frac{1}{18}\left[n(n-1)(2n+5) - \sum_{h=1}^{g} t_h(t_h - 1)(2t_h + 5)\right], \tag{7.4}$$

where *n* denotes the number of data points, stands for a number of tied groups (tied groups are set of data samples with the same value), while the represents a number of ties of extent *h.* Mann–Kendall Statistics (*S*) is then computed in a standardized statistics form (*Z*) as given below (Eq. 7.5):

$$Z = \begin{bmatrix} \frac{S-1}{\sqrt{\mathrm{Var}(S)}} \text{ if } S > 0 \\ 0 \text{ if } S = 0 \\ \frac{S+1}{\sqrt{\mathrm{Var}(S)}} \text{ if } S < 0 \end{bmatrix}, \tag{7.5}$$

Modified Mann–Kendall (MMK) Trend Test

When data is not random and influenced by autocorrelation, the modified Mann–Kendall tests may be used for trend detection studies (Krishan et al. 2015). Hamed and Rao (1998) have proposed a variance correction approach to address the issue of serial correlation in trend analysis. Data are initially detrended and the effective sample size is calculated using the ranks of significant serial correlation coefficients which are then used to correct the inflated (or deflated) variance of the test statistic (Blain 2013; Patakamuri and O'Brien 2020).

In this method, the effect of all significant autocorrelation coefficients is removed from the data set as proposed by Hamed and Rao (1998). For this purpose, a modified variance of *S*, designated as Var(*S*)*, was adopted as follows (Eq. 7.6):

$$\mathrm{Var}(S)^* = V(S)\frac{n}{n^*} \tag{7.6}$$

where n^* = effective sample size. The n/n^* ratio was computed directly from the equation proposed by Hamed and Rao (1998) as follows (Eq. 7.7):

$$\frac{n}{n^*} = 1 + \frac{2}{n(n-1)(n-2)}\sum_{i=1}^{n-1}(n-1)(n-i-1)(n-i-2)r_i \tag{7.7}$$

where n = actual number of observations and r = lag - i significant autocorrelation coefficient of rank i of time series. Once Var(*S*) * was computed from Eq. (7.7), then it was substituted for Var(*S*) in Eq. (7.6). At 90, 95, and 99% confidence levels, the null hypothesis (H_0) is rejected, and the alternative hypothesis (H_A) of a trend is accepted if Z is greater than 1.645, 1.96, and 2.33, respectively, a positive or negative value of Z indicates an upward or downward trend, respectively, in hydro-meteorological variables.

Sen's Slope Estimator

Sen's slope estimator, which is also non-parametric, is used to estimate the true slope overtime of an existing linear trend in hydro-meteorological variables. This is necessary because these variables do not follow a particular pattern in their fluctuation or deviation from the normal distribution over time (Sen 1968). Some of the studies which applied this estimator include Aditya et al (2021), who used it for rainfall trend analysis in West Kalimantan; Aswad et al. (2020), for time series rainfall of 70 years in Iraq; Agbo and Ekpo, (2021) for trend analysis of the variations of ambient temperature in Calabar, Southern Nigeria; and Ghalhari et al. (2022), in trend analysis of Humidex as a heat discomfort index in Iran. With the assumption that the trend is linear, Sen's slope can be computed as follows (Eq. 7.8):

$$f(t) = Qt + B \tag{7.8}$$

where Q is the slope and B is a constant. To obtain the slope estimate Q in Eq. 7.8, the slopes of all the data value pairs are calculated as (Eq. 7.9):

$$Q_i = \frac{x_j - x_k}{j - k} \tag{7.9}$$

where x_j and x_k denote the data measurements at time and k, j is subsequent time after time k ($j > k$). If there are values of x_j in the data time series, we get as many as $N = n(n - 1)/2$ slope estimates Qi. The Sen's estimator of the slope is the median of these N values of Qi. The N values of Qi are ranked from the smallest to the largest, and the Sen's estimator is the median of all slopes, Sen's estimator is computed as $Q_{\text{med}} = T(N + 1)/2$ if N is odd, and it is considered as $Q_{\text{med}} = [T(N/2) + T(N + 2)/2]/2$ if N is even. In the end, Q_{med} is calculated by a two-sided test at 100 $(1 - \alpha)\%$ confidence interval, and then a true slope can be obtained by the non-parametric test. A positive value of Q_i indicates an increasing or upward trend, while a negative value denotes a decreasing or downward trend in the time series.

Rainfall Variability

The city of Lagos is the hub of economic activities in southwestern Nigeria, and it is very prone to flooding; hence, the study of the variability of rainfall will help decision-makers come up with policies for water resources management as well as reduce loss of lives and economic goods, so many indices exist for the study of rainfall variability: Standard Precipitation Index (SPI), Coefficient of Variance (CV), Precipitation Concentration Index (PCI), Rainfall Variability Index (RVI), in the present study, PCI (e.g., Adegun and Balogun 2015; Ezenwaji et al. 2017), and RVI (e.g., Ogunrinde et al. 2019; Wagaye and Antensay 2020) were applied in studying rainfall variability at seasonal and annual time scales.

Precipitation Concentration Index (PCI)

Precipitation Concentration Index (PCI) is a powerful temporal rainfall distribution indicator that can be used for seasonal and annual time scales, according to Zhang et al. (2019). PCI proposed by Oliver (1980) and further developed by De Luis et al. (2011) was calculated at seasonal and annual time scales as indicated by Eqs. (7.10–7.12). PCI values below 10 represent low precipitation distribution (uniform monthly precipitation distribution during the season/year), values from 11 to 15 signify moderate precipitation distribution, and values ranging 16 and 20 denote irregular precipitation distribution, while high precipitation concentration (strong irregularity of precipitation) is denoted by values above 20 (Zamani et al. 2018; Ezenwaji et al. 2017; Oliver 1980) (Eqs. 7.10, 7.11 and 7.12).

$$\mathrm{PCI_{wet}} = \frac{\sum_{i=1}^{12} P_i^2}{\left(\sum_{i=1}^{7} P_i\right)^2} \times 58 \tag{7.10}$$

$$\mathrm{PCI_{dry}} = \frac{\sum_{i=1}^{12} P_i^2}{\left(\sum_{i=1}^{5} P_i\right)^2} \times 42 \tag{7.11}$$

$$\mathrm{PCI_{annual}} = \frac{\sum_{i=1}^{12} P_i^2}{\left(\sum_{i=1}^{12} P_i\right)^2} \times 100 \tag{7.12}$$

where P_i represents monthly precipitation of any given month i.

Rainfall Variability Index (RVI)

Rainfall Variability Index is used for the assessment of the frequency and severity of drought by separating wet and dry seasons/years from available rainfall time series, RVI is computed as standardized rainfall anomaly using Eqs. (7.13–7.15), and the result is classified into different climatic regimes such as extreme drought (δ <– 1.65), severe drought (–1.28 > δ >– 1.65), moderate drought (–0.84 > δ >– 1.28), and no drought (δ >– 0.84) according to Agnew and Chapel (1999) (Eqs. 7.13, 7.14 and 7.15).

$$\delta_w = \frac{(P_{\mathrm{w}} - \mu)}{\sigma} \tag{7.13}$$

$$\delta_d = \frac{(P_d - \mu)}{\sigma} \tag{7.14}$$

$$\delta_i = \frac{(P_i - \mu)}{\sigma} \tag{7.15}$$

where δ_w, δ_d, andδ_i indicate the wet season, dry season, and annual RVI; P_w, P_d, and P_i represent the wet season, dry season, and annual rainfall; *and* σ and μ denote standard deviation and mean in the wet season, dry season, and annual rainfall.

Results and Discussion

Rainfall Climatology: Monthly, Seasonal, and Annual

The monthly mean of rainfall amount in Lagos (Fig. 7.2) for the period (1984–2013) shows that rainfall in Lagos is bimodal with the first peak in June for Ikeja (284.26 mm) and Lagos-Roof (383.76 mm) stations while the second peak is in September (188.42 mm) and October (213.59 mm) for Ikeja and Lagos-Roof stations, respectively; minimum mean monthly rainfall was recorded in January (16.61 mm) for Ikeja station and in December (16.41 mm) for Lagos Roof station. The maximum cumulative monthly rainfall observed were 619.5 mm (June 1997) and 753.5 mm (May 2003) in Ikeja and Lagos-Roof stations.

The seasonal accumulated amount of rainfall depicted by Fig. 7.3 shows that Ikeja station in the wet season recorded values ranging from 871.2 mm (1986) to 1766.4 mm (1997), whereas Lagos roof station recorded values ranging from 775.7 mm (2008) to 2301 mm (1996). Dry season rainfall observed in Ikeja station ranges from 77.9 mm (2000) to 588 mm (2013), while Lagos Roof station observed values ranging from 28 mm (1989) to 352.9 mm (2004).

The annual rainfall amount depicted by Fig. 7.4 shows that Ikeja station recorded values ranging from 1045 mm to 2019.9 mm, while Lagos Roof recorded values ranging from 939.2 mm to 2535.9 mm. The descriptive statistics of the annual rainfall in the two weather stations is presented in Table 7.2. The monthly, seasonal, and annual trends of rainfall for the period of study (1984–2013) is presented in Table 7.3.

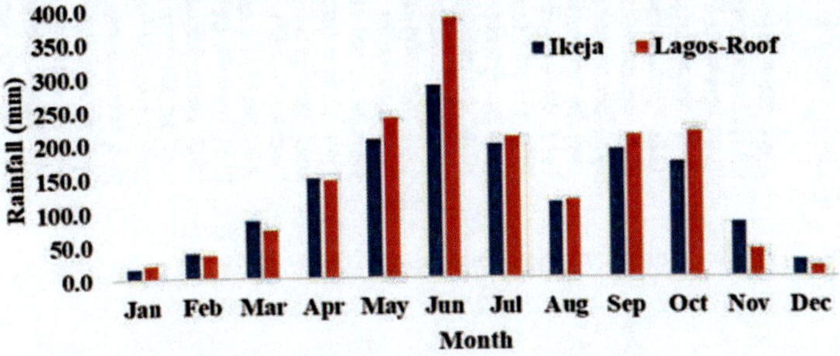

Fig. 7.2 Mean monthly amount of rainfall for the period (1984–2013)

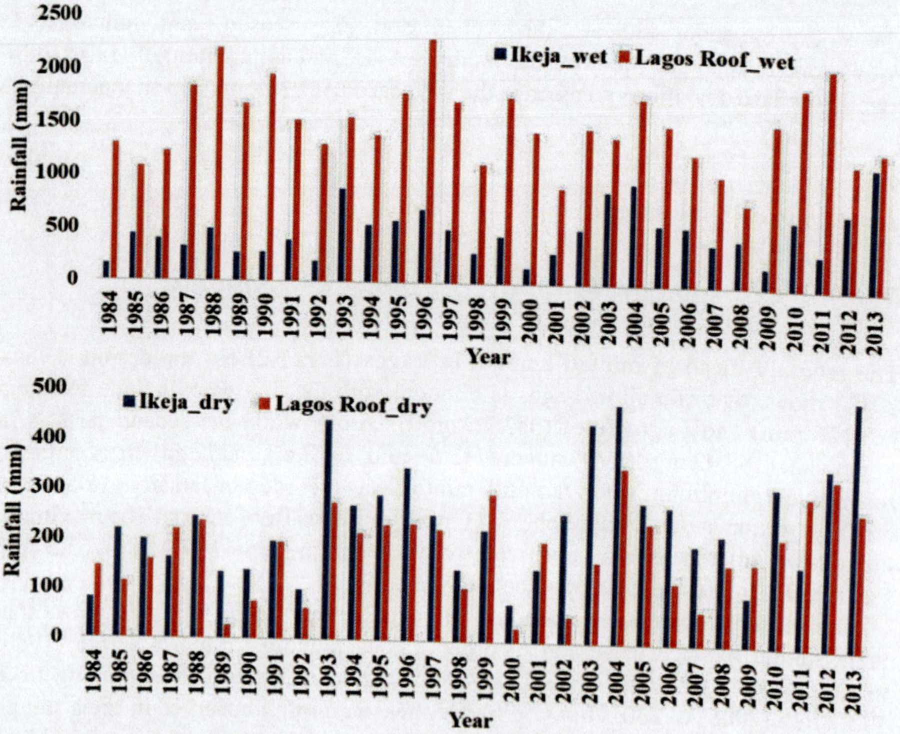

Fig. 7.3 Seasonal amount of rainfall for the period (1984–2013)

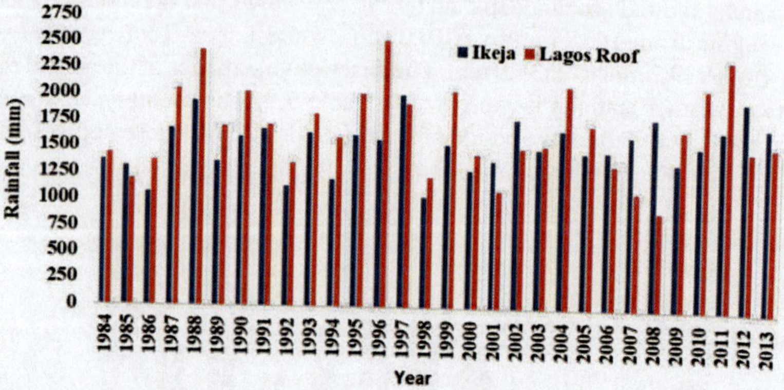

Fig. 7.4 Annual rainfall amount for the period (1984–2013)

Table 7.2 Descriptive statistics of weather stations for annual rainfall (1984–2013)

Station	Min	Max	Mean	First Quartile	Median	Third Quartile	SD	C.V	SE
Ikeja	1045	2019.90	1546.8	1378.7	1572.55	1689.60	254.40	0.16	46.4
Lagos Roof	939.2	2535.90	1689.9	1393.2	1652.45	2029.0	414.45	0.25	75.7

Min = Minimum; Max = Maximum; SD = Standard Deviation; C.V = Coefficient of Variance; S.E = Standard Error

Table 7.3 Monthly, seasonal, and annual trend for the period (1984–2013)

Ikeja					Lagos-Roof			
Period	MM-K test (Z)	Sen's slope (Q)	Trend	Sig	MM-K test (Z)	Sen's slope (Q)	Trend	Sig
Jan	2.24	0.13	↑	**	0.25	0.10	↑	
Feb	1.50	0.68	↑		0.82	0.54	↑	
Mar	0.30	0.25	↑		0.74	0.80	↑	
Apr	0.32	0.32	↑		0.64	1.30	↑	
May	0.92	1.68	↑		0.71	1.28	↑	
June	2.17	6.31	↑	**	−0.21	−0.88	↓	
July	−0.64	−1.79	↓		−0.24	−0.86	↓	
Aug	0.07	0.17	↑		−0.78	−1.26	↓	
Sep	0.50	0.8	↑		−1.85	3.42	↓	*
Oct	1.03	1.37	↑		−1.21	−2.26	↓	
Nov	1.62	2.06	↑		0.80	0.54	↑	
Dec	0.70	0.06	↑		0.67	0.03	↑	
Wet season	0.67	3.48	↑		−0.71	−7.8	↓	
Dry season	2.10	4.89	↑	**	1.64	3.4	↑	*
Annual	1.81	10.1	↑	*	−0.21	−2.01	↓	

↑ = upward; ↓ = downward; Sig. = Significance; * = at 90%; ** = at 95%

Rainfall Trends

See Table 7.3.

Monthly Rainfall Trend

Ikeja and Lagos Roof weather stations exhibit a positive trend from January to May and November to December, and both stations also observe a decreasing trend in July. However, a contrasting trend for both stations is observed from June and August to October, while rainfall trend is positive in Ikeja from June and August to October;

Lagos Roof depicts a negative rainfall trend in the same period. Ikeja observed a 95% confidence level of positive trend in January and June, respectively, Lagos Roof showed a 90% confidence level of negative trend in September. The magnitude of the positive trend for Ikeja and Lagos-Roof ranges from 0.13 mm/year to 6.31 mm/year and 0.10/year to 1.30 mm/year, respectively. July is the only month Ikeja station observed a negative trend, and it was in the magnitude of –1.79 mm/year; the rate of decrease in monthly rainfall observed in Lagos Roof ranges from –0.88 mm/year to –3.42 mm/year.

Seasonal Rainfall Trend

Dry season rainfall is observed to be at increased rates of 4.89 mm/year and 3.40 mm/year for Ikeja and Lagos Roof stations, respectively; Ikeja station shows significance at 95% confidence level, while Lagos Roof station depicts 90% confidence level significance. Wet season recorded contrasting observations for both stations. Ikeja station shows an upward trend at a rate of 3.48 mm/year, while Lagos Roof station depicted a downward trend in the magnitude of –7.8 mm/year; wet season trend for both stations was not significant.

Annual Rainfall Trend

Ikeja station observed an increase in annual rainfall at a magnitude of 10.1 mm/year, while Lagos Roof recorded a decrease in annual rainfall at a rate of –2.01 mm/year. The upward trend in Ikeja shows significance at 90% confidence level, but no significance was observed in the downward trend in Lagos Roof station.

Rainfall Variability

Precipitation Concentration Index (PCI)

Seasonal PCI for Ikeja and Lagos Roof stations depicted in Fig. 7.5 shows that there is higher uniformity in wet season rainfall than dry season rainfall. Wet season PCI for both stations ranges between 9 and 20, and Ikeja station recorded 9 and 13 as lowest and highest PCI values, respectively, while Lagos Roof recorded 9 and 20 as its lowest and highest PCI values in wet season, respectively; 37 and 63% of the wet season in the period under review witnessed low and moderate PCI categories in Ikeja station, respectively, while 13, 80, and 7% of the wet season in Lagos Roof observed low, moderate, and irregular PCI categories, respectively. Dry season PCI for both stations ranges from 9 to 42, and Ikeja recorded 10 and 41 as the minimum and maximum values, while Lagos Roof observed 9 and 42 as the minimum and maximum values, respectively; 3, 47, 33, and 17% of the study duration recorded

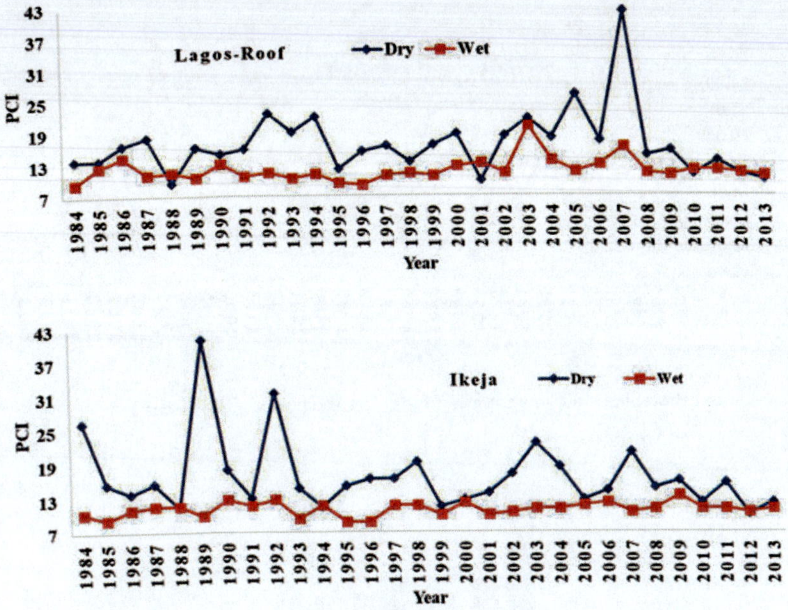

Fig. 7.5 Seasonal Precipitation Concentration Index for the period (1984–2013)

PCI categories of low, moderate, irregular, and strong irregularity of precipitation in Ikeja, while 10, 37, 36, and 17% of the period observed PCI categories of low, moderate, irregular, and strong irregularity of precipitation in Lagos Roof station.

Annual PCI for Ikeja and Lagos Roof stations (Fig. 7.6) ranges from 11 to 28. Ikeja station observed the lowest and highest annual PCI values of 11 and 20 in 1993 and 2009, while Lagos Roof station recorded its lowest and highest annual PCI values of 13 and 28 in 1996 and 2003, respectively. Ikeja station recorded moderate and irregular precipitation in 63 and 37% of the period examined, while Lagos Roof station observed moderate, irregular, and strong irregularity of precipitation in 37, 50, and 13% of the study duration (Fig. 7.7).

Rainfall Variability Index (RVI)

Seasonal RVI: Rainfall wet season standardized anomaly (Fig. 7.8) shows that Ikeja station recorded 57% negative to 43% positive anomaly tendency, while Lagos Roof station recorded 50% negative to 50% positive tendency. Categories of climate regimes for Ikeja were 80% no drought, 10% moderate drought, 3% severe drought, and 7% extreme drought, while 77% no drought, 13% moderate drought, 7% severe drought, and 3% extreme drought for the study period were recorded in Lagos Roof station. Dry season standardized anomaly (Fig. 7.9) indicates that Ikeja station recorded 53% negative to 47% positive tendency, while Lagos Roof station recorded

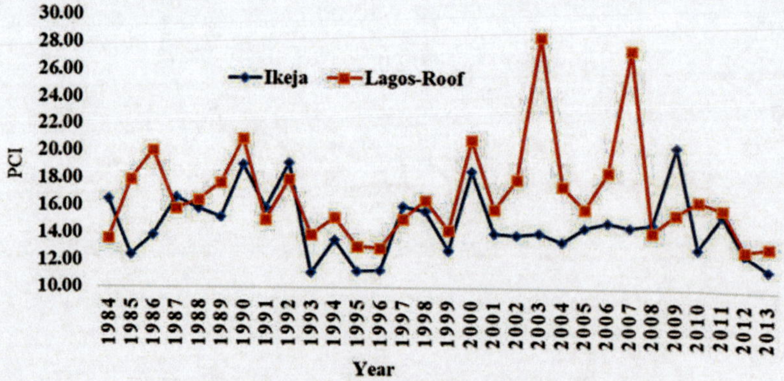

Fig. 7.6 Annual Precipitation Concentration Index for the period (1984–2013)

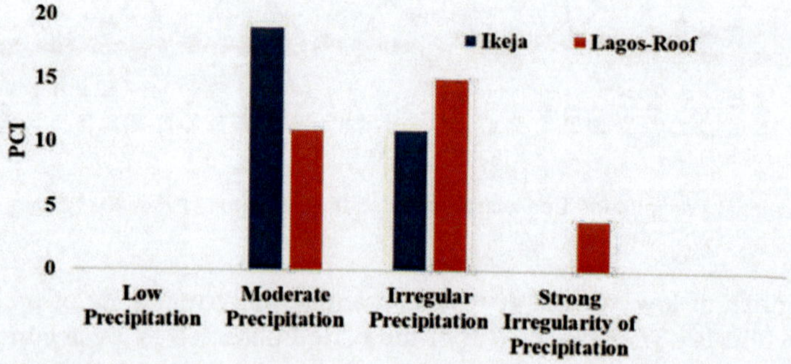

Fig. 7.7 Frequency occurrence of PCI categories for the period (1984–2013)

47% negative to 53% positive tendency. Classes of climate regimes for Ikeja were 73% no drought, 20% moderate drought, and 7% severe drought, while Lagos Roof station recorded 77% no drought, 7% moderate drought, 13% severe drought, and 3% extreme drought periods.

The annual standardized anomaly depicted by Fig. 7.10 shows a 43% negative to 57% positive tendency for Ikeja station and a 50% negative to 50% positive tendency for Lagos Roof station. Climate regimes were 80, 7, 7, and 6% of no drought, moderate drought, severe drought, and extreme drought years in Ikeja station, while 73, 17, 7, and 3% of no drought, moderate drought, severe drought, and extreme drought years in Lagos Roof station, respectively. About 37, 33, and 30% of the study period show uniformity in positive anomaly (1987, 1988, 1990, 1991, 1993, 1995, 1996, 1997, 2004, 2010 and 2011), consistency in negative anomaly (1984, 1985, 1986, 1992, 1994, 1998, 2000, 2001, 2003 and 2006), and contrasting anomaly (1989, 1999, 2002, 2005, 2007, 2008, 2009, 2012 and 2013) annually, respectively,

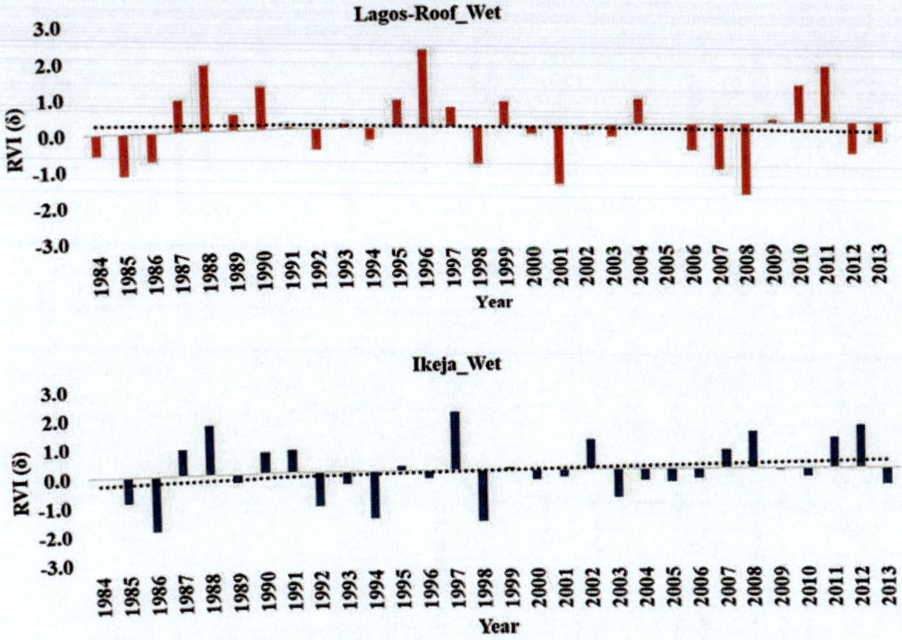

Fig. 7.8 Wet season Rainfall Variability Index (RVI) for the period (1984–2013)

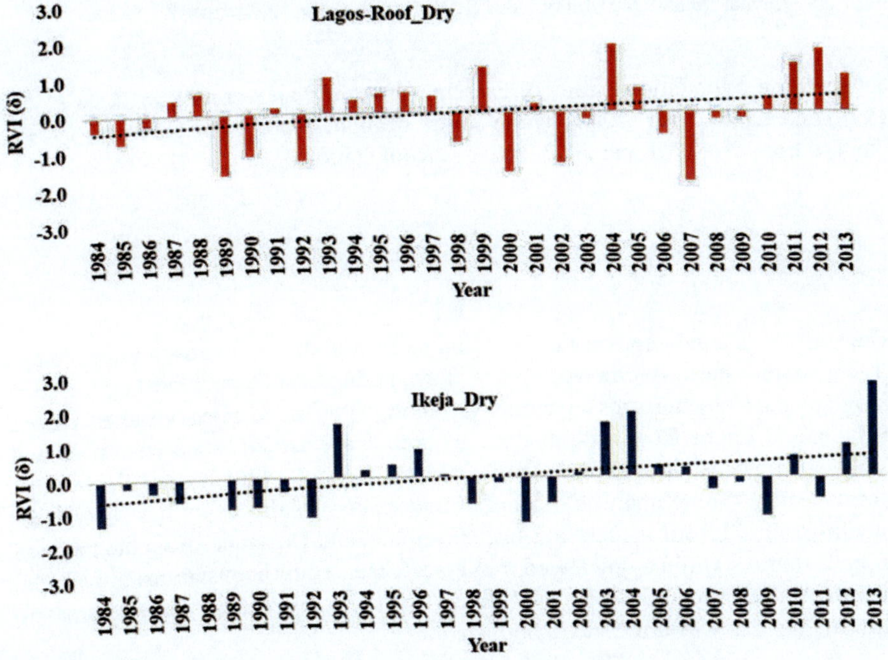

Fig. 7.9 Dry season Rainfall Variability Index (RVI) for the period (1984–2013)

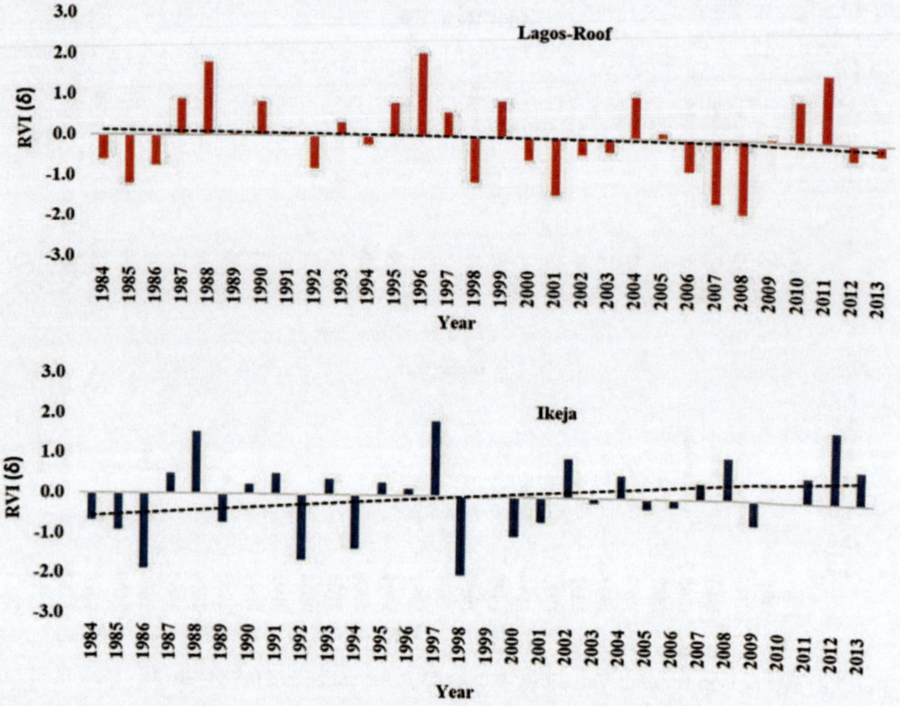

Fig. 7.10 Annual Rainfall Variability Index (RVI) for the period (1984–2013)

between the two stations studied. Extreme drought was recorded in Ikeja (1986, 1998) and Lagos Roof (2008), while severe drought occurred in 1992 and 1994 in Ikeja station, and 2001 and 2007 in Lagos Roof station.

Conclusion

The study was conducted on rainfall trend and variability in Lagos, a littoral mega city in Southeastern Nigeria with a view of evaluating the potential local influence of geographical characteristics on rainfall variability. Two independent weather stations at Ikeja and Lagos Roof with distinct geographical characteristics were used and analysis was done through the modified Mann–Kendall (MM-K) and Sen's slope estimator (Q). Trend analysis of rainfall undertaken by this study has shown that monthly rainfall from January to May, November, and December is on the increase in both stations while a downward trend was observed in both stations only in the month of July, even as the months of June, August, and September show contrasting upward and a downward trend in both stations.

Seasonally, both stations observed an upward trend in dry season rainfall, which signals the strength of other forcing functions apart from the African Easterly Jet (AEJ) that may be responsible for rainfall during the wet season (Omotosho and Abiodun 2007), and contrary trend in the wet season. On the annual scale, rainfall is increasing in Ikeja station, which is in agreement with the findings of Akinsanola and Ogunjobi (2015) and decreasing in Lagos Roof station for the period under investigation; this disparity between closely sited stations could be attributed to the influence of local climate forcing and climate change effect especially in a densely populated city like Lagos that has witnessed very rapid urbanization in the past few decades. This is congruent with Trenberth (2008) and Ogou et al. (2021) who observed that there is a direct influence of global warming on changes in precipitation and heavy rains and that the patterns of where it rains also have been observed to change, with dry areas becoming drier throughout the subtropics and wet areas becoming wetter, especially in mid to high latitudes. In India, Nair et al. (2018) observed that there are different remote and local climate forcings to seasonal and regional inhomogeneity in rainfall in India among which are the Eastern and Central Pacific El Niño Southern Oscillation, equatorial zonal winds, Atlantic zonal mode and surface temperatures of the Arabian Sea and Bay of Bengal, as well as the sea surface temperature of the North Atlantic and extratropical oceans. Simulations by Vera et al. (2019) also confirm that anthropogenic forcing is a necessary condition to explain the positive trends detected in the co-variability between tropical SST and regional U200 anomalies in rainfall variability in the South American Altiplano.

Rainfall variability study conducted herein indicated high levels of variability within Lagos across different time scales in conformity with the findings of Sojobi et al. (2016). Urbanization has been observed to increase extreme rainfall events over or downwind of cities (Jia et al. 2019). PCI is more uniform in the wet season than in the dry season, which is in agreement with the findings of Ezenwaji et al. (2017) in Awka, which is a tropical city in Nigeria with the same climate features (Omotosho and Abiodun 2007). Wet season PCI for both stations varies from 9 to 20 while dry season PCI for both stations varies from 9 to 42; low and moderate rainfall dominated wet season, while strong and irregular rainfall dominated the dry season. Annually, PCI in Ikeja station is dominated by 63% moderate rainfall in contrast to 37% moderate rainfall recorded in Lagos Roof station. RVI study shows that Lagos Roof station recorded a higher positive tendency than Ikeja station in both wet and dry seasons as well as high percentages of no drought characteristics in both seasons (77 and 80% for the wet season while 77 and 73% for the dry season for Lagos Roof and Ikeja stations, respectively). Annual RVI shows that Ikeja station has 57% positive tendency compared to 50% positive tendency recorded in Lagos Roof station, some level of consistency was recorded between both stations in the first and second decade. However, contrasting anomaly between the stations was more visible in the last decade of the study, which could be likened to the effect of a changing climate.

Limitations, Perspectives for Future Research, and Recommendations

As can be seen from the study, finding out what is happening to rainfall trends in mega cities is vital due to urbanization and the impact of climate change and variability which has become rampant in recent years. This is not only important for accuracy and precision in rainfall prediction but also in water resource planning and environmental management for a sustainable society and environment. These efforts provide a potential data source to address variability on heavy rainfall, in terms of their intensities and frequencies, within and across different climate regions.

Lagos is an economic city that is occasionally prone to heavy flooding annually; therefore, there is a need for policymakers to conceptualize a framework for mitigation and adaptation. The result of this study also calls for more research using hourly, daily, and weekly data with a longer time frame as this study is limited to monthly data series of 30 years (1984–2013).

Acknowledgements Our appreciation goes to the management of the Nigerian Meteorological Agency for providing the data for this study.

References

Abdul Aziz OI, Burn DH (2006) Trend and variability in the hydrological regime of the Mackenzie River Basin. J Hydrol 319:282–294

Abiodun BJ, Lawal KA, Salami AT, Abatan AA (2012) Potential influences of global warming on future climate and extreme events in Nigeria. Reg Environ Change. https://doi.org/10.1007/s10113-012-0381-7

Adegun O, Balogun I (2015) Analysis of rainfall distribution in Owerri and Enugu, Nigeria using Precipitation Concentration Index. Ethiop J Environ Stud Manag 8(4):408–422. https://doi.org/10.4314/ejesm.v8i4.6

Adegun O, Odunuga S, Ajayi OS (2015) Climate, runoff and landuse trends in the Owo River Catchment in Nigeria. Published by Copernicus Publications on behalf of the International Association of Hydrological Sciences, Proc IAHS, vol 371, pp 137–142. https://doi.org/10.5194/piahs-371-137-2015

Aditya R, Gusmayanti E, Sudrajat J (2021) Rainfall trend analysis using Mann-Kendall and Sen's slope estimator test in West Kalimantan. In: 2nd international conference on tropical meteorology and atmospheric sciences, IOP conference series: Earth and environmental science, vol 893, p 012006. https://doi.org/10.1088/1755-1315/893/1/012006

Agbo EP, Ekpo CM (2021).Trend analysis of the variations of ambient temperature using Mann-Kendall test and Sen's estimate in Calabar, southern Nigeria. In: International conference on recent trends in applied research (ICoRTAR) 2020. Journal of physics: conference series, vol 1734, p 012016. https://doi.org/10.1088/1742-6596/1734/1/012016

Agnew CT, Chappel A (1999) Drought in the Sahel. GeoJournal 48:299–311

Akinsanola AA, Ogunjobi KO (2014) Analysis of rainfall and temperature variability over Nigeria. Glob J Hum-Soc Sci (B) XIV(III). Version I 2014 Online ISSN: 2249-460x & Print ISSN: 0975-587

Akinsanola AA, Ogunjobi KO (2015) Recent homogeneity analysis and long-term Spatio-temporal rainfall trends in Nigeria. Theor Appl Climatol. https://doi.org/10.1007/s00704-015-1701-x

Animashaun IM, Oguntunde PG, Akinwumiju AS, Olubanjo OO (2020) Rainfall analysis over the Niger central hydrological area, Nigeria: variability, trend, and changepoint detection. Sci Afr 8:e00419. www.elsevier.com/locate/sciaf

Aswad FKH, Yousif AA, Ibrahim SA (2020) Trend analysis using Mann-Kendall and Sen's slope estimator test for annual and monthly rainfall for Sinjar district, Iraq. Journal of University of Duhok, vol 32, no 2 (Pure and Eng. Sciences), pp 501–508, 3232 (Special Issue). 3rd international conference on recent innovations in engineering (ICRIE) Duhok, 10 Sep 202

Bello NJ (1998) A study of evidence of climate change based on rainfall seasonality and the reliability of rainfall regime in Nigeria. Proc Sustain Afr 4:30–32

Blain GC (2013) The modified Mann-Kendall test: on the performance of three variance correction approaches. Bragantia Camp 72(4):416–425. https://doi.org/10.1590/brag.2013.045

Daramola MT, Eresanya EO, Erhabor SC (2017) Analysis of rainfall and temperature over climatic zones in Nigeria. J Geogr Environ Earth Sci Int 11(2):1–14. Article no. JGEESI.35304. ISSN: 2454-7352

De Luis M, González-Hidalgo JC, Brunetti M, Longares LA (2011) Precipitation concentration changes in Spain 1946–2005. Nat Hazard 11(5):1259–1265

Ezenwaji EE, Nzoiwu CP, Chima GN (2017) Analysis of precipitation concentration index (PCI) for Awka Urban Area, Nigeria. Hydrol Current Res 8:287. https://doi.org/10.4172/2157-7587.1000287

Ghalhari GF, Dehghan SF, Asghari M (2022) Trend analysis of Humidex as a heat discomfort index using Mann-Kendall and Sen's slope estimator statistical tests. Environ Health Eng Manag J 9(2):165–176. https://doi.org/10.34172/EHEM.2022.18

Hamed KH, Rao AR (1998) A modified Mann-Kendall trend test for autocorrelated data. J Hydrol 204:182–196

Hamed KH (2008a) Discussion on "To prewhiten or not to prewhiten in trend analysis. Hydrol Sci–J–des Sciences Hydrologiques 53(3)

Hamed KH (2008b) Trend detection in hydrologic data: the Mann-Kendall trend test under the scaling hypothesis. J Hydrol 2008(349):350–363

Hamed KH (2009) Enhancing the effectiveness of prewhitening in trend analysis of hydrologic data. J Hydrol 368(2009):143–155

Ifabiyi IP, Ojoye S (2013). Rainfall trends in the Sudano-Sahelian ecological zone of Nigeria. Earth Sci Res 2(2). ISSN 1927-0542 E-ISSN 19270550. https://doi.org/10.5539/esr.v2n2p194

Jia G, Shevliakova E, Artaxo P, De Noblet-Ducoudré N, Houghton R, House J, Kitajima K, Lennard C, Popp A, Sirin A, Sukumar R, Verchot L (2019). Land–climate interactions. In: Shukla PR, Skea J, Calvo Buendia E, Masson-Delmotte V, Pörtner H-O, Roberts DC, Zhai P, Slade R, Connors S, van Diemen R, Ferrat M, Haughey E, Luz S, Neogi S, Pathak M, Petzold J, Portugal Pereira J, Vyas P, Huntley E, Kissick K, Belkacemi M, Malley J (eds) Climate change and land: an IPCC special report on climate change, desertification, land degradation, sustainable land management, food security, and greenhouse gas fluxes in terrestrial ecosystems. IPCC

Kendall MG (1975) Rank correlation methods, 4th edn. Charles Griffin, London, U.K

Krishan G, Chandniha SK, Lohani AK (2015) Rainfall trend analysis of Punjab, India using statistical non-parametric test. Curr World Environ 10(3)

Longe EO, Malomo S, Olorunniwo MA (1987) Hydrogeology of Lagos metropolis. J Afr Earth Sc 6(2):163–174

Mann HB (1945) Non-parametric test against trend. Econometrica 13:245–259

Nair PJ, Chakraborty A, Varikoden H, Francis PA, Kuttippurath J (2018) The local and global climate forcings induced inhomogeneity of Indian rainfall. Sci Rep 2018(8):6026. https://doi.org/10.1038/s41598-018-24021-x

Odjugo PAO (2010) General overview of climate change impacts in Nigeria. J Hum Ecol 29:47–55

Odunuga S, Oyeande L (2007) Change detection and hydrological implications in the Lower Ogun flood plain, SW Nigeria. Remote sensing for environmental monitoring and change detection. In: Proceedings of symposium HS3007 at IUGG2007, Perugia, July 2007. IAHS Publ 316, pp 91–99

Ogou FK, Ojeh VN, Naabil E, Mbah I (2021) Hydro-climatic and water availability changes and its relationship with NDVI in northern Sub-Saharan Africa. Earth Syst Environ. https://doi.org/10.1007/s41748-021-00260-3

Ogungbenro SB, Morakinyo TE (2014) Rainfall distribution and change detection across climatic zones in Nigeria. J Weather Clim Extrem 5–6:1–6. https://doi.org/10.1016/j.wace.2014.10.002

Ogunrinde AT, Oguntunde PG, Akinwumiju AS, Fasinmirin JT (2019) Analysis of recent changes in rainfall and drought indices in Nigeria, 1981–2015. Hydrol Sci J. https://doi.org/10.1080/02626667.2019.1673396

Oguntunde PG, Abiodun BJ, Lischeid G (2011a) Rainfall trends in Nigeria, 1901–2000. J Hydrol 411:207–218

Oguntunde PG, Abiodun BJ, Olukunle OJ, Olufayo AA (2011b) Trends and variability in pan evaporation and other climatic variables at Ibadan, Nigeria, 1973–2008. Meteorol Appl. https://doi.org/10.1002/met.281

Oliver JE (1980) Monthly precipitation distribution: a comparative index. Prof Geogr 32:300–309

Oloruntade AJ, Mohammad TA, Wayayo A (2017) Rainfall trends in the Niger South Basin, Nigeria, 1948–2008, Pertanika J Sci Technol 25(2):476–496

Omotosho JB (1988) Spatial variation of rainfall in Nigeria during the Little Dry Season. Atmos Res 2:137–147

Omotosho JB, Abiodun BJ (2007) A numerical study of moisture build-up and rainfall over West Africa. Meteorol Appl 14(3):209–225

Patakamuri SK, O'Brien N (2020) Modified versions of Mann Kendall and Spearman's Rho trend tests version 1.5.0. Technical Report, March 17, 2020. https://doi.org/10.13140/RG.2.2.20621.56806

Salami AW, Ikpee OD, Ibitoye AB, Oritola SF (2016) Trend analysis of hydro-meteorological variables in the coastal area of Lagos using standard anomaly index methods. J Appl Sci Environ Manag 20(3):797–808

Sen PK (1968) Estimates of the regression coefficient based on Kendall's Tau. J Am Stat Assoc 63(324):1379–1389. https://doi.org/10.2307/2285891

Sojobi AO, Balogun II, Salami AW (2016) Climate change in Lagos: what really happened? Environ Monit Assess 188:556. https://doi.org/10.1007/s10661-016-5549-z

Vera CS, Díaz LB, Saurral RI (2019) Influence of anthropogenically-forced global warming and natural climate variability in the rainfall changes observed over the south American altiplano. Front Environ Sci 7:87. https://doi.org/10.3389/fenvs.2019.00087

Wagaye BA, Antensay M (2020). Rainfall variability and trends over central Ethiopia. Int J Environ Sci Nat Res 24(4):556144. https://doi.org/10.19080/IJESNR.2020.24.556144. 0217

You S, Wang CH (2004) The Mann-Kendall test modified by effective sample size to detect trend in serially correlated hydrological series. Water Resour Manag 18:201–218

Zamani R, Mirabbasi R, Nazeri M, Meshram SG, Ahmadi F (2018) Spatio-temporal analysis of daily, seasonal and annual precipitation concentration in Jharkhand state, India. Stoch Env Res Risk Assess 32(4):1085–1097

Zhang K, Yao Y, Qian X, Wang J (2019) Various characteristics of precipitation concentration index and its cause analysis in China between 1960 and 2016. Int J Climatol 2019:1–11. https://doi.org/10.1002/joc.6092

Chapter 8
Effect of Climate Change on Soil Organic Carbon Storage in Four Land Use Types in Abakaliki, South Eastern Nigeria

J. E. Orji, C. A. Igwe, and P. I. Ezeaku

Abstract Soil organic carbon (SOC) has been found to negatively impact the environment due to its emission of carbon (iv) oxide (CO_2) a greenhouse gas with a serious effect on global climate change. A study was conducted to determine the effect of climate change on soil organic carbon storage in four land use types in the Abakaliki, south eastern Nigeria. The four land use types selected include managed forest plantation (MF), fallow land, grass land and continuously cultivated soil (CCS). Undisturbed and core samples were collected at three soil depths (0–20, 20–40 and 40–60 cm) and replicated five times for bulk density, total porosity, saturated hydraulic conductivity, particle size distribution, pH, organic carbon, total nitrogen available phosphorus and exchangeable acidity. The results showed that the soils in the area were predominantly sandy loam and sandy clay loam in texture. Bulk density values were lowest in the fallow land and highest in the continuously cultivated soil across the three depths. Saturated hydraulic conductivity values were highest in MF and lowest in CCS soil across the three depths. The pH of the soils was slightly acidic. Fallow land recorded the highest soil organic carbon (SOC), total nitrogen and available phosphorus while continuously cultivated soil had the lowest values for the soil nutrients at 0–20 cm. The result on total organic C storage showed that MF recorded the highest value (340.67 Mg C ha^{-1}), while CCS recorded the lowest value (151.49 Mg C ha^{-1}) across the three soil depths.

Keywords Climate change · Organic carbon · Storage · Land use · Soil

J. E. Orji (✉)
Department of Agriculture, Alex Ekwueme Federal University, Ndufu Alike, Ebonyi state, Nigeria
e-mail: orji.jephter@funai.edu.ng

C. A. Igwe · P. I. Ezeaku
Department of Soil Science, University of Nigeria, Nsukka, Nigeria
e-mail: vc.unn@unn.edu.ng

P. I. Ezeaku
e-mail: peter.ezeaku@unn.edu.ng

J. C. Egbueri et al. (eds.), *Climate Change Impacts on Nigeria*, Springer Climate,
https://doi.org/10.1007/978-3-031-21007-5_8

Introduction

Soil organic matter (SOM) is known as a major component of soil that affects its physical, chemical and biological properties, and contributes greatly to its proper functioning on which human societies depend. SOM is made of organic compounds that are highly enriched in carbon (Murphy 2015; Chokor and Egborah 2018). Soil organic carbon, an integral part of soil organic matter is the component of soil obtained from dead living matter decomposed by soil microorganisms. It is a known source of nutrients for plants and energy for soil microorganisms. Soil organic carbon (SOC) levels are directly related to the amount of organic matter contained in the soil and SOC is often how organic matter is measured in soils (D'Amato et al. 2011). SOC levels result from the interactions of several ecosystem processes, of which photosynthesis, respiration and decomposition are key. Photosynthesis is the fixation of atmospheric CO_2 into the plant biomass (Powers et al. 2011).

The land use type, changes in land use and management practices influence the amount and rate of soil carbon losses (Guggenberger et al., 1995). Land use types affect the amount of soil carbon emission. Han et al. (2011) observed increase in soil organic carbon when arable crops were converted to grass land in China. Soil organic carbon plays an important role in supplying plant nutrients (N, P, S) enhancing cation exchange capacity, improving soil water retention and supporting biological activities (Dudal and Deckers 1993).

Soil comprises the largest pool of terrestrial C and through soil organic matter (SOM) cycling; it represents either an important sink in atmospheric CO_2 or a possible source of greenhouse gases. The concentration of carbon (iv) oxide in the atmosphere which encourages climate change is greatly influenced by the balance between soil organic carbon (SOC) inputs and outputs (IPCC 2007). Normally, the carbon pool in the soils is approximately twice as that in the atmosphere. The amount of soil organic carbon in the soil is controlled by several factors like climate (rainfall, temperature, etc.), parent materials and land use types (Akpan et al. 2018; Vose et al. 2012).

Forests, fallow and grass land vegetation contributions to the soil organic carbon pool include leaf litter, coarse woody debris, roots exudates and dissolved organic matters leached from the top (litter) layer (Nave et al. 2010; D'Amore and Kane 2016). The stabilization of these organic inputs in the mineral soil matrix (storage) or biodegraded and returned to the atmosphere as carbon (IV) oxide (CO_2) depends on complex, fine-scale interactions involving soil minerals, plants and soil organisms and organic components, and all these interactions are influenced by several factors such as climate, soil management practices and land use types (Johnson et al. 2009).

Soil organic carbon is lost and released to the atmosphere when trees are felled and soils ploughed (tillage). Among the emitted greenhouse gases which encourage global warming that results to climate change, carbon (iv) oxide contributes 72% of the total gases emitted (IPCC 2007). Regular disturbances on soils through tillage operations during crop production result to loss of soil organic carbon to the atmosphere as CO_2. Therefore, to reduce atmospheric carbon (iv) oxide in order to mitigate the effect of climate change, the existing atmospheric carbon must be captured and stored in soils and plant biomass. However, relatively little is known about the effect of climate change on carbon storage in different land use types in the Abakaliki, South eastern Nigeria. The objective of the study, therefore, is to determine the effect of climate change on carbon storage in four land use types in the Abakaliki, South eastern Nigeria.

Materials and Methods

Site Description

The study was carried out at four locations with a particular land use type situated in the Abakaliki Ebonyi State, South Eastern Nigeria in 2019 cropping season. The Abakaliki in Ebonyi State lies between Latitude 06° 25 N and longitude 08° 3′ E and altitude of 170 m. The rainfall ranges from 1700 to 2000 mm with mean annual rainfall of 1800 mm (ODRNI 1989). The mean annual temperature ranges from 27 to 31 °C throughout the year. The relative humidity is high 80% during rainy season but declines to 65% in the dry season (ODRNI 1989).

The relief of the area is generally undulating and no location exceeds 200 m above sea level. Geologically, the soil of the area is underlain by sedimentary rocks derived from successive marine deposits. According to FDALR (1987), the Abakaliki agricultural zone lies within Asu River and is associated with brown olive shales, fine grained sandstone and mudstone. The soils of the area being basically clayey loam underlain with laterite are particularly suitable for the cultivation of rice, yam, cassava, maize, etc. The soil is shallow with unconsolidated parent material up to 1 m depth belonging to the order ultisol classified as Typic Haplustult (FDALR 1987). The major vegetation of the area is derived savanna characterized by growth of shrubs, herbs, dispersed large trees and common tropical grasses.

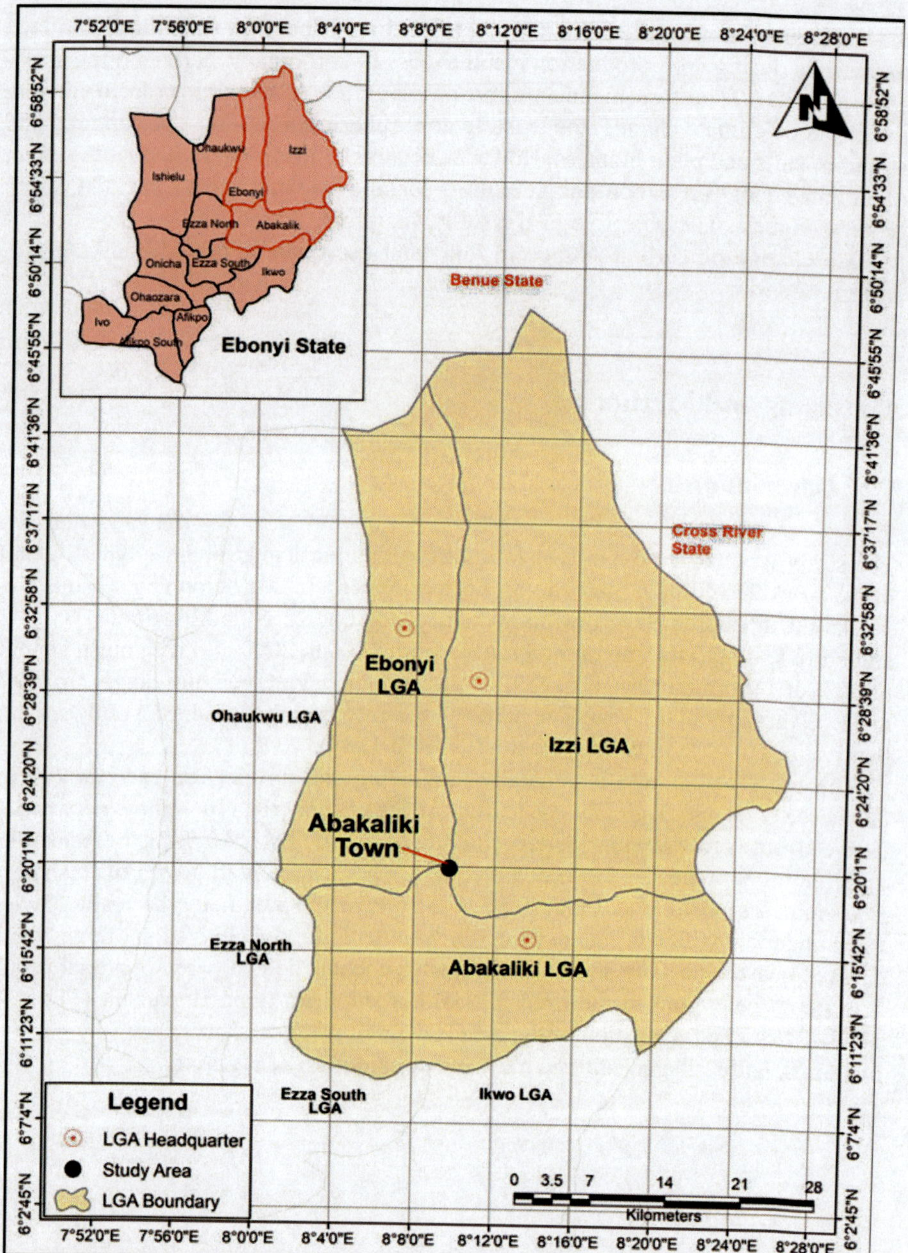

The map of Ebonyi state, showing Abakaliki the study area (the Abakaliki area covers three local government areas viz*: Abakaliki, Ebonyi and Izzi).*

The four land use types and their land use history are described below as follows;

Managed Forest Plantation (MF): Artificial *Gmelina arborea* was established at Azugwu Abakaliki during the colonial era for preservation of forestry resources spanning for 50 years (Nwite et al. 2018). Other species of trees and shrubs are also found growing in the plantation. Hunting and farming activities are prohibited although poaching is not completely eliminated as people still encroach in the forest for farming related activities. The land area for the study was measured 100 × 100 m which is equivalent to 1 ha.

Fallow Land (FL): The fallow is located behind All Saints Cathedral, Abakaliki near Ogoja road. It lies between latitude 06° 19′ 17N and longitude 08°06′ 23E (Google map 2019). The fallow has lasted for more than 10 years. There are different types of vegetation covering the land but among major ones are scattered trees, herbs, shrubs and grasses like guinea grass (*Panicum maximum),* wire grass (*Sporobulus pyramidalis*), goat weed (*Ageratum conyzoides*), stubborn weed *(Sida acuta)*. The land area of 100 × 100 m equivalent to 1 ha was measured for the study.

Grass Land (GL) is located at Bishop Otubelu College Abakaliki. The grass land is periodically undercut especially the area that serves as field for games, some parts of the land is used to graze animals. The grass land has existed over 20 years. The land is grown with herbs but dominated by grasses such as *Tridax*, giant star grass, guinea grass, wire grass, goat weed *and* stubborn weed. The land area for the purpose of the study is measured 100 × 100 m equivalent to 1 ha.

Continuously Cultivated Soil (CCS) otherwise known as arable land is located behind Law Faculty of Ebonyi State University Abakaliki. The land is the experimental farm of the Department of Soil Science and Environmental Management, Faculty of Agriculture, Ebonyi State University, Abakaliki. The area has been under yearly cultivation for more than 18 years by students and staff of the department. It has been subjected to conventional tillage operations, use of chemicals (like fertilizers, herbicides and pesticides), organic amendments and other cultural practices. Common crops grown in rotation include yam (*Dioscorea spp*), cassava (*Manihot spp*), maize (*Zea mays* L), vegetables, cucumber and leguminous crops. The land area measured out for the purpose of the research is 100 × 100 m equivalent to 1 ha.

Soil Sampling Procedure

Soil samples were collected from the four (4) selected land use types. Undisturbed samples were randomly collected from all the four land use types at three different depths (0–20, 20–40 and 40–60 cm) using open-faced coring tube of 6 cm height and 5 cm diameter for determination of physical properties of the soil. Within each land use type, two (2) core and an auger samples of the soils were collected from each of the three depths and replicated five times for physicochemical properties determination of soils of the land use types.

The experiment was laid out in a randomized complete block design (RCBD) with four treatments (the four land use types viz: Continuously cultivated soil, fallow land, grass land and managed Gmelina plantation), three sub treatment was the soil depths (0–20, 20–40 and 40–60 cm) and replicated five times.

Laboratory Methods

The disturbed soil samples were air dried and sieved through 2 mm mesh. The undisturbed core samples were used for bulk density, total porosity and hydraulic conductivity determinations only. The samples <2 mm were used for the determination of particles size distribution (in water and calgon), soil pH, soil organic carbon, total Nitrogen, available phosphorus, exchangeable acidity according to the standard analytical procedures. Particle size distribution was determined using Bouyoucous hydrometer method as described by Gee and Bauder (1986). The textural class of the soil was determined using the textural triangle.

Bulk density was determined using core sampling method after oven drying the soil samples to a constant weight at temperature of 105 °C for 24 h ((Blake and Hartage 1986; Grossman and Reinsch 2002). Total porosity was calculated from soil bulk density using as assumed particle density of 2.70 gcm^{-3}. Saturated hydraulic conductivity (Ksat) was determined by core method as described by Klute and Dirksen (1986). Soil pH was determined both in water and 0.1 N potassium chloride solution at the ratio 1:2.5 soil/water suspension using standardized glass electrode pH metre (Udo et al. 2009).

Soil organic carbon content of the soil and aggregate-associated carbon was quantified by Walkley and Black wet oxidation method as described by Nelson and Sommers (1998). Total nitrogen content of the soil (<2.00 mm) was determined by the Micro Kjeldahl digestion method using $CuSO_4$ (Copper sulphate) and Na_2SO_4 (Sodium sulphate) catalyst mixture (Bremner and Mulvanney 1982). Available phosphorus was determined using Bray II bicarbonate extraction method (Bray and Kurtz 1945) using 0.03 N ammonium fluoride with 0.1NHCl. The phosphorus in the extract was determined with a photo-electric colorimeter. Exchangeable acidity (EA) was calculated by summing the values of exchangeable aluminium and hydrogen; EA = $Al^{3+} + H^{+}$.

Soil carbon storage ($MgCha^{-1}$) was quantified as the product of fractional mass of carbon, soil depth (m), soil bulk density (mg/m^3) and land area/ha (m^2/ha) thus.

$MgC\ ha^{-1} = [\%\ C \times BD \times d \times 10^4 m^2]/100$ (Lal et al. 1998).

where $MgC\ ha^{-1}$ = mega gram carbon per hectare (1 Mg = 10^6 g).

%C = percentage of carbon given by laboratory result.

Bd = soil bulk density (gcm^{-3}).

d = depth in metres.

Data Analysis

An analysis of variance (ANOVA) of each soil properties within each depth and land use type was performed on the soil data generated from laboratory using the Genstat Discovery Software, edition 4. Mean separation was done using the Fisher's least significant difference (FLSD) at 5% level of probability. Descriptive statistics (mean, range and coefficient of variation) was used to assess the variability in soil physicochemical properties across the land use types. Coefficients of variation were ranked according to Aweto (1980) as follows: CV < 20% low variation, CV = 20–50% moderate variation and CV = 50% and above high variation.

Results and Discussion

Physical Properties of Soils of the Different Land Use Types

The results in Table 8.1 show the physical properties of the soils of the four land use types studied. The result revealed that the sand fraction was significantly different (P < 0.05) among the land use types and varied from 620 to 740 gkg^{-1} with coefficient of variation (CV) of 5.0% at 0–20 cm depth, 580 to 640 gkg^{-1} with CV of 6.0% at 20–40 cm depth and, 580 to 680 gkg^{-1} (CV: 7.0%) at 40–60 cm. The sand fraction decreased generally with depth in all land use types except at 40–60 cm depth in FL which could be attributed to the selective movement of soil particles from one horizon to another by erosion and variations in weathering intensity (Kolo 2019). The sand fractions are similar in variation irrespective of different land use types as evidenced by their CV values (CV ≤ 7.0%). The general decrease in sand fraction with depth was reported by Thomas and Oke (2018). The silt fraction was found to be significantly different (P < 0.05) among the four land use types. Silt fraction varied from 90 to 190 gkg^{-1} with CV of 11.0%, 70 to 170 gkg^{-1} with CV of 39.0% and 70 to 150 gkg^{-1} with CV of 39.0% for 0–20, 20–40 and 40–60 cm depths, respectively (Table 8.1). The distribution of silt fraction was irregular with depth across the land use types. Although in CCS, FL and GL land use types, silt fractions decreased with increase in depths but the reverse was the case with MF which could be as a result of loessial deposit on soils of managed Gmelina plantation (Maniyunda and Gwari 2016; Maniyunda 2018). According to Sharu et al. (2013) the development of soils in-situ is responsible for the irregular distribution of silt down the soil depths.

The clay fraction varied from 130 to 230 gkg^{-1} (CV: 30.0%) at 0–20 cm depth, 250 to 270 gkg^{-1} (CV: 4.0%) at 20–40 cm depth and 230 to 310 gkg^{-1} with CV of 13% at 40–60 cm depth. The clay fraction was found to be significantly different (P < 0.05) among the four land use types. According to the result, clay fraction increased with depth at 0–20 and 40 cm but varied at 40–60 cm depth. Continuously cultivated soil (CCS), fallow land (FL) and grass land (GL) land use types recorded higher values of clay fractions at the sub soil (20–40 cm and 40–60 cm) depths compared to the top

Table 8.1 Physical properties of soils of the four land use types

Land use	Sand (gkg^{-1})	Silt (gkg^{-1})	Clay (gkg^{-1})	TC	Bd (gcm^{-3})	TP (%)	Ksat (cmh^{-1})
0–20 cm							
CCS	680	190	130	SL	1.75	36.41	15.52
FL	740	110	150	SL	1.51	46.50	25.75
GL	740	130	130	SL	1.69	36.60	1854
MF	680	90	230	SCL	1.58	43.49	2453
MEAN	710	130	160		1.63	40.75	21.09
CV (%)	5.0	11.0	30.0		6.0	12.0	23.0
FLSD (0.05)	6.29	58.35	6.29		0.15	2.26	8.73
20–40 cm							
CCS	580	170	250	SCL	1.65	36.79	13.05
FL	640	90	270	SCL	1.35	50.00	16.92
GL	660	70	270	SCL	1.68	36.60	13.74
MF	640	110	250	SCL	1.68	36.60	28.41
MEAN	630	110	260		1.59	40.00	18.04
CV (%)	6.0	39.0	4.0		9.0	17.0	39.0
FLSD (0.05)	25.94	34.09	50.71		0.89	15.73	1.89
40–60 cm							
CCS	580	150	270	SCL	1.62	38.52	5.80
FL	680	70	250	SCL	1.33	50.97	7.09
GL	620	70	310	SCL	1.64	38.30	17.10
MF	640	130	230	SCL	1.65	38.35	14.20
MEAN	630	105	265		1.56	41.54	11.05
CV (%)	7.0	39.0	13.0		9.0	15.0	50.0
FLSD (0.05)	15.41	62.90	3.15		0.99	5.03	4.40

TC—Textural class Bd—Bulk density, TP—Total porosity, Ksat—Saturated hydraulic conductivity, CCS = Continuously cultivated soil, FL = Fallow land, GL = Grass land, MF = Managed Gmelina plantation. CV = Coefficient of variation

soils (0–20 cm) (Table 8.1). This might be attributed to the selective transport of finer soil particles into lower soil depths by eluviation/illuviation processes (Akamigbo et al. 2001; Mustapha et al. 2003; Osuaku et al. 2014; Amuyou and Kontigo 2015). The higher clay on the upper depth (20–40 cm) of MF land use type might be attributed to clay accumulation from the selective dissolution of more soluble minerals (Boul et al. 2003).

The particle size distribution result is in tandem with the general trend in particle size distribution sand > clay > silt, this was also observed by Asadu and Akamigbo

(1983) in most soils of southeastern, Nigeria. The dominance of sand fraction in the soils may be attributed to high content of quartz mineral in the parent material. Lawal et al. (2013) and Osujicke et al. (2016) reported that sand dominated the mineral fraction due to quartz rich parent material, clay migration, surface erosion and geological sorting of soil materials by biological activities in their field of study. Low silt fraction observed in the soils could be attributed to leaching and nature of the parent material (Nsor 2017). The textural class of the soils of the different land use types varied from sandy loam to sand clay loam (Table 8.1). The textural variation across the land use types might be attributed to the selective removal of soil particles from one location to another by agents of soil erosion and variation in weathering intensity (Kolo 2019).

Across the soil depths (0–60 cm), the value of bulk density was significantly different ($P < 0.05$) among the land use types and ranged from 1.33 to 1.75 gcm^{-3}. The highest (1.75 gcm^{-3}) and lowest (1.51 gcm^{-3}) bulk density values at 0–20 cmsoil depth were recorded at continuously cultivated soil (CCS) and fallow land (FL), respectively. The values of bulk density ranged between 1.33 and 1.65 gcm^{-3} with CV of 9.0% (mean = 1.63 gcm^{-3}) at 0–20 cm depth, 1.35 and 1.68 gcm^{-3} (CV 9.0%; mean = 1.59 gcm^{-3}) at 20–40 cm depth and 1.51 and 1.75 gcm^{-3} with CV of 6.0% (mean = 1.56 gcm^{-3}) at 40–60 cm depth (Table 8.1). Fallow land recorded the lowest bulk density among the land use types across the depths which ranged from 1.33 to 1.50 gcm^{-3} which has advantage of increased porosity (Brady and Weil 2010; Vogelmann et al. 2010) with accompanied adequate aeration. The increase in bulk density with depth may be attributed to compaction caused by the weight of the overlying horizons and decrease in organic matter with depth. This agrees with the reports of Chaudhari et al. (2013) and Fedaku et al. (2018) as they reported increasing bulk density with depth due to changes in organic matter content, porosity and compaction.

The mean values of total porosity ranged from 40.00 to 41.54% for the different land use types and across the depths defying a particular trend. Total porosity values of the soils were significantly different ($P < 0.05$) among the land use types. Total porosity varied from 38.30 to 50.90% with CV of 15% at 0–20 cm depth, 36.60 to 50.00% with CV of 17% at 20–40 cm depth and 36. 41 to 46.50% with CV of 12% at 40–60 cm depth. The trend in enhancement of total porosity by the different land use types at 0–20 cm depth is FL > MF > GL > CCS. Increase in bulk density reduces pore spaces (Okolo et al. 2015). Also, continuous traffic and agricultural activities on land (Akamigbo et al. 2001). The decrease in total porosities with depth might be attributed to the decrease in organic matter content with depth (Hassan and Shuaibu 2006).

The values of hydraulic conductivity of the soils were significantly different ($P < 0.05$) among the land use types The result showed that Ksat ranged from 15.52 to 25.75 cmh^{-1} with CV of 39.0% at 0–20 cm, 13.05 to 28.41 $cmhr^{-1}$ (CV: 39.0%) at 20–40 cm and 5.80 to 17.10$cmhr^{-1}$ (CV: 50.0%) at 40–60 cm. The wide variation (23.0, 39.0 and 50.0%) observed in Ksat might be due to variation in particle size distribution and moisture content (Shelton 2003; Obalum et al. 2011). Hydraulic conductivity increases with decrease in bulk density, increase in porosity and organic matter content (Anikwe 2000). The values of Ksat were found to be lower in sub soil when compared with the top soils suggesting low rate of water transmission in the former due to accumulation of clay (Ezeaku et al. 2015). At 0–20 cm depth, fallow land recorded the highest value (25.75 cmh^{-1}) when compared with other land used types. In all, 0–20 cm recorded the highest mean value when compared with the other depths and the trend of the decrease in the mean values was 0–20 > 20–40 < 40–60 cm (Table 8.2). Antonio et al. (2001) and Bagarello and Lovino (2003) reported high Ksat on top soils due to higher water content and smaller pores.

Table 8.2 Chemical properties of soils of the land use types

Land use	pH (H_2O)	pH (KCl)	OC	TN	Av.P ($mgkg^{-1}$)	H^+	Al^{3+} $cmolkg^{-1}$	EA
0–20 cm								
CCS	6.20	5.00	4.74	0.98	4.89	1.60	0.20	1.80
FL	6.40	5.20	11.37	1.68	11.99	1.40	0.60	2.00
GL	6.90	5.60	6.58	1.40	5.60	1.20	0.00	1.20
MP	5.60	4.20	10.27	1.54	9.40	1.20	1.00	2.20
Mean	5.38	5.00	8.24	1.40	7.97	1.45	0.55	1.90
CV (%)	10.0	12.0	38.0	21.0	38.0	14.0	113.0	26.0
FLSD (0.05)	0.06	0.37	0.07	0.26	0.72	0.37	0.27	0.39
20–40 cm								
CCS	5.60	4.30	3.29	0.70	3.86	1.60	0.80	2.40
FL	5.90	4.70	5.69	1.12	10.93	1.60	1.00	2.60
GL	6.40	5.20	5.53	0.84	4.20	1.00	0.80	1.80
MP	5.50	4.30	7.11	0.98	8.80	1.40	1.20	2.60
Mean	5.85	4.63	5.41	0.91	6.95	1.40	1.05	2.35
CV (%	7.0	9.0	29.0	19.0	32.0	20.0	35.0	9.0
FLSD (0.05)	0.19	0.17	0.02	0.13	0.14	0.52	0.32	0.09
40–60 cm								
CCS	5.60	4.30	2.37	0.70	3.30	1.40	1.00	2.60
FL	6.00	5.00	4.19	0.98	9.60	1.40	1.20	2.60
GL	6.80	5.70	3.56	0.70	3.58	1.20	0.20	1.00
MP	5.60	4.30	6.72	0.84	7.18	1.60	0.20	1.80
Mean	6.00	4.83	4.21	0.81	5.92	1.35	1.10	2.45
CV (%)	9.0	14.0	47.0	17.0	40.0	7.0	35.0	12.0
FLSD (0.05)	0.41	0.43	0.06	0.09	0.25	0.17	0.14	0.34

OM: organic matter; TN: total nitrogen; Av. P available phosphorus; H^+: exchangeable hydrogen; Al^{3+}: exchangeable aluminium; CV: coefficient of variation. CCS = Continuously cultivated soil, FL = Fallow land, GL = Grass land, MF = Managed Gmelina plantation

Chemical Properties of Soils of Various Land Use Type

The chemical properties of the soils of the various land use types are presented in Table 8.2. The soil pH in H_2O and KCl ranged from 5.60 to 6.90 and 4.20 to 5.20 with CV of 10.0 and 9.0%, respectively, 5.50–6.40 and 4.30–5.20 with CV of 7 and 9%, respectively, at 20–40 cm. At 40–60 cm depth, pH in H_2O and KCl ranged from 5.60 to 6.80 and 4.30 to 5.7 with CV of 10 and 14%, respectively. The soil pH was significantly different ($P < 0.05$) among the land use types. Generally, the pH values both in H_2O and KCl were rated moderately acidic to slightly acidic according to the ratings of Enwezor et al. (1989), Fasina et al. (2000), Raji and Mohammed (2007) and Adeputu et al. (2014). This level of soil reaction may be attributed to the influence of the parent materials, continuous cultivation, application commercial fertilizers and loss of exchangeable bases through leaching (Brady and Weil 2002; Havlin et al. 2006; Maniyunda 2018).

Soil organic carbon (SOC) was significantly different ($P < 0.05$) among land use types across the depths. The highest (11.37 gkg^{-1}) and lowest (4.74 gkg^{-1}) values of SOC were recorded at fallow land (FL) and continuously cultivated soil (CCS) at 0–20 cm. Soil organic carbon varied from 4.74 to 11.37 gkg^{-1} with CV of 38.0% (mean = 8.24 gkg^{-1}) at 0–20 cm, for soils at 20–40 cm, SOC varied from 3.29 to 7.11 gkg^{-1} with CV of 29.0% (mean = 5.41 gkg^{-1}) and 2.37 to 6.72 gkg^{-1} with CV of 43.0% (mean = 4.21 gkg^{-1}) at 40–60 cm depths. Generally, SOC contents in the soil under different land use types significantly decreased with depths giving the least values at 40–60 cm depths. The trend of soil organic carbon content under different land use types at 0–20 cm depths is FL > MF > GL > CCS and in the depth intervals the trend is as follows: 0–20 > 20–40 > 40–60 cm (Table 8.2).

Low values of SOC recorded in soils of the various land use types might be attributed to the low content of organic matter in the soils. Mbah and Idike (2011) observed that organic manure like poultry manure has the capacity of improving the nutrient condition of soils. Soil organic carbon had the highest values at the depths of 0–20 cm, with fallow land recording the highest value (11.37 gkg^{-1}) while the lowest (4.74 gkg^{-1}) was obtained in continuously cultivated soils. These variations might be due to high level of humification of organic matter in the top soils as reported by Nsor et al. (2014). Vegetation cover was observed to encourage high organic carbon and organic matter contents of the soils as observed by Chokor and Egborah (2018). The studied soils had generally low content of organic carbon since the SOC of the soils were below the critical value (<17.0 gkg^{-1}) (as reported by Loveland and Webb 2003 and Emmanuel et al. 2018). This could be attributed to leaching and washing effects of high erosivity in the region (Igwe 2005).

Total nitrogen (TN) showed significant difference ($P < 0.05$) among the land use types. Total N at CCS (0.98 gkg^{-1}); FL (1.68 gkg^{-1}); GL (1.40 gkg^{-1}) and MF (1.54 gkg^{-1}) at 0–20 cm. On the other hand, TN recorded 0.70, 1.12, 0.84 and 0.98 $gkg^{-1}$for CCS, FL, GL and MF, respectively, at 20–40 cm soil depth. Similarly, at 40–60 cm the values of TN across the various LUTs are 0.70, 0.98, 0.70 and 0.84 $gkg^{-1}$for CCS, FL, GL and MF, respectively. Total N at 0–20, 20–40 and 40–60 cm varied from 0.98 to 1.68 gkg^{-1} with CV of 21.0%, 0.70 to 1.12 gkg^{-1} with CV of 19.0% and 0.70 to 0.98 gkg^{-1} with CV of 17.0%, respectively. Total nitrogen was generally low, (Maniyunda 2018) which according to Enwezor et al. (1989) could be rated very low to low. The content of TN recorded low variation in its distribution across the land use types with CV ranged from 17 to 21%. The highest TN value (1.68 gkg$-^{1}$) and the lowest value (0.98 gkg^{-1}) were obtained at 0–20 cm in fallow land and continuously cultivated soil, respectively. The low values of total nitrogen obtained might be attributed to low organic matter content of the soils as OM accounts for 93 to 97% of total nitrogen (Meysner et al. 2006). Total nitrogen of the soils of various land use types decreased with increase in depths with 0–20 cm recording the highest values. Njoku (2012) and Uzoho et al. (2014) attributed low TN to losses through run-off, low organic matter, high N mineralization and crop removal. The higher TN content values in surface soils when compared with the sub-surface soils of the different land use types might be due to nutrient recycling through litter fall (Egbuchua and Ojobor 2011).

The results of available phosphorus showed that the values ranged between 4.89 and 11.99 $mgkg^{-1}$ with CV of 3.0% at 0–20 cm; 3.86 to 10.93 $mgkg^{-1}$ 32.0% at 20–40 cm and 3.30 to 9.60 $mgkg^{-1}$ with CV of 44.0% at 40–60 cm (Table 8.2). Available phosphorus values were rated low to high with content values between 4.89 and 11.99 $mgkg^{-1}$ when compared to the critical value of 10–16 $mgkg^{-1}$ as reported by Adeoye and Agboola (1985); Chokor and Egborah (2018). The CV values of available phosphorus varied between 32 and 40% and considered to be moderately high (Tabi and Ogunkule 2007; Maniyunda 2018). The value of available P at 0–20 cm depths of the soils of FL land use type was higher than 10 $mgkg^{-1}$ critical limit recommended for most commonly cultivated crops whereas others are below. This agrees with the observations of Uponi and Adeoye (2000), Aduayi et al. (2002) and Obigbesan (2009) that reported that the critical limit of available P recommended for most commonly cultivated crops is 10 $mgkg^{-1}$.

Available P decreased with increasing soil depths in all the soils of the land use types with 40–60 cm recording the lowest values. At 20–40 cm FL recorded the highest values (10.93 $mgkg^{-1}$) above the critical limit for available P (10 $mgkg^{-1}$) when compared with other land use types which recorded low values that fall below the critical limit of 10 $mgkg^{-1}$ (Table 8.2). The low values of available P recorded at the sub soils might partly be due to the nature of the parent material and as well due to the ease with which phosphorus undergoes fixation with iron and aluminium oxides under well drained acidic medium (Onyekwere et al. 2001; Nuga et al. 2006). According to Amhaklian and Osemuota (2012), tropical soils are generally low in available P due to low apatite content of the soil forming materials.

Exchangeable acidity (EA) was significantly different ($P < 0.05$) among the various land use types. The values of EA in CCS = 1.80 $cmolkg^{-1}$, FL = 2.00 $cmolkg^{-1}$, GL = 1.20 $cmolkg^{-1}$ and 2.20 $cmolkg^{-1}$ at 0–20 cm depth. At 20–40 cm the values of EA include 2.40, 2.60, 1.80 and 2.60 $cmolkg^{-1}$ for CCS, FL, GL and MF, respectively. Similarly, at 40–60 cm the values of EA include 2.60, 2.60, 1.00 and 2.60 $cmolkg^{-1}$ for CCS, FL, GL and MF, respectively. The exchangeable acidity ranged between 1.40 and 2.60 $cmolkg^{-1}$ with CV of 26.0% at 0–20 cm, 2.20 and

2.60 cmolkg^{-1} with CV of 9.0% at 20–40 cm and 2.20 and 2.80 cmolkg-^{1} with CV of 12.0% at 40–60 cmsoil depths.

Exchangeable hydrogen was significantly different ($P < 0.05$) among the land use types and ranged from 1.20 to 1.60 cmolkg^{-1} with CV of 14.0%. At 20–40 cm H^+ varied from 1.00 to 1.60 cmolkg^{-1} with CV of 20.0%. Exchangeable H^+ varied from 1.20 to 1.40 cmolkg^{-1} with CV of 7.0% at 40–60 cm soil depth. Exchangeable aluminium ranged between 0.00 and 1.4 cmolkg^{-1} with CV of 113.0% at 0–20 cmsoil depth. At 20–40 cm Al^{3+} ranged between 0.80 and 1.20 cmolkg^{-1} with CV of 35.0%. Exchangeable Al^{3+} ranged from 0.20 to 1.20 cmolkg^{-1} with CV of 35% at 40–60 cm soil depths. Exchangeable acidity values were rated low to relatively high according to the rating of Enwezor et al. (1989). Higher values (2.60 cmolkg^{-1} each) were observed in soils of the CCS and FL land use types at 40–60 cm. This might be attributed to the presence of iron and aluminium oxides at the sub soils. The high values of EA could be attributed to the application commercial fertilizer, uptake of basic cations and leaching (Orji Uzor and Obasi 2015; Mohammed 2017). Mustapha et al. (2003) and Fasina et al. (2005) attributed higher values of EA on subsoil to the solubility and movement of exchangeable acidity down to the soil from surface soils. The variation of EA across the soil depth is low with a CV range between 15.0 and 24.0% according to Aweto (1980).

Carbon Storage on Soils of the Four Land Use Types Across the Three Soil Depths

The results of carbon storage across the land use types are presented in Table 8.3. The result showed that carbon storage ranged between 29.28 and 126.88 MgCha^{-1} at 0–20 cmsoil depth, 47.41 and 135.48 MgCha^{-1} at 20–40 cm soil depth and 42.54 and 78.32 MgCha^{-1} across the land use types (Table 8.3). The total carbon storage across the depths showed that 0–20 cm recorded the highest value (352.18 MgCha^{-1}) whereas 40–60 cm recorded the lowest value (265.97 MgCha^{-1}) which implies that carbon storage decreased with increase in depth. The highest value (126.88 MgCha^{-1}) for carbon storage was obtained in managed forest plantation, and the lowest value (29.28 MgCha^{-1}) was obtained in continuously cultivated soil at 0–20 cm soil depth. The mean values of carbon storage across the soil depths ranged from 66.49 to 88.05 MgCha^{-1} whereas the coefficient of variation varied between 25.0 and 60.0%.

The total carbon storage across the land use types is as follows; CCS = 151.49 MgCha^{-1}, FL = 217.08 MgCha^{-1}, GL = 195.92 MgCha^{-1} and MF = 340.67 MgCha^{-1} (Table 8.3). The trend for increase in total carbon storage across the land use types is MF > FL > GL > CCS. This could be attributed to the ability of forest trees to sequester and store carbon in soil through the litter fall which increases SOM and in turn encourages carbon storage. Many researchers (Nowak et al. 2010; Mbah and Idike 2011; Nowak and Greenfield 2012) have reported that

Table 8.3 Carbon storage on soils of the four land use types across the three soil depths

Carbon storage ($MgCha^{-1}$)				
Land use	0–20 cm	20–40 cm	40–60 cm	Total
CCS	29.28	47.41	74.81	151.49
FL	95.05	51.72	70.31	217.08
GL	100.98	52.41	42.54	195.92
MF	126.88	135.48	78.32	340.67
TOTAL	352.18	287.01	265.97	905.16
Mean	88.05	71.76	66.49	226.29
CV (%)	48.0	60.0	25.0	

CCS = Continuously cultivated soil, FL = Fallow land, GL = Grass land, MF = Managed forest plantation, CV = Coefficient of variation.

natural undisturbed forests store high quantity of carbon. On the other hand, the low carbon storage observed in CCS during the period of study could be as a result of tillage operation, run-off, leaching and nutrient uptake. The observation corroborates with the reports of previous studies that tillage adversely affects carbon storage in soils (Anikwe 2006; Mbah and Idike 2011). Similarly, Bationo et al. (2007) in their study of soil organic carbon dynamics, functions and management in West African agro-ecosystems observed rapid decline of SOC levels with continuous cultivation.

Conclusion

The study showed that physical and chemical properties of soils of the various land use types including carbon storage were significantly higher in managed forest plantation, fallow land and grass land than in continuously cultivated soil due to the effect of climate change. The trend in carbon storage was as follows: managed forest plantation > fallow land > grassland > continuously cultivated soil. The soil physicochemical properties including total carbon storage were significantly higher at 0–20 cm soil depth than 20–40 and 40–60 cm soil depths. In all the land use types, managed forest plantation had the highest carbon storage while continuously cultivated soil had the lowest. Therefore, in the face of climate change, the land uses such as managed forest plantation, fallow land and grassland should be encouraged in order to capture and store the existing atmospheric carbon into the soil and biomass to mitigate the effect of climate change while continuous cultivation should be discouraged since it encourages carbon emission into the atmosphere which promotes climate change. It is, therefore, recommended that continuous cultivation which exposes the soil and hinders carbon storage should be discouraged especially now climate change is ravaging the global ecosystem rather managed forest plantation, fallow land and grass land should be encouraged in order to mitigate the effect of climate change.

Acknowledgements My profound gratitude goes to Profs Charles A. Igwe and Peter I. Ezeaku for their encouragement and support. I also acknowledge Johnbosco C. Egbueri, Joshua O. Ighalo and Chaitanya Pende the organizers of this work.

References

Adeoye GO, Agboola AA (1985) Critical levels of soil Ph, avaliable P, K Zn and Mn in sedimentary soil of south west Nigeria. Fertil Res 6:65–71

Adeputu JA, Adetunji MT, Ige DV (2014) Soil fertility and crop nutrition. Fact Affect Soil Acid 108–218

Aduayi EA, Chude VO, Adebusuyi BA, Olayiwola SO (2002) Fertilizer use and management practices for crops in Nigeria. Fed Minis Agric Rural Dev, Abuja, Nigeria. 9:63–65

Akamigbo FOR, Ezedimma FOC, Igwe CA (2001) Properties and classification of some Fadama soils in Bauchi State, Nigeria. In: Proceeding of the 27th annual conference of soil science society of nigeria held between November 4–9th. Univ. of Calabar, Nigeria, pp 45–48

Akpan US, Udofia GN, Edem SO (2018) Influence of parent material and land use on Organic Carbon associated with particle size in soils of Akwabom state In: Oluwatosin e tal (Eds). Proceedings of the 42nd annual conference of oil science society of Nigeria held on 12th–16th March, 2018 at Institute of Agricultural Research and Training Obatemi Awolowo. University Moor plantation, Ibadan, pp 52–62

Amhakhian SO, Osemwota IO (2012) Physical and chemical properties of soils in Kogi State, Guinea savanna zone of Nigeria. Nigeria J Soil Sci 22(1):44–52

Amuyou UA, Kontigo KE (2015) Toposequence analysis of soil properties of an agricultural field in Obudu mountain slopes, Cross River State, Nigeria. Eur J Phys Agric Sci 3(1):1–11

Anikwe MAN (2000) Amelioration of heavy clay loam soil with rice dust and its effect on Soil physical properties and maize yield. Biores Technol 74:169–173

Anikwe MAN (2006) Soil quality assessment and monitory: A review of current research efforts. New Generation Books, New Generation Ventures Ltd, Enugu, Southeastern Nigeria p, p 208

Antonio CDA, Angulo-Jaramilo R, deSouza ES, Netto AM, Cameiro CJG, Montenegro A (2001) Determinants of hydraulic conductivity and sorptivity of soils using a disk infitrometer. Br J Agric Environ Eng 10(4):1802–1899

Asadu CLA, Akamigbo FOR (1983) Influence of parent materials on the soils of Southeastern Nigeria. East Afric Agric For J 48:81–91

Aweto AO (1980) Variability of upper slope soil development on sandstone in South Western Nigeria. Nigeria Geograph J 25:27–37

Bagarello V, Lovino M (2003) Field testing parameter sensitivity of the two-term infiltration equation using differentiated linearization. Soil Sci Soc Am J 2:358–367

Bationo A, Kihars J, Van lauwa B, Waswa B, Kimetu J (2007) Soil organic carbon Dynamics, functions and management West African agro-ecosystem. Agric Syst 94:13–25

Blake GR, Hartage KH (1986) Bulk density. In: Klute A (ed) Methods of soil analysis part 1, vol 409. American Society of Agronomy, Madison Wisconsin, pp 364–375

Boul SW, Southard RJ, Graham RC, McDaniel PA (2003) Genesis and classification, 5th edn. Iowa State Press, A Blackwell Publishing Company, pp 3–34

Brady NC, Weil RR (2002) The Nature and properties of soils, 4th edn. Prentice Hall, New Jessey, U.S.A

Brady NC, Weil RR (2010) Elements of the natural and properties of soil, 2nd edn. Prentice Hall, Upper Saddle River, N J

Bray RH, Kurtz LT (1945) Determination of Total Organic and available forms of phosphorus in soils. Soil Sci 59:39–45

Bremner JM, Mulvanney CS (1982) Total nitrogen. In: Page A, et al (eds) Methods of soil analysis Part 2 ASA. Madison, Wisconsin, pp 595–624

Chaudhari PR, Ahire DV, Ahire VD, Chkravarty M, Maity S (2013) Soil bulk density as related to soil texture, organic matter content and available total nutrients of Coinboire soils. Int J Sci Res Publ 3(2):1–8

Chokor JU, Egborah JO (2018) Effect of Land Use types on selected soil physical and chemical properties in the Univeristy of Benin Nigeria. J Soil Sci 28(1): 276–286

D'Amato AW, Bradford JB, Fraver S, Palik BJ (2011) Forest management for mitigation and adaptation to climate changes. Insights from long term silviculture experiments. For Ecol Manage 262:803–816

D'Amore D, Kane E (2016) Climate change and forest sol carbon. US Department of Agriculture, forest service, climate change resource centre

Dudal R, Deckers J (1993) Soil organic matter in relation to soil productivity. In: Mulongoy K, Merckx R (eds) Soil organic matter dynamics and suitability of tropical agriculture.John Wiley and son, West Sussex, United Kingdom, p 02

Egbuchua CN, Ojobor S (2011) Characterization of some hydric soils in the niger-delta region of Nigeria for land evaluation purposes. Int J Adv Biol Res 1:77–82

Emmanuel OC, Akintola OA, Ezekiel-Adewoyin DT, Tetteh FM (2018) Grain and biomass yield of Cowpea as influenced by NPK Fertilizer application in the Sudan Savanna zone of Ghana. In: Proceedings of the 42nd Annual Conference of the Soil Sciene of Nigeria. pp 360–365

Enwezor WO, Udo EJ, Usoro NJ, Ayotade KA, Adepetu JA, Chude VO, Udegbu CI (1989) Fertilizer use and management practices for crops in Nigeria. Series 2. Federal Procurement and Distribution Division, Federal Ministry of Agriculture, Water Resources and Rural Development, Lagos, pp 119–121

Ezeaku PI, Eze FU, Oku EJ (2015) Profile distribution and degradation of soil properties of an Ultisol in Nsukka Semi-Humid Area of Nigeria. Afric J Agric Res 10(11):1306–1311

Fasina AS, Omolayo FO, Ajayi OS, Falodun AA (2000) Influence of land use on soil properties of three mapping units in Southwestern Nigeria: implications for sustainable soil management. Res J Appl Sci 2(8):879–883

Fasina AS, Aruleba JO, Omolayo FO, Shittu OS, Okusami YA (2005) Properties and classification of five soils formed on granatic parent materials of humid Southwest Nigeria. Nigeria J Soil Sci 15(2):21–29

Federal Department of Agriculture and Land Resoruces FDALR (1987) Reconnaissance soil map of Nigeria. Federal Ministry of Agriculture and Rural development Abuja

Fekadu A, Kibebew KA, Bobe BA, Asmare MB (2018) Characterization and classification of soils of Yikalo Subwatershed in Lay Gayint District, Northwestern Highlands of Ethiopia Endalkachew. Eurasian J Soil Sci 7(2):151–166

Gee GW, Bander JW (1986) Particle size analysis by hydrometer method, a simphfied method of routine textural analysis and sensitivity test on mineral parameters, soil SC SOC Am J. 43 1004–1007

Google map (2019) https://www.google.com

Grossman RB, Reinsch TG (2002) Bulk density and linear Extensibility core method. In Dare JH, Topp GC (eds) Methods of soil analysis. Part 4. Physical methods. SSSA incorporated. Guinea Savanna of Nigeria. Int J Agric Biosyst Eng 7(5):330–334

Guggenberger G, Zech W, Haumaier L, Christensen BT (1995) Land use effects on the compostion of organic matter in particle ion 13C NMR analysis. Eur J Soil Sci 46:147–158

Han WX, Fang JY, Reich PB, Ian Woodward F, Wang ZH (2011) Biogeography andvariability of eleven mineral elements in plant leaves across gradients of climate, soil and plant functional type in China. Ecol Lett 14:788–796

Hassan AM, Shuaibu M (2006) Physico-chemical properties of the soils at Walba-Daba, Gombe State. In: Proceeding of the 24th annual conference of horticultural society of Nigeria held between 17th–22nd November, 2006. Gombe State University, Gombe, Nigeria, pp 163–168

Havlin JL, Beaton JD, Tisdale S, Nelson WL (2006) Soil fertility and fertilizers. An introduction to nutrient management. 7th edn. Prentice Hall of India private limited, New Delhi, 110001, p 633
Igwe CA (2005) Erodibility in relation to water-dispersible clay forsomesoils of eastern Nigeria. LandDegrad Dev 16:87–96
IPCC (2007) Climate change 2007. Impacts adaptation and vulnerability. Contribution of working group II to fourth assessment report of the international panel on climate. In: Parry ML, Canziani OF, Paulutikof JP, Vander Linden PJ, Hanson CE (eds). Cambridge University press.
Johnson K, Scatena FN, Johnson AH, Pan Y (2009) Controls on soil organic matter content within a northern hard wood forest. Geoderma 148:346–356
JUO ASR (1979) Selected methods for soil and plant analysis. International institute of Tropical Agriculture (IITA), Ibadan, Nigeria
Klute A, Dirksen C (1986) Hydraulic conductivity and diffusivity: laboratory methods. In: Klute A (eds) Methods of soil analysis part 1: Physical and mineralogical methods, 2nd edn. Agronomy 9, America Society of Agronomy, Madison, WI, pp 199–224
Kolo J (2019) Investigation of topographic influence on soils supporting Cannibis sativa and their alternative use for rice production in Southwestern Nigeria. Unpublished PHD Thesis, Soil Science Department, University of Nigeria, Nsukka, p 87
Lal R, Kimble J, Follelt RF, Cole CJ (1998) The potential of U.S crop land to sequester carbon and mitigate the green house effect. Ann. Arbor. Press, Chelsea, USA
Lawal BA, Ojanuga AG, Tsado PA, Mohammed A (2013) Characterization, classification and agricultural potentials of soils on a toposequence in Southern 59
Loveland P, Webb J (2003) Is there a critical level of organic matter in the agricultural soils of Temperate regions. A review. Soil Tillage Res 70:1–18
Maniyunda LM, Gwari MG (2016) Soil suitability assessment of Haplustalfs for maize and groundnut production in a sub-humid environment of Nigeria. Int J Dev Sustain 393–403
Maniyunda LM (2018) Lithosequence effect on selected properties of Soils of sub-humid high plain in Northwestern Nigeria. Nigeria J Soil Sci 28(1):133–143. Pub.SSSN
Mbah CN, Idike FI (2011) Carbon storage in tropical agricultural soils of South Eastern under different management practices. Int R J Agric SC Soil SC 1(2):053–057
Meysner T, Szajdak L, Kus J (2006) Impact of the farming systems on the content of biologically active substances and the forms of nitrogen in the soils. Agro Res 4: 531–542
Mohammed YM (2017) Impact of landuse and management practice on soil physical and chemical quality indicators of vertisols at Pawe North Western Ethiopia. Afri J Agric Res 12(15):1306–1313
Murphy BW (2015) Impacts of soil organic matter on soil properties—a review with emphasis on Australian soils. In: Soil Research, vol 53. Department of the Environment, Canberra, Australia, 605–635. https://doi.org/10.1071/SR14246
Mustapha S, Jonah K, Voncir N (2003) Topogenic variability in some selected physico-chemical properties of soils at Abakukar Tefewa Belewa University (Yelma Campus) Bauchi. Agric Bus Technol J 1(1):1–12
Nave LE, Vance ED, Swanston CW, Curtis PS (2010) Harvest impacts on soil carbon storage in temperate forests. For Ecol Manage 259:857–866
Nelson DW, Sommers LE (1998) Total organic carbon and organic matter. In: Spaus DL, Page AL, Helmke PA, Leoppert RH, Soltanpour PN, Tahatabai MA, Johnson CI (eds) Methods of soil analysis, 3rd edn. Amenrican society of Agronomy, Madison, WI, USA, pp 961–1010
Njoku JD (2012) Application of geographical information system technology in soil fertility variation analysis in Southeastern Nigeria. Int J Sci Technol 1:47–59
Nowak DJ, Greenfield EJ (2012) Tree and imperious cover change in US cities. For Urban Green 11(1):21–30
Nowak DJ, Noble MH, Sisinni SM, Dwyer JF (2010) People and trees: assessing the U.S carbon forest resources. J For Washington 99(3):37–42
Nsor ME (2017) Pedogenesis, classification and management implications of soils derived from Nanka sands of false bedded sandstones lithology in Orumba North L.G.A, Southeastern, Nigeria.

In: Proceeding of the 41st annual conference of soil science society of Nigeria held between 13–17th March, 2017 at ATBU, Bauchi, Nigeria, pp 195–204

Nsor ME, Lifu MI, Otu EJ (2014) Influence of cocoa pods husks on the morphological and physico-chemical characteristics of Rhodic Kandiudults derived from basalts in central Cross River state. J Sci Agric Food Technol Environ 13:17

Nuga BO, Eluwa NC, Abimbola GE, Wokocha CC (2006) Characterization and classification of soils along a toposequence in Ikwano local government area of Abia State, Nigeria. Agric J 1(2):192–197

Nwite JN, Orji JE, Okolo C (2018) Effect of different land use systems on soil carbon storage and structured land use systems on soil carbon storage and structured indices in Abakaliki, Nigeria. India J Ecol 45(3):322–327

Obalum SE, Nwite JC, Oppong J, Igwe CA, Wakastuki T (2011) Variations in selected soil physical properties with land forms and slopes within an inland valley ecosystem in Ashanti Region of Ghana. Soil Water Res 6(2):73–82

Obigbesan GO (2009) Impact of Liebig's research on the development of agriculture in Africa. In: Proceedings 33rd conference of the soil science society of Nigeria held at Ado-Ekiti. p 51

ODRIN (1989) Nigeria profile of agricultural potential. Overseas Development Natural Resources Institute, United Kingdom, p 15

Okolo CC, Nwite JN, Ezeaku PI (2015) Evaluation of heavy metal concentration in soils at mechanic village, Abakaliki, southeast Nigeria and the implication for agricultural land use planning. Soil contamination and conservation discovery publishing House PVT Ltd. New Delli India 2015:1–11

Onyekwere IN, Akpan-idiok AU, Amalu UC, Asawalam DO, Eze PC (2001) Constraints and opportunities in agricultural utilization of some wetland soils in Akwa-Ibom. In: Proceeding of the 27th annual conference of soil science society of Nigeria held between November 4–9th. Calabar, Nigeria, pp 139–146

Orji Uzor EN, Obasi SN (2015) Properties and classification of erosion prone soils of Ukpo, Nnewi South L.G.A, Anambra State, Nigeria. Int J Agric Sci Rural Dev 15(2):1079–1084

Osuaku SK, Ukaegbu EP, Poly-Mbah CP, Nnawuhe CO, Okoro GO, Osuaku HE (2014) Variability in some physical and chemical properties of soils along a toposequence in Alvan Ikoku federal college of education, Owerri, Imo State, Nigeria. Int J Res Appl Nat a Soil Sci 2(11):17–24

Osujieke DN, Onweremadu EU, Ahukaemere CM, Ndukwu BN (2016) Classification of soils of a toposequence underlain by coastal plain sand in Southeast Nigeria. Nigerian J Soil Environ Res 14:256–263. Plant growth–Modeling the interaction between nitrogen and phosphorus. New Phytol 194:953–960

Powers M, Kolka R, Palik B, McDonald R, Jurgensen M (2011) Long term management impacts on carbon storage in lake states forests. For Ecol Manage 262:424–431

Raji BA, Mohammed K (2007) The nature of acidity in Nigeria Savanna soil. Samaru J Agric Res 16:15–24

Sharu MB, Yakubu M, Noma SS, Tsafe IA (2013) Characterization and classification of soils on an agricultural landscape in Dingyadi District, Sokoto State, Nigeria. Nigerian J Basic Appl Sci 21:137–147

Shelton IJ (2003) Soil erosion: causes and effects. Facts Sheet. ISSN. 1198-712X. Ontario Ministry of Agriculture, Food and Rural Affairs

Tabi FO, Ogunkule AO (2007) Spatial variation of some soil physicochemical properties of Alfisol in south western, Nigeria. J Soil Environ Res

Thomas EY, Oke OS (2018) Assessment of some trace elements from selected soil along the top sequence in Ibadan, South western Nigeria. In: Proceedings of the 42nd annual conf. of the soil science society of Nigeria, pp 434–447

Udo EJ, Ibia TO, Ogunwale JA, Anuo AO, Esu IE (2009) Manual of soil, plant and water analysis, 1st edn. Sidon Book Ltd, Lagos, Nigeria, pp 17–76

Uponi JI, Adeoye GO (2000) Soil testing and plant analysis: an overview. Agronomy in Nigeria, p 177

Uzoho BU, Ekpe II, Ahukaemere CM, Ndukwu BN, Okoli NH, Osis FA, Chris-Emenyeonu CM (2014) Nitrogen status of soils of selected land uses of cropping systems in the humid tropical rainforest, Southeastern Nigeria. Adv Life Sci Technol 25:24–33

Vogelmann ES, Reichert JM, Reinert DJ, Mentges MI, Vieira DA, Peixoto de Barros CA, Fasinmirin JT (2010) Water repellency in soils of humid subtropical climate of Rio Grande do Sul, Brazil. Soil Tillage Res 110(1):126–133. ISSN 0167-1987. https://doi.org/10.1016/j.still.2010.07.006

Vose JM, Peterson DL, Patel–Weyhand T (2012) Effects of climatic variability and change on forest ecosystems: a comprehensive science synthesis for the US forest sector. Portland, or: US Department of Agriculture, forest service, Pacitic Northwest Research station, p 265

Chapter 9
The Impact of Climate Change on the State of Carbon Footprint in Nigeria

Sodiq Solagbade Oguntade, Abdullahi Tunde Aborode, Oluwatoni Honour Afinjuomo, Victor Akolade Kayode, Toyeeb Abidemi Atanda, Oluwayomi Dapo Amupitan, Olatomide Blessing Ojajune, Abeeb Babajide Ajagbe, and Michael Olamilekan Omonitan

Abstract Carbon emission in Nigeria is among the highest globally, and the country suffers from the impacts of climate change. The Nigerian government, international, and national non-governmental organisations have made efforts to curtail this scourge through laws, policies, programmes and projects. However, the country's carbon footprint continues to increase, with over 82 million tonnes of CO_2 emissions and CO_2 emissions per capita of 0.44 tonnes per person in 2016. These carbon emissions adversely affect essential sectors in Nigeria, such as agriculture, health, power generation and security. The rainfall and temperature patterns have changed geographically in Nigeria due to climate change induced by carbon emissions, with some areas receiving excess rainfall and heat and vice versa. Therefore, all stakeholders must take holistic measures to reduce the carbon footprint, especially the government. A major challenge for this research is the non-availability of an up-to-date carbon profile of

S. S. Oguntade (✉)
IHE Delft Institute for Water Education, Delft, Netherlands
e-mail: sodiqoguntade@yahoo.com

A. T. Aborode
Healthy Africans Platform, Research and Development Hub, Ibadan, Nigeria

O. H. Afinjuomo
University of Ilorin, Ilorin, Nigeria

V. A. Kayode
Kogi State University, Anyigba, Nigeria

T. A. Atanda
Institute of Environmental Sciences, Jagiellonian University, Kraków, Poland

O. D. Amupitan · O. B. Ojajune
Federal University of Technology Akure, Akure, Nigeria

A. B. Ajagbe
Obafemi Awolowo University, Ile-Ife, Nigeria

M. O. Omonitan
Federal University of Agriculture Abeokuta, Abeokuta, Nigeria

J. C. Egbueri et al. (eds.), *Climate Change Impacts on Nigeria*, Springer Climate,
https://doi.org/10.1007/978-3-031-21007-5_9

Nigeria to validate the progress made so far in reducing the carbon footprint, which should be the focus of subsequent research. Nevertheless, this chapter discusses the impact of climate change on the state of carbon footprint in Nigeria, the challenges, the implications, the current interventions, and sustainable recommendations.

Keywords Climate change · Impact · Carbon footprint · Nigeria

Introduction

Continuous human activities such as transportation, food, and electrical consumption are the major determinants of a person's carbon footprint (WWF 2018). However, in most developing countries, including Nigeria, over-population, poverty, and urbanisation have resulted in high carbon footprints with dire consequences on their environmental sustainability, climate change and planetary health (Ali et al. 2016).

Ali et al. (2016), Rafindadi (2016), Riti and Shu (2016), Ali et al. (2019), Adejumo and Asongu (2020), and Adejumo (2020) have previously looked at how biological, environmental degradation in Nigeria is affected by the amount of productive land needed to fulfil human needs, such as food, housing, woods, and the sequestrated carbon from fossil fuel burning.

In 1970, the population of Nigeria was around 56 million, with only 17.8% inhabiting urban residences. There was a substantial improvement in the population existing and surviving in the nation, as 186 million people were present in 2016, a 324% increase in 46 years (WPP 2019). As people migrated to cities and towns for better economic and livelihood opportunities, population growth coincided with the explosion of the oil industry. As a result, the effects of climate change increase through GHG emissions and carbon footprints. Out of a total population of 212 million, over 102 million people live in cities (National Bureau of Statistics 2021).

Furthermore, the population increase and high level of carbon emission at the different phases of agriculture production, consumption, industrialisation, and other economic activities are the carbon footprints that are mainly determined by the amount of energy consumed (Adejumo 2020). However, poverty and food security are major problems in developing countries like Nigeria (Omotayo et al. 2018). Infrastructure and lifestyle differences are significant factors affecting Nigeria's carbon footprints, as are differences in urban and rural infrastructure and lifestyle preferences (Ali et al. 2019).

From 1961 to 2013, data from the global footprint network was used to compute each country's ecological footprint, which ranked Nigeria 165th out of 180th based on a 1.16-hectare ecological footprint per capita and a 0.7-hectare ecological footprint capacity (Razack and Ludin 2014). Nigeria should reduce climate change activities that contribute to carbon footprints as a result of this observed concern regarding environmental sustainability, such as transportation, electricity generation, deforestation, and begin to embrace a circular economy which involves the utilisation of environmental resources in a manner that will ensure reduction in waste and pollution, the longer a product or material is used, the better off it is.

As a result, the direct impact of human activities on the environment results in climate change which in turn modify to different effects such as carbon emissions and water pollution. Among the causes of carbon emission in Nigeria include oil spillage, industrial fumes, biological chemicals, etc. (Ali et al. 2019). In addition, Nigeria's carbon emissions exceed its pledge at the climate action summit in 2019 to reverse the harmful outcome of climate change due to the consistent impact and activities of humans and industries on the environment, while there have been low actions to reduce these activities. Instead, it keeps increasing the harmful effects of climate change in Nigeria. Broader discussion and progressive actions in reducing carbon footprint status in Nigeria are needed because it is one of the required recipes for achieving sustainable development. However, few literatures have identified how climate change results in carbon emissions in Nigeria and how harmful it is to the well-being of Nigerians. Therefore, understanding the impact of climate change on the state of carbon footprint in Nigeria is necessary for implementing policies on alleviating climate change towards achieving sustainable development in Nigeria by 2030.

Sources and Causative Factors of Carbon Emissions in Nigeria

Carbon emissions in Nigeria come from various sectors such as the power industry, transport, buildings, and other sectors. According to (Worldometer n.d.), the transport sector accounts for the highest (31.9%), followed by the non-combustion sector (27.5%), other industrial combustion (17.2%), power industry (14.8%), and buildings (8.7%). Nigeria's CO_2 emissions per capita are approximately 0.44 tonnes per person as of 2016 population (185, 960, 241), accounting for 0.23% of global CO_2 emissions (Worldometer n.d.).

Many factors have contributed to the country's increase in $C0_2$ emissions over the years. For example, despite being the largest economy in Africa, a significant percentage (40%) of the country's population does not have access to an electricity source and largely depends on wood and charcoal for cooking. In addition, more than 50% of the country's power generation is from fossil fuels (Fig. 9.1). Furthermore, with a population growth projected at 3%, by 2030, electricity demand will increase by 16.8% (Ministry of Foreign Affairs of Netherlands 2021). The number of generators has increased in cities' population, coupled with epileptic electricity supply by the national grid. Nigerians spend about \$22 billion (~5% of Nigeria's GDP) annually on fuel for generators (Ministry of Foreign Affairs of Netherlands 2021). Also, land-use change and forestry (LUCF) due to increased urbanisation and population growth are major drivers of carbon emissions in Nigeria. In 2014, forest land

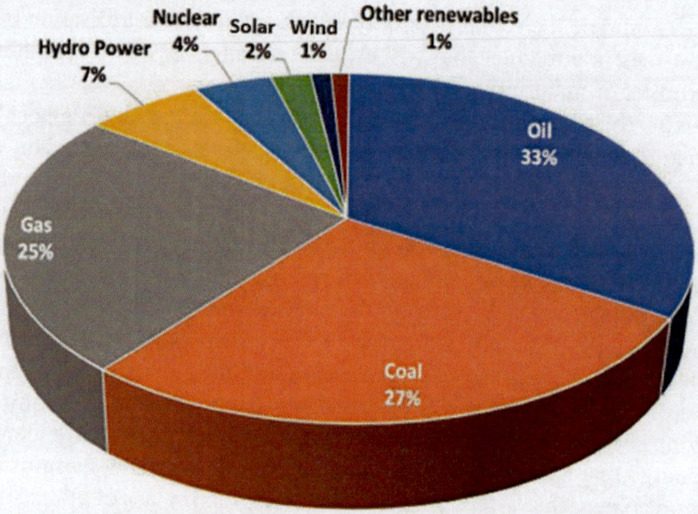

Fig. 9.1 Sources of energy generation in Nigeria (Ministry of Foreign Affairs of Netherlands 2021)

accounted for the majority of LUCF emissions, with only a minor contribution from burning biomass (USAID 2016). The non-revolutionary situation in Nigeria's transportation sector is also a major factor in the increased carbon emissions. Currently, there are no operational electric intra-city rail lines or trams in Nigeria, which means that many Nigerians rely heavily on public transportation that runs on fossil fuels. Similarly, the state governments' public transportation buses run on fossil fuels.

The Implication of not Reducing Carbon Footprint

Global warming results from the high rate of carbon footprint without reduction, so global warming is driving rapid climate changes and weather shifts. Humans are releasing heat-trapping greenhouse gases (GHG) into the atmosphere rapidly, so these changes are happening rapidly (Alhorr et al. 2014). The most important anthropogenic greenhouse gas is carbon dioxide (CO_2), which is abundant in the atmosphere and can remain there for thousands of years (Mardiana and Riffat 2015). Carbon dioxide emissions can have severe repercussions and impacts on humans and the environment. CO_2 emissions act as a blanket, trapping heat in the atmosphere and significantly warming the planet (Klufallah et al. 2014). This segment protects the universe from cooling, thus raising global temperatures.

The Fifth Assessment Report compiled by the Intergovernmental Panel on Climate Change (IPCC) revealed that the major factor causing global warming is greenhouse gases (GHG) emitted from human activity; the organisation was formed in the mid-twentieth century. Agribusiness and related land-use changes accounted for 17% of the world's anthropogenic GHG emissions in 2010 (Edenhofer et al. 2015).

Climate change will alter all aspects of the hydrological cycle ranging from evaporation through precipitation, run off and discharge (McGuire et al. 2002). Nigeria as a country is already facing impacts such as high temperatures, variability in rainfall, sea-level rise, threats to food security, flooding, low agricultural yield, drought, desertification or desert encroachment, land degradation, extreme weather events, access to freshwater, diseases, and loss of biodiversity, which is prominently related to uncontrolled and increasing carbon footprint (Elisha et al. 2017; Ebele and Emodi 2016; Olaniyi et al. 2013). Observations have shown that there have been fluctuations between two well-defined seasons (dry and rainy season) in Nigeria over the last few years; the start date for the rainy season in Southwest Nigeria lasts mainly from March to September, but with the adverse effect of carbon emission on the environment. Most farmers have been lamenting about late and inadequate precipitation, which is already harming agricultural exportation from the country, and also, local consumption has been greatly affected, thereby skyrocketing the prices of farm produce (Haider 2019).

In 2019, the Nigerian government under the Muhammed Buhari-led Administration pledged to reduce greenhouse emissions by 20% in 2020 and a further 45% depending on international support (Energy Capital & Power 2022). Despite these commitments by the federal government under the current administration, no significant progress has been recorded to help slow down carbon emissions in the country.

As a result of the failure of the Federal Government to initiate policies to monitor carbon footprints, the problem has continued to persist and will continue for a long time; if necessary, actions are not taken to cushion the effect of carbon emissions in the country. The following subsections explain the environmental factors that have been greatly influenced by the impact of carbon emissions on the environment.

Temperature and Rainfall

Carbon emission affects the average air temperature, which has an impact on rainfall distribution. The average earth's temperature rises due to carbon dioxide from carbon emissions. Particles in the earth's atmosphere circulate more quickly at extremely high temperatures. The atmosphere of the earth hence gives off more energy. This energy from the atmosphere is transported worldwide, causing variation in the atmospheric condition. The severity of the atmospheric condition increases with atmospheric energy levels.

The temperature in Nigeria has risen significantly since the 1980s (Enete 2014; Federal Ministry of Environment 2014). The Northern part of Nigeria has been experiencing drought due to the increased temperature. A good example is an extremely scorching sun. The increased temperature dries up the water, which results in a

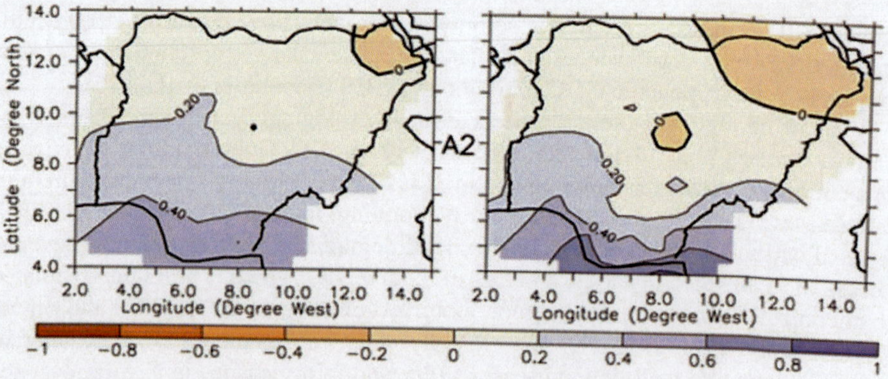

Fig. 9.2 Projected changes in average daily rainfall in Nigeria (BNRCC 2011)

drought. Lake Chad and other lakes in Nigeria are shrinking due to climate change and are at risk of disappearing (Dioha and Emodi 2018; Elisha et al. 2017), thus, causing tension with other neighbouring countries and people depending on the lakes.

In the coastal areas and southern parts of Nigeria, there is variability and increment in rainfall patterns and sea level, leading to flooding, which causes unpredictable weather patterns, displacement of water bodies, loss of lives and properties, and increased population pressure (Matemilola 2019). Increased rainfall and sea-level rise will exacerbate flooding and sweep away coastal areas (Akande et al. 2017; Ebele and Emodi 2016). Climate projections for the next years have shown a substantial increase in temperature in ecological zones (Akande et al. 2017).

Precipitation forecasts between 2046 and 2065 revealed that there would be a 0.4 mm per day increase in the southern part of the country and a 0.2 mm per day decrease in the Northern region (Fig. 9.2). Similarly, temperature forecasts indicated that the number of extremely hot days (≥38 °C) could increase by 7 days in mangrove (South), 23 days in rain forest (South), 41 days in tall grass savanna, and 88 days in short grass savanna annually (North) as shown in Fig. 9.3.

Geographic Impacts

Geographical location plays an essential role in the amount of carbon emission released into the atmosphere. Different regions have their domestic needs. Densely populated regions release high concentrations of carbon emission because of the population's energy needs. The need to accommodate the population has also led to the expansion of infrastructural facilities, which have greatly affected the natural cycle. Large, dispersed regions produce greater carbon dioxide emissions than smaller regions (Ribeiro et al. 2019). According to a study conducted in the United States, the bigger the city, the greater the impact of changing its population on its carbon output (Fig. 9.4).

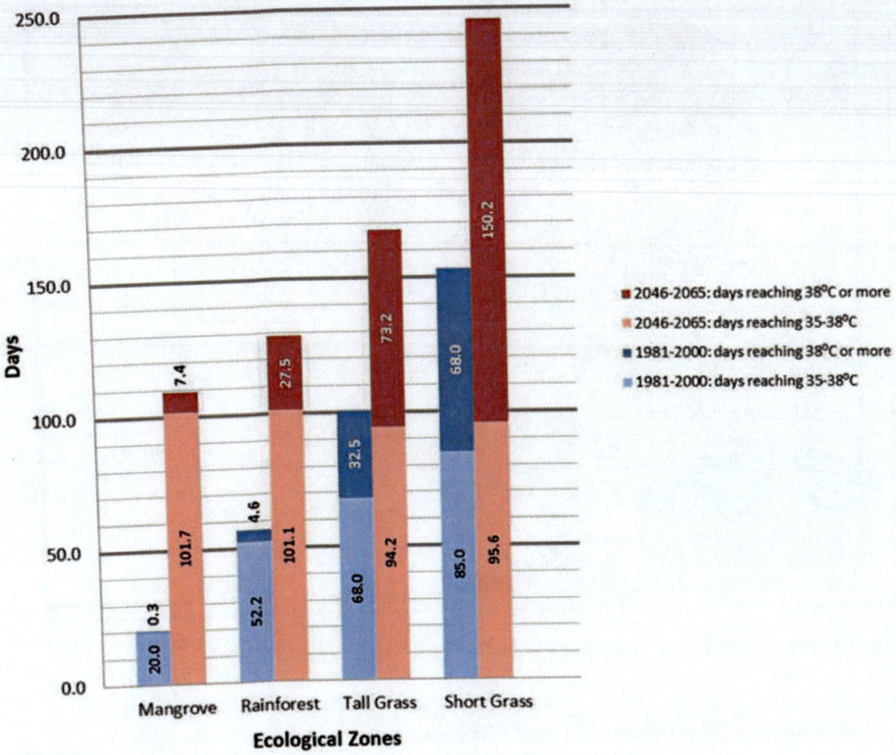

Fig. 9.3 Nigeria's annual severe heat days by ecological zone (BNRCC 2011)

Fig. 9.4 Strong positive relationship between CO_2 emissions and population (Ribeiro et al. 2019)

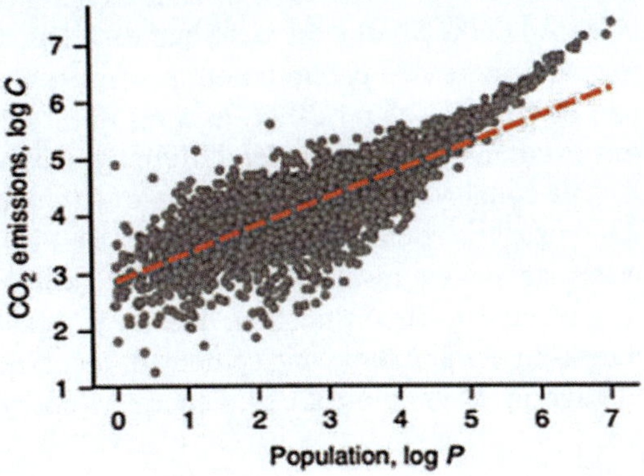

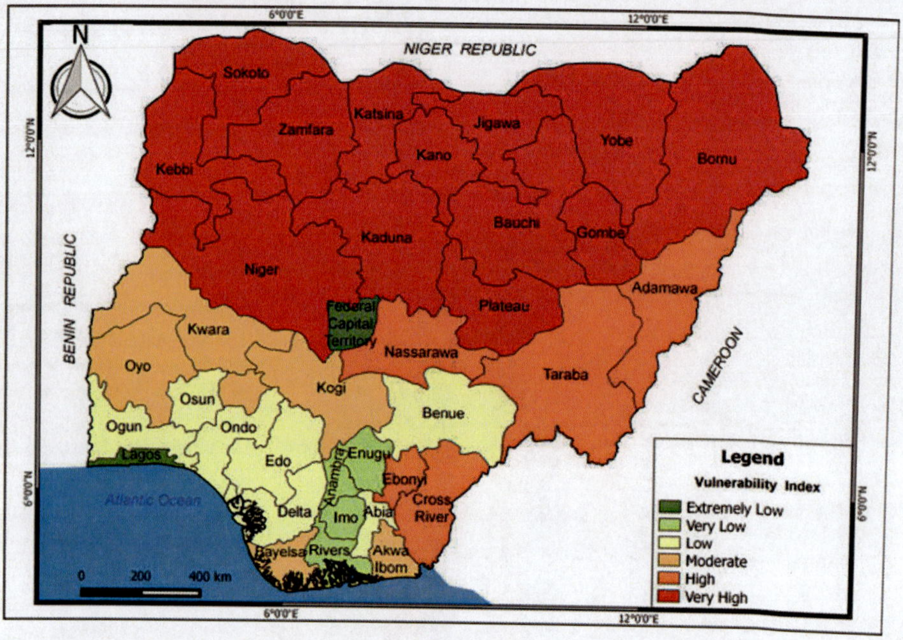

Fig. 9.5 Patterns of vulnerability to climate change in Nigeria (Madu 2016)

An analysis of the vulnerability of Northern states to climate change revealed that they are more vulnerable than their Southern counterparts due to the increasing desert encroachments and droughts (Madu 2016; Federal Ministry of Environment 2014), as shown in Fig. 9.5. It is most vulnerable in the Northeast and Northwest, while it is less so in the Southwest and Southeast. Due to coastal erosion, flooding, sea-level rise, and increased precipitation, several settlements are displaced in the southern part of Nigeria (Niger Delta) as a result of coastal erosion, flooding, and sea-level rise (Matemilola 2019; Federal Ministry of Environment 2014; Sayne 2011).

The country has a tropical climate with two precipitation regimes: high precipitation for the southwest and southeast and low precipitation for the North. Due to water stress as a result of climate change induced by carbon emission, the North is subjected to desertification. However, coastal erosion, flooding, sea-level rise, increased precipitation, and settlement displacement are typical in the southern part of Nigeria (Matemilola, 2019; Federal Ministry of Environment 2014; Sayne 2011).

Food Security and Water

The pattern of vulnerability to climate change exacerbated by carbon emissions also affects food security and agricultural activities (Madu 2016). Factors such as overgrazing, cutting of trees (deforestation), the combination of rising heat and less rain

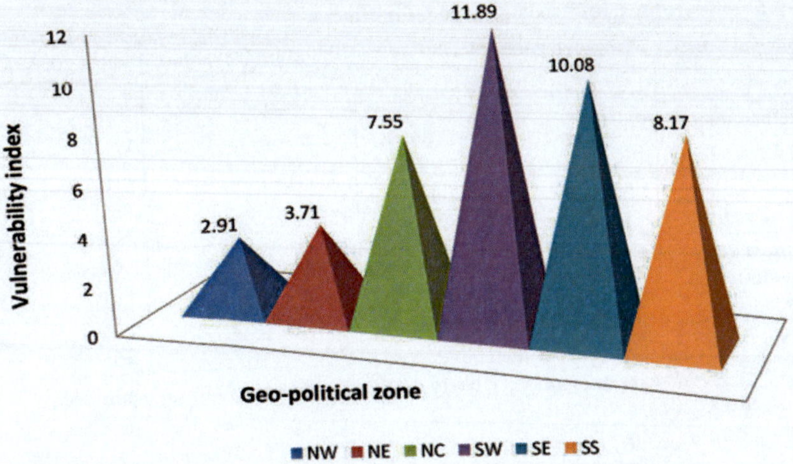

Fig. 9.6 Nigerian geopolitical zones' climate change vulnerability (Ignatius 2016)

have fastened desertification, loss of the wetlands, and an intense reduction in the volume of surface water, reduced flora and fauna, and biotic resources (Abdulkadir et al. 2017; Akande et al. 2017; Ebele and Emodi 2016; Federal Ministry of Environment 2014). However, the impacts of climate change on food and water depend on the region; some regions are more vulnerable than others (Fig. 9.6). A lower value indicates greater vulnerability.

There have been a lot of cases of deficit in the availability of food across the country as a result of the low level of food production due to the increasing variability in rainfall, excessive heat, and lack of water (Abdulkadir et al. 2017; Elum and Momodu 2017; Ebele and Emodi 2016; Enete 2014). For example, Lake Chad's shrinkage has affected socio-economic activities in the Northeast and Northwest parts of the country. Due to the critical level of climate variability in the Northern part of Nigeria, there is a ripple effect on the other parts of the country (Sayne 2011). In the Southwest part of the country, the risk of flooding due to climate variability has affected food production (Haider 2019). A comparison of the impacts of heat stress as a result of climate change on poultry farming in the North and South in 2017 (Kaduna and Oyo States) revealed that 10% of the farmers suffered losses in production (Liverpool-Tasie et al. 2019).

Forestry

Forestry is also affected by the non-reduction of the carbon footprints of Nigeria. Carbon emissions heighten climate change effects such as flooding and adverse weather conditions that result in extreme wind, which can destroy trees and other forest vegetations (Ogbuabor and Egwuchukwu 2017). In 2003, Nigeria's cost of

Table 9.1 Projected trends of ecological zones in Nigeria

Climate variables	Mangrove zone	Rain forest	Tall grass (savanna)	Short grass (Sahel)
Temperature	↑	↑	↑	↑
Rainfall amount	↑	↑	↓	↓
Rainfall variability	↑	↑	↑	↑
Extreme rainfall events-droughts	Likely	Likely	↑	↑
Extreme rainfall events-storms and floods	↑	↑	Likely	Likely
Sea level rise	↑	NA	NA	NA

Legend: ↑ likely increase or increase; ↓ likely decrease or decrease; NA not applicable
Source (BNRCC 2011)

deforestation and non-timber forest product losses was estimated to be around 1.7% of the country's GDP (Ogbuabor and Egwuchukwu 2017; Abdulkadir et al. 2017). Moreover, there is high dependence on forest products such as timber, wood, and charcoal for cooking, emitting carbon due to the increasing population (BNRCC 2011).

The mangrove forest in the Niger Delta, the largest mangroves in Africa and one of the largest mangrove forests in the world, has suffered a reduction due to emissions from oil and gas exploration in the region. The size has reduced from 9,990 km^2 in 1980 to 7,386 km^2 in 2006 (World Rainforest Movement n.d.). In addition, the rainforest in southwest Nigeria has been impacted by climate change due to variability in rainfall and temperature between 1953 and 2012 (Adedoyin et al. 2019). Similarly, in the savannah agro-ecological zones of Nigeria, climate change has affected the restoration of parkland trees (Abdullahi 2020). Table 9.1 gives Nigeria's projected trends in ecological zones due to climate change.

Health

In Nigeria, climate change has numerous negative effects on the health and well-being of residents. For example, extreme weather events such as heat waves directly affect human health in a direct manner (BNRCC 2011). Climate change indirectly impacts the human health by causing malnutrition, typhoid fever, cholera, and excess heat, which can lead to meningitis, a result of a shortage of food production and increased air pollution (Abdulkadir et al. 2017; Osuafor and Nnorom 2014; BNRCC 2011). Drought and high temperatures have reduced fresh water availability, increased heat stress, affected hygiene, and increased the chances of diarrhoea and many other water-borne diseases, including typhoid, cholera, and river blindness in the country's savanna and Sahel regions (Nkechi et al. 2016; BNRCC 2011). In the north-western part of Nigeria, cerebrospinal meningitis has been projected to increase due to climate change by 32 to 38% between 2020 and 2035 and 43 to 91% between 2060 and 2075 (Abdussalam et al. 2014).

Furthermore, in coastal areas of Nigeria, extreme winds and rainfall, rising sea levels and floods make people vulnerable to injuries, drowning, death, and serious physical and mental trauma (Abdulkadir et al. 2017; Amanchukwu et al. 2015). Diseases such as hepatitis and malaria have been on the rise in south-western Nigeria due to increased rainfall intensity, flooding, stagnant water, and contaminated groundwater (Ebele and Emodi 2016; Osuafor and Nnorom 2014; BNRCC 2011). In addition, sea-level rise in Southern Nigeria has resulted in aquifer salinization as well as pollution with sediment and sewage, which is a threat to public health (Sayne 2011). Generally, a rise in the CO_2 activity in the blood is also associated with cancer because it creates a hostile and low pH environment that suits the growth of cancer cells (Hulikova et al. 2014).

Economy

Agriculture is a significant part of Nigeria's economy, employing many Nigerians. There is a climate variation in Nigeria's agricultural pattern. For example, cattle, guinea fowls, and turkeys are dominant in Northern Nigeria, where rainfall and humidity are lower, and the presence of natural grasses for grazing is very scarce and heavily reliant on rainfall. However, the southern part of Nigeria receives more rainfall, and grazing areas are moister, with more parasites and endemic diseases, because high temperatures and humidity promote parasite growth outside their host. Climate stress caused by carbon emissions thus reduces fodder, water intake, grazing period, and thus the growth rate and economic output (Idowu et al. 2011). High temperatures have reduced livestock production by reducing fertility, lowering meat and milk outputs, and increasing livestock mortalities, particularly in piggery, poultry, and rodentary, by at least 15% per year. Also, pests and diseases due to climate change impacts have reduced investment profit by more than 20% per year (Idowu et al. 2011).

Furthermore, crop production in Nigeria depends highly on rainfall (Haider 2019). As a result, the erratic rainfall pattern caused by climate change induced by carbon emissions makes it hard for farmers to schedule their operations and minimise the farming season and growing days duration (Anabaraonye et al. 2019; BNRCC 2011). In addition, floods and destroyed farmlands drive pest and disease migration because of climate change, whereas high temperatures damage crops (Idowu et al. 2011). These variations in rainfall and sunshine hours have reduced the harvests of annual crops such as maize, cassava, and yam by more than 2.5% per annum. Similarly, cash crops such as cocoa, kola nut, and oil palm have annually decreased by 5.5 metric tonnes/hectare (Idowu et al. 2011). Thus, climatic conditions that negatively affect crop and animal yields will have a substantial adverse effect on the nation's Gross Domestic Product (Anabaraonye et al. 2019; Solomon and Edet 2018; Ogbuabor and

Egwuchukwu 2017; Ebele and Emodi 2016). Additionally, extreme weather events such as floods lead to the destruction of properties and infrastructures, resulting in an extraordinary amount of money being spent to restore them (Federal Government of Nigeria 2013).

Energy

Non-reduction of the carbon footprints of Nigeria affects hydroelectric and thermal generation in the country (Ebele and Emodi 2016). There are fewer trees in the northern part of Nigeria due to the low amount of rainfall; this affects hydroelectric output due to the unavailability of biomass for fuel (Ebele and Emodi 2016; BNRCC 2011). Alternatively, the high precipitation level in rainforest zones also affects power generation because of damage to the transmission lines and substation equipment. These events are caused by extreme weather events that follow climate change (Ebele and Emodi 2016; BNRCC 2011). Other power plants in the country are also at risk from extreme winds and heavy storms (BNRCC 2011).

The Implications of Low Reduction of Carbon Footprint

Carbon emission and climate change are interlinked, making carbon footprint a severe issue to be evaluated. This is because the impacts of climate change on all aspects of human endeavour are more worrisome daily. A carbon footprint is the amount of greenhouse gas emitted by an entity (Bergeler 2012). Carbon footprint may occur through natural processes or human activities. The individual's carbon footprint is measured by the environmental impacts resulting from the burning of fossil fuels. It is estimated as the tonnes of CO_2 emitted in a year. Human carbon dioxide and other greenhouse gas emissions are the main cause of climate change and one of the world's biggest challenges (Collins et al. 2013).

The generated carbon emissions in Nigeria were about 82,634,214 tonnes in 2016 (Worldometer n.d.). The dependence of the country's population on forest wood for energy sources has increased the carbon emission ratio, thereby contributing to climate change. This is evident in the research conducted by Global Forest Watch that Nigeria has lost a total of 214,240.37 acres of forest land between 2010 and 2019, contributing to about 19.72 Metric tonnes of carbon dioxide (Daisy 2020).

In addition, it is known that the primary element causing global warming is a result of the gas released when fossil fuels are burnt (carbon dioxide). There is an increase in the earth's temperature, resulting in melting polar ice caps, flooding of low-lying areas, and rising sea levels due to excess release of the gaseous elements. These are the threats to the climate, and unfortunately, they are fast becoming a reality and might no longer be regarded as threats if there is a non-reduction of the country's carbon footprint. Hence, in oil-producing countries like Nigeria, greenhouse gas emissions have an increased rate of climate- weather change. Therefore, the high rate of fossil fuel utilisation contributes to great risk vis-à-vis climate change impacts Nigeria's future (Anomohanran 2017; Adekomaya et al. 2016).

Furthermore, the agricultural sector is crucial to Nigeria's economy. According to the World Bank GDP data (2016), nearly 78% of the country's total land mass, representing 708,000 km^2, is under agricultural cultivation. Of these, 48.0% constitute arable lands, 42.8% are under permanent meadows and pastures, and the remaining 9.2% are under permanent crop production. The agriculture sector is also the country's largest employer and accounts for 24.4% of GDP (2016). Available evidence shows that climate change is global likewise its impacts, but the most adverse effects will be felt mainly by developing countries, especially those in Africa, due to their low level of coping capabilities (Nwafor 2007; Jagtap 2007). Nigeria is one of these developing countries (Akpodiogaga and Odjugo 2010). As the planet warms from greenhouse gases emission, rainfall patterns shift, and extreme events such as droughts, floods, and forest fires become more frequent (Zoellick 2009), which results in poor and unpredictable yields, thereby making farmers more vulnerable, particularly in Africa (UNFCCC 2007). Climate variability and trends in Nigeria include rising temperatures, increased duration of dry spells, the intensity of heavy rain events, increased aridity and drought threatening the country's agricultural sector as a result of limited infrastructure, lack of access to inputs, and finance and substantial post-harvest losses. Given the importance of the agricultural sector to livelihoods and the economy in Nigeria, extreme weather events, such as drought, cyclones, tornadoes, and floods, can affect economic growth through production costs, infrastructure damages, and the need for spending out of budget (Federal Government of Nigeria 2013). It is projected that crop yield in Africa may fall by 10–20% by 2050 or up to 50% due to climate change (Jones and Thornton 2002). For example, cassava, a commonly grown crop in Nigeria, is relatively well adapted to hot and dry conditions; however, it is susceptible to water logging, and heavy rainfall events could impact production yields. The First Biennial Update Report of the Federal Republic of Nigeria under the UNFCCC, Nigeria, in 2018 estimated that just 1% of Nigerian agriculture is irrigated. With most of the country's agriculture reliant upon rain-fed agriculture and smallholder farmers using traditional methods, slight variations in rainfall would make it extremely difficult for the farmers to plan for planting and other operations. In the southern production region, flooding, erosion, and soil loss are of great concern. In the North, a traditional livestock production zone, decreasing precipitation, and increased temperatures are of significant concern. Thus, leading to a reduction in productivity (Anabaraonye et al. 2019; BNRCC 2011).

According to statistics, if there is no adaptation, climate change could contribute to a loss of 6 to 30% of Nigeria's GDP by the year 2050. This projection was arrived at due to the wide range of climate change impacts affecting all Nigerian sectors (BNRCC 2011). This makes it imperative and essential to reduce the country's carbon footprint.

The country's carbon footprint increments can also largely affect the health sector. This would cause climate change and, in return, affect the health of humans by exposing them to diseases. There is every likelihood of water-borne diseases such as malaria spreading when there is an increase in rainfall. Malaria accounts for 45% of outpatients in Nigeria, and about 50% of Nigerians have a minimum of one malaria sickness in a year (Ayeni 2011). Heatwaves and hurricanes are also created due to an increase in global warming. This occurs when the ocean warms up and heats the surrounding air. Heatwaves have caused more deaths due to heat-related illnesses and cardiovascular failure in recent years than in the last sixty years (United Nations 2021). The exhaust from public transport in Nigeria, the use of generators as a means of electricity in households, and its usage in industries result in air pollution caused by the carbon dioxide concentration in the environment, which increases respiratory diseases like Asthma.

The energy sector of Nigeria has also been affected by climate change impacts. For example, hydroelectric and thermal power productions are expected to negatively affect the current low electricity supply in Nigeria supply of electricity in Nigeria (Ebele and Emodi 2016). The reason for this is the reduced precipitation in northern Nigeria, which has affected the growth of trees and biomass for fuel, thus impacting hydroelectric power production (Ebele and Emodi 2016; BNRCC 2011). The reduction in river discharge and increased temperature has affected the capability of thermal electric production; an increased temperature reduces the capacity to transmit electricity. Higher evapotranspiration due to drought has also affected the water volume in dams.

An example is the Kanji dam hydroelectric station in Niger State, the largest completed hydroelectric power project in Nigeria, with an installed capacity of 760 megawatts. Its capacity has continued to decline due to drought, thus, resulting in lower power generation. A negative reservoir inflow trend was recorded between 1985 and 2014 (Salami et al. 2011, Adegbehin et al. 2016), as shown in Fig. 9.7. However, the electricity generated depends on the reservoir inflow at the Kanji dam, which receives precipitation and flows from over 4,000 km and nine countries in West Africa. The research by (Adegbehin et al. 2016) confirmed that a strong positive correlation (0.92) exists between the reservoir inflow and electricity generation at Kanji Dam.

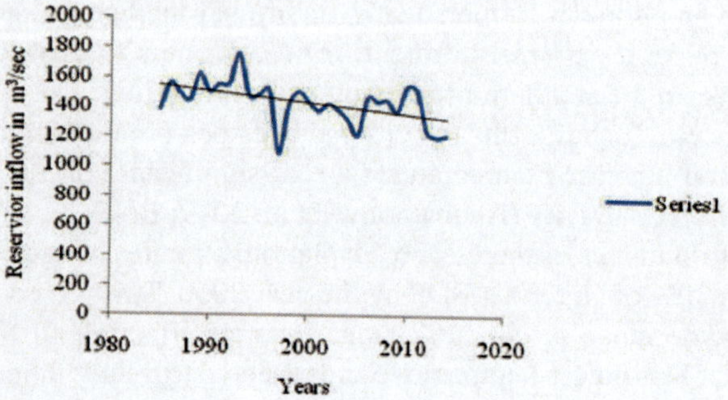

Fig. 9.7 Trend of reservoir inflow in Kanji Hydroelectric power station between 1985 and 2014 (Adegbehin et al. 2016)

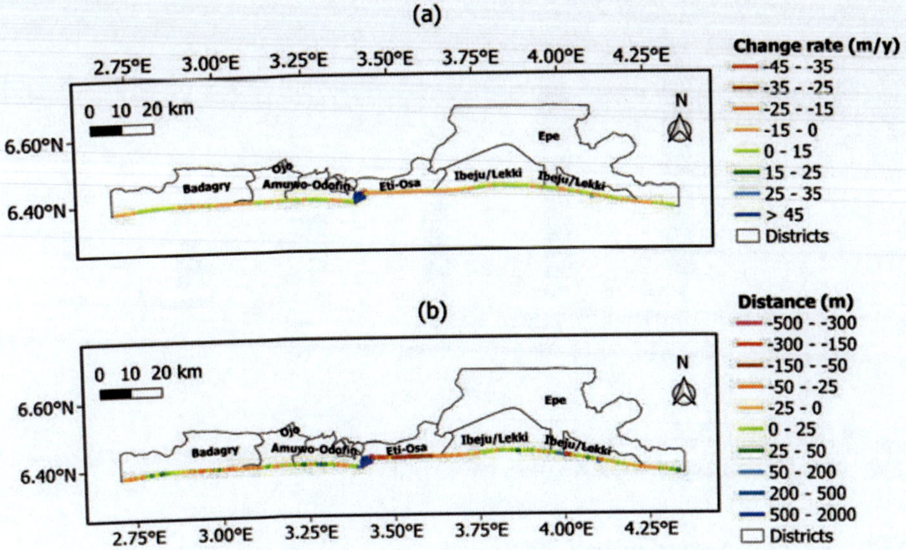

Fig. 9.8 **a** change rate (m/year) **b** distance change (m) (Adeaga et al. 2021)

Transmission lines and substation equipment have also been damaged in areas that experienced intense rainfall, such as the coastal and rainforest areas, which will have an adverse effect on the generation of power (Ebele and Emodi 2016; BNRCC 2011).

Sectors such as tourism and manufacturing have also experienced the impacts of climate change in Nigeria. For example, beach-based tourism such as the Lagos bar beach and Lekki Island has been affected due to sea-level rise that washes away the shores. Also, this sea-level rise may destroy industrial layouts due to flooding. This will then reduce the productivity and efficiency of the manufacturing sector (Ebele and Emodi 2016). Figure 9.8 shows the Lagos shoreline dynamics between 2001 and 2020.

Climate change has also threatened the country's security due to the increased struggle for natural resources such as water, food, and land. An example of this is the increased farmers-herders conflict in Nigeria due to desert encroachment in the Northern part of the country, which forced the herders to migrate in search of greener pastures for their cattle. According to data obtained from the Armed Conflict Location and Event Data Project (ACLED), which is a data collection, assessment, and conflict tracking project with its headquarters in the United States, at least 8,343 people have been killed as a direct result of the farmer-herder conflicts that have been occurring in Nigeria between 2005 and 2021. The data further showed that Benue State had been the target of the most attacks, followed by Plateau State, Kaduna State, Taraba State, and Nasarawa State in that order (Fig. 9.9).

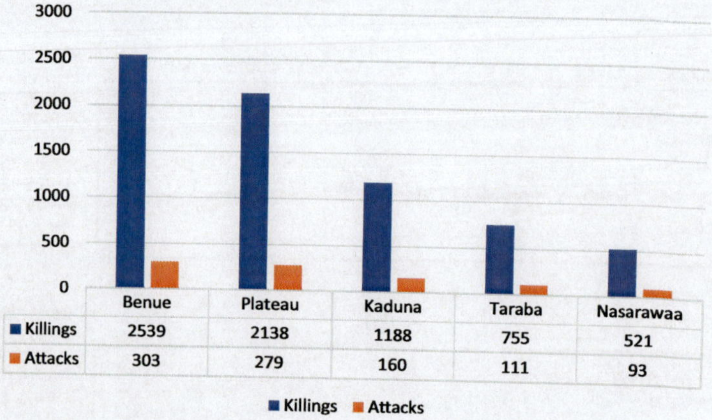

Fig. 9.9 Number of attacks and fatalities as a result of farmer-herder clashes (2005–2021). *Source* (The Armed Conflict Location & Event Data Project, n.d.)

Efforts in Understanding, Mitigating, and Adapting to Climate Change Impact on Carbon Emission in Nigeria

Climate change is one of the greatest threats to Nigeria's society, and its harmful impacts are well known, such as flooding, depletion of soil nutrients, unemployment, increased crime rates, hunger, poverty, and many others (Ebele and Emodi 2016). Climate change has been the topmost threat to Nigeria's development; therefore, several efforts have been made to understand, adapt, and mitigate it (Ebele and Emodi 2016).

Nigeria has put in place a lot of policies and strategies for adaptation and mitigation of the consequences of climate change. Some of these laws and policies include the Climate Change bill signed by President Muhammadu Buhari on 18 November 2021, on the heels of Nigeria's commitment at COP26 held in Glasgow. The climate change bill allows for establishment of a National Climate Change Council and advances the country's goal of becoming an emissions-free nation by 2050 (*TheCable* 2021). Furthermore, there is a national policy on climate change aimed at adopting climate-smart practices and fostering sustainable development. Also, the Climate Change Department was created in the Federal Ministry of Environment to implement the Climate Change Convention, the Kyoto Protocol and the Paris Agreement. The department provides information on Nigeria's Nationally determined contributions (NDCs), green bonds in Nigeria, vulnerability assessments and climate change adaptation strategies (Department of Climate Change Federal Ministry of Environment n.d.). Table 9.2 shows some green bonds funded projects.

Furthermore, the National Policy on Drought and Desertification and National Forest Policy was signed in 2017 with the aim of improving the sustainable management of forest resources in Nigeria. The objectives of the national drought plan include implementing holistic drought monitoring and detection systems and

Table 9.2 Projects funded with Nigeria's green bonds

Project	Ministry	Description
Renewable energy micro utilities in 45 communities	Federal ministry of power, works and housing	An initiative to provide access to electricity to 45 communities across the country employing mini-grids with distributed loads of between 33 and 50 KW per community
Energising education	Federal ministry of power, works, and housing	A rural electrification initiative that seeks to develop clean off-grid independent power plant
Afforestation programme	Ministry of environment	Afforestation programme to increase forest coverage through the plantation of seedlings to cover 131,000 hectares of land

Source (Federal Government of Nigeria, n.d.)

completing vulnerability assessments for all drought-vulnerable sectors, populations, and zones (Federal Ministry of Environment 2018). Other policies include the 2015 National Renewable Energy and Energy Efficiency Policy (NREEP), approved by the Federal Executive Council (FEC) for the electricity sector. The policy focuses on improving hydropower, solar, biomass, geothermal, wave, and tidal energy power plants as well as co-generation power plants to reduce the reliance on fuel wood which accounts for over 50% of domestic energy usage in Nigeria (Federal Ministry of Power 2015). In 2020, the National Adaptation Plan Framework was established to reduce the vulnerability to climate change through building adaptive capacity and resilience (Federal Ministry of Environment 2020).

Apart from the Nigerian Governments' efforts, several international and national organisations have contributed to efforts to reduce the effects of carbon emissions and climate change impacts in Nigeria. Wastesmart, a social innovation organisation, for example, has a project called Youth Environmental Empowerment Programme (YEPP) that will last between 2017 and 2024 to train public servants and young Nigerians with the skills and competence to drive climate change threats into the potential for advancement and national development (Youth Environmental Empowerment Program n.d.). Similarly, the European Union (UE) under the EU Global Climate Change Alliance Plus Initiative (GCCA +) has a 7.5 million Euros project in Nigeria known as the Nigeria Climate Change Response Programme (NCCRP) that will last between 2020 and 2024. The aim is to strengthen Nigeria's greenhouse gas emission profile to identify mitigation actions that will reduce energy use and costs in the furtherance of renewable energy and more eco-friendly solutions (Nigeria Climate Change Response Programme n.d.).

National non-governmental organisations such as Eleven Eleven Twelve Foundation, Plogging Nigeria, and Elomgreenie have all championed the education of young Nigerians, especially in primary and secondary schools, as well as non-educated

Nigerians at markets and other public places on climate change mitigation and adaptation strategies. Moreover, the Great Green Wall Initiative and the Nigerian Conservation Foundation have contributed to climate change mitigation by planting trees to fight deforestation, land degradation, and desert encroachment.

Conclusion

Human activities such as car exhaust, industrial fumes, and energy generation have significantly intensified climate change in the form of carbon footprint. Furthermore, global changes such as the increase in population, industrialisation, agricultural production, consumption, urbanisation, and other economic activities have all constituted to increase in carbon emissions. Nigeria is a developing country that relies on oil revenue as its main foreign exchange earnings and main sources of income; over the years, there has been a large increase in greenhouse gas emissions, especially CO_2 emissions, which directly induced climate change. As highlighted, this increase in Nigeria's carbon footprint will make the country more vulnerable to climate change impacts.

The impact or implication of a low carbon footprint reduction can be destructive, and the country is currently facing these challenges. Nigeria has witnessed an increase in intensities of rainfall in the southern region of the country and the high temperature in the northern part of the country, which consequently has led to the desertification of forest areas, reduction in agricultural productivity, adverse effects on energy generation and transmission, increased incidence of climate change induced diseases and security challenges as a result of climate change, linked to a high level of carbon emission. Therefore, the country must take measures to reduce its carbon footprints for sustainable development. Although the government has taken measures to reduce carbon emissions with the support of national and international non-governmental organisations, more actions are needed if the country wants to achieve its target of a net-zero economy by 2050.

Recommendation

This chapter showed that reducing carbon footprints is expedient for eluding the impacts of climate change in Nigeria. Carbon emission has plagued critical sectors in Nigeria and has retarded the country's growth due to its adverse effects on power generation, health, agriculture, and security. However, measures such as planting trees will help the country lower its carbon footprints. Trees planting is an inexpensive method, and it requires collective efforts of Nigerians across the board to nurture them in sequestrating these carbon emissions. Furthermore, simple modifications in how farmers maintain their land can also be a low-cost and effective approach to reducing carbon emissions from the atmosphere in Nigeria. For example, planting crops with less tillage and adding biochar, a charcoal made from plant matter, to the soil. Through terrestrial carbon sequestration, biochar is important for removing CO_2 from the atmosphere.

Another essential measure to reduce carbon emissions in Nigeria is the investment in renewable energy; as discussed, the energy generation in Nigeria relies heavily on fossil fuels. But there is a lot of potential for renewable energy in Nigeria. Solar and wind are Nigeria's under-utilised energy sources that offer tremendous advantages. First, job creation in the form of technicians and other stakeholders. Second, the two will boost the agricultural sector through enhanced post-harvest storage, a major problem in the agricultural food chain in Nigeria. Also, a shift to renewable energy will foster international trade, especially with European countries, which are frontiers in the new world order on clean energy. Lastly, if the two sources are well-utilised, especially in the Northern part of the country, which has more potential, Nigeria could be exporting excess energy to its neighbouring countries and generating additional income, thus, increasing the GDP.

The major limitation of this research is the absence of up-to-date data on the carbon profile of Nigeria, which made it difficult to ascertain the country's progress so far. The earliest and most reliable data available was in 2016. Also, there are no metrics available to measure the progress of the policies and laws that have been made on climate change mitigation and adaptation in Nigeria. Therefore, in a subsequent study, the focus should be on critically evaluating the most recent progress the country has made in reducing carbon emissions.

References

Abdulkadir A, Maryam LA, Muhammad TI (2017) Climate change and its implications on human existence in Nigeria: a review. Bayero J Pure Appl Sci 10(2):152–158. https://www.ajol.info/index.php/bajopas/article/viewFile/170772/160195

Abdullahi IN (2020) The impact of climate change on the management and regeneration of parkland trees in the Savannah zones of Northern Nigeria. Bangor University (United Kingdom)

Abdussalam AF, Monaghan AJ, Steinhoff DF, Dukic VM, Hayden MH, Hopson TM, Thornes JE, Leckebusch GC (2014) The impact of climate change on meningitis in Northwest Nigeria: an assessment using CMIP5 climate model simulations. Weath Clim Soc 6(3):371–379

Adeaga O, Folorunsho R, Foli BAK, Akinbaloye O (2021) Assessment of shoreline change along the coast of Lagos, Nigeria. Remote Sens Earth Syst Sci 4(3):186–198. https://doi.org/10.1007/s41976-021-00059-w

Adedoyin LB, Balogun L, Ayodele O (2019) Analysis of climate change in rainforest ecosystems of Southwestern Nigeria. Sci Res Ann 10:220–231

Adegbehin AB, Yusuf YO, Iguisi EO, Zubairu I (2016) Reservoir inflow pattern and its effects on hydroelectric power generation at the Kainji Dam, Niger State, Nigeria. Environ Impact III 1:233–244. https://doi.org/10.2495/eid160211

Adejumo AV, Asongu SA (2020) Foreign direct investment, domestic investment and green growth in Nigeria: any spillovers? In: International Business, Trade and Institutional Sustainability, pp 839–861. Springer, Cham

Adejumo OO (2020) Openness and greenness: pay-offs or trade-offs for the Nigerian economy. In: International business, trade and institutional sustainability, pp 863–886. Springer, Cham. https://doi.org/10.1007/978-3-030-26759-9_51

Adekomaya O, Jamiru T, Sadiku R, Huan Z (2016) Sustaining the shelf life of fresh food in cold chain-a burden on the environment. Alex Eng J 55(2):1359–1365

Akande A, Costa AC, Mateu J, Henriques R (2017) Geospatial analysis of extreme weather events in Nigeria (1985–2015) using self-organising maps. Adv Meteorol https://doi.org/10.1155/2017/8576150

Akpodiogaga P, Odjugo O (2010) General overview of climate change impacts in Nigeria. J Hum Ecol 29(1):47–55

Alhorr Y, Eliskandarani E, Elsarrag E (2014) Approaches to reducing carbon dioxide emissions in the built environment: low carbon cities. Int J Sustain Built Environ 3(2):167–178

Ali HS, Law SH, Zannah TI (2016) Dynamic impact of urbanisation, economic growth, energy consumption, and trade openness on CO 2 emissions in Nigeria. Environ Sci Pollut Res 23(12):12435–12443

Ali HS, Law SH, Lin WL, Yusop Z, Chin L, Bare UAA (2019) Financial development and carbon dioxide emissions in Nigeria: evidence from the ARDL bounds approach. GeoJournal 84(3):641–655

Amanchukwu RN, Amadi-Ali TG, Ololube NP (2015) Climate change education in Nigeria: the role of curriculum review. Education 5(3):71–79. https://www.researchgate.net/profile/Nwachukwu_Prince_Ololube/publication/283081778_Climate_Change_Education_in_Nigeria_The_Role_of_Curriculum_Review/links/56292c5508ae518e347c967b/Climate-Change-Education-in-Nigeria-The-Role-of-Curriculum-Review.pdf

Anabaraonye B, Okafor CJ, Ikuelogbon OJ (2019) Educating farmers and fishermen in rural areas in Nigeria on climate change mitigation and adaptation for global sustainability. Int J Sci Eng Res 10(4):1391–1398. https://www.ijser.org/researchpaper/Educating-Farmers-and-Fishermen-in-Rural-Areas-inNigeria-on-Climate-Change-Mitigation-and-Adaptation-for-Global-Sustainability.pdfd

Anomohanran O (2017) Bridging environmental impact of fossil fuel energy: the contributing role of alternative energy. J Eng Stud Res 23(2):22–26

Ayeni AO (2011) Malaria morbidity in Akure, Southwest, Nigeria: a temporal observation in a climate change scenario. Trends Appl Sci Res 6(5):488

Bergeler E (2012) What is the carbon footprint and why is it bad? Science niblets The Science behind everyday topics. http://www.scienceniblets.org/everyday-science/what-is-the-carbon-foot-print-and-why-is-it-bad.html

BNRCC (Building Nigeria's Response to Climate Change) (2011) National adaptation strategy and plan of action on climate change for Nigeria (NASPA-CCN). Prepared for the Federal Ministry of Environment Special Climate Change Unit. http://csdevnet.org/wp-content/uploads/NATIONAL-ADAPTATION-STRATEGY-AND-PLAN-OF-ACTION.pdf

Collins M, Knutti R, Arblaster J, Dufresne J-L, Fichefet T, Friedlingstein P, Gao X, Gutowski WJ, Johns T, Krinner G (2013) Long-term climate change: projections, commitments and irreversibility. In: Climate change 2013-the physical science basis: contribution of working group I to the fifth assessment report of the intergovernmental panel on climate change, pp 1029–1136. Cambridge University Press

Daisy D (2020) The carbon brief profile: Nigeria. https://www.carbonbrief.org/the-carbon-brief-profile-nigeria/

Department of Climate Change Federal Ministry of Environment. (n.d.). Department of Climate Change. https://climatechange.gov.ng/mandate/. Accessed 26 June 2022

Dioha MO, Emodi NV (2018) Energy-climate dilemma in Nigeria: Options for the future. IAEE Energy Forum 2:29–32. https://www.researchgate.net/publication/323830158_Energy-Climate_Dilemma_in_Nigeria_Options_for_the_Future

Ebele NE, Emodi NV (2016) Climate change and its impact in Nigeria economy. J Sci Res Rep 10(6):1–13. http://www.journaljsrr.com/index.php/JSRR/article/view/21917/40737

Edenhofer O, Pichs-Madruga R, Sokona Y, Farahani E, Kadner S, Kadner K, Seyboth A, Adler I, Baum S, Myhre G (2015) Climate change 2014: mitigation of climate change; working group III contribution to the IPCC fifth assessment report. Cambridge University Press, Cambridge, UK

Elisha I, Sawa BA, Ejeh UL (2017) Evidence of climate change and adaptation strategies among grain farmers in Sokoto State, Nigeria. IOSR J Environ Sci Toxicol Food Technol (IOSR-JESTFT) 11(3):1–7. https://www.iosrjournals.org/iosr-jestft/papers/vol11-issue%203/Version-2/A1103020107.pdf

Elum ZA, Momodu AS (2017) Climate change mitigation and renewable energy for sustainable development in Nigeria: a discourse approach. Renew Sustain Energy Rev 76:72–80. https://doi.org/10.1016/j.rser.2017.03.040

Energy Capital & Power (2022) Nigeria commits to reducing greenhouse gas emissions with climate pledge. https://energycapitalpower.com/nigeria-commits-to-reducing-greenhouse-gas-emissions-with-climate-pledge/

Enete IC (2014) Impacts of climate change on agricultural production in Enugu State, Nigeria. J Earth Sci Clim Change 5(9):234. https://www.omicsonline.org/open-access/impacts-of-climate-change-on-agricultural-production-in-enugu-state-nigeria-2157-7617.1000234.php?aid=32633

Federal Government of Nigeria (2013) Nigeria post-disaster needs assessment–2012 floods. https://www.gfdrr.org/sites/default/files/publication/NIGERIA_PDNA_PRINT_05_29_2013_WEB.pdf

Federal Government of Nigeria (n.d.) Federal government of Nigeria green bonds. https://www.chapelhilldenham.com/public/assets/campaign/Green%20Bonds%20FAQ%2006022019.pdf

Federal Ministry of Environment (2014) United nations climate change Nigeria. National Communication (NC). NC 2. 2014. https://unfccc.int/sites/default/files/resource/nganc2.pdf

Federal Ministry of Environment (2018) National drought plan. pp 1–103

Federal Ministry of Environment (2020) Nigeria's national adaptation plan framework (NAP). pp 1–60

Federal Ministry of Power (2015) National renewable energy and energy efficiency policy (NREEP)

Haider H (2019) Climate change in Nigeria: impacts and responses. K4D Helpdesk Report, pp 1–38. http://www.rockfound.org/initiatives/climate/climate_change.shtml%0Awww.iied.org/HS/publications.html.%0AHOW%0Ahttps://assets.publishing.service.gov.uk/media/5dcd7a1aed915d0719bf4542/675_Climate_Change_in_Nigeria.pdf

Hulikova A, Aveyard N, Harris AL, Vaughan-Jones RD, Swietach P (2014) Intracellular carbonic anhydrase activity sensitises cancer cell pH signaling to dynamic changes in CO2 partial pressure. J Biol Chem 289(37):25418–25430

Idowu AA, Ayoola SO, Opele AI, Ikenweiwe NB (2011) Impact of climate change in Nigeria. Iranica J Energy Environ 2(2):145–152. https://www.researchgate.net/publication/228459699

Ignatius AM (2016) Rurality and climate change vulnerability in Nigeria: assessment towards evidence based even rural development policy

Jagtap S (2007) Managing vulnerability to extreme weather and climate events: implications for agriculture and food security in Africa. In: Proceedings of the international conference on climate change and economic sustainability. Nnamdi Azikiwe University, Enugu, Nigeria, pp 12–14

Jones PG, Thornton PK (2002) Croppers to livestockers: livelihood transition to 2010 in Africa due to climate change. Global Environmental Change, World Health Organization, Geneva, Switzerland

Klufallah MM, Nuruddin MF, Khamidi MF, Jamaludin N (2014) Assessment of carbon emission reduction for buildings projects in Malaysia-a comparative analysis, vol 3. In: E3S Web of Conferences; EDP Sciences: Bangi, Malaysia

Liverpool-Tasie LSO, Sanou A, Tambo JA (2019) Climate change adaptation among poultry farmers: evidence from Nigeria. Clim Change 157(3):527–544. https://doi.org/10.1007/s10584-019-02574-8

Madu IA (2016) Rurality and climate change vulnerability in Nigeria: assessment towards evidence based even rural development policy. In: 2016 Berlin Conference on Global Environmental Change Held from 23–24 May 2016. Freie Universität Berlin, pp 1–23. https://pdfs.semanticscholar.org/508b/94cab07b84a703b44eca1089326cc98d7495.pdf?_ga=2.154518008.112403230.1572433568-162569160.1557482164

Mardiana A, Riffat SB (2015) Building energy consumption and carbon-dioxide emissions: threat to climate change. J Earth Sci Clim Change S3:1

Matemilola S (2019) Mainstreaming climate change into the EIA process in Nigeria: perspectives from projects in the Niger Delta Region. Climate 7(2):29. https://doi.org/10.3390/cli7020029

McGuire B, Mason IM, Mason I, Kilburn CRJ, Kilburn C, Killburn C (2002) Natural hazards and environmental change. Oxford University Press

Ministry of Foreign Affairs of Netherlands (2021) Solar Report Nigeria. 65

National Bureau of Statistics (2021) Annual abstracts of statistics, Abuja, Nigeria

Nigeria Climate Change Response Programme (n.d.) Nigeria climate change response programme-NCCRP. https://www.gcca.eu/programmes/nigeria-climate-change-response-programme-nccrp. Accessed 26 June 2022

Nkechi OG, Alphonsus NA, Eze E (2016) Mitigating climate change in Nigeria: African traditional religious values in focus. Medit J Soc Sci 7(6):299–308. https://www.mcser.org/journal/index.php/mjss/article/view/9612

Nwafor JC (2007) Global climate change: the driver of multiple causes of flood intensity in Sub-Saharan Africa. In: International conference on climate change and economic sustainability. Nnamdi Azikiwe University, Enugu, Nigeria, pp 12–14

Ogbuabor JE, Egwuchukwu EI (2017) The impact of climate change on the Nigerian economy. Int J Energy Econ Policy 7(2):217–223. https://dergipark.org.tr/en/download/article-file/361739

Olaniyi OA, Ojekunle ZO, Amujo BT (2013) Review of climate change and its effect on Nigeria ecosystem. Int J Afr Asian Stud-An Open Access Int J 1(1):55–65. http://dx.doi.org/10.22161/ijreh.3.3.3

Omotayo AO, Ogunniyi AI, Tchereni BH, Nkonki-Mandleni B (2018) Understanding the link between households' poverty and food security in South-West Nigeria. J Dev Areas 52(3):27–38. https://doi.org/10.1353/jda.2018.0034

Osuafor AM, Nnorom NR (2014) Impact of climate change on food security in Nigeria. AFRREV STECH 3(1). https://www.ajol.info/index.php/stech/article/view/103132

Rafindadi AA (2016) Does the need for economic growth influence energy consumption and CO2 emissions in Nigeria? Evidence from the innovation accounting test. Renew Sustain Energy Rev 62:1209–1225

Razack NTAA, Ludin ANM (2014) Ecological footprint and food consumption in Minna, Nigeria. In: IOP conference series: earth and environment science, vol 18. IOP Publishing, p 012179

Ribeiro HV, Rybski D, Kropp JP (2019) Effects of changing population or density on urban carbon dioxide emissions. Nat Commun 10(1):3204. https://doi.org/10.1038/s41467-019-11184-y

Riti JS, Shu Y (2016) Renewable energy, energy efficiency, and ecofriendly environment (RE 5) in Nigeria. Energy Sustain Soc 6(1):13

Salami AW, Sule BF, Okeola OG (2011) Assessment of climate variability on Kainji hydropower reservoir. In: Annual conference of the national association of hydrological sciences. Abuja, pp 18–21

Sayne A (2011) Climate change adaptation and conflict in Nigeria. USIP, Washington, DC. https://www.usip.org/sites/default/files/Climate_Change_Nigeria.pdf

Solomon E, Edet OG (2018) Determinants of climate change adaptation strategies among farm households in Delta State, Nigeria. Curr Invest Agric Curr Res 5(3). http://dx.doi.org/10.32474/CIACR.2018.05.000213

The Armed Conflict Location & Event Data Project (n.d.) The armed conflict location & event data project. https://acleddata.com/#/dashboard. Accessed 26 June 2022

The Cable (2021) https://www.thecable.ng/breaking-buhari-signs-climate-change-bill-into-law

UNFCCC (2007) Climate change country parties discuss approaches to reduce emissions from deforestation in developing countries. https://unfccc.int/press/news_room/newsletter/in_focus/0Aitems/3944.php

United Nations (2021) The sustainable development goals report 2021. United Nations Publication Issued by the Department of Economic and Social Affairs, pp 1–64

USAID (2016) Greenhouse Gas Emissions in Nigeria. 2014:10–11

World Rainforest Movement (n.d.) Niger deltas mangrove communities threatened by continued gas flaring, map alert action in africa, in mangroves are disappearing and with them, the livelihoods, vol 14

Worldometer (n.d.) Nigeria CO2 emissions–worldometer. https://www.worldometers.info/co2-emissions/nigeria-co2-emissions/. Accessed 26 June 2022

WPP (2019) World Population Prospects 2019: Highlights. United Nations, Department of Economic and Social Affairs, Population Division. ST/ESA/SER.A/423

WWF (2018) Living planet report: aiming higher. In: Grooten M, Almond RE (eds). WWF, Gland

Youth Environmental Empowerment Program (n.d.) Project YEPP. https://www.climate-chance.org/en/best-pratices/project-yepp-youth-environmental-enpowerment-program/. Accessed 26 June 2022

Zoellick RB (2009) A climate smart future. Washington, DC, World Bank

Chapter 10
Coppicing Capacity of *Pycnanthus angolensis* for Sustainable Forestry Techniques in the Climate Change

Olunike Adedoyin Bello

Abstract Demographic pressure and climate change are two of the major challenges of the twenty-first century in which climate change is possibly the greatest threat, as it is leading to extreme weather conditions, increasing areas of drought, and species extinction. *Pycnanthus angolensis* is one of the overexploited indigenous tree species used in phytomedicine that is threatened by environmental factors such as climate change. Therefore, for sustainable utilisation of this species, coppicing capacity of decapitated seedlings of *Pycnanthus angolensis* (False African Nutmeg) was investigated to determine the effects of location and climate change. Five seedlings each from Osun (Gbongan, Ajaba), Ekiti (Otun, Ayetoro), and Oyo (Idito, Adewumi) at 12 weeks after transplanting were decapitated from diameter (0.9 to 2.1 cm) and heights (4.50 to 20.50 cm) (from the first four nodes of the plant) and laid out in an open nursery in two replicates in a complete randomised design (CRD) to evaluate new shoot height (SH), plant height (PH), number of leaves (NL), and number of nodes (NN) fortnightly for 14 weeks. The data were analysed with observed means subjected to Tukey Post Hoc. The SH ranged from 1.55 ± 1.01 to13.40 ± 1.01 cm, NN ranged from 2.30 ± 1.00 to 18.00 ± 1.00, and NL ranged from 1.35 ± 0.81 to 11.10 ± 0.81 cm. The NB ranged from 0.10 ± 0.62 to 3.40 ± 0.62 while the PH ranged from 8.90 ± 1.05 cm to 25.20 ± 1.05 cm. It was observed that *Pycnanthus angolensis* sourced from Gbongan responded positively to coppicing in a strategic way of conservation. The present study confirmed that *Pycnanthus angolensis* could be coppiced well at plant height from 4.50 to 20.50 cm. There was steady increment in coppicing capacity of *Pycnanthus angolensis* and proved to be site-dependent. Coppicing today remains a common forest management system for sustainable livelihood in climate change scenario.

Keywords Climate change · Sustainable agriculture · Coppicing capacity · *Pycnanthus angolensis* · Seedling · Decapitation

O. A. Bello (✉)
Department of Biology, Osun State College Of Education, Ila-Orangun, Nigeria
e-mail: nikebello73@gmail.com

J. C. Egbueri et al. (eds.), *Climate Change Impacts on Nigeria*, Springer Climate,
https://doi.org/10.1007/978-3-031-21007-5_10

Introduction

The developments of anthropogenic factors caused by an increase in population and consequently, by increased production and growth of the agrosphere and climate change are the major challenges of the twenty-first century. The worldwide population continues growing exponentially, and it is expected to reach ~9:8 × 109 in 2050 (Hunter et al. 2007). Climate change generated by greenhouse gas emissions is possibly the greatest threat, as it is leading to extreme weather conditions, increasing areas of drought, and species extinction, among others (Carnicer et al. 2011; Wernberg et al. 2013). In this adverse context, food production needs to be increased significantly. Increasing food production is not enough though. Breeding programmes in which coppice is one of the breeding methods should also consider food safety and environmental care among their objectives (Portirio et al. 2018). The rising demand for tree products and expanding population pressure resulted in the decline of forest cover and consequently the demand for tree products exceeded the supply. There is much concern about the sustainability of intensive cutting of trees for firewood and other economic values in Nigeria (Neke 2004). The supply of wood and woody biomass products in Nigeria comes from different forest and vegetation types and production systems, including natural forests, woodlands, bush lands, community woodlots, and farm forestry. The share of the total domestic energy is fuel wood and tree residues 70%, dung 8%, agricultural residues 7%, and the rest comes from other sources.

Coppicing is one of the oldest forestry systems known from many countries worldwide (Fujimori 2001). Coppicing is a common response strategy of trees to damage, enabling survival after severe disturbances such as cutting. It was frequently used as a means of firewood until the second half of the nineteenth century (Peterken 1993). Since then, the conversion of coppices to high forests has been the principal trend especially in central and northwestern Europe (Peterken 1993), particularly due to demand for higher-quality timber (Hédl et al. 2010). Willows and black locust trees coppice easily. Other trees that were traditionally coppiced include ash maple, oak, chestnut, and elder berry. The practice of coppice of trees has been practised for thousands years.

Coppicing today remains a common forest management system in countries of south-eastern Europe (Maděra et al. 2014) to provide firewood on the local scale. The importance of coppices increases in the field of nature protection where there is an increased growth (Mason and MacDonald 2002), and as a natural habitat for the reproduction and life of birds, small mammals, and invertebrates (Spitzer et al. 2008). Nowadays, with the growing interest in biomass as an alternative energy source, a growing interest in the potential renewal of coppicing can also be seen in countries of central and Western Europe (Šplíchalová et al. 2012).

Decapitation is the removal of apical dominance in plant. Experiment shows that decapitation is a dwarfism process because it reduces the total size of a tree (Wilson

2000). Decapitation and canopy management is important crop architectural intervention, which helps the formation of branches, induce abundant and healthy inflorescence, which leads to good setting and seed yield (Gour et al. 2007). Decapitation can serve as tool to study defoliation, which may be as a result of herbivores or insects eating up plants leaves as well as the delayed senescence phenomena that appear after defoliation (Yordanov et al. 2008). Although the interest in coppicing is increasing, there is still a lack of information regarding the effect of coppices on *Pycnanthus angolensis* as a tree species.

Pycnanthus angolensis, (Akomu) an evergreen tree from the family Myristicaceae mainly grows in the West and Central African countries like; Ghana, Sierra Leone, and Cameroon (Fig. 10.1). This plant is generally known as False African nutmeg with a large, small crown of branches at right angles to the bole. *Pycnanthus angolensis* has some common names aside of False African nutmeg such as Akomu in Yoruba, Walele in Ivory Coast, Eteng in Cameroon, Calabo in Spanish, and Lolako in Zaire (USDA 2015). *Pycnanthus angolensis* is medically active plant harvested from the wild worldwide. The species are found in upland and wet evergreen forest and semi-deciduous forest with more than 1,600 mm rainfall (Orwa et al. 2009). Its outer bark is greyish brown, with orange brown exudates. The leaves are alternate, simple and entire, without stipules. The fruit is oblong and yellowish-orange when ripe. Seeds of *Pycnanthus angolensis* are recalcitrant with germination duration of 16–36 days after planting.

Extracts of *Pycnanthus angolensis* are effective for wound healing (Agyare et al. 2009) and it has antioxidant activities that are good for stomach ulcer treatment due to its anti-adhesive activity against *helicobacter pylori* on human stomach cells (Ancolio et al. 2002). The leaf juice is used for the prevention of white tongue thrush and the latex is applied to wounds as an anti-coagulant. It has also been reported to be useful for the treatment of female sterility, gonorrhoea, general infertility, rhino pharyngeal and broncho-pneumonia, eye troubles such as cataract, skin infection, as well as antidote for poison (Rahaman 2011).

Its genetic resources are now under threat of total loss or endangered due to excessive harvest and poor environmental conditions (IUCN 2011). The conventional propagation methods through seeds stem and root cuttings have limitations including low seed viability and germination rate, as well as delayed rooting of seedlings and plantlets.

This chapter in the book seeks to determine the coppicing capacity of *Pycnanthus angolensis* for sustainable forestry techniques in the climate change.

Materials and Methods

This study was carried out in the tissue culture laboratory of the Department of Agronomy, University of Ibadan, Oyo State. One indigenous tree species, known for its relative morphological traits and preferable by the farmers, namely *Pycnanthus angolensis* was selected for evaluating its coppice potential.

Fig. 10.1 Habit and mature fruits of *Pycnanthus angolensis*

Study Areas of Pycnanthus Angolensis

Three states, Oyo, Ekiti, and Osun states were purposively selected due to the availability and abundance of the plant (Fig. 10.2). **Ekiti state** is situated on latitudes 7° 25′ and 80° 5′ East of Greenwich Meridian and longitude 4° 45′ and 5′ 46° North of the Equator (Adegboyega and Bolanle 2021). The state is surrounded to the North by Kwara and Kogi states, bordered by Osun state to the west, Edo to the East, and Ondo to the South. The state enjoys a hot weather with two distinct seasons: rainy season (April to October) and dry season (November to March) and the temperature ranges from 21 to 28 °C (Oluwasusi and Tijani 2013) with high moisture. The total land area of the state is about 6,353 km^2 and it has a community of 3,930,212 (National Population Commission (NPC) 2006) with more than 60 per cent residing in rural areas. In **Osun State,** Sofoluwe et al. (2011) opined that Ekiti state is situated on latitude 7.0° and 9.0° N and longitude 2.8° and 6.8° E, with a total land area of 8,602 km^2. In **Oyo State,** Adejumo et al. (2016) reported that it is geographically situated on latitude 7° 02′ N and 9° 10′ N and longitude 2°04′ E and 4° 30′ E, enclosed by Ogun state, Kwara state, partly surrounded by Ogun state and partially by the Republic of Benin, while Osun State on the eastern part. The state enjoys a hot weather with prominent wet and dry seasons (Adejumo et al. 2016).

Coppicing capacity of decapitated seedlings of *Pycnanthus angolensis* (False African Nutmeg) was investigated to determine the effect of location with the climate change. Five seedlings each from Osun (Gbongan, Ajaba), Ekiti (Otun, Ayetoro), and Oyo (Idito, Adewumi) at 12 weeks after transplanting were decapitated from diameter (from 0.9 to 2.1 cm) and heights (from 4.50 to 20.50 cm) (from the first four nodes of the plant) and laid out in an open nursery in two replicates in a complete randomised design (CRD) to evaluate new shoot height (SH), plant height (PH), number of leaves

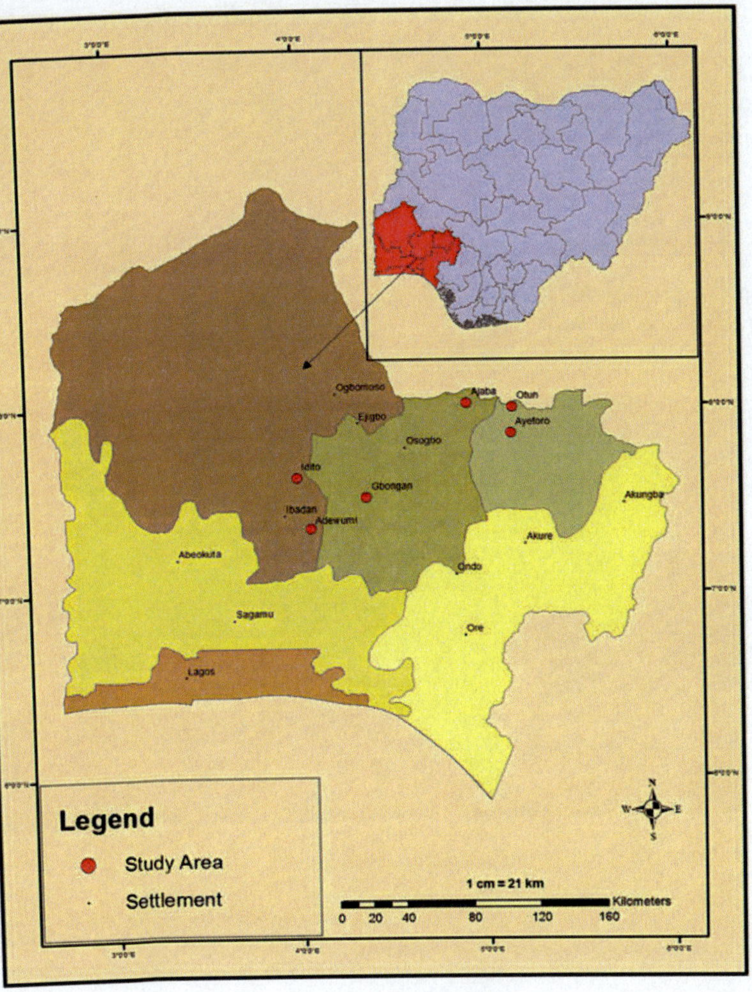

Fig. 10.2 Map showing the study areas of *Pycnanthus angolensis*

(NL), and number of nodes (NN) were measured fortnightly for 14 weeks. The data were analysed with observed means subjected to Tukey Post Hoc. The experiment was laid in 6 × 3 factorial experiments in a complete randomised design (CRD). Factor A is locations while factor B is states (Fig. 10.3a, b). Watering was done to field capacity once daily in the morning.

Fig. 10.3 **a** Decapitation of *Pycnanthus angolensis*. **b** Seedling development of *Pycnanthus angolensis* after decapitation

Fig. 10.4 Coppiced seedlings of *Pycnanthus angolensis*

Data Collection and Analysis

The following growth traits were measured to evaluate the new shoot height (SH), plant height (PH), number of leaves (NL), and number of nodes (NN). The numbers of leaves (NL) and number of nodes (NN) were counted visually, while new shoot height (SH) and plant height were measured with metre ruler fortnightly for nine weeks. The data taken were analysed. The observed means were subjected to Tukey Post Hoc to show mean separation.

Results and Discussions

Climate change generated by greenhouse gas emissions is possibly the greatest threat, as it is leading to extreme weather conditions, increasing areas of drought, and species extinction, among others (Carnicer et al. 2011; Wernberg et al. 2013). In this adverse context, breeding programmes should consider coppice as one of the breeding methods. Decapitation and canopy management is important crop/tree architectural intervention, which help the formation of branches, induce abundant and healthy inflorescence, which leads to good setting and seed yield (Gour et al. 2007).

Shoot Height (SH)

From the first cutting time height ranging from 4.50 to 20.50 cm after planting, shoot height ranged from 5.03 ± 1.01 cm (Adewumi) to 11.20 ± 1.01 cm (Idito). Summary of the marginal means of the observations, together with their respective standard error, for each time point (measured in weeks) for SH is presented on Table 10.1. From the table, the least observed decapitation value was observed on seedling from Ajaba (with value 1.55 ± 1.01) in the first week while the highest observation occurred in the eleventh week on seedling from Gbongan (with value 13.40 ± 1.01). This could be due to genetic makeup of the plant from different locations. The coppicing of a tree species is strong-minded in part by genetic factors (Tang et al. 1999).

The analysis of variance for the observations on the seedlings from the respective locations is presented on Table 10.2. From the table, location and period of observation (that is, week) had significant effect on decapitation, at five per cent (5%) alpha level (that is, 95% confidence interval. The results obtained showed that decapitation had significant effect on SH from the base to the decapitated points and were short throughout the period of experiment as compared to previous study (no data) The reduction in SH observed on all the decapitated sources confirmed that decapitation is a dwarfing process.

Table 10.1 Summary of marginal mean for SH per location per time point after decapitation

Week	Location					
	Ajaba	Otun	Gbongan	Idito	Ayetoro	Adewumi
1	1.55 ± 1.01	1.80 ± 1.01	1.78 ± 1.01	1.60 ± 1.01	2.08 ± 1.01	1.70 ± 1.01
2	2.37 ± 1.01	2.17 ± 1.01	3.01 ± 1.01	2.75 ± 1.01	2.40 ± 1.01	2.80 ± 1.01
3	3.55 ± 1.01	2.88 ± 1.01	4.29 ± 1.01	3.64 ± 1.01	2.85 ± 1.01	2.90 ± 1.01
4	3.70 ± 1.01	3.31 ± 1.01	4.94 ± 1.01	4.50 ± 1.01	3.31 ± 1.01	4.11 ± 1.01
5	4.01 ± 1.01	3.61 ± 1.01	5.46 ± 1.01	5.06 ± 1.01	3.84 ± 1.01	4.55 ± 1.01
6	4.80 ± 1.01	3.95 ± 1.01	6.16 ± 1.01	5.70 ± 1.01	4.50 ± 1.01	5.03 ± 1.01
7	5.25 ± 1.01	4.47 ± 1.01	6.55 ± 1.01	5.83 ± 1.01	4.66 ± 1.01	4.86 ± 1.01
8	5.95 ± 1.01	4.70 ± 1.01	6.51 ± 1.01	6.42 ± 1.01	4.65 ± 1.01	5.40 ± 1.01
9	6.25 ± 1.01	5.33 ± 1.01	7.35 ± 1.01	6.77 ± 1.01	4.85 ± 1.01	5.81 ± 1.01
10	6.95 ± 1.01	7.20 ± 1.01	12.10 ± 1.01	8.70 ± 1.01	5.30 ± 1.01	6.15 ± 1.01
11	6.05 ± 1.01	6.80 ± 1.01	13.40 ± 1.01	9.49 ± 1.01	6.05 ± 1.01	5.75 ± 1.01
12	3.53 ± 1.01	3.53 ± 1.01	3.53 ± 1.01	3.53 ± 1.01	3.53 ± 1.01	3.53 ± 1.01
13	7.30 ± 1.01	6.32 ± 1.01	9.01 ± 1.01	10.20 ± 1.01	6.70 ± 1.01	4.63 ± 1.01
14	5.65 ± 1.01	10.75 ± 1.01	9.00 ± 1.01	11.20 ± 1.01	5.83 ± 1.01	5.03 ± 1.01

(N.B.: Each cell contains mean ± standard error)

Table 10.2 Analysis of variance for shoot height

Source	Degree of freedom	Sum of square	Mean square	F	p-value	Remark
Location	5	643.921	128.784	12.524	<0.001	Significant
Week	13	3229.065	2487.39	24.156	<0.001	Significant
Location * week	65	891.741	13.719	1.334	0.045	Significant
Error	756	7773.875	10.283			
Total	839	12,538.6				

Number of Nodes (NN)

Table 10.3 presents the decapitation marginal means for the seedlings observed from the six sources for NN. From the table, Adewumi was observed to present the least observation (with value of 2.30 ± 1.00) on the first week while seedlings from Gbongan were observed to have the highest observation (on the fourteenth week, i.e. 18.00 ± 1.00). Table 10.4 is the ANOVA for the observations made on NN. From the table, location was observed to have significant effect on decapitation while the readings over the week were also observed to significantly vary from one another. This might be due to differences in stored reserves and ability to restore and regrowth.

Table 10.3 Summary of marginal mean for NN per location per time point after decapitation

Week	Location					
	Ajaba	Otun	Gbongan	Idito	Ayetoro	Adewumi
1	3.20 ± 1.00	3.00 ± 1.00	2.80 ± 1.00	2.90 ± 1.00	2.60 ± 1.00	2.30 ± 1.00
2	3.60 ± 1.00	3.70 ± 1.00	3.90 ± 1.00	4.30 ± 1.00	3.70 ± 1.00	3.70 ± 1.00
3	5.70 ± 1.00	6.70 ± 1.00	7.70 ± 1.00	6.10 ± 1.00	5.70 ± 1.00	7.00 ± 1.00
4	8.50 ± 1.00	7.10 ± 1.00	9.20 ± 1.00	6.90 ± 1.00	6.00 ± 1.00	7.80 ± 1.00
5	8.90 ± 1.00	8.40 ± 1.00	9.60 ± 1.00	8.50 ± 1.00	7.60 ± 1.00	8.70 ± 1.00
6	9.20 ± 1.00	8.60 ± 1.00	8.90 ± 1.00	9.80 ± 1.00	8.00 ± 1.00	8.50 ± 1.00
7	9.70 ± 1.00	9.60 ± 1.00	10.50 ± 1.00	8.70 ± 1.00	8.30 ± 1.00	9.10 ± 1.00
8	10.80 ± 1.00	9.20 ± 1.00	10.40 ± 1.00	10.50 ± 1.00	9.90 ± 1.00	10.20 ± 1.00
9	11.40 ± 1.00	10.40 ± 1.00	13.60 ± 1.00	11.10 ± 1.00	10.50 ± 1.00	10.70 ± 1.00
10	9.70 ± 1.00	9.60 ± 1.00	14.10 ± 1.00	11.10 ± 1.00	11.00 ± 1.00	11.30 ± 1.00
11	10.40 ± 1.00	13.50 ± 1.00	16.60 ± 1.00	12.70 ± 1.00	12.20 ± 1.00	9.40 ± 1.00
12	1.90 ± 1.00	1.90 ± 1.00	1.90 ± 1.00	1.90 ± 1.00	1.90 ± 1.00	1.90 ± 1.00
13	10.80 ± 1.00	11.00 ± 1.00	16.60 ± 1.00	12.90 ± 1.00	13.30 ± 1.00	9.70 ± 1.00
14	8.60 ± 1.00	9.90 ± 1.00	18.00 ± 1.00	13.00 ± 1.00	13.10 ± 1.00	8.10 ± 1.00

(N.B.: Each cell contains mean ± standard error)

Table 10.4 Analysis of variance for NN

Source	Degree of freedom	Sum of square	Mean square	F	p-value	Remark
Location	5	599.852	119.97	11.966	<0.001	Significant
Week	13	9763.395	751.03	74.909	<0.001	Significant
Location*week	65	1114.348	17.144	1.71	0.001	Significant
Error	756	7579.6	10.026			
Total	839	19,057.2				

This is supported by Van der Heyden and Stock (1996) that carbon is important to support growth and recovery of a plant after a disturbance.

Number of Leaves (NL)

Table 10.5 presents the marginal means of decapitation using NL over the fourteen time points. From the table, the least 1.35 ± 0.81 reading was observed from seedlings from Adewumi in the first week while the highest 11.10 ± 0.81 occurred from seedlings gotten from Gbongan on the fourteenth week. It is important to note, however, that readings for SH, NN and NL on the twelfth week seemed to be at par

Table 10.5 Summary of marginal mean for NL per location per time after decapitation

Week	Location					
	Ajaba	Otun	Gbongan	Idito	Ayetoro	Adewumi
1	2.00 ± 0.81	1.90 ± 0.81	2.40 ± 0.81	2.20 ± 0.81	1.40 ± 0.81	1.35 ± 0.81
2	2.60 ± 0.81	3.20 ± 0.81	3.90 ± 0.81	3.10 ± 0.81	2.70 ± 0.81	2.60 ± 0.81
3	5.40 ± 0.81	4.10 ± 0.81	6.30 ± 0.81	4.90 ± 0.81	4.60 ± 0.81	5.10 ± 0.81
4	6.20 ± 0.81	5.40 ± 0.81	7.70 ± 0.81	5.70 ± 0.81	4.70 ± 0.81	6.10 ± 0.81
5	7.40 ± 0.81	6.40 ± 0.81	8.10 ± 0.81	5.80 ± 0.81	5.40 ± 0.81	5.90 ± 0.81
6	8.20 ± 0.81	7.20 ± 0.81	7.20 ± 0.81	6.60 ± 0.81	5.50 ± 0.81	6.30 ± 0.81
7	8.70 ± 0.81	7.30 ± 0.81	8.30 ± 0.81	6.70 ± 0.81	6.00 ± 0.81	6.80 ± 0.81
8	8.50 ± 0.81	6.90 ± 0.81	8.00 ± 0.81	7.00 ± 0.81	6.20 ± 0.81	7.20 ± 0.81
9	8.60 ± 0.81	7.00 ± 0.81	8.50 ± 0.81	7.10 ± 0.81	6.80 ± 0.81	6.90 ± 0.81
10	4.00 ± 0.81	6.90 ± 0.81	10.40 ± 0.81	6.30 ± 0.81	4.70 ± 0.81	1.80 ± 0.81
11	4.50 ± 0.81	7.00 ± 0.81	10.00 ± 0.81	6.70 ± 0.81	4.60 ± 0.81	2.30 ± 0.81
12	1.80 ± 0.81	1.80 ± 0.81	1.80 ± 0.81	1.80 ± 0.81	1.80 ± 0.81	1.80 ± 0.81
13	5.20 ± 0.81	6.50 ± 0.81	10.70 ± 0.81	7.30 ± 0.81	5.60 ± 0.81	3.10 ± 0.81
14	6.80 ± 0.81	7.00 ± 0.81	11.10 ± 0.81	7.40 ± 0.81	5.80 ± 0.81	3.70 ± 0.81

(N.B.: Each cell contains mean ± standard error)

Table 10.6 Analysis of variance for NL

Source	Degree of freedom	Sum of square	Mean square	F	p-value	Remark
Location	5	811.224	162.245	24.521	<0.001	Significant
Week	13	2966.99	228.23	34.493	<0.001	Significant
Location*week	65	904.71	13.919	2.104	<0.001	Significant
Error	756	5002.2	6.617			
Total	839	9685.124				

with readings in the first week. From the ANOVA (Table 10.6) for NL observations, there is significant difference in the observations from the different locations and there is significant difference in the readings across the different weeks of observation.

Number of Branches (NB)

Table 10.7 presents the marginal means for the NB observations. From the above table, the highest (3.40 ± 0.62) NB was observed on seedlings from Ayetoro in the second week of assessment, Idito had the least (0.10 ± 0.62) on the sixth week. Table 10.8 is the ANOVA table for the observations on NB. The table showed a significant difference for seedlings from the different locations for the NB values. Also, there is significant difference in the weekly observations. Decapitation greatly influenced branching, as number of branches increased rapidly after decapitation in each location. This may be due to auxins in the apical meristem inhibiting the rapid growth and development of branches might have been removed. Branching is a function of apical dominance and correlative inhibition of axillary buds by apical shoots and apical control suppresses growth of younger shoots (Leakey and Longman 1986).

Table 10.7 Summary of marginal mean for NB per location per time point

Week	Location					
	Ajaba	Otun	Gbongan	Idito	Ayetoro	Adewumi
1	1.70 ± 0.62	1.40 ± 0.62	2.10 ± 0.62	0.40 ± 0.62	1.20 ± 0.62	0.90 ± 0.62
2	1.80 ± 0.62	1.60 ± 0.62	1.90 ± 0.62	0.40 ± 0.62	3.40 ± 0.62	0.60 ± 0.62
3	1.50 ± 0.62	1.70 ± 0.62	1.70 ± 0.62	0.40 ± 0.62	1.20 ± 0.62	0.60 ± 0.62
4	1.50 ± 0.62	1.60 ± 0.62	1.90 ± 0.62	0.20 ± 0.62	1.00 ± 0.62	0.60 ± 0.62
5	1.50 ± 0.62	1.80 ± 0.62	1.60 ± 0.62	0.20 ± 0.62	1.40 ± 0.62	0.90 ± 0.62
6	1.50 ± 0.62	2.20 ± 0.62	2.20 ± 0.62	0.10 ± 0.62	0.90 ± 0.62	0.90 ± 0.62
7	2.20 ± 0.62	3.10 ± 0.62	1.50 ± 0.62	0.20 ± 0.62	1.40 ± 0.62	0.90 ± 0.62
8	2.10 ± 0.62	1.40 ± 0.62	1.40 ± 0.62	0.40 ± 0.62	1.30 ± 0.62	0.80 ± 0.62
9	1.50 ± 0.62	2.50 ± 0.62	1.90 ± 0.62	0.50 ± 0.62	1.70 ± 0.62	0.80 ± 0.62
10	0.60 ± 0.62	1.40 ± 0.62	1.40 ± 0.62	0.80 ± 0.62	2.50 ± 0.62	1.20 ± 0.62
11	1.70 ± 0.62	2.20 ± 0.62	2.10 ± 0.62	2.30 ± 0.62	2.50 ± 0.62	2.20 ± 0.62
12	0.00 ± 0.62	0.00 ± 0.62	0.00 ± 0.62	0.00 ± 0.62	0.00 ± 0.62	0.00 ± 0.62
13	1.70 ± 0.62	3.20 ± 0.62	2.30 ± 0.62	1.30 ± 0.62	2.20 ± 0.62	1.20 ± 0.62
14	1.70 ± 0.62	3.35 ± 0.62	2.30 ± 0.62	1.30 ± 0.62	2.40 ± 0.62	1.80 ± 0.62

(N.B.: Each cell contains mean ± standard error)

Table 10.8 Analysis of variance for NB

Source	Degree of freedom	Sum of square	Mean square	F	p-value	Remark
Location	5	185.406	37.081	9.581	<0.001	Significant
Week	13	223.118	17.163	4.434	<0.001	Significant
Location*week	65	153.098	2.355	0.609	0.993	Not significant
Error	756	2926.025	3.87			
Total	839	3487.647				

Table 10.9 Summary of marginal mean for PH per location per time after decapitation

Week	Location					
	Ajaba	Otun	Gbongan	Idito	Ayetoro	Adewumi
1	11.06 ± 1.05	8.90 ± 1.05	13.03 ± 1.05	17.40 ± 1.05	10.63 ± 1.05	13.10 ± 1.05
2	12.35 ± 1.05	10.10 ± 1.05	14.70 ± 1.05	17.65 ± 1.05	11.28 ± 1.05	13.58 ± 1.05
3	12.70 ± 1.05	10.30 ± 1.05	16.02 ± 1.05	18.52 ± 1.05	11.35 ± 1.05	13.46 ± 1.05
4	13.21 ± 1.05	10.30 ± 1.05	16.35 ± 1.05	19.39 ± 1.05	11.91 ± 1.05	14.85 ± 1.05
5	13.52 ± 1.05	10.73 ± 1.05	16.50 ± 1.05	19.71 ± 1.05	12.18 ± 1.05	15.37 ± 1.05
6	14.16 ± 1.05	11.20 ± 1.05	17.25 ± 1.05	20.25 ± 1.05	12.48 ± 1.05	15.20 ± 1.05
7	14.68 ± 1.05	10.90 ± 1.05	15.66 ± 1.05	21.67 ± 1.05	12.44 ± 1.05	15.36 ± 1.05
8	15.20 ± 1.05	11.51 ± 1.05	18.23 ± 1.05	18.93 ± 1.05	12.81 ± 1.05	16.65 ± 1.05
9	15.55 ± 1.05	11.30 ± 1.05	18.90 ± 1.05	21.73 ± 1.05	13.16 ± 1.05	17.25 ± 1.05
10	16.45 ± 1.05	12.65 ± 1.05	23.75 ± 1.05	23.00 ± 1.05	13.60 ± 1.05	16.65 ± 1.05
11	12.70 ± 1.05	11.85 ± 1.05	16.75 ± 1.05	23.95 ± 1.05	13.40 ± 1.05	15.85 ± 1.05
12	0.00 ± 1.05	0.00 ± 1.05	0.00 ± 1.05	0.00 ± 1.05	0.00 ± 1.05	0.00 ± 1.05
13	11.30 ± 1.05	13.25 ± 1.05	21.28 ± 1.05	24.40 ± 1.05	13.70 ± 1.05	12.75 ± 1.05
14	12.85 ± 1.05	15.25 ± 1.05	21.45 ± 1.05	25.20 ± 1.05	14.13 ± 1.05	13.05 ± 1.05

(N.B.: Each cell contains mean ± standard error)

Plant Height (PH) of Pycnanthus Angolensis

From the first cutting time at the height ranging from 4.50 to 20.50 cm after planting, re-sprouted plant height ranged from 12.85 ± 1.05 cm (Ajaba) to 25.20 ± 1.05 cm (Idito) at the fourteenth week (Fig. 10.3). Table 10.9 presents the marginal means for PH values. From the table, the highest PH value (25.20 ± 1.05 cm) was observed on seedlings from Idito on the fourteenth week of observation, the least (8.90 ± 1.05 cm) from Otun. From Table 10.10, there is significant difference in the PH values from the different locations. Also, there is significant difference in the weekly observations. However, the present study confirmed that *Pycnanthus angolensis* could be coppiced well at plant height from 4.50 to 20.50 cm. There was steady increment in coppicing capacity of *Pycnanthus angolensis*.

Effect of Seedling Source on Growth Parameters

The rising demand for tree products and expanding population pressure resulted in the decline of forest cover and consequently the demand for tree products exceeded the supply. There is much concern about the sustainability of intensive cutting of trees for firewood and other economic values in Nigeria (Neke et al. 2006) food safety and environmental care among their objective The growth and recovery of

Table 10.10 Analysis of variance for PH

Source	Degree of freedom	Sum of square	Mean square	F	*p*-valuc	Remark
Location	5	7625.03	1525.006	115.422	<0.001*	Significant
Week	13	14,403.78	1107.983	83.859	<0.001*	Significant
Location*week	65	1912.897	29.429	2.227	<0.001*	Significant
Error	756	9988.563	13.212			
Total	839	33,930.27				

Significant*

trees from disturbances such as decapitation or cuttings is determined by the intensity of the disorder and the accessibility of resources which positively influence climate change. Coppicing today remains a common forest management system for sustainable livelihood in climate change. Table 10.11 presents the Post Hoc test for the effect of location on decapitation observed on the ANOVA Tables 10.2, 10.4, 10.6, 10.8, and 10.10). From Table 10.11, it was observed observations for SH readings from Ayetoro, Adewumi, Otun, and Ajaba were not statistically significantly different from one another at the 95% confidence interval. However, seedlings from these four locations were observed to differ significantly from seedlings from Idito and Gbongan. Also, the least SH mean value 4.325 cm was observed to occur on re-sprout seedlings from Ayetoro, while the highest (6.649 cm) was found on re-sprout seedlings from Gbongan.

Similarly, observations from variable PH for Ayetoro and Otun were found to produce similar, or non-significant observations, to each other while observations from Ayetoro and Ajaba were also observed to be similar to each other but not to Otun. From the PH row on Table 10.11, the highest PH value (19.416) for the seedlings was observed to occur at Idito, while the least (10.588) occurred in seedlings from Otun.

The NN from Ayetoro, Adewumi, Otun, Ajaba, and Idito were not statistically significantly different from one another at the 95% confidence interval. However,

Table 10.11 Effect of seedling sources on growth parameters of *Pycnanthus angolensis*

Variable	Location					
	Ayetoro	Adewumi	Otun	Ajaba	Idito	Gbongan
SH	4.325^a	4.446^a	4.773^a	4.779^a	6.099^b	6.649^b
NN	8.129^a	7.743^a	8.043^a	8.029^a	8.600^a	10.271^b
NL	4.700^a	4.364^a	5.614^b	5.707^b	5.614^b	7.457^c
NB	1.650^c	0.957^{ab}	1.961^c	1.500^{bc}	0.607^a	1.736^c
PH	11.648^{ab}	13.794^c	10.588^a	12.552^b	19.416^e	16.419^d

Cells with the same letter is not significantly different from each other; the order of the letter corresponds to increase in measurement according to Tukey's HSD Post HOC seedlings from these five locations were observed to differ significantly from seedlings from Gbongan. The highest NN (10.271) was recorded from Gbongan and least (7.743) from Adewumi.

The NL from Ayetoro and Adewumi, Otun, Ajaba, and Idito were not statistically significantly different from one another at the 95% confidence interval but the Gbongan had the highest (7.457) and the least NL (4.364) from Adewumi.

Furthermore, the NB from Ayetoro, Otun, Ajaba, and Gbongan were not statistically significantly different from one another at the 95% confidence interval. However, seedlings from Otun had the highest NB (1.961) and the least (0.607) from Idito. All the changes observed from each location might be due differences in stored reserves and ability to restore and regrowth. This is supported by Van der Heyden and Stock (1996) and Canadell and Lopez-Soria (1998) that carbon is important to support growth and recovery of a plant after a disturbance. Growth and yields of *Pycnanthus angolensis* are site-dependent. This is similar to the report on coppicing ability of teak (*Tectona grandis*) after thinning (Thaiutsa et al. 1999).

Time Against Variable of Interest

The Post HOC table for time of observation for the observations is presented in Table 10.12. The second column of Table 10.12 (that is, SH) presents a comparative analysis of readings on SH value on the seedlings over a period of fourteen weeks. Readings on the SH column showed a progressive increase from the first week up to the fourteenth week. The table also showed that no statistically significant result was observed generally for SH from the point of observation to the third week. However, a significant change in SH value from the first week was observed on the fourth week. Similarly, effect of change was observed to be stationary from the ninth week to the fourteenth week for this variable. Shoot increased on weekly basis and this might be due to recruitment of earlier accumulated reserves of either nutrients or starch (Latt et al. 2000; Garcia et al. 2001). Branching of the plant was induced due to decapitation and this was because branching in trees is a function of apical dominance. This is similar to Noggle and Fritz (1987).

This study showed that coppicing *Pycnanthus angolensis* at different locations or sources affect shoot height, number of leaves, number of nodes, leaf production, number of branches, and the total plant height. This is similar to research from several scientists that among several factors that contribute to the production variability, climate (Barros and Novais 1990), and physical (Melo 1994), chemical (Santana 1994), and biological (Facelli and Pickett 1991) properties of soil stand out. *Pycnanthus angolensis* responded positively to coppicing strategy way of conservation. It is, therefore, recommended that *Pycnanthus angolensis* from the first cutting time at the height ranging from 4.50 to 20.50 cm after planting, re-sprouted to plant height ranged from 12.85 ± 1.05 cm (Ajaba) to 25.20 ± 1.05 cm (Idito) at the fourteenth week. This is contrary to the report of Chibinga et al. (2016) on the effect of coppicing levels on the growth of *Julbernardia globiflora* that cutting must not

Table 10.12 Post HOC test for effect of time on each of the variable under consideration

Time (Week)	Variable of interest				
	SH	NN	NL	NB	PH
1	1.752^{a}	2.800^{a}	1.900^{a}	1.283^{b}	12.353^{b}
2	2.583^{ab}	3.817^{a}	3.017^{a}	1.617^{b}	13.277^{bc}
3	3.352^{abc}	6.483^{b}	5.067^{b}	1.183^{ab}	13.725^{bcd}
4	3.978^{bcd}	7.583^{bc}	5.967^{bcde}	1.133^{ab}	14.335^{bcde}
5	4.422^{bcde}	8.617^{cd}	6.500^{bcde}	1.233^{b}	14.668^{cde}
6	5.023^{cde}	8.833^{cd}	6.833^{cde}	1.300^{b}	15.090^{cdef}
7	5.270^{cde}	9.317^{cde}	7.300^{de}	1.550^{b}	15.118^{cdef}
8	5.605^{def}	10.167^{def}	7.300^{de}	1.233^{b}	15.555^{defg}
9	6.060^{efg}	11.283^{fg}	7.483^{e}	1.483^{b}	16.315^{efg}
10	7.733^{g}	11.133^{efg}	5.683^{bc}	1.317^{b}	17.683^{g}
11	7.923^{g}	12.467^{g}	5.850^{bcd}	2.167^{b}	15.750^{defg}
12	3.530^{abc}	1.900^{a}	1.800^{a}	0.000^{a}	0.000^{a}
13	7.360^{fg}	12.383^{g}	6.400^{bcde}	1.983^{b}	16.113^{efg}
14	7.910^{g}	11.783^{fg}	6.967^{cde}	2.142^{b}	16.992^{fg}

Cells with the same letter is not significantly different from each other; the order of the letter corresponds to increase in measurement according to Tukey's HSD Post HOC
SH = Shoot Height. NN = Number of Nodes, NB = Number of Branches, and PH = Plant Height

less than 20 cm from the ground. Therefore, coppicing below 20 cm is good for tree regeneration in the study.

Conclusion

This study hereby reported that *Pycnanthus angolensis* is capable of regenerating quickly to improve afforestation and as well reduce the effect of climate change in our locality and environment at large. However, Gbongan thrived well in all the growth parameters in a steady and progressive manner. This plant coppiced well on weekly basis on shoot height, number of leaves, number of nodes, leaf production, number of branches, and the total plant height. Coppicing is a silvicultural practice for sustainable forestry techniques because it uses nature's capacity. Coppicing is generally the primary regeneration mechanism after a woody plant has been cut or severely damaged, where stem and roots remain in place (Forrester et al. 2003). Furthermore, the re-sprouting ability of a plant can have major impacts on tree recovery and re-establishment to reduce ecosystem dependence on seed production and germination for population maintenance and growth, because seedling establishment is not a reliable means of re-establishment, especially where resources such as water are limited (Bond and Midgley 2001).

Recommendation

A current bottleneck in plant breeding programmes is climate change under different environmental conditions. However, coppicing is a silvicultural practice for sustainable forestry techniques because it uses nature's capacity to alleviate the problem of climate change. As a management tool, it has been proposed as a method for minimising damage to indigenous vegetation through sustainable fuelwood production (Kennedy 1998).

References

Adegboyega ER, Bolanle AO (2021) Using forest resources for economic development and poverty reduction in Ekiti state. Int J Appl Agric Sci 7(4):183

Adejumo AA, Adekoya AE, Sangotegbe NS (2016) Perceived effect of waste generation on the climate among rural households in Oyo state, Nigeria. J Environ Pollut Res 4(31):50–60

Agyare C, Asase A, Lechtenberg M, Niehues M, Deters A, Hensel A (2009) An ethnopharmacological survey and in vitro confirmation of ethnopharmacological use of medicinal plants used for wound healing in Bosomtwi-Atwima-Kwanwoma area, Ghana. J Ethnopharmacol 125(3):393–403

Ancolio CM, Azes V, Oliver E (2002) Antimalarial activities of extracts and alkaloids isolated from six plants used in traditional medicine in Mali and Sao Tome. Phytother Res 16:646–649

Barros NF, Novais RF (1990) Relação soloeucalipto. Universidade Federal de Viçosa, Viçosa, p 330p

Bond WJ, Midgley JJ (2001) Ecology of sprouting in woody plants: the persistence niche. Trends Ecol Evol 16(1):45–51

Canadell J, Lopez-Soria L (1998) Lignotuber reserves support regrowth following clipping of two mediterranean shrubs. Funct Ecol 12:31–38

Carnicer J, Coll M, Ninyerola M, Pons X, Sánchez G, Peñuelas J (2011) Widespread crown condition decline, food web disruption, and amplified tree mortality with increased climate change-type drought. Proc Natl Acad Sci USA 108(4):1474–1478

Chibinga OC, Nyangito MM, Musimba NRK, Simbaya J, Daura MT (2016) Effect of the coppicing levels on the re-growth of Julbernardia globiflora. Livest Res Rural Dev 28:42

Facelli JM, Pickett STA (1991) Plant litter: its dynamics and effects on plant community structure. Bot Rev 57:1–32

Forrester D, Bauhus J, Connell M (2003) Competition in thinned Silvertop Ash (*Eucalyptus sieberi* L. Johnson) stands from early coppice growth. For Ecol Manag 174(1–3):459–475

Fujimori T (2001) Ecological and silvicultural strategies for sustainable forest management. Elsevier, Amsterdam, p 398

Garcia H, Nygren P, Desfontaines L (2001) Dynamics of non-structural carbohydrates and biomass yield in a fodder legume tree at different harvest intensities. Tree Physiol 21:523–531

Gour VS, Sharma SK, Emmanuel CJSK, Kant T (2007) Stomata and chlorophyll content as marker traits for hardening of in vitro raised Balanites aegyptiaca (L.) Del. plantlets. Natl Acad Sci Lett 30:45–47

Hédl R, Kopecký M, Komárek J (2010) Half a century of succession in a temperate Oakwood: from species-rich community to mesic forest. Divers Distrib 16:267–276

Hunter MC, Smith RG, Schipanski ME, Atwood LW, Mortensen DA (2007) Agriculture in 2050: recalibrating Plant Phenomics 11 targets for sustainable intensification. Bioscience 67(4):386–391

International Union for Conservation of Nature (IUCN) (2011). Red list of threatened species version 2011. http://www.iucnredlist.org
Kennedy AD (1998) Coppicing of Tarconanthus camphorates (Compositae) as a source of sustainable fuelwood production: an example from the Laikipia Plateau, Kenya. African J Ecol 36:148–158
Latt CR, Nair PKR, Kang BT (2000) Interactions among cutting frequency, reserve carbohydrates, and post-cutting biomass production in Gliricidia sepium and Leucaena leucocephala. Agrofor Syst 50:27–46
Leakey RRB, Longman KA (1986) Physiological, environmental and genetic variation in apical dominance as determined by decapitation in Triplochiton scleroxylon K Schum. Tree Physiol 1:193–207
Maděra P, Habrová H, Šenfeldr M, Kholová I, Lvončík S, Ehrenbergerová L, Pavliš J et al (2019) Growth dynamics of endemic *Dracaena cinnabari* Balf. f. of Socotra Island suggest essential elements for a conservation strategy. Biologia 74:339–349
Mason CF, MacDonald SM (2002) Responses of ground flora to coppice management in english woodland—a study using permanent quadrats. Biodivers Conserv 11:1773–1789
Melo VF (1994) Relação entre a reserva mineral dos solos e a produção de nutrientes em povoamentos puros de Eucalyptus saligna Smith no Rio Grande do Sul. 145 f. Dissertação (Mestrado), Universidade Federal de Viçosa, Viçosa
National Population Commission NPC (2006) Population figure, National Population Commission, Abuja, Nigeria, p 26
Neke KS (2004) The regeneration ecology of savanna woodlands in relation to human utilisation. University of the Witwatersrand, Johannesburg
Noggle GR, Fritz GJ (1987) Introductory plant physiology, 2nd edn. Prentice-Hall, Inc., Englewood Cliffs, N.J., USA
Oluwasusi JO, Tijani SA (2013) Farmers' adaptation strategies to the effect of climate variation on yam production: a case study in Ekiti state Nigeria. Agrosearch 13(2):20–31. https://doi.org/10.4314/agrosh.v13i2.3
Orwa C, Mutua A, Kindt R, Jamnadass R, Simons A (2009) Agroforestree database: a tree reference and selection guide version 4.0. World Agro forestry Centre ICRAF, p 09. https://worldagroforestry.org/publication/agroforestree-database-tree-reference-and-selection-guide-version-40
Peterken GF (1993) Woodland conservation and management. Chapman and Hall, London, p 314
Porfirio LL, Newth D, Finnigan JJ, Cai Y (2018) Economic shifts in agricultural production and trade due to climate change. Palgrave Commun 4(1)
Rahaman O (2011) A review of medicinal uses of *Pycnanthus angolensis* (Akomu) in African Traditional medicine. www.pub.acs.org/doi/full/
Santana RC (1994) Crescimento e eficiência nutricional de procedências de E. grandis e E. saligna em sítios do estado de São Paulo. 73 f. Dissertação (Mestrado em Agronomia), Universidade Federal de Viçosa, Viçosa
Sofoluwe NA, Tijani AA, Baruwa OI (2011) Farmers' perception and adaptation to climate change in Osun State, Nigeria. Afr J Agric Res 6(20):4789–4794
Spitzer L, Konvicka M, Benes J, Tropek R, Tuf IH, Tufova J (2008) Does closure of traditionally managed open woodlands threaten epigeic invertebrates? Effects of coppicing and high deer densities. Biol Cons 141:827–837
Šplíchalová M, Adamec Z, Kadavý J et al (2012) Probability model of sessile oak (Quercus petraea (Matt.) Liebl.) stump sprouting in the Czech Republic. Eur J Forest Res 131:1611–1618
Tang Z, Chambers JL, Guddanti S, Yu S, Barnett JP (1999) Seasonal shoot and needle growth of loblolly pine responds to thinning, fertilization, and crown position. For Ecol Manage 120:117–130
Thaiutsa B, Puangchit L, Yarwudhi C, Wacharinrat C, Kobayashi S (1999) (Coppicing Ability of Teak (*Tectona grandis*) after Thinning
Usda N (2015) The plants database. National Plant Data Team, Greensboro. http://plants.usda.gov. Accessed May 2011

Van der Heyden F, Stock WD (1996) Regrowth of a semiarid shrub following simulated browsing: the role of reserve carbon. Funct Ecol 10:647–653

Wernberg T, Smale DA, Tuya F, et al (2013) An extreme climatic event alters marine ecosystem structure in a global biodiversity hotspot. Nat Clim Change 3(1):78–82

Wilson BF (2000) Apical control of branch growth and angle in woody plants. Am J Bot 87(5):601–607

Yordanov I, Goltsev V, Stefanov D, Chernev P, Zaharieva I, Kirova M, Strasser RJ et al (2008) Preservation of photosynthetic electron transport from senescence-induced inactivation in primary leaves after decapitation and defoliation of bean plants. J Plant Physiol 165(18):1954–1963

Chapter 11
Assessment of Tree Growth Competition Indices for Biodiversity Conservation in IITA Forest Ibadan, Nigeria

P. O. Ige and O. O. Komolafe

Abstract Forest growth assessment is a key tool for sustainable forest management. Understanding relative contributions of competition in the growth of the forest is of great important which determines the forest structure and also gives an insight to various influences of tree response to climate. There is dearth of information on forest growth using Competition Indices (CI). Hence, this study assessed CI effects on stand growth in International Institute of Tropical Agriculture (IITA) Forest, Ibadan, Nigeria, towards improving the forest health status and biodiversity conservation. Data were collected from the forest using four systematic line transect (270 m each) at 200 m apart for plot demarcation. Sixteen sample plots of 25 m × 25 m were alternately laid to collect data. All trees with diameter at breast height (DBH) ≥ 10 cm were estimated. Characterizing the joint influence of tree size, climate and competition in each plot, overtopped trees were considered subject trees and 10 m search radius was used in identification of competitor's tree for distance dependent (DD). Measurement of influence of neighbouring trees for distance independent (DI) was based on plot-centred. Eight Competition Indices were assessed (CI_1-CI_8). Best DD and DI were adapted each into Basal Area Increment model (BAI) before and after adding competition measures. Best model was selected using Root Mean Square Error (RMSE), Coefficient of Determination (R^2), Akaike Information Criterion (AIC) and Bayesian Information Criterion (BIC). Data were analysed using descriptive statistic and regression at $\alpha_{0.05}$. The stand comprises 389 stem ha^{-1}. The diameter at breast height (DBH), tree total height (THt), numbers of tree per hectares (N/ha) and volume (V) ranged from 25.12 ± 1.023 cm, 18.548 ± 0.324 m, 442 and 1.035 ± 0.136m^3, respectively. The computational analysis shows that basal area increment (BAI) model is a function of neighbourhood interactions and the best spatial indices were better growth predictors than the best non-spatial indices. The best CI growth model was: BAI = exp (−3.769 + 0.026DBH + 0.012C_6) (RMSE = 0.064, AIC = −774.031, BIC = −759.324 and R^2 = 0.912). This implied that DD CI predicted the growth predictability well compared to DI indices.

P. O. Ige (✉) · O. O. Komolafe
Department of Social and Environmental Forestry, University of Ibadan, Ibadan, Nigeria
e-mail: igepetero@gmail.com

J. C. Egbueri et al. (eds.), *Climate Change Impacts on Nigeria*, Springer Climate,
https://doi.org/10.1007/978-3-031-21007-5_11

Keywords Tree growth characteristics · Inter-tree-competition · Stand health indices · Biodiversity competition

Introduction

Nigeria tropical rain forest has a large numbers of species, which are been represented by few tree and their growth pattern and rates varies (Aigbe et al. 2013). Growth is an irreversible process which takes place in all living things. Tree growth simply means the increase in magnitude and quantity of the vegetative structures. As trees grow in the forest, competition sets in for photosynthesis, space and resources. Competition is an interaction between individual for survival for limited resources resulting to decrease for survivorship, reproduction and growth of the competing individual (Ige and Adesoye 2017). However, it was asserted by Ige (2017), that tree height and tree diameter within a forest will be constrained by the pressure of adjacent trees. Competition or growth rate in the forest often determines the shape and the structures of the forest stand (Coomes and Allen 2007). Individual tree growth competition is also an essential environmental process that plays substantial roles in population dynamics, survival, growth and species replacement on forest composition and stand structure (Amiri and Naghdi 2016; Ige 2017). However, trees growing in a given population usually exhibit large variation in growth. Coomes and Allen (2007) emphasized on the need for understanding the different variation in growth which is the basis for forest structures and biomass and also noted that tree growth declined with altitude. It was ascertain by Pelemo et al. (2011) that some trees grows poorly in the forest not as a result of competition but due to the influence of some other disturbances such as floods, windstorms, fire and human inflicted damages which make the forest to be instable and make the tree less favourable to grow properly. Various attempt of predicting the tree growth as accurate and precisely basically brought out the study of effect of competition on individual tree, two general methods are widely used for tree growth competitor indices which are the distance-independent indices and distance-dependent indices (Tome and Burkhart 1989; Amiri and Naghdi 2016).

Distance-independent indices or non-spatial indices generally measure and portray the competition status of trees in the stand which requires not the trees coordinate or the relative location of the neighbouring trees with the dimension of the subject tree not required as ascertained by (Tome and Burkhart 1989; Contreras et al. 2011). Obtaining distance-independent indices variables are relatively easy, and the calculation is less demanding in terms of data and time.

Distance-dependent indices try to explain a tree's competitive status based on the direct conditions of their neighbouring tree (Contreras et al. 2011). This generally measures the zones of influence of the neighbouring trees which best improve estimates of individual tree growth (Ige 2017). In estimating the tree growth competition using distance-dependent and distance-independent indices, strong positive correlation has been proven to exist between tree growth and basal area. Basal areal basically deals with the average amount of an area occupied by tree stem, thus making diameter at breast height (Dbh) a good predictors of forest dynamics which also improve the dependability of timber volume, growth and yield models (Brooks et al. 2008; Onyekachi and Osho 2018).

Materials and Method

Study Area

This study was carried out in International Institute of Tropical Agriculture (IITA) Forest (Fig. 11.1). International Institute of Tropical Agriculture forest is geographically located in Akinyele Local Government Area of Oyo State Nigeria. It lies between latitudes 7° 30′ 5.1264" and 7° 28′ 55.52" North and longitudes 3° 54′ 47.50" and 3° 52′ 44.49" East in the city of Ibadan. IITA forest has a humid tropical climate with well-known wet and dry seasons, with the wet season commencing from March and ends in October and dry season that lasts from November to February, and it has an average daily temperature of about 21–23 °C and the maximum temperature ranges from 28–34 °C. The forest used to experience bimodal rainfall pattern between 1300 and 1500 mm, which falls between the month of May and September. The mean daily relative humidity ranges between 64 and 83% (Ariyo et al. 2012). The forest reserve has a low lying and gentle undulating topography with an elevation ranges between 243 and 292 m. The parent rock materials of the soil are been forms through the underlying crystalline and banded gneiss which weathers to form site-specific soils. In the upland areas clay, quartz gravel and sand are predominant soil types, while the bottom of the valley has poorly drained clay and sandy soils (Oluyinka 2020). Some part of IITA forest has a highly diverse plant species. The vegetation of this area could be classified as tropical semi-deciduous forest with diverse of vegetation types ranging from derived savanna, secondary forest and riparian types (Osunsina et al. 2012).

Sampling Techniques and Data Collection

Reconnaissance was carried out so as to assess the forest stand and see the different changes that are currently taken place at the reserve. The survey carried out revealed that there was no evidence of logging in the forest, though the forest is a secondary forest that is currently undergoing reservation phase for biodiversity conservation. The sampling procedure used for the research work was adopted after the visitation to the study area. Simple systematic line transect was adopted for this study for plot laying and data collection. A total of 16 temporary sample plots were used for this research work. In laying of plots for data collection, simple systematic line transect

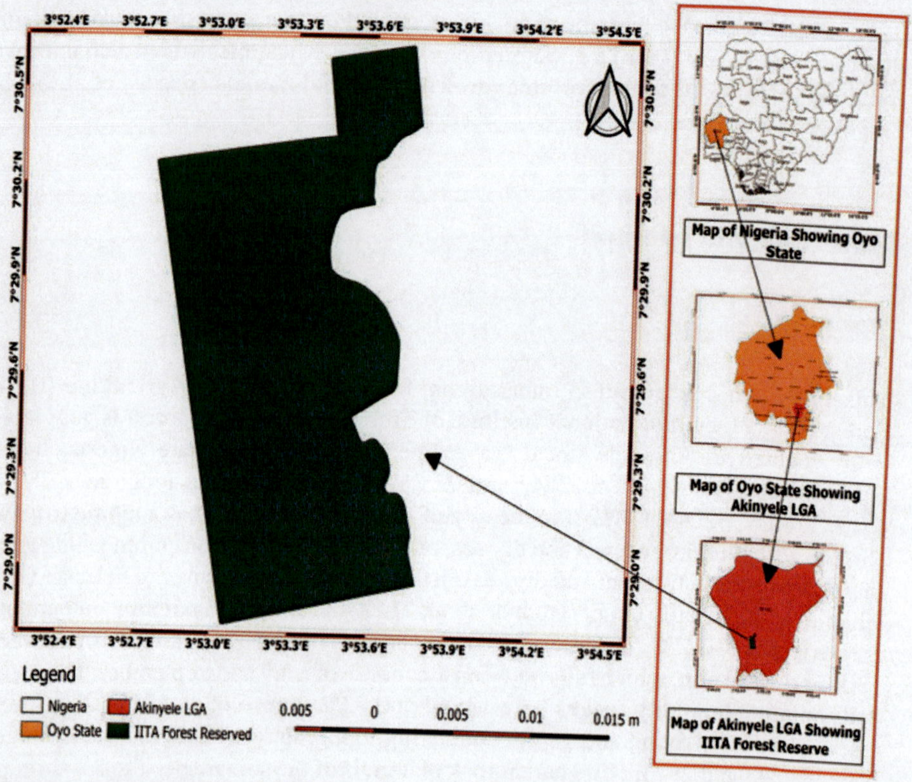

Fig. 11.1 Map of IITA forest reserve

has used by Adekunle et al. (2013) was adopted and modified for plot laying, and four parallel transects of equal distance (300 m) were delineated at 200 m apart for this study. A total number of 4 sample plot of equal size (25 m × 25 m) were laid alternatively on each transect and 50 m interval distance offset away from each sample plot was observed so as to decrease replication of tree species. To minimize the edge effect, 20 m offset was measured at the beginning of each transect (Fig. 11.2).

Data Collection

On each sample plot, all trees with diameter at breast height (DBH) ≥ 10 cm variables were identified and measured as used by Adekunle et al. (2013). To estimate volume per stand, the diameters at the base, middle and top were measured using Spiegel relaskop, while merchantable height and the total height of all the tree were measured using Haga Altimeter. Competitors tree was been identified by weighing the dimension of the subject tree and its neighbouring tree. Tree that are completely

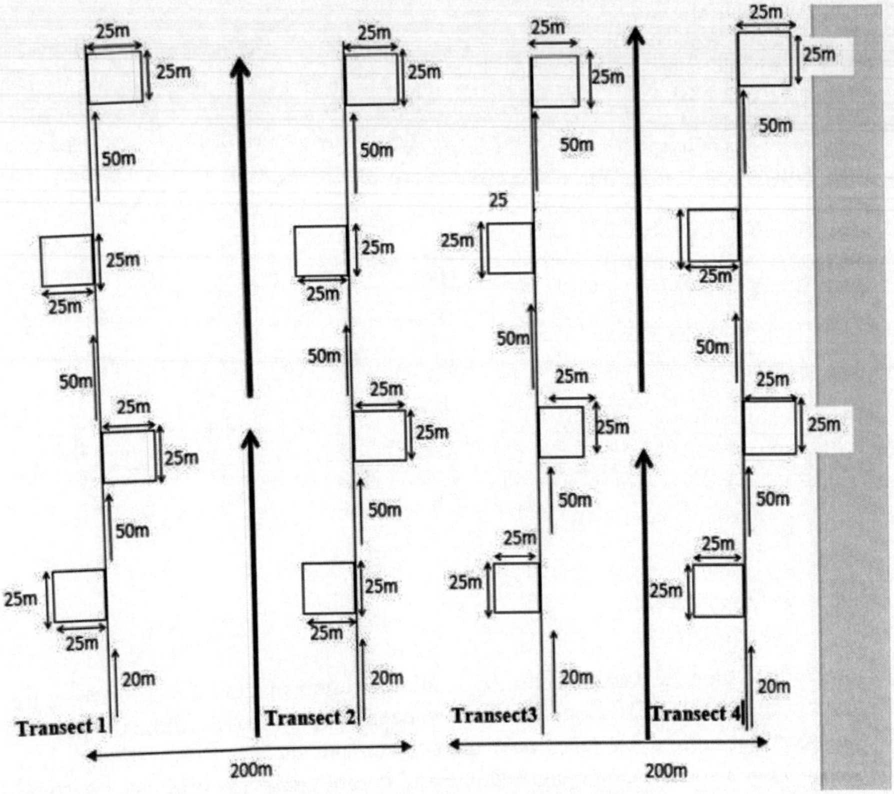

Fig. 11.2 Systematic line transects sampling technique for Plot layout

suppressed are considered subject tree. All the relevant information for computational evaluation of the competition indices of each subject tree and its neighbouring trees such as the horizontal distance, height and diameter of the entire subject tree to its competitors within the search radius of 10 m were measured and recorded.

Tree Growth Competition Indices

All the tree growth variables were assessed. Distance-dependent and distance-independent indices that are generally used were adopted to examining the tree growth competition indices.

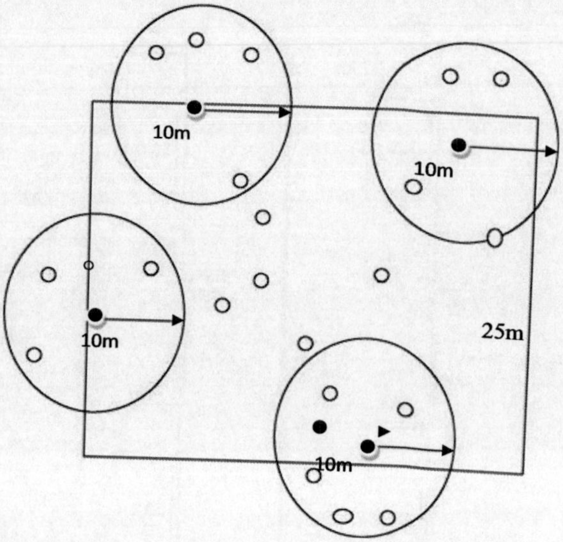

Fig. 11.3 Sample plot with four subject trees (solid dots) with various numbers of neighbours (open dots) within their respective competition plots of fixed radius of 10 m

Distance Dependent

Distance dependent was carried out by spatial location of the affected subject tree for their computations. Diverse method have been adopted to determine the pressure of the potential competitor trees over the subject tree such as fixed radius, crown-influence-zone overlap, DBH angle-gauge and height gauge method (Ige 2017). On the sample plot trees that are completely overtopped were considered as the core tree, and fixed radius of 10 m was used to measure the dimension of trees considered as neighbour trees (Figs. 11.3 and 11.4).

Distance Independent

Measurements were based on plot-centred rather than tree-centred neighbourhood data as used in distance-dependent indices.

Competition of each subject tree was quantified using four distance-independent competition indices (CI 1–4) and four distance-independent competition indices (CI 5–9) as given in Table 11.1 were used respectively for the competition indices. The indices used for this study were selected from the literature, taking into consideration the availability of tree variables for this study with their simplicity to describe the competition situation of a tree.

Relative spacing was computed using this equation:

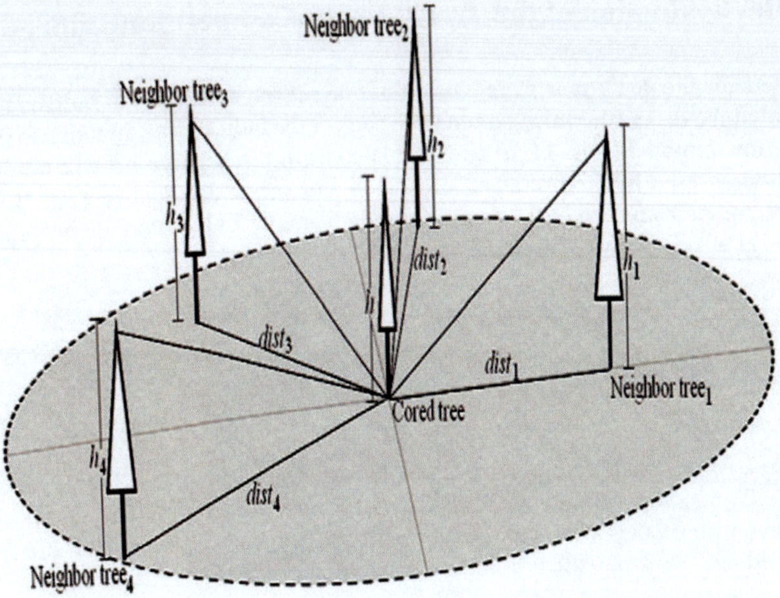

Fig. 11.4 Measurement of distance of each competitor trees to the core or subject tree

Table 11.1 Competition indices evaluated in this study

Equation no	Competition indices	Source
Distance-independent competition indices		
11.1	$\frac{\sum_{j=1}^{n} d_j{}^2}{d_i{}^2}$	Corona and Ferrara (1989)
11.2	$\frac{\left(1-\left[1-\left(\frac{BAL}{G}\right)\right]\right)}{RS}$	Schröder and Gadow (1999)
11.3	$\frac{\sum_{d_i<j\neq i}^{n}(g_j)}{S}$	Wykoff et al. (1982)
11.4	$\sum_{j\neq i}^{n}\left(g_{j;d_j>d_i}\right)/G$	Daniel (1976)
Distance-dependent competition indices		
11.5	$\sum_{i=1}^{n} h_i\, arctan\left(\frac{d_i}{dist_i}\right)$	Rouvinen and Kuuluvainen (1997)
11.6	$\sum_{i=1}^{n} h_i\left(\frac{h_j}{dist_i}\right)$	Braathe (1980)
11.7	$\sum_{i=1}^{n} d_i\left(\frac{d_j}{dist_i}\right)$	Daniel (1976)
11.8	$\sum_{j=1}^{n}\frac{d_j}{d_i\,(l_{ij}+1)}$	Daniel (1976), Ige and Adesoye (2017)

Where: n = number of neighbours within the 10 m radius competition plot; BAL basal area of neighbour trees larger than the cored tree (m^2h^{-1}); G is total basal area of the trees within plot (m^2h^{-1}); g_j is basal area of competitor tree ; $dist_i$ is the horizontal distance from the ith neighbour tree to the subject tree (m); h_i height of the subject tree (m); h is height of the competitor tree (m); l_{ij}, distance between competitor (j) and subject (i) tree (m), S is plot area

Model Evaluation of the Tree Growth Competition Indices

Mathematical calculations were carried out on all the competition indices (CI 1–8) specified above, and the value gotten from each of the indices was included in the basal area growth model (Eqs. 11.10 and 11.11) so as to predict tree growth competition function as used by Contreras et al. (2011) and Ige and Adesoye (2017). All the values gotten from each of the competition indices (CI 1–10) were incorporated into the basal area increment model.

$$BAI = \exp(b_0 + b_1 \ln(DBH) + b_2 CI) \tag{11.9}$$

$$\ln BAI = b_0 + b_1 DBH + b_2 H + b_3 CI \tag{11.10}$$

where: BAI = basal area increment, DBH = diameter at breast height, H = tree height, CI = competition indices, and b_0, b_1, b_2 and b_3 = regression parameters.

The entire models were assessed based on graphical and numerical analysis of the residuals. Four statistical fit indices used to evaluate the model are:

Coefficient of Determination (R^2): The (R^2) measures the proportion of variation in the dependent variable based on the behaviour of the independent variable. R^2 value must be higher for the model to be considered valid. The formula below was used to compute it.

$$R^2 = \frac{SSregression}{SSTotal} \times 100 \tag{11.11}$$

Root Mean Square Error (RMSE): This is the sum square of the vertical distances between the data point and its corresponding data point on the regression line. The RMSE must be relatively small for the model to be considered valid.

$$\text{Rmse} = \sqrt{\frac{\sum (Y_i - Y')^2}{N}} \tag{11.12}$$

Bayesian Information Criterion (BIC): For a model to be considered valid, its BIC must be relatively low.

$$\text{BIC} = n \ln\left(\frac{rss}{n}\right) + p \ln n \tag{11.13}$$

Akaike Information Criteria (AIC): This estimates the amount of information lost by a model, thereby estimating the quality of the model. For a model to be considered good fit, its AIC value must be relatively low.

$$AIC = n \ln\left(\frac{rss}{n}\right) + 2p$$

where SS = Sum square, rss = residual sum square, n = sample size, p = number of model fixed parameters, Y_i = the observed value and Y^i = the theoretical value predicted by the model.

Model Validation

Validating of the model was based on the qualitative assessment of the model outputs compared with the data set kept aside for validation. The 25% data set kept aside for model validation was used for this purpose. This was done by examining the significant different between the predicted value and the actual value using two sample t-tests for paired samples. Confidence level of 5% alpha was also used to test for statistical significance. There is no significant different between the observed value and the predicted value when the p-value is greater than 0.05 ($p > 0.05$), and the model is considered acceptable.

Tree Growth Competition Indices

Table 11.2 gives the statistical summary of the growth characteristics obtained for this study. The diameter at breast height (DBH) for the study area ranges from 10 to 170 cm with mean value of 25.12 ± 1.03 cm. The tree height ranges from 7.70 to 38.10 m with a mean value of 18.55 ± 0.32 m.The numbers of tree per hectare in a sample plot ranges from 96 to 704 with a mean value of 442. The mean volume and basal area were1.04 ± 0.14m^3 and 0.08 ± 0.01 m^2, respectively, with minimum and maximum values of 0.003 and 24.676 m^3; and 0.007 and 2.270m^2. The crown diameter had a mean value of 5.89 ± 0.08 m with respective minimum and maximum values of 3 and 13.7 m. The crown length and crown ratio had respective mean value of 2.97 ± 0.06 and 0.17 ± 0.004 with their minimum and maximum values of 1 and 7.8 m; and 0.05and 0.51.

Table 11.3 gives the correlation matrix between the basal area and the various competition indices. There was a strong positive correlation between basal area and competition indices 3(CI_3), and this was as a result of the similarities in the indices formulation and the association among the input variables (the basal area in the CI_3) and the sample plot (S). All competition of each subject tree in the study area was quantified using four (4) spatial (dependent) and non-spatial (independent) indices. The results of the competition indices estimated in this study area are presented in Table 11.4. The value of CI_1 ranges from 0.0203 ± 0.0183 to 1.4972 ± 0.0183. Indices estimated with CI_2 ranges from 0.00006 ± 0 0.0002 to 0.0435 ± 0 0.0002, CI_3 had a value range of 0.1087 ± 0.1575 to 36.3215 ± 0.1575, CI_4 and CI_5 had

Table 11.2 Statistical summary of the tree growth characteristics

Stand growth variable	Mean	Count	MIN	MAX
DBH (cm)	25.123 ± 1.026	389	10	170
THT (m)	18.548 ± 0.324	389	7.7	38.1
MHT (m)	15.038 ± 0.336	389	3.8	36
VOL (m^3)	1.035 ± 0.136	389	0.003	24.676
BAL (m^2)	0.083 ± 0.010	389	0.007	2.270
CL (m)	2.9688 ± 0.059	389	1	7.8
CR (m)	0.174 ± 0.004	389	0.051	0.506
SC	93 0.140 ± 1.896	389	22	225
N/ha	442	389	96	704
Dq (cm)	31.488 ± 0.406	389	18.047	52.655
CD (m)	5.894 ± 0.083	389	3	13.7

where: DBH = diameter at breast height, THT = Tree Total Height, MHT = Merchantable Height, VOL = Volume, BAL = Basal Area, CL = Crown Length, CR = Crown Ratio, SC = slenderness Coefficient, N/ha = numbers of tree per hectare, Dq = Quadratic Mean Diameter, CD = Crown Diameter

a range value 0.55726 ± 0.0058–0.9960 ± 0.0058 and 0.110604 ± 0.0044–0.4325 ± 0.0044, respectively. There was a change pattern in the estimate of competition indices 6 (CI_6), where the range value was higher compared to other competition indices the range value is between 22.425 ± 0.8775 and 87.7998 ± 0.8775, while CI_7 and CI_8 had their values ranges from 0.002 ± 0.0039–0.3704 ± 0.0039 and 0.1075 ± 0.0393–2.7669 ± 0.0393, respectively. In order to asses and compare the competition indices equation for tree growth competition in the study area, a basal area model was used. The best model was been selected using four goodness of fit which are Root Mean Square Error (RMSE), Akaike Information Criteria (AIC), Bayesian Information Criteria (BIC) and Coefficient of Determination (R^2). Table 11.5 gives the summary of the models fit indices and the rank of all the fit indices.

In Table 11.6, the model selected for distance-independent competition indices for Model $BAI = \exp(b_0 + b_1(DBH) + b_2CI)$ is the model with competition indices 3 (CI_3) while for the distance dependent was model with CI_6. The second model which the dependent variable is the natural logarithm of BAI (lnBAI) the model selected for distance-independent indices is model with CI_3 and for distance dependent is model with CI_7 ($\ln BAI = b_0 + b_1 DBH + b_2 H + b_3 CI_7$). Further analysis which included model verification and residual analysis was carried out to select the best from the four selected CI model from the two BAI and lnBAI models used (selected on basis of distance-dependent competition indices and distance-independent CI indices).

It was observed that all the models had a p-value greater than the alpha level, which thus indicates that the models are not significant. Figures 11.5, 11.6, 11.7, and 11.8 show the residual analysis carried out on the selected competition indices models. It was revealed that only Fig. 11.5 met the assumption of homoscedasticity. The error where randomly and evenly distributed with zero means and constant variance along

Table 11.3 Correlation matrix between basal area increment and various competition indices

	DBH	THt	lnBAI	BAI	C1	C2	C3	C4	*C5*	*C6*	*C7*	*C8*
DBH	1											
THt	0.595	1										
lnBAI	0.925	0.659	1									
BAI	0.926	0.441	0.731	1								
C1	0.108	0.072	0.127	0.081	1							
C2	0.826	0.460	0.691	0.828	-0.137	1						
C3	0.926	0.4409	0.731	1	0.081	0.828	1					
C4	0.210	0.024	0.243	0.149	0.510	0.076	0.149	1				
C5	0.139	0.131	0.208	0.055	−0.005	0.041	0.055	0.214	1			
C6	0.154	0.181	0.209	0.071	0.149	0.066	0.071	0.341	0.356	1		
C7	0.057	0.089	0.055	0.0372	−0.042	0.087	0.037	0.238	−0.104	0.553	1	
C8	0.101	0.005	0.103	0.075	0.616	−0.033	0.075	0.367	−0.104	0.328	−0.087	1

C1–C4 are distance-independent competition indices (non-spatial), while C5–C8 are distance-dependent competition indices (spatial)

Table 11.4 Estimated mean for the competition indices

	Mean	Min	Max
CI_1	0.36867 ± 0.0183	0.0203 ± 0.0183	1.4972 ± 0.0183
CI_2	0.0017 ± 0 0.0002	0.00006 ± 0 0.0002	0.0435 ± 0 0.0002
CI_3	1.3266 ± 0.1575	0.1087 ± 0.1575	36.3215 ± 0.1575
CI_4	0.8917 ± 0.0058	0.55726 ± 0.0058	0.9960 ± 0.0058
CI_5	0.2958 ± 0.0044	0.110604 ± 0.0044	0.4325 ± 0.0044
CI_6	50.3021 ± 0.8775	22.425 ± 0.8775	87.7998 ± 0.8775
CI_7	0.0334 ± 0.0039	0.002 ± 0.0039	0.3704 ± 0.0039
CI_8	0.7408 ± 0.0393	0.1075 ± 0.0393	2.7669 ± 0.0393

Where CI = competition indices, ± Standard error

Table 11.5 Models parameters and fit Indices

Basal Area Model + CI_i	B_0	B_1	B_2	B_3	*RMSE*	*AIC*	*BIC*	R^2	*Rank*
	$BAI = exp(b_0 + b_1(DBH) + b_2(CI)$								
CI_1	−3.046	0.023	0.072	–	0.072	−703.245	−688.538	0.8874	369.068
CI_2	−3.014	0.0212	14.75	–	0.069	−724.873	−710.1664	0.895	368.373
CI_3	−4.173	0.055	−0.120	–	0.021	−1437.97	−1423.263	0.991	343.164
CI_4	−9.382	0.024	6.738	–	0.063	−780.688	−765.981	0.914	385.087
CI_5	−3.830	0.025	2.314	–	0.066	−751.016	−736.309	0.904	365.620
CI_6	−3.769	0.026	0.012	–	0.064	−774.031	−759.324	0.912	364.759
CI_7	−3.077	0.024	1.063	–	0.071	−715.026	−700.319	0.892	367.896
CI_8	−3.208	0.024	0.172	–	0.069	−731.929	−717.222	0.898	367.322
No CI	−3.0141	0.0234			0.072	−704.455	−693.425	0.887	368.041
	$lnBAI = b_0 + b_1DBH + b_2H + b_3CI$								
CI_1	−5.131	0.0455	0.032	0.004	0.431	342.536	360.919	0.868	773.653
CI_2	−5.254	0.055	0.031	−66.011	0.403	303.149	321.533	0.85	773.747
CI_3	−5.448	0.094	0.004	−0.306	0.228	−29.005	10.622	0.963	750.423
CI_4	−5.763	0.043	0.035	0.699	0.424	333.043	351.426	0.872	774.543
CI_5	−5.471	0.044	0.030	1.367	0.415	320.839	339.223	0.877	769.365
CI_6	−5.275	0.044	0.031	0.004	0.426	336.639	355.022	0.870	783.513
CI_7	−5.126	0.045	0.032	−0.234	0.430	342.025	360.408	0.868	772.314
CI_8	−5.158	0.044	0.032	0.036	0.429	341.285	359.669	0.868	777.869
No CI	−5.130	0.045	0.032		0.867	344.539	361.246	0.868	772.443

Note CI = Competition indices, C_1–C_8 are the competition indices equation numbers, C_1–C_4 are the distance-independent indices, while C_5–C_8 are the distance-dependent indices

Table 11.6 Fitted model validation for the selected models with competition indices incorporated

Models form	CI	Types of indices	Mean of predicted value	Mean of observed value	Validation T-Value	P-value	Shapiro wilk
$BAI = \exp(b_0 + b_1(DBH) + b_2CI)$ $BAI = \exp(-5.448 + 0.094DBH + 0.004$	C3	DD	0.067583	0.07400361	−0.40072	0.6891	2.2×10^{-16}
$BAI = \exp(b_0 + b_1(DBH) + b_2CI)$ $BAI = \exp(-3.7687 + 0.0255DBH + 0.0117CI$	C6	DD	0.06758328	0.092105	−1.7371	0.08424	2.2×10^{-16}
$\ln BAI = b_0 + b_1DBH + b_2H + b_3CI$ $\ln BAI = -5.448 + 0.094DBH + 0.004H - 0.306$	C3	DID	−3.432653	−3.436474	0.024624	0.9804	4.672×10^{-11}
$\ln BAI = b_0 + b_1DBH + b_2H + b_3CI$ $\ln BAI = -5.448 + 0.045DBH + 0.032H - 0.306CI$	C7	DID	−3.432653	−3.462242	0.20646	0.8367	1.397×10^{-9}

where CI = competition indices, C3, C6 and C7 are the competition indices equation numbers, DD = distance-dependent indices, DID = distance-independent indices

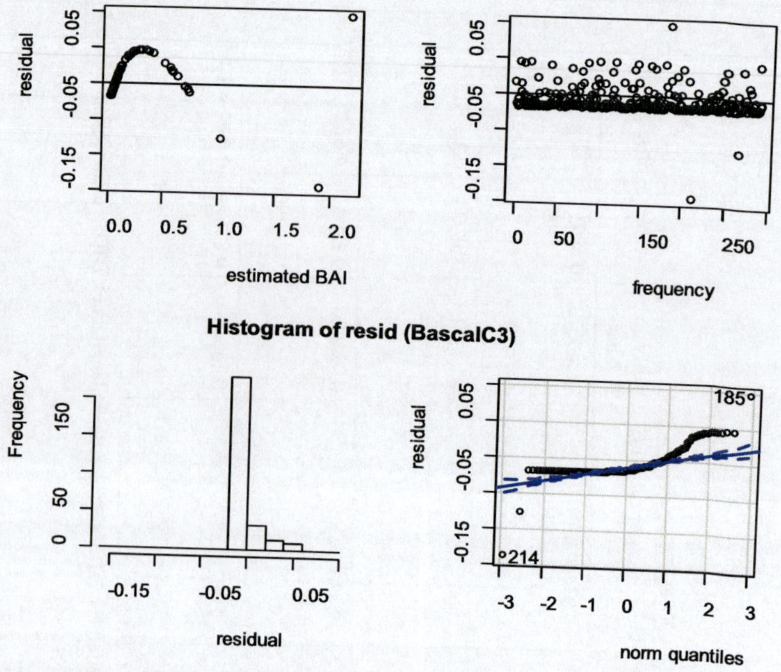

Fig. 11.5 Residual analysis for model 1 distance-independent CI_3

the positive and negative region of the X-axis. The distribution pattern was uniform and this thus implies that Model 1 distance-dependent competition indices CI_6 is good and superior to other models used in the study area.

Discussion

Model is now a daily routine used in forestry for predicting growth and yield, modelling diameter distributions, basal area model and tree crown model and many more (Ogana et al. 2015). Models are simply used for prediction and projection. Tree growth competition model was developed for this study in order to examine the competitive effect on each tree. Several studies had opined that decision of the management of the forest is often predetermined on information about current and future resources condition. As such, this study has directed effort in obtaining prediction models on tree growth competition with special focus on incorporation of spatial and non-spatial indices in the basal area growth model. The distance dependent involves spatially location of the subject tree to the competitors tree, while the distance independent examined the effect of the subject trees in relative to the stand measured at the centre of the plot. Studies have shown that adding of competition indices to tree growth improves the predictability of the model due to inclusion of

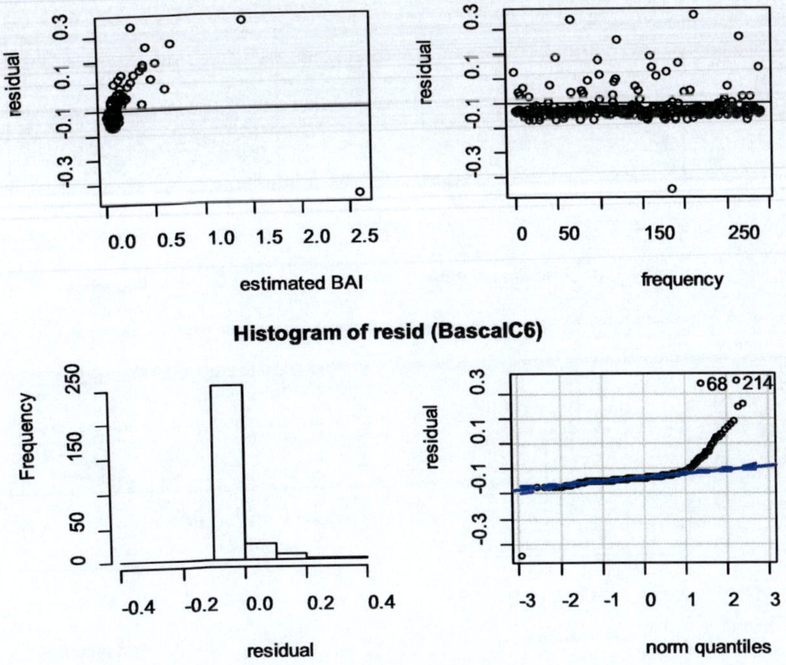

Fig. 11.6 Residual analysis for model 1 distance-dependent CI_6

trees variables in the competition indices (Contreras et al. 2011; Maleki et al. 2015; Ige 2017). For the study area, it was observed that distance-dependent competition indices C6 gave better estimation of tree growth competition and its effect on the growth of neighbouring trees. This study was in contrast with what was reported by Biging and Dobbertin (1992) that estimation of crown parameter improved the performance of distance depend indices measure, because competition indices that performed best for this study only uses height and distance in is computational competition index. However, Fraver et al. (2014) noted that inter-tree competition significantly affects growth rates as observed in better performance of model with competition indices when compared to models with no competition indices. Examining the effect of competition on tree growth, basal area model as used by Contreras et al. (2011) was used with four goodness of fit being used in selecting the best model for predicting the tree growth competition indices for the study area. For distance-dependent competition indices model, CI_6 gave better estimates for tree growth competition, while CI_3 gave better estimate for tree growth competition using the distance independent. Further analysis was however carried out on the DD CI_6 and DID C_3model. The residual analysis carried out on CI_6 and CI_3 shows that the spread residual above and below the zero line for model with CI_6 follows the law of homoscedasticity this indicates that the distance-dependent indices are more superior to distance-independent indices. Values gotten for the Root Mean Square Error of all the models are very small and high values gotten from the adjusted coefficients

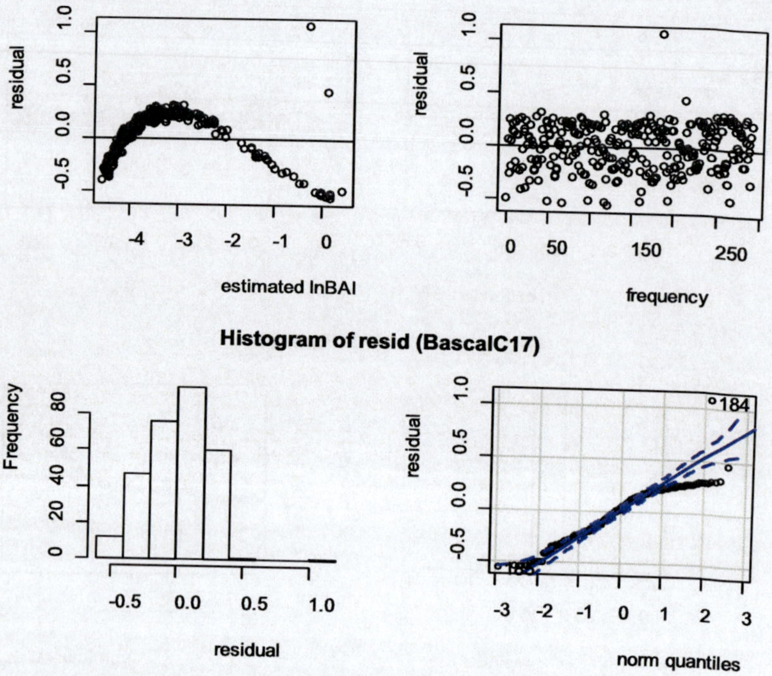

Fig. 11.7 Residual analysis for model 2 distance-independent CI_3

of determination shows that the model form is well adapted and biological realistic which is an indication that one of the major factors contributing to individual tree growth variation is the inter-tree competition as such, this finding also agrees with what was reported by Ige and Adesoye (2017).

Conclusions and Recommendations

The result of this study revealed the present assessment of stand growth characteristics and evaluation of tree growth competition indices in the study area. The study area has an estimated number of 389 tree stem per hectare which compares well with what is been observed in tropical forest ecosystem. Tree growth competition indices are not often address in many natural forests. Tree growth competition evaluated for this study involves using eight measures of tree competition index examined in terms of their effectiveness as growth predictor for the study area. This study demonstrated that one of the factors that influence forest processes and structure is competition. The inclusion of distance-dependent indices described the effects of tree neighbourhood

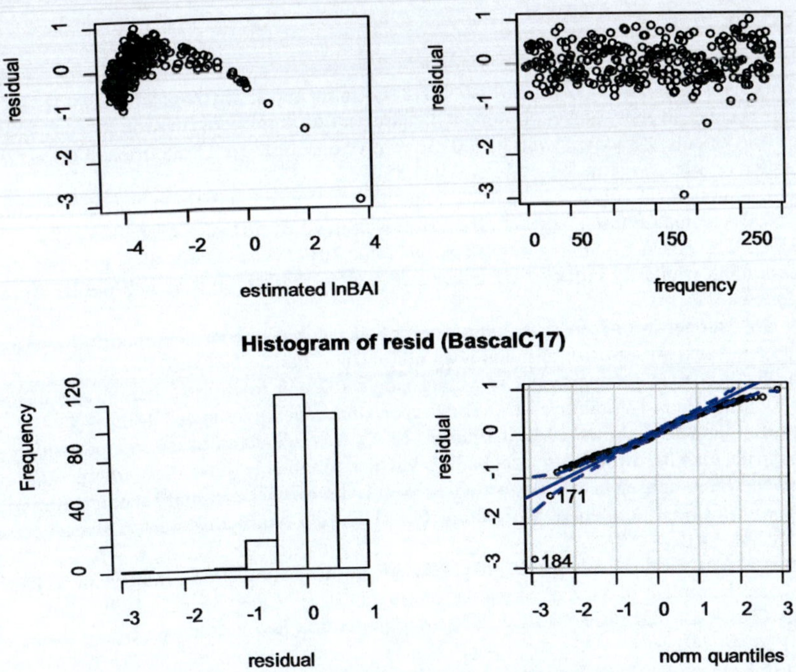

Fig. 11.8 Residual analysis for model 2 distance dependent with CI_7

maintained in the complex stand structure compared to distance-independent indices in the study area. One major constrain to use of distance dependent is the acquisition of tree attributes such as location and distance measurement which are time consuming and labour intensive. A positive strong correlation was found between two competition indices and tree growth variables, and this is an indicator that competition exists between trees. The growth model examined in this study revealed that model form of BAI $=$ exp $(-3.7687 + 0.0255\text{DBH} + 0.0117\text{CI}_6$ of distance-dependent competition indices predicted the growth predictability well compared to distance-independent indices. Therefore, for basal area increment growth model, competition index computed with the sum of ratio of height of subject tree multiplies with the ratio of height of competitor tree to the distance of the subject tree in a plot that are recommended so as to increase tree growth predictability for the study area.

Acknowledgements We appreciate the Management and Staff of International Institute of Tropical Agriculture, Ibadan, for kind approval and hospitality during the data collection.

References

Adekunle VAJ, Olagoke AO, Ogundare LF (2013) Logging impacts in tropical lowland humid forest on tree species diversity and environmental conservation. Appl Ecol Environ Res 11(3):491–511

Aigbe HI, Akindele SO, Onyekwelu JC (2013) Volume equations for Oban tropical Forest Reserve, Cross River State, Nigeria. Niger J For 43(1):53–67

Amiri M, Naghdi R (2016) Assessment of competition indices of an unlogged oriental beech mixed stand in Hyrcanian forests, Northern Iran. Biodiversitas J Biol Divers 17(1):306–314

Ariyo OC, Oluwalana SA, Faleyimu OI, Ariyo MO (2012) Assessment of vegetation structural diversity and similarity index of IITA forest reserve in Ibadan, Oyo State Nigeria. Agrosearch 12(2):136–158

Biging GS, Dobbertin M (1992) A comparison of distance-dependent competition measures for height and basal area growth of individual conifer trees. For Sci 38(3):695–720

Braathe P (1980) Height increment of young single trees in relation to height and distance of neighbouring trees. Mitteilungen Der Forstlichen Bundesversuchsanstalt 130:43–47

Brooks JR, Jiang L, Perkowski M, Sharma B (2008) A whole stand basal area projection model for Appalachian hardwoods. In: Jacobs DF, Michler, Charles H (eds) Proceedings, 16th central hardwood forest conference; 2008 April 8–9; West Lafayette, IN. General Technical Report NRS-P-24, vol 24. US Department of Agriculture, Forest Service, Northern Research Station, Newtown Square, PA, pp 324–330

Contreras MA, Affleck D, Chung W (2011) Evaluating tree competition indices as predictors of basal area increment in western Montana forests. For Ecol Manag 262(11):1939–1949

Coomes DA, Allen RB (2007) Effects of size, competition and altitude on tree growth. J Ecol 95(5):1084–1097

Corona P, Ferrara A (1989) Individual competition indices for conifer plantations. Agr Ecosyst Environ 27(1–4):429–437

Daniels RF (1976) Simple competition indices and their correlation with annual loblolly pine tree growth. For Sci 22(4):454–456

Fraver S, D'Amato AW, Bradford JB, Jonsson BG, Jönsson M, Esseen PA (2014) Tree growth and competition in an old-growth P icea abies forest of boreal S weden: influence of tree spatial patterning. J Veg Sci 25(2):374–385

Ige PO (2017) Development of models for assessing the slenderness of *Triplochiton Scleroxylon* K. Schum Stands in Oniganbari Forest Reserve, Nigeria. Niger J For 47(2):44–50

Ige PO, Adesoye PO (2017) Assessment of non-spatially explicit competition indices effects on Diameter Growth of Gmelina arborea Roxb. Stands in Omo Forest Reserve, Nigeria. J For For Prod 10:106–118

Maleki K, Kiviste A, Korjus H (2015) Analysis of individual tree competition on diameter growth of silver birch in Estonia. For Syst 24(2):8

Ogana FN, Osho JSA, Gorgoso-Varela JJ (2015) Comparison of Beta, Gamma Weibull distributions for characterising tree diameter in Oluwa Forest Reserve, Ondo State, Nigeria. J Nat Sci Res 5(4):28–36

Oluyinka Christopher A (2020). Comparative analyses of diversity and similarity indices of west bank forest and block a forest of the international institute of tropical agriculture (IITA) Ibadan, Oyo State, Nigeria. Int J For Res

Onyekachi C, Osho JS (2018) Basal area-stump diameter models for Tectona grandis Linn. f. stands in Omo Forest Reserve, Nigeria. J For Environ Sci 34(2):119–125

Osunsina IOO, Inah EI, Ogunjinmi AA, Onadeko SA, Osunsina O (2012) Distribution and diversity of flora and fauna in international institute of tropical agriculture (IITA) forest and nature reserve, Ibadan, Oyo State Nigeria. J Agric For Soc Sci (JOAFSS) 10(2):289–302

Pelemo OJ, Adeofun CO, Osudiala CS, Adetogun AC (2011) Assessment of growth dynamics of tree species in SNR2, Akure forest reserve, Nigeria. J Res for Wildl Environ 3(2):39–45

Rouvinen S, Kuuluvainen T (1997) Structure and asymmetry of tree crowns in relation to local competition in a natural mature Scots pine forest. Can J For Res 27(6):890–902

Schröder J, Gadow KV (1999) Testing a new competition index for Maritime pine in northwestern Spain. Can J For Res 29(2):280–283

Tome M, Burkhart HE (1989) Distance-dependent competition measures for predicting growth of individual trees. For Sci 35(3):816–831

Wykoff WR, Crookston NL, Stage AR (1982) User's guide to the stand prognosis model. General Technical Report INT-133, vol 112. Department of Agriculture, Forest Service, Intermountain Forest and Range Experiment Station, Ogden, UT. U.S

Chapter 12
Review on Climate Change Impacts on Air Quality in Nigeria

Nwanneka N. Onwudiwe

Abstract Climate change and air pollution pose the greatest environmental challenges that are affecting the whole nations in the world. These environmental challenges have not been effectively managed or controlled by environmental policies in Nigeria because of poor coping capacity. There is an ample evidence that climate change impacts in Nigeria result from climate related factors such as increased temperatures, increased rainfall, sea-level rise, extreme weather events, and, especially, increased health risk. Climate change has an impact on air quality because it is highly dependent on weather. Human activities have added more and more greenhouse gases into the atmosphere. The air quality we breathe in Nigeria is mainly affected by the release of air contaminants via production of petroleum products and bush burning, which has affected lives and health condition of many people in environment and the emission of greenhouse gases (GHGs) into the atmosphere is confirmed to deplete the activities of the ozone layer which protects us from harmful ultraviolet radiation. The impact of the change will be difficult to handle and it will be potentially very long lasting. However, this study reviews causes of climate change in Nigeria, its impact on air quality in the environments and useful policies to enforce to achieve sustainability of lives.

Keywords Climate change · Greenhouse gases · Air quality · Policies

Introduction

Climate change is regarded as the differences in the mean daily weather factors such as precipitation, wind, humidity, sunlight, and degree in temperature of a place over a given time. An alteration to the chemistry of the global atmosphere brought about by human activities could also be described as a change of weather conditions

N. N. Onwudiwe (✉)
Natural Science Unit, University of Nigeria, Nsukka, Nigeria
e-mail: diego.onwudiwe25@gmail.com

J. C. Egbueri et al. (eds.), *Climate Change Impacts on Nigeria*, Springer Climate,
https://doi.org/10.1007/978-3-031-21007-5_12

(Adejuwon 2006). Climate change causes environmental problems such as desertification, debasement of land, water scarcity, food security, and air and water contamination. Increasing alterations in weather designs and predicted maximum climate events may threaten people's subsistence if no action is taken. According to scientific proof, climate change and global warming are undeniably causing severe and potentially damaging effects. Climate change and air pollution are the most critical health and sustainability issues facing the world. The two are very much related. Besides its effects on human health, atmospheric gases (CO_2) influence the climate of the Earth by absorbing energy and blocking heat from escaping, causing climate variability (Akinola et al. 2014). Till date, with a little collaboration, policymaking on climate change and air quality goes on parallel ways and the strategy is designed in a way that it will meet both present and future needs without any compromise. Virtually all the nations in the world with this design are developing solutions to minimize climate change and air pollution impacts in our environments (Ajiere and Nwaerema 2020). The effect of climate change has outspread on both human and natural ecosystem beyond the global. In other parts of the world, its impact is evident, but in Africa, its impact is considered endangered since the ecosystems and economy is fragile (Ajiere and Nwaerema 2020). Policymaking on climate and air quality is integrated to use tools to assess greenhouse gas emissions and air pollutants together. This allows multiple impacts on climate, health, and crops to be quantified. A number of Nigeria's policies and strategic initiatives can act as climate change mitigation measures, if properly implemented. Some of the policy initiatives like "oases rehabilitation in the National Action to Combat Desertification and the National Policy on Drought and Desertification" could be adapted as measure and plans and modify into policy options for climate change response in the Nation (Obioh 2008). This strategic response will enable policies to be translated into purposeful intersectional pursuit for the maintenance of the environments (Obioh 2008).

Climate change was identified by the World Health Organization as one of the top ten threats to global health in 2019; Nigeria is among the top 10 countries most vulnerable to climate change, with approximately 6% of the country's land area are affected by extreme weather conditions (World Bank 2019). Furthermore, the rural and urban migration stream flow on the significant increase and decrease of other effects of climate change in Nigeria have been documented (Cattaneo and Massetti 2019; Akinwumi et al. 2020). As a result of climate change, Nigeria's agriculture and food security, water resources, public health, and settlements have suffered significantly. Climate change is likely to cause about 250,000 deaths worldwide within the year 2030 and 2050 resulting from malnutrition, malaria, diarrhea, and heat stress. This was estimated by the World Health Organization (WHO) and the evaluation of this report was done to express the cost of annual damage to health (WHO 2021a). Moreover, record from the National Emergency Management Agency (NEMA) stated that flood displaced 2 millions of people in Nigeria as at 2012 another record of about the number 100000 displaced in 2015, 92,000 in 2016, 250,000 in 2017, and roughly 1.9 million in 2018 were injured.

The impact of climate change is not only vulnerable to agriculture and environment but it also poses a risk to people and have various unpleasant consequences on health and social activities (Pittock 2005). A rise in sea-level of about 2 m to 6 m in the coastal region of the world could lead to the migration of millions of people from their home (Rowley et al. 2007; Nicholls et al. 2011). IPCC (2007a) revealed that by 2022, some 75 and 50 million people in Africa will be at risk of severe impacts due to weather extremes. It has also been reported that 27–53 million Nigerians living in the coastal region were forced out of their homes as a result of the rise in sea-level (USAID 2012). Increased recurring outbreak of vector and water borne diseases that is associated with climate change could lead to morbidity and mortality (WHO 2020).

The developed nations encounter less effects of climate change more than the developing nations (Maikasuwa 2013). The fact is that the developed nations have progressed a lot in research and technology despite being the major contributors of climate change to discover working policies for adaptive measures that could help to minimize the outcomes of climate change impacts in their nation (Jagtap 2007; Enete and Amusa 2010; Ebele and Emodi 2016; Elum and Momodu 2017). Such an advanced adaptive response and technologies are very slim in developing country like Nigeria. However, the African countries especially are more vulnerable to climate change effect and lack of willingness to enhance durability of adaptive response are very low (ND-GAIN 2021).

Air pollution is the release of pollutants into our environment. It can be defined as the presence of quantities of biological materials or particulate pollutants in the outdoor or indoor atmosphere and of time duration such as to be harmful to human and other living organisms (Fagorite et al. 2021). Air pollution is an outcome of reduced air quality as a result of emission of natural substances and human pursuit in the form of gases into the atmosphere which has direct or indirect harmful impacts on human health and environment (IPCC 2014). The emission of contaminated substance into the atmospheric environment has been a serious problem affecting all nations in various ways. The emission of these polluted substances are the main reason the air is contaminated they include carbon monoxides (CO), nitric oxides (NOx) volatile organic compounds (VOCs) and sulfur dioxide (SO_2). These air pollutants have posed a great threat to the most sustainable programs in Nigeria and have tremendous effect on the life of citizens. The major sources of gas emission in some parts of Nigeria, such as the Southwest particularly Lagos, are the landfills, combustion of waste materials, and generator set (Akinola et al. 2014) meanwhile, in the oil producing Niger region, the air quality is poor and happened to be among the top 10 most polluted areas in the world, this is because of the menace of black soot and other hazardous chemical which are being emitted into the atmosphere which have become a great concern to people living in the state (Weil 2014; Kuenzer et al. 2014; Ede and Edokpa 2015).

According to the World Health Organization (WHO) regulations, air pollution affects health even at low levels and the average concentrations of $PM_{2.5}$ should not go beyond 5 $\mu g/m^3$ per year, and the average daily exposure of particulate matter of air should not go beyond 25 $\mu g/m^3$ in a year. Its yearly exposure was reported to cause about 7 millions of premature deaths (WHO 2021b). The cause of air pollution in our environments has risen so greatly with transformation of work, population, and commercial development (Grutter et al. 2014), and the sources of these particulate substances are mainly caused by release of gases, waste, and toxicants from the commercial industries during productions, rise in movement of vehicles in the city due to continuous congestions, road and building construction activities, and domestic burning activities (Fort Air Partnership (FAP) 2014). This study is aim to review causes of climate change and the impact on air quality in Nigeria.

General Effects of Climate Change in Nigeria

The challenges existing with climate change vary beyond the nation. Nigeria's climate consists of two seasons, the dry and rainy (wet) seasons. This has led to seasons of drought and excess flood, which affected agricultural activities and caused a loss of shelter (Akande et al. 2017). These seasons differ in lengths depending on the geographical location. Studies on climate change susceptibility demonstrated that states in the north are more vulnerable to climate change than the southern states (Madu 2016; Federal Ministry of Environment 2014). For instant, South usually has a longer period of rainy season which starts between the months of March to November than the Northern part; theirs is short and happens between the months of May to September. The Northern region experience frequent dry season which is usually hot and their average value of temperature reaches about 38 °C per month. The North Central is the least susceptibility to climate change, which could be related to steady precipitations and irrigation which occurs yearly due to the disposal of lakes in the region, which helps to make up for lower than anticipated rains (Federal Ministry of Environment 2014). Although the North Central area also experience climate change toward dryness, as well as varying decrease in rainfall, these factors are a growing concern about the region and a downward spiral affecting its economic performance and agriculture output, and a falling drift in precipitation has been observed in Plateau, Nasarawa, and the Federal Capital Territory (Ideki and Weli 2019). The most vulnerable regions are the Northeast and Northwest, and the results of high level of heat and less rainfalls has quickened desertification, destroying wetlands and reducing the quantity of surface water and biodiversity resources on the land (Abdulkadir et al. 2017; Akande et al. 2017; Ebele and Emodi 2016; Federal Ministry of Environment 2014).

Generally, the Harmattan is a hot and dry wind, which is felt more in the North than in the South. The mean temperature in south is around 32–33 °C. The vulnerability features vary beyond the country and the most vulnerable area is the Northeast. One of the exposed factors which affect the North is decreased rainfall which jeopardizes rain fed agriculture and affects water supply. Meanwhile, the Southeast is exposed to high level of water decline because of soil porosity (Federal Ministry of Environment 2014). There is less vulnerability in the Southeast and Southwest than in the North. The South-South (Niger Delta region) is highly exposed; this is because of rise in rainfall and overflowing of water on the land which has caused displacement of many communities (Sayne 2011; Federal Ministry of Environment 2014; Saheed et al. 2019). The effects of climate change on human health in Nigeria are numerous and indirect. Already, Nigeria's residents are experiencing poor nutritional imbalances and climate change will only exacerbate this problem. There could be a shift in the prevalence of vector-borne diseases, such as malaria (mosquitoes) and sleeping sickness (tsetse flies). The changes in patterns of rainfall and temperature have caused mosquitoes to thrive in areas that have water logging and poor drainage (Akpomi and Vipene 2016). Climate change often appears very esoteric but in Nigeria, it's real. We already have an increasing incidence of disease, declining agricultural productivity, and a rising number of heat waves. There is clear evidence that climate change is reality and it affects almost every aspect of our lives. Our everyday lives involve adapting to change and mitigating its effects. However, we must first understand what climate change is, how we contribute to it, in order to adapt and reduce our vulnerabilities (Olaniyi et al. 2013).

Causes of Climate Change in Nigeria

The natural processes (biogeographical) and human activities known as (anthropogenic) are the factors that cause climate change (Odjugo 2010). The natural processes consist of astronomical and extraterrestrial features. Among the Earth's astronomical features are the eccentricity of its orbit all over the Sun, the orbit's changes over time, changes in volcanic activity, and changes in orbital trajectories, while extraterrestrial factors include the quantity and quality of solar radiation (IPCC 2007b; Odjugo 2010). In addition to external changes in the climate system, such as ocean currents or atmospheric circulation, climate variability can also result from internal changes (IPCC 2005; NRC 2011). Spore (2008) outlined how weather patterns can provide an easy understanding of climate change. He claims that sunlight is emitted onto Earth's surface by the Sun every day. Part of the heat is absorbed by the Earth; another part is reflected into the atmosphere, and a third part is radiated as infrared rays. Under normal conditions, the Earth's temperature is stabilized by cloud cover and water vapor absorbing these rays. As a result of human activity, GHG concentrations have increased significantly, and our planet is facing an increasingly dangerous situation. These gases absorb solar radiation and radiate it back to Earth, resulting in a general rise in temperatures known as global warming. Human activities is a major component of anthropogenic factor which involve the release of large quantity of greenhouse gases (GHGs) into the atmosphere which deplete the ozone layer and create a reduction in the amount of carbon in the atmosphere (IPCC 2007b). According to Chukwu and Asiegbu (2011), other important factor contributing to greenhouse gas emission into our environment is the growing number of rearing farm animals whose digestive processes continuously produce methane and nitrous oxide. It has been established that human activities, directly or indirectly, knowingly or inadvertently, to degrade our environment is irrefutably responsible for the climate change that we experience, in addition to natural variability seen over a similar timeframe (Chukwu 2008; Olaniyi et al. 2013; Dipeolu 2015). Although some agricultural practices and desertification release GHG gases into the atmosphere, these gases originate from the combustion of fossil fuels. The greenhouse gases that cause global warming are carbon dioxide (CO_2), methane (CH_4), water vapor, and hydrofluorocarbon (HFCs). The fact is that, CO_2 is the primary gas that causes climate change, but its impact is very low. One molecule of CO_2 can damage ozone layer for only a few years, but other greenhouse gases can damage the ozone layer for decades to thousands of years (Odjugo 2013). The emission of CO_2 by human activities may contribute less to the greenhouse effect than many other gases and heat radiation from the Sun to the Earth's surface can be effectively absorbed by greenhouse gases. It works by keeping the atmosphere warmer than it should be. Based on current GHG emission trends, the Earth is projected to experience temperature increase between 1.5 and 4.5 °C by 2030 (Nebedum and Emodi 2016). As reported by Ogunniran (2018), climate change due to greenhouse effects has been a threat to people, natural systems, and the environment for more than a decade due to high temperatures, acidic rain, flooding, and susceptibility to pests and disease outbreaks, which makes life more difficult. There is no doubt that climate change is global phenomenon, as are its impacts. However, the developing countries will be hit more severely than the developed countries due to their limited coping abilities (Nwafor 2007; Jagtap 2007) and Nigeria is one of those countries. Nigeria has already been affected by various ecological problems caused by changing or variable climate (Odjugo 2001; Odjugo and Ikhuoria 2003) (Fig. 12.1).

Causes of Air Pollution in Nigeria

In this day and age, air pollution is the greatest plague, affecting not only our climate but also the public and our individual health, resulting in increased morbidity and mortality (Komolafe et al. 2014). The two major factors that cause air pollution are natural and anthropogenic activities. Even so, the anthropogenic sources produce most of the pollutants in urban areas; most of them are emitted by human activities (Abaje et al. 2020). The natural phenomenon results from natural processes that are not induced by human activity. Air pollution and urban climate are interconnected in several ways. Different studies in other region of the world have explored the relationship between meteorological parameters and pollutants such as ozone, SO_2, and NO_2 (Broomfield et al. 1996; Davis et al. 1998; Perez and Trier 2001; Krupa et al. 2003), but no studies as reported by Odekanle et al. (2017), have been conducted on the effects of meteorological parameters on air pollution during transport in Nigeria. People who live in large urban areas are mostly affected by air pollution, as road emissions contribute the most to poor air quality. Industrial accidents also exist; however, this can lead to the spread of a toxic fog that is deadly to the local populations. Many factors play a role in the dispersion of pollutants, including atmospheric stability, wind direction, and wind speed (Kelishadi and Poursafa 2010). Some other sources of pollution include dust, animal waste, combustion of fossil fuel, vegetation,

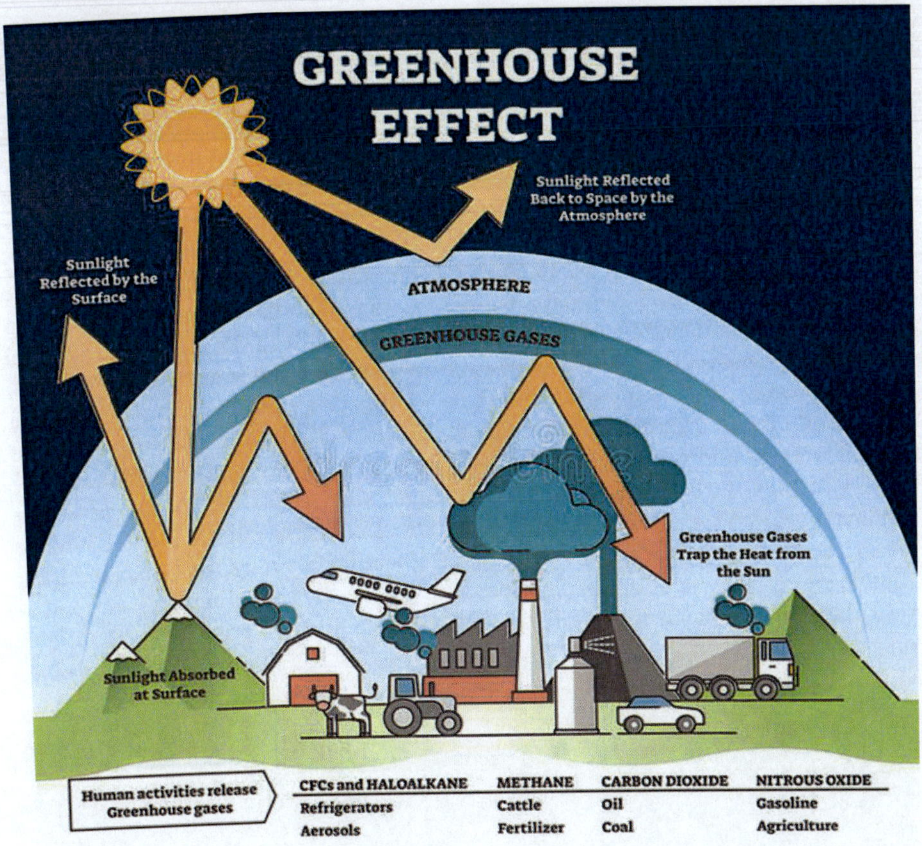

Fig. 12.1 Greenhouse effect and climate change from global warming outline concept. *Source* Dreamstime.com

and anthropogenic causes. These sources follow a natural cycle, which occur more common under certain conditions and less frequently under others. Dust storms are often cause by wind, especially in areas where there is little or no vegetation as a result of lack of rainfall (Fagorite et al. 2021). When this pollutant is present in the air, it can warm the atmosphere and pose a health hazard for living organism, it can also interfere with photosynthesis (Brook et al. 2010). Beside dust; other cause of air pollution is digestion from animals, especially from cattle which help in the emission of gas known as methane. When it is warmer, number of flowers and trees do release large volume of volatile compounds. A number of these compounds react with nitrogen oxides, sulfur dioxide, and carbon compounds to produce ground-level recurrent hazes made of ozone (Tawari and Abowei 2012). Among the sources of carbon emissions are transportation, manufacturing processes, the burning of bushes, aerosol usage, and the combustion of fossil fuels in other for power supply in the production companies and residential areas (Yakubu 2017). Furthermore, large the portion of the electricity required for other services and air conditioning is also consumed, which results in more emissions of pollutants in the environment. All gases emitted by burning fossil fuel and combustion vehicles are part of human activities which is anthropogenic cause and has pose a great contribution in global warming of the Earth (Olaniyi et al. 2013). In addition, air can also be polluted through building up of waste and agriculture practices (Ekiyor et al. 2019). The height of distress on air pollution on people living in the environment of these mega cities in Nigeria (Lagos, Kano, Port Harcourt, Ibadan and Warri) through refineries and other industrial activities is very pathetic and on assumption that the nation is affected majorly by the pollutants from all these industrial practices and the inhabitants of these cities are to be informed about the effect of air pollution to their health so to take precaution and measures (Ibemere 2019). The use of some construction materials like the cement, asbestos roofing, and ceiling tiles contributes to indoor air pollution and can lead to some health issues indirectly. According to the central pollution control board, the cement industry ranks among the 17 most polluting industries, and cement manufacture creates dust during its production (Ogedengbe and Oke 2011). The amount of air pollutants created via industrial and mobile sources affect the environmental outdoor air space, and the transportation of these air contaminants over long distance have significant consequences on the environment. Generally, air pollution in Nigeria is very alarming and havoc to the citizen (Fig. 12.2).

Fig. 12.2 How air pollution is silently choking life out of Nigerians in major cities. *Source* Bussinessday.ng

Impacts of Climate Change on Air in Nigeria

The effects of climate change on human health, as well as the environment, are already evident in Nigeria. The Earth's temperature increases as a result of introducing carbon dioxide into the atmosphere via combustion of fossil fuel such as coal and gas and this increase in temperature makes the Earth to become much like a large blanket of heat around us throughout the year. This heat trapped in the environment disrupts many interconnected systems. The air quality of an environment is tied to the atmospheric processes of the location, on measures of how human actions affect the environment (Ezeonyejiaku and Obiakor 2013) and poor air quality is when the emissions of nitrogen oxides, methane, and other harmful gases are above levels acceptable for human health (Ajiere and Nwaerema 2020). The decrease in the quality of air as a result of climate change via the emission of pollutants can affect human health. It has been estimated that over seven million people die every year from diseases caused by air pollution (Ezeonyejiaku et al. 2022). Moreover, it elevates particulate matter and ozone levels in the air, which has detrimental effects on health. Nigeria, especially in the Niger Delta region, has been struggling with the issue of pollution of the air quality (Tawari and Abowei 2012), especially after the appearance of carbon black (black soot) in the city of Port-Harcourt atmosphere. Nevertheless, the adverse health effects of poor air quality due to pollution in the city have become a source of concern in recent years (Yakubu 2017; Ugbebor et al. 2019). According to some reports, the soot pollution is attributed to the incomplete combustion of fossil fuels, biomass, tires, and illegal artisanal refineries. Among its serious risks are a decrease in human health, a decrease in agriculture, and a reduction in climate change (Akinfolarin et al. 2018).

Air pollutants compromises air quality of the environment (Ibe et al. 2017). As pollutants, air pollution is incredibly hazardous to human health and the effect of much exposure of it could lead to development of diseases such as asthma, bronchitis, coughing, breathing difficulties, or decrease lung function. The adverse effects of these conditions of climate change can lead to more medication usage, frequenting hospitals for consultation and can also lead to untimely death (Omoruyi and Onafalujo 2011; WHO 2014; Ibe et al. 2017). It is conceivable that changes in climate conditions will extend the transmission seasons of important vector-borne diseases and change the geographical distribution of these diseases. There is evidence that malaria is caused by climate conditions, a unicellular organism known as Plasmodium, and transmitted through the bite of an infected female Anopheles mosquito. A survey shows that malaria accounts for over 45% of all outpatient visits and almost 50% of Nigerians experience malaria at least once a year (Ayeni 2011). Wildfires in Nigeria are caused by human activities and lightening fires are very rare because of the hot and dry Harmattan winds flowing from the north to south between December and March. The area situated above latitude 10°N in Nigeria experiences more wildfire, mainly because this is the savanna ecozones with a shorter duration of rainfall and a longer period of dryness (IFFN 2006). Therefore, wildfires is very much experience when there is increase in drought and increased wildfires spread due to climate change can decrease air quality and lead to some aliments such as respiratory and cardiovascular diseases. Much exposure of it could affect organs in our body and worsen existing health conditions that are linked with climate change. Obviously, people react to air pollutants, but the interaction depends on the characteristics of pollutant they are exposed to, the level of exposure, as well as genetic makeup (Duflo et al. 2008). Those who are sensitive will experience greater allergic reactions when they are exposed to both allergens and air pollutants and this could lead to increased risk of respiratory illness (APHA 2016).

When the concentration of toxic gases, nitrogen oxides, and methane is higher than what is considered safe for human health, the air of that place has poor quality. At high concentrations, ground-level ozone is a toxic pollutant that is formed by combining other gases under favorable conditions. It can affect both human and crop, as well as damage the environments. Similarly, a third form of ozone is found in the stratosphere, and it acts as a barrier against UV radiation from the Sun. Thus, climate change is expected to result in greater concentrations of grand level ozone when environmental conditions change, such as increased temperature (Kath 2010).

Assessing population exposure to air pollutants requires air quality monitoring. Public health effects are influenced by the type, severity, frequency, duration, and toxicity of the pollutant (Zhang et al. 2019; Manisalidis et al. 2020). Nonetheless, the air quality index has become a widely used indicator for assessing and indexing the state of an air environment over a place or a region, it is still considered a matrix in its own time. Most developed countries use air quality indexes for routine and daily activities, which are available on digital platforms. But some developing countries lack access to such platforms, such as Nigeria, where air quality monitoring is rarely measured (Ezeonyejiaku et al. 2022).

As reported by Ede and Edokpa (2015), the south-south (Niger delta region) of Nigeria has poor air quality which affects emissions circulation. There is also evidence from another study that the air quality in the North-eastern Nigeria that is, the same region of the country Nigeria particularly Port Harcourt is very unhealthy and is a risk health factor to the exposed people and can disrupt the respiratory and cardiovascular systems after a long exposure (Ugbebor et al. 2019).

Recent research recorded by Ezeonyejiaku et al. (2022), on air quality measure in some areas (Awka, Ekwulobia and Nnewi) of Anambra state located in the eastern part of Nigeria, indicated that oxides of carbon (CO_2 and CO) showed highest mean concentrations across towns within the residential and commercial areas. Meanwhile, methane (CH_4) concentrations in all the towns examined, excluding Onitsha, were higher and exceeded WHO limits.

Industrial developments in Nigeria process both wet and dry wet, and each of these activities entails related environmental impacts and contributes significantly to climate change. The paper, cement, iron, and steel industries that emit pollutant or toxicants into the atmosphere are the most vulnerable to climate regulation. As a consequence, manufacturing companies in Nigeria rely on hydropower, which is epileptic, requiring the use of more generating sets, which in turn leads to more carbon emissions (Saka 2020; Etim et al. 2021).

Policy that Can be Use to Improve Air Quality in Nigeria

Several rules and regulations accompany the Nigeria climate policy intended for agencies and individuals to adhere in order to attain its objective. A major objective of the policy is to address variety of human-caused causes of climate change in so many of ways. It aims to mitigate climate change's negative impact. In order to mitigate climate change, Nigeria should implement a national climate change policy and enact laws that prohibit the release of toxic gases from industrial areas. Mitigation measures of the effect of climate change could include the following:

Climate change impacts and its vulnerability effect on populations should be brought into the spotlight by agencies in control. This can be accomplished by providing basic training, workshops, and awareness programs to raise awareness and promote climate change adaptation (Madu 2016). In addition, these actions would provide benefits to global health in the long run (Borokinni 2017). However, climate change threatens air quality in Nigeria, but air pollution can be minimized by improving access to clean household energy, which in turn would involve using refined cooking stoves without hydrocarbon and solar light (Ajiere and Nwaerema 2020).

To the industry, we must provide clean technologies that can minimize smokestack emissions and improve on urban solid waste management.

In the field of transportation, it is necessary to reduce the number of large polluting long-distance vehicles such as trucks and lorries that release polluting gases to the environment and to reintroduce rail transportation systems, electric cars, reposition heavy-duty vehicles to cleaner models, and use high-efficiency vehicles and fuels (Suleiman 2013). Some developed economies have successfully evolved effective climate change measures for their countries. The political leadership in this country should seek the support of developed economies that have succeeded in implementing such measures in their countries. By providing such aid, the crisis could be rapidly addressed in the areas of health, waste management, water quality, and food shortages. In order to build environmental stability, soil fertility, and greater food production, reduce environmental-induced migration, and stabilize the political arena; the government must establish a nationwide program of reforestation, tree planting, and windbreaks in the northern areas (Urim et al. 2013).

Conclusion

Children, as well as adults in Nigeria, are affected by the issues associated with climate change and air quality studies face a lot of challenges. The most prominent issues in "Nigeria" has been deficiencies in equipment, infrastructure, and lack of expertise and weak policy frameworks. In addition to these and many other factors, the country has been unable to achieve their long term desire objectives, the ecosystem quality and safety of public health it is pursuing. In Nigeria, air quality can be improved by implementing mitigation measures in several sectors, including transportation, industry, waste, oil and gas, agriculture, power, cooking, and lighting households.

Recommendation

In light of the fact that the effects of climate change can be felt in a wide variety of ways, including air quality, human settlements, health, wetlands, water resources, business, commerce, financial services, food security, land degradation, forestry, and biodiversity, it is recommended that Nigerian citizens become aware of the issues surrounding climate change. A sustainable way to achieve this objective is through education and capacity building as a means of integrating adaptation and mitigation.

The government should also put more of their efforts to plant trees, as this will help clean up the environment. Government should abate illegal refining of crude oil and finance into low-carbon fuels and solar products in Nigeria. It is also very important for the government to take, support, and enforce policies to combat air pollution to ensure quality environment. However, there should also be joint efforts between the citizens and the government to address climate change on air quality in every community in Nigeria.

The government can help to produce energy for the society by assembling plants and crushing them in a modern bio-power. In the presence of ultraviolet light, free electrons are released from material, causing free radicals to form, which are responsible for oxidizing VOCs and NOx emissions. Titanium dioxide can be use; it has been reported to be effective in reducing air pollution by scientists and has been found by researchers to have this property.

Research on the impact of climate change on the quality of air and its mitigation in the society needs to be in-depth.

Acknowledgements I want to thank my Colleague Emmanuel Akubunwa who brought the information and topic to my interest. This also motivated me to write. I will also remain grateful to my beloved husband, Mr Uzochukwu Obiegbu for his financial supports, care, advice, and encouragement in writing this work. I sincerely appreciate the editors for their responsibilities toward this work. Above all, I thank God Almighty whose grace has sustained me all through.

References

Abaje IB, Bello Y, Ahmad SA (2020) A review of air quality and concentrations of air pollutants in Nigeria. J Appl Sci Environ Manag 24(2):373–379

Abdulkadir A, Maryam LA, Muhammad TI (2017) Climate change and its implications on human existence in Nigeria: a review. Bayero J Pure Appl Sci 10(2):152–158

Adejuwon JO (2006) Food crop production in Nigeria II. Potential effects of climate change. Climate Res 32:229–245

Ajiere S, Nwaerema P (2020) Climate change and air pollution: implication for human health and environment in rivers state. Int J Res Innov Soc Sci 4(5):276–279
Akande A, Ana Cristina C, Jorge M, Roberto H (2017) Geospatial analysis of extreme weather events in Nigeria (1985–2015) using self-organizing maps. Adv Meteorol 2:1–11
Akinfolarin OM, Obunwo CC, Boisa N (2018) Air quality characteristics of emerging industrial areas in Port Harcourt. Niger J Chem Soc Niger 43(1):7–14
Akinola AK, Suleiman AA, Adeleye YBA, Francis OA, Dauda RA (2014) Air pollution and climate change in Lagos, Nigeria: needs for proactive approaches to risk management and adaptation. Am J Environ Sci 10(4):412–423
Akinwumi AM, Adewumi JR, Obiora-Okeke OA (2020) Impact of climate change on the streamflow of Ala River, Akure, Nigeria. Sust Water Res Manag 7(1):1–11
Akpomi ME, Vipene J (2016) Promoting knowledge of climate change (CC) amongst Nigerians: implications for education managers. J Educ Pract 7:32. ISSN 2222-1735
American Public Health Association (APHA) (2016) Climate change decreases the quality of the air we breathe. https://www.apha.org/~/media/files/pdf/factsheets/climate/airquality.ashx. Accessed 6 Dec 2021
Ayeni AO (2011) Malaria morbidity in Akure, Southwest, Nigeria: a temporal observation in a climate change scenario. Trends Appl Sci Res 6:488–494
Borokinni J (2017) Climate change and health impacts. The nation newspaper. https://nationonlien.ng.net?climate-change-health-impact/. Accessed 19 July 2022
Brook RD, Rajagopalan S, Pope CA, Brook JR, Bhatnagar A (2010) Particulate matter air pollution and cardiovascular disease: an update to the scientific statement from the American Heart Association. Circulation 121:2331–2378
Broomfield P, Royle JA, Steinberg LJ, Yang Q (1996) Accounting for meteorological effects in measuring urban ozone levels and trends. Atmos B Environ 17:3067–3077
Cattaneo C, Massetti E (2019) Does harmful climate increase or decrease migration? Evidence from rural household Nigeria. Clim Change Econ 10(4):1950013
Chukwu GU (2008) Poverty-driven causes and effects of Environmental degradation in Nigeria. Pac J Sci Technol 9(2):599–602
Chukwu GU, Asiegbu AD (2011) Global warming and climate change; impact and implications in Nigeria. Int J Curr Res 3(6):112–115
Davis JM, Eder BK, Nychka D, Yang Q (1998) Modeling the effects of meteorology on ozone in Houston using cluster analysis and generalized additive models. Atmos Environ 32:2505–2520
Dipeolu AA (2015) Climate change and challenges of environmental sustainability in Lagos State, Nigeria. J Environ Earth Sci 5:17
Duflo E, Greenstone M, Hanna R (2008) Indoor air pollution, health and economic well-being. S.A.P.I.EN.S, 1(1). Sapiens.revues.org. Accessed 27 Dec 2022
Ebele NE, Emodi NV (2016) Climate change and its impact in Nigerian economy. J Sci Res Rep 10(6):1–13
Ede PN, Edokpa DO (2015a) Regional air quality of the Nigeria's Niger Delta. Open J Air Pollut 4(1):4236–54348
Ekiyor TH, Horsfall IT, Kalagbor AI, Egbara-Dedua (2019) Evaluating the effects of air pollution on the agricultural yield of selected crops in eleme, Nigeria. J Appl Sci Environ Manage 23(10):1771–1778
Elum ZA, Momodu AS (2017) Climate change mitigation and renewable energy for sustainable development in Nigeria: a discourse approach. Renew Sustain Energy Rev 76:72–80
Enete AA, Amusa TA (2010) Challenges of agricultural adaptation to climate change in Nigeria: a synthesis from the literature. Field action science reports. J Field Action 4:1–11
Etim MA, Babaremu KO, Justin LD, Omole DO (2021) Health risk and environmental assessment of cement production in Nigeria. Atmosphere 12(9):1111
Ezeonyejiaku CD, Obiakor MO (2013) Spatial biomonitoring of trace element contamination and atmospheric quality assessment. Am J Life Sci 1(2):67–73

Ezeonyejiaku CD, Okoye CO, Ezeonyejiaku NJ, Obiakor MO (2022) Air quality in Nigerian urban environments: a comprehensive assessment of gaseous pollutants and particle concentrations. Curr Appl Sci Technol 22:5. https://doi.org/10.55003/cast.2022.05.22.011
Fagorite VI, Anifowose FA, Chiokwe VN (2021) Air pollution; causes, effects and remediation in Nigeria. Int J Adv Acad Res 7(1):13–30
Federal Ministry of Environment (2014) United nations climate change Nigeria. National Communication (NC). NC 2. https://unfccc.int/sites/default/files/resource/nganc2.pdf. Accessed 17 Nov 2021
Fort Air Partnership (FAP) (2014) How is air quality monitored? Albrta environment, government of alberta. www.fortair.org. Accessed 27 Dec 2022
Grutter M, Arellano J, Bezanilla H, Friedrich M, Plaza E, Rivera C, Stremme W (2014) Characterization of air pollution in Mexico city by remote sensing. Geophys Res Abstr 16:1
Ibe F, Opara A, Chu N, Ikech A (2017) Ambient air quality assessment of Orlu, Southeastern. Niger J Appl Sci 17:441–457
Ibemere D (2019) How air pollution is silently choking life out of Nigerians in major cities, 22 September 2019. https://businessday.ng/features/article/how-air-pollution-is-silently-choking-life-out-of-nigerians-in-major-cities/. Accessed 15th Jul 2022
Ideki O, Weli V (2019) Assessment of drought vulnerability and occurrence zones in north central Nigeria. Atmos Climate Sci 09(03):298–309
Intergovernmental Panel on Climate Change (IPCC) (2005) Global warming early signs on climate change. https://www.climatehotmap.org. Accessed 19 Jun 2022
International Forest Fire News (IFFN) (2006) Fire situation in Nigeria. https://gfmc.online/wpcontent/uploads/12-IFFN-34-Nigeria-1.pdf. Accessed 21 Jul 2022
Intergovernmental Panel on Climate (IPCC) (2007a) Climate change 2007: AR4 synthesis report. Contribution of working groups I, II and III to the fourth assessment report of the intergovernmental panel on climate change core writing team. In: Pachauri RK, Reisinger A (eds) IPCC, Geneva, Switzerland, pp 104–112
Intergovernmental panel on climate change (IPCC) (2007b) Climate change impact, adaptation and vulnerability: contribution of working group II to the forth assessment reports of the IPCC. J Environ Q Madison 37(6):2407
Intergovernmental panel on climate change (IPCC) (2014) AR5 Climate change: synthesis report. Contribution of working groups I, II and III to the fifth assessment report of the intergovernmental panel on climate change core writing team. In: Pachauri RK, Reisinger A (eds) IPCC, Geneva, Switzerlaand, pp 151–169
Jagtap S (2007) Managing vulnerability to extreme weather and climate events: implications for agriculture and food security in Africa. In: Proceedings of the inter-national conference on climate change and economic sustainability held at Nnamdi Azikiwe University, Awka, Nigeria, pp 42–52
Kath L (2010) Air Pollution and climate change. Sci Environ Policy 24:1–3
Kelishadi R, Poursafa P (2010) Air pollution and non-respiratory health hazards for children. Arch Med Sci 6:483–495
Komolafe AA, Adegboyega SA, Anifowose AYB, Akinluyi FO, Awoniran DR (2014) Air pollution and climate change in Lagos, Nigeria: needs for proactive approaches to risk management and adaptation. Am J Environ Sci 10(4):412–423
Krupa S, Nosal M, Ferdinand JA, Stevenson RE, Skelly JM (2003) A multivariate statistical model integrating passive sampler and meteorology data to predict the frequency distributions of hourly ambient ozone (O_3) concentrations. Environ Pollu 124:173–178
Kuenzer C, Van B, Gessner U, Dech S (2014) Land surface dynamics and environmental challenges of the Niger Delta, Africa: remote sensing-based analysis spanning three decades (1986–2013). Appl Geogr 53:354–368
Madu IA (2016) Rurality and climate change vulnerability in Nigeria: assessment towards evidence based even rural development policy. Paper presented at the 2016 Berlin Conference on Global Environmental Change, 23–24 May 2016 at Freie Universität Berlin. https://pdfs.semanticscholar. Accessed 19 July 2022

Maikasuwa SA (2013) Climate change and development countries: issues and policy implication. J Res Dev 187(941):1–10

Manisalidis I, Stavropoulou E, Stavropoulos A, Bezirtzoglou E (2020) Environmental and health impacts of air pollution: a review. Front Public Health 8(14). https://doi.org/10.3389/fpubh.2020.0001

National research Council (NRC) (2011) Climate stabilization targets: emissions, concentrations and impacts over decades to Millennia. The National Academies Press, Washington, DC, p 298. https://doi.org/10.17226/12877. Accessed 12 Nov 2021

Nd-Gain (2021) Ranking of country index. https://gain.nd.edu/our-work/country-index/rankings/. Accessed 19 Oct 2021

Nebedum EE, Emodi NV (2016) Climate change and Its Impact in Nigerian economy. J Sci Res Rep 10(6):1–13

Nicholls RJ, Marinova N, Lowe JA, Brown S, Vellinga P, De Gusmao D (2011) Phil Trans Math Phys Eng Sci 369:161–181

Nwafor JC (2007) Global climate change: the driver of multiple causes of flood intensity in Sub-Saharan Africa. Paper presented at the International Conference on Climate Change and Economic Sustain-ability held at Nnamdi Azikiwe University, Enugu, Nigeria, pp 12–14

Obioh IB (2008) Proposed climate change policy and the framework for integration into national planning in Nigeria. Centre for energy research and development, OAU, IFE, Nigeria. https://climatechange.gov.ng/wp-content/uploads/2020/09/national-climate-change-policy-1-1.pdf. Accessed 7 Oct 2021

Odekanle EL, Adeyeye MA, Akeredolu FA, Sonibare J (2017) Variability of meteorological factors on in-cabin and pedestrians exposures to CO and VOC in South-west Nigeria. J Atmos Pollut 5(1):1–8

Odjugo PAO (2001) A global warming and food production: a global and regional analysis. Afr J Environ Stud 2(2):85–91

Odjugo PAO, Ikhuoria AI (2003) The impact of climate change and anthropogenic factors on desertification in the semi-arid region of Nigeria. Global J Environ Sci 2(2):118–126

Odjugo PAO (2005a) An analysis of rainfall pattern in Nigeria. Global J Environ Sci 4(2):139–145

Odjugo PAO (2005b) General overview of climate change impacts in Nigeria. J Hum Ecol 29(1):47–55

Odjugo PAO (2010) General overview of climate change impacts in Nigeria. J Hum Ecol 29(1):47–55

Ogedengbe 1, Oke AO (2011) Pollution impact of cement production on air, soil and water in a production location in Nigeria. J Sci Technol 31(2):46–56

Odjugo PAO (2013) Analysis of climate change awareness in Nigeria. Acad J 8(26):1203–1211

Olaniyi OA, Ojekunle ZO, Amujo BT (2013) Review of climate change and its effect on Nigeria ecosystem. Int J Afr Asian Stud Open Access Int J 1:2013. https://core.ac.uk/download/pdf/234689728.pdf. Accessed 2 Nov 2021

Ogunniran BI (2018) Ozone layer depletion and climate change in Nigeria-problems and prospects; a review. Global J Res Rev 5(1):2

Omoruyi EP, Onafalujo AK (2011) Effects of climate change on health risks in Nigeria. Asian J Bus Manag Sci 1(1):204–215

Perez P, Trier A (2001) Prediction of NO and NO_2 concentrations near a street with heavy traffic in Santiago, Chile. Atmos Environ 35:1783–1789

Pittock AB (2005) Climate change: turning up the heat, Ist. Earthscan, London, p 316

Rowley RJ, Kostelnick D, Braaten X, Li X, Meisel J (2007) Risk of rising sea level to production and land area. Eos Trans Am Geophys Union 88(9):105–107

Saka L (2020) State, industries and regulation in Nigeria: Ife iron and steel company and environmental pollution. J Manag Soc Sci 9(1):750–768

Saheed M, Oludare H, Adedeji IE, Fatima K (2019) Mainstreaming climate change into the EIA process in Nigeria: perspectives from projects in the Niger Delta Region. https://doi.org/10.3390/cli7020029. Accessed 4 Jan 2022

Sayne A (2011) Climate change adaptation and conflict in Nigeria. USIP, Washington, DC. https://www.usip.org/sites/default/files/Climate_Change_Nigeria.pdf. Accessed 18 July 2022

Spore (2008) Climate change. Spore special issue, Aug 2008. CTA, Wageningen, pp 1–23

Suleiman IL (2013) Examining air pollution and control measures in urban centers of Nigeria. Int J Environ Eng Manag 4(6):621–628

Tawari CC, Abowei JFN (2012) Air pollution in the niger delta area of Nigeria. Int J Fish Aquat Sci 1(2):94–117

Ugbebor JN, Yorkor B, Amadi G (2019) Assessment of air quality and its health implications on Abuja campus residence, University of Port Harcourt, Nigeria. J Sci Technol Environ Inf 7(1):500–509

Urim UM, Imhonopi D, Egharevba ME (2013) Advancing solutions for climate change in Nigeria: an appraisal. In: A panoply of readings in social sciences: lessons for and from Nigeria. Department of Sociology, Covenant University, Ota, Ogun State, pp 233–247

USAID (2012) Nigeria climate vulnerability profile. https://www.climatelinks.org/sites/default/files/asset/document/nigeria_climate_vulnerability_profile_jan2013.pdf. Accessed 3 Nov 2021

Weli VE (2014) Atmospheric concentration of particulate pollutants and its implications for respiratory health hazard management in Port Harcourt Metropolis, Nigeria. Civil Environ Res 6(23):11–17

WHO (2014) 7 million premature deaths annually linked to air pollution. https://www.who.int/news/item/25-03-2014-7-million-premature-deaths-annually-linked-to-air-pollution. Accessed 7 Jan 2022

World Bank (2019) Building climate resilience: experience from Nigeria. https://www.worldbank.org/en/results/2019/04/18/building-climate-resilience-experience-from-nigeria. Accessed 3 Nov 2021

World Health Organization (WHO). (2021a). Climate change. https:// www.who.int/heli/risks/climate/climatechange/en/. Accessed 3 Nov 2021a

World Health Organization (WHO) (2021b) WHO Air quality guidelines. https://www.c40knowledgehub.org/s/article/WHO-Air-Quality-Guidelines?language=enUS. Accessed 5 Nov 2021b

World health organization Africa (WHO) (2020) Nigeria strengthens capacity to address impact of climate change on health. https://www.afro.who.int/news/nigeria-strengthens-capacity-address-impact-climate-change-health. Accessed 3 Nov 2021

Yakubu O (2017) Particle (soot) pollution in Port Harcourt Rivers State, Nigeria-double air pollution burden understanding and tracking potential environmental public health impacts. Environments 5(1):2–8

Zhang JJ, Adcock IM, Bai Z, Chung KF, Duan X, Fang Z (2019) Health effects of air pollution: what we need to know and to do in the next decade. J Thorac Dis 11(4):1727–1730

Chapter 13
Sustaining Livestock Production Under the Changing Climate: Africa Scenario for Nigeria Resilience and Adaptation Actions

Akeem Babatunde Sikiru, Silpa Mullakkalparambil Velayyudhan, Mini R. Reshma Nair, Sejian Veerasamy, and John Olayinka Makinde

Abstract Climate change is a global millennial challenge affecting humans in all facets of life. In Africa, livestock production is one of the sectors most affected by climate change because livestock production is extensive, smallholders operated, and low external input operation. The paper gives an insight for sustaining Africa's livestock production in resilience to climate change. The study obtained secondary meteorological and livestock production data for the continent over forty years, and the data were analysed for the understanding of changes over the years and prediction of future scenario. The result revealed that by the year 2050, the average surface temperature in Africa will rise by 2.2 °C, while the livestock population will continue to grow. The result also indicated that the scenario revealed in the study is an Africa-wide condition, but the West Africa subregion of Africa could be the most vulnerable to the climate change impact of temperature rise. The study also indicated that small ruminant animals including sheep and goats could be more

A. B. Sikiru (✉)
Department of Animal Production, Federal University of Technology, Minna, Niger 920211, Nigeria
e-mail: akeembaba01@gmail.com

S. M. Velayyudhan · M. R. R. Nair · S. Veerasamy
ICAR-National Institute of Animal Nutrition and Physiology, Adugodi, Hosur Road, Bangalore, Karnataka 560030, India

S. M. Velayyudhan
Institute of Animal Breeding and Genetics, Justus-Liebig-Universität Gießen, 35390 Giessen, Germany

M. R. R. Nair
Academy of Climate Change Education and Research, Kerala Agricultural University, Vellanikkara, Thrissur, Kerala 680656, India

A. B. Sikiru
Federal University of Agriculture, Zuru 872101, Kebbi, Nigeria

J. O. Makinde
Federal University Gashua, Gashua 671105, Yobe, Nigeria

J. C. Egbueri et al. (eds.), *Climate Change Impacts on Nigeria*, Springer Climate,
https://doi.org/10.1007/978-3-031-21007-5_13

tolerant to the future changing climate scenarios in Africa compared with other livestock because their population keeps increasing over the years despite the climate change threats. Finally, future projections of climate conditions indicated that Africa could be a hub of future livestock production, but climate conditions would affect optimal animal performances and production. Therefore, several strategies to sustain livestock production were suggested as resilient means for adapting to the changing climate conditions.

Keywords Adaptation · Africa · Amelioration · Climate · Heat stress · Livestock · Resilience

Introduction

The rising human population has led to an increase in global demand for livestock and agricultural products (Naqvi and Sejian 2011). However, erratic climatic conditions have posed a threat across the world—a situation which necessitated the convocation of the Intergovernmental Panel on Climate Change (IPCC), which reported that the mean global surface temperature in 2100 could increase between 0.3 and 4.8 °C higher (IPCC 2014a). This rising temperature is a matter of concern, especially in Africa, which has been identified as the most vulnerable region to climate change in the world (IPCC 2014b). There are varying risk factors of climate which have been assessed and found to have a potential negative impact on the Africa, such as the reduction in water resources, reduced crop productivity, and increased incidence of vector-borne and waterborne diseases; these are critical to livestock production and its consequence on the environment (IPCC 2014b).

Livestock production contributes immensely to the Africa economy and contributing to the standard of living of poor and marginal farmers while ensuring food security (AU-IBAR 2015). The sector was also reported to be contributing between 30 and 80% to the entire agricultural gross domestic product (GDP) of most countries in the sub-Saharan Africa region (AU-IBAR 2015). Therefore, the livestock production sector could be confidently described as a high potential for delivering sustainable development goals and socio-economic transformation in Africa, according to the release of the African Union (AU) since about a decade ago (AU 2014). This has placed African policymakers and stakeholders in the increasing awareness creation about livestock production as well as exploiting the livestock production to meet up the rising demand for protein and promotion of economic growth (AU-IBAR 2015).

In attestation of the foregoing facts, Pica-Ciamarra et al. (2013) stated that the demand for livestock and their products are expected to increase two to eightfold by 2050 in Africa. Meanwhile, based on the current issues in the livestock sector in Africa coupled with low investment and growth, meeting up with the projected demand is nearly impossible (Herrero et al. 2014). This is reflected in the yearly increase in the import of animal products to the continent, leading to a drastic increase in livestock product import bills and loss of foreign exchange (AU-IBAR 2015). These are ancillary factors to a series of negative effects on the economy, exacerbation of food insecurity, unemployment, and poor trading (Akokpari 2007).

Although Africa is a blessed continent in terms of extensive range land, water resources, and others that can boost livestock production, the climatic change challenges have always threatened the livestock production sector directly or indirectly with severe consequence on animals growth, production, and resistance to diseases (Otte et al. 2012; Yaméogo et al. 2014; AU-IBAR 2015). Therefore, overcoming the constraint of climate change is highly desirable for the Africa livestock production sector because it would help the livestock farmers to sustain production of animals to produce animal proteins for feeding the teeming growing population in the continent. Hence, in this chapter, data on climate and livestock production in Africa were analysed, and relevant information on various adverse effects of climate change on the African livestock population and production in future under climate change scenarios was reported. Furthermore, the chapter also suggested various mitigation strategies for sustainable livestock production in the face of changing climate conditions and a specific framework for resilience and adaptation in Nigeria.

Methodology

Sources of the Data

Climate Data

The monthly surface temperature data for the periods between the year 1980 and 2019 in Africa were retrieved from the database of the National Oceanic and Atmospheric Administration (NOAA) of the USA (https://www.climate.gov). While data on relative humidity were retrieved from the public datasets of the Copernicus Climate Data Store (https://cds.climate.copernicus.eu/#!/home) and the United Nations (UN) data explorer platform (http://data.un.org/Explorer.aspx?d=CLINO).

Livestock Production Data

There was a retrieval of livestock production data for Africa between 1980 and 2019 from the Food and Agriculture data (FAOSTAT 2019). The data obtained include populations of goat, sheep, and cattle; milk and meat production output from cattle and the small ruminant, respectively.

Data Analyses

The data retrieved for both the climate and livestock production were organized into ten year groups, including 1980–1989, 1990–1999, 2000–2009, and 2010–2019; and in each ten years group, a year was divided into four quarters (3 months units) giving a total of 16 quarters in the 40 year period which represented the unit time used in the analysis. The data were then subjected to descriptive statistical analysis and time series modelling using Microsoft Excel with the aim of understanding patterns of changes over the years and to identify contributions of seasonal, irregular, and trend components to the livestock production The time series model used for the study simple time series model is described by Sikiru et al. (2020). The model takes y (t) as the time series value of a parameter over a given period of time, S (t) is the seasonal component associated with the parameter changes over a given period of time, I (t) is the irregular component associated with the parameter changes over a given period of time, and T (t) is the trend component associated with the parameter changes over a given period of time (Pollock et al. 1999).

Results and Discussion

The average increase in the surface temperature recorded between 1980 and 2019 was 1.3 °C, but by the year 2050, the increase in surface temperature of Africa could increase up to 2.2 °C and this indicates that temperature increase in Africa is real and almost impossible (Fig. 13.1a–c). The relationship between increase in temperature changes, temperature-humidity index (THI), and population of livestock and products, outputs indicated an inverse relationships whereby, as the temperature and temperature-humidity index (THI) increases, there is decrease in population of the livestock in a perfect inverse relationship of $P < 0.01$ and $R^2 = 0.72$ (Fig. 13.1).

The temperature-humidity index follows similar patterns as the surface temperature changes which indicated that there could be a future increase in the THI and consequential non-favourable climate conditions for livestock production in Africa. Meanwhile, for both the temperature and THI increases, West Africa subregion is the major driver as the bulk of the increase may be found in the West Africa and these may lead to reduced livestock products supply compared with the Africa-wide average (Fig. 13.1). Finally, the projections indicated potential livestock population and production increase, West Africa could be the hub of the future production except for milk, and the climate conditions in the region are the least suitable for optimal animal performances (Fig. 13.1).

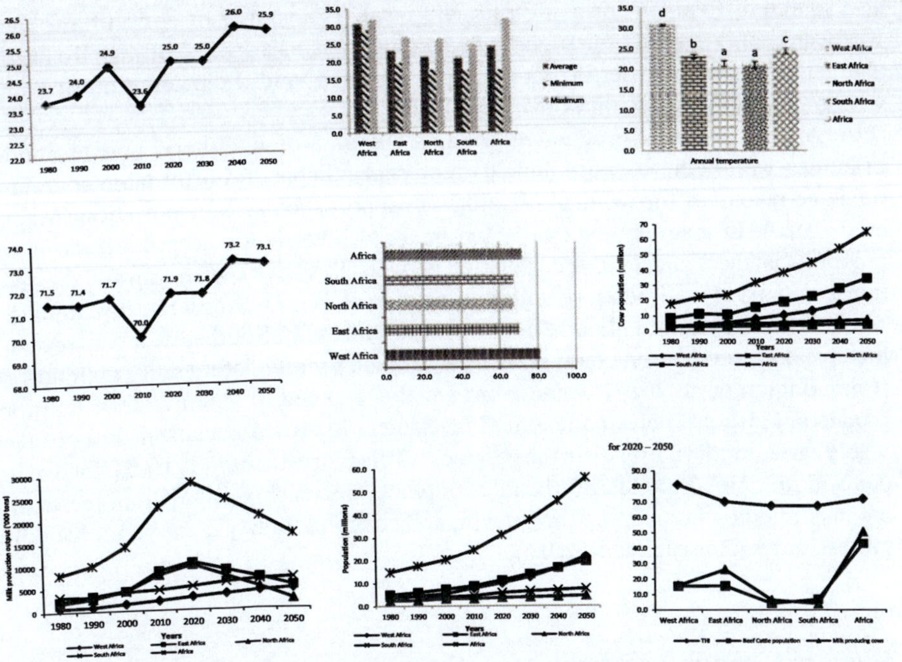

Fig. 13.1 Various consequences of climate change on African continent with specific emphasis on livestock production across the sub region. **a** Average annual recorded and projected temperature for Africa (°C), **b** Minimum and maximum projected temperature (°C) for Africa by regions 2020–2050, **c** Projected variations in regional annual temperatures (°C) for Africa 2020 to 2050, **d** Average annual calculated and projected temperature-humidity-index for Africa (°C), **e** Regional projected Temperature-Humidity-Index for Africa 2020–2050, **f** Regional populations and trends of milk producing cows distribution in Africa, **g** Regional past and future trends of milk production output ('000 tons) for Africa, **h** Regional populations and trends of cattle slaughtered for beef production in Africa, **i** Relationships between temperature-humidity-index and regional population trends of beef and milk producing cattle in Africa for 2020–2050

Discussion

Significance of Livestock Production to the African Continent

The benefits of livestock production in Africa are huge and as such, the analysed continental climate data of Africa concerning livestock population as demonstrated in this study established and increasing trend which could be hampered by climate variable specifically surface temperature and relative humidity (Fig. 13.1). These outcomes described possible consequences of climate change on livestock production in Africa and how it could compromise the significance of livestock production in Africa. In Africa, livestock production is a principal component of human survival, and it is a vital source of livelihood of people, especially in the rural areas (Raheem et al. 2019). This is because livestock farming provides multiple benefits to the farmer as a source of foods in forms of milk and meat, as a source of agricultural power where animals are used as draught in land cultivation, as a source of manure used as alternative to chemical fertilizers in crop production, and the promotion of human welfare via supply of incomes (Meltzer 1995).

In Africa, livestock production remains a source of livelihood and promotion of human well-being because animals represent reliable and affordable sources of food and resources for work and livelihood support such as transportation (Mapiye et al. 2020). In a survey data collected on 1624 households across Africa, it was reported that there is a positive correlation between livestock ownership and the nutritional status and well-being of children (Ruel et al. 2017). Similarly, investigations across Africa, spanning Nigeria in the west to Malawi in Southeast, it was revealed that livestock production could serve as an avenue for alleviating child malnutrition (Hetherington et al. 2017). Apart from benefits accrued to smallholders, livestock production also promotes commercial investment in Africa because it is a growing viable agricultural enterprise in the continent. This is driven majorly by the increasing demand for animal products, urban development, increasing population, increasing income, middle-class population growth, and changing lifestyles of young working professionals (Roessler et al. 2016).

Climate Change and Economic Consequences in the African Livestock Sector

In an economy and economic development terms, climate change could have strong negative implications for the economy of the African livestock sector; this was earlier opined in a report by Hendrix (2017). The direct impacts of climate change on animals such as heat stress, production decline, impaired reproductive efficiency, and increased incidence of animal diseases could severely affect the continent's economy since huge incomes are associated with livestock productivity (Ogenga et al. 2018). Climate change could also pose several indirect consequences like animal feed and fodder shortage, water scarcity, alarming multiplication, and spread of vectors transmitting diseases, all of which can contribute directly and indirectly to unemployment and hamper economic growth in Africa (Hopkins and Prado 2007; Thornton et al. 2009). For instance, climate change has been reported affected Sudan, a country in the Northeast Africa and projection of economic loss if unabated, could result in an approximate loss of up to 105.5 billion USD by the year 2050 (Borgomeo et al. 2018; Siddig et al. 2020).

Another indirect consequence of climate change that could negatively affect the economy is the reduction in forage yield and insufficient availability of water for animal (Escarcha et al. 2018a, b). Although these could be indirectly, it could be confounded with a reduction in revenue generation from animal sales as well as directly aggravating competition for land resources and land uses between crop farmers and livestock herders (Thornton and Herrero 2010a). This situation is the leading factor continually forcing livestock keepers to migrate in search of greener pasture and, in recent years, has created economic and security challenges. This is a concern in countries such as Nigeria, where food security is presently at the risk of food security promoted by ethnoreligious cynicism (Sikiru 2016, 2020).

The economic implications were simulated in a climate-livestock model which indicated that higher atmospheric temperature corresponds to the lower profit maximization in a livestock enterprise (Seo and Mendelsohn 2008a). In application of this model to the African scenario, it revealed that increasing temperature in Africa will continuously be a leading factor responsible for reducing production and population of cattle and chicken in Africa. However, the model proposed the possibility of sustainable production of sheep and goats despite the temperature increase. Taking all factors into account, and looking at the simulated model, climate change could cause between \$9 and \$12 billion loss in Africa by 2030 if current conditions are left unabated (Seo and Mendelsohn 2008b).

Climate Impacts on Livestock Production Performances and Management

The impact of climate change on livestock performances and management cannot be overlooked, and some of the possible scenarios in Africa and their potential impacts were presented in this study following an elucidation of the relationships between the increasing surface temperature, high temperature-humidity index, and livestock performances (Fig. 13.2). Changes in these climatic parameters to the high extreme levels can negatively affect livestock production and management as well as increase the risk of the vulnerability of livestock rearing communities to natural disasters such as drought (Zougmoré et al. 2016). The extreme frequencies of these parameters and their interaction with other factors have changed the livestock production systems among African livestock producers which has consequently led to loss of productivity (IPCC 2014b; Zougmoré et al. 2016).

Changing climate conditions could also affect livestock production at both the species and breed levels and in addition could differently hamper production performances in different animals and breeds (Hoffmann 2010). For example, it has been established through prediction modelling that the demand for beef and cattle and dairy compared with goats and sheep will decrease significantly because of temperature rises (Kurukulasuriya et al. 2006; Rust and Rust 2013). Meanwhile, goats and chicken as well as sheep preferred wetter agro-climatic zones over beef and dairy cattle (Seo and Mendelsohn 2006). Furthermore, livestock net revenues in Africa are observed to be highly sensitive to climate, in areas with moderate climate conditions, livestock keepers were reported to generate higher revenue and vice versa compared with livestock keepers in the harsh climates (Kurukulasuriya et al. 2006).

The direct and indirect impacts of climate change on livestock could also impair their reproductive efficiency and hence have been identified to significantly reduce the

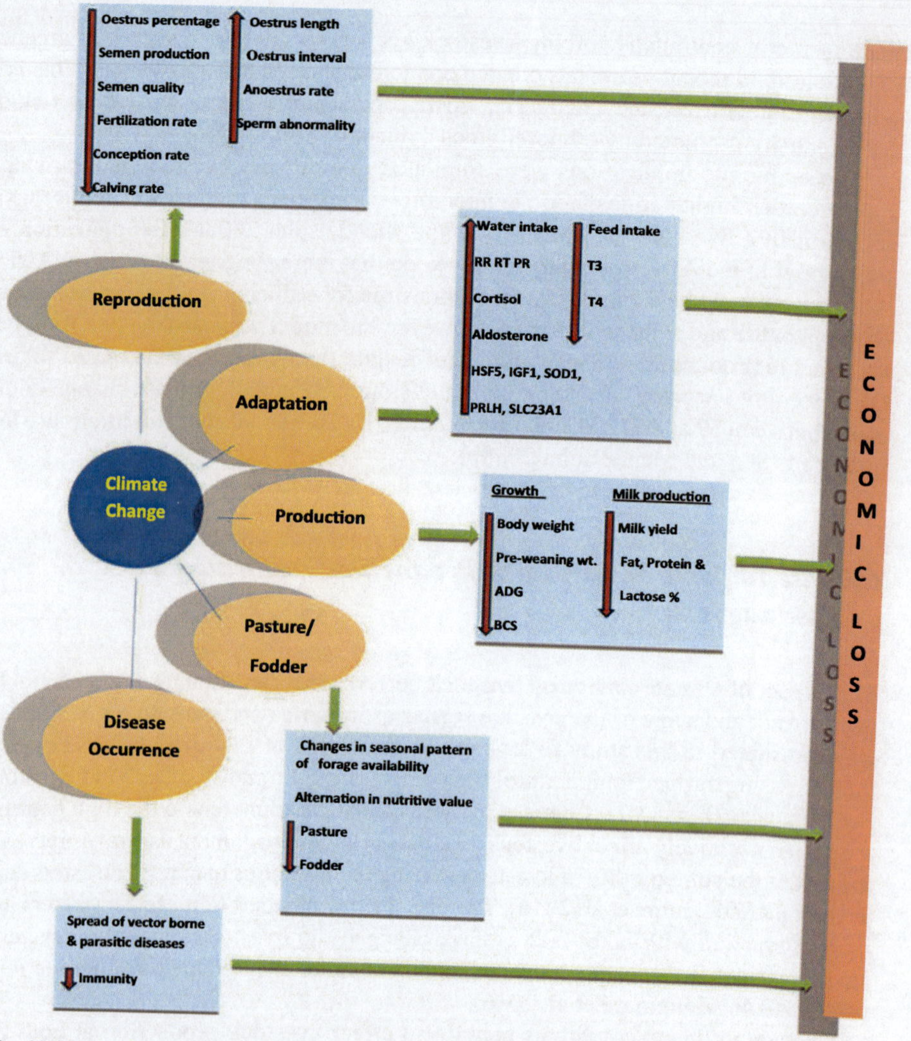

Fig. 13.2 Impacts of climate change variabilities on livestock production and management in relation to economic loss arising from poor performance characteristics of animals in Africa. The figure indicated negative effects of climate variations on decreasing reproductive performances, reducing growth, milk production and increased spread of diseases due to decreased immune competence

animal production performances in terms of fertility and reproduction outcomes (Rust and Rust 2013). The indigenous African cattle constitute the highest percentage of the meat-producing animals in Africa, and they are mainly reared on natural pastures, with their finishing custom feedlots. These animals could be severely affected by the negative effects of varying climatic conditions, and as a result, production management could be affected, leading to low animal productivity (Scholtz and Theunissen 2010; Rust and Rust 2013). This was also reported by Furstenberg and Scholtz (2008), who anticipated that global warming can lead to a 30% decline in land productivity in Africa, a situation which was identified as a consequence of the reduction in the rangelands and forests in Africa since the production of animals will become more difficult and less profitable.

In a confirmation of these possibilities, Scholtz et al. (2018) reported that the effect of climate on pre- and post-weaning performances of different breeds of cattle in South Africa was reduced due to increased summer temperature. It was also reported in the same study that there was a reduction in the average daily weight gain of the calves as a result of the heat waves increase during summer season. Similarly, Okouruwa (2015) studied the effects of heat stress on the live weight of West African Dwarf (WAD) goats in Nigeria in a study where goats were grouped into three types of housing production systems: T1 (confined to their pens throughout the day), T2 (allowed in the open yard from 8.00 am and 1.00 pm), and T3 (allowed in the open yard from 1.00 and 6.00 pm). The body weights records of the animals were obtained before and after the exposure of the animals to different housing confinement. The analysis of variance of the obtained records of weight changes of the goats indicated a significant reduction in the average body weight of the animals subjected to the open yard grazing from 1.00 and 6.00 pm which was attributed to the negative impacts of heat stress on the growth and production of the animals.

Milk production is another economic activity of livestock sector that could be adversely affected by heat stress occasioned by climate change as posited by Kekana et al. (2018). These researchers evaluated the effect of high thermal stress on milk production traits in lactating Jersey cattle reared in two communal regions of South Africa. The summer thermal-humidity index (THI) of the two regions was THI-1 (72–83: extreme caution) and THI-2 (75–87: danger). It was observed that there was a significant reduction in milk yield, urea nitrogen, fat, and protein and lactose percentage of lactating Jersey cows subjected to higher thermal stress (THI-2). In another study, Salem and Bouraoui (2009) also reported a 10% reduction in the milk yield per cow during summer heat stress conditions in Tunisia which is also within the continent of Africa. The foregoing paragraphs and projected increase in temperature and THI in Africa submitted that climatic changes could negatively affect production performances and management of animals in Africa. Also, the health of the animals could get compromised due to climate change since favourable environment for the survival of pathogens could be created to enhance the spread of vector-borne and parasitic diseases as well as reducing immunity as the underlying promoters of reduced animal productivity (Rust and Rust 2013).

Impact of Climate Change on Livestock Reproduction and Fertility

In addition to the direct and or indirect impacts of harsh climate on livestock production performances, climate change can also impact livestock reproduction and fertility. This is possible through the exertion of heat stress, which could have serious implications on animal reproduction (Sikiru et al. 2020). Changing climatic conditions could modulate the hypothalamus-hypophyseal-pituitary-ovarian axis of the animals, which could reduce the secretion of gonadotropin-releasing hormones. This could lead to complex malfunctioning of the primary reproductive hormones oestrogen, testosterone, follicle-stimulating hormone, and luteinizing hormone, thereby making the animal unproductive (Kebede 2016).

In female animals, climate change-related stress shortens the duration of receptivity for mating (Ekesbo and Gunnarsson 2018); in assisted reproductive technology practices such as artificial insemination, heat stress could induce conception failures in animals (Hansen 2014). Furthermore, elevated temperatures due to climate change could induce oestrogen malfunction and non-viability of mature follicles via modulation of the PGF-2α secretion. In addition to conception losses, heat stress could also prevent or extend an animal's returning to breeding after the lactation (Bilby et al. 2008). In male animals, when elevated temperatures cause heat stress, there is an increase in intra-testicular temperature, which could result in the production of poor quality semen, reduced secretion of reproductive hormones, and overall reproductive failures (Bhakat et al. 2014).

Being a tropical continent, Africa is more vulnerable to heat stress and exposure of livestock to extended nutritional crisis as an indirect impact of heat stress can impair endocrine functions which could cumulatively lead to reduced reproductive output (Habeeb et al. 2018). According to Gwazdauskas (1985), climate change modifies seasonal features regulating the animal's reproductive performances, and this could be linked with differences in feeds available based on season.

Impact of Climate Change on Pasture/Fodder Availability in Africa

Livestock production in Africa is primarily an extensive system wherein the animals are left in the open grasslands for grazing. It is smallholders operated especially the nomadic smallholders who actively engage in this system of rearing. These nomads often move from one location to another along with their herd in search of green pasture (McIntire et al. 1992). However, the pattern is changing in the recent years from pure nomadic pastoralism to agro-pastoralism because of the reduced availability of forages and fodder which could be attributed reduced rainfall due to climate change. This changing pattern of rainfall negatively affects pastoralists movement and migration search for greener pastures (Palmer and Bennett 2013; Hoffman and Vogel 2008).

Therefore, while responding to climate change-induced forage shortages and poor rangeland productivity, livestock keepers now engage in the production of food crops and other arable plants as a coping strategy. This has led to what could be described as a crop-livestock integration commonly found across different parts of Africa. In

Nigeria, for example, smallholders livestock keepers in the Southern and Northern Guinea Savannah agro-ecological zones are now agro-pastoralists instead of being purely pastoral nomads because they are now engaged in the cultivation of common staple crops in addition to the rearing of animals (Baiyeri et al. 2019). This practice seems to benefit principles of sustainable agriculture, but it must be stated that it has an indirect implication on food security because it shifts the focus of livestock keepers away from animal production. In relation to the feed shortage, climate change also harms soil quality, thereby affecting fodder production and yield (Swanepoel and Tshuma 2017).

Since climate change causes reduced pasture and rangeland productivity through shortage of water or precipitation, drought remains one of the greatest limitations affecting animal production within the complex of the rangeland productivity. For instance, it was reported that in South Africa, even years after a ravaging drought, which caused huge loss of pasture on extensive rangeland, close to half and quarter populations of cattle and goats were, respectively, lost in KwaZulu Natal which is a major livestock production hub in the country as results of the cumulative consequences of drought (Vetter et al. 2020).

Similarly, it was reported that in Burkina Faso which is a landlocked country where almost half of its population depends on traditional livestock production, extreme drought once led to the loss of 87% of the livestock population. This loss was attributed to the non-availability of feed resources from both the natural rangelands and non-availability of residues from crop processing (Traore and Owiyo 2013). The implications of this type of climate effect usually extend beyond livestock production systems as it usually affects economic conditions of the people who depend on the animals for food supply and income generation. Furthermore, while describing climate change impact on forage and fodder crop production in Nigeria, Shiawoya and Tsado (2011) identified rainfall instability as a major cause of forage and fodder shortage in Nigeria. Hence, it was concluded that to overcome the problem of climate change effects of rainfall on grassland productivity, deliberate efforts are required for the promotion of better feed resource utilization.

It could be deduced from the foregoing submissions that livestock production in Africa is climate dependent because predominant animal keepers are smallholders who solely depend on natural grassland, but unfortunately, the productivity of the natural grassland has reduced in recent years. Therefore, it becomes necessary to investigate the extent of climate impacts on feed resources used by smallholders with the adaptation of the animals as this could give a better understanding of ways to promote the optimum use of scarce available pasture/fodders.

Climate Change Effects on Livestock Disease Outbreaks in Africa

The protection of livestock against climate vulnerabilities in Africa requires evaluation of climate change impacts on the livestock disease occurrence, distribution, outbreak, and or biology of some disease vectors and related pathogenic organisms (Grace et al. 2015). This becomes highly important because livestock diseases of economic importance are threats to the sector, even in other parts of the world with stable climatic conditions and production management compared with Africa. It was reported that flooding increased prevalence and transmission of pathogenic organisms through faecal-oral routes in Great Britain (Gale et al. 2009).

Meanwhile, flooding is almost one of the most common climate change consequences ravaging African countries, and most of these locations have livestock production as their economic mainstay. A specific example of these scenarios is the Jigawa state of Nigeria, which faces huge economic losses annually due to the inevitable flood occurrence along the flood plain of the Hadejia–Nguru Wetland, where livestock production and other agricultural activities are negatively affected virtually every year (Kaugama and Ahmed 2014).

Climate change has been implicated in the outbreak of some livestock and human diseases of economic importance across Africa. In East Africa, the outbreak of the Rift Valley Fever virus has been linked with increasing temperature, which enhanced transmission of the diseases by many biting insects found in the area (Freeman et al. 2007). Similarly, climate change has been identified as a major factor promoting the existence and occurrence of trypanosomiasis in Africa. Unfortunately, the tsetse fly, which transmits the protozoa, adapts to every changing climate condition in Africa from riverside to savannah grassland (McDermott et al. 2002). This further aggravates the transmission of diseases across regions in Africa.

Furthermore, increasing temperature and relative humidity promotes the occurrence and spread of other diseases allied to livestock production such as helminthiasis caused by roundworms, tapeworms, and trematodes. Apart from these, human diseases related to livestock production are also influenced by climate change results of the change in animal and crop production practices, notable among these diseases include malaria, schistosomiasis, and lymphatic filariasis (Patz et al. 2005).

Ameliorative Strategies for Livestock Production in Response to Climate Change in Africa

Improved Livestock Management Practices for Reducing the Animal's Response to Heat Stress

The management practices for reducing heat stress in animals involve multiple approaches aimed at reducing most metabolic losses that the animal would have incurred in response to climate pressures (Sarangi 2018; Sejian et al. 2018). Therefore, the management strategies that can be adopted for livestock management should primarily aim at looking into the modification of housing and feeding practices of the animals. Designing and redesigning of the animal shelter to provide proper ventilation and enabling cooling facilities for the amelioration of the heat load on the livestock are important (Sejian et al. 2018). The provision of shade is one of the easiest and cheapest methods suitable for decreasing the adverse effect of solar radiation (Sarangi 2018). Studies conducted by Muller et al. (1994a, b) also revealed that animals subjected to heat stress and reared in dry lots containing shade showed lower signs of distress compared to those reared without shade.

In pasture-based animal production systems, the provision of shade for animals on rangeland and increasing proximity to water resources for the alleviation of heat stress are significant strategies that can be employed by local pastoralists in protecting grazing cattle. Farmers can also graze their cattle near dams thereby giving them access to ad libitum drinking water (Katiyatiya et al. 2014). These are because animals having better access to shade have been reported to have higher weight gain, milk production, and reproductive performance (Berger et al. 2004; Al-Dawood 2017). Furthermore, according to Hammadi et al. (2012), it was observed that there was a reduction in rectal temperature and respiration rate in goats that were provided shade during grazing periods.

Feeding and nutritional management could also play a crucial role in livestock physiological modulation in response to the amelioration of heat stress (Al-Dawood 2017). Feeding animals during cooler periods of the day and also at more frequent intervals is advised during the summer. This maintains normal feed intake of the animal and prevents the co-occurrence of peak metabolic and climatic heat load (Mader and Davis 2004). Also, increasing the frequency helps in minimizing the diurnal fluctuation in rumen metabolites and increases feed utilization efficiency in the gut (Soto-Navarro et al. 2000).

Deliberate exposure of livestock to changing climate conditions without hampering the performances of the animals is the main aim of production management (Seré et al. 2007). Although several attempts and strategies of management have been reported from different parts of the world, there is a scarcity of experimented production management strategies reported from Africa. However, there is indigenous knowledge of livestock production management existing among the pastoralists and agro-pastoralists in different parts of Africa. Indigenous knowledge comprises a wide variety of traditional and innovative technologies such as rainwater harvesting systems whereby water harvested during the wet season is kept for animals' consumption during the dry season (Ajani et al. 2013).

Implementation of Strategic Nutritional Intervention

Animals depend on an adequate supply of nutrients for their normal functioning and production performances. Impacts of climate change on animal nutrition have been documented and reported to include poor intake of feed and poor digestibility due to heat stress (Rojas-Downing et al. 2017a). These reports have opened up the potential application of nutritional manipulation for adaptation of livestock to changing climate scenarios. Management of feeding practices has been suggested as a strategy for improving livestock production and protecting climate change. According to earlier reports and recommendations for hot climate areas as is found in Africa, feeding management practices including manipulation of animal dietary composition in such a way that animals derive maximum benefits of nutrients per quantity of feeds provided without generation of obnoxious gases was recommended by Renaudeau et al. (2012). In a different view by Thornton and Herrero (2010b), since livestock production in the tropics, including Africa, is predominantly extensive grazing, it was recommended that livestock producers inculcate agro-forestry as part of their activities and plant species from such practices should be incorporated into the formulation of the animals' diets.

Nutritional strategy for adapting to changing climate scenarios could also be in the form of manipulating the animals' digestive physiology for the utilization of lesser use animal feed resources (Makkar 2016). For example, the ruminant animals could be supplemented with crop residues in addition to their extensive grazing during the dry season when natural rangelands become non-productive (Alphonce 2017). This has been demonstrated from studies and reports across African countries where animals were supplemented with both rich concentrate and poor feed residue to cushion the negative effects of climate impacts on rangeland productivity.

According to reports of Mengesha (2016), supplementation of protein-rich leaves as strategic feed resources in Ethiopian highland sheep led to an improvement of dry matter intakes, body weight changes, and nutrient digestibility. The study concluded that such a protein supplement improved the low quality of the sheep feed during the dry season, which led to improved performance of the sheep. Similarly, it was also reported by Sikiru et al. (2018) that cassava bran and fish processing waste are potential dry season feed resources suitable for growing lambs since it has no negative impact on the health of the animals and it promotes growth and performance of the animals. Furthermore, Sikiru and Makinde (2018) reported that low-cost supplemental feeding practices in lactating goats under extension production have significant impacts on the improvement of pruning weight gain of goat kids. Similarly, it was reported that an indigenous Nigerian goat (Red Sokoto) made appreciable weight gains in the long dry season when supplemented with crop residue-based diets (Malau-Aduli et al. 2003).

Finally, while conducting a survey on livestock production and management in the Republic of South Africa, Mapiye et al. (2009) recommended that nutritional management should be considered as a major factor in the management of pedo-climatic conditions threatening the productivity of indigenous livestock in the country. It could be inferred from the above submissions that nutritional manipulations stand as potential ameliorative strategies for improving animal efficiencies in Africa under climate conditions. This strategy mainly focuses on two thematic areas vis-a-vis strategic supplemental feeding for inhibiting the negative effects of forage and fodder shortages as well as manipulation of functional digestive physiology of the ruminant animals.

Adaptation Strategies for Livestock Production in Response to Climate Change

This present study and similar other previous studies suggested that African countries could be most affected by climate change due to limited skills and equipment for disaster management, inadequate financial resources for climate management, weak institutional capacity, and high dependence on rain-fed agriculture (Bagamba et al. 2012). Therefore, implementing breeding of thermo-tolerant animal breeds, developing early warning systems, improved pasture management, better use of water resources, empowering women towards livestock rearing, and upgrading capacity building programmes for professionals working with farmers and conducting research on livestock production are the main strategies for sustainable livestock production in the African continent. In Fig. 13.3 of this chapter, some of these strategies for sustaining African livestock production in the changing climate scenario were presented.

Breeding Policy for Thermo-Tolerance Breeds of Livestock

Africans host many indigenous species of animals that tolerate heat stress and disease; hence, preparing breeding policies for exploitation of the traits in these indigenous breeds would aid in combating the losses incurred through climate change in livestock production (Dzama 2016). Although these breeds may have lower production rates when compared to the exotic breeds, their ability to thrive in harsh environmental conditions with low feed intake provides a basis for adaptation to climate change. Also, the cost of expenditure on maintenance and treatment of such adapted breeds would be lower than the high producing exotic breeds. Thus, preparing suitable breeding policies that would encourage the use of the indigenous germplasm and also their crosses with exotic animal breed may be considered.

However, it is necessary to understand the agro-climatic condition prevailing in a region before introducing any specific breed into the breeding policy of the area (Zougmoré et al. 2016). Meanwhile, there are a few projects and policies on the use of indigenous livestock of Africa, and most breeding project focus on conservation rather than utilization of the beneficial traits (Katiyatiya et al. 2014; Dzama 2016). Hence, there arises a need to scrutinize the existing policies and amend them as well as encouraging of governmental intervention such as subsidies and tax incentives to indigenous animal breeders and breeding companies to promote the use of indigenous germplasm in animal breeding.

The adoption of suitable breed-species mix with indigenous African livestock breeds known for their thermo-tolerance and disease resistance should be the primary objective in the livestock agriculture policies (Dzama 2016). According to Katiyatiya et al. (2014) who surveyed the farmers of Eastern Cape Province of South Africa, to evaluate their perceptions and knowledge on heat stress and tick resistance in cattle. They observed that most of the farmers in this region preferred rearing indigenous

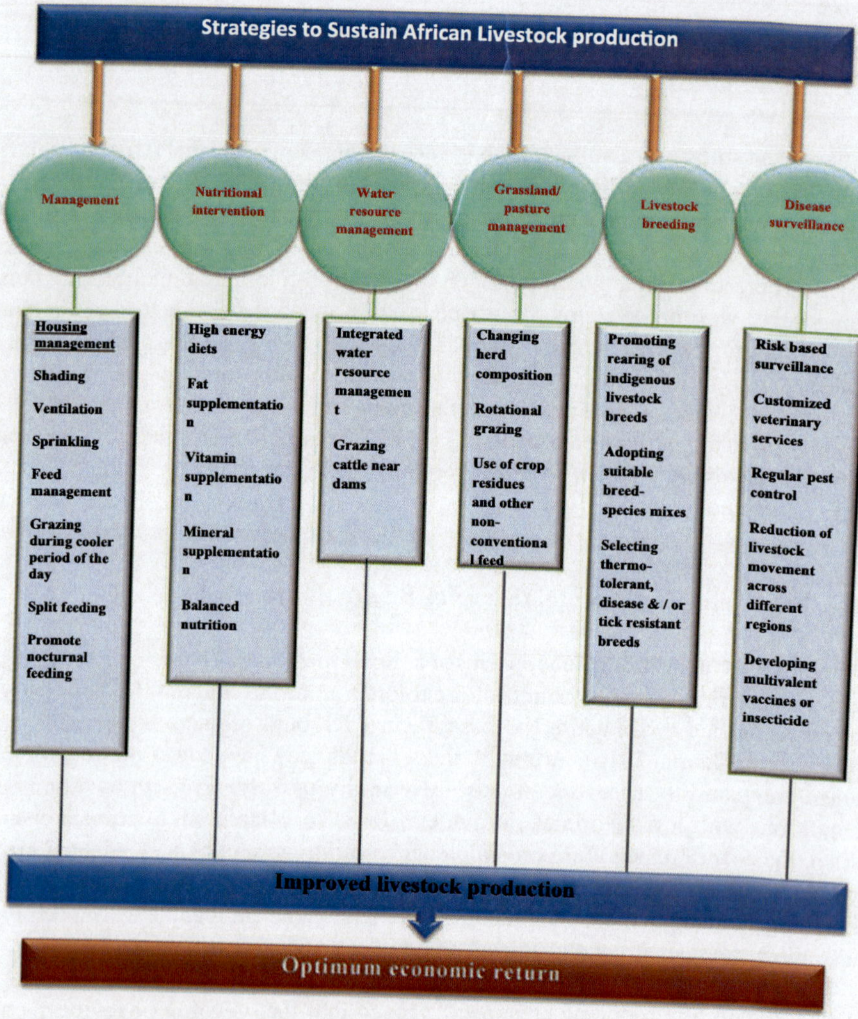

Fig. 13.3 Different strategies to sustain African livestock production in the changing climate scenario

Ngumi cattle and other non-descript cattle, which are again crossed with Ngumi and exotic cattle breeds. This is because the Ngumi cattle are known for their adaptability in low grazing areas and also for possessing good heat and tick tolerance (Musemwa et al. 2010). Despite having comparatively lower productivity than the other exotic cattle breeds, the farmers in this region preferred rearing Ngumi cattle as they provided sustainable farming. Similar trends of farmers shifting towards indigenous Sanga breeds of cattle like Nguni, Tuli, and Tswana were also reported among the commercial and smallholder sector in South Africa (Dzama 2016).

Furthermore, a significant rise in poultry rearing also projects in this region due to the increasing demand for chicken. This sector is again reported to be dominated by small-scale farmers who engaged in the rearing of indigenous flocks that are well adapted to produce well and survive under harsh environmental conditions (Dzama 2016). These indicated that enhancing livestock rearing using indigenous breeds can guarantee the farmers with a sustained production and income (Zougmoré et al. 2016).

In an effort to promote the use of indigenous animals breeding as adaptation to climate change, Onzima et al. (2018) conducted a genome-wide characterization study on six Ugandan goat breeds and identified several thermo-tolerance-related genes. Some of the genes are HOXC12, HOXC13, PPP1R36, HSPA2, DNAJC24, DNAJC13, KPNA4, MTOR, SH2B1, MAPK3, and IGF1, which are involved either directly or indirectly in molecular responses to heat stress highlighting the significance of indigenous goat breeds. Meanwhile, across each country in Africa, there are indigenous breeds which are well adapted to the environmental conditions persisting in Africa such as the N'Dama, Boran, Ogaden, Kenana, Ankole, East African shorthorn Zebu, Fogera breeds of cattle; WAD goats, Kigezi, Mubende, Small East African breeds of the goat; WAD sheep, Horro sheep, Red Maasai, and Nguni sheep (Adedeji 2012; Okourwa 2015; Taye et al. 2017; Onzima et al. 2018). Also, Srikanth et al. (2019) reported some differentially expressed heat stress-related genes in indigenous breeds of Kenyan chicken reared in both the lowland and highland agro-climatic zones of the country, and some of the genes include CCNB2, Crb2, CHST9, SESN1, NR4A3, COMMD4, TTC32, H1F0, ACYP1, and RPS28. The potential biomarkers generated across these studies may serve to develop appropriate future breeding strategies for chicken in Africa.

Implementation of Early Warning Systems for Climate Change Preparedness and Response

Timely forecasting of future weather events in a relevant manner could improve the planning and preparedness against undesirable climatic conditions (Nhamo et al. 2019). Such early warning systems can play a role in disseminating information on droughts, floods, and also on disease outbreaks to farmers. Though Africa has numerous early warning systems (EWSs), limited reports are available on their effectiveness. Lumbroso (2018) stated that EWSs in Uganda and the rest of Africa have to build a strong scientific and technical basis to aid people at risk. Also, sufficient funding should be provided to enable better response and interventions.

Improved Management of Water and Water Resources

Climate change adaptation practices at the farm level in Africa can have significant development outcomes in addition to reducing exposure to weather risks (Rojas-Downing et al. 2017b). It was projected that by the year 2025, about 64% of the world may be living under water stress conditions due to water scarcity and depletion of water sources; Africa is no exception in this projection (Rosegrant et al. 2002). The implication on livestock production is that there will be less water for animal consumption; hence, there is the need for livestock systems requiring less water use and development of livestock adaptation capacity to water shortage as an ameliorative coping strategy. Research investigating the effects of water shortage in livestock production, however, is scanty. Therefore, it is important to consider it as an appropriate mitigation strategy for sustainable livestock production (Thornton et al. 2009).

Water consumption is the most limiting resource for grazing animals, whether during drought or not in the tropics and subtropical countries of Africa because during the dry season, there is always a water shortage coupled with high temperature which puts livestock under stress conditions (Gaughan 2012). Therefore, assessment of animal endurance capacity under water shortage without limiting physiological and productive performance is worthy of exploration to ameliorate climate change effects on livestock production in Africa (Casamassima et al. 2008). Water availability and quality are the main challenges climate change posed on animal production, especially in an extensive livestock system. Therefore, the integrative adaptive approach is a very crucial part of climate change mitigation needs (Sejian et al, 2015). Livestock adaptation to water shortage could serve the entire world, including African countries as a climate resilience strategy for fighting food insecurity (Belay et al. 2017).

Improved Management of Grazing/Pasture Land Management

Changing the pattern of grazing management, such as reducing the number of animals per pasture land area, the use of crop residues and feeding of animals with fodder crops which are usually in abundance compared with the grasses and legumes for animals' consumption are leading adaptation strategies employed by smallholder agro-pastoralists in Africa. In confirmation of these strategies as farmers-driven adaptation approaches, Woldeamlak et al. (2015) stated that smallholder farmers across Africa have made adjustments in livestock production in response to climate change that includes feeding crop residues and feeding fodder crops known as kinchib (*Euphorbia tirucalli*), cactus (Ficus carica) to goats, and camels. In a similar report, Ducrotoy et al. (2017) stated that changing of herd size by reducing the animal population in a particular place for a given time at a prominent livestock production centre in Nigeria significantly contributed to better grazing land management and improved animal performances.

Therefore, since these are farmer-driven strategies, their acceptance by farmers as adaptation strategies could be easier when they become institutionalized by governments and other relevant organizations involved in livestock production in Africa. These forms of adaptive measures could be easily adopted across the continent through the engagement of community-based organizations and farmers' cooperatives in grazing management (Escarcha et al. 2018a, b). This will involve educating and informing their fellow community members about the need to contribute to climate change adaptation through reduction of grazing animals per time within a given area of land to encourage pasture land conservation, increasing forage growth and development as well as the establishment of community fodder banks (Davies et al. 2020).

Continuous Surveillance of Livestock Disease Occurrences and Vector Migration Patterns

Disease surveillance—an information-based activity—involves the collection and analysis of information on disease occurrence. Well-functioning surveillance systems along with timely responses are predicted to reduce the cost of outbreaks by 95% (Grace 2014). The very first step towards this would be to reduce the knowledge gaps on pests and diseases to enrich the data on epidemiological diseases (Bett et al. 2017). This would also help in developing risk maps to assess the climatic influence on such diseases. Risk-based (targeted) surveillance must be encouraged which focuses on diseases prevalent across different sectors, subpopulations, and areas (Grace 2014).

Participatory disease surveillance involving local communities has helped early detection and reporting of diseases such as rinderpest and avian influenza in Africa (Grace et al. 2015). Such surveillances have wider coverage and are cost effective when compared to the traditional surveillance measures which follow the assumption that the probability of disease occurrence is constant throughout the reference population (Grace et al. 2015). Having such a relevant and informative data will then aid in planning customized veterinary services, like vaccination and deworming schedules, for every region. Pest protection is also an important disease surveillance measure as vectors transmit infectious livestock diseases (Bett et al. 2017). Practising regular pest control measures will help to keep check on the vectors.

The movement of livestock across different regions and national boundaries, for the trade of nomadic pastoralism, can further aggravate the spread of disease, which led to a coordinated approach and the establishment of the Intergovernmental Authority on Development, in Africa. This was aimed at harmonizing the surveillance and control of livestock diseases by developing standard methods and procedures (Bett et al. 2017). Technological advancements aiming towards quicker detection, response, and effective management are being practised. The development of safer vaccines for Rift Valley Fever by Warimwe et al. (2016) is one such example. As some of the climate-sensitive infectious diseases co-occur in the same ecology, future researches are targeted towards tackling multiple diseases together. Developing multivalent vaccines or insecticides to control multiple diseases and vectors would be cost effective and useful (Bett et al. 2017).

Implementation of Livestock Production with an Emphasis on Women Empowerment

Women are more capable of managing ecosystem services and food security than men; they possess immense leadership qualities and play a potent role in handling various issues like energy consumption, deforestation, population growth, economic growth, and management, developing scientific research and technologies, contributing towards policy making, and many more. Thus, it should be of prime importance to involve them actively in livestock rearing (Sejian et al. 2015). Several activities associated with livestock production such as fodder collection, feeding, watering, and health care, management, milking, and household-level processing, value addition, and marketing are performed by women (Patel et al. 2016). Therefore, implementation of livestock production systems with an emphasis on women empowerment can be an adaptation strategy to climate change in Africa.

Implementation of Livestock Extension, Education, and Capacity Building Programmes

Farmers across Africa adopt various measures to combat the deleterious effects of climate change within the scope of their agricultural knowledge. However, they are ignorant of the other possible adaptive strategies available in other parts of the world such as India, where a capacity building programme referred to as the *Krishi Vigyan Kendra* (KVK) is bringing out entrepreneurship prowess in dairy keepers following training on knowledge gained via capacity building cum livestock extension programme for dairy development (Kumar et al. 2011). Adaptation strategies would be successful only when the knowledge and inputs are transferred down to the ultimate target groups who are the poor farmers.

Developing a suitable capacity building programme (CBP) to enhance the knowledge and understanding of livestock owners and local, commercial market stakeholders towards the impact of climate change on livestock production would increase their awareness of the global changes in climate and its effects on livestock production. For a successful implementation of adaptation strategies, CBP should focus on the basic understanding of climate change, its impacts on biodiversity and ecosystems, and its role towards conservation and sustenance of eco-services and also strategies for assessing vulnerability and adaptation (Sejian et al. 2015). This will stimulate active participation by all sectors including governmental, non-governmental, private sector, and volunteers to support livestock farmers and besides support timely upgrading of technologies, policies, and frameworks which has to be made to ensure effective use of feasible adaptation strategies.

Conclusion and Recommendations

Nigeria can be regarded as the centre point of animal products consumption and production in West Africa. The population of the country also keeps growing while serving as the flourishing market for neighbouring Sahel countries that raise livestock in the region. The country also drives beef consumption in the West Africa region because 50% of beef consumption in the ECOWAS region are from Nigeria (Bénard et al. 2010). Unfortunately, livestock production in Nigeria could be at the risk of climate change because it is largely pasture-based and extensive grazing management (Ducrotoy et al. 2018).

Meanwhile, available data, records, and submissions on livestock production in Africa and its concerns with climate change examined in this present study considered West Africa as the driver of the future demand for livestock products as well as the most vulnerable region to climate change consequences in Africa. Hence, since Nigeria is the nerve cord of livestock production in West Africa, a working hypothesis for action as follows proposed for livestock production under climate change for the promotion of food security and human well-being in Africa, West Africa, and Nigeria is as follows:

> In Africa, increasing ambient temperature, relative humidity, damaging wind speed, extreme thermal radiation and reducing precipitation are environmental stressors that could arise from climate change and they could have negative impacts on livestock production, which could serve as potential hindrances affecting animals' health, growth, productivity, and reproduction; and these can increase the risks of poverty and food insecurity in the continent.

Therefore, this study suggests the need for scientific and institutional experimental testing of the hypothesis stated above for generating empirical data that can be used for developing models that can provide resilience and adaptation measures as critical need for sustainable livestock production in Africa for promotion of food security and human well-being under changing climate conditions.

Climate changes have severe impacts on the livestock production system in many parts of the tropical and subtropical countries of Africa; but despite the importance of the livestock production sector to African farmers and the magnitude of climate change consequences, they may likely face in future, little or no hope exists to overcome these challenges. This is because the intersections between climate change and livestock production are a relatively neglected area of research in Africa.

There is need for carrying out empirical data generated from experimental studies focusing on biological, productive, and other climate change effects on animal production and performances. This is because the trends of climate change variables (especially temperature increase) generated in this study indicated that among the various subregions of Africa, the West African subregion could be highly vulnerable to the devastating climate consequences concerning livestock production.

Furthermore, available reports via socio-economic assessment revealed that climate change has led to serious repercussions affecting animal production and reproductive efficiency as well as disease outbreaks. These climate change consequences indirectly affected animal production in Africa as a result of accompanying reduced fodder and water resources. Therefore, several solutions, as well as adaptation strategies proposed in this chapter if implemented, could sustain livestock production in the face of changing climatic conditions in the Africa.

Africa is undoubtedly in need of researchers working in the areas of livestock production and climate change to sustain the continents' livestock production and promotion of global food security, considering the cofounding effects of Africa population growth and migration on other parts of the world. Future research efforts are needed to develop cost-effective, easy-to-use technologies catering to the needs of African livestock farmers to mitigate and adapt to climate change realities as it affects their production systems. Apart from scientific research investigations, there is a need for institutional policies that can enhance farmers' adaptation to climate change, and this includes but not limited to refinements of existing animal breeding policies for the development of more climate-resilient livestock breeds specific to different agro-ecological zones of Africa. Finally, special initiatives are required for the development of more adaptation and coping strategies in future through international collaborations of governmental, non-governmental, and developmental agencies for conservation of water resources and pasture land management to ensure optimum livestock production in Africa.

References

Adedeji TA (2012) Effect of some qualitative traits and non-genetic factors on heat tolerance attributes of extensively reared West African Dwarf (WAD) goats. Int J Appl Agric Apicul Res 8:68–81

African Union AU (2014) Malabo declaration on accelerated agricultural growth and transformation for shared prosperity and improved livelihoods

African Union Interafrican Bureau for Animal Resources AU-IBAR (2015) Livestock Development Strategy for Africa (LiDeSA), pp 2015–2035

Ajani EN, Mgbenka RN, Okeke MN (2013) Use of indigenous knowledge as a strategy for climate change adaptation among farmers in sub-Saharan Africa: policy implications. Asian J Agric Ext Econ Sociol 23–40

Akokpari J (2007) The political economy of human insecurity in Sub-Saharan Africa. Institute of Developing Economies

Al-Dawood A (2017) Towards heat stress management in small ruminants–a review. Ann Anim Sci 17(1):59–88

Alphonce HI (2017) Assessment of the extent of utilizing crop residues as ruminant feed in crop–livestock farming systems in Babati district, Tanzania. Doctoral dissertation, Sokoine University of Agriculture

Bagamba F, Bashaasha B, Claessens L, Antle J (2012) Assessing climate change impacts and adaptation strategies for smallholder agricultural systems in Uganda. Afr Crop Sci J 20(2):303–316

Baiyeri PK, Foleng HN, Machebe NS, Nwobodo CE (2019) Crop-livestock interaction for sustainable agriculture. In: Innovations in sustainable agriculture. Springer, Cham, pp 557–582

Belay A, Recha J, Woldeamanuel T, Morton JF (2017). Smallholder farmers adaptation to climate change and determinants of their adaptation in the central rift valley of Ethiopia. In: Proceedings of the impact of El Niño on biodiversity, agriculture, and food security, p 63. (23 Feb 2017)

Bénard C, Bonnet B, Guibert B (2010) Demand for farm animal products in Nigeria: an opportunity for Sahel countries. Grain De Sel 51(1):14–15

Berger Y, Billon P, Bocquier F, Caja G, Cannas A, Mc KB, Marnet PG, Thomas D (2004) Principles of sheep dairying in North America. Cooperative Extension Publishing, A3767. University of Wisconsin-Madison, USA, p 156

Bett B, Kiunga P, Gachohi J, Sindato C, Mbotha D, Robinson T, Lindahl J, Grace D (2017) Effects of climate change on the occurrence and distribution of livestock diseases. Prev Vet Med 137:119–129

Bhakat M, Mohanty TK, Gupta AK, Abdullah M (2014) Effect of season on semen quality of crossbred (Karan Fries) bulls. Adv Anim Vet Sci 2(11):632–637
Bilby TR, Baumgard LH, Collier RJ, Zimbelman RB, Rhoads ML (2008) Heat stress effects on fertility: consequences and possible solutions. In: The proceedings of the 2008 South Western nutritional conference 2008
Borgomeo E, Vadheim B, Woldeyes FB, Alamirew T, Tamru S, Charles KJ, Kebede S, Walker O (2018) The distributional and multi-sectoral impacts of rainfall shocks: evidence from computable general equilibrium modelling for the Awash Basin, Ethiopia. Ecol Econ 146:621–632
Casamassima D, Pizzo R, Palazzo M, D'alessandro AG, Martemucci G (2008) Effect of water restriction on productive performance and blood parameters in comisana sheep reared under intensive condition. Small Rumin Res 78(1–3):169–175
Davies JE, Spear D, Ziervogel G, Hegga S, Ndapewa Angula M, Kunamwene I, Togarepi C (2020) Avenues of understanding: mapping the intersecting barriers to adaptation in Namibia. Climate Dev 12(3):268–280
Ducrotoy MJ, Revie CW, Shaw AP, Musa UB, Bertu WJ, Gusi AM, Ocholi RA, Majekodunmi AO, Welburn SC (2017) Wealth, household heterogeneity, and livelihood diversification of Fulani pastoralists in the Kachia Grazing Reserve, Northern Nigeria, during a period of social transition. PloS one 12(3)
Ducrotoy MJ, Majekodunmi AO, Shaw AP, Bagulo H, Bertu WJ, Gusi AM, Ocholi R, Welburn SC (2018) Patterns of passage into protected areas: drivers and outcomes of Fulani immigration, settlement and integration into the Kachia Grazing Reserve, Northwest Nigeria. Pastoralism 8(1):1–16
Dzama K (2016) Is the livestock sector in southern Africa prepared for climate change? South African Institute of International Affairs. Policy Briefing, p 153
Ekesbo I, Gunnarsson S (2018) Farm animal behaviour: characteristics for assessment of health and welfare. CABI
Escarcha JF, Lassa JA, Zander KK (2018a) Livestock under climate change: a systematic review of impacts and adaptation. Climate 6(3):54
Escarcha JF, Lassa JA, Palacpac EP, Zander KK (2018b) Understanding climate change impacts on water buffalo production through farmers' perceptions. Clim Risk Manag 20:50–63
FAOSTAT (2019). Food agriculture and organization (FAOSTAT). http://www.fao.org/faostat/en/#data/QC
Freeman A, Herrero M, Mwai O, Thornton P (2007) Vulnerability, climate change and livestock-Research opportunities and challenges for poverty alleviation. An open Access Journal published by ICRISAT, vol 4, no 1
Furstenburg D, Scholtz MM (2008) Global climate change and animal production in Southern Africa: a short review. Poster presented at 10th World Congress on Animal Production, Cape Town, South Africa
Gale P, Drew T, Phipps LP, David G, Wooldridge M (2009) The effect of climate change on the occurrence and prevalence of livestock diseases in Great Britain: a review. J Appl Microbiol 106(5):1409–1423
Gaughan JB (2012) Basic principles involved in adaption of livestock to climate change. In: Sejian V, Naqvi SMK, Ezeji T, Lakritz J, Lal R (eds) Environmental stress and amelioration in livestock production. Springer-Verlag GMbH Publisher, Germany, Berlin, Heidelberg, pp 245–261
Grace D, Bett B, Lindahl J, Robinson T (2015) Climate and livestock disease: assessing the vulnerability of agricultural systems to livestock pests under climate change scenarios. CCAFS Working Paper no. 116. Copenhagen, Denmark. CGIAR Research Program on Climate Change, Agriculture and Food Security (CCAFS). www.ccafs.cgiar.org
Grace D (2014) The business case for one health. Onderstepoort J Vet Sci 81:2
Gwazdauskas FC (1985) Effects of climate on reproduction in cattle. J Dairy Sci 68(6):1568–1578
Habeeb AA, Gad AE, Atta AM (2018) Temperature-humidity indices as indicators to heat stress of climatic conditions with relation to production and reproduction of farm animals. Int J Biotech Recent Adv 1(2):35–50

Hammadi M, Fehem A, Harrabi H, Ayeb N, Khorchani T, Salama AAK, Casals R, Such X, Caja G (2012) Shading effects on respiratory rate and rectal temperature in Tunisian local goat kids during summer season. In: Proceedings of XI international conference on goats, Gran Canaria, Spain, 23–27 Sept 2012, p 127

Hansen PJ (2014) Current and future assisted reproductive technologies for mammalian farm animals. In: Current and future reproductive technologies and world food production. Springer, New York, NY, pp 1–22

Hendrix CS (2017) The streetlight effect in climate change research on Africa. Glob Environ Chang 43:137–147

Herrero M, Havlík P, McIntire J, Palazzo A, Valin H (2014) African Livestock Futures: Realizing the potential of livestock for food security, poverty reduction and the environment in Sub-Saharan Africa. Office of the Special Representative of the UN Secretary General for Food Security and Nutrition and the United Nations System Influenza Coordination. (UNSIC) Geneva, Switzerland

Hetherington JB, Wiethoelter AK, Negin J, Mor SM (2017) Livestock ownership, animal source foods and child nutritional outcomes in seven rural village clusters in Sub-Saharan Africa. Agric Food Secur 6(1):9

Hoffman T, Vogel C (2008) Climate change impacts on African rangelands. Rangelands 30(3):12–17

Hoffmann I (2010) Climate change and the characterization, breeding and conservation of animal genetic resources. Anim Genet 41:32–46

Hopkins A, Del Prado A (2007) Implications of climate change for grassland in Europe: impacts, adaptations and mitigation options: a review. Grass Forage Sci 62(2):118–126

IPCC (2014b) Summary for policymakers. In: Climate change 2014: impacts, adaptation, and vulnerability. Part A: global and sectoral aspects. Contribution of working group II to the fifth assessment report of the intergovernmental panel on climate change. In: Field CB, Barros VR, Dokken DJ, Mach KJ, Mastrandrea MD, Bilir TE, Chatterjee M, Ebi KL, Estrada YO, Genova RC, Girma B, Kissel ES, Levy AN, MacCracken S, Mastrandrea PR, White LL (eds). Cambridge University Press, Cambridge, United Kingdom and New York, NY, USA, pp 1–32

IPCC (2014a) Mitigation of climate change. In: Edenhofer O, Pichs-Madruga R, Sokona Y, Farahani E, Kadner S, Seyboth K, Adler A, Baum I, Brunner S, Eickemeier P, Kriemann B, Savolainen J, Schlömer S, von Stechow C, Zwickel T, Minx JC (eds.) Climate change 2014a: mitigation of climate change. Contribution of working group III to the fifth assessment report of the intergovernmental panel on climate change. Cambridge University Press, Cambridge, United Kingdom and New York, NY, USA

Katiyatiya CLF, Muchenje V, Mushunje A (2014) Farmers' perceptions and knowledge of cattle adaptation to heat stress and tick resistance in the Eastern Cape, South Africa. Asian Australas J Anim Sci 27(11):1663–1670

Kaugama HH, Ahmed BA (2014) Prospect and challenges of farming along the Hadejia-Nguru Wetland in Jigawa State Nigeria. Int J Acad Res Econ Manag Sci 3(6):43

Kebede D (2016) Impact of climate change on livestock productive and reproductive performance. Livestock Res Rural Dev 28:227

Kekana TW, Nherera-Chokuda FV, Muya MC, Manyama KM, Lehloenya KC (2018) Milk production and blood metabolites of dairy cattle as influenced by thermal-humidity index. Trop Anim Health Prod 50:921–924

Kumar S, Kumar B, Hindustani S, Sankhala G (2011) Knowledge level of dairy farmers in operational area of Krishi Vigyan Kendra about improved dairy farming practices. Res J Agric Sci 2(1):122–124

Kurukulasuriya P, Mendelsohn R, Hassan R, Benhin J, Deressa T, Diop M, Dinar A et al (2006) Will African agriculture survive climate change?. World Bank Econ Rev 20(3):367–388

Lumbroso D (2018) How can policy makers in sub-Saharan Africa make early warning systems more effective? The case of Uganda. Int J Disaster Risk Reduct 27:530–540

Mader TL, Davis MS (2004) Effect of management strategies on reducing heat stress of feedlot cattle: feed and water intake. J Anim Sci 82:3077–3087

Makkar HP (2016) Smart livestock feeding strategies for harvesting triple gain–the desired outcomes in planet, people and profit dimensions: a developing country perspective. Anim Prod Sci 56(3):519–534

Malau-Aduli BS, Eduvie LO, Lakpini CAM, Malau-Aduli AEO (2003) Variations in liveweight gains, milk yield and composition of Red Sokoto goats fed crop-residue-based supplements in the subhumid zone of Nigeria. Livest Prod Sci 83(1):63–71

Mapiye C, Chimonyo M, Dzama K, Raats JG, Mapekula M (2009) Opportunities for improving Nguni cattle production in the smallholder farming systems of South Africa. Livest Sci 124(1–3):196–204

Mapiye O, Chikwanha OC, Makombe G, Dzama K, Mapiye C (2020) Livelihood, food and nutrition security in Southern Africa: What role do indigenous cattle genetic resources play? Diversity 12(2):74

McDermott JJ, Kristjanson PM, Kruska RL, Reid RS, Robinson TP, Coleman PG, Jones PG, Thornton PK (2002) Effects of climate, human population and socio-economic changes on tsetse-transmitted trypanosomiasis to 2050. In: The African trypanosomes. Springer, Boston, MA, pp 25–38

McIntire J, Bourzat D, Prabhu P (1992) Crop-livestock interaction in sub-Saharan Africa. World Bank, Washington, DC

Meltzer MI (1995) Livestock in Africa: the economics of ownership and production, and the potential for improvement. Agric Hum Values 12(2):4–18

Mengesha M (2016) Effect of graded levels of Tagasaste (Chamaecytisus palmensis) leaves supplementation on performance of yearling Menz sheep in Ethiopian Highlands. MSc thesis in Animal and Range Sciences (Specialization: Animal Nutrition). Hawassa University, Hawassa, Ethiopia

Muller CJC, Botha JA, Coetzer WA, Smith WW (1994a) Effect of shade on various parameters of Friesian cows in a Mediterranean climate in South Africa. 2. Physiological responses. South Afric J Anim Sci 24:56–60

Muller CJC, Botha JA, Smith WW (1994b) Effect of shade on various parameters of Friesian cows in a Mediterranean climate in South Africa. 2. Physiological responses. South Afric J Anim Sci 24:61–66

Musemwa L, Mushunje A, Chimonyo M, Mapiye C (2010) Low cattle market off-take rates in communal production systems of South Africa: causes and mitigation strategies. J Sustain Dev Afr 12:209–226

Naqvi SMK, Sejian V (2011) Global climate change: role of livestock. Asian J Agric Sci 3(1):19–25

Nhamo L, Mabhaudhi T, Modi AT (2019) Preparedness or repeated short-term relief aid? Build Drought Resil through Early Warn South Afr 45(1):75–85

Ogenga JO, Mugalavai EM, Nyandiko NO (2018) Impact of rainfall variability on food production under rainfed agriculture in Homa Bay County, Kenya. Int J Res Sci Publ 861

Okouruwa MI (2015) Effect of coat characteristics on physiological traits and heat tolerance of West African Dwarf sheep in Southern Nigeria. Open J Anim Sci 5:351–357

Onzima RB, Upadhyay MR, Doekes HP, Brito LF, Bosse M, Kanis E, Groenen MAM, Crooijmans RPMA (2018) Genome-wide characterization of selection signatures and runs of homozygosity in Ugandan goat breeds. Front Genet 9:1–13. https://doi.org/10.3389/fgene.2018.00318

Otte J, Costales A, Dijkman J, Pica-Ciamarra U, Robinson T, Ahuja V, Ly C, Roland-Holst D (2012) Livestock sector development for poverty reduction: an economic and policy perspective–livestock's many virtues. FAO (Food and Agriculture Organization of the United Nations), Rome, Italy, pp 161

Palmer AR, Bennett JE (2013) Degradation of communal rangelands in South Africa: towards an improved understanding to inform policy. Afr J Range Forage Sci 30(1–2):57–63

Patel SJ, Patel MD, Patel JH, Patel AS, Gelani RN (2016) Role of women gender in livestock sector: a review. J Livest Sci 7:92–96

Patz JA, Confalonieri UE, Amerasinghe FP, Chua KB, Daszak P, Hyatt AD, Molyneux D, Thomson M, Yameogo L, Lazaro MM, Vasconcelos P (2005) Human health: ecosystem regulation of infectious diseases. In: Ecosystems and human well-being: current state and trends: findings of

the condition and trends working group of the millennium ecosystem assessment, vol 14, pp 391-415. (Dec 2005)
Pica-Ciamarra U, Baker D, Morgan N, Ly C, Nouala S (2013) Investing in African livestock: business opportunities in 2030–2050. Livestock data innovation in Africa Project. A joint initiative of the World Bank, FAO, ILRI, AU-IBAR with support from the Bill and Melinda Gates Foundation, Seattle, WA, p 1–14
Pollock DSG, Green RC, Nguyen T (1999) Handbook of time series analysis, signal processing and dynamics. Elsevier, New York, USA, p 848. ISBN-13: 9780080507873
Raheem D, Carrascosa C, Oluwole OB, Nieuwland M, Saraiva A, Millán R, Raposo A (2019) Traditional consumption of and rearing edible insects in Africa, Asia and Europe. Crit Rev Food Sci Nutr 59(14):2169–2188
Renaudeau D, Collin A, Yahav S, De Basilio V, Gourdine JL, Collier RJ (2012) Adaptation to hot climate and strategies to alleviate heat stress in livestock production. Animal 6(5):707–728
Roessler R, Mpouam S, Muchemwa T, Schlecht E (2016) Emerging development pathways of urban livestock production in rapidly growing West Africa cities. Sustainability 8(11):1199
Rojas-Downing MM, Nejadhashemi AP, Harrigan T, Woznicki SA (2017a) Climate change and livestock: Impacts, adaptation, and mitigation. Clim Risk Manag 16:145–63. (1 Jan 2017a)
Rojas-Downing MM, Nejadhashemi AP, Harrigan T, Woznicki SA (2017b) Climate change and livestock: impacts, adaptation, and mitigation. Clim Risk Manag 16:145–163
Rosegrant MW, Cai X, Cline SA (2002) World water and food to 2025: dealing with scarcity. Int Food Policy Res Inst
Ruel MT, Quisumbing AR, Balagamwala M (2017) Nutrition-sensitive agriculture: What have we learned and where do we go from here? Int Food Policy Res Inst 1681
Rust JM, Rust T (2013) Climate change and livestock production: a review with emphasis on Africa. S Afr J Anim Sci 43(3):256–267
Ryan SJ, McNally A, Johnson LR, Mordecai EA, Ben-Horin T, Paaijmans K, Lafferty KD (2015) Mapping physiological suitability limits for malaria in Africa under climate change. Vector-Borne Zoonotic Dis 15(12):718–725
Salem BM, Bouraoui R (2009) Heat stress in Tunisia, effects on dairy cows and potential means of alleviating it. S Afr J Anim Sci. 39:256–259
Sarangi S (2018) Adaptability of goats to heat stress: a review. Pharma Innov J 7(4):1114–1126
Scholtz M, Pyoos G, Seshoka M, Theunissen A (2018) The effect of climate on the pre- and post-weaning performance of different beef cattle genotypes. J Anim Sci 96:521. (suppl Supplement 3; Champaign)
Scholtz MM, Theunissen A (2010) The use of indigenous cattle in terminal cross-breeding to improve beef cattle production in Sub-Saharan Africa. Anim Genet Res 46:33–39
Sejian V, Samal L, Haque N, Bagath M, Hyder I, Maurya VP, Bhatta R, Ravindra JP, Prasad CS, Lal R (2015) Overview on adaptation, mitigation and amelioration strategies to improve livestock production under the changing climatic scenario. In: Sejian V, Gaughan J, Baumgard L, Prasad C (eds) Climate change impact on livestock: adaptation and mitigation. Springer, New Delhi
Sejian V, Bhatta R, Gaughan JB, Dunshea FR, Lacetera N (2018) Review: adaptation of animals to heat stress. Animal 12:S2:s431–s444
Seo SN, Mendelsohn R (2008a) Measuring impacts and adaptations to climate change: a structural Ricardian model of African livestock management. Agric Econ 38(2):151–65
Seo SN, Mendelsohn R (2008b) Animal husbandry in Africa: climate change impacts and adaptations. Afr J Agric Resour Econ 2(311-2016-5520):65–82
Seo SN, Mendelsohn R (2006) Climate change adaptation in Africa: a microeconomic analysis of livestock choice. CEEPA Discussion Paper No. 19, Centre for Environmental Economics and Policy in Africa, University of Pretoria, South Africa
Seré C, Van der Zijpp AJ, Persley G, Rege E (2007) Dynamics of livestock production systems, drivers of change and prospects for animal genetic resources. In: Proceedings of the first international technical conference on animal genetic resources for food and agriculture, 3–7 Sept 2007, Interlaken, Switzerland, pp 4–24

Shiawoya EL, Tsado DN (2011) Forage and fodder crop production in Nigeria: problems and prospects. World J Life Sci Med Res 1(4):88–93

Siddig K, Stepanyan D, Wiebelt M, Grethe H, Zhu T (2020) Climate change and agriculture in the Sudan: impact pathways beyond changes in mean rainfall and temperature. Ecol Econ 169:106566

Sikiru AB (2016) Assessment of feed resources utilization for livestock production by Agro Pastoralists in Tafa Local Government Area of Nigeria. J Rangel Sci 6(1):43–52. (1 Jan 2016)

Sikiru AB, Makinde OJ (2018) Effect of low-cost supplemental feeding practices on pre-weaning weight gain of goat kids in Tafa LGA Nigeria. Int J Agric Res Sustain Food Suffic (IJARSFS) 5(1):237–247. (16 Feb 2018)

Sikiru AB, Yousuf MB, Ademola SG (2018) Cassava Bran-Fish processing waste as dry season feed resources for sheep in Nigeria Southern Guinea Savannah. J Rangel Sci 8(1):11–22

Sikiru AB (2020) Livestock production at the nexus of resources competition and ethnoreligious cynicism in Nigeria-implicative analysis on food security. J Rangel Sci 10(3):267–281

Sikiru AB, Alemede IC, Arangasamy A, Egena SSA, Ijaiya AT (2020) Temperature elevation effects on livestock production-a study and forecast of Bengaluru climate data. Ecologia 10:1–8

Soto-Navarro SA, Krehbie CR, Duff GC, Galyean ML, Brown MS, Steiner RL (2000) Influence of feed intake fluctuation and frequency of feeding on nutrient digestion, digesta kinetics, and ruminal fermentation profiles in limit-fed steers. J Anim Sci 78:2215–2222

Srikanth K, Kumar H, Park W, Byun M, Lim D, Kemp S, tePas MFW, Kim J-M, Park J-E (2019) Cardiac and skeletal muscle transcriptome response to heat stress in Kenyan chicken ecotypes adapted to low and high altitudes reveal differences in thermal tolerance and stress response. Front Genet 10:993

Swanepoel PA, Tshuma F (2017) Soil quality effects on regeneration of annual Medicago pastures in the Swartland of South Africa. Afr J Range Forage Sci 34(4):201–208

Taye M, Lee W, Caetano-Anolles K, Dessie T, Hanotte O, Mwai OA, Kemp S, Cho S, Oh SJ, Lee H, Kim H (2017) Whole genome detection of signature of positive selection in African cattle reveals selection for thermotolerance. Anim Sci J 88:1889–1901

Thornton PK, van de Steeg J, Notenbaert A, Herrero M (2009) The impacts of climate change on livestock and livestock systems in developing countries: A review of what we know and what we need to know. Agric Syst 101(3):113–127

Thornton PK, Herrero M (2010a) Potential for reduced methane and carbon dioxide emissions from livestock and pasture management in the tropics. Proc Natl Acad Sci 107(46):19667–19672

Thornton PK, Herrero M (2010b) The inter-linkages between rapid growth in livestock production, climate change, and the impacts on water resources, land use, and deforestation. The World Bank

Traore S, Owiyo T (2013) Dirty droughts causing loss and damage in Northern Burkina Faso. Int J Global Warm 5(4):498–513

Vetter S, Goodall VL, Alcock R (2020) Effect of drought on communal livestock farmers in KwaZulu-Natal, South Africa. Afr J Range Forage Sci 37(1):93–106

Warimwe GM, Gesharisha J, Carr BV, Otieno S, Otingah K, Wright D, Charleston B, Okoth E, Elena L-G, Lorenzo G, Ayman E-B, Alharbi NK, Al-Dubaib MA, Brun A, Gilbert SC, Nene V, Hill AVS (2016) Chimpanzee adenovirus vaccine provides multispecies protection against Rift Valley fever. Sci Rep 6:20617. https://doi.org/10.1038/srep20617

Woldeamlak B, Radeny M, Mungai C (2015) Agricultural adaptation and institutional responses to climate change vulnerability in Ethiopia. CCAFS Working Paper no. 106. CGIAR Research Program on Climate Change, Agriculture and Food Security (CCAFS). Copenhagen, Denmark. www.ccafs.cgiar.org

Yaméogo ND, Nabassaga T, Ncube M (2014) Diversification and sophistication of livestock products: the case of African countries. Food Policy 49(2):398–407

Zougmoré R, Partey S, Ouédraogo M, Omitoyin B, Thomas T, Ayantunde A, Ericksen P, Said M, Jalloh A (2016) Toward climate-smart agriculture in West Africa: a review of climate change impacts, adaptation strategies and policy developments for the livestock, fishery and crop production sectors. Agric Food Secur 5(26). https://doi.org/10.1186/s40066-016-0075-3

Chapter 14
Mapping and Assessing the Seasonal Dynamics of Surface Urban Heat Intensity Using LandSAT-8 OLI/TIRS Images

Oluibukun Gbenga Ajayi, Tosin Sandra Kolade, and Mahmud Baba

Abstract Increased heat intensity in urban climate has serious implications on human health, contributing to urban liveability and vitality. As a way of mitigating the effect of excessive heat temperature in the urban area, it is imperative to examine the level of surface temperature in urban areas over time so that the urban heat intensity and its attendant consequences can be put into consideration when undertaking sustainable urban planning. This study examined the spatiotemporal dynamics of surface urban heat intensity in Bosso Local Government Area of Niger State using remotely sensed images. Landsat-8 OLI/TIRS images of the year 2015, 2017, 2019, and 2021 for both dry and wet seasons were used to determine the study area's Normalized Difference Vegetation Index (NDVI), surface emissivity, land surface temperature (LST), and Normalized Difference Built-up Index (NDBI), using ArcGIS 10.8 software. The result showed that a rise in built-up density, surface emissivity, and a decrease in vegetation density yields an increase in LST, while vegetation density proved to be of little effect in dry season when compared to the rainy season because most vegetation experiences draught at this time of the year. The result also showed that LST is higher in rainy season than it was in dry season because the wind, which decreases the effect of LST, is weak at this season of the year. The least value for surface emissivity in dry season was recorded to be 0.98605 while that of rainy is 0.98698, which implies that the emissivity of materials in the study area was observed to be higher in the rainy season than dry season. Furthermore, the result affirmed that a rise in urbanization gives rise to LST, likewise an increase in vegetation density of an area will lead to a decrease in the area's urban heat intensity.

O. G. Ajayi (✉)
Department of Land and Spatial Sciences, Namibia University of Science and Technology, Windhoek, Namibia
e-mail: oajayi@nust.na; gbenga.ajayi@futminna.edu.ng

O. G. Ajayi · T. S. Kolade · M. Baba
Department of Surveying and Geoinformatics, Federal University of Technology, Minna, Nigeria
e-mail: tosin.kolade4@gmail.com

M. Baba
e-mail: Mahmud.baba@futminna.edu.ng

J. C. Egbueri et al. (eds.), *Climate Change Impacts on Nigeria*, Springer Climate,
https://doi.org/10.1007/978-3-031-21007-5_14

The results also proved that wet periods can be hotter than dry periods of the year due to the presence of weak winds.

Keywords Normalized difference vegetation index · Land surface temperature · Surface emissivity · Normalized difference built-up index

Introduction

The atmospheric systems and energy balance of the earth are gradually being altered as a result of chaotic urbanization, which has a direct impact on human thermal discomfort. Such issues are exacerbated in the cities, since the urban environment is the object of man's most arbitrary landscape-modifying actions (Gomes and Caracristi 2021). In the light of the challenges of global warming and its characterizing dynamics related to earth's surface alterations, such as agricultural expansion, desertification, urban development, and so on, examining the surface urban heat intensity is critical. In this regard, a lot of effort has gone into determining land surface temperature using remote sensing data (Garouani et al. 2021). The removal of vegetation within urban areas, changes in urban thermal and physical properties of construction materials, building, morphology, surface roughness, urbanization and anthropogenic heat sources, all modifies, alters, or affects local energy and leads to increase in atmospheric temperature in urban areas compared to their surroundings (Ayanlade et al. 2021). Adequate and accurate information about the status of the land surface temperature (LST) of specific areas of interest is required for successful geo-environmental management, which involves the monitoring and modelling of the environment (Agbor and Makinde 2018). There are many natural and anthropogenic factors responsible for the increase or decrease in LST, while the degree of LST is seasonal and location dependent. Climate change and urbanization have been reported as one of the critically significant factors responsible for the change in land use and LST (Argueso et al. 2015; Elhadi et al. 2020).

The heat intensity effect of solar radiation varies significantly across urban and rural areas. It has been observed to be higher in urban or metropolitan areas than in rural areas. The term urban heat island (UHI) is used to describe this phenomenon, which is primarily impacted by the amount of plant and water pervious surfaces present in an urbanized area. Because water pervious surfaces and vegetation have been replaced by impervious surfaces in urban environments, there is less evaporation to reduce LST (Michael et al. 2012).

In an urban environment, natural vegetation is eliminated and replaced by non-transpiring and non-evaporating surfaces that have low capacity for solar reflectivity and high capacity for heat absorption such as concrete, asphalt, and metals in most cases, resulting in a significant modification of the earth's surface (Andrew 2012; Ridwan et al. 2021). This change eventually causes incoming solar energy to be redistributed, resulting in the rural–urban disparity in air temperatures and surface radiance (Guiling et al. 2008).

The influence of urbanization is tremendous, and it affects or alters the natural ecosystem; therefore, understanding UHI is vital for a variety of applications in earth and physical sciences as well as environmental management techniques (Aneeqa et al. 2016). The demand for agricultural production, food, and shelter is increasing as the global population grows. As a result of anthropogenic activities, land cover characteristics are shifting to satisfy rising population need and replacing vegetated areas with impermeable surfaces, inadvertently leading to climate change (Imran et al. 2021). Increased heat intensity in the urban climate has major consequences on human health and the usage of outdoor areas, as well as many activities that contribute to the liveability and vitality of cities. It causes different multifaceted issues such as skin cancer and greater energy consumption because air conditioners are often required (Michael et al. 2012; Naserikia et al. 2019).

LST is defined by how hot the "surface" of the earth would feel when touched in a particular region (Przyborski 2021) while the surface in this context and as used in satellite remote sensing refers to whatever a satellite observes as its signal pierces through the atmosphere to the earth. Surface heat fluxes, which are affected by urbanization, influence the LST in an area (Dousset and Gourmelon 2003). Therefore, understanding the spatiotemporal distribution of LST will aid in deciphering its mechanism and determining possible mitigation techniques (Sun et al. 2009).

Apart from the LST, other indices that have been reported to contribute to the urban heat island of a location include vegetation which is often measured using the Normalized Difference Vegetative Indices (NDVI), urbanization or built-up areas which is often measured using the Normalized Difference Building Indices (NDBI), surface emissivity, etc.

The NDVI is an index used to detect and ascertain the existence or presence of live green vegetation. Most visible light (0.4–0.7 m) is absorbed by healthy vegetation, whereas most near-infrared light (0.7–1.1 m) is reflected. In contrast, unhealthy or sparse vegetation will reflect less near-infrared light and more visible light (Weirer and Herring 2010). As a result, greater radiation that is reflected in the near-infrared wavelengths than the visible wavelengths indicates the existence of green vegetation, while minimal variation in intensity between the two wavelengths suggests the existence of either non-vegetated surfaces or sparse vegetation (Weirer and Herring 2010). The built-up density for each area is described by the NDBI, which is synonymous to the vegetation density described by the NDVI. The ratio of short red infrared (SWIR) to near infrared (NIR) is calculated as NDBI, with indices ranging from -1 to 1 (Kshetri 2018).

The impact of the relationship or connection between the NDVI and LST, especially in locations where the urban heat intensity phenomenon is more prevalent and mitigating efforts are required, cannot be overemphasized. This is primarily because denser vegetation lowers LST by ensuring the transfer of latent heat to the atmosphere from the surface via evapotranspiration. The NDVI is used to investigate this relationship and, as a result, provides insight into how plants or vegetation's natural cooling mechanism can be exploited to improve urban thermal settings. In general, it is expected that lower LSTs are recorded or observed in locations or places with a high NDVI, implying that the two have an indirect relationship. However, surface evapotranspiration and soil moisture levels may significantly alter or modify the dynamics of this relationship (Yuan and Bauer 2007).

A property or attribute of a surface that determines the volume of energy that is emitted by an object at a particular temperature when compared to a blackbody at the same temperature is known as surface emissivity (EXERGEN 2021). It can also be described as the ability of a surface or an object to convert heat energy it receives into radiant energy (Sekertekin and Bonafoni 2020). The emissivity is gotten from NDVI after the fractional vegetation (Pv) cover has been estimated and then calculated from the reflectance values of the materials on the earth surface based on the results of the NDVI. The emissivity of materials on the earth's surface reflects how well they absorb all incident radiation and convert it to internal energy before emitting (re-radiating) the received energy at the highest rate feasible per unit area (Isa et al. 2016).

One of the important features or characteristics that can be observed by satellite remote sensing is the surface temperature of an area. This data (surface temperature) has a wide range of applications in ecology, environmental studies as well as spatial data modelling, which is frequently employed in web-based GIS applications (Sameen and Al Kubaisy 2014).

Recently, the connection and correlation between LST and other factors or indices have received significant research attention (Peng et al. 2020). Researchers frequently look into the individual interaction between LST, vegetation, surface emissivity, and water, as well as the impact of urban land growth on temperature change, while little known effort has been invested in the combined effects of some of these interactions, a gap this research seeks to fill. The fundamental goal of this study is to investigate the combined effect and interdependence of vegetation (using the Normalized Difference Vegetation Index (NDVI)), built-up area (using Normalized Difference Built-up Index (NDBI)), surface emissivity, and the LST in the assessment of urban heat intensity using Bosso Local Government Area of Niger State, Nigeria, as a case study.

Materials and Methods

Bosso, which is the project site of this study, is one of the Local Government Areas (LGA) in Niger State, Nigeria (see Fig. 14.1). With its administrative headquarters situated in Maikunkele, it covers an area of about 1592 km^2 and a population of about 147,359 according to the 2006 census. Bosso LGA has a high heat flow (Mohammed et al. 2019) which could be as a result of increased urban activities, hence, making it a suitable location for this study.

The study area experiences both wet (rainy) and dry seasons annually, and an attempt was made to assess the effect of seasonal variation of the LST or the heat intensity of the study area. Therefore, two images were downloaded for each year: one at dry season (February) and the second at rainy season (September), for the years 2015, 2017, 2019, and 2021, making a total of eight images. The properties of the Landsat-8 images, generated from their metadata, are presented in Table 14.1.

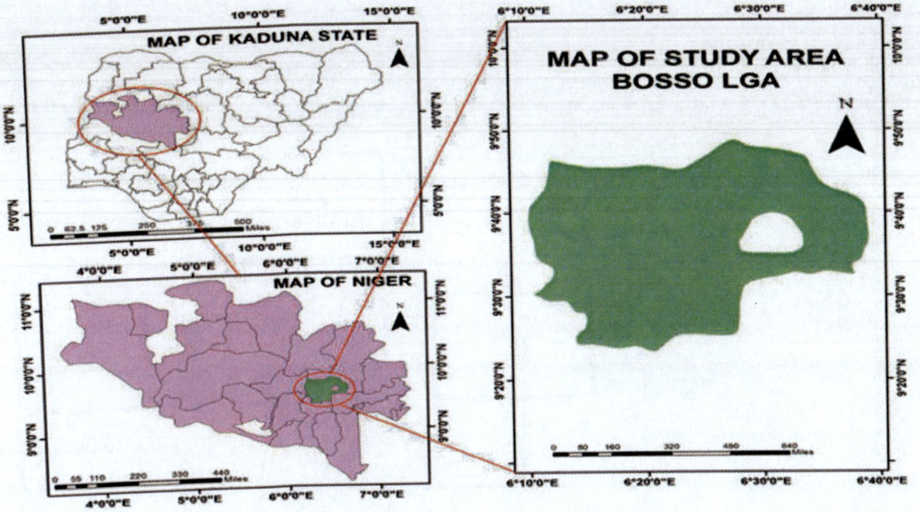

Fig. 14.1 Cartographic description of the study area

Table 14.1 Characteristics of the used Landsat-8 OLI images

Data	Sensor	Bands	Resolution/grid cell size	Season	Date obtained	Year
Landsat-8	OLI	11	30 m	Dry	25-02-2015	2015
Landsat-8	OLI	11	30 m	Rainy	21-09-2015	2015
Landsat-8	OLI	11	30 m	Dry	14-02-2017	2017
Landsat-8	OLI	11	30 m	Rainy	26-09-2017	2017
Landsat-8	OLI	11	30 m	Dry	20-02-2019	2019
Landsat-8	OLI	11	30 m	Rainy	27-05-2019	2019
Landsat-8	OLI	11	30 m	Dry	25-02-2021	2021
Landsat-8	OLI	11	30 m	Rainy	21-09-2021	2021

OLI = Operational land imager

The step-by-step or detailed procedure adopted for the execution of this study is presented in Fig. 14.2. The used Landsat-8 OLI/TIRS C2 L1 band images for 2015, 2017, 2019, and 2021 were downloaded from https://earthexplorer.usgs.gov/.

Radiometric and Atmospheric Correction

On satellite images, radiometric and atmospheric correction is frequently used to reduce the atmosphere's absorption and scattering effects. As the electromagnetic (EM) energy travels from the sun to the earth and back to the sensor, through the

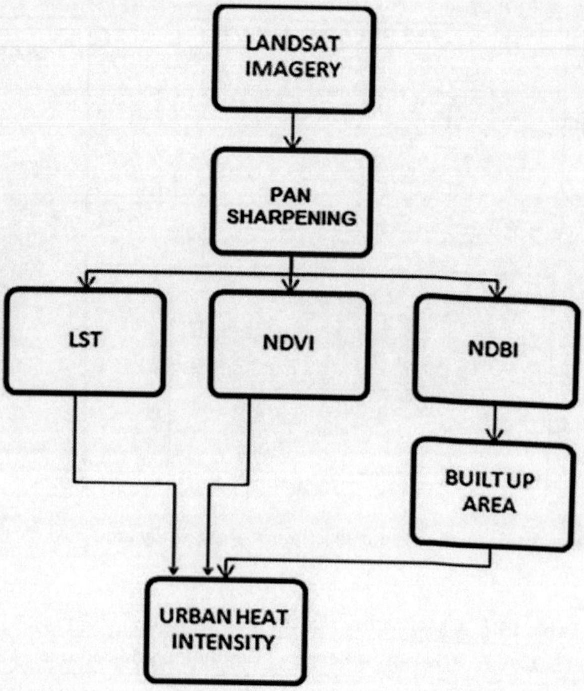

Fig. 14.2 Methodology workflow of estimating urban heat intensity from Landsat images

atmosphere, absorption diminishes the intensity of EM energy resulting in haziness, while the energy is redirected in the atmosphere by scattering resulting in an adjacent effect in which neighbouring pixels are shared and thus affecting image quality (GISgeography 2021). Radiometric and atmospheric correction removes the effect of sensor influence and atmospheric effect on the reflectance value of satellite images. Radiometric and atmospheric correction was done by computing the Top of Atmospheric (TOA) spectral reflectance (Eq. 14.1), followed by the correction of sun angle (Eq. 14.2) for all the satellite images downloaded for this study.

Procedure for Obtaining LST

LST can be estimated using the Landsat-8 OLI/TIRS C2 L1 thermal bands by applying Eqs. (14.1)–(14.8) (Rosado et al. 2020) presented in the simplified procedure described by the following five (5) steps.

I. Top of Atmospheric (TOA) spectral reflectance.

On a given surface, the ratio of reflected solar radiation to incident solar radiation is often referred to as the ratio of TOA radiance (Eq. 14.1), which is a unitless measurement. The mean solar spectral irradiance and the solar zenith angle derived from satellite-measured spectral radiance are used for the estimation of TOA (Rosado et al. 2020).

$$TOA(L) = ML * Qcal + AL \tag{14.1}$$

where M_L is the band-specific multiplicative rescaling factor from the metadata, Q_{cal} corresponds to band 10, and A_L is the band-specific additive rescaling factor from the metadata.

II. Calculation of Brightness Temperature

The thermal band detectors record TOA's brightness temperature in the form of digital numbers (DNs), which is then converted to surface temperature using the single channel algorithm. Surface temperatures obtained with Eq. (14.3) are deemed to be highly accurate (Michael et al. 2012).

$$BT = (K2/(ln(K1/L) + 1)) - 273.15 \tag{14.2}$$

where K_1 and K_2 are the band-specific thermal conversion constant from the metadata and L is the TOA. In order to obtain the results in degree Celsius, absolute zero is added to the radiant temperature as presented in Eq. (14.3).

$$BT = \left(1321.0789/Ln\left((777.8853/''\%TOA\%'') + 1\right)\right) - 273.15 \tag{14.3}$$

III. Extracting the NDVI

Reflectance data of Landsat-8 images, i.e. the visible red (Red) and the near-infrared (NIR) bands (bands 4 and 5, respectively) were used to extract the NDVI.

The NDVI was extracted using the expression presented in Eqs. (14.4) and (14.5).

$$\text{NDVI} = (\text{Band } 5 - \text{Band } 4)/(\text{Band } 5 + \text{Band } 4) \tag{14.4}$$

It should be noted that calculating the NDVI is crucial since it is necessary to estimate the proportion of vegetation (Pv), which is related to the NDVI, and emissivity (ε), which is related to the proportion of vegetation.

$$\text{NDVI} = \text{Float}(\text{Band } 5 - \text{Band } 4)/\text{Float}(\text{Band } 5 + \text{Band } 4) \tag{14.5}$$

IV. Estimating vegetation proportion (Pν)

The proportion of vegetation was estimated using Eq. (14.6) (Carlson and Ripley 1997).

$$Pv = \text{Square}((\text{NDVI} - \text{NDVImin})/(\text{NDVImax} - \text{NDVImin})) \tag{14.6}$$

The minimum and maximum values of the NDVI are gotten from the properties of NDVI under the source tab or the highest and lowest value range in ArcGIS 10.8.

V. Estimating Surface Emissivity (ε)

The effectiveness of a material's surface in emitting energy as thermal radiation is known as its emissivity. Thermal radiation is electromagnetic radiation that includes both visible (light) and infrared (infrared) wavelengths that are invisible to the human eye (Sobrino et al. 2013). The mathematical expression used for the estimation of emissivity is presented in Eq. (14.7).

$$\varepsilon = m * Pv + n \tag{14.7}$$

where m = emissivity of vegetation (0.004), Pv = percentage of vegetation, and n = soil emissivity value (0.986).

VI Estimating the Land Surface Temperature

Equation (14.8) presents the mathematical expression used for the estimation of LST.

$$LST = (BT/(1 + (0.00115 * BT/1.4388) * Ln(\varepsilon))) \tag{14.8}$$

where BT = Brightness temperature and ε = Emissivity.

Estimating the NDBI

The NDBI was extracted using the reflectance data (short red infrared (SWIR) and near infrared (NIR) bands) of Landsat-8 images. The NDBI is estimated as the ratio between short red infrared (SWIR) and near infrared (NIR) and has indices ranging from −1 to 1. Generally, the mathematical expression for estimating the NDBI for an area is shown in Eq. (14.9), while Eq. (14.10) presents the equation used to extract the NDBI specifically from Landsat-8 image.

$$NDBI = (SWIR - NIR)/(SWIR + NIR) \tag{14.9}$$

$$NDBI = float(band\ 6 - band\ 5)/float(band\ 6 + band\ 5) \tag{14.10}$$

where band 6 is a SWIR band and band 5 is a NIR band. Built-up areas were extracted from the built-up density in the form of point features for clear depiction of urbanization.

Interpretation of NDBI and NDVI

Generally, the value of the NDBI and NDVI calculation ranges from −1 to 1. The representation of the values within the range for NDVI is presented in Table 14.2,

Table 14.2 NDVI value range and feature represented (Kshteri 2018)

NDVI value range	Feature represented
−1–0	Water bodies
−0.1–0.1	Snow, barren rocks, or sand
0.2–0.5	Senescing crops, grasslands, or shrubs
0.6–1.0	Tropical rainforest or dense vegetation

Table 14.3 NDVI value range and the represented state of plant's health

NDVI value range	State of plant
−1–0	Inanimate objects or plants that are dead
0–0.33	Plants that are unhealthy
0.33–0.66	Plants that are moderately healthy
0.66–1	Plants that are very healthy

while Table 14.3 presents the range of index values representing the state of plant health (EOS 2019). For the NDBI, while negative values represent non-urban land areas, urban land areas are represented by positive values.

Results and Discussion

For clearer presentation, the results were presented and discussed in two subsections. While the results obtained in the seasonal dynamics of the surface heat intensity of the study area in the dry season were presented in subsection "Surface Heat Intensity of the Area in the Dry Season", the result obtained for the same analysis in the rainy season was presented in subsection "Surface Heat Intensity of the Area in the Rainy Season". The relationships and effects of the combined indices (NDVI, NDBI, and LST) on the surface urban heat intensity are also presented.

Surface Heat Intensity of the Area in the Dry Season

Maps depicting the spatial distribution of NDVI, NDBI, and the spatial variation of LST values of the study area for the dry season are presented in Figs. 14.3, 14.4 and 14.5, respectively, while Table 14.4 contains the estimated maximum values of the NDBI, NDVI, and LST obtained for 2015, 2017, 2019, and 2021 for the dry season. From the results, it was observed that in dry season, the highest maximum value of LST was recorded in the year 2017 (51 °C) which also yielded the highest NDVI and NDBI values of 0.388 and 0.577, respectively, while the lowest maximum value of LST was recorded in the year 2015 (33 °C).

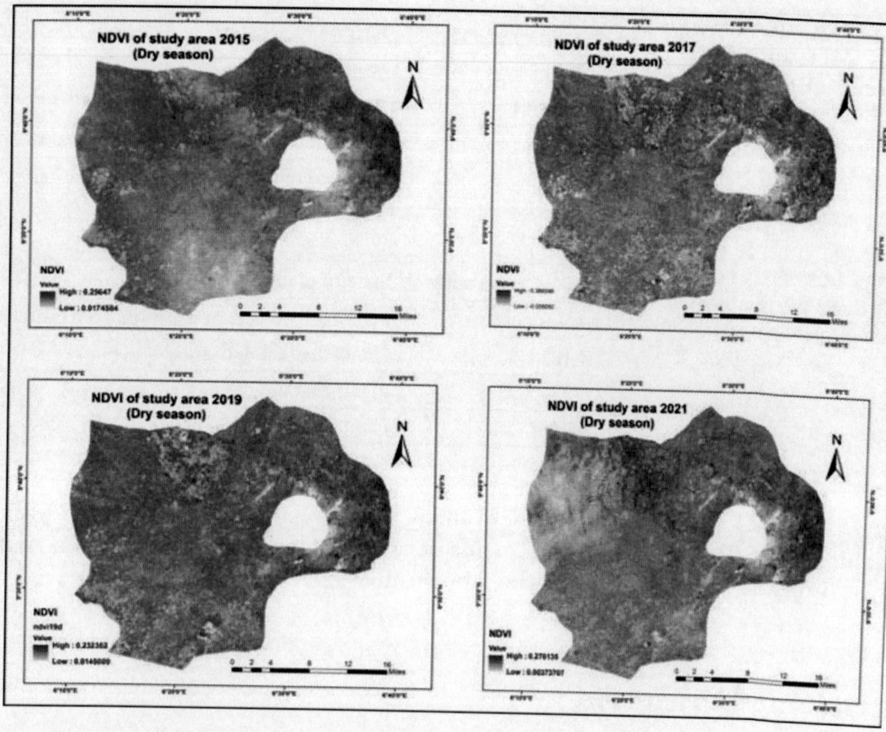

Fig. 14.3 Spatial distribution of NDVI in dry season

Also, the value for LST is lowest in the year 2015, but its value for NDBI (0.425) is higher than the values recorded for the years 2019 and 2021 which are 0.414 and 0.320, respectively, while the vegetative index value for the year 2021 (0.278) is higher than that of the year 2015 (0.257). Therefore, the lowest LST in the dry season over the years understudied is justifiably expected to be reported in the year 2021 since vegetation reduces the effect of LST, and the value of the NDBI recorded for the year 2021 is the least compared to other years. Also, since NDVI value below 0.5 indicates the presence of shrubs, rocks, sand, unhealthy plants, grasses (unhealthy or moderately healthy), etc., and the highest maximum value for NDVI in dry season is 0.388 as recorded in the year 2017, it evidently shows that most plants within the study area have already dried up at this time of the year; therefore, the effect of NDVI in dry season is minimal.

Also, since surface emissivity contributes to surface temperature in an area along with built-up index, the surface emissivity of the study area was examined for each of the years understudied for the dry season, and the obtained result is presented in Fig. 14.6. The maximum surface emissivity recorded in the years 2017 and 2021 illustrates the extent to which various materials, such as bare land or soils, engineering structures, metal, concrete, and tar, absorb all incident radiation completely, and convert it to internal energy. The absorbed energy is then emitted (re-radiated) into

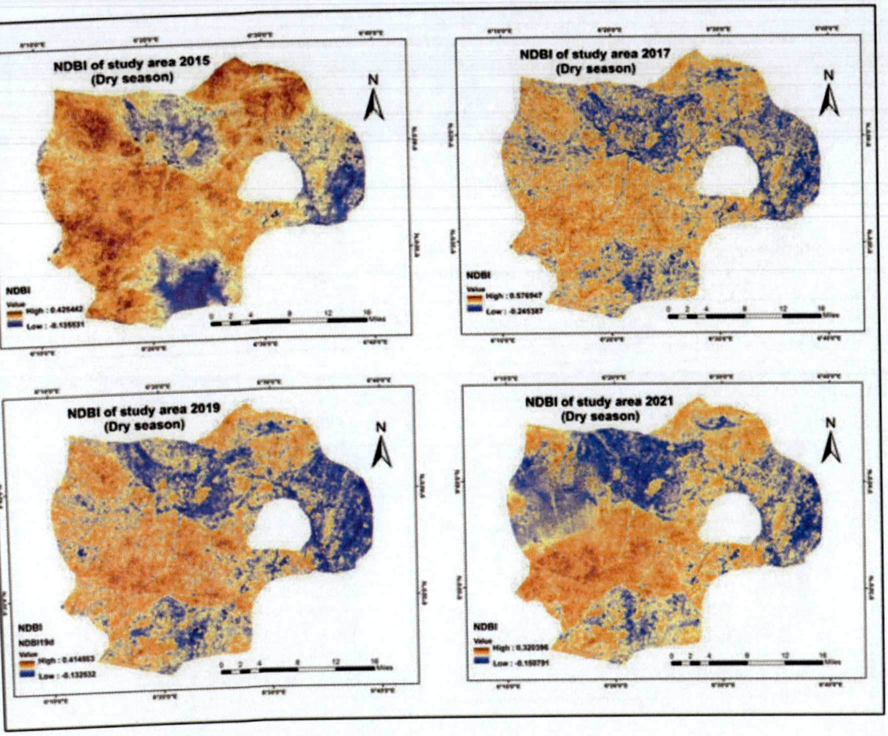

Fig. 14.4 Spatial distribution of NDBI in dry season

the atmosphere and thereby contributing to LST. The recorded surface emissivity in the years 2017 and 2021 is higher than that of the year 2015 for the dry season which explains why the least value of LST for dry season was recorded in the year 2015 and not in 2021. It also explains why LST in the year 2021 is higher than what it was in the year 2019.

Surface Heat Intensity of the Area in the Rainy Season

Figures 14.7, 14.8 and 14.9 present the graphical description of the spatial distribution of NDVI, NDBI, and the spatial variation of LST of the study area for the rainy season, respectively. The estimated maximum values of the NDBI, NDVI, and LST obtained for the years 2015, 2017, 2019, and 2021 for the rainy season are presented in Table 14.5. Similar to what was observed in the results obtained for the dry season, the lowest value of LST was recorded in the year 2021, but its value for built-up index (0.267) is higher than NDBI values for the years 2017 and 2015 which are 0.259 and 0.243, respectively. Also, the year 2019 had the highest maximum value for LST (70 °C), while the year also recorded the lowest NDVI value of 0.544 and the highest

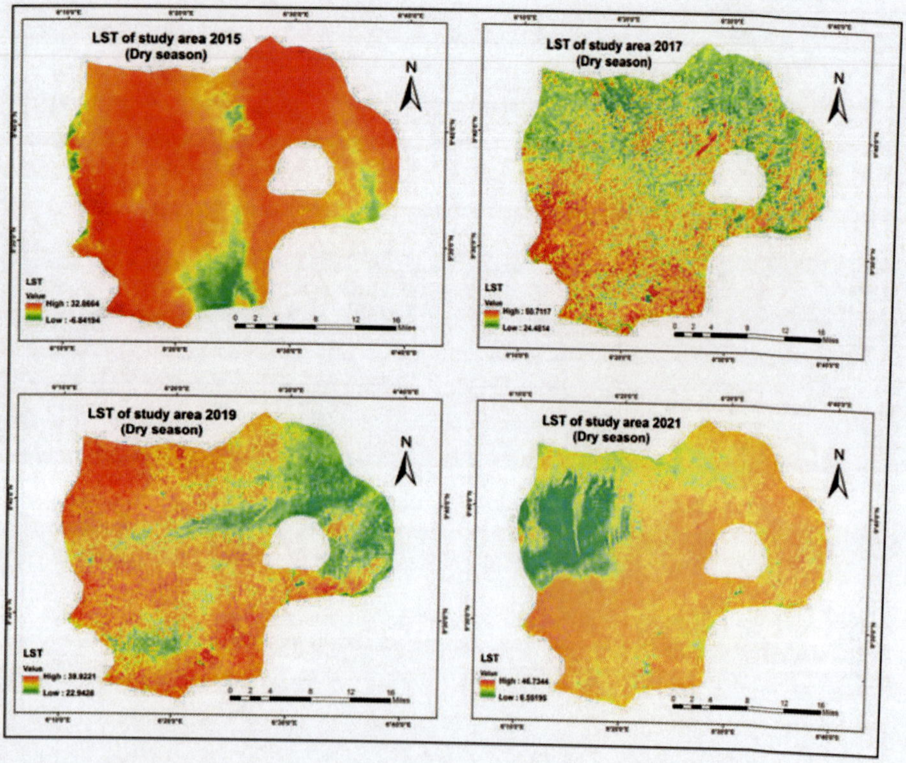

Fig. 14.5 Spatial variation of LST in dry season

Table 14.4 Estimated NDVI, NDBI, and LST maximum values for the dry seasons

Year	NDVI	NDBI	LST (°C)
2015	0.257	0.425	33
2017	0.388	0.577	51
2019	0.232	0.414	40
2021	0.278	0.320	47

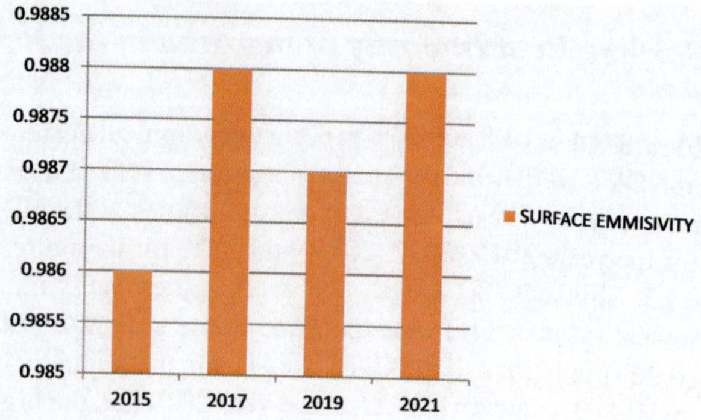

Fig. 14.6 Surface emissivity in dry season

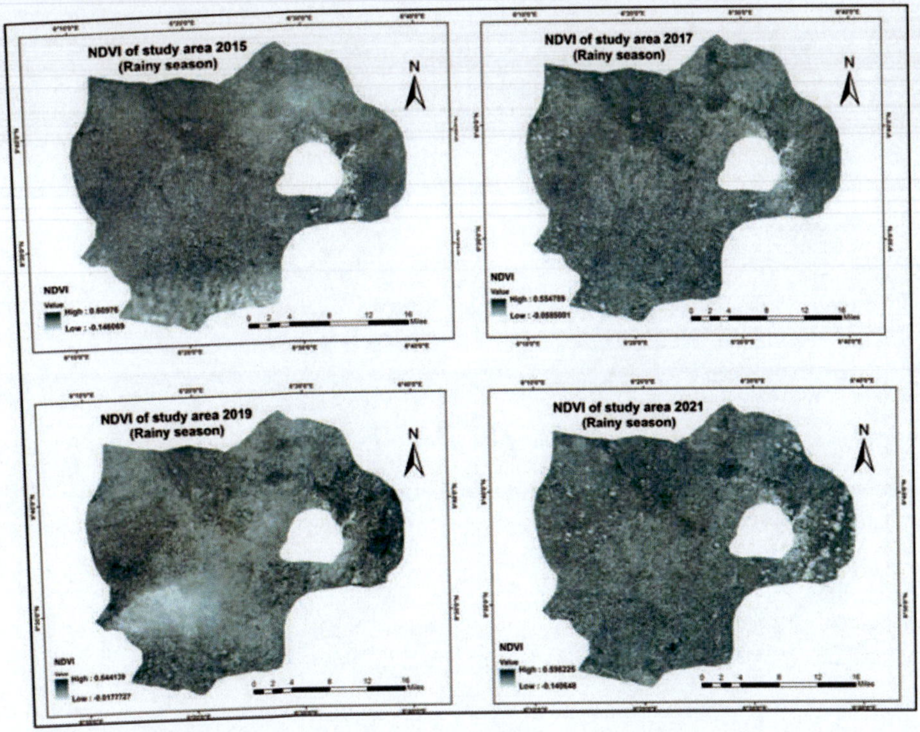

Fig. 14.7 Spatial distribution of NDVI in rainy season

NDBI value of 0.421. The implication of this is that in the year 2019, urbanization rate was very high which had a negative effect on vegetation thereby increasing the urban heat intensity of the study area in the rainy season. In contrast, for the year 2021, there was a very low heat intensity judging by the recorded lowest maximum value for LST of 29 °C. This can be attributed to the high NDVI of 0.598 and low NDBI of 0.267, which implies an improvement in vegetation growth and reduced urban growth, respectively.

Although the vegetation index for the years 2021 and 2015 is very close, the lowest value of LST recorded in the rainy season of the year 2015 is justified by the lowest value for built-up index recorded in the same season and the highest value of the NDVI recorded in the dry season over the studied time epochs. This suggests that while there are low or reduced urbanization activities, there were high vegetated activities which led to the recorded low LST within the period of study.

Also, the surface emissivity of the study area in the rainy season was examined over the studied years and the result is presented in Fig. 14.10. Similar to the surface emissivity of the study area in the dry season, the surface emissivity for the years 2015 and 2017 was observed to be considerably higher than what it was in the years 2019 and 2021 in the rainy season, which explains why the least value of LST for the rainy season was recorded in the year 2021 and not in year 2015.

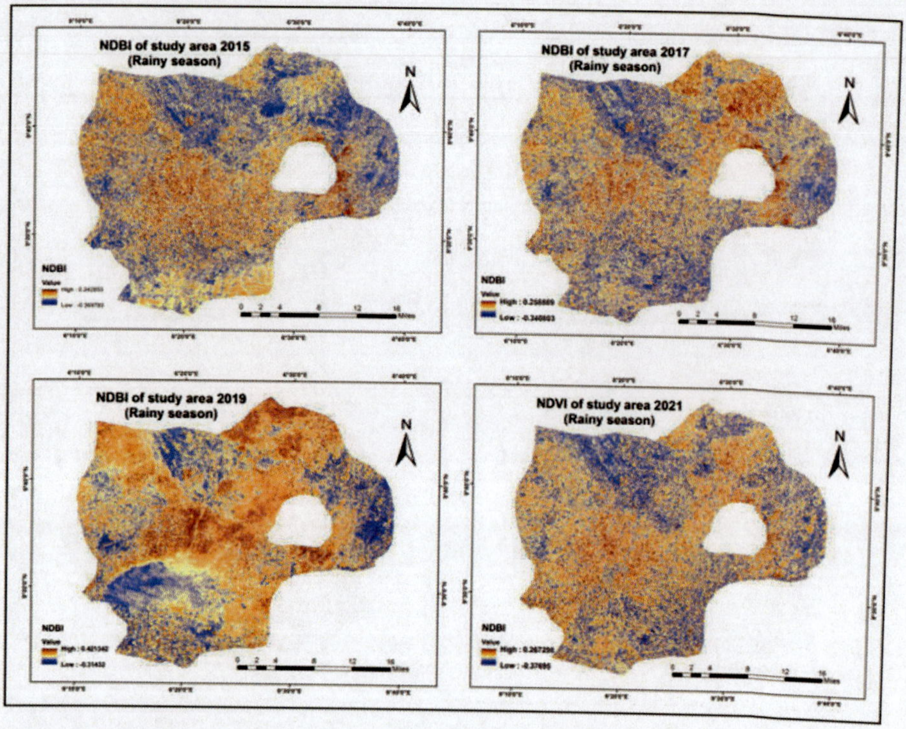

Fig. 14.8 Spatial distribution of NDBI in rainy season

Generally, it was observed from the analysis that the surface emissivity values of the study area obtained in the rainy season were significantly higher than the surface emissivity values of the study area obtained for dry season across all the years, except for the results obtained for the year 2019. This is because the strength of wind in the study area is weak at this season of the year.

Conclusion

This research shows the potential of using Geographic Information System (GIS) and remote sensing techniques in assessing the surface heat intensity of any location. The findings of this study affirmed that a rise in surface urban heat intensity is resulted from an increase in urbanization and a decrease in vegetation. The effect of increased built-up density and surface emissivity which resulted to an increase in LST proves that increase in urbanization leads to an increase in surface heat intensity. The study also proved that a rise in vegetation density leads to a decrease in urban heat intensity and that the effect is more significant in the rainy season than dry season as vegetation is usually dry at this season of the year, where soil moisture content is low. In as

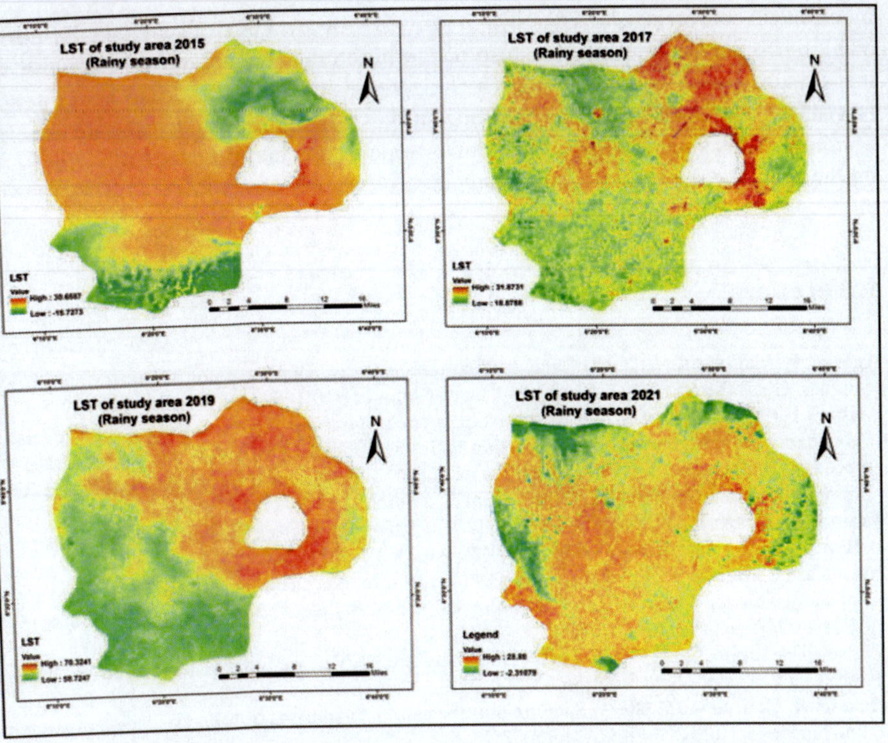

Fig. 14.9 Spatial variation of LST in rainy season

Table 14.5 Estimated NDVI, NDBI, and LST maximum values for the rainy seasons

Year	NDVI	NDBI	LST (°C)
2015	0.610	0.243	31
2017	0.555	0.259	32
2019	0.544	0.421	70
2021	0.598	0.267	29

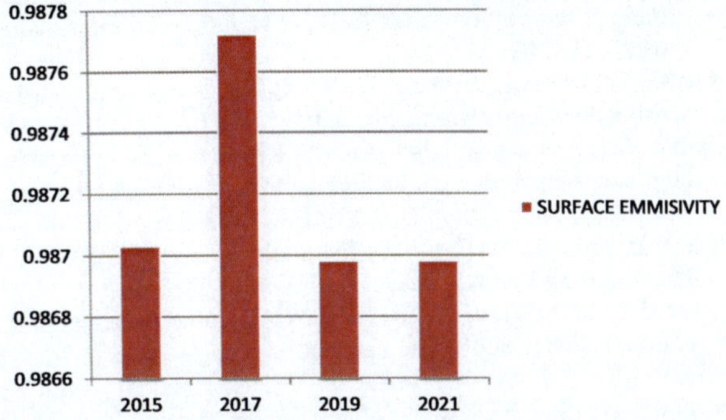

Fig. 14.10 Surface emissivity in rainy season

much as vegetation is more effective in rainy season, wet periods are usually periods of high urban heat intensity because of the high emissivity values associated with this season. This is because wind, which decreases the effect of urban heat intensity, is weak at this season of the year. Further studies will attempt to investigate the effect of other factors such as terrain elevation or topography and presence of water bodies on the urban heat intensity of the study area.

References

Agbor CF, Makinde EO (2018) Land surface temperature mapping using Geoinformatics techniques. Geoinform FCE CTU 17(1):2–17. https://doi.org/10.14311/gi.17.1.2

Andrew F (2012) Exploring the relationship between land surface temperature and vegetation abundance for urban heat island mitigation in Seville, Spain. LUMA-GIS Thesis Nr 15:1–58

Aneeqa A, Muhammad AB, Khadija N, Huma A (2016) Assessment of Urban Heat Island (UHI) using remote sensing and GIS. Glob J Hum-Soc Sci 2(2):1–9

Argueso D, Evans JP, Pitman AJ, Luca A (2015) Effects of city expansion on heat stress under climate change conditions. PLoS One 10(2), Article ID:e0117066

Ayanlade A, Aigbiremolen M, Oladosu O (2021) Variations in urban land surface temperature intensity over four cities in different ecological zones. Sci Rep. https://doi.org/10.1038/s41598-021-99693-z

Carlson TN, Ripley DA (1997) On the relation between NDVI, fractional vegetation cover, and leaf area index. Remote Sens Environ 62:241–252

Dousset B, Gourmelon F (2003) Satellite multi-sensor data analysis of urban surface temperatures and landcover. ISPRS J Photogramm 58:43–54

EXERGEN Global industrial sales. (2021) What is emissivity? https://www.exergenglobal.com/tech-notes/31-what-is-emissivity

Elhadi KM, Yungang C, Guoxiang L, Moshbeh RK, Ashraf AB, Fawzi Z, Mohammed S (2020) Study for predicting land surface temperature (LST) using Landsat data: a comparison of four algorithms. Adv Civil Eng. https://doi.org/10.1155/2020/7363546

Garouani M, Amyay M, Lahrach A, Oulidi HA (2021) Land surface temperature in response to land use/cover change based on remote sensing data and GIS techniques: application to Saïss Plain, Morocco. J Ecol Eng 22(7):100–112

GIS Geography (2021) What is atmospheric correction in remote sensing? https://gisgeography.com/atmospheric-correction/

Gomes BY, Caracristi I (2021) Seasonal analysis on land surface temperature (LST) and normalized difference vegetation index (NDVI) variations in the Iguatu semi-arid hinterland, Ceará. Int J Hydrol 5(6):289–294

Guiling W, Weimei J, Ming W (2008) An assessment of urban heat island effect using remote sensing data. http://hdl.handle.net/1834/5876

Imran HM, Hossain A, Islam AKMS, Rahman A, Bhuiyan AE, Paul S, Alam A (2021) Impact of land cover changes on land surface temperature and human thermal comfort in Dhaka City of Bangladesh. https://doi.org/10.1007/s41748-021-00243-4

Isa Z, Baba I, Ayuba Z (2016) Estimation of Land Surface Temperature of Kaduna Metropolis, Nigeria using Landsat images. Sci World J 11(3):1–7

Kshetri T (2018) NDVI, NDBI & NDWI Calculation using LandSAT 7 and 8. https://www.linkedin.com/pulse/ndvi-ndbi-ndwi-calculation-using-landsat-7-8-tek-bahadur-kshteri

Michael SA, Saviour M, NaaDedei T (2012) Mapping land surface temperature and land cover to detect urban heat island effect: a case study of Tarkwa, South West Ghana. Department of Geomatics Engineering, University of Mines and Technology, P.O. Box 237, Tarkwa, Ghana

Mohammed A, Adewumi T, Kazeem AS, Abdulwaheed R, Adetona AA, Usman A (2019) Assessment of geothermal potentials in some parts of upper Benue trough Northeast Nigeria using Aeromagnetic data. J Geosci Eng Environ Technol 4(1). https://journal.uir.ac.id/index.php/JGEET/article/view/2090

Naserikia M, Shamsabadi EA, Filho WL (2019) The Urban Heat Island in an urban context: a case study of Mashhad, Iran. https://www.ncbi.nlm.nih.gov/pmc/articles/PMC6388183

Peng X, Wu W, Zheng Y, Sun J, Hu T, Wang P (2020) Correlation analysis of land surface temperature and topographic elements in Hangzhou, China. Scientific Rep. https://www.nature.com/articles/s41598-020-67423-6

Przyborski P (2021) Land surface temperature. Earth observatory. https://earthobservatory.nasa.gov/global-maps/MOD_LSTD_M

Ridwan, Rasyidi ES, Syafri, Rahman R, Okviyani N, Jumadil, Ma'rief AA (2021) Assessment of the relationship between building density and urban heat island using Landsat images in Makassar city. In: International conference on research collaboration of environmental science. https://doi.org/10.1088/1755-1315/802/1/012042

Rosado MRG, Guzman MEA, Lopez CJE, Molina MW, Garcia HLC, Yedra EL (2020) Mapping the LST (Land Surface Temperature) with satellite information and software ArcGis. In: International conference on research collaboration of environmental science. https://doi.org/10.1088/1757-899X/811/1/012045

Sameen IM, Al Kubaisy AM (2014) Automatic surface temperature mapping in ArcGIS using LANDSAT-8 TIRS and ENVI tools case study: Al Habbaniyah Lake. J Environ Earth Sci 12(4):1–7

Sekertekin A, Bonafoni S (2020) Land surface temperature. Retrieval from Landsat 5, 7, and 8 over Rural Areas: Assessment of Different Retrieval Algorithms and Emissivity Models and Toolbox Implementation. Retrieved from https://www.mdpi.com/2072-4292/12/2/294/htm

Sobrino J, Wang N, Wu H, Li ZL (2013) Land surface emissivity retrieval from satellite data. Int J Remote Sens 34(9–10):3084–3127. https://doi.org/10.1080/01431161.2012.716540

Sun Q, Tan J, Xu Y (2009) An ERDAS image processing method for retrieving LST and describing urban heat evolution: a case study in the Pearl River Delta Region in South China. Environ Earth Sci 1047–1055

Weirer J, Herring D (2010) Measuring vegetation (NDVI & EVI) (NASA Earth Observatory). http://earthobservatory.nasa.gov/Features/MeasuringVegetation/measuring_vegetation_2.php

Yuan F, Bauer ME (2007) Comparison of impervious surface area and normalized difference vegetation index as indicators of surface urban heat island effects in Landsat imagery. Remote Sens Environ 106:375–386

Chapter 15
Agroecology as a Response to Sustainable Development Under Climate Change in Southeast Nigeria

Chukwuma Otum Ume, Tochukwu Linda Onah, Adaku Bridget Ezeibe, Ogochukwu Gabriella Onah, Patience Ifeyinwa Opata, Ezinne Orie Idika, and Kalu Uche Felix

Abstract Nations of the world have seen unprecedented changes in climate variables in recent decades. But it is unclear to what extent climate change has impacted and will impact food systems in some developing regions, and how policymakers can frame an approach to encouraging adaptation and advancing climate-smart agriculture. Many studies attempting to link agroecology to climate change adaptation do so without understanding the potential of Agroecology not only to mitigate climate change—which is the weak response—but also to reverse its impact and "climate proof" our food systems. By modeling the near and far future impacts of climate change on crop production, we showed how climate will impact crop production under two crop production systems (agroecology and non-agroecology production systems). The overarching aim is to derive sustainable development strategies and lessons for policymakers and climate researchers—essential components of environment and Agricultural development. Using case studies from Nigeria, we observed that transitioning to agroecology, even at the farm level also transforms farm designs, thereby affecting their overall food and nutrition status. The result showed that the use of agroecology management practices not only reduces the impact of climate change in the near future but will also lead to increased crop yield in the future. The finding suggests that to feed over 400 million projected population of Nigeria by 2050, the use of agroecological practices will be a better alternative to conventional farming methods. To advance the use of agroecological farming methods, governments at every level in Nigeria need to mainstream organic agriculture in national government policies. This is important as it will not only address climate change impacts, but also hunger and poverty.

Keywords Agroecology · Bio-economic farm models · Crop syst · Aqua crop · Organic farming · Sustainable development

---

C. O. Ume · T. L. Onah · A. B. Ezeibe · O. G. Onah (✉) · P. I. Opata · K. U. Felix
Department of Agricultural Economics, University of Nigeria, Nsukka, Nigeria
e-mail: ogochi.onah@unn.edu.ng

E. O. Idika
Department of Agricultural Education, University of Nigeria, Nsukka, Nigeria

J. C. Egbueri et al. (eds.), *Climate Change Impacts on Nigeria*, Springer Climate,
https://doi.org/10.1007/978-3-031-21007-5_15

Introduction

The impact of climate change undermines efforts in achieving Sustainable Development Goals. According to the Intergovernmental Panel on Climate Change, climate change is the alteration in global climatic patterns, which can be observed by the changes in weather properties over a period of time (IPCC 2007a). The impact of climate change has become evident, not only in the threats it poses to the environment, but also in the fight toward poverty eradication, disease control, and zero hunger. These effects are traceable to their direct and indirect impact on the agricultural sector, especially in crop production(Ume 2018). Although substantial efforts have been directed toward climate change mitigation (i.e., addressing the causes of climate change), there is also a critical need to build resilience capacities within the agricultural sector and national food systems. As stated in IPCC (2007a, b), such a climate resilience framework and adaptive capacities will help grapple with the current and future impacts of climate change. According to the IPCC (2007a, b, p. 3), climate change adaptation is the "adjustment in natural or human systems to a new or changing environment". This means that efforts in climate change adaptation among farmers will create a system capable of cushioning present and future impacts from Climate Change thereby enabling farmers to cope with future climate change outcomes.

Furthermore, the climate change and food security report of the Food and Agricultural Organization (FAO) indicated that changes in climate properties will aggravate the challenges confronting food crop systems (Food and Agriculture Organization 2018). According to the report, climate change will have a direct and indirect impact on crop growth, market flows, and livelihood assets of the rural populace. In recent decades, progress in climate research has improved the understanding of how climate change influences agricultural productivity, triggers biodiversity losses, and undermines sustainable development efforts (Kessete, Moges, and Steenhuis 2019; Mamoon and Rahman 2019; Mengistu et al. 2019). As recent changes in climate properties have been more rapid compared to any other time in human history, literature has called for innovative research direction that critically projects climate impacts to establish targeted and more coordinated adaptation decisions (Mengistu et al. 2019). Key to the ability of farmers to make an informed decision and adapt to climate change and variability will be access to relevant information and knowledge (Challinor et al. 2005).

In Nigeria, due to high poverty levels, population expansion, and heavy dependence on rain-fed agriculture, climate change impact on the national economy of most nations in the region is expected to be exacerbated as agriculture plays a dominant role in supporting economic growth and rural livelihoods (Boko 2007). According to Challinor et al. (2005) crop farmers in Africa are vulnerable to climate change impacts due to three critical reasons: (1) vulnerability of crops to variability in climate properties, (2) low adaptive capacity of farmers, and (3) institutional failure in facilitating climate change adaptation. The foregoing suggests that, in Africa, there exists a triple burden of high poverty levels, population expansion, and the growing demand to safeguard the environment (Ume 2017). These challenges place the demand on farmers at all levels to increase productivity in other to meet the food security needs of the population. The need for achieving food and nutritional security and at the same time some level of environmental stewardship among farmers in Africa have called for a rapid shift from conventional agricultural practices to a more sustainable production method. This call transcends a shift in farming practices that employs external inputs, which has adverse effects on the soil and people's health, to a sustainable organic farming system that simultaneously supports environmental sustainability and food productivity, while promoting rural livelihoods. Agroecology and organic agriculture have been advanced as sustainable farming practice that can improve the above challenges, as it has been proven to be efficient, productive, and resilient (Altieri and Toledo 2011; Altieri, Funes-Monzote and Petersen 2012; Wezel et al. 2014).

According to FAO, climate change impact first affects food systems and livelihood groups with a higher level of vulnerability. Ume, Opata et al. (2021) noted that among farmers in Nigeria, female smallholder farmers are expected to be of low adaptive capacity and high exposure and sensitivity to climate impacts due to socioeconomic and institutional factors that undermine their adaptation efforts. There is, therefore, a growing focus on the need for "transformational adaptation" (Eakin and Wehbe 2009), that is the re-evaluation of institutional and socioeconomic factors or relations, established over time, which determine limits the adaptive capacity of smallholder farmers. Institutional transformation is therefore imperative if meaningful adaptation effort is to be achieved in overcoming the impact of climate change as smallholder farmers contribute over 70% of food and labor in Nigeria. Most of the governments in the continent still assume that maximizing food production will automatically lead to a reduction in food insecurity and malnutrition. It is also widely assumed that commercialization in agriculture and large-scale agribusiness leads to better food systems, with better-nourished food system members. Research has shown that around 80% of the Africa's poor derive their livelihoods from production-based entitlement and not market based entitlements (Thompson 2015). Given this fact, food system initiatives built narrowly on strengthening green revolution strategies may ignore the bulk of the continent's poorest individuals. We need to create a food system in which smallholder farmers can become empowered actors.

This research intends to provide context-specific evidence on the need for agroecology and sustainable agriculture in counteracting the impact of climate change in the agricultural sector in Nigeria. The overarching aim is to study the climate change implications and sustainability of organic farming compared to conventional farming, thereby deriving evidence and lessons for policymakers, and eliciting informed actions and procedures on how to fashion agroecological strategies. This will help fast-track the development of Agroecosystem practices in the food systems of Africa. To achieve the research goal, this research will seek to provide answers to the following questions:

(i) What is the impact of climate change on crop productivity in Nigeria?
(ii) What are the different agroecology measures employed by smallholder farmers in adapting to the impact of climate change on crop productivity?
(iii) How effective are the adaptation measures adopted by farmers in different agroecological regions?

Agroecology and Farming Systems in Nigeria

By definition, a farm that is not more than 5 ha is categorized as a small scale (CGIAR 2013). Based on this classification, over 80% of farmers in Nigeria are smallholder farmers (Mgbenka and Mbah 2016), as over 80% of farmers in Nigeria farm below 5 ha. This set of farmers produces over 98% of the food consumed in Nigeria apart from wheat and about 99% of the total crop output (Mgbenka and Mbah 2016), suggesting that they play the dominant role in the agricultural sector of the economy. This also points to the fact that a typical characteristic of the production system in the country is that a disproportionately greater portion of the farming output rests in the hands of smallholder farmers. Therefore, one can reasonably submit that a typical farming community in Nigeria comprise of smallholder farmers, producing food (crop and animal), not just for family consumption but for commercial purposes as well. According to Adewumi and Omoresho (2002), it is the progress of these farming communities that will, to a great extent, determine the progress of the agricultural sector.

Nigeria is a federation comprising 36 states with six geopolitical zones and seven vegetation types: Mangrove Swamp and Coastal Vegetation, Lowland Rain Forest, Freshwater Swamp Forest, Derived Savanna, Sudan Savanna, Guinea Savanna, and Sahel Savanna. In all these zones, farming is predominantly small-scale. Despite the small-scale nature of farming in the country, Nigeria is the highest producer of maize and cassava globally followed by Brazil (Opata et al. 2021), and the highest producer of rice in Africa (Akpoti et al. 2021). The southeast region of Nigeria has seen unprecedented changes in climate variables for more than two decades (Chukwuemeka et al. 2018). But it is unclear to what extent climate change has impacted and will impact major food crops in southeast Nigeria, and how policymakers can frame an approach to encouraging adaptation and advancing climate-smart agriculture. This study seeks to examine these concerns and understand how and to what extent changes in climatic conditions will affect crop farmers' productivity in the short and long run. The overarching aim is to derive evidence and lessons for policymakers and climate researchers—essential components of climate change and international development.

Small-scale farmers produce a greater percentage of these three crops. According to the Animal Science Association of Nigeria, large-scale farmers are mostly concentrated within the animal sub-sector (Finelib 2019). Although the impact of climate change on food and nutritional security and environmental sustainability is continuously gaining attention across Nigeria, the Southeast region is, however, complicated as it is also burdened with environmental issues such as soil erosion (Okorafor et al. 2017). However, because the technology employed in the agricultural revolution for climate change adaptation in required capital, only a large few farmers could maximize production under climate change.

In some other parts of Africa, such as Kenya and Tanzania, we have seen how effective Agroecology and indigenous knowledge can be in fostering sustainable agriculture among smallholder farmers through soil fertility improvement (Simon et al. 2020), drought control (Botelho et al. 2016), and platform for social interactions among farmer (Silici 2014). The emphasis should now shift from the big-scale transformation approach to the small-scale improvement strategy approach, which is attuned to Nigerian age-long farm practice. Innovative approaches to increase productivity enhance the resilience of agricultural systems, and give back control of the food system to the smallholders are therefore needed. No published research covering the knowledge and practice of agroecology by farmers in southeast Nigeria based on detailed and consistent field data is available. Rural empowerment and inadequate research in the area of indigenous knowledge use are listed as a constraint in the newly adopted Nigeria's Agriculture Transformation Agenda (Federal Ministry of Agriculture and Rural Development 2017).

According to Wezel et al. (2009), the word "agroecology" was first mentioned by Bensin (1928). Since then, the term has increasingly attracted the interest of policymakers, advocacy groups, and researchers from different disciplines. Alexander and Jauneau attribute this rise in popularity to the need for agriculture to respond to the many sustainability challenges including food security, biodiversity conservation, and rural development. Over the years, due to the multidisciplinary/cross-disciplinary nature of the term "agroecology", diverse definitions and conceptualizations have emerged. Table 15.1 presents the different definitions and conceptualizations starting from Bensin (1928) when the world was first used.

The definitions presented show that the concept of agroecology has been an evolving concept both in research and in practice. Between 1928 and 2000, agroecology was conceptualized as a method of applying principles of ecology in climate and land management to increase farm production with reduced immediate ecological consequences production (Bensin 1928; Azzi 1956; Gliessman 1998). From the year 2000, the food system component was added to the definition. Here, agroecology as a concept is discussed beyond farm-level analysis and immediate biophysical impacts at the farm and field to include how all the elements of a food production–distribution-consumption system come together and interact with one another (Dalgaard et al. 2003; Francis et al. 2003; Clements and Shrestha 2004; United States Department of Agriculture (USDA) 2007). Subsequently, the emphasis shifted to food sovereignty and the right to food, without losing the sustainable production and food system components (De Schutter 2011; FAO 2014; Edwards 2017).

Ever since the concept of agroecology has continuously evolved and has been re-formed. In fact, because of the diversity of ways, the term agroecology is understood and approached, the Food and Agricultural Organization (FAO) has maintained a database of definitions of agroecology from published documents, authored by researchers, academia, civil society, legal documents, governments, and policies

Table 15.1 Model scenarios indicate the mean annual precipitation and temperature changes with respect to the baseline scenario

	A2 (2010–2039)		A2 (2070–2100)	
	Temperature (°C)	Precipitation (mm)	Temperature (°C)	Precipitation (mm)
Aba	27.9	60.86	30.1	102.3
Umuahia	28.9	99.73	32.0	112.5
Bende	28.9[a]	66.23	31.1	96.2
Okigwe	28.2	70.86	30.4	152.8
Orlu	28.1	114.08	29.2	124.2
Owerri	28.2	75.01	32.1	98.2
Enugu	28.2	72.0	30.5	63.2
Agbani	29.9	67.2	33.2	61.5
Udi	28.1	77.0	30.5	61.0
Awgu	28.9	85.7	32.1	98.1
Nsukka	30	114.5	33.5	122.2
Enugu	28.9[a]	87.2	32.6	74.8
Ezike	28.8	103.5	31.1	115.0
Ebonyi North	30.1	121.1	33.9	121.8
Ebonyi South	29.1	148.5	32.1	135.8
Ebonyi Central	30.1	121.1	32.1	115.0
Aguata	28.9[a]	144.0	31.5	122.0
Anambra	28.9[a]	150.0	31.6	135.2
Awka	28.9[a]	132.0	31.5	111.9
Onitsha	28.9[a]	122.0	31.5	129.0

[a]Signifies extreme variations

(FAO 2019c). In fact, the FAO (2019a, b, c) database currently has 401 definitions, in English, French, and Spanish which shows the variety of ways in which agroecology is approached or conceptualized. However, in general, the underlining characteristic of agroecology as a practice is the need to minimize the use of external chemicals in crop or animal production, in that regard, for this study, the categorization of farmers into agroecology and non-agroecology is based on the use and non-use of chemical in crop production.

Materials and Methods

Model Choice

There are three kinds of quantitative methods of climate impact investigation; these are the agronomic models, the Ricardian models, and the agroecological zoning studies (Kessete et al. 2019). Whereas the Ricardian models depend on long-term time-series data, it flaws in terms of forecasting agricultural productivity under changing climates (Reinsborough 2003). To capture the effect of climate change on crop productivity, agronomic models such as CropSyst and DSSAT are usually employed to capture the complexity involved in crop yield modeling (Challinor et al. 2005). In this regard, such models could uncover the sensitivity of the impact of climate change on crop yields, management practices, and the biophysical environment. Several studies have employed this model to predict yields as well as to conduct climate change impact assessments at the farm-level impact of climate change (Stöckle et al. 2003, 2014; Abi Saab et al. 2015). Hence, their usefulness in simulation has been proven largely and well employed in climate change impact modeling. The disadvantage of this technique, however, is that they ignore an important aspect of crop production which is the decision makers' adaptation behavior (Schönhart et al. 2011). However, it is generally acknowledged that the decision makers' adaptation behavior is endogenous in understanding the impact of climate change on crop production, especially in the prioritization of adaptation options.

The above flaws necessitated the need to adopt the well-known integrated models known as the bio-economic farm models (BEFM) in this study (Linderhof et al. 2019; Pahmeyer et al. 2021). In this study, we simultaneously accounted for biophysical changes, as well as agroecology management decisions, for different farming systems. This means that the BEFM allows for accessing the different levels of agroecology practices that a farmer decides to adopt on the farm. Because data for the analysis was based on the same agroecological zone with similar soil type and fertility, it becomes easy to isolate the impact of farmers' decision or choice of use of agroecology practices. This makes the bio-economic farm models appropriate for ex-ante assessment of different climate, policy, and technological scenarios even with limited availability of data (Linderhof et al. 2019).

Farm Survey and Modeling Approach

We identified 19 agricultural zones based on the agroecological diversity of the zones (Morgan 2019). We calibrated the BEFM for the agricultural zones in Southeast Nigeria (Aba, Umuahia, Bende, Okigwe, Orlu, Owerri, Enugu, Agbani, Udi, Awgu, Nsukka, Enugu Ezike, Ebonyi North, Ebonyi South, Ebonyi Central, Aguata, Anambra, Awka, and Onitsha). We differentiated these zones based on similarities

Fig. 15.1 Flow chart of our modeling system

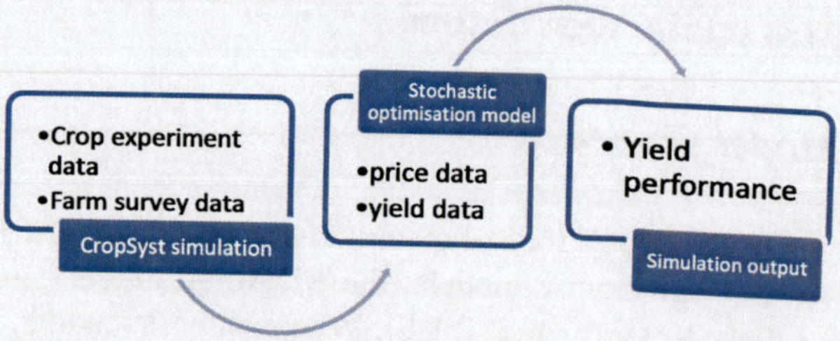

in soil characteristics. We surveyed 1221 farm households from different agricultural zones during the years 2020–2021. We used the sampling procedure of IMEA (2010) in the selection of farm households. We used the BEFM taking into account the risk component. The aim is to consider the impact of climate change on the three main crops cultivated in the region (maize, rice, and cassava). As a first step in the modeling process, we spatially downscaled the climate change scenarios to the local level. Downscaling is a process of deriving regional or local climate information from a general circulation model. The downscaled scenarios are applied to the crop model. With the combination, we can estimate the impact of climate change on the productivity of the crops under consideration. Figure 15.1 shows a schematic representation of the process. To calibrate the simulation of the crop model, we used the actual farm management practices as well as crop experiment data collected from farm surveys. The outcome of the simulation model (in this case, the crop yield) is used in the farm-level stochastic optimization model to show the impact of climate change (yield performance).

Climate Change Scenarios

We employed the A2 greenhouse gas emission scenarios of the Intergovernmental Panel on Climate Change (IPCC) in this analysis (IPCC 2007b). The A2 scenario was preferred as it is the higher end of the Special Report on Emissions Scenarios (SRES) emissions. From an adaptation and impact point of view, a farm system that adapts to a larger climate change scenario can as well adapt to a smaller climate change scenario. The scenario assumes that there will be an increase in growth, the population will peak by the mid-century, and there will be an increase in new technologies. From the 23 General Circulation Models (GCM), we downscaled the temperature and precipitation for future periods. This gave rise to a monthly temperature deviation (ΔT) and monthly precipitation deviation (ΔP) from the historic data supplied. The result is presented in Table 15.1. The Fig. 15.2 as presented in Table 15.1 show that in all the agricultural zones precipitation and temperature increased but the variation differs among the agricultural zones.

To determine the variations in yield amount under different climate change scenarios, we apply the obtained climate change scenarios to the crop simulation

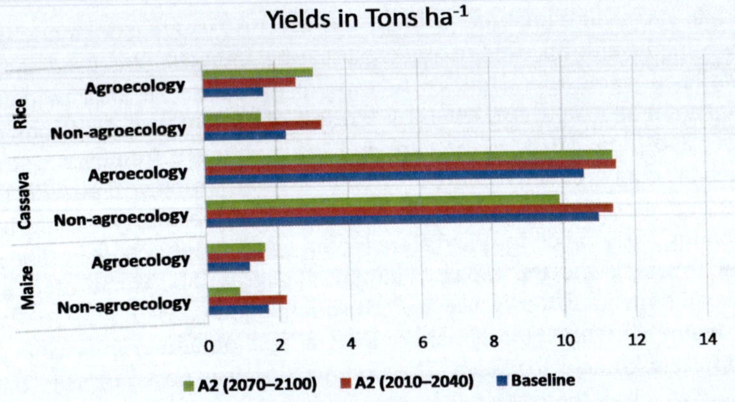

Fig. 15.2 Crop yield volatilities under different climate change scenarios and management options

models. We employed stochastic weather generators (WGs) to estimate the daily data since the crop model will require daily time step data.

Crop Model

We analyzed the crop yields under the near-future scenario (2011–2040) and far-future scenario (2071–2100) with the help of Cropping Systems Simulation (CropSyst) and AquaCrop models. Calibration of the CropSyst and AquaCrop model was carried out based on conservative parameters stipulated in (Hsiao et al. 2009). The models were validated using data collected in the 2020 and 2021 cropping seasons. The production of cassava was simulated by CropSyst, while the production of maize and rice was simulated by AquaCrop. The models were calibrated with experimental data based on input using Agroecology management (use of organic fertilizer, push-and-pull technology, and zero tillage) and Non-agroecology management (inorganic fertilizers, pesticide use, and soil tilling practices). These were carried out for three years of weather and growth records. The input use of agroecology and non-agroecology farm management patterns were used to indicate the two-management option of interest in the study. The models were then utilized for the above-mentioned sceneries and periods.

Estimation of Value-Variance

Our interest is not only to investigate the yield variance as a result of climate change, but to also understand the covariance between them. This is important in order not to produce a wrong estimation of impact. For example, in the case of a mixed or multiple cropping situation, without the use of a stochastic modeling framework, we might end up over or underestimating impacts. We used the expected value-variance framework strategy to capture the effect of risk associated with crop allocation and input use levels as a result of agricultural decisions employed (Hardaker and Lien 2010). In the expected value-variance framework strategy, we determine the choice of activities or crop combination, which will provide the farmer with the highest utility by taking into consideration the utility variability and its corresponding covariance for the different crop combination choices. The approach is utilized to assess what the impact of climate change will be as a result of a farmer's responses in terms of resource use and crop combinations. Using the BEFM model, we can determine what the optimal cropping allocation and crop combination will be for different climate sceneries. The optimization was based on certainty equivalent (CE) which is the expected income and variance of income of the different combinations (Riddel 2012). The expected certainty equivalent is given by

$$maxEU = E(Y) - \frac{\lambda}{2}V(Y)$$

$$Subject\ to\ Bj \geq \sum_{i=1}^{n} aijHi$$

where *E(Y)* = total expected income.

V(Y) = Income variance.

λ = Absolute risk aversion.

Bj = availability of *j*th resource.

aji = input use coefficient for crop *i*.

Hi is the area under each crop.

The expected certainty equivalent can be used to estimate the risk-free income such that options with higher expected certainty equivalent are preferred over the option with lower expected certainty equivalent. We used the survey data to measure the availabilities and constraints employed in the equation. We used the long-term prices as compiled in FAOSTAT, while the farm-level price, variable and fixed costs, depreciation and tax were from our survey data.

Results

How Will Climate Change Impact the Yield of Major Crops in Nigeria?

The result, as presented in Table 15.2, shows that in the near future, climate change will lead to an increase in maize production by 0.5 tons per ha for non-agroecology management methods, corresponding to a 29.4% increase in yield. For the agroecology management methods, the increase in yield was found to be about 0.4 tons per ha corresponding to a 33% increase in yield. In terms of cassava production, for Non-agroecology management practices, cassava yield will increase by 0.4 tons per ha corresponding to a 3.6% increase in yield. For the agroecology management practices, we observed a decrease in yield in the near future. The decrease was found to be about 0.1 tons per ha, corresponding to a 0.9% decrease in yield. This finding corresponds to the reports of Adhikari, Nejadhashemi, and Woznicki (2015) and Calleja-Cabrera et al. (2020), who indicated that climate change will lead to a decrease in yield in root crop production but contradict the finding made by Chukwuma et al. (2020) that in the near-future agroecological methods will boost yield. Finally, in terms of rice production, we found that in the near future, climate change would lead to an increase in yield both for the agroecological and non-agroecological management farming systems. For the non-agroecology farming system, we observed an increase of about 1 ton per ha, and for the agroecology farming system, we observed a similar amount of increase (0.9 ton per ha).

In the long run, we observed more variation between the two farming systems. For the non-agroecology farming system, we observed a decrease in yield across the three crops under consideration. Starting with the maize crop, we found that in the long run there would be a substantial decrease in the yield of maize for the non-agroecology farming systems. While climate change will lead to 0.8 tons per ha decrease in yield under the non-agroecology farming system, we observed that climate change would not have any significant impact on maize production using agroecological practices. The result in Table 15.2 shows that in the long run under climate change, there will be an increase in production by 0.43 tons per ha. For the cassava crop, we observed that under climate change, there would be a reduction in

Table 15.2 Projected gross margins under (USD ha^{-1})

Crop	Management option	Baseline	A2 (2010–2040)	A2 (2070–2100)
Maize	Non-agroecology	102.3	122.8	117.2
	Agroecology	72.0	102.6	153.2
Cassava	Non-agroecology	116.2	146.5	111.3
	Agroecology	105.3	122.2	120.4
Rice	Non-agroecology	338.5	129.2	−94.8
	Agroecology	313.1	243	270.1

yield of about 1.1 tons per ha in the long run under the non-agroecology farming system. Conversely, in the long run under climate change, there will be an increase in cassava production by 0.8 tons per ha. Similarly, for rice production, a similar trend was observed. While climate change will lead to a decrease in rice yields by 0.7 tons per ha in the far future, we observed that in the far future under climate change, rice yield would increase by 1.4 tons per ha under the agroecology cropping system.

How Will Climate Change Impact Farmers' Income?

The expected gross margins under different climate change scenarios and management options in the agricultural zones of southeast Nigeria are presented in Table 15.2. Even though there was an increase in rice yield under the non-agroecological system, we observed a negative gross margin. This could be explained by the emphasis on rice production in the area (UNEP 2015) as introduced in the model, which might lead to a glut. The returns from rice production might become very low and even negative in the far future. In addition, in the far future climate scenario, the agroecological system appears to be more suitable for small-scale farmers than the conventional farming system. Apart from cassava where we observed a 1.8 USD reduction per haectare, there is a consistent increase in gross margin for all the three crops under consideration. In the near future, we observe an increase in welfare for small-scale farmers producing maize and cassava but a reduction in gross margin for farmers producing rice. This is the case for both agroecology and non-agroecology farmers.

Discussion

To provide meaningful insight into the result, we discuss the findings following the research questions stipulated earlier.

What is the Impact of Climate Change at the Farm Level in Nigeria

Climate change will have a positive impact on the production of basic crops in Nigeria in the short run. However, in the long run, there will be a negative impact depending on the management practices employed. It is interesting to point out that in the near future, we did not observe any significant reduction in yield for the three crops under consideration. This is in line with a report by Haider (2019) that climate change impact will be much more observable in the far future.

To explain the impact of climate change at the farm level, we discuss the dynamic soil factors vulnerable to climate impact. The soil quality can depend on a number of factors. Quality soil has the robust ability to support crop needs. The implication is that adaptation options that are targeted for the far future will be necessary for others to ensure sustainable development and food security. It is expected that the population of Nigeria will spike to about 400 million people by 2050 (World Population Review 2019). This shows that the farming system required to feed this population has to ensure continuous yield in the long run. The soil, air, and water temperature affect crop maturity. This change is induced by the change in atmospheric temperature and precipitation. An increase in temperature due to climate change might lead to the heating up of the soil water, leading to the evaporation of water from the soil. Although this impact can be reduced through irrigation, in places where water is scarce, irrigation might not be possible. Over a long period, the impact of climate change might lead to drought, salination, and desertification. This will lead to the loss of peat soils (Boko 2007), which further impact the ability of soil crop needs.

To further understand the impact of climate change on the soil, we need to understand the dynamic properties of soil that are non-static. These dynamic properties include the organic matter content of the soil and the alteration of the soil pH levels.

The soil organic matter in the soil determines several soil functions such as the soil nutrient-supplying power, water-holding capacity of the soil, the soil structure, and the food for the huge numbers of organisms that live in the soil. According to Palm et al. (2001), it is the soil organic matter that has the highest potential to be impacted by climate change. With the increase in the temperature across the agricultural zones, there will be a net release of soil organic matter due to increased decomposition and mineralization of the organic matter content in the soil. With this release over time, the soil organic matter content will be substantially reduced (Palm et al. 2001). Indirectly, with the rapid decrease in the soil organic matter, major soil properties such as the water-holding capacity, aggregate formation, and stability, cation exchange capacity, and the general soil nutrient content will be reduced.

Apart from the soil organic matter, the soil acidity, or pH, are highly sensitive to climate change impact. At the farm level, the impact of climate change can be observed in the effect it has on the soils and the alteration of the functions that the soil performs for crop production. The important factors that determine the soil pH are the rate of weathering of parent material and vegetation types (Baveye et al. 2020). This means that one of the direct effects of climate change will be the change in pH levels. An increase in precipitation will lead to an increase in leaching, which will lead to an increase in the level of soil acidity. Our result showed that in all the agricultural zones, the precipitation will increase over the near and far future. This increase in precipitation will create more humid environments which are known to decrease soil pH over time due to soil acidification as a result of leaching occasioned by the high amounts of rainfall. In Okigwe and Ebonyi north agricultural zones where precipitation increase was observed to be up to 20%, it suggests the possibility of increased pH levels in these areas.

Another impact of climate change on the soil pH level is the increased CO_2 concentration in soil water. The dissolving of CO_2 in water has been found to reduce soil pH levels (Zhang et al. 2021). With the dissolution of carbon dioxide in water, there will higher release of hydrogen ions, which leads to a decrease in pH level. Indirectly, the organic matter decomposition and decay occur as a result of increased temperature. The soil organic matters contain the element carbon (C) (Zhang et al. 2021), and as they decompose, the carbon is released into the water. Because carbon is unstable, they easily react with oxygen to form CO_2.

In terms of greenhouse gas emissions, Baveye et al. (2020) showed that nitrous oxides are one of the major agricultural emissions from farming. According to Menšík et al. (2019), nitrous oxide gas possesses over 300 times the potential to cause global warming than carbon dioxide. Since nitrogen forms the primary nutrient in fertilizers, when farmers introduce it into the soil, the unused nitrogen is released into the atmosphere as nitrous oxide. The use of synthetic nitrogen fertilizers by conventional farmers has been known to release unused nitrous oxides, while natural nitrogen from nitrogen-fixing crops like legumes, compost, and manure does not release nitrous oxide (Menšík et al. 2019).

What Are the Different Agroecology Measures Employed by Smallholder Farmers in Adapting to the Impact of Climate Change

Climate change will have a positive impact on the production of basic crops in Nigeria in the short run. However, in the long run, there will be a negative impact depending on the management practices employed. Using information from the farm household survey, we identified three agroecological practices employed by smallholder farmers in coping with climate change impact. The agroecological practice includes the use of organic fertilization, the use of biological insect control measures, and the adoption of reduced or zero tillage. We present the agroecological practices according to the analytical framework in Wezel et al. (2014) corresponding to three categories: increase efficiency in the use of inputs, the direct substitution of inputs, and total redesign (Fig. 15.3). By increase in efficiency, we refer to agroecological practices adopted by farmers to decrease the consumption of inputs, e.g., Splitting Nitrogen fertilizer application. At the same time, not compromise crop productivity. Substitution practices refer to the direct replacement of external chemical inputs such as pesticides and fertilizers by adopting the use of organic fertilizers, and push-and-pull technology. Lastly, redesign refers to changes in the farming system such as moving from tillage to agroforestry (Wezel et al. 2014).

Splitting Nitrogen Fertilizer Application

According to the farmers surveyed, who employ the splitting nitrogen fertilizer application technique, the approach helps them to supply optimum nitrogen requirements for the crops without incurring waste. The idea is that nitrogen is supplied to the crop only when it is needed. By matching the supply with the demand, excess nitrogen

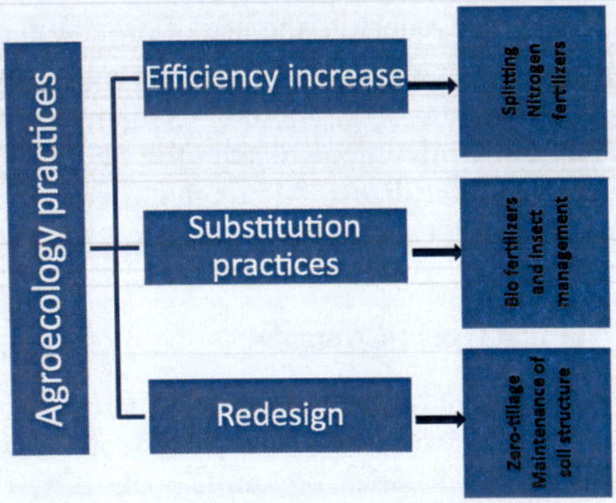

Fig. 15.3 Different levels of agroecology practices. *Source* Wezel et al. (2014)

is left in the soil, thereby reducing the evaporation of nitrous gases. In this way, agroecological practices not only improve adaptation by ensuring that plants receive adequate nutrient, but also aids in the mitigation of climate change. To understand the uptake and allocation dynamics, farmers base the idea on the inverse relationship between crop nitrogen requirement and plant growth. This phenomenon applies to all crops (Lemaire et al. 2008). According to the farmers, the nitrogen requirement for crops decreases monotonically with crop growth. This is based on the fact that the ontogenetic or visible morphological characteristics of crops, especially in leaf area per unit of plant mass declines with growth. In addition, their mobilization of Nitrogen from shaded leaves up to top leaves also declines with nitrogen deficiency. This is also evident in yellowish leaves at the top relative to lower leaves. With these indicators, farmers can determine when nitrogen supply will be needed (Lemaire et al. 2008).

Organic and Bio-fertilizers

Utilization of bio-fertilizers is another way to improve nutrient availability without the use of chemical fertilizers. Bio-fertilizers are microorganisms that have the ability to colonize rhizosphere, which has the ability to supply primary nutrients to the soil, thereby promoting the growth of the host plant (Gautam et al. 2021). The arbuscular mycorrhizal fungi (AMF), Plant growth-promoting rhizobacteria (PGPR), and nitrogen-fixing rhizobia are the three major groups of microorganisms that are considered bio-fertilizers (Wezel et al. 2014): the most common bio-fertilizer is the nitrogen-fixing rhizobia which have been in existence for centuries. Recently, there have been efforts in the commercialization of PGPR and AMF inoculants, especially in Europe and America (Gautam et al. 2021). The bio-fertilizers have been reported to be very effective in the uptake of N, growth promotion, and maintaining crop yield in the face of climate change (Wezel et al. 2014; Dasgupta et al. 2021; Maan and Garcha 2021).

Organic fertilization entails substituting the use of inorganic fertilizers. It also entails fertilization efficiency as it helps in the improvement of general soil fertility.

In terms of redesign, the use of organic fertilization can be a critical factor. This is because, to ensure that there is sufficient inorganic manure for the soil, there will be a need for the integration of crop and animal farming systems, hence a redesign of the farm. Soil biological activities and mineralization increased with the application of organic fertilizers. Although it involved higher labor and energy demand, organic fertilization can lead to sufficient utilization of soil Nitrogen (Maan and Garcha 2021).

Natural Insect Control

Biological and botanical insect control is an agroecological practice that aims to replace the use of synthetic pesticides. Botanical insect control is natural pesticides, botanical pesticides, or simply botanicals. These natural pesticides are used to counteract the associated negative effects of synthetic pesticides while efficiently serving the purpose of pesticides. Although they have not been effectively used on large scale, farmers who employ the use of these natural pesticides in Nigeria on the small scale observe positive effects. Our interview with the farmers showed that the use of these botanicals is of immense benefit to the farmers, especially when they cannot afford the cost of chemical pesticides. Studies have shown that climate change can lead to an increase in pest infestation (Grünig et al. 2020; Skendžić et al. 2021). This is due to increased temperature and humidity (Skendžić et al. 2021). With the use of natural insect control, farmers can address the challenge of pest infestation naturally without attracting additional costs. This might explains the increase in the gross margin under climate change for the agroecological farming system. Natural pesticides are derived from several sources. These include those derived from the seeds of trees, pyrethrum extracted from flowers, crude aqueous extracts of plants, and those that are based on plant essential oils (Nemet 2009; Wezel et al. 2014; Kassie et al. 2020). For traditional agriculture among small-scale farmers, and for organized organic farming where synthetic pesticides are not allowed, the use of botanicals has been very effective (Picco et al. 2016).

Biological pest control as an agroecological practice replaces synthetic pesticides. The bio-organisms release natural enemies into the agroecosystems which have been proven effective in pest control (Fenibo, Ijoma, and Matambo 2021). The so-called biopesticides involve the use of AMF inoculants, bacteria, or other fungi to control organisms that harm crops (Fenibo, Ijoma, and Matambo 2021). According to Fenibo, Ijoma, and Matambo (2021), biopesticides eliminates pest in three ways (i) through a process known as antibiosis (antagonistic association between two organisms), (ii) through competition, and (iii) through the inactivation of pathogen germination (Using pheromones to disturb sexual reproduction of targeted insect pests is another biological control option). Another importance of natural pesticides and biological pest control is that it reduces the risk risks to human health as water pollution is avoided.

Redesign

Practices that are considered redesign practices are those practices that affect a large part or even the whole of the cropping system (Wezel et al. 2014). This means that the farmers adopting these practices engage in a holistic shift and rethinking of their cropping system in response to climate change. Most of the efficiency and substitution practices can also be classified as redesigns (e.g., organic farming). Another important redesign is the adoption of zero-tillage agroforestry practice. While zero tillage entails that the farm does disturb the soil in terms of tilling before planting seeds, agroforestry means an intentional integration of shrubs and trees into the farming systems that will generate environmental benefits for the farmer (Brown et al. 2018). The trees and shrubs are grown among or around the crops. Most times, agroforestry practices are natural in that farmers intentionally plant in forest areas. Other times, the farmers develop an artificial forest planting trees and shrubs among the crops.

The agroforestry redesign bridges the gap by integrating agriculture and forestry into an integrated system that is capable of addressing the environmental problem of climate change. It improves resilience in the cropping systems and even lessens the impacts of climate change. Small-scale farm household surveyed is of the opinion that adopting the agroforestry redesign has helped them in improving crop yield as it increases soil fertility and reduces heat from the sun while providing other benefits to human welfare.

How Effective Are the Agroecology Practices in Counteracting Climate Change Impacts

Recent studies suggest that is likely impossible to meet the Paris Agreement's goal of reducing the impact of climate change without major reductions in emissions from food and agriculture (Farrelly 2016; FAO 2019a; Poppe et al. 2021). The premise was supported by the World Resources Institute study, which reported several industries that emissions from the agricultural sector need to be reduced by 39% by the year 2050. Critics have argued that moving to organic farming will lead to a reduction in yields (Elizabeth Amechi and Caleb 2015). They argue that the use of organic farming will mean greater land use. For instance, a study by Xu et al. (2018) suggested that 100% adoption of organic farming will increase the lands needed to produce the same amount of food production. The study also argued that a conventional production system would yield the same food under climate changes and soil degradation. However, just as elicited in this study, the findings only hold for the near future. However. In addition, controlled trials at the Rodale Institute have shown it is possible to get the same yields in some organic systems over the long term, and more research may help more farmers achieve those yields in the fields of the future. "Conventional yields are high because there have been decades of research and billions of dollars invested," said Jessica Shade, the director of science programs at the Organic Center. "Organic is a pretty nascent field, and it gets a pretty limited amount of funds. Even with that small amount of funding, we've seen dramatic increases in yield."

Studies have shown global evidence of soil mismanagement either through obvious mismanagement or ignorance. the conventional farming system, although it produces increased crop yields in the near future, only succeeds in putting increased pressure on the soil (FAO 2019a) leading to soil degradation. Our finding suggests that for the three crops assessed in this study, the adoption of agroecology practice will provide a better adaptation option to cope with the impact of climate change in the far future. These findings can be explained as follows.

Firstly, at the farm level, farmers' decisions in terms of the production systems and soil management practices employed play a significant role in determining yield under climate change. Although the use of inorganic fertilizer and other green revolution strategies might lead to improved food production in the near future, over a long period of turn, the yield will gradually reduce below what it would be if sustainable production patterns were followed. This is in line with the law of diminishing returns (Shephard 1974; Shi et al. 2020) which suggests that there will be a decrease in marginal yield which a consistent use of external inputs, assuming all other factors remain constant. Our findings suggest that agroecology provides an alternative cropping system that might be effective in grappling with diminishing returns.

Secondly, it has also been suggested by Elizabeth Amechi and Caleb (2015) that organic farming increases water infiltration capacity at the same time reduces surface runoff. This has the effect of preventing flooding of agricultural fields and reducing soil erosion. These two advantages make it possible for agroecological farmlands to robustly adapt to climate change impacts, thereby increasing yields.

Thirdly, the practice of organic agriculture and zero tillage can also help combat the impact of climate change by improving the amount of carbon in the soil (Baveye et al. 2020). Improvement of carbon storage in the soil over a long time has the potential of raising productivity under climate change scenarios. By implication, the use of these agroecological practices lower carbon emissions to the atmosphere by storing more carbon in the soil, thereby not only an adaptation strategy but also a mitigation response. Many similarly, the use of push-and-pull technology practices have shown to be critical in returning crop nitrogen and residues to the soil which has been found to enhance productivity and favors carbon storage (Baveye et al. 2020).

Fourthly, organic farming and zero tillage have been found to increase the water-holding capacity of the soil. With the increased temperature in the near and far future, soils rich in organic matter content will be better in their water-holding capacity properties compared to conventional soils. additionally, with an increase in precipitation, erosion and flooding are highly controlled in organic soils compared to conventional soils (FAO 2019b).

Finally, the push-and-pull technology has been shown to be instrumental in increasing the conservation of soil moisture in farms where the technology is utilized. The desmodium plants serve as cover crops that help to reduce the intensity of the sun's heat on the soil, thereby preventing excessive evaporation from the soil. In this way, soil water content loss is minimized. Furthermore, the nitrogen fixation property of the desmodium plants helps to fix nitrogen in the soil, thereby eliminating the need for artificial nitrogen application which has been found to deplete soil fertility in the far future (Nemet 2009).

Limitation of the Study

The study assumes that all the farm households surveyed are commercial farms, hence did not take into account the associated household consumption demands. This means that commercial and subsistence farm households were not differentiated. Secondly, the changes in yield as determined do not take into account the possibility of diseases and pest incidences. This is also the case for the crop type, as the study did not take into account the possibility of improved varieties and technological changes.

Conclusion

This study investigated the role of agroecology in counteracting the negative impact of climate change on crop production in Nigeria. By modeling the near and far future impacts of climate change on crop production, we showed how climate will impact crop production under two crop production systems (agroecology and non-agroecology production systems). Our climate and crop models show the climate change impact on cassava, maize, and rice production in Nigeria. These three crops were selected as they are the three most common staple crops produced in Nigeria. The result showed that the use of agroecology management practices not only reduces the impact of climate change in the near future, but will also lead to increased crop yield in the future. The finding suggests that to feed over 400 million projected population of Nigeria by 2050, the use of agroecological practices will be a better alternative to conventional farming methods. To advance the use of agroecological farming methods, governments at every level in Nigeria need to mainstream organic agriculture in national government policies. This is important as it will not only address climate change impacts but also hunger and poverty.

Beyond adaptation, governmental agricultural policies in the country should also acknowledge organic farming as a potent strategy for mitigation and reducing greenhouse gas emissions due to the potential of organic farming in sequestering carbon from the atmosphere. The government at all levels should begin to help farmers adapt to climate change by promoting agroecological practices. This can be done by investing in climate research and extension service delivery. Green evolution strategies, in developing nations such as Nigeria, should incorporate initiatives that are based on the principles of agroecology (Emeana et al. 2019) into the Nationally Intended Mitigation Contribution (NIMC).

References

Abi Saab MT, Todorovic M, Albrizio R (2015) Comparing AquaCrop and CropSyst models in simulating barley growth and yield under different water and nitrogen regimes. Does calibration year influence the performance of crop growth models? Agric Water Manag 147:21–33. https://doi.org/10.1016/j.agwat.2014.08.001

Adewumi M, Omorisho A (2002) An analysis of production objectives of small-scale rural farming households in kwara state, nigeria. J Rural Dev 25(1):201–211. https://ageconsearch.umn.edu/record/288176/files/5.pdf

Adhikari U, Nejadhashemi AP, Woznicki SA (2015) Climate change and eastern Africa: a review of impact on major crops. Food Energy Secur 4(2):110–132. https://doi.org/10.1002/fes3.61

Akpoti K et al (2021) The potential for expansion of irrigated rice under alternate wetting and drying in Burkina Faso. Agric Water Manag 247:106758. https://doi.org/10.1016/j.agwat.2021.106758

Altieri MA, Toledo VM (2011) The agroecological revolution in Latin America: rescuing nature, ensuring food sovereignty and empowering peasants. J Peasant Stud 38(3):587–612

Altieri MA, Funes-Monzote FR, Petersen P (2012) Agroecologically efficient agricultural systems for smallholder farmers: contributions to food sovereignty. Agron Sustain Dev 32(1):1–13. https://doi.org/10.1007/s13593-011-0065-6

Azzi G (1956) Agricultural ecology. Constable & Company, London

Baveye PC et al (2020) Soil organic matter research and climate change: merely re-storing carbon versus restoring soil functions. Front Environ Sci 8. https://doi.org/10.3389/fenvs.2020.579904

Bensin BM (1928) Agroecological characteristics description and classification of the local corn varieties chorotypes. Unknown. https://scholar.google.com/scholar?q=BensinB.M.%281928%29

Bensin BM (1930) Possibilities for international cooperation in agroecological investigations. Int Rev Agr Mo Bull Agr Sci 21:277–284. http://scholar.google.com/scholar_lookup?title=PossibilitiesforinternationalcooperationinagroecologicalInvestigations&author=B.M.Bensin&journal=Int.Rev.Agr.Mo.Bull.Agr.Sci.Pract.%28Rome%29&volume=21&pages=277-284&publication_year=1930

Boko M (2007) Executive summary. In: Parry ML et al (eds) Chapter 9: Africa. .. Retrieved .', In: Parry ML (ed) Climate change 2007: impacts. Cambridge University Press (CUP), London. Print version: CUP, Cambridge, UK

Botelho MIV, Cardoso IM, Otsuki K (2016) "I made a pact with God, with nature, and with myself": exploring deep agroecology. Agroecol Sustain Food Syst 40(2):116–131. https://doi.org/10.1080/21683565.2015.1115798

Brown SE et al (2018) Evidence for the impacts of agroforestry on agricultural productivity, ecosystem services, and human well-being in high-income countries: a systematic map protocol. Environ Evid 7(1):24. https://doi.org/10.1186/s13750-018-0136-0

Calleja-Cabrera J et al (2020) Root growth adaptation to climate change in crops. Front Plant Sci 11. https://doi.org/10.3389/fpls.2020.00544

CGIAR (2013) African agriculture: does farm size really matter?, Water, land and ecosystems. https://wle.cgiar.org/thrive/big-questions/does-farm-size-really-matter-africa/african-agriculture-does-farm-size-really

Challinor A et al (2005) Simulation of the impact of high temperature stress on annual crop yields. Agric Meteorol 135:180–189. https://doi.org/10.1016/j.agrformet.2005.11.015

Chukwuemeka SU, Alaezi K, Ume C (2018) Climate change vulnerability analysis of smallholder farmers in Enugu state Nigeria: Gender sensitive approach. J Aridland Agric, 1–6. https://doi.org/10.25081/jaa.2018.v4.3374

Chukwuma U, Stephanie D, Nuppenau E-A (2020) Linking agency to food and nutrition security (FNS): the role of agroecology. In: Food policy and governance: examining linkages and food security outcomes, p 303

Clements DR, Shrestha A (2004) New dimensions in agroecology for developing a biological approach to crop production. J Crop Improv 11(1):1–20. https://doi.org/10.1300/J411v11n01_01

Dalgaard T, Hutchings NJ, Porter JR (2003) Agroecology, scaling and interdisciplinarity. Agric Ecosyst Environ 100(1):39–51. https://www.sciencedirect.com/science/article/pii/S016788090300152X
Dasgupta D et al (2021) Chapter 1—Microbial biofertilizers: recent trends and future outlook. In: De Mandal S, Passari AKBT-RA, in MB (eds) Academic Press, pp 1–26. https://doi.org/10.1016/B978-0-12-822098-6.00001-X
De Schutter O (2011) Agro-ecology and the right to food, United Nations Human Rights Office of the High Commissioner
Eakin HC, Wehbe MB (2009) Linking local vulnerability to system sustainability in a resilience framework: two cases from Latin America. Climatic change. Springer, pp 355–377. https://doi.org/10.1007/s10584-008-9514-x
Edwards CA (2017) Agroecology: a transdisciplinary, participatory and action-oriented approach (Ed., VE Méndez et al.). Taylor & Francis, London
Elizabeth Amechi O, Caleb E (2015) Organic farming in Nigeria: need for popularization and policy. World J Agric Sci 11(4):346–355. https://doi.org/10.5829/idosi.wjas.2015.11.6.1885
Emeana E et al (2019) Evaluating the role of public agricultural extension and advisory services in promoting agro-ecology transition in Southeast Nigeria. Agroecology and sustainable food systems. Taylor and Francis Inc., pp 123–144. https://doi.org/10.1080/21683565.2018.1509410
FAO (2014) International symposium on agroecology for food security and nutrition, food and agriculture organization of the United Nations. http://www.fao.org/about/meetings/afns/en/
FAO (2019a) Agroecological and other innovative approaches for sustainable agriculture and food systems that enhance food security and nutrition (Summary)
FAO (2019b) Climate change|Food and agriculture organization of the United Nations. http://www.fao.org/climate-change/en/. Accessed 23 April 2020
FAO (2019c) Definitions: agroecology knowledge hub, food and agriculture organization of the United Nations. https://www.fao.org/agroecology/knowledge/definitions/en/?page=2&ipp=6&no_cache=1&tx_dynalist_pi1[par]=YToxOntzOjE6IkwiO3M6MToiMCI7fQ==
Farrelly M (2016) Agroecology contributes to the sustainable development goals. Farm Matt 32(3):32–34
Federal Ministry of Agriculture and Rural Development (2017) Agricultural transformation Agenda
Fenibo EO, Ijoma GN, Matambo T (2021) Biopesticides in sustainable agriculture: a critical sustainable development driver governed by green chemistry principles. Front Sustain Food Syst 5. https://doi.org/10.3389/fsufs.2021.619058
Finelib (2019) List of Nigeria poultry farms. https://www.finelib.com/agricultuFinelibre/poultry
Food and Agriculture Organization (2018) Climate change and food security, forestry unit. http://www.fao.org/forestry/15538-079b31d45081fe9c3dbc6ff34de4807e4.pdf
Francis C et al (2003) Agroecology: the ecology of food systems. J Sustain Agric 22(3):99–118. https://doi.org/10.1300/J064v22n03
Gautam K et al (2021) Chapter 1—Microbial biofertilizer: types, applications, and current challenges for sustainable agricultural production. In: Rakshit A et al (eds). Woodhead Publishing, pp 3–19. https://doi.org/10.1016/B978-0-12-821667-5.00014-2
Gliessman S (1998) Agroecology: ecological processes in sustainable agriculture. Ann Arbor Press, Chelsea, Michigan
Grünig M et al (2020) Crop and forest pest metawebs shift towards increased linkage and suitability overlap under climate change. Commun Biol 3(1):1–10. https://doi.org/10.1038/s42003-020-0962-9
Haider H (2019) Climate change in Nigeria: impacts and responses. https://opendocs.ids.ac.uk/opendocs/handle/20.500.12413/14761
Hardaker JB, Lien G (2010) Stochastic efficiency analysis with risk aversion bounds: a comment. Austr J Agric Resour Econ 54(3):379–383. https://doi.org/10.1111/j.1467-8489.2010.00498.x
Hsiao TC et al (2009) AquaCrop—the FAO crop model to simulate yield response to water: III. Parameterization and testing for maize. Agron J 101(3):448–459. https://doi.org/10.2134/agronj2008.0218s

IMEA (2010) Macroregion methodology report of Mato Grosso, Instituto Mato-Grossense de Economia Agropecuária

IPCC (2007a) Annex B. Glossary of terms, intergovernmental panel on climate change. https://archive.ipcc.ch/pdf/glossary/tar-ipcc-terms-en.pdf. Accessed 22 Apr 2020

IPCC (2007b) The physical science basis. contribution of working group I to the fourth assessment report of the intergovernmental panel on climate change summary for policymakers. Cambridge University Press, Cambridge, UK and New York, USA

Kassie M et al (2020) 'Women's empowerment boosts the gains in dietary diversity from agricultural technology adoption in rural Kenya'. Food Policy, November 2019, p 101957. https://doi.org/10.1016/j.foodpol.2020.101957

Kessete N, Moges MA, Steenhuis TS (2019) Chapter 2—Evaluating the applicability and scalability of bias corrected CFSR climate data for hydrological modeling in upper Blue Nile basin, Ethiopia. In: Melesse AM, Abtew W, Senay GBT-EH, and CV (eds). Elsevier, pp 11–22. https://doi.org/10.1016/B978-0-12-815998-9.00002-6

Lemaire G, Jeuffroy M-H, Gastal F (2008) Diagnosis tool for plant and crop N status in vegetative stage: theory and practice for crop N management. Eur J Agron 28:614–624. https://doi.org/10.1016/j.eja.2008.01.005

Linderhof V, Janssen V, Achterbosch T (2019) Does agricultural commercialization affect food security: the case of crop-producing households in the regions of post-reform Vietnam? Sustainability 11(5). https://doi.org/10.3390/su11051263

Maan PK, Garcha S (2021) Chapter 3—Production technology, properties, and quality management. In: Rakshit A et al (eds). Woodhead Publishing, pp 31–43. https://doi.org/10.1016/B978-0-12-821667-5.00013-0

Mamoon AA, Rahman A (2019) Chapter 4—Uncertainty analysis in design rainfall estimation due to limited data length: a case study in Qatar. In: Melesse AM, Abtew W, Senay GBT-EH, and CV (eds). Elsevier, pp 37–45. https://doi.org/10.1016/B978-0-12-815998-9.00004-X

Mengistu S et al (2019) Chapter 3—Evaluation of long-term satellite-based retrieved precipitation estimates and spatiotemporal rainfall variability: the case study of Awash basin, Ethiopia. In: Melesse AM, Abtew W, Senay GBT-EH and CV (eds) Elsevier, pp. 23–35. https://doi.org/10.1016/B978-0-12-815998-9.00003-8

Menšík L, Hlisnikovský L, Kunzová E (2019) The state of the soil organic matter and nutrients in the long-term field experiments with application of organic and mineral fertilizers in different soil-climate conditions in the view of expecting climate change. In: Larramendy M, Soloneski S (eds) Organic fertilizers. IntechOpen, Rijeka. https://doi.org/10.5772/intechopen.86716

Mgbenka RN, Mbah EN (2016) A review of smallholder farming in Nigeria: need for transformation. Int J Agric Extens Rural Dev Stud 3(2):43–54. http://www.eajournals.org/wp-content/uploads/A-Review-of-Smallholder-Farming-In-Nigeria.pdf

Morgan WB (2019) Agriculture in Southern Nigeria, excluding the cameroons, 35(2):138–150

Nemet GF (2009) Demand-pull, technology-push, and government-led incentives for non-incremental technical change. Res Policy 38(5):700–709. https://doi.org/10.1016/j.respol.2009.01.004

Okorafor OO, Akinbile CO, Adeyemo AJ (2017) Soil erosion in South Eastern Nigeria: a review. Sci Res J (SCIRJ) V(IX):30–37

Opata PI et al (2021) Adoption of improved cassava varieties in Nigeria: Insights from DNA fingerprinting versus self-reporting varietal identification approaches. Outlook Agric [Preprint]. https://doi.org/10.1177/0030727021990032

Pahmeyer C, Kuhn T, Britz W (2021) Single plots or shares of land—How modeling of crop choices in bio-economic farm models influences simulation results

Palm CA, Gachengo CN, Delve RJ, Cadisch G, Giller KE (2001) Organic inputs for soil fertility management in tropical agroecosystems: application of an organic resource database. Agric Ecosyst Environ 83(1–2):27–42. https://www.sciencedirect.com/science/article/pii/S016788090000267X

Picco S et al (2016) We are IntechOpen, the world's leading publisher of Open Access books Built by scientists, for scientists TOP 1%. Intech, (tourism), p 13

Poppe K, Vrolijk H, van Dijk R (2021) Design of a system for information transfer to reduce administrative burdens in the agrifood sector. Int J Food Syst Dyn 12(4):301–313. https://doi.org/10.18461/ijfsd.v12i4.92

Reinsborough MJ (2003) A ricardian model of climate change in Canada published by: Wiley on behalf of the Canadian Economics Association Stable. https://www.jstor.org/stable/3131913. Linked references are available on JSTOR for this article: All use subject to https://www.abo, 36(1):21–40

Riddel M (2012) Comparing risk preferences over financial and environmental lotteries. J Risk Uncert 45(2):135–157. http://www.jstor.org/stable/43550154

Schönhart M, Schmid E, Schneider UA (2011) CropRota—a crop rotation model to support integrated land use assessments. Eur J Agron 34(4):263–277. https://doi.org/10.1016/j.eja.2011.02.004

Shephard RW (1974) Eichhorn W et al (eds) Production Theory © Springer, Berlin, Heidelberg, pp 287–288

Shi P et al (2020) Does the law of diminishing returns in leaf scaling apply to vines?—evidence from 12 species of climbing plants. Glob Ecol Conserv 21:e00830. https://doi.org/10.1016/j.gecco.2019.e00830

Silici L (2014) Agroecology. What it is and what it has to offer. International Institute for Environment and Development, London, (June), pp 1–27

Simon X, Montero M, Bermudez Ó (2020) Advancing food security through agroecological technologies: the implementation of the biointensive method in the dry corridor of Nicaragua. Sustainability (Switzerland) 12(3). https://doi.org/10.3390/su12030844

Skendžić S et al (2021) The impact of climate change on agricultural insect pests. Insects 12(5):440. https://doi.org/10.3390/insects12050440

Stöckle CO et al (2014) CropSyst model evolution: from field to regional to global scales and from research to decision support systems. Environ Model Softw 62:361–369. https://doi.org/10.1016/j.envsoft.2014.09.006

Stöckle CO, Donatelli M, Nelson R (2003) CropSyst, a cropping systems simulation model. Eur J Agron 18(3):289–307. https://doi.org/10.1016/S1161-0301(02)00109-0

Thompson PB (2015) From world hunger to food sovereignty: food ethics and human development. J Glob Ethics 11(3):336–350. https://doi.org/10.1080/17449626.2015.1100651

Ume C (2017) Critical perspective on climate change adaptation among farmers in developing nations: unpacking divergent approaches. Mod Conc Dev Agron 1(1). https://doi.org/10.31031/mcda.2017.01.000504

Ume C (2018) Critical perspective on climate change adaptation among farmers in developing nations: unpacking divergent approaches. Mod Conc Dev Agron 1(1):1–6. https://doi.org/10.31031/mcda.2017.01.000504

Ume CO, Opata PI, Onyekuru ANJ (2021) Gender and climate change adaptation among rural households in Nigeria. Afric Handb Clim Change Adapt, 1–17. https://doi.org/10.1007/978-3-030-42091-8_182-1

UNEP (2015) Integrated assessment of the impact of trade liberalization. Environment [Preprint]

United States Department of Agriculture (USDA) (2007) Sustainable agriculture: definitions and terms. Related terms: alternative farming systems. https://www.nal.usda.gov/afsic/sustainable-agriculture-definitions-and-terms-related-terms#term1

Wezel A et al (2014) Agroecological practices for sustainable agriculture. A review. Agron Sustain Dev 34(1):1–20. https://doi.org/10.1007/s13593-013-0180-7

Wezel A, Bellon S, Doré T (2009) Agroecology as a science, a movement and a practice. A review. Agron Sustain Dev 29(503):503–515. https://doi.org/10.1051/agro/2009004
Wezel A, Jauneau J-C (2011) Agroecology—interpretations, approaches and their links to nature conservation, rural development and ecotourism. Integr AgricConserv Ecotour: Exam Field 1(11):324. https://doi.org/10.1007/978-94-007-1309-3
World Population Review (2019) *Nigeria Population 2019*. Available at: http://worldpopulationreview.com/countries/nigeria-population/. Accessed 5 Mar 2019
Xu Q et al (2018) Why do farmers not convert to organic farming? Modeling conversion to organic farming as a major change. Nat Resour Model 31(3):e12171. https://doi.org/10.1111/nrm.12171
Zhang J et al (2021) Rapid clinical progression and its correlates among acute HIV infected men who have sex with men in China: findings from a 5-year multicenter prospective cohort study. Front Immunol 12. https://doi.org/10.3389/fimmu.2021.712802

Chapter 16
Climate Change Impact on Nigerian Ecology, Vegetation/Forest, Carbon and Biomass Management

R. E. Ita and F. O. Ogbemudia

Abstract Nigeria is currently facing diverse environmental glitches from climate change which are detrimental to the vegetation ecology and ecosystem functioning. Hence, exploring the responses of vegetation to climate change is a necessity to promote management techniques towards their protection and sustenance. This study assessed the impacts of climate change on Nigerian ecology, vegetation as well as carbon and biomass management. Climate change in different Nigerian zones results in variabilities in temperature and rainfall, thereby affecting Nigerian ecology/vegetation. High rainfall in the southern parts of Nigeria causes flooding of vegetation and forests, while high temperature and heat in the Northern parts of Nigeria result in wildfires, especially in savanna vegetation. Flooding issues from sea level rise in the south-south regions, lead to poor regeneration of vegetation, decline in vegetation cover, extinction of plant species and low forest productivity. Fire outbreaks in vegetation due to high temperature, especially in the northern zones, result in land exposure to extreme winds, erosion and nutrient loss. Due to desertification, large area of cropland and vegetation, especially in the Sahel region are lost. Fluctuations in rainfall patterns cause changes in soil moisture regimes which directly hamper vegetation growth. In northern regions of Nigeria, extreme temperature limits soil moisture thereby reducing carbon fixation and inputs in biomass. Drought in northern states of Nigeria causes a decline in vegetation and soil carbon stocks. From the foregoing, there is a need for the adoption of appropriate management techniques towards the management, protection and sustenance of vegetation amidst this environmental crisis.

Keywords Climate change · Vegetation · Biomass · Carbon storage · Flooding · Fire outbreaks · Drought · Desertification

R. E. Ita (✉)
Department of Biological Sciences, Ritman University, Ikot Ekpene, Akwa Ibom State, Nigeria
e-mail: alwaizfwesh247@yahoo.com

F. O. Ogbemudia
Department of Botany and Ecological Studies, University of Uyo, Uyo, Akwa Ibom State, Nigeria

J. C. Egbueri et al. (eds.), *Climate Change Impacts on Nigeria*, Springer Climate,
https://doi.org/10.1007/978-3-031-21007-5_16

Introduction

Irrefutable scientific justifications exist concerning the continuous warming of the earth and changing of the climate as a result of diverse anthropogenic activities. These two scenarios have been reported to be associated with devastating consequences on humans, animals, plants and environment as a whole (IPCC 2014). These consequences are felt globally, particularly in the tropical regions (Williams et al. 2018) where several biological and physical changes in the environment have been elicited (IPCC 2014). It is worth noting that, even though countries with low and middle income, especially Africa, are facing higher vulnerability to climate change, their inclination to enhance resilience in these countries is low (ND-GAIN 2021).

Nigeria is one such country in Africa plagued with a high prevalence of climate change. As reported by World Bank (2019), Nigeria ranks amongst the first ten countries in Africa with high susceptibility to climate change effects, with approximately 6% of total land mass exposed to extremities in weather conditions. Various incidences stemming from climate change have been documented in Nigeria ranging from flooding, drought, deforestation, irregularity in patterns of rainfall and encroachment by desert (Audu et al. 2013; Onyekuru and Marchant 2016; Olaniyi et al. 2019). As reported by Jibrillah et al. (2019), Olaniyi et al. (2019) and Giri et al. (2021), these have impacted negatively on the health of humans, land cover and use, as well as the livelihood of the country.

The effects of climate change on Nigerian ecology, vegetation, carbon and biomass is one topical issue that cannot be overlooked and underestimated. The climate of Nigeria over the years has faced variabilities in rainfall and temperature patterns and these have posed impacts that are detrimental to the vegetation ecology and ecosystem functioning. For instance, the increased intensities and durations of rainfall have resulted in runoff and consistent flooding in various places (Enete 2014). In the southern parts of Nigeria, sea level rise is expected due to increased precipitation which further exacerbates issues relating to coastal land submersion and flooding (Ebele and Emodi 2016; Akande et al. 2017). The drought in the northern parts of Nigeria has continued to be a constant scenario due to low or absence of rainfall and temperature rise (Amanchukwu et al. 2015). This causes water bodies to dry up and be at risk of disappearing (Elisha et al. 2017; Dioha and Emodi 2018).

These impacts posed by fluctuations in temperature and rainfall patterns also affect the vegetation ecology since biomass management, cycling and availability of nutrients, microbial and physiological activities are dependent on them. For instance, slight changes in rainfall pattern may be consequential on moisture availability and can trigger erosion as well as increased acidity and salinity levels in the soil. Fluctuation in temperature and rainfall may also cause disturbances such as high winds, floods, wildfires, landslides, avalanches, etc. Since these factors are connected to the growth and productivity of vegetation, they can cause modifications in the composition and structure of vegetation in the long run. Within this purview, this review aims at examining the impacts of climate change on Nigerian ecology, vegetation/forest, carbon and biomass management so as to understand the dynamism of vegetation in response to changes in climate as this will help in better and future management of the ecosystems.

Impacts of Climate Change

Impacts on Nigerian Ecology

Nigeria as a country has a land mass of 923, 768 km^2 and lies between longitudes 2° 2′ and 14° 30′ E and latitudes 4° 1′ and 13° 9′ N. Nigeria consists of six ecological regions namely; rainforest, mangrove swamp, Guinea savanna, Derived savanna, Sahel savanna and Sudan savannah (Fig. 16.1). In the rainforest, mangrove swamps and some areas of the derived savanna, the rainfall pattern is bimodal, while in the Sahel, Sudan and Guinea savannas, rainfall distribution is unimodal. Rainfall ranges from 500 mm (Sahel savanna) to approximately 3000 mm (mangrove swamp) annually. In the past years, the ecology of Nigeria has faced several alterations due to changes in climate as a function of anthropogenic upheavals. This is evidenced in the form of temperature and rainfall variabilities.

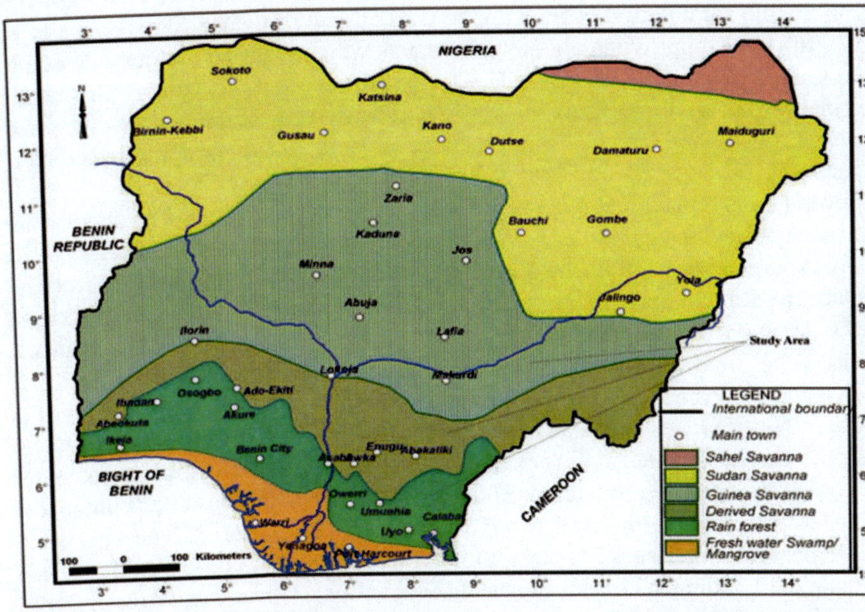

Fig. 16.1 Map of Nigeria showing the ecological zones. *Source* Odebode and Eniola (2019)

An increase in temperature and a decrease in rainfall across the globe are the highest impacts posed by climate change (Hengeveld et al. 2002). Temperature elevation has resulted in the melting of glaciers and ice caps. Melting of the ice caps and glaciers in temperate regions, rainfall increase in some regions of the world and ocean expansion resulting from the warming of water have led to a rise in sea level, erosion and coastal flooding (Hengeveld et al. 2002). Globally, the current rise in sea level is estimated at 0.2 m. This value according to Hengeveld and Whitewood (2005), is predicted to upsurge by 1 m in 2100. This implies that the current estimate of rise in sea level (0.2 m) has flooded about 3400 km^2 of Nigerian coastal areas, and if the projected 1 m rise in sea level is attained in 2100, about 18, 400 km^2 of Nigerian coastal areas will be faced with flooding issues (NEST 2003).

Coastal regions in Nigeria with <10 m above sea level such as Lagos, Calabar, Forcados, Warri, Bonny and Port Harcourt, would face threats arising from sea level rise (Ibrahim et al. 2018). The intrusion of the sea as a result of rise in sea level enhances the infiltration of salt water into freshwater which destroys coastal wetlands, beaches and mangroves. The flooding and erosion of Nigerian coastal areas pose serious environmental glitches in areas like Akwa Ibom, Delta, Lagos, Cross River, Bayelsa and Rivers States. Abu (2007) reported that as a result of rise in sea level, nearly 14 million inhabitants were displaced in Nigerian coastal regions. In Sokoto and Jigawa States 2 million and 40, 000 inhabitants were displaced resulting from flood incidence that took place in the year 2010, respectively, (Elisha et al. 2017). The flood incidence also led to the closure of Usman Dan Fodio University in Sokoto for a period of time due to the collapse of their bridge (Elisha et al. 2017). In Lagos, approximately 689 inhabitants were displaced in Ajegunle due to flood issues (Yekken 2011). BNRCC (2011) also reported that heavy rains which took place in Ikot Ibom Itam, Akwa Ibom State, caused the disappearance of approximately 4 streams and subjected many houses and landscapes to flooding and erosion, respectively.

Several transitions from Sudan to Sahel savanna and Guinea to Sudan savanna are noticeable at high intensities in the Northern parts of Nigeria (Gadzama and Ayuba 2016). A large expansion of the Sahara desert is observed trying to take over and displace the Sahel region. This expansion is estimated at 1–10 km annually (Yaqub 2007). Arising from this, the Northern part of the Nigerian States such as Katsina, Borno, Yobe, Jigawa and Sokoto are threatened severely by sand dunes (Fig. 16.2) typifying encroachment of desert (Odjugo and Ikhuoria 2003). These sand dunes in Northern Nigeria have displaced many agricultural lands leading to low production of crops. This has elicited a mass exodus of people to settlements that are not under the susceptibility of desertification. This emigration of people has often culminated in clashes between the emigrants and the inhabitants (Yaqub 2007). It also results in intense urbanization issues (Akonga 2001). These sand dunes have also threatened the main road leading to Damasak town of Mobbar Local Government Area of Borno State, Nigeria (BNRCC 2012).

Changes in climate also cause an alteration in the water cycle (Mcquire et al. 2002). Irregular rainfall patterns, rainfall decrease and temperature increase, particularly in Africa, will recharge groundwater sources (lakes, wells, rivers and springs) minimally

Fig. 16.2 Encroachment of sand dunes in Yusufari Local Government Area, Yobe State, Nigeria. *Source* Abdullahi (2022)

and this will bring forth scarcity of water and other related problems. In Nigeria, water resources like rivers have been lost due to drying up (Yugunda 2002). The southern region of Lake Chad which occupies some part of Northeastern region of Nigeria (Yobe and Borno States), also dried up due to climate change (Beyioku 2016). In the last 60 years, about 90% of Lake Chad has undergone shrinking after the long-lasting drought incidence in the early 1970s (Beyioku 2016). In 1963, the total surface area of the Chad was over 40, 000 km^2, but presently, it has <1300 km^2 (Beyioku 2016). This puts the Chad together with other northern rivers in Nigeria at the risk of disappearing. The scarcity of water arising from drought issues poses severe tension and threat to the left-over limited water resources, as everybody is dependent on them. This condition triggers the contamination and pollution of such water bodies culminating in the transmission of waterborne related diseases such as typhoid fever, cholera, river blindness and infection by guinea worm (BNRCC 2011).

Climate change also brings about a high spread of pathogenic insects in areas susceptible to this environmental menace. For instance, high rainfall in coastal and southern regions of Nigeria will bring about a high prevalence of mosquito and malaria sickness in the region. Also, increasing temperature will result in massive mosquito migration from the northern regions of Nigeria to a favourable clime.

Impacts on Forestry and Vegetation

Vegetation and forests play pivotal roles in our ecosystems such as modification of climate, protection of soil, primary productivity, and regulation of flood (Adegboyega et al. 2016). In Nigeria, these resources provide numerous benefits to man ranging from being sources of food, fuel wood, raw materials and shelter (Adegboyega et al. 2016). However, in the wake of climate change, forests and vegetation are constantly pressurized and threatened in Nigeria. As a result of high winds and erosion, produce from forests such as timber, cane and wood are on the decline in many forest ecosystems across the country, especially in the southern regions (Ogbuabor and Egwuchukwu 2017). This not only leads to a reduction in the nation's income, but also results in a high cost of furniture and materials for building (Ogbuabor and Egwuchukwu 2017).

Change in climate has caused many native tropical forests housing diverse flora and fauna species to be lost in Nigeria (Federal Ministry of Environment 2014). For example, Beyioku (2016) reported that forests surrounding Oyo have been totally decimated to grassland due to climate change issues. Constant water logging and flooding incidences have led to tree species mortality, making planting trees and regeneration of forests an uphill task (Ammer 2019). The persistent rise in sea level especially in the south-south region sharing proximity with the ocean, has resulted in land cover deterioration, degradation of mangroves, biodiversity loss, destruction of aquifers and catchment making water availability a serious problem in the region. The rise in sea level around coastal regions like in the Niger Delta, leads to the intrusion of salt water into plant communities growing in low-lying niches which can damage plants or vegetation with low tolerance to salinity, causing a wide-scale destruction of wetlands (Sayne 2011). This introduction of saline water in vegetation that is low lying, affects and alters the chemical and physical properties of the soil which, in turn, inhibit vegetation growth by preventing the root of plants from absorbing soil water and lowering the amount of available water to plants (Audu et al. 2013). This enhances moisture stress in vegetation.

Flooding events can be very detrimental to terrestrial and aquatic plants (Fig. 16.3). Anoxic conditions in soil caused by flooding around coastal areas in Nigeria can destroy the roots of plants and other important soil microorganisms. Flooding and runoff of storm due to the rise in sea level water can introduce several impurities into aquatic ecosystems. This makes the water to lose its clarity, making it prejudicial to aquatic plants as decreased availability of sunlight hinders their ability to photosynthesize.

In the northern parts of Nigeria dominated by the savanna biome, a slight change or rainfall reduction can affect the ecosystem through habitat degradation. According to the Federal Ministry of Environment (2014), desertification together with excessive exploitation of marginal lands in the northern regions has resulted in severe environmental degradation. Trees are facing damages arising from the wildfires which are pronounced in the northern and arid parts of Nigeria (Federal Ministry of Environment 2014). Low precipitation and high temperature culminating in issues of drought in the north, negatively affect vegetation cover and its productivity (Ammer 2019; Jibrilah et al. 2019). High temperature in northern states has turned surviving plants/vegetation into scrubs due to its negative influence on vegetation physiology, transpiration, translocation of nutrients, tropism and respiration (Audu

Fig. 16.3 Destruction of farmlands and crops due to flooding incidences. *Source* Hansen (2020)

et al. 2013). For example, under high temperature, there is a rapid loss of water from plants/vegetation by transpiration resulting in high water demand by plants. This causes imbalances in vegetation with regard to water requirements.

Odjugo and Ikhuoria (2003) had reported that high temperature and delayed rainfall in Northern areas like Katsina, Kano and Sokoto, led to high evapotranspiration rates in vegetation which culminated in wilting and subsequent death of plant species. High temperatures wither and weaken vegetation, exacerbating poor photosynthetic and productivity rates. For instance, Edem et al. (2016) reported a low yield of maize due to early cessation and delay in rainfall in the Gassol Local Government Area of Taraba State, Nigeria. The production of rice in Niger State, Nigeria, was reported to have reduced by 17% due to an increase in relative humidity by 1%, while an increase in temperature by 1%, resulted in an increase in 52.3% production of rice (Ayinde et al. 2013). A one percent temperature increase significantly reduced the yield of groundnut in Nigeria.

In the south-south regions of Nigeria, especially the Niger Delta where high rainfall is predominant, vegetation is struggling due to variabilities in rainfall patterns (Okon et al. 2018). In the northern parts of Nigeria dominated by grasslands, intense heat and high temperature are seriously affecting the vegetation of the place due to frequent wildfires (Fig. 16.4). Okon et al. (2018). Also, constant drought and high temperature in the Northern states (Bauchi, Gombe, Adamawa, Borno, Katsina, Jigawa, Yobe, Kebbi, Kano, Zamfara and Sokoto) have led to desertification, whereas encroachment by deserts (Fig. 16.5) has led to the loss of agricultural and grazing terrains to sand dunes which hamper vegetation growth greatly. As reported by Federal Ministry of Environment (2012), areas in these regions that were formerly occupied by bushes, grasses and scattered trees have undergone a transition to sand-like terrains. For example, in Tosha town of Yobe State, Nigeria, the yields of farmlands in uplands have been on a gradual decrease in the past years due to the emergence and formation of sand dunes (Gadzama and Ayuba 2016). Farmlands with proximity to Sansan local Government Area are also being threatened by these sand dunes (Gadzama and Ayuba 2016). Onah et al. (2016) also asserted that the loss of grazing terrains due to desertification in northern parts of Nigeria has forced pastoral nomads to migrate and encroach the southern regions for pasture, causing the destruction of farmlands. The desert which formerly occupied around 35% of the land mass of Nigeria, is progressing to about 0.6 km annually with deforestation rates being placed at 3.5% yearly (Yugunda 2002). This has caused an advancement in the desert belt from Kano, Kebbi and Maiduguri to regions like Jos, New Bussa, Sheleng and Kaduna, with savannas interfacing between the deserts and forest ecosystems along Kogi, Oyo, Benue and Osun States (Ibrahim et al. 2018). Aliyu (2013) reported that about 50% of lands used for agricultural activities in Gudu Local Government Area of Sokoto State was lost as a result of desertification and intrusion by sand dunes. Ragatoa et al. (2019) also reported that around 30 ha of land used for cropping were lost annually as a result of desertification in the Sahel areas of Nigeria. This low productivity of plants, on the other hand, affects wildlife sustainability since they are dependent on plants for food. Hence, this brings about displacement and a decline in wildlife populations.

Alteration of vegetation physiognomy is also prominent in Nigeria emerging from climate change. For instance, the persistence of temperature increases in the northern regions, causes changes in moisture availability in the soil, hence making vegetation shift as a response mechanism. Trees are compelled to move to altitudes that are higher for suitable and convivial climates or habitats for their existence and survival.

Fig. 16.4 Wildfire outbreak in Gulani Local Government, Yobe State, Nigeria destroying vegetation. *Source* Usman (2021)

Fig. 16.5 Degradation of farmlands and vegetation in Northern Nigeria due to desertification. *Source* The New Humanitarian (2017)

In Nigeria, Odujugo (2010) reported a shift in crops (guinea corn) that were predominant and grown in the extreme north of Nigeria to the guinea savanna as a result of delay or low rainfall and early rain cessation. Likewise, in southern regions of Zamfara, millet steadily replaced guinea corn (Odujugo 2010). Climate change has caused so many crops which preferred the Sahelo–Sudan niches for their growth to migrate to the guinea savanna, with those of the guinea savanna moving to occupy the northern forest zones (Odujugo 2010).

Impacts on Biomass and Carbon Management

The biomass of forests serves as important reservoirs for carbon by constantly exchanging CO_2 with the atmosphere, resulting from anthropogenic and natural processes. They have high carbon sequestration and storage in comparison to other ecosystems (Popo-Ola et al. 2012). They serve as a "braking mechanism", especially in the era of climate change. This is because, they act as sinks for carbon, sequestering atmospheric carbon and storing them in various pools like litter (leaves, roots, stems) soil and plants. However, the sequestration and storage of carbon by forests differ depending on the age of the forest. According to Binyam (2012), a young growing forest can sequester tonnes of carbon which is proportional to its biomass growth. An old forest, even when it is devoid of net growth, can serve as a pool by storing a large amount of carbon in its biomass (Binyam 2012). Hence, less amount of carbon is held in young growing forests having high carbon sequestration ability, while more carbon is stored in old forests with low carbon sequestration abilities

(Binyam 2012). Globally, IPCC (2000) reported that plants and soil store 19 and 81% of carbon in the biosphere. In all tropical, boreal and temperate forests, about 31.3 and 69% of carbon is stored in biomass and soils, respectively (IPCC 2000). About 50% of carbon is stored in soils and biomass of tropical forests (IPCC 2000). One of the principal greenhouse gases contributing to global warming and climate change is CO_2. During photosynthesis, vegetation helps greatly in the removal of this gas from the atmosphere and stores them in their various biomass. Resulting climate change, many forested areas have been lost permanently culminating in a decline in the earth's ability to store carbon.

The cycling of carbon in forests is greatly influenced by sets of climatic variables and CO_2 concentrations in the atmosphere. A slight change in these variables will certainly pose severe impacts on forests, especially in the area of carbon storage and management. Deforestation, natural perturbations and changes in rainfall and temperature patterns can be very critical in the development of forest stands (Keeton et al. 2007). This, according to McNulty (2002) and Zuidema et al. (2013), also affects the accumulation and storage of carbon. The impacts of climate change on vegetation or forest carbon storage and biomass may be positive and negative, although the negatives outweigh the positives. Positively, changes in climate have been reported to enhance water use efficiency, Net Primary Productivity and growth of forests under high CO_2 concentrations (Peters et al. 2013). Negatively, under high CO_2 concentrations or global warming, the rates of putrefaction of organic matter and heterotrophic respiration are bound to increase causing a decrease in the effectiveness of this sink to store carbon, thereby making forest ecosystem or vegetation generally, a CO_2 source. In the northern regions of Nigeria with dryness orchestrated by extreme temperature, soil moisture is limited, which, in turn, reduces carbon fixation and inputs in biomass of plants through a reduction in Net Primary Productivity. Drought resulting in the mortality of trees in northern states of Nigeria causes a decline in the carbon stock of above-ground biomass. Drought also contributes to increased loss of soil carbon to the atmosphere from the litter pools (Cleveland et al. 2010). Aside from CO_2, an elevation in the ozone concentration can result in a decrease in the production of biomass in forest ecosystems. Wildfires, especially in the extreme north due to high temperature, expedite the loss of stored organic carbon in soil and vegetation. Loss of vegetation cover and wildfires due to variabilities in rainfall and temperature, exposes the soil to prolonged erosion and leaching incidences leading to organic carbon losses from forest ecosystems. Under high temperature, the root biomass of vegetation is likely to decrease.

Conclusion and Recommendations

It is evidenced from this study that changes in climate pose several impacts on the ecology, forestry, vegetation, carbon and biomass management in Nigeria. These impacts manifest as a result of incidences such as increased temperature, variabilities in rainfall patterns, droughts, flooding and wildfires across various zones of the country. In the northern states, characterized by dryness and drought (due to increased temperature and low rainfall), vegetation cover and forests are disappearing at geometric rates giving room to encroachment by deserts and sand dunes. In the south-south region, with high rainfall, vegetation and forests are lost due to sea level rise and flooding scenarios. The net primary productivity, biomass and carbon stock in various pools of forests are also affected by these extremities in climate across the different Nigerian zones. From the foregoing, it is, therefore, imperative for actions to be taken by environmental stakeholders in Nigeria towards controlling and regulating anthropogenic activities that will lead to the release of greenhouse gases responsible for climate change. Since vegetation plays vital roles in ecosystem productivity and sustainability, there is a need for appropriate management techniques to be adopted for the management, protection and sustenance of vegetation amidst this environmental crisis.

References

Abdullahi M (2022) Great green wall: a glance at Nigeria's line of defence against the Sahara. https://humanglemedia.com/great-green-wall-a-glance-at-nigerias-line-of-defence-against-the-sahara/. Accessed 9 June 2022

Abu B (2007) Sea level rise and the Niger Delta of Nigeria. J Wetland 3(1):44–52

Adegboyega SA, Olajuyigbe AE, Balogun I, Olatoye O (2016) Monitoring drought and effects on vegetation in Sokoto State, Nigeria using statistical and geospatial techniques. Ethiop J Environ Stud Manag 9(1):56–69. https://doi.org/10.4314/ejesm.v9i1.6

Akande A, Costa AC, Mateu J, Henriques R (2017) Geospatial analysis of extreme weather events in Nigeria (1985–2015) using self-organizing maps. Adv Meteorol. https://doi.org/10.1155/2017/8576150

Akonga AZ (2001) The causes and impacts of drought. ANIS Monograph 3:1–16

Aliyu MM (2013) Assessment of the loss of agricultural farmland using remote sensing techniques in Gudu local government area of Sokoto State. J Educ Soc Res 3(8):145–150. https://doi.org/10.5901/jesr.2013.v3n8p145

Amanchukwu RN, Amadi-Ali TG, Ololube NP (2015) Climate change education in Nigeria: the role of curriculum review. Education 5(3):71–79. https://doi.org/10.5923/j.edu.20150503.01

Ammer C (2019) Diversity and forest productivity in a changing climate. New Phytol 221:50–66. https://doi.org/10.1111/nph.15263

Audu EB, Audu HO, Binbol NL, Gana JN (2013) Climate Change and its implication on agriculture in Nigeria. Abuja J Geogr Dev 3(2):1–15

Ayinde OE, Ojehomon VET, Daramola FS, Falaki AA (2013) Evaluation of the effects of climate change on rice production in Niger State, Nigeria. Ethiop. J Environ Stud Manag 6(6):763–773. https://doi.org/10.4314/ejesm.v6i6.7S

Beyioku J (2016) Climate change in Nigeria: a brief review of causes, effects and solutions. https://fmic.gov.ng/climate-change-nigeria-brief-review-causes-effects-solution/. Accessed 14 Apr 2022

Binyam AY (2012) Carbon stock potentials of woodlands and land use and land cover changes in North Western Lowlands of Ethiopia. M.Sc Dissertation. Hawassa University, Wondo Genet, Ethiopia, 129 p

BNRCC (Building Nigeria's Response to Climate Change) (2011) National adaptation strategy and plan of action on climate change for Nigeria (NASPA-CCN). Prepared for the Federal Ministry of Environment Special Climate Change Unit. http://csdevnet.org/wpcontent/uploads/NATIONAL-ADAPTATION-STRATEGY-AND-PLAN-OF-ACTION.pdf

BNRCC (Building Nigeria's Response to Climate Change) (2012). Improving livelihoods and stabilizing sand dunes in two communities in the sahel of Northeastern Nigeria. University of Maiduguri Press, Maiduguri

Cleveland CC, Wieder WR, Reed SC, Townsend AR (2010) Experimental drought in a tropical rain forest increases soil carbon dioxide losses to the atmosphere. Ecology 91:2313–2323. https://doi.org/10.1890/09-1582.1

Dioha MO, Emodi NV (2018) Energy-climate dilemma in Nigeria: options for the future. IAEE Energy Forum

Ebele NE, Emodi NV (2016) Climate change and its impact in Nigerian economy. J Sci Res Rep 10(6):1–13. https://doi.org/10.9734/JSRR/2016/25162

Edem O, Ahmed YM, Gambo MN, Tukura E (2016) Effect of rainfall variability on maize yield in Gassol LGA, Taraba State, Nigeria. J Agric Biotechnol 1(1):1–8. https://doi.org/10.20936/JAB/160101

Elisha I, Sawa BA, Ejeh UD (2017) Evidence of climate change and adaptation strategies among grain farmers in Sokoto State, Nigeria. IOSR J Environ Sci Toxicol Food Technol 11(3):1–7. https://doi.org/10.9790/2402-1103020107

Enete IC (2014) Impacts of climate change on agricultural production in Enugu State, Nigeria. J Earth Sci Clim Change 5(9):234. https://doi.org/10.4172/2157-7617.1000234

Federal Ministry of Environment (2012) Great Green Wall for the Sahara and Sahel Initiative, National Strategic Action Plan. https://landportal.org/library/resources/lex-faoc151052/great-green-wall-sahara-and-sahel-initiative-national-strategic. Accessed 12 Apr 2022

Federal Ministry of Environment (2014) United Nations Climate Change Nigeria. National Communication (NC). NC 2. 2014. https://unfccc.int/sites/default/files/resource/nganc2.pdf. Accessed 12 Apr 2022

Gadzama NM, Ayuba HK (2016) On major environmental problem of desertification in Northern Nigeria with sustainable efforts to managing it. World J Sci Technol Sustain Dev 13(1):18–30. https://doi.org/10.1108/WJSTSD-06-2015-0035

Giri M, Bista G, Singh PK, Pandey R (2021) Climate change vulnerability assessment of urban informal settlers in Nepal, a least developed country. J Clean Prod 307:127213. https://doi.org/10.1016/j.jclepro.2021.127213

Hansen P (2020) Nigeria: Floods, droughts, and a lack of plan. https://www.climatescorecard.or/2020/11/nigeria-flood-droughts-and-a-lack-of-a-plan/. Accessed 9 June 2022

Hengeveld H, Whitewood B (2005) Understanding climate change 2005: a synthesis of climate change science. https://publications.gc.ca/site/eng/272648/publication.html. Accessed 20 Apr 2022. Environment Canada, Canada, pp 31–35

Hengeveld GH, Bush E, Edwards P (2002) Frequently asked questions about climate change science. Environment Canada, Canada. https://publications.gc.ca/site/fra/9.616210/publication.html. Accessed 30 Mar 2022

Ibrahim MA, Abubakar BY, Balarabe ML (2018) Sequestrated carbon content in tree species and diurnal temperature influence for adaptive climate change resilience in Nigeria. In: Leal Filho W (ed) Handbook of climate change resilience. Springer Nature, Switzerland, pp 1–25. https://doi.org/10.1007/978-3-319-71025-9_4-1

IPCC (2000) Land use, land-use change, and forestry special report. Cambridge University Press, Cambridge, p 377p

IPCC (Intergovernmental Panel on Climate Change) (2014) Synthesis Report. Contribution of Working Groups I, II and III to the Fifth Assessment Report of the Intergovernmental Panel on Climate Change [Core Writing Team, Pachauri RK, Meyer LA (eds)]. IPCC, Geneva, Switzerland, p. 151

Jibrillah AM, Ja'afar M, Choy LK (2019) Monitoring vegetation change in the dryland ecosystem of Sokoto, northwestern Nigeria using geoinformatics. Indonesian J Geogr 51(1):9–17. https://doi.org/10.22146/ijg.33207

Keeton WS, Kraft CE, Warren DR (2007) Mature and old-growth riparian forests: structure, dynamics, and effects on Adirondack stream habitats. Ecol Appl 17:852–868. https://doi.org/10.1890/06-1172

McNulty SG (2002) Hurricane impacts on US forest carbon sequestration. Environ Pollut 116:17–24. https://doi.org/10.1016/s0269-7491(01)00242-1

Mcquire B, Mason I, Kilburn C (2002) Natural hazards and environmental change. Arnold, London, pp 53–63

ND-GAIN (2021) Ranking of country index. Available at. https://gain.nd.edu/our-work/country-index/rankings/. Accessed 14 June 2021 and 5 Apr 2022

NEST (Nigerian Environmental Study/Action Team) (2003) Climate change in Nigeria. A communication guide for reporters and educators. NEST, Ibadan, pp 5–16

Odebode OS, Eniola PO (2019) Health effects of charcoal production as perceived by the rural dwellers in rainforest and guinea savannah agro-ecological zones of Nigeria. J Sci Res Rep 22(5):1–12

Odujugo PAO (2010) Shift in crops production as a means of adaptation to climate change in the semi-arid region of Nigeria. J Meteorol Clim Sci 8(1):4

Odjugo PAO, Khuoria AI (2003) The impact of climate change and anthropogenic factors on desertification in the semi-arid region of Nigeria. Glob J Environ Sci 2(2):118–126. https://doi.org/10.4314/gjes.v2i2.2418

Ogbuabor JE, Egwuchukwu EI (2017) The impact of climate change on the Nigerian economy. Int J Energy Econ Policy 7(2):217–223

Okon EM, Falana BM, Solaja SO, Yakubu SO, Alabi OO, Okikiola BT, Awe TE, Adesina BT, Tokula BE, Kipchumba AK, Edeme AB (2018) Systematic review of climate change impact research in Nigeria: implication for sustainable development. Heliyon 7(9):1–21. https://doi.org/10.1016/j.heliyon.2021.e07941

Olaniyi OA, Olutimehin IO, Funmilayo OA (2019) Review of climate change and its effect on Nigeria ecosystem. Int J Rural Dev Environ Health Res 3(3):92–100. https://doi.org/10.22161/ijreh.3.3.3

Onah NG, Ali AN, Eze E (2016) Mitigating climate change in Nigeria: African traditional religious values in focus. Mediterr J Soc Sci 7(6):299–308

Onyekuru NA, Marchant R (2016) Assessing the economic impact of climate change on forest resource use in Nigeria: a Ricardian approach. Agric for Meteorol 220:10–20. https://doi.org/10.1016/j.agrformet.2016.01.001

Peters EB, Wythers KR, Zhang SX, Bradford JB, Reich PB (2013) Potential climate change impacts on temperate forest ecosystem processes. Can J for Res 43:939–950. https://doi.org/10.1139/cjfr-2013-0013

Popo-Ola FS, Aiyeloja AA, Adedeji GA (2012) Sustaining carbon sink potentials in tropical forest. J Agric Soc Res 12(1):64–73

Ragatoa DS, Ogunjobi KO, Klutse NAB, Okhimamhe AA, Eichie JO (2019) A change comparison of heat wave aspects in climatic zones of Nigeria. Environ Earth Sci 78(111):1–16. https://doi.org/10.1007/s12665-019-8112-8

Sayne A (2011) Climate change adaptation and conflict in Nigeria. USIP, Washington, DC. https://www.usip.org/sites/default/files/Climate_Change_Nigeria.pdf. Accessed 20 Apr 2022

The New Humanitarian (2017) Nigerian farmers can't fight desertification alone. https://naijafarmers.com/land-degradation-nigeria/. Accessed 9 June 2022

Usman S (2021) One dies as bushfire destroys over 90 farmlands in Yobe. https://www.google.com/amp/s/dailypost.ng/2021/12/02/one-dies-as-bushfire-destroys-over-90-farmlands-in-yobe-/%3famp=1. Accessed 9 June 2022

Williams PA, Crespo O, Abu M, Simpson NP (2018) A systematic review of how vulnerability of smallholder agricultural systems to changing climate is assessed in Africa. Environ Res Lett 13(10):103004. https://doi.org/10.1088/1748-9326/aae026

World Bank (2019) Building climate resilience: experience from Nigeria. https://www.worldbank.org/en/results/2019/04/18/building-climate-resilience-experience-from-nigeria. Accessed 15 Apr 2022

Yaqub CN (2007) Desert encroachment in Africa: extent, causes and impacts. J Arid Environ 4(1):14–20

Yekken S (2011) Innovative spatial planning in mitigating climate change-related vulnerability in Nigerian urban. A tutorial review. In: Proceedings of Federal University of Technology, Minna, Nigeria, pp 2–10

Yugunda BS (2002) Socio-economic and cultural impacts of desert encroachment in Nigeria. J Environ Dyn 5(2):19–30

Zuidema PA, Baker PJ, Peter G, Schippers PV, der Sleen P, Vlam M, Sterck F (2013) Tropical forests and global change: filling knowledge gaps. Trends Plant Sci 18:413–419. https://doi.org/10.1016/j.tplants.2013.05.006

Chapter 17
Impacts of Climate Change on Sustainable Development in Nigeria

Ifeyinwa Ijeoma Obianyo, Sylvia Echezona Kelechi, and Azikiwe Peter Onwualu

Abstract Economic, social, and environmental issues associated with climate change pose a serious threat to sustainable development globally. In Nigeria and indeed other developing countries, the devastating effects of climate change are felt on the livelihood and health of the people more than the developed economies where there is advanced knowledge and understanding of the issues and where efforts have been made toward mitigation of the impacts and adaptation of systems to cope with the impacts. This chapter presents a deeper understanding and knowledge of the concepts of sustainable development and climate change and the nexus between the two. The status of sustainable development and challenges is discussed. The causes of climate change in Nigeria are presented, followed by the impact of climate change on sustainable development with emphasis on how the impact affects urban and rural dwellers as well as gender and socio-economic dimensions of the impact. The chapter ends with suggestions on sustainable strategies for mitigating the impacts, strengthening the relevant institutions and thus fostering the achievement of the SDGs in Nigeria. Data used for the development of the chapter were obtained through structured questionnaires as well as desk study using a comprehensive literature review approach. Samples and case studies were drawn from rural and urban dwellers. Recommendations were provided on a sustainable strategy for mitigating the impacts based on building sustainable resilient and adaptation initiatives, strong institutions, and investments that will minimize the impacts of climate change on sustainable development in Nigeria.

I. I. Obianyo (✉) · A. P. Onwualu
Department of Civil Engineering, African University of Science and Technology, Abuja, Nigeria
e-mail: iobianyo@aust.edu.ng

S. E. Kelechi
Department of Civil Engineering, Bayero University, PMB, Kano State 3011, Nigeria

Department of Mechanical and Civil Engineering, Purdue University, Northwest, Hammond, IN 46323, USA

A. P. Onwualu
Department of Materials Science and Engineering, African University of Science and Technology, Abuja, Nigeria

J. C. Egbueri et al. (eds.), *Climate Change Impacts on Nigeria*, Springer Climate,
https://doi.org/10.1007/978-3-031-21007-5_17

Keywords Climate change · Sustainable development · Mitigation · Environmental sustainability · Resilience · Adaptation

Introduction

In developing countries like Nigeria, the impacts of climate change on sustainable development have not been given the needed attention. Globally, some measures have been taken to curtail the devastating effects of climate change on the sustainable development of society. These efforts are yet to yield the desired results as industrialization is on the rise. Industrialization among other factors has contributed immensely to global warming across the globe (Ahmed Ali et al. 2022; Mgbemene et al. 2016). There is a need for industries to begin to think of green and sustainable manufacturing processes that will aid the reduction of greenhouse gases emitted into the atmosphere via production processes. As the rate of industrialization and technological advancements increases, the resultant effects which include environmental pollution and global warming/climate change will escalate. Majority of the greenhouse gases that causes ozone layer depletion are generated by industries across the globe (Heger et al. 2019; Manisalidis et al. 2020). In as much as industries are needed to drive the developmental agendas of nations, leaders need to strike a balance between achieving developments and preserving the environment. Production processes that contribute to climate change need to be banned to compel industries to opt for sustainable production processes. In other words, both local and foreign investors should be tasked with the responsibility of ensuring that their investments are economically, socially, and environmentally sustainable. This will not only be beneficial to their investments and their host communities, but it will also help to preserve our planet.

Since Nigeria's independence in 1960, different regimes have come up with different developmental plans and policies for the country. However, the sustainable development of Nigeria has been affected by different factors such as lack of political will, corruption, brain drain, poor management of resources, climate change, and so on. Research on the effect of some of these factors on sustainable development has been conducted by different researchers. Olarinde et al. explored the impact of corruption on sustainable development in Nigeria and their study asserted that sustainable growth and development will be impossible to achieve in a corruption-infested society (Olarinde and Jonathan 2021). A study on the effect of poor management of resources on sustainable development in Nigeria recommended the equitable distribution of revenue from resources and enforcing of laws aimed at the development of capital projects and reduction of pollution (Mayomi and Omachoko 2017). Effective resource allocation and utilization were identified as the missing link hindering Nigeria from achieving sustainable development by Ehiorobo (2018). The result of their findings indicated that resource allocation and utilization in Nigeria have been carried out with gross inefficiency due to several factors that include a lack of accountability and ethnic prejudices. The Nigerian legal perspective on the impact of climate change on sustainable development was explored by Orie (2015). The study revealed the failure of the Nigerian government to understand the interconnection between climate change and sustainable development which will inform the formulation of mitigation strategies/plans for climate change hindering the achievement of the sustainable development. Despite previous studies on the impacts of several factors on sustainable development (Ejumudo 2015; Ndubuisi-okolo and Anekwe 2018; Olaniyi et al. 2013; Osakede et al. 2016), studies on the impact of climate change on the sustainable development of Nigeria are yet to be elaborately explored with the view of providing adaptation and mitigation strategies for climate change issues. This chapter X-ray's the connection between climate change and sustainable development in Nigeria by analyzing the impacts on climate change on sustainable development as well as providing adaptation and mitigation measures for tackling the issues of climate change.

Conceptual Issues

National Development

Education, poverty, health, employment, safety and security, social cohesion, and good governance are the metrics used to measure the development of a country (USAID 2020). However, some aspects of development cannot be measured and they include freedom, conservation, and sustainability (Mensah 2019). Classification of countries as developed, developing, and underdeveloped is based on different parameters. These parameters include economic growth, political stability, human resources, social amenities, industrialization, and infrastructural advancement (Chete et al. 1999). Although Nigeria is blessed with both natural and human resources, it is still recognized as a developing country due to a lack of proper management of these resources by the leaders of the nation. The political will to drive developmental projects is largely missing in the nation's leadership. Basic needs such as water, electricity, and shelter are hardly provided by the government. This incompetency of the Nigerian leadership does not only affect the development of the country but critical issue concerns such as environmental pollution and climate change are not given adequate attention.

Development Plans and Policies in Nigeria

Every country has their developmental plans and policies aimed at facilitating the achievement of the country's developmental goals. Implementation of these plans and policies has been an issue of challenge in developing countries like Nigeria (Emmanuel 2019). Since independence, different administrations in Nigeria have formulated great development plans and policies to facilitate the sustainable development of the nation. Some of these policies and laws that relate to climate change and sustainable development include National Policy on Climate Change and Climate Change Policy Response and Strategy 2021–2030 (2021), National Integrated Infrastructure Master Plan (2021), Medium Term National Development Plan 2021–2025 (2021), National Action Plan on Gender and Climate Change for Nigeria (2020), National Forest Policy (2020), Nigeria Economic Sustainability Plan (2020), National Adaptation Plan Framework (2020), Nigeria's National Action Plan to reduce short-lived climate pollutants (2019), The flare gas (prevention of waste and pollution) regulations (2018), Economic Recovery and Growth Plan 2017–2020 (2017), Nigeria's green bonds program (2017), National Gas Policy (2017), National policy on the environment (2016), National Renewable Energy and Energy Efficiency Policy (2015), National Environmental (Desertification Control and Drought Mitigation) Regulations (2016), Nigeria Vision 2020 (2010), and Nigerian Biofuel Policy and Incentives (2007). Unfortunately, these plans, policies, and laws were not fully implemented after the expiration of the tenures of the regimes of the government that initiated them because of the lack of continuum in governance in Nigeria. One of the best ways to ensure the achievement of these developmental plans and policies in Nigeria is by effective monitoring and evaluation of the phases of implementation (Obianyo et al. 2021). This calls for the establishment of strong institutions for monitoring and evaluating the developmental plans and policy implementation in Nigeria.

Millennium Development Goals in Nigeria: A Post-mortem

An appraisal of the achievements of the Millennium Development Goals (MDGs) in Nigeria will provide insights into its limitations which will guide decision-makers in making suitable policies. These policies will help to overcome the challenges encountered in MDGs implementation while working to achieve SDGs. The MDGs were set up to achieve eight substantive goals in September 2000. Implementation of MDGs in Nigeria was faced with various setbacks such as policy issues, lack of political will, corruption, and political instability (Okoroafor and Anuforo 2012). The year 2015 was set for the achievement of the Millennium Development Goals (MDGs). However, some countries lagged in achieving the MDGs (Fehling et al. 2015). There was a lack of synergies between governments, industries, NGOs, and individuals in the pursuit of MDGs fulfilment (Lomazzi et al. 2014). Ayobami reported in his study on the lack of partnership and participation of the local government and recommended full collaboration with local government to facilitate the effective implementation of the MDGs (Ayobami 2018). The study conducted by Okon and Ukwayi revealed a lack of transparency and accountability for MDGs projects' fund approval and disbursement (Okon and Ukwayi 2012).

Sustainability and Sustainable Development

Consideration of the future outcomes of the decisions and actions of today promotes sustainability. Taking proactive steps toward the preservation of society and its environment is key to achieving sustainability. Sustainability, therefore, entails a holistic approach targeted at protecting our planet for future generations while utilizing the available resources of the day. According to Obianyo et al. sustainability is harnessing the resources of today without harming the opportunity of future generations to harness theirs. Therefore, sustainable development needs concerted efforts in ensuring that the development of today will not have negative effects on the life of future generations (Obianyo et al. 2020). For a development to be sustainable, its impacts on the environment and the society at large must be positive.

Sustainable Development Goals and Climate Change

In the year 2015, the Member States of the United Nations adopted the 2030 Agenda for Sustainable Development which provides a shared blueprint for peace and prosperity for people and the planet, now and into the future. The 17 Sustainable Development Goals (SDGs) were developed to aid the achievement of the 2030 Agenda for Sustainable Development in global partnership (Morton et al. 2017). It is expected that by the year 2030, all countries must have taken urgent actions that will result in the achievements of the SDGs. The Division for Sustainable Development Goals in the United Nations Department of Economic and Social Affairs was created to deliver functional support for the achievement of all the thematic areas of the SDGs and to conduct an evaluation of activities for the implementation of the 2030 Agenda by the United Nations (2018).

Climate change is a global issue with different levels of impact across the nations of the world. The impacts depend on various factors including human activities and natural phenomena. The nexus between climate change and human activities can be felt in different areas of life such as farming, weather, health, and the environment. Harmful activities of humans have largely contributed to global warming (Nwankwoala 2015; Trenberth 2018). These activities include cutting down trees, open/bush burning, illegal mining, fossil fuel emissions, and industrialization. The recently revised National Climate Change Policy Response and Strategy (NCCPRS) was formulated to address the increasing issues of climate change in Nigeria. This well-detailed policy document contains a critical component of climate change mitigation and adaptation. In order to effectively implement the NCCPRS, the establishment of a national climate change law targeted at mitigating the impacts of climate change on Nigeria must be put in place (Orie 2015). The national climate change law must comply with the United Nations Framework Convention on Climate Change and its Protocol which need to be integrated with existing community-based sustainable development policies in Nigeria. Several public and private institutions have been established to deal with sustainable development and climate change issues in Nigeria. Strengthening these institutions is key to ensuring the effectiveness of their operations in the area of sustainable development.

Materials and Methods

Study Area

Abuja Municipal Area Council (AMAC) and Gwagwalada Area Council (GAC) of Federal Capital Territory (FCT), Abuja were used as the urban and rural study areas for this research, respectively. AMAC, situated in the heart of Abuja, has major infrastructure, unlike the GAC which is situated at the outskirt of Abuja and is faced with so many infrastructural challenges. The choice of both Area Councils comprising of urban and rural settlements was done to ascertain the impacts of climate change on both settlements to assess whether climate change affects people living in urban and rural places differently. The location map of the study area is presented in Fig. 17.1. Abuja has a tropical climate with precipitation, that is, consists of mostly rainfall and the pattern of rainfall is seasonal (Itiowe et al. 2019). The city of Abuja has an average daily temperature, atmospheric pressure, wind speed, and relative humidity of 32.49 ° C, 968.32 hpa, 0.76 m/s, and 55.51%, respectively (Ayegba et al. 2019).

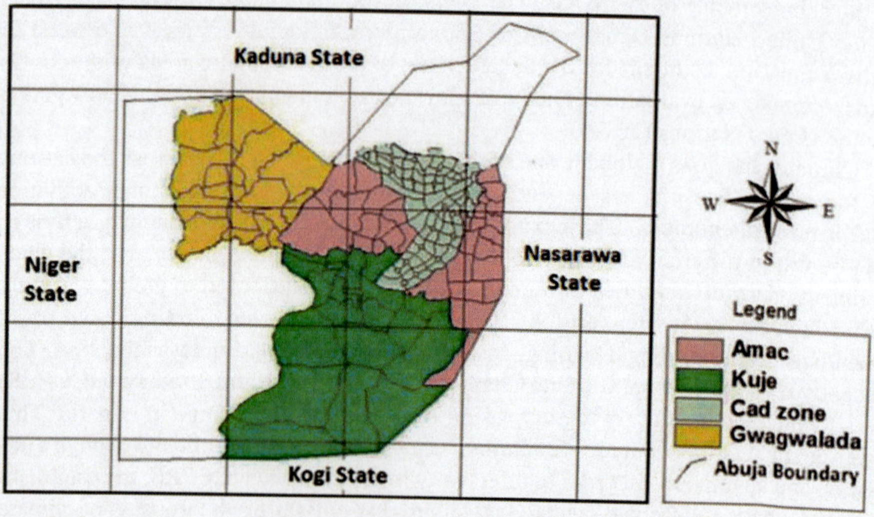

Fig. 17.1 Location map of the study area (Hassan and Okobia 2008)

Data Collection and Analysis

Data used for this study were collected using a structured questionnaire survey. The structured questionnaire survey sampling was conducted in urban and rural settlements. A total of 120 participants comprising 60 males (30 males from urban areas and 30 males from rural areas) and 60 females (30 females from urban areas and 30 females from rural areas) were used for the study. The data analysis of this study was carried out using Minitab and Excel software.

Results and Discussion

Demographic Information of the Respondents

The age distribution of the study participants is presented in Fig. 17.2. The majority of the participants fall within the economic/working-age group (18–69) with 20, 47, and 27% for the age groups of 18–30 years, 31–45 years, and 46–69 years, respectively. Only 6% of the participants have aged 70 years and above. This age distribution of the participants implies that the outcomes of this study would be very important in the decision-making of Nigeria since the views are majorly from the workforce of the nation.

The employment status of the study participants surveyed was identified concerning gender and presented in Fig. 17.3. Among the 60 males and 60 females

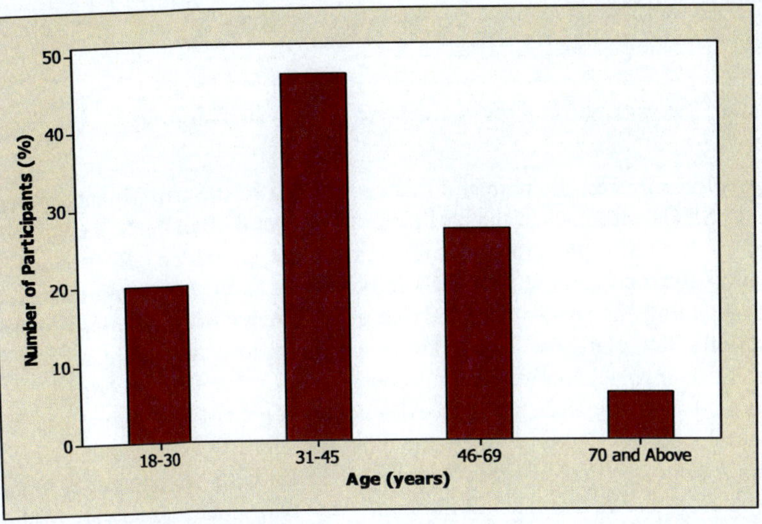

Fig. 17.2 Age distribution of participants

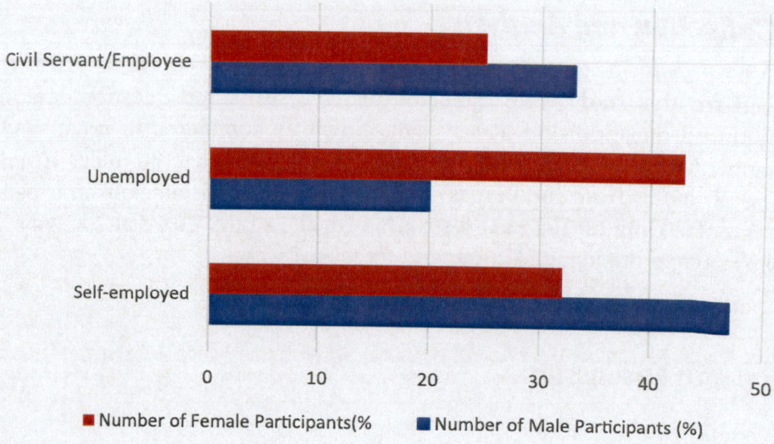

Fig. 17.3 Employment status of participants

sampled in this study, the percentage of unemployed women was higher than the percentage of men who are unemployed. This implies that women are likely to suffer the impact of climate change issues more than men. Therefore, more resources should be allocated to economically empower women. The results also indicated that more men are self-employed and are employed by the government than women. In order to bridge this gap, gender equity programs, such as special business grants for women, engaging more qualified women in government positions, scholarships for women, tax waivers for women in business, and provision of job offers to women upon graduation from the higher institution and free skill training for women should be implemented rigorously.

Status of Sustainable Development Goals in Nigeria

The questionnaire results indicated that only a few of the participants know about all the 17 SDGs with 93% of the participants not aware of all the SDGs as presented in Fig. 17.4. The implication of this is that more concerted efforts are required in creating awareness about the SDGs in both rural and urban communities. Stakeholders including the government and non-governmental organizations need to come up with effective plans for SDGs' awareness programs targeted at sensitizing the society on their roles required for the achievement of the SDGs in Nigeria.

This study revealed that sustainable development goals 4 and 1 which are "Quality Education" and "No Poverty" were the most familiar SDGs as shown in Fig. 17.5. Climate Action was ranked 4th among the 17 SDGs with 8% popularity among the participants. It shows that a small number of the participants are aware of this goal such that awareness campaigns are needed to get everyone involved in tackling the global issue of climate change. It is important to note that "Quality Education", "No

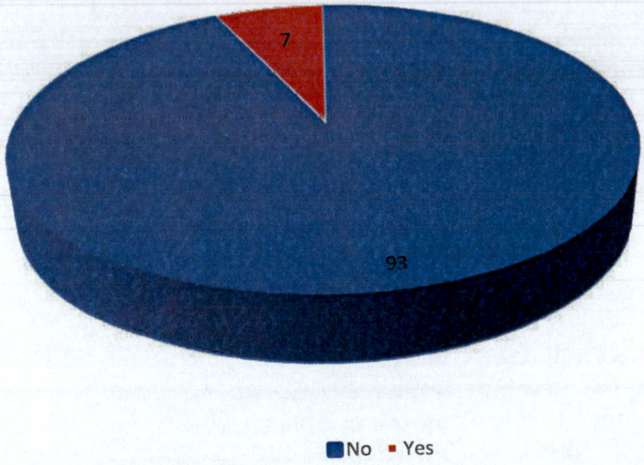

Fig. 17.4 Participants' knowledge of all the 17 SDGs (%)

Poverty", and "Gender Equality" goals with 15%, 12%, and 10%, respectively, were the goals that the majority of the respondents are familiar with and this is traceable to the more awareness campaigns carried out on these goals by the government, Civil Society Organizations, and Voluntary Organizations (Malaolu and Ogbuabor 2017). SDGs such as "Life below Water", "Life on Land", "Decent Work and Economic Growth", and "Responsible Consumption and Production" have very low popularity due to poor awareness campaigns on them. A robust campaign program on SDGs in different major languages in Nigeria will increase the knowledge of SDGs in Nigeria (Okwudili and Opeyemi 2019). The more people know about the SDGs, the more they will likely get involved in the collective actions targeted at achieving the SDGs.

This study X-rayed the order of the importance of the SDGs as perceived by Nigerians. The results presented in Fig. 17.6 indicated that "Quality Education" was perceived as the most important SDG in Nigeria. This is followed by "Good Health and Well-being" and "No Poverty". Climate Action was perceived to be the fourth most important SDG in Nigeria and this implies that conscious actions are required to be taken to mitigate the causes of climate change. The results also revealed the peculiarity of the SDGs to Nigerian society as shown in Fig. 17.6. This will assist the government and stakeholders in decision-making by channelling more resources and efforts to the goals that are perceived to be more important by Nigerians.

Challenges of Achieving Sustainable Development in Nigeria

Efforts targeted at achieving SDGs in Nigeria have been truncated by some obstacles (Lawrence et al. 2020; Obute 2021; Okwudili and Opeyemi 2019). Eight years to the full achievement of the SDGs by 2030, the implementation of these goals in Nigeria

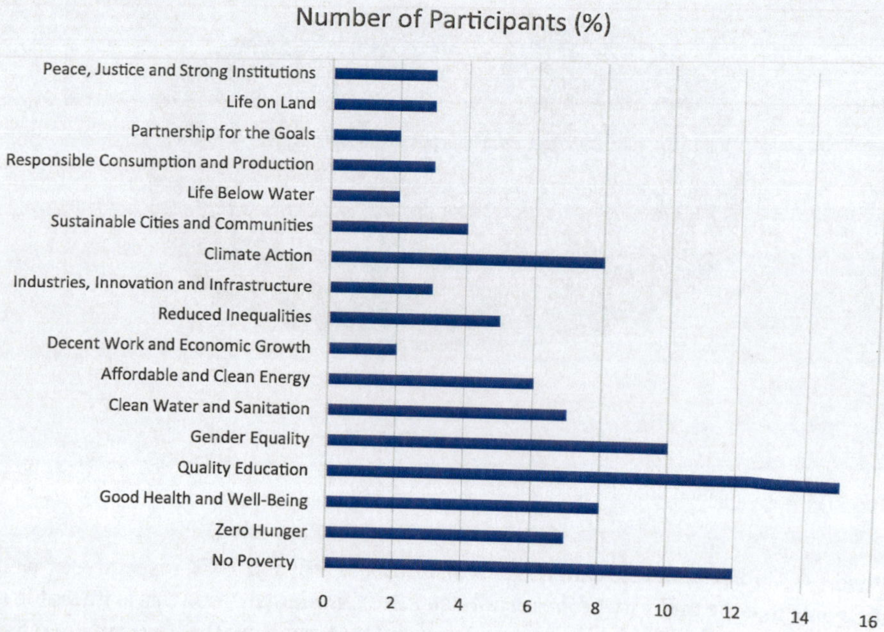

Fig. 17.5 The popularity of the SDGs in Nigeria

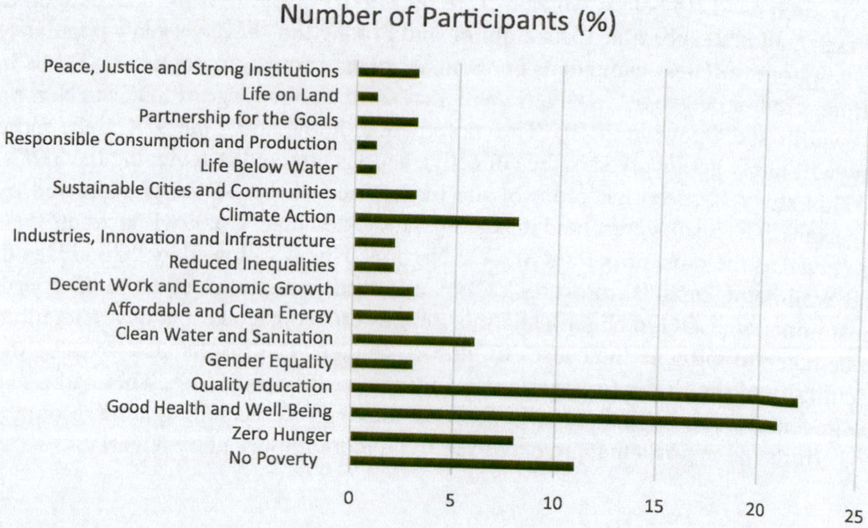

Fig. 17.6 Order of importance of the SDGs in Nigeria as perceived by participants

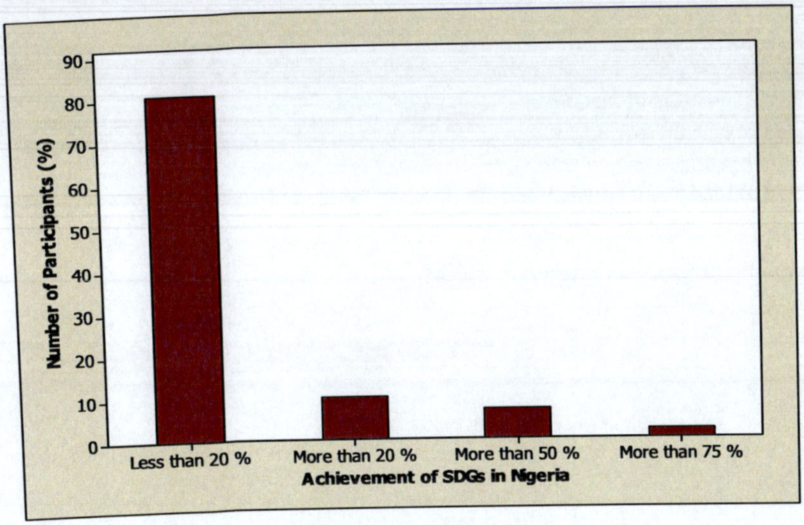

Fig. 17.7 Status of the achievement of SDGs in Nigeria

is far-fetched. Participants were asked to rate the percentage of the achievement of the SDGs in this study. An overwhelming 81% of the respondents indicated that only 20% of the SDGs have been achieved in Nigeria as shown in Fig. 17.7. This is not surprising considering the myriads of socio-economic and environmental challenges plaguing Nigeria (Ojo et al. 2014).

Obstacles to the full actualization of sustainable development in Nigeria were carefully considered with the aim of identifying the most pressing factor. Insecurity was identified as the major hindrance to the achievement of SDGs in Nigeria as indicated in Fig. 17.8. The dimensions of insecurity in Nigeria are manifold and complicated. Despite the efforts of the Nigerian security agencies to make the nation safe, the scope of insecurity in Nigeria has been on the increase and this has affected the sustainable development of the nation (Achumba et al. 2013). Insecurity in Nigeria has affected every aspect of the country including food, livelihood, and economy. Farmers have been adversely affected resulting in the high cost of food supplies in Nigeria. This has led to an increase in the level of poverty and hunger as people cannot afford the prices of food items in the market. Improper management of revenues from the nation's resources, poor implementation strategies for the SDGs, and lack of political will by the leaders to drive SDGs' actions and programs were also identified as setbacks to the achievements of SDGs in Nigeria.

Different groups and individual have contributed their efforts and resources to the sustainable development of Nigeria since the inception of the nation. Participants in this study were asked to identify the group among the stakeholders that have the highest responsibility for championing sustainable development in Nigeria as the government as shown in Fig. 17.9. The results of the survey also indicated the need for the collective efforts of individuals in achieving the SDGs in Nigeria. Contributions

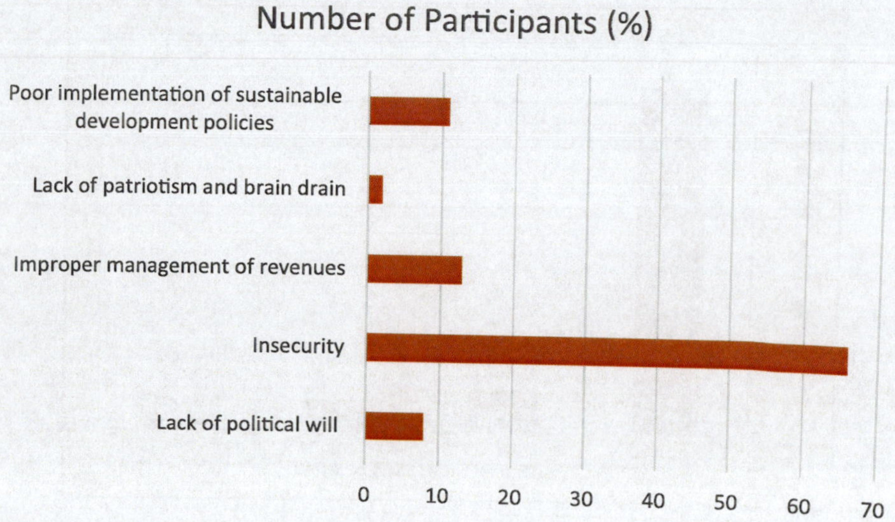

Fig. 17.8 Challenges of sustainable development in Nigeria

of the financial institutions, non-governmental organizations, research institutions, and industries will help facilitate the timely achievement of SDGs in Nigeria.

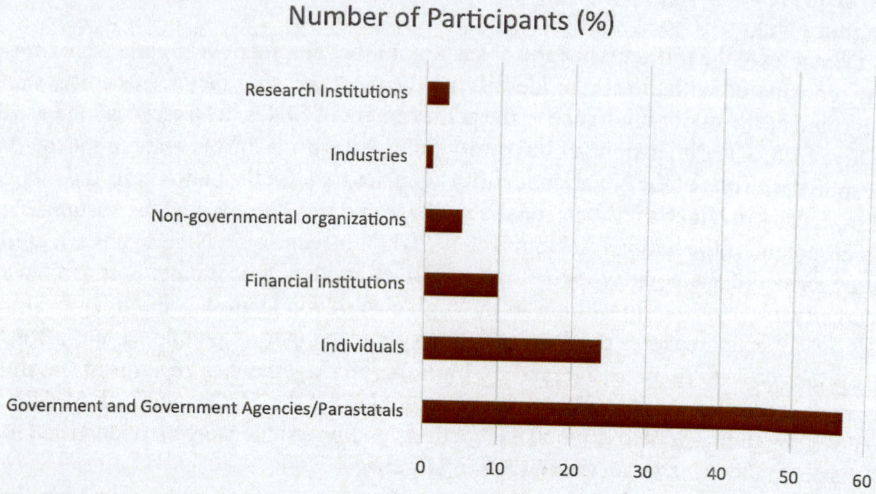

Fig. 17.9 Responsibility for achieving SDGs in Nigeria

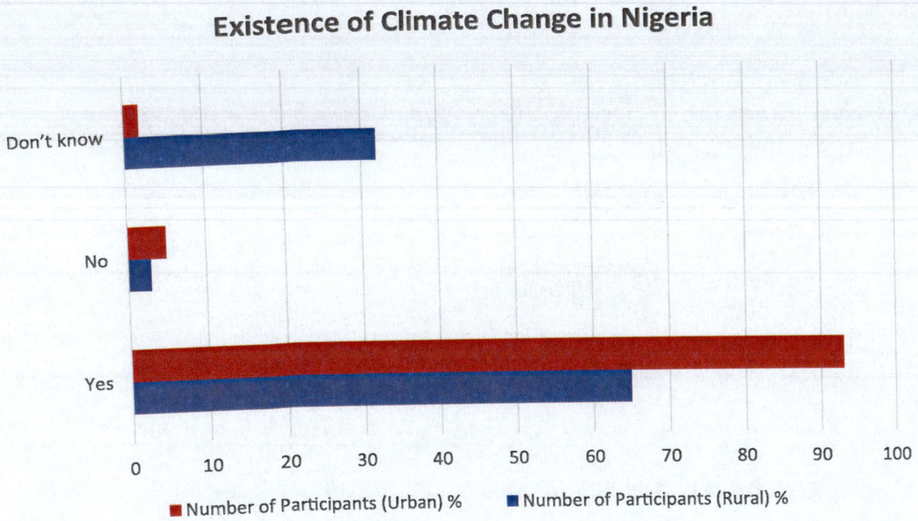

Fig. 17.10 Knowledge of the existence of climate change in Nigeria

Climate Change in Nigeria

Unlike the developed economies, Nigeria is yet to channel reasonable resources toward achieving the SDGs 11. The knowledge of the existence of climate change in the rural and urban settlements of Nigeria was unravelled in this study. Over 30% of rural participants are indifferent to the existence of climate change in Nigeria as shown in Fig. 17.10. The results indicated that urban residents are more aware of the existence of climate change in Nigeria than rural dwellers. This implies that more efforts are required in the dissemination of relevant climate change information to the rural part of Nigeria.

Causes of Climate Change in Nigeria

Rural and urban participants were asked to identify the causes of climate change in Nigeria. A similarity exists in the views of both the rural and urban residents on factors that cause climate change as presented in Fig. 17.11. Both rural and urban dwellers sampled acknowledged environmental pollution, urbanization, deforestation, and industrialization as factors that cause climate change. Among these factors, environmental pollution and industrialization were judged to be the major causes of climate change in Nigeria. Environmental and industrial human activities such as open burning, gas flaring, oil spillage, improper waste disposals, and release of carbon dioxide into the atmosphere have contributed to the climate change issues of ozone layer depletion and flooding (Abdulkareem et al. 2012; Gimah et al. 2019;

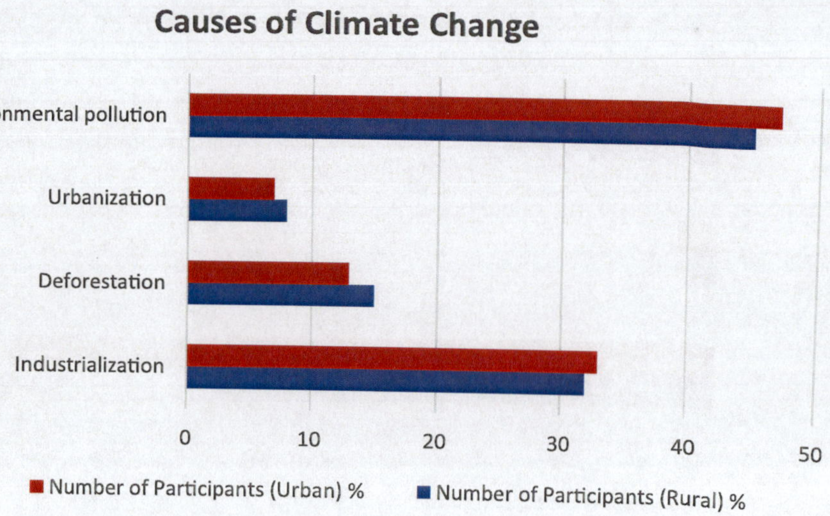

Fig. 17.11 Causes of climate change in Nigeria

Mafimisebi 2016). Enactment and implementation of policies and laws to end these devastating human activities require the concerted efforts of all the stakeholders.

Impact of Climate Change on Sustainable Development

The intensity of the impacts of climate change on sustainable development was explored in this study. Participants were asked to identify the causes of climate change as well as the areas where climate change affect their life and sustainable development. The impacts of climate change on sustainable development concerning gender, as well as the rural and urban settlements of Nigeria, were studied. According to the survey results, climate change affects rural and urban dwellers differently. Poverty and hunger/high cost of living was identified as the major impacts of climate change on rural dwellers while poor health condition and global warming mostly affect the urban residents as shown in Fig. 17.12a. Unpredictable weather, flooding, and lack of clean water affect the rural and urban dwellers similarly. The effect of the unpredictable weather conditions on farming contributes to the increased hunger and high cost of living in Nigeria (Matemilola and Elegbede 2017; Oti et al. 2021). Gender impacts of climate change were also explored in this study. The survey results showed the disparity between the ways climate change affects male and female. Overall, the impact of climate change is felt more by women than men, especially in the areas of poor health conditions and poverty as shown in Fig. 17.12b. However, more men indicated that they are affected by global warming than women.

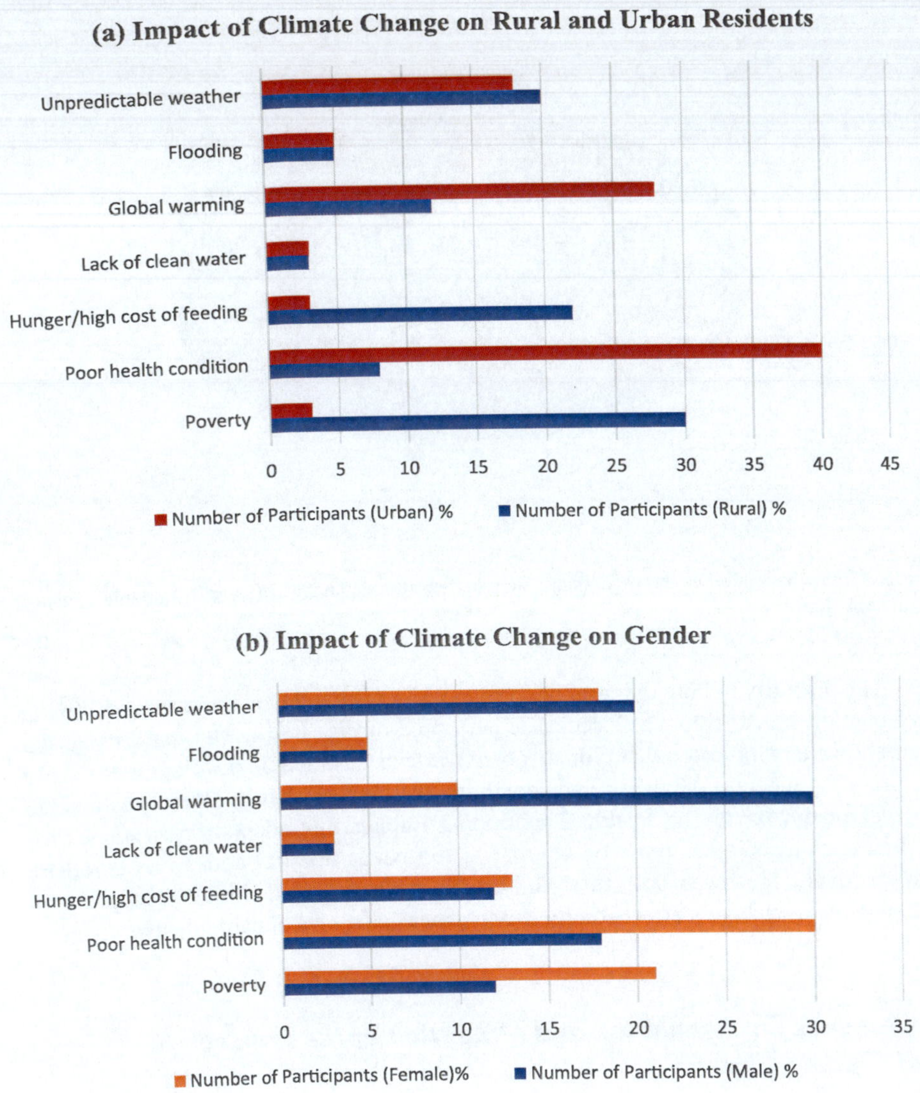

Fig. 17.12 Impact of climate change on Nigeria on: **a** Rural and urban residents and **b** Gender impact

Participants' opinions were sought on whether climate change affects sustainable development in Nigeria. Only 1% of the respondents perceived that climate change does not affect sustainable development whereas 3% were indifferent as presented in Fig. 17.13. The majority of the participants (96%) asserted that climate change affects sustainable development in Nigeria. This implies that tackling climate change issues should be taken seriously for the sustainable development of Nigeria to be achieved.

Fig. 17.13 Perception of the participants on whether climate change affects sustainable development in Nigeria

The disparity in how the rural and urban residents perceived the effect of climate change on sustainable development is presented in Fig. 17.14. The majority of the rural dwellers indicated that climate change affects sustainable development mostly in farming whereas the urban dwellers think that climate change affects sustainable development mostly on health. This finding implies that adaptation strategies for climate change issues must be tailored to the needs of rural and urban dwellers, respectively. However, both rural and urban dwellers have a similar perception on how climate change's effects on the environment affect sustainable development.

Strategies for Adaptation and Mitigation of the Impacts of Climate Change

Alleviation of the impacts of climate change can be achieved when every stakeholder takes decisive actions aimed at preventing the causes of climate change. Global collaborative efforts are necessary for tackling the issues of climate change (Tando et al. 2020). The interdependence of the SDGs implies that achieving a goal in the SDGs can result in achieving other SDGs. Hence, strategies for mitigating climate change will facilitate the achievement of improved health and well-being, sustainable cities and communities, responsible consumption and production, clean water and sanitation, zero hunger, as well as ensure the alleviation of poverty.

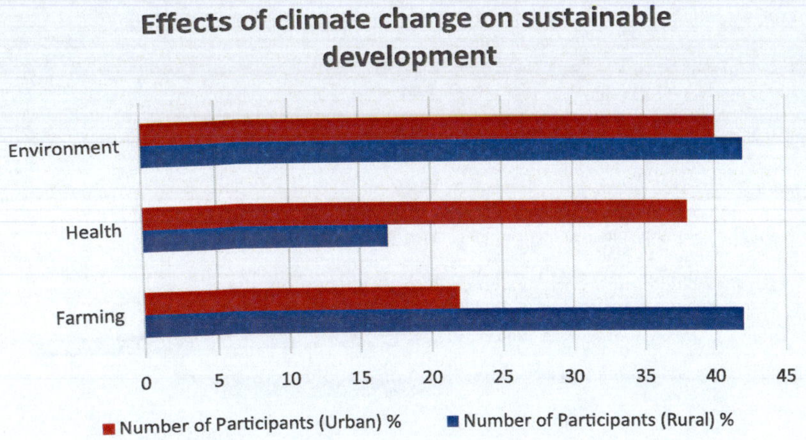

Fig. 17.14 Effects of climate change on sustainable development

Climate change is real and its impacts can be devastating to the economic and social life of the society (Dryzek et al. 2022). This study explores the adaptation strategies required to cope with the impact of climate change on Nigeria. From Fig. 17.15, among other suggested strategies, a decrease in pollution by industries and provision of aid/support to rural areas were the most voted adaptation strategies for climate change issues.

An effective way to address global issues of concern is by finding sustainable ways of preventing the cause of the issues. In a bid to address the root causes of climate

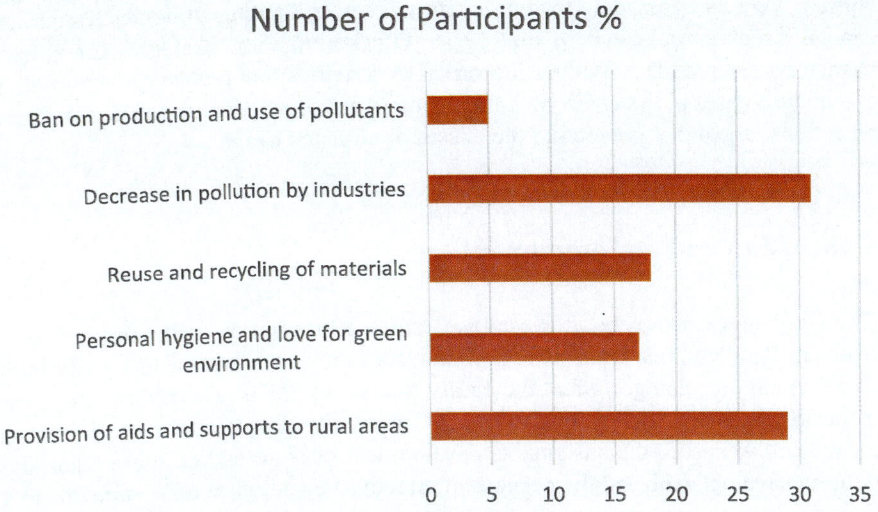

Fig. 17.15 Adaptation strategies for climate change issues

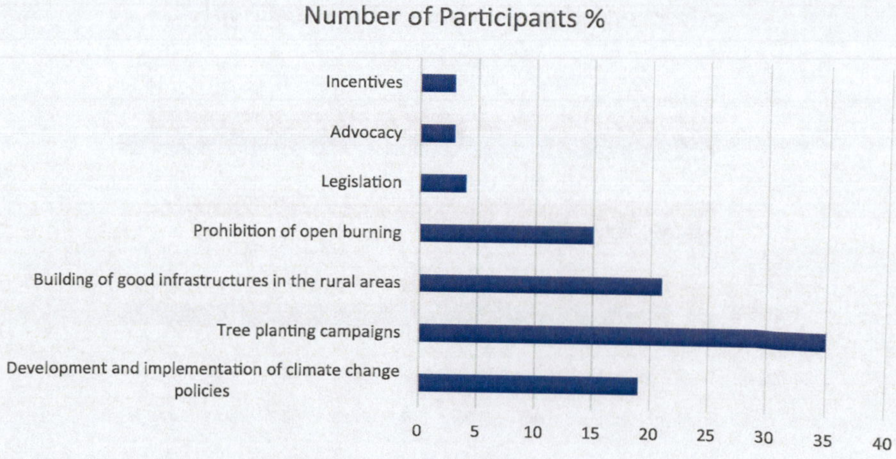

Fig. 17.16 Mitigation strategies for climate change issues

change, participants were asked to identify the mitigation strategies that could solve the challenges of climate change. Tree planting campaigns were selected as the best mitigation strategy for curbing the issues of climate change. A green environment will help to absorb more greenhouse gases while releasing oxygen into the environment and this will ensure that ozone layer depletion is averted. Tree planting is one of the ways to prevent the adverse weather conditions caused by climate change (Hopkins et al. 2021). Although tree-planting campaigns have been initiated in some parts of Nigeria, the use of wood and charcoal for cooking has been on the rise (Gobir et al. 2021). Building good infrastructure in the rural areas and prohibition of open burning were recognized by the participants as key mitigating strategies for climate change issues as indicated in Fig. 17.16. Other mitigation strategies for climate change issues include providing incentives to activities that reduce the emission of greenhouse gases to the environment, advocacy, and legislation of policies and laws on actions targeted at preventing the causes of climate change.

Conclusion and Recommendation

Climate change can only be addressed by decisive efforts/actions of the entire society. It is not a government affair, it is everyone's affair. The results of this study indicate that climate changes affect the ability to achieve SDGs in Nigeria in the areas of health, farming, and the environment. This study identified plans and policies formulated to ensure the sustainable development of Nigeria. Yet, these plans and policies have not achieved the purpose of effective development of Nigeria due to a lack of adequate implementation of them. Comparisons of the results obtained from the rural and urban dwellers were carried out to assess the degree of the impacts of climate change on the sustainable development of both. It was found that climate change impacts on sustainable development were more severe on the rural dwellers than urban dwellers. However, in terms of health, the impacts were felt more by urban dwellers compared to rural dwellers. This implies that there is a need for decision-makers including public and private organizations to channel more resources for the mitigation of these impacts to rural areas while not forgetting urban dwellers. There was also gender disparity in the ability to cope with the impacts of climate change among the subjects studied. Based on the results obtained from this study, the following recommendations are necessary for the adaptation and mitigation of the issues of climate change which greatly impact the sustainable development of Nigeria.

- Conducting a campaign on tree planting in schools and organizations.
- Development of relevant knowledge through research that can inform policy initiatives toward overcoming the challenge of climate change.
- Discouraging the deforestation caused by bush burning and cutting down of trees for cooking purposes.
- Encouraging the reuse and recycling of materials by setting up recycling plants in every local government area in Nigeria.
- Prevention of gas flaring, oil spillage, and deforestation is expedient to safeguarding our environments.
- Reduction in the rapid rate of urbanization by building infrastructures such as good road networks, water, electricity, and other social amenities.
- Educate the children about climate change because they are the future generations and are the ones most affected by its impacts.
- Enactment of regulations and laws for waste management and disposal to mitigate the issues of climate change.

References

Abdulkareem AS, Afolabi AS, Abdulfatai J, Uthman H, Odigure JO (2012) Oil exploration and climate change : a case study of heat radiation from gas flaring in the niger delta area of Nigeria. In: Sustainable development - authoritative and leading edge content for environmental management, pp 1–28

Achumba IC, Ighomereho OS, Akpor-Robaro MOM (2013) Security challenges in Nigeria and the implications for business activities and sustainable development. J Econ Sustain Develop 4(2):79–99

Ahmed F, Ali I, Kousar S, Ahmed S (2022) The environmental impact of industrialization and foreign direct investment: empirical evidence from Asia-Pacific region. Environ Sci Pollut Res 29(20):29778–29792. https://doi.org/10.1007/s11356-021-17560-w

Ayegba A, Yusuf Y, Simeon AU, Justin IO, Augustine OE, Suleiman AO, Chioma OR (2019) A study of the daily weather variation of Abuja, North Central, Nigeria. Int J Eng Appl Sci (IJEAS) 6(8):76–79. Accessed from www.ijeas.org

Ayobami E (2018) Local government and the attainment of sustainable development goals in Nigeria: Lessons from the millennium development goals the concept of development. African J Govern Develop 7(1):50–62

Chete LN, Adeoti JO, Adeyinka FM, Ogundele O (1999) Industrial Development and growth in Nigeria. British Med J 318(7187):822–823. Accessed from http://data.worldbank.org/indicator/SP.POP.TOTL?locations=NG

Dryzek JS, Norgaard RB, Schlosberg D (2022). Climate Change and Society: Approaches and Responses. https://doi.org/10.1093/oxfordhb/9780199566600.003.0001

Ehiorobo AO (2018) Efficient resource allocation and utilization: the missing link in Nigeria's quest for sustainable development. Econ Business 0394:264–275. https://doi.org/10.2478/eb-2018-0020

Ejumudo BKO (2015) Sustainable development in Nigeria: the policy gap and action dilemma. J Econ Sustain Develop 6(23):78–85

Emmanuel U (2019) Development plans and policies in Nigeria: observed impediments and practical best alternatives. Int J Res Sci Innov (IJRSI) VI(VII)

Fehling M, Nelson BD, Venkatapuram S (2015) Limitations of the millennium development goals: a literature review. Glob Public Health 8(10):1109–1122. https://doi.org/10.1080/17441692.2013.845676

Gimah BG, Bodo T (2019) Curbing human activities that degrade the environment: the relevance of environmental adult education. Earth Environ Sci Res Rev 2(5):1–7

Gobir AA, Aliyu AA, Abubakar AA, Esekhaigbe C, Joshua IA, Adagba KO, Ag N (2021) Climate change awareness and related tree planting practices in a rural community in North-Western Nigeria. J Commun Med Primary Health Care 33(1):41–49. https://doi.org/10.4314/jcmphc.v33i1.4

Hassan SM, Okobia LE (2008) Survey of ambient air quality in some parts of the federal capital territory, Abuja, Nigeria. Abuja J Geography Develop 2(2):0–17. Accessed from https://www.researchgate.net/profile/Efegbidiki_Okobia/publication/276278771_Survey_of_Ambient_Air_Quality_in_Some_Parts_of_The_Federal_Capital_Territory_Abuja_Nigeria/links/55552a0308ae980ca60ad862.pdf

Heger GC, Brönnimann S, Cowan T, Friedman AR, Hawkins E, Iles C, Undorf S (2019) Causes of climate change over the historical record. Environ Res Lett 14. https://doi.org/10.1088/1748-9326/ab4557

Hopkins LP, Campos LA, January-bevers DJ, Caton EK (2021) A simple tree planting framework to improve climate, air pollution , health , and urban heat in vulnerable locations using non-traditional partners. Plants People Planet 1–15. https://doi.org/10.1002/ppp3.10245

Itiowe T, Hassan SM, Agidi VA (2019) Analysis of rainfall trends and patterns in Abuja, Nigeria. Current J Appl Sci Technol 1–7. https://doi.org/10.9734/cjast/2019/v34i430139

Lawrence AW, Ihebuzor N, Lawrence DO (2020) Some challenges militating against developing countries achieving SDG 4 on targets: Nigeria as case study. Mod Econ 11(07):1307–1328. https://doi.org/10.4236/me.2020.117093

Lomazzi M, Borisch B, Laaser U (2014) The millennium development goals: experiences, achievements and what's next. Global Health Action 7(1). https://doi.org/10.3402/gha.v7.23695

Mafimisebi OP (2016) Environmental risk of gas flaring In Nigeria: lessons from Chevron Nigeria and Ilaje crisis. J Environ Earth Sci 6(3):180–204

Malaolu VA, Ogbuabor JE (2017) Towards achieving sustainable development goals in Nigeria: Role of CSOs and VOPEs. Saudi J Busin Manag Stud 2(5A):482–487. https://doi.org/10.21276/sjbms

Manisalidis I, Stavropoulou E, Stavropoulos A, Bezirtzoglou E (2020) Environmental and health impacts of air pollution: a review. Front Public Health 8(February):1–13. https://doi.org/10.3389/fpubh.2020.00014

Matemilola S, Elegbede I (2017) The challenges of food security in Nigeria. Open Access Library J 4:1–22. https://doi.org/10.4236/oalib.1104185

Mayomi AO, Omachoko A (2017) Resource management as panacea for sustainable national economic development and national unity. Int J Sci Technol Res 6(01):160–164

Mensah J (2019) Sustainable development: meaning, history, principles, pillars, and implications for human action: Literature review. Cogent Soc Sci 5(1). https://doi.org/10.1080/23311886.2019.1653531

Mgbemene CA, Nnaji CC, Nwozor C (2016) Review article industrialization and its backlash: focus on climate change and its consequences. J Environ Sci Technol 9(4):301–316. https://doi.org/10.3923/jest.2016.301.316

Morton S, Pencheon D, Squires N (2017) Sustainable development goals (SDGs), and their implementation. Br Med Bull 124(1):81–90. https://doi.org/10.1093/bmb/ldx031

Ndubuisi-okolo PU, Anekwe RI (2018) Strategies for achieving sustainable development in Nigeria: the nexus. Int J Acad Res Busin Soc Sci 8(6):176–187. https://doi.org/10.6007/IJARBSS/v8-i6/4195

Nwankwoala HN (2015) Causes of climate and environmental changes: the need for environmental-friendly education policy in Nigeria. J Educ Pract 6:224–234

Obianyo II, Anosike-francis E, Odochi G, Geng Y, Jin R, Peter A, Soboyejo ABO (2020) Multivariate regression models for predicting the compressive strength of bone ash stabilized lateritic soil for sustainable building. Constr Build Mater 263:120677. https://doi.org/10.1016/j.conbuildmat.2020.120677

Obianyo II, Ihekweme GO, Mahamat AA, Chibuzor Onyelowe K, Onwualu AP, Soboyejo ABO (2021) Overcoming the obstacles to sustainable housing and urban development in Nigeria: the role of research and innovation. Cleaner Eng Technol 4:2666–7908. https://doi.org/10.1016/j.clet.2021.100226

Obute MA (2021) Socio-economic challenges of sustainable development in nigeria and the need for human development, social inclusion and government policies/programs in actualizing sustainable development goals by the year. IOSR J Human Soc Sci (IOSR-JHSS) 26(11):26–39. https://doi.org/10.9790/0837-2611022639

Ojo PO, Aworawo F, Ifedayo TE (2014) Governance and the challenge of socio-economic development in Nigeria. Afe Babalola Univ: J Sustain Develop Law Policy 3(1):132–148

Okon DE, Ukwayi JK (2012) Challenges and prospects of the millennium development goals (MDGS) in Nigeria. Glob J Soc Sci 11(2):119–132

Okoroafor EC, Anuforo EE (2012) The millennium development goals (MDGS) and the problem of policy implementation in Nigeria. Int J Develop Manage Rev (INJODEMAR) 7:63–73

Okwudili UE, Opeyemi OP (2019) Nigeria and the sustainable development goals: why the indigenous languages count. Europ Sci J ESJ 15(5):52–68. https://doi.org/10.19044/esj.2019.v15n5p52

Olaniyi OA, Ojekunle ZO, Amujo BT (2013) Review of climate change and its effect on Nigeria ecosystem. Int J African Asian Stud 1:57–65

Olarinde OM, Jonathan JM (2021) The impact of corruption on economic growth in Nigeria. J Asian Develop 7(2):27–50. https://doi.org/10.52941/jad.v7i2.2

Orie EG (2015) Climate change and sustainable development: the Nigerian legal experience. NOUN Current Issues Nigerian Law 5:68–114

Osakede KO, Ijimakinwa SO, Adesanya TO (2016) Climate change and its impacts on the development. Int J Account Res (IJAR) 2(9):1–12

Oti OG, Okoye CU, Obasi IO (2021) Awareness and perception of climate change among farmers in Nigeria: implications for food security. Niger Agric J 52(2):164–173

Tando CE, Sudarmo, Haryanti RH (2020) Collaborative governance in new era for problem solving: a literature review. IOP Conf Ser: Earth Environ Sci 423(012023). https://doi.org/10.1088/1755-1315/423/1/012023

Trenberth K (2018) Climate change caused by human activities is happening and it already has major consequences. J Energy Natl Resour Law 1–6. https://doi.org/10.1002/2017EF000709

United Nations (2018) The 2030 agenda and the sustainable development goals an opportunity for Latin America and the Caribbean thank you for your interest in this ECLAC publication. United Nations publication

USAID (2020) Country Development Cooperation Strategy (CDCS), 1–39

Chapter 18
Landslide Occurrences in Southeastern Nigeria: A Literature Analysis on the Impact of Rainfall

Chinanu O. Unigwe, Johnbosco C. Egbueri, Michael E. Omeka, Daniel A. Ayejoto, Johnson C. Agbasi, and Vincent E. Nwazelibe

Abstract Climate changes have been noted as a factor impacting slope instability and triggering landslides all around the world. The type, extent, degree, and direction of changes in the stability conditions in response to anticipated climatic changes, however, are unclear due to the dynamic character of the system. Landslides have one of the worst effects on people and the environment, leading to economic downturns, population displacement, and the loss of life, among other things, in southeastern Nigeria, which has experienced several environmental disasters. It is believed that the recent climate changes that have occurred all over the world have had an impact on the occurrence of landslides in the study location. Over the last decade, the earth's temperature has risen by 0.08 °C, and the rate of warming has doubled in the last two decades. This study also found that rainfall in southeast Nigeria has increased in intensity while decreasing in total yearly precipitation when opposed to past decades. Nevertheless, a greater amount of rain falling on fewer wet days is a definite indication that weather-related problems would most likely arise (like flooding, landslides, and erosion). Along with climate change (increase in temperature and precipitation), landslides in this region are also significantly facilitated by several other factors, including geology, slope features, groundwater conditions, vegetation, and human

C. O. Unigwe (✉)
Boone Pickens School of Geology, Oklahoma State University, Stillwater, OK 74078, USA
e-mail: unigwe.chinanu@funai.edu.ng

J. C. Egbueri · J. C. Agbasi
Department of Geology, Chukwuemeka Odumegwu Ojukwu University, Anambra State, Uli, Nigeria

M. E. Omeka
Department of Geology, University of Calabar, Cross-River State, Calabar, Nigeria

D. A. Ayejoto
Department of Environmental and Sustainability Sciences, Texas Christian University, Fort Worth, Texas, United States

V. E. Nwazelibe
Albert Ludwig University of Freibury, Freiburg Im Breisgau, Germany

J. C. Egbueri et al. (eds.), *Climate Change Impacts on Nigeria*, Springer Climate,
https://doi.org/10.1007/978-3-031-21007-5_18

development activities. Although landslides are a natural occurrence that cannot be completely halted, they can be controlled and reduced. As a result, effective management techniques are strongly advised.

Keywords Landslides · Rainfall · Southeastern Nigeria · Climatic change · Human-induced activities · Weather-related hazards

Introduction

One of the most critical elements in evaluating the overall impact of climate change will be changes in rainfall patterns and other forms of precipitation. Global warming refers to long-term changes in the climate triggered by both natural or human-induced factors. This change termed "climatic change" is already altering the severity and frequency of many weather-related hazards, also increasing the risk and undermining the adaptability of poor and marginalized people who rely on acreage, water, and seasonal mean weather patterns. Landslides are one of the most prevalent geoenvironmental disasters experienced all over the globe. It is a form of mass wasting process that occurs on both natural and engineered slopes and involves the movement of a mass of rock, debris, or dirt down a slope under gravity's effect (Cruden and Varnes 1996; Hungr et al. 2013; Gariano and Guzzetti 2016). Landslides' failure mechanism involves flowing, sliding, toppling, falling, or spreading, and many landslides demonstrate a combination of these actions at the same moment or throughout the landslide's existence (Gariano and Guzzetti 2016).

The frequency of devastating landslides, as well as the number of studies on landslides, increased dramatically in the second half of the twentieth century (Ulakpa et al. 2020a, b, c, d). Nigeria, like many other developing African countries, has been susceptible to a wide range of natural disasters including landslides. The impact of natural disasters on Nigeria is increasing due to rapid urbanization/industrialization, proliferation of settlements, and lifelines over high-risk areas. Developing countries have always struggled to afford the enormous expenses of mitigating natural disasters through massive engineering projects and efficient land-use planning. Ulakpa et al. (2020a) noted, however, that many industrialized societies are reluctant to engage in systemic measures to minimize natural risks due to the economic downturn. Even if the incidence of environmental disasters remained constant, economic and environmental considerations show that the damage inflicted by the cataclysmic events may be excessive even for industrialized societies.

Several authors have researched extensively the major factors that lead to the occurrence of landslides, which include geology, geomorphology, soil type, landuse land cover, hydrologic conditions, and so on (Emeh and Igwe 2017; Ozioko and Igwe (2019; Nebeokike et al. 2020; Egbueri et al. 2021). However, Igwe et al. (2013) reported that rainfall is the most critical element in slope instability among other landslide conditioning factors. Landslides are widespread in many parts of Nigeria during prolonged periods of heavy rainfall when the slopes are partial to fully saturated with water. Generally, rainfall-induced landslides are obvious in cut-hillslopes and involve a large volume of soil mass moving rapidly. According to a study by Igwe et al. (2013), rainfall-induced landslides are one of the primary drivers of environmental annihilation and deflation, causing damage worth millions of dollars per year all over the world. Nonetheless, landslide origins, impacts, and processes have been under-reported in most West African countries such as Nigeria, where landslides are frequent and have a heavy effect on the economy (Igwe et al. 2013; Egbueri et al. 2021).

In southeastern Nigeria, there has been a limited number of studies on landslides compared to the widespread phenomenon in the region. However, studies have shown that steep slopes, geologic conditions, increasing population pressure, inadequate vegetation cover, and excessive rainfall subject most regions in southeastern Nigeria vulnerable to landslides (Okagbue and Ezechi 1988; Igwe et al. 2015; Egbueri et al. 2021). Okagbue (1988, 1992) further said that landslides in southeastern Nigeria are triggered by a range of factors, including poor geotechnical characteristics of the soil, topography, and human activity. However, it is obvious that landslides have caused damage in various regions of the study area, but their assessment, prediction, and management remain a mystery, as existing findings show that erosion is one of the causes that initiate landslides in the region (Igwe et al. 2015). Igwe and Egbueri (2018) also reported that most of the studied gullies within southeastern Nigeria are associated with landslides. Furthermore, rainstorms are another major triggering factor of landslides in these areas, also human activities, geologic attributes, and processes equally play a role in compounding the intensity of these occurrences, and the possibility of a landslide of similar magnitudes persisting in these areas is high (Igwe et al. 2015). A recent study by Ozioko and Igwe (2019) reported that lithology (geology), slope, distance to drainage, and land cover have the highest influence on the landslide incidence within the Iva Valley in southeastern Nigeria. Following reports of landslides in southeast Nigeria, it is necessary to ascertain how climate change may be influencing the occurrence of landslides in the area under study. To comprehend the influence and magnitude of meteorological indicators on the incidence of landslides in the research region, this study aims to conduct a comprehensive review of the existing literature. Additionally, it is thought that this study will serve as a guide for the disaster management plan.

Study Area

This study concentrated on the southeastern states of Nigeria (Enugu, Anambra, Imo, Abia, and the Ebonyi States), which are situated between the wet south and semi-arid north of the country and are part of the Guinea savanna vegetation zone. Geographically, it is situated between latitudes 04° 47' and 07° 44' N and longitudes 07° 54' and 08° 27' E, with a surface area of approximately 29,095 km^2 (Fig. 18.1) (Ogbuene 2010). In southeastern Nigeria, temperature and rainfall in one are mostly seasonal and change greatly each year. The region experiences annual rainfall ranging

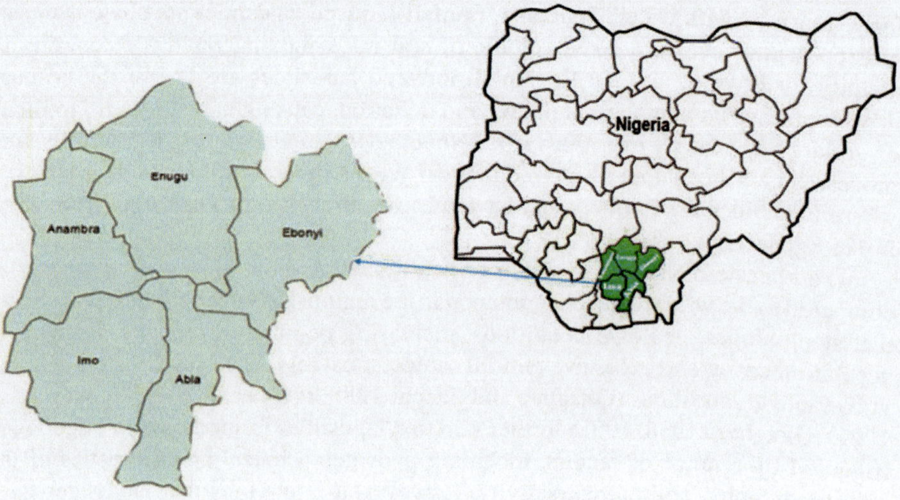

Fig. 18.1 Map of Nigeria showing the five southeastern states (Egboka et al. 2019)

from 1500 to 2500 mm, with mean annual temperatures often between 21 and 32 °C (Ogbuene 2010; Igwe 2014). The wet and dry seasons alternately occur in the area. The dry season lasts for roughly 6 to 7 months, from October to March or April, whereas the wet season lasts for roughly 5 to 6 months, from May to September. However, because of the current global climate change, which also causes several other weather-related risks, these meteorological conditions vary greatly.

Based on geology, southeastern Nigeria as shown in Fig. 18.2 falls within three Nigeria sedimentary basins successively from the oldest to the youngest; the Aptian-Santonian Abakaliki Basin comprising the Asu-River Group, Odukapani Formation, Eze-aku Formation and Agwu Shale, followed by the Campanian-Early Eocene Anambra Basin comprising of the Nkporo Group, Mamu Formation, Ajali Formation, Nsukka Formation, which is the followed by the Late Eocene-Pliocene Niger Delta Basin comprising of the Imo Shale, Ameki Group, and Ogwashi-Asaba Formation (Nwajide 1980; Arua 1980; Anyanwu and Arua 1990; Oboh-Ikuenobe et al. 2005; Ene et al. 2018).

Landslides, despite being a significant environmental threat in southeastern Nigeria, there haven't been many studies in the region. Following previous works done on landslides within southeastern Nigeria. It has been observed that landslides are majorly experienced in Enugu and Anambra States with little literature highlighting the Orlu Area in Imo State. The slopes are within a range of 2.34° to 90°, with the geologic formations Ajali and Nanka as the most prone formations for landslides within southeastern Nigeria (Table 18.1).

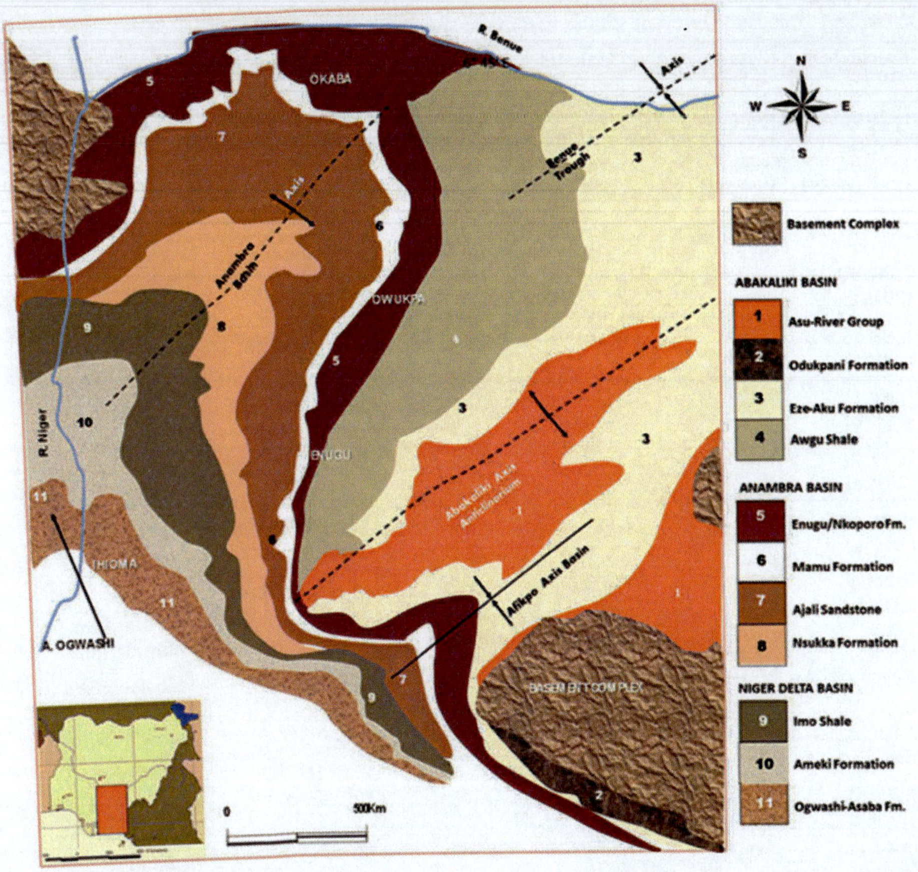

Fig. 18.2 Map showing the geologic formations in southeastern Nigeria (after Akande et al. 2007; Dim et al. 2017; Ene et al. 2018)

Climatic Indicators

Rainfall Variability as an Important Climatic Indicator

According to a paper by Akinsanola and Ogunjobi (2014), Nigeria's mean monthly rainfall patterns from 1971 to 2000 for January through June and July through December, respectively, show that rainfall declines in all seasons from the coast to the Sahel. This was in line with the findings by Nicholson (1994) that observed rainfall in West Africa often declines with latitude, virtually following zonal isohyets. However, from January to December, the amount of rainfall at the lower latitude was almost two times greater than at the higher latitude. Additionally, it was noted that the rainfall pattern below latitude 10°N is bimodal, with the major peak occurring in June–July and the sequel peak arising in September with a brief hiatus in

Table 18.1 Landslide relevant studies within the study region

Researcher	State	Geologic formation	Slope (^{0})	Landslide nature	Method of study
Igwe et al. (2013)	Enugu	Ajali	> 42	Landslide only	Geotechnical analysis
Igwe (2014)	Enugu	Ajali	–	Landslide only	Geotechnical analysis
Igwe and Fukuoke (2014)	Enugu	Ajali	10–60	Landslide only	Geotechnical analysis
Igwe (2015a)	Enugu	Ajali	> 35	Landslide only	Geotechnical and Geospatial
Igwe (2015b)	Enugu	Ajali	–	Landslide only	Geotechnical analysis
Igwe (2015c)	Enugu	Ajali	–	Landslide only	Geotechnical analysis
Igwe (2015d)	Enugu	Ajali	–	Landslide only	Geotechnical analysis
Ige et al. (2016)	Anambra	Nanka	–	In association with gully	Liquid equilibrium stimulation
Igwe (2017)	Anambra	Nanka	65–90	Landslide only	Geotechnical and Geophysical
Emeh and Igwe (2017)	Enugu	Ajali	–	In association with gully	Geotechnical and Geochemistry
Igwe and Chukwu (2018a)	Ebonyi	Asu-River Group	30–60	Mine site	Liquid equilibrium stimulation
Igwe and Chukwu (2018b)	Ebonyi	Asu-River Group	30–60	Mine site	Geotechnical analysis
Egboka et al. (2019)	All southeastern states	Ajali, Ameki, Nanka	> 45	Both	Geotechnical analysis
Ozioko and Igwe (2019)	Enugu	Ajali	5–45	Landslide only	Heuristic and Bivariate analysis
Igwe and Una (2019)	Anambra	Nanka	36–80	In association with gully	Geotechnical analysis
Nebeokike et al. (2020)	Enugu	Ajali	39–60	Landslide only	Liquid equilibrium stimulation

(continued)

Table 18.1 (continued)

Researcher	Statc	Geologic formation	Slope (⁰)	Landslide nature	Method of study
Ibeh (2020)	Anambra	Nanka	–	In association with gully	Liquid equilibrium stimulation
Ulakpa et al. (2020a)	Enugu, Anambra, Imo	Ajali, Nanka, Benin	2.34–66.5	In association with gully	Geospatial modeling
Ulakpa et al. (2020b)	Enugu, Anambra, Imo	Ajali, Nanka, Benin	–	In association with gully	Geospatial modeling
Ulakpa et al. (2020c)	Anambra	Nanka	–	Both	Statistical approach
Ulakpa et al. (2020d)	Enugu, Anambra, Imo	Ajali, Nanka, Benin	–	Both	Geospatial modeling
Egbueri et al. (2021)	Anambra	Nanka, Ogwashi, Benin	18–85	In association with gully	Liquid equilibrium stimulation

August (known as the "August Break") due to the absence of the Africa Easterly Jet (Omotosho 2007). Okechukwu and Mbajiorgu's (2020) study on southeastern Nigeria from 1993 to 2013 reveals that between April and September, there was a high average monthly rainfall that varied from 370 to 126 mm in Abia State and Enugu State, respectively. On the other hand, between December and March, the dry season experienced low average monthly rainfall in the states of Ebonyi and Abia State, ranging from 9 to 123 mm, respectively (Fig. 18.3). According to a study by Omoja et al. (2021) conducted in Awka, Anambra States, between 2011 and 2020, it was observed that the year 2019 has the highest annual rainfall, while the year 2013 has the lowest. Additionally, it was noted that rainfall patterns changed from 2012 to 2014, dropping, and then abruptly rising, from 2016 to 2019 (Fig. 18.4). The state-wide average monthly rainfall followed the same pattern across the five states of the study area and their slight variations in values are shown in Fig. 18.3. In decreasing order from the north, the state-wide average annual rainfall was 2203, 2058, 1885, 1764, and 1663 mm for Abia, Imo, Ebonyi, Anambra, and the Enugu States, respectively. The long-term average annual rainfall trend showed a gradual increase over the years of study, from 1993 to 2013. The average annual rainfall in the study area was found to be 1914 mm, which is close to 2000 mm as reported by Okonkwo and Mbajiorgu (2010).

For Imo State, a comparable study on the trend analysis of rainfall data from the National Root Crop Research Institute, Umudike indicates variability in rainfall with

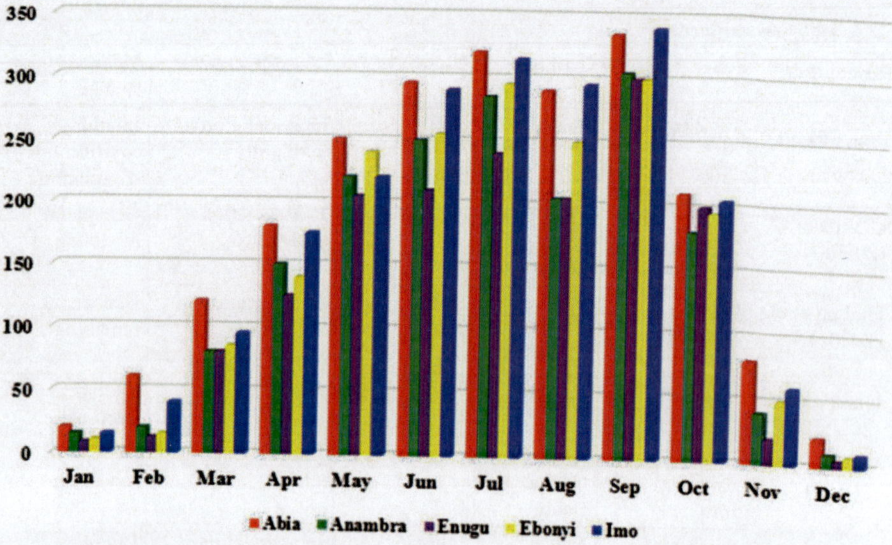

Fig. 18.3 Distribution of average monthly rainfall within the states of the study area from 1993 to 2013 (modified after Okechukwu and Mbajiorgu 2020)

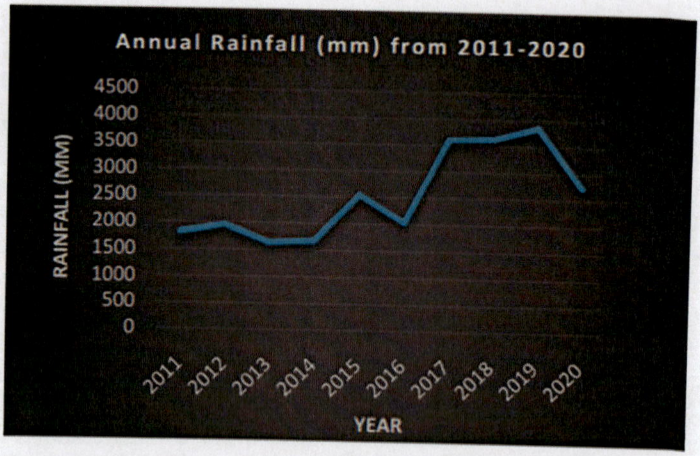

Fig. 18.4 Trend of annual rainfall from 2011 to 2020 in Awka, Anambra State (Omoja et al. 2021)

greater amounts experienced in 1974, 1984, 1996, and 2000 (Fig. 18.5) (Chidiebere-Mark et al. 2014). It was noted that between 2002 and 2012, there was a consistent decline in rainfall (Chidiebere-Mark et al. 2014). In southeastern Nigeria's Imo State, Okorie et al. (2019) report on the impact of rainfall variability over 30 years on crop production. According to research done over three decades (1988–1997, 1998–2007, and 2008–2017), the years 1993 (first decade), 1998 (second decade), and 2013 (third decade) had the lowest mean annual rainfall, with values of 181.9 mm, 136.7 mm, and

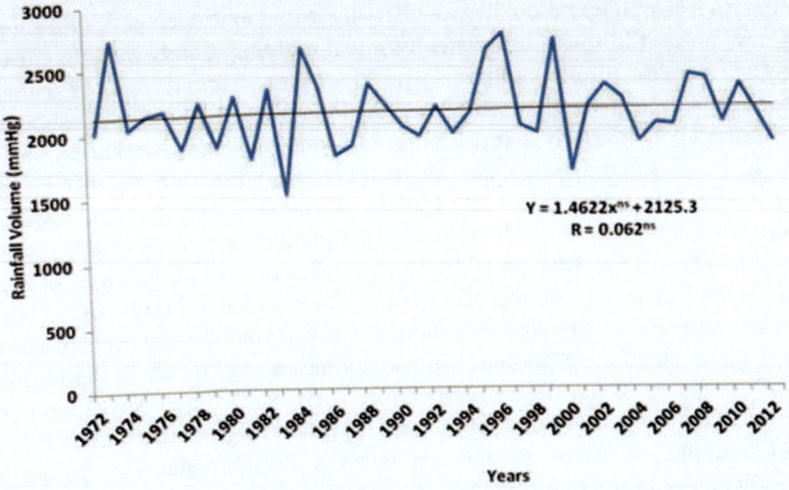

Fig. 18.5 Trend result of rainfall volume from 1972 to 2012 in Imo State (Chidiebere-Mark et al. 2014)

164.6 mm, respectively. However, for the first (1988–1997), second (1998–2007), and third (2008–2017) decades, the highest mean annual rainfalls of 246.8 mm, 267.4 mm, and 236.2 mm, respectively, were recorded in 1990, 2002, and 2012, respectively.

According to Osugiri et al. (2019), rainfall volume in southeast Nigeria varied significantly across years from 1972 to 2014. Consequently, there is extremely little (0.049) relationship between the volume of rainfall and time. Although there was no particularly noticeable trend, the amount of rainfall in the area increased at an increasing rate. The results were found to be consistent with research by Onyeneke and Mmagu (2015) and Babatunde et al. (2011), according to which the total amount of rainfall in the rainforest and coastal regions of Nigeria will not change; however, the intensity and pattern will. According to research conducted by Abiodun et al. (2011), rainfall intensity in Nigeria's rainforest region is showing a clear pattern toward declining. Therefore, the likelihood of weather-related risks like landslides and erosion to occur is strongly indicated by an increase in rainfall over a period of time with fewer rainy days.

Temperature Variability as an Important Climatic Indicator

One of the markers of climate change is a rise in temperature (Asfaw et al. 2018; Ekwueme and Agunwamba 2021). In all of Nigeria's major agro-ecological zones, according to Onyeneke et al. (2020), the temperature has been steadily rising over time. In the Savannah regions of Guinea and Sudan, temperatures have increased, according to research by Freduah et al. (2019). Similar evidence of declining precipitation was present in these agrometeorological regions, which was also characterized by an increase in temperature. Using the Mann–Kendall test, Ekwueme and Agunwamba (2021) recently studied air temperature and found a consistent trend throughout all southeastern states. Enugu, Umuahia, Owerri, Abakaliki, and Awka all had a considerable rise in mean air temperature, with increase in air temperature of 0.04698, 0.01600, 0.004153, 0.003295, and 0.01859 °C/year, respectively. The statistical significance of all the states' temperature trends ($p < 0.05$) was high. According to Figs. 18.6, 18.7, 18.8, 18.9 and 18.10, a maximum value of 32.2, 27.21, 28.1, 27.8, and 28.2 °C/year was recorded at Enugu, Umuahia, Owerri, Abakaliki, and Awka, respectively. The statistics show that all the States' temperatures are rising, which is the most striking finding. More evaporation may occur in the region as a result of the rate at which the air temperature is rising. An analysis of temperature data from 1972 to 2012 for Imo Sate, acquired from the Agrometeorological Station, NRCRI Umudike, by Ekwueme and Agunwamba (2021), reveals an upward trend. The coefficient of correlation between temperature and time was found to be 0.768 and is statistically significant, indicating that there is a strong positive association between the two variables. As a result, a key element in temperature variations is time. This proves that localized warming is substantial and real. This conclusion was reinforced by Okorie et al. (2019), who noted that a consistent rise in surface temperature is indicative of climate fluctuations in the Imo State. In southeastern Nigeria, a study on forest cover and the effects of several changing climatic factors found a consistent increase in decadal surface air temperature trends, with a mean of 32.3 for Owerri, Imo State (Njoku 2012). Long-term variability of rainfall and temperature in Nigeria exhibits observable indications of climate change, according to Abiodun et al. (2011) and Ekwueme and Agunwamba (2021). This implies that when Nigeria's temperature goes up, heat waves or protracted hot days may become more frequent (Abiodun et al. 2011).

Review of Some Landslide Causative Factors in Southeastern Nigeria

There have been some other prevalent risk factors that, in addition to climatic indicators (rainfall and temperature), have a significant impact on the occurrence of landslides in southeast Nigeria. These factors are detailed below and were noted by the previous researcher in southeastern Nigeria.

Geologic Condition and Steep Slope

In landslide activity, soil type and elevation are important factors. However, research has indicated that these two elements increase the risk of landslides in southeastern

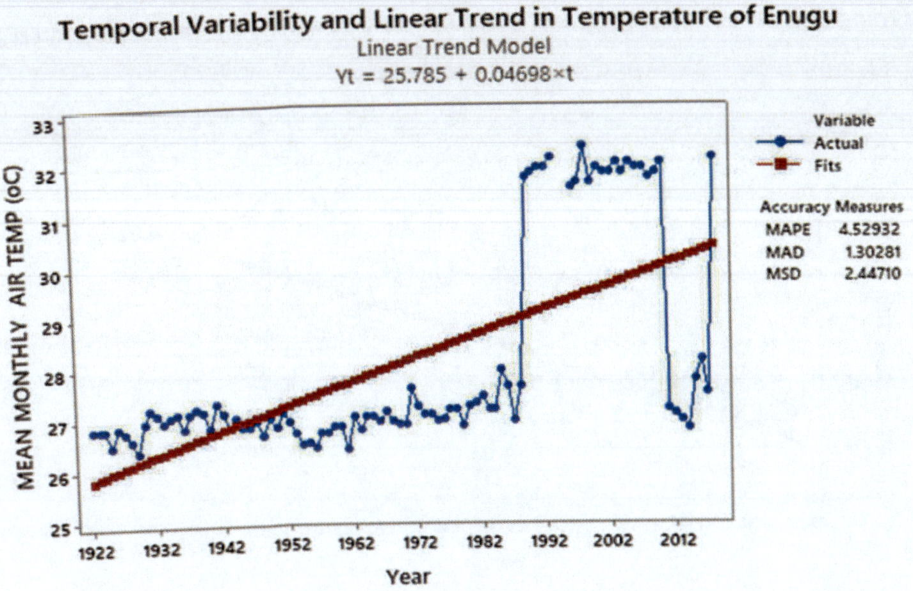

Fig. 18.6 Linear nature of Trend Line of Air Temperature in Enugu, Enugu State (Ekwueme and Agunwamba 2021)

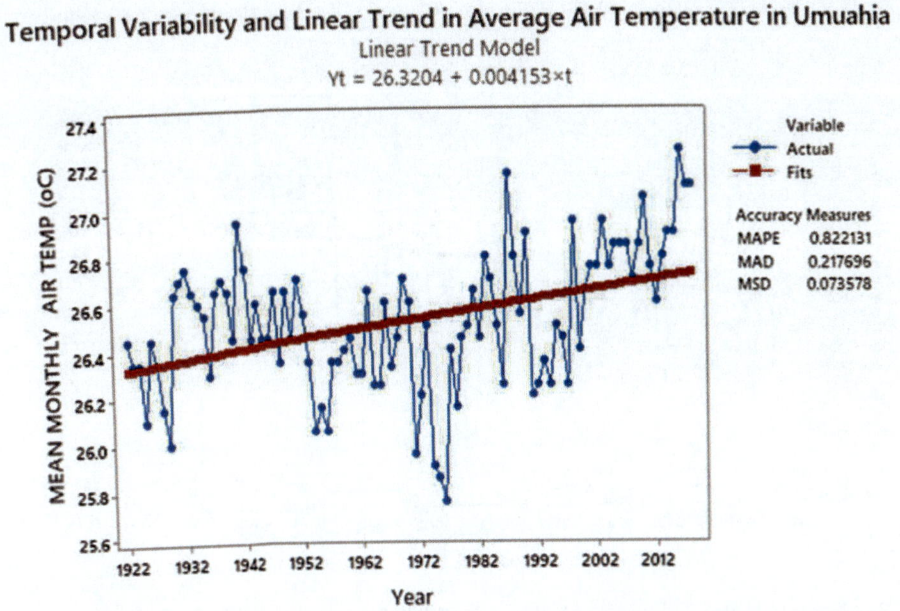

Fig. 18.7 Linear nature of Trend Line of Air Temperature in Umuahia, Abia State (Ekwueme and Agunwamba 2021)

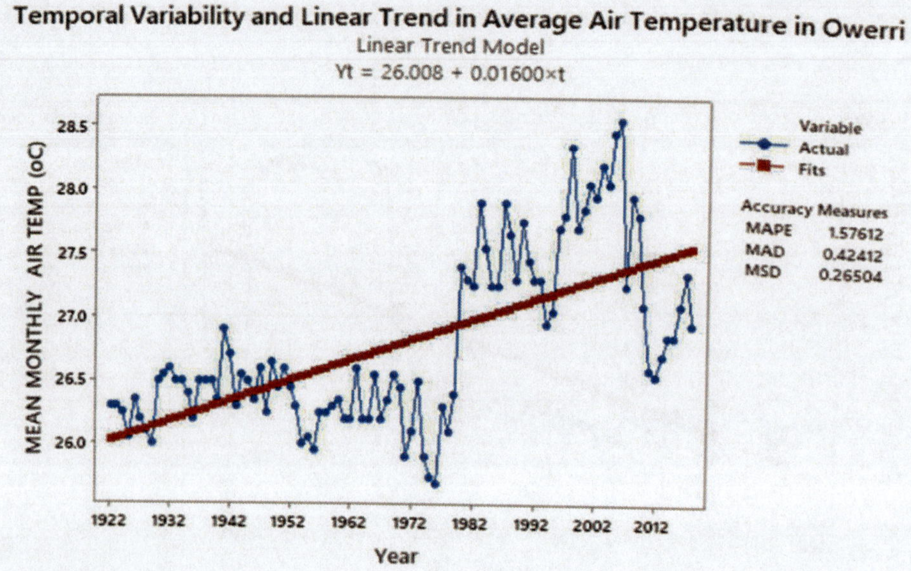

Fig. 18.8 Linear nature of Trend Line of Air Temperature in Owerri, Imo State (Ekwueme and Agunwamba 2021)

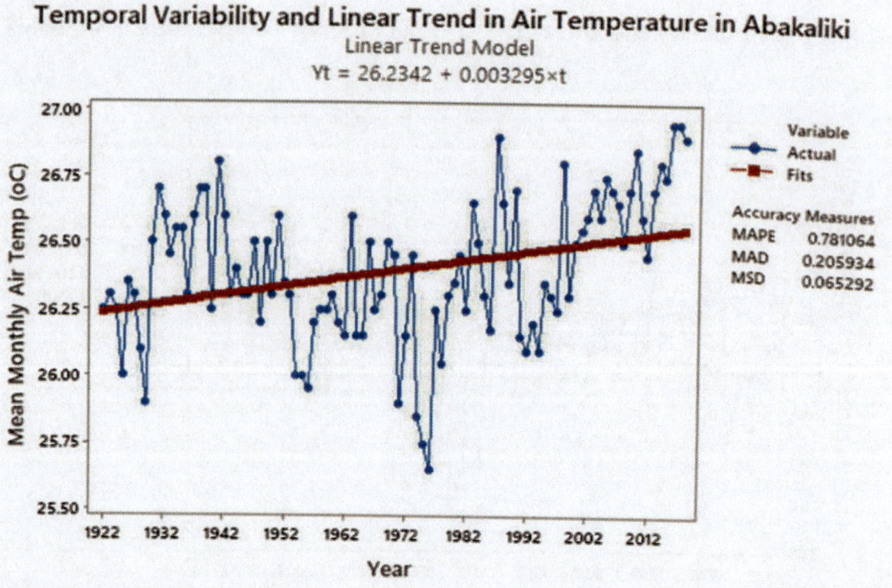

Fig. 18.9 Linear nature of Trend Line of Air Temperature in Abakaliki, Ebonyi State (Ekwueme and Agunwamba 2021)

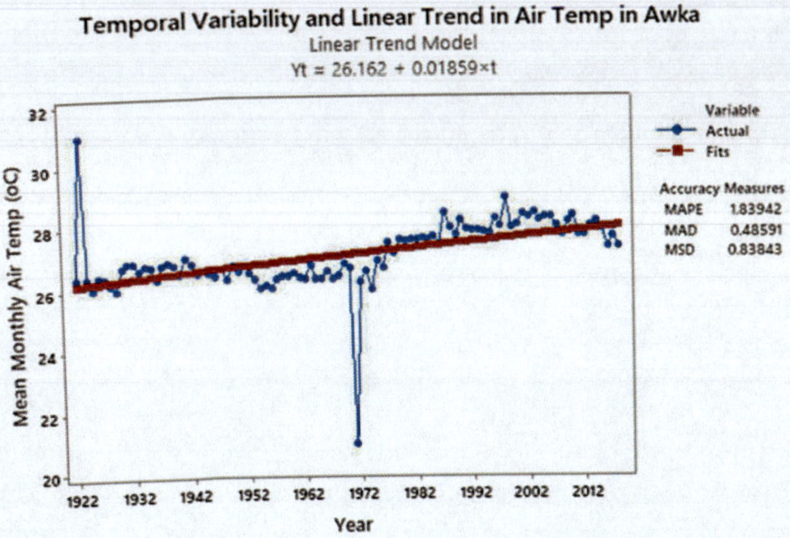

Fig. 18.10 Linear nature of Trend Line of Air Temperature in Awka, Anambra State (Ekwueme and Agunwamba 2021)

Nigeria (Ozioko and Igwe 2019). Environmental hazards such as landslides and gullying are frequent in southeastern Nigeria. These two hazards are particularly prevalent in the states of Enugu and Anambra. According to a report by Igbokwe et al. (2008), the two states (Enugu and Anambra) have the highest number of gully and landslide scars among all southeastern Nigerian states, which was attributed to the area's underlying geology. The youngest geologic deposit underlies Enugu and Anambra States, and among all, the Ajali and Nanka have been dubbed notorious for gullying and landslides by various authors (Nwajide 2013; Egboka 2019; Egbueri et al. 2021). Landslides, on the other hand, have been observed in the Ajali and Nanka formations in all of the analyzed studies in this research (Table 18.1). According to investigations, both formations are loose and friable with poor geotechnical properties, which causes the soil texture prone to precipitation due to human activities such as mining and quarrying (Igwe et al. 2013; Egboka et al. 2019). Both formations are devoid of clayey material, according to the particle size distribution analysis (Emeh and Igwe 2017; Igwe and Una 2019; Nebeokike et al. 2020; Egbueri et al. 2021). Clayey materials, by their nature and composition, aid in serving as a natural binder in soil mass and aid in resisting erosive forces caused by water or any other mechanical action (Egbueri and Igwe 2020). Studies have reported that soils with little or no clay content are vulnerable to failure (Nebeokike et al. 2020; Egbueri et al. 2021).

Steep slopes are widespread in road cuts and gully sites in southeastern Nigeria. In southeastern Nigeria, a steep slope contributes to the occurrence of landslides. The presence of numerous erosion scars, which is a result of physical weathering, exposes several locations to a high slope (Fig. 18.11). The majority of the observed slopes have quite high slope angles, according to the studies examined in this research

Fig. 18.11 Image area showing numerous landslides at steep slopes (Emeh and Igwe 2017)

(Ulakpa et al. 2020a,b; Ozioko and Igwe 2019; Nebeokike et al. 2020; Egbueri et al. 2021). A high slope angle combined with heavy rain will result in a landslide or slope failure (Igwe and Una 2019). Human activities including vibrations emitted by heavy equipment and machinery will also contribute to the collapse of steep slopes.

Groundwater Condition

The most well-known natural dynamic drivers of landslides are changes in groundwater (Brönnimann 2011). Depending on the situation, groundwater can be both a resource and a problem. Kemper (2004) reported that pumping water from the groundwater reservoir can provide water for the inhabitants while also increasing the stability of the landmass on which they live in communities with limited access to potable water. A change in the infiltration of water into a slope (from rainfall), a volumetric change in porosity or a change in the slope's drainage conditions induces groundwater fluctuations (e.g., changes in seepage pressures or flooding at the base) (Ibeh 2020). Variations in groundwater pressure affect pore-water pressures, which alter effective normal stresses and, as a result, the shear strength of potential failure surfaces (Puzrin and Randolph 2015). In southeastern Nigeria, studies by Nebeokike et al. (2020) and Egbueri et al. (2021) on liquid equilibrium stimulation for pore-water pressure reveal that there is a higher build-up of pore-water pressure during the rainy season than during the dry season. A rise in groundwater level was attributed to the high build-up of water pressure. Egbueri et al. (2021) reported that water is seen dripping from some of the slopes within the Nanka at the peak of the rainy season. However, according to Igwe et al. (2013), the majority of landslides in the area occur around the start of the rainy season because infiltrating water reduces the suction in the slopes.

Land Use Pattern and Human Activities

Changes in land cover and land use can either raise or decrease the likelihood of landslides in a given area. Many areas in southeastern Nigeria have shed vegetative cover as a result of growing urbanization and industrialization, as people look for habitat, farmland, or sources of income through lumbering. Trees are being cut down without being replaced, which not only harms our environment but also puts our citizens at risk by providing pathways for environmental disasters such as gullies and landslides in this present era of climatic changes. Most locations in southeastern Nigeria are classed by build-up land cover type, according to studies by Ozioko and Igwe (2019) and Ulakpa et al. (2020a, b). This indicates that habitats in these places are threatened by the loss of plant cover and that such sites are more likely to experience gradual soil erosion or landslides whenever there is torrential rain.

Landslides are triggered by a combination of factors, including population growth and the intensity of human activity. In southeastern Nigeria, construction activities, land transition, extractive industries, urbanization, industrialization, excessive use of agrochemicals, mining and quarrying activities, and other human activities are among the landslide triggering events in the studied area (Igwe 2015a, b, c, d; Nebeokike et al. 2020; Egbueri et al. 2021). Moving vehicles, heavy machinery, and explosives in mining activities can create abrupt ground accelerations, causing instantaneous shear stresses exceeding the mobilized shear strength of the soil mass. In addition, human-induced activities such as bush cutting, lumbering, overgrazing, and bush burning in southeastern Nigeria affect the soil's vegetative cover (Ulakpa et al. 2020a, b; Ogbuene 2010). Weather-related hazards such as landslides, soil erosion, and flooding harm the human environment in high practice or the occurrence of such events.

Landslide Occurrence in Southeastern Nigeria

Many landslides have been caused by storms that produce intense rainfall in a short period, such as some hours, or have a more moderate intensity that can last several days (Jibson 1989). Shallow landslides in soils and weathered rock are frequently formed on steep slopes during the more intense phases of a storm, according to studies, and thresholds of combined severity and length may be required to cause them (Gryta and Bartholomew 1989; Jacobson et al. 1989; Jibson 1989). In the southeastern area of Nigeria, annual rainfall averages around 2000 mm. The yearly

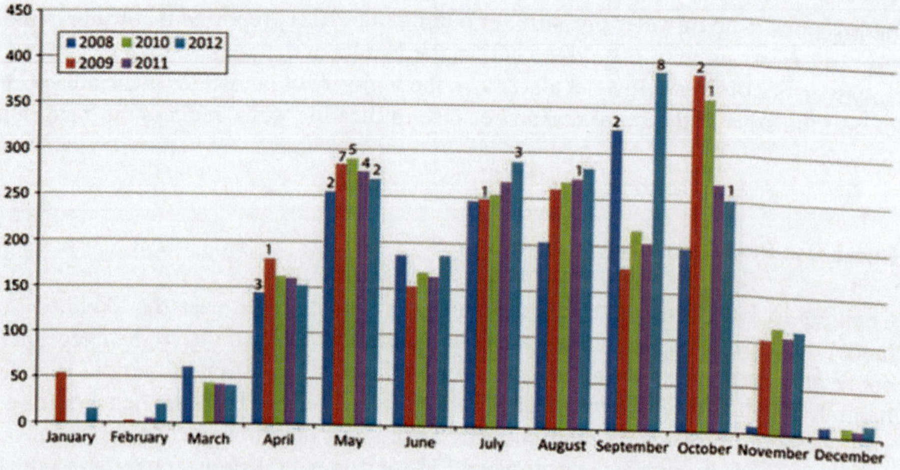

Fig. 18.12 Compared monthly rainfall and the timings of the 43 landslide events at Iva Valley (Igwe et al. 2013)

heavy rainfall has a significant negative influence on the landscape and landform. Deep gullies are formed by such a scenario, which cut deep into the soil and into the subsurface. Most shallow landslides are thought to be caused by rapid rainfall infiltration, which causes soil saturation and a transient rise in pore-water pressures. Igwe et al. (2013) reported that landslides occur majorly during rainfall events. A comparison chart between monthly rainfall and the timing of 43 landslide events within the Iva Valley area in Enugu State is shown in Fig. 18.12. It reveals that the rainfall amount in May appears to support the idea that intense/short-duration rainfall is the main factor in landslides' trigger. Furthermore, southeastern Nigerian studies on slope stability analysis also found that slopes are more stable in the dry season than in the rainy season (Ige et al. 2016; Chukwu and Igwe 2018a; Ibeh 2020; Nebeokike et al. 2020; Egbueri et al. 2021). This could be due to an increase in pore-water pressure during the rainy season's peak. Continuous change in wet and dry seasons, on the other hand, weakens the soil matrix and exposes it to failure. Although rainfall intensity is the most common natural cause of landslides, human activities such as deforestation, fossil fuel burning, and indiscriminate construction are known to alter rainfall intensity indirectly over time.

Impact and Management of Landslide in Southeastern Nigeria

Economic disruption, population displacement, and detrimental effects on the natural environment have all been reported to be caused by slope instability combined with severe storms or torrential rain and floods (Igwe 2017; Igwe and Una 2019; Ozioko and Igwe 2019; Egbueri et al. 2021). However, landslides have a detrimental influence on the environment in the research area, resulting in the loss of arable land, which will result in a decline in farming activities, and food productivity, and a negative impact on the economy. It also results in the loss of vegetation cover, a reduction in available land for the construction of both residential and industrial structures, vandalism of transportation and communication systems, migration of people from their natural habitats, and the destruction of homes and human property, all of which cost millions of Naira.

Landslides are frequently characterized as local problems, but their effects and cost implications may drive them to become State or national issues (Highland and Bobrowsky 2008). For a long time, researchers and government agencies have worked to reduce the damage caused by landslides around the world (Yalcin 2007). According to research, landslides in southeastern Nigeria have progressed beyond the interests of local communities and are now a national issue (Igwe and Una 2019). However, the Federal Government of Nigeria has recently spent a huge sum of money to repair the damage and prevent future landslides in this area. The majority of the measures used to mitigate this risk have failed (Fig. 18.13); yet, there is a need for a need to be innovation.

Fig. 18.13 Evidence of failed mitigation measure at Nanka community in Anambra State (Igwe and Una 2019)

Conclusions and Recommendations

The major effects of climate change in most regions of the world are an increase in temperature (global warming) and a decrease in precipitation. In many regions of the world, these have either detrimental or beneficial ecological effects. All over the globe, increased land-based, ice fragility, and melting have been brought on by the rising temperature. On the other hand, sea-level rise, coastal flooding, landslides, and erosion have been impacted by the warming of the Arctic, warm and freezing temperate ice, increasing rainfall in some areas of the world, and ocean expansion as the water warms.

Nigeria, and particularly the southeast, has demonstrated significant climatic change effects. Studies on several climate indicators, such as rainfall and temperature, have revealed an increase in weather-related risks such as landslides, erosions, and flooding. However, research has shown that when compared to earlier decades, rainfall in southeast Nigeria has increased in intensity but total annual rainfall has decreased. But an increasingly large amount of rain falling on fewer wet days is a clear sign that weather-related problems are likely to arise (like flooding, landslides, and erosion). Additionally, taking into account a few natural and anthropogenic features in southeast Nigeria, including the geology, topography, groundwater levels, land-use patterns, and human developmental activities. These elements contribute to landslide hazards in southeastern Nigeria in one way or another through the effects of climate change. For instance, the fact that southeastern Nigeria is majorly dominated by sedimentary rocks of which many of which have poor geotechnical properties, owing to the uneven landforms, altered land-use patterns and human development activities (quarrying for construction purposes and population growth), it is however believed that the area will be readily accessible for hazards like erosion and landslides. Furthermore, since the hazard of concern in this study, "landslides," is a natural occurrence, it is obvious that, if not appropriately managed and controlled to a certain extent, southeastern Nigeria will continue to be in peril from this dreadful disaster. In the note of this, the following management strategies are recommended.

i. Following a study by Igwe and Una (2019), it was discovered that some areas of the study region do not have a proper drainage system based on an understanding of geology, geomorphology, hydrogeology, and geotechnical characteristics of the area. They did advise, however, that a properly constructed, interconnected drainage system would reduce the quantity of moisture entering the soils, increasing shear strength and slope stability. This is thought to tackle the problem of erosion-caused hill cuts. Yalcin (2007) proposed a similar method in the East Black Sea region of Turkey. The well-designed drainage system will minimize not only sliding but also reactivation of movements in the event of a sudden failure. Highland and Bobrowsky (2008) found that proper drainage is required to prevent sliding or, in the case of an existing failure, to prevent the movement from reactivating.
ii. According to Gray and Leiser (1982), biotechnical stabilization uses mechanical and biological factors to prevent and stop slope failures and erosion. Soil

bioengineering stabilization, according to Gray and Sotir (1996), is a specific subset of biotechnical stabilization in which live plant components function as mechanical elements in the slope protection system. Individuals and the government should take responsibility for planting various stabilization plants that can stabilize the slopes, according to these reports. In many different environments, plants like vetiver and bamboo trees function effectively in stabilizing slopes against erosion and landslide hazards (Highland and Bobrowsky 2008).

iii. Public awareness programs: The general public should be aware of the risks and precautions. Farmers should be taught about the need of planting trees and grasses that will aid stabilize slopes, and landslide hazard information should be incorporated into long-term plans for areas prone to landslides. Furthermore, it urged the implementation of a proper land-use policy to effectively mitigate issues such as afforestation and overgrazing in some regions.

References

Anyanwu NP, Arua I (1990) Ichnofossils from the Imo Formation and their paleoenvironmental significance. J Min Geol 26(1):1–4

Ajetomobi J, Abiodun A, Hassan R (2011) Impacts of climate change on rice agriculture in Nigeria. Trop Subtrop Agroecosyst 14(2):613–22

Akande SO, Ogunmoyero IB, Petersen HI, Nytoft HP (2007) Source rock evaluation of coals from the lower Maastrichtian Mamu formation SE Nigeria. J Pet Geol 30(4):303–324

Akinsanola AA, Ogunjobi KO (2014) Analysis of rainfall and temperature variability over Nigeria. Glob J Human-Soc Sci: B Geography, Geosci, Environ Disaster Manage 14(3):1–7

Arua I (1980) Palaeocene macrofossils from the Imo Shale in Anambra state, Nigeria. J Min Geol 17(1):81–4

Asfaw A, Simane B, Hassen A, Bantider A (2018) Variability and time series trend analysis of rainfall and temperature in northcentral Ethiopia: a case study in Woleka Sub-Basin. Weather Clim Extrem 19:29–41. https://doi.org/10.1016/j.wace.2017.12.002

Babatunde JA, Salami AT, Tadross M (2011) Developing climate change scenarios, biophysical impacts and adaptation strategies in Nigeria. A final report submitted to Nigerian Environmental Study/Action Team (NEST) as part of the Building Nigeria Response to Climate Change (BNRCC) Project

Brönnimann CS (2011) Effect of groundwater on landslide triggering. EPFL

Chidiebere-Mark NM, Eze CC, Nwankwo OO, Onyeneke RU, Ukwandu JC, Ejike RD, Enyia CO (2014) Perception and adaptation of food crop farmers to climate change in Imo State, Nigeria. In: 14th annual national conference of the Nigerian association of agricultural economists

Cruden DM, Varnes DJ (1996) Landslide types and processes. In: Turner AK, Schuster RL (eds) Landslides, investigation and mitigation, special report 247. Transportation Research Board, Washington DC, pp 36–75. ISSN: 0360–859X, ISBN:030906208X

Dim CIP, Onuoha KM, Okeugo CG, Ozumba BM (2017) Petroleum system elements within the late cretaceous and early paleogene sediments of Nigeria's inland basins: an integrated sequence stratigraphic approach. J Afr Earth Sci 130:76–86

Egboka BCE, Orji AE, Nwankwoala HO (2019) Gully erosion and landslides in Southeastern Nigeria: causes, consequences and control measures. Glob J Eng Sci. https://doi.org/10.33552/GJES.2019.02.000541

Egbueri JC, Igwe O (2020) The impact of hydrogeomorphological characteristics on gullying processes in erosion-prone geological units in parts of southeast Nigeria. Geol Ecol Landscape. https://doi.org/10.1080/24749508.2020.1711637

Egbueri JC, Igwe O, Unigwe CO (2021) Gully slope distribution characteristics and stability analysis for soil erosion risk ranking in parts of Southeastern Nigeria: a case study. Environ Earth Sci. https://doi.org/10.1007/s12665-021-09605-7

Ekwueme BN, Agunwamba JC (2021) Trend analysis and variability of air temperature and rainfall in regional river basins. Civ Eng J 1(7):816–26

Emeh C, Igwe O (2017) Effect of environmental pollution on susceptibility of sesquioxide-rich soils to water erosion. Geol Ecol Landscape. https://doi.org/10.1080/24749508.2018.1452484

Ene EG, Dim CI, Chidubem E (2018) Assessment of some pristine sand deposit from Imo Formation (Southeastern Nigeria) as construction aggregate and industrial material. Environ Earth Sci. https://doi.org/10.1007/s12665-018-7957-6

Freduah G, Fidelman P, Smith TF (2019) Adaptive capacity of small-scale coastal fishers to climate and non-climate stressors in the Western region of Ghana. Geogr J 185(1):96–110

Gariano SL, Guzzetti F (2016) Landslides in a changing climate. Earth Sci Rev. https://doi.org/10.1016/j.earscirev.2016.08.011

Gray DH, Leiser AT (1982) Biotechnical slope protection and Erosion control, 271. Van Nostrand Reinhold Company, New York

Gray DH, Sotir RB (1996) Biotechnical and soil bioengineering slope stabilization–A practical guide for Erosion control, 378. Wiley, New York

Gryta JJ, Bartholomew MJ (1989) Debris-avalanche type features in Watauga County, North Carolina. In: Lewis SE (ed) Geologic investigations in the Blue Ridge of Northwestern North Carolina, Carolina geological society guide-book, North Carolina Division of Land Resources, Boone, Article 5, p 22

Highland LM, Bobrowsky P (2008) The landslide handbook—Aguideto understanding landslides: Reston, Virginia, U.S. Geological Survey circular 1325, 129

Hungr O, Leroueil S, Picarelli L (2013) The Varnes classification of landslide types, an update. Landslides 11(2):167–194. https://doi.org/10.1007/s10346-013-0436-y

Ibeh CU (2020) Effect of changing groundwater level on shallow landslide at the basin scale: a case study in the Odo basin of south eastern Nigeria. J Afr Earth Sci. https://doi.org/10.1016/j.jafrearsci.2020.103773

Igbokwe JI, Akinyede JOB, Dang BT, Alaga TMN, Ono MN, Nnodu VC, Anike LO (2008) Mapping and monitoring of the impact of gully erosion in south-eastern Nigeria with satellite remote sensing and geographic information system. The international archives of the photogrammetry. Remote Sens Spatial Inf Sci 37:865–871

Ige OO, Oyeleke TA, Baiyegunhi C, Oloniniyi TL, Sigabi L (2016) Liquefaction, landslide and slope stability analyses of soils: A case study of soils from part of Kwara. Kogi and Anambra States of Nigeria Natural Hazards and Earth System Sciences Discussion. https://doi.org/10.5194/nhess-2016-297

Igwe O (2014) The compressibility and shear characteristics of soils associated with landslides in geologically different localities—case examples from Nigeria. Arab J Geosci. https://doi.org/10.1007/s12517-014-1616-3

Igwe O (2015a) Predisposing factors and the mechanisms of rainfall-induced slope movements in Ugwueme South-East Nigeria. Bull Eng Geol Environ. https://doi.org/10.1007/s10064-015-0767-0

Igwe O (2015d) The mechanisms and geotechnical characteristics of slope failures at a mining district. Journal geological society of India, Southeast Nigeria. https://doi.org/10.1007/s12594-015-0239-y

Igwe O (2017) The hydrogeological attributes and mechanisms of a receding sedimentary terrain in the Anambra Basin. Southern Nigeria. Environ Earth Sci 76(1):1–22

Igwe O (2018) The mechanisms of enlargement and hydrogeological characteristics of Nanka Gully, Southeast Nigeria. J Geol Soc India 92:227–234

Igwe O, Chukwu C (2018a) Evaluation of the mechanical properties and critical slope parameters of mine tailings at Enyigba. South Eastern, Nigeria, Geo-Engineering. https://doi.org/10.1186/s40703-018-0075-1

Igwe O, Chukwu C (2018b) Slope stability analysis of mine waste dumps at a mine site in Southeastern Nigeria. Bull Eng Geol Environ. https://doi.org/10.1007/s10064-018-1304-8

Igwe O, Una CO (2019) Landslide impacts and management in Nanka area. Southeast Nigeria. Geoenviron Disast 6:5. https://doi.org/10.1186/s40677-019-0122-z

Igwe O, Fukuoka H (2014) The effect of water-saturation on the stability of problematic slopes at the Iva Valley area, Southeast Nigeria. Arabian J Geosci. https://doi.org/10.1007/s12517-014-1398-7

Igwe O, Mode W, Nnebedum O, Okonkwo I, Oha I (2013) The analysis of rainfall-induced slope failures at Iva Valley area of Enugu State, Nigeria. Environmental Earth Sciences. https://doi.org/10.1007/s12665-013-2647-x

Igwe O (2015b) The study of the factors controlling rainfall-induced landslides at a failure-prone catchment area in Enugu. South-eastern Nigeria using remote sensing data. Landslides. https://doi.org/10.1007/s10346-015-0627-9

Igwe O (2015a) The geotechnical characteristics of landslides on the sedimentary and metamorphic terrains of South-East Nigeria, West Africa. Geoenvironmental Disasters. https://doi.org/10.1186/s40677-014-0008-z

Jacobson RB, Cron ED, McGeehin JP (1989) Slope movements triggered by heavy rainfall, November 3–5, 1985, in Virginia and West Virginia, U.S.A. In: Schultz AP, Jibson RW (eds) Landslide processes of the Eastern United States and Puerto Rico, Geological Society of America Special Paper 236, pp 1–14

Jibson RW (1989) Debris flows in Southern Puerto Rico. In: Schultz AP, Jibson RW (eds) Landslide processes of the Eastern United States and Puerto Rico, Geological Society of America Special Paper 236, pp 29–56

Kemper KE (2004) Groundwater—from development to management. Hydrogeol J 12(1):3–5

Nebeokike UC, Igwe O, Egbueri JC, Ifediegwu SI (2020) Erodibility characteristics and slope stability analysis of geological units prone to erosion in Udi area, southeast Nigeria. Model Earth Syst Environ. https://doi.org/10.1007/s40808-020-00741-w

Nicholson SE (1994) Recent rainfall fluctuations in Africa and their relationship to past conditions over the continent. Holocene 4:121–131

Njoku JD (2006) Analysis of the effect of global warming on forest of southeastern Nigeria using remotely sensed data. A PhD thesis: Imo State University, Owerri, Nigeria

Nwajide CS (1980) Eocene tidal sedimentation in the Anambra Basin, southern Nigeria. Sediment Geol 1–25(3):189–207

Nwajide CS (2013) Geology of Nigeria's sedimentary basins. CSS Bookshops Limited, Nigeria, p 565

Oboh-Ikuenobe FE, Obi CG, Jaramillo CA (2005) Lithofacies, palynofacies, and sequence stratigraphy of Palaeogene strata in Southeastern Nigeria. J Afr Earth Sci 1–41(1–2):79–101

Ogbuene EB (2010) Environmental consequences of rainfall variability and deforestation in Southeastern Nigeria. Int J Water Soil Resourc Res 1(1–3):100–115

Okagbue CO (1988) A landslide in a quasi-stable slope. Eng Geolol 25:69–82

Okagbue CO (1992) The 1988 Nanka landslide, Anambra State, Nigeria. Bull Int Assoc Eng Geol 46(1):79–87

Okagbue CO, Ezechi JC (1988) Geotechnical characteristics of soils susceptible to severe gullying in eastern Nigeria. Bull Int Assoc Eng Geol 38:111–119

Okechukwu ME, Mbajiorgu CC (2020) Spatial distribution of rainfall and reference evapotranspiration in southeast Nigeria. Agric Eng Int 22(1)

Okonkwo GI, Mbajiorgu CC (2010) Rainfall intensity-duration-frequency analysis for Southeastern Nigeria. Agricult Eng Int: CIGR J 8:12(1)

Okorie FC, Njoku JD, Onweremadu EU, Iwuji MC (2019) Climate change and food crop production in Imo state Nigeria: an analysis of 30 year period. African J Soc Behav Sci 9(2)

Omoja UC, Okpalaku BN, Uchechukwu UN, Obiekezie TN (2021) Variability of rainfall in Awka, Anambra state Nigeria. IOSR J Appl Phys 13(4):50–56

Omotosho JB (2007) Pre-rainy season moisture build-up and storm precipitation delivery in the West Africa Sahel. Int J Climatol 28:937–946. https://doi.org/10.1002/joc.1548

Onyeneke RU, Emenekwe CC, Munonye JO, Nwajiuba CA, Uwazie UI, Amadi MU, Izuogu CU, Njoku CL, Onyeneke LU (2020) Progress in climate–agricultural vulnerability assessment in Nigeria. Atmosphere 11:190. https://doi.org/10.3390/atmos11020190

Onyeneke RU, Mmagu CJ (2015) Crop farmers' understanding of climate change, and adaptation practices to climate change in southeast Nigeria. In: Proceedings of the Conference in honour of Professor Chinedum Uzoma Nwajiuba held at the Oak of Mamre Library, Imo State, Nigeria, July 19 and 20

Osugiri II, Ben-Chendo GN, Okwara MO, Anyanwu GU, Utazi CO (2019) Analysis of the trends of climatic variables in Southeast Nigeria. Int J Current Res 11:3910–3913. https://doi.org/10.24941/ijcr.31970.05.2019

Ozioko OH, Igwe O (2019) GIS-based landslide susceptibility mapping using heuristic and bivariate statistical methods for Iva Valley and environs, Southeast Nigeria. Environ Monit Assess. https://doi.org/10.1007/s10661-019-7951-9

Puzrin AM, Randolph MF (2015) Effects of pore water pressure dissipation on rate dependency of shear strength in localised failure of soils. Int J Numer Anal Meth Geomech 39(10):1045–1062

Ulakpa ROE, Okwu VUD, Chukwu KE (2020a) An evaluation of the effect of landslide using statistical approach in Nanka, Anambra state. African Res J Environ. https://doi.org/10.36265/arejoen.2020.030106

Ulakpa ROE, Okwu VUD, Chukwu KE, Eyankware MO (2020b) Landslide susceptibility modelling in selected States Across SE. Nigeria. Environ Ecosyst Sci. https://doi.org/10.26480/ees.01.2020.23.27

Ulakpa ROE, Okwu VUD, Chukwu KE, Eyankware MO (2020c) A GIS-based approach for long-term prediction of landslide; a case study of three states in SE. Nigeria. Environ Contamin Rev. https://doi.org/10.26480/ecr.02.2020.66.70

Ulakpa ROE, Okwu VUD, Chukwu KE, Iwueke NT (2020d) A GIS-Based approach for long-term prediction of landslide; a case study of three states in SE. Nigeria Environ Contamin Rev. https://doi.org/10.26480/ecr.02.2020.45.49

Yalcin A (2007) Environmental impacts of landslides: a case study from East Black Sea region. Turkey. Environ Eng Sci 24(6):821–833

Chapter 19
Climate Change and Drought in the Dryland Areas of Nigeria

S. U. Usman, Nura Umar, and A. I. Abdulhamid

Abstract Current rapid changes in global, regional and local weather and climate, which are believed to have been caused by human-induced changes in atmospheric chemistry, are impacting different parts of the world differently. Some dry areas are becoming drier while some humid areas are becoming wetter. This study assesses the impact of climate change on drought events and vulnerability to drought in the dryland areas of Nigeria (Borno, Jigawa, Kano, Katsina, Kebbi, Sokoto, Yobe and Zamfara States). Objectives of the study include determining; (i) past and future drought intensity, (ii) past and future drought frequency, and (iii) people's vulnerability to drought in the study area. Historical (1901–2020) and projected (2021–2100) rainfall data (CMIP6–SSP1, 3 and 5) for the study area were downloaded from the World Bank Climate Change Knowledge Portal and Standardised Precipitation Index was used in analysing the data. The results indicate alternating drought and non-drought conditions with higher drought occurrences in the 1970s and 1980s and low occurrences in the last two decades. On one hand, SSP1 indicates an increase in drought in the near future (2021–2050) and a decrease in the distant future. On the other hand, SSP3 and 5 indicate the opposite. Inhabitants of the area are found to be one of the most vulnerable to drought in the country. A holistic approach to drought adaptation, especially addressing factors influencing drought vulnerability, should be vigorously pursued in the area. Flood monitoring should be part of the environmental plan for the area.

Keywords Climate change · Drought · Drought vulnerability · Dryland areas · Standardised precipitation index

---

S. U. Usman (✉) · N. Umar · A. I. Abdulhamid
Department of Environmental Sciences, Federal University Dutse, Jigawa State, Dutse, Nigeria
e-mail: sani.u@fud.edu.ng

N. Umar
e-mail: nuraumar@fud.edu.ng

A. I. Abdulhamid
e-mail: abdulhamidaz@fud.edu.ng

J. C. Egbueri et al. (eds.), *Climate Change Impacts on Nigeria*, Springer Climate,
https://doi.org/10.1007/978-3-031-21007-5_19

Introduction

Climate change entails changes in many weather and climate processes, which emanate as responses to energy fluxes within the climate system. The Earth's history is chequered by changes in the climate. However, the current rapid changes especially global warming, which is likely to be 1.5 °C above the pre-industrial level by 2030, are believed to be human induced (Intergovernmental Panel on Climate Change [IPCC] 2018). Global warming has varied climatic implications across the globe. Thermodynamically, it gives rise to an increase in atmospheric water vapour by a 7% per degree increase in temperature, in accordance with the Clausius–Clapeyron relationship (Trenberth 2011). This enhances the hydrological cycle, which implies a global distribution of atmospheric moisture through an increase in the latitudinal exchange of heat and moisture (Soldatenko 2019). However, it has been argued that a commensurate increase in precipitation should not be expected as it will be just about 2–3% due to the suppressive effects of adjustments in atmospheric energy budgets in response to fluxes in greenhouse gases and absorbing aerosols (Allan et al. 2020).

Flood and drought are among the prominent features of climate change. Drought being a product of ocean, land and atmosphere coupling is bound to be affected by the warming of the climate in terms of its frequency and intensity. Rainfall over continental areas is strongly dependent on the transport of water vapour from the ocean (Gimeno et al. 2012) and low-level jets are the major media of the transport (Gimeno et al. 2016). Such is the case in Nigeria's dryland areas, located within the West African Sahel region. The source of moisture for this area is water vapour advected from the Gulf of Guinea and the tropical Atlantic Ocean via the West African Monsoon (WAM) (Pomposi et al. 2020; Dey et al. 2021; Sori et al. 2019). The WAM consists of the African Easterly Jet (AEJ), the African Easterly Waves (AEWs), the Tropical Easterly Jet (TEJ), the Saharan heat low (SHL) and the Intertropical Convergence Zone (ITCZ) (Raj et al. 2019). It has been described as one of the most complex monsoon systems on Earth, as land, ocean and atmosphere are highly coupled (Niang et al. 2020). This is in addition to being highly sensitive to the warming of the climate.

On one hand, southward displacement of the ITCZ in the Sahelian region is associated with drought (Giannini et al. 2003; Schneider et al. 2014; Thomas and Nigam 2018). This anomaly has been ascribed to comparative warming of the Southern Hemisphere (Haywood et al. 2013), enhanced warming of the equatorial surface (Zhou et al. 2019), meridional sea surface temperature contrast (Mischell and Lee 2022) and human sulphate aerosol emission (Biasutti and Giannini 2006; Allen et al. 2015). On the other hand, warming of the Northern Hemisphere results in a northward shift of the ITCZ and consequently rainfall surplus over the Sahel (Bonfils et al. 2020). Both the shift and narrowing of the ITCZ are reported to be in response to climate change (Byrne and Schneider 2016; Watt-Mayer et al. 2019), and they will have implications for tropical regions (Byrne et al. 2018; Mamalakis et al. 2021).

SHL affects rainfall conditions in the study area in different ways and is also sensitive to the warming of the climate. Its northward shift, which has been reported to be in response to global warming, will lead to the northward displacement of the WAM and is expected to bring more rainfall into the study area (Monerie et al. 2020). Conversely, its southward displacement inhibits the northward shift of the WAM, which results in drought over the Sahel (Biasutti et al. 2009; Harsma et al. 2005). Intensification of SHL increases thermal contrast between the Sahara and Gulf of Guinea, which brings more moisture to the Sahel (Lavaysse et al. 2016).

AEJ and TEJ are also instrumental in moisture transport to the Sahel region and negative anomalies in their strength and location result in drought in the region. King (2016) reported that the position of the tropical rain belt coincides with the position of AEJ and TEJ. The study also revealed strong correlations ($r = 0.50$ to $r = 0.70$) between TEJ strength and rainfall. In the same vein, Bercos-Hickey et al. (2020) reported a negative relationship between Saharan mineral dust and northward displacement of AEJ, which is related to drought in the Sahel. Intensification of the Saharan heat low also influences the AEJ and AEWs (Lavaysse et al. 2010).

The semi-arid part of Nigeria's dryland areas, which is made up of the states (Borno, Jigawa, Kano, Katsina, Kebbi, Sokoto, Yobe and Zamfara) that are adjacent to the Niger Republic, is the one that is the most vulnerable to drought in the country. This is due to its environmental and socio-demographic characteristics. It is also the one that has been hit the hardest by the past drought episodes, which include causing acute poverty and starvation among the populace (Mortimore 2009). The area also has the highest drought frequency in the country. Adejuwon and Dada (2021) reported that the frequency of drought occurrence in this area ranges from 12 to 16 years out of the 31 years covered by the study. In Yobe State, Gana (2018) reported that a drought episode caused losses of about 70–80% of harvests and livestock. Thus, drought can be considered one of the foremost environmental challenges in the area.

Having considered the aforementioned points, it is evident that further warming of the atmosphere is bound to have a profound effect on drought frequency and intensity in the study area. This will wreak more havoc and inflict untold hardship on the populace, as it will be detrimental to the food security and livelihood of the majority. It is also pertinent to note that for effective investigation of future occurrences of drought in this area, climate models that adequately involve all the major processes controlling rainfall in the area have to be used. One of such models is the World Climate Research Programme's Coupled Model Intercomparison Project Phase 6 (CMIP6), which consists of Shared Socio-economic Pathways (SSPs) 1.9, 2.6, 4.5 7.0 and 8.5 (representing low-to-very high CO_2 emission scenarios) involving 21 ensembles (Eyring et al. 2015). Therefore, this study aimed at assessing the historical and future occurrences of drought in the study area in the context of CMIP6 climate change scenarios. The objectives of the study are to (i) examine historical and future drought intensity in the study area, (ii) assess historical and future drought frequency in the study area and (iii) examine the vulnerability of people to drought in the study area. The findings of this study will be useful as a basis for food security, water supply and even human security planning for the area.

Materials and Methods

The Study Area

The study area is located in the extreme northern part of Nigeria, roughly between latitudes 10°N to 14°N and 3°30'E to 14°45'E as shown in Fig. 19.1. It consists of Borno, Jigawa, Kano, Katsina, Kebbi, Sokoto, Yobe and Zamfara States. It covers 289,803 km^2, representing 31.8% of Nigeria's total land area. The estimated combined population of these states as of 2018 was 47, 309, 759 (NBS 2018), representing 25.7% of the total population of Nigeria. The largest proportion of the adult population in this area is engaged in small-scale agricultural (crop and livestock) production ranging from 70% in Borno and Kano States (FAO and ICRISAT 2019; Rigaud et al. 2021), to 80% in Katsina State (World Bank 2020) and 90% in Jigawa State (Sanusi et al. 2013). As such, states with the highest population of cattle and sheep (Zamfara) as well as goats (Katsina) in the country are found in this area (National Agricultural Extension and Research Liaison Services [NAERLS] and Federal Ministry of Agriculture and Rural Development [FMARD] 2020). The presence of rivers such as the Niger River (in Kebbi State) and its tributaries like Rima and Sokoto Rivers (in Sokoto and Zamfara States), Komadugu-Yobe River (in Borno and Yobe States) and its tributaries like Kano River (in Kano State), Hadejia and Jama'are Rivers (in Jigawa State) as well as the Lake Chad, which bounded Borno State to the east allow for extensive irrigation agriculture and fishing in the area. About half of the study area (Borno, Jigawa, Kano and Yobe States) falls within the Lake Chad Basin while the other half (Katsina, Kebbi, Sokoto and Zamfara) falls within the Niger River Basin. The area has tropical continental (Aw) type of climate characterised by wet (June–September) and dry (October–May) seasons. It is also characterised by a severe rate of desertification (Olagunju 2015). Vegetation of the area ranges from Sudan to Sahel Savanna types.

Drought Computations

Observed (1901–2020) and projected CMIP6 (2021–2100) rainfall data for the coordinates of the synoptic stations in the states involved were downloaded from the World Bank Group Climate Change Knowledge Portal. Standardised Precipitation Index (SPI) was used to characterise the rainfall condition of the study period. Its calculation involves fitting the annual rainfall data to a gamma distribution (McKee et al. 1993) with a probability density function defined as

$$g(x) = \frac{1}{\beta^{\alpha}\Gamma(\alpha)} x^{\alpha-1} e^{-x/\beta} \quad (19.1)$$

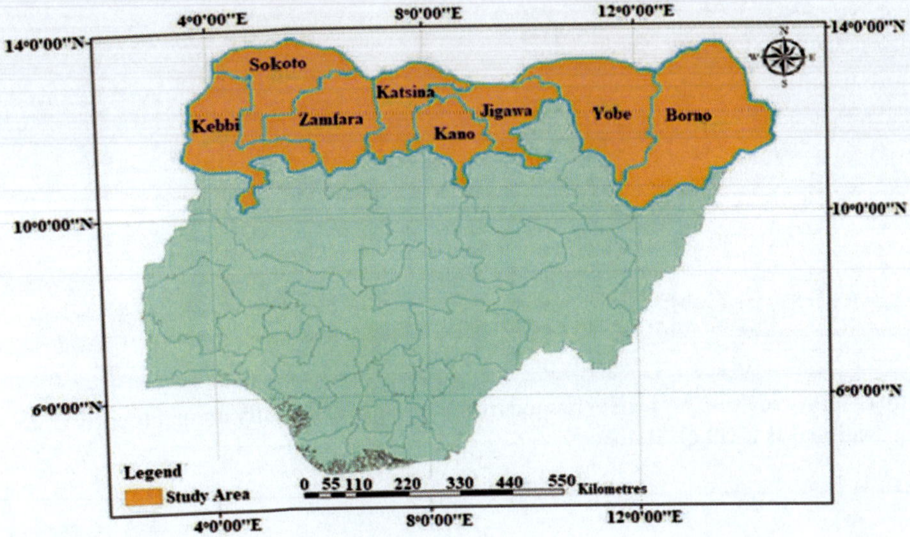

Fig. 19.1 Map of Nigeria showing the study area

where $x > 0$ is the annual rainfall and $\Gamma(\alpha)$ is the gamma function and in which the shape (α) and scale (β) parameters were estimated using the method of maximum likelihood:

$$\hat{\alpha} = \frac{1}{4A}\left(1 + \sqrt{1 + \frac{4A}{3}}\right) \tag{19.2}$$

$$\hat{\beta} = \frac{\overline{x}}{\hat{\alpha}} \tag{19.3}$$

where for n observations

$$\mathrm{A} = \ln(\overline{x}) - \frac{\sum \ln(x)}{n} \tag{19.4}$$

Table 19.1 is used to define rainfall conditions from the results of SPI computations. Drought frequency (F) is calculated as

$$\mathrm{F} = \frac{n}{N} \times 100 \tag{19.5}$$

where N is the number of years involved and n is the number of drought occurrences during the years involved. GIS was used in depicting the frequency of drought for both the observed and projected periods in the study area. Indicators of people's

Table 19.1 Drought classification by SPI value

SPI value	Category
≥2.00	Extremely wet
1.50 to 1.99	Very wet
1.00 to 1.49	Moderately wet
−0.99 to 0.99	Near normal
−1.00 to −1.49	Moderate drought
−1.50 to −1.99	Severe drought
≤−2.0	Extreme drought

Source McKee et al. (1993)

vulnerability to drought were extracted from literature especially annual reports from the National Bureau of Statistics.

Results and Discussion

Historical Drought Intensity

The result of SPI computations involving Eqs. 19.1–19.4 for both the observed period (1901–2020) and the projected period (2021–2100) is presented in Fig. 19.2a–h. Drought in this area is characterised by high spatial and temporal variability, with attendant losses of human and animal lives and acute malnutrition among the populace. The 1913/1914, 1949, 1972/1973 and 1983 to 1987 droughts, which are regional in nature, affected the whole study area. They led to waves of migration and immigration within the area and from the neighbouring Sahelian countries, respectively (Watts 1977; Weiss 2003). These led to unplanned urban growth, increases in criminal activities (Weiss 2003) and the number of beggars in the urban areas (Mortimore and Batterbury 2013).

In response to the 1972/1973 drought, River Basin Development Authorities were established all over Nigeria, which led to the construction of several dams in order to boost irrigation agriculture. For instance, over 20 dams were constructed on the tributaries of the Komadugu-Yobe River (Lake Chad Basin Commission 2000), which empties its water into Lake Chad. Construction of such dams and subsequent droughts within its basin resulted in the shrinkage of Lake Chad by about 90–95% (African Organisation of Supreme Audit Institutions [AFROSAI] 2015) and the deterioration of important wetlands and floodplains in the area. Apart from being Ramsar Sites, Baturiya and Hadejia-Nguru wetlands (in Jigawa and Yobe States) serve as a source of livelihood to over 1.5 million farmers, pastoralists and fishermen (United Nations Office for Project Services [UNOPS] 2014) and also as habitats to migratory birds. However, persistent droughts and attendant widespread construction of dams in the upstream areas have led to their shrinkage by two-thirds (Werz and

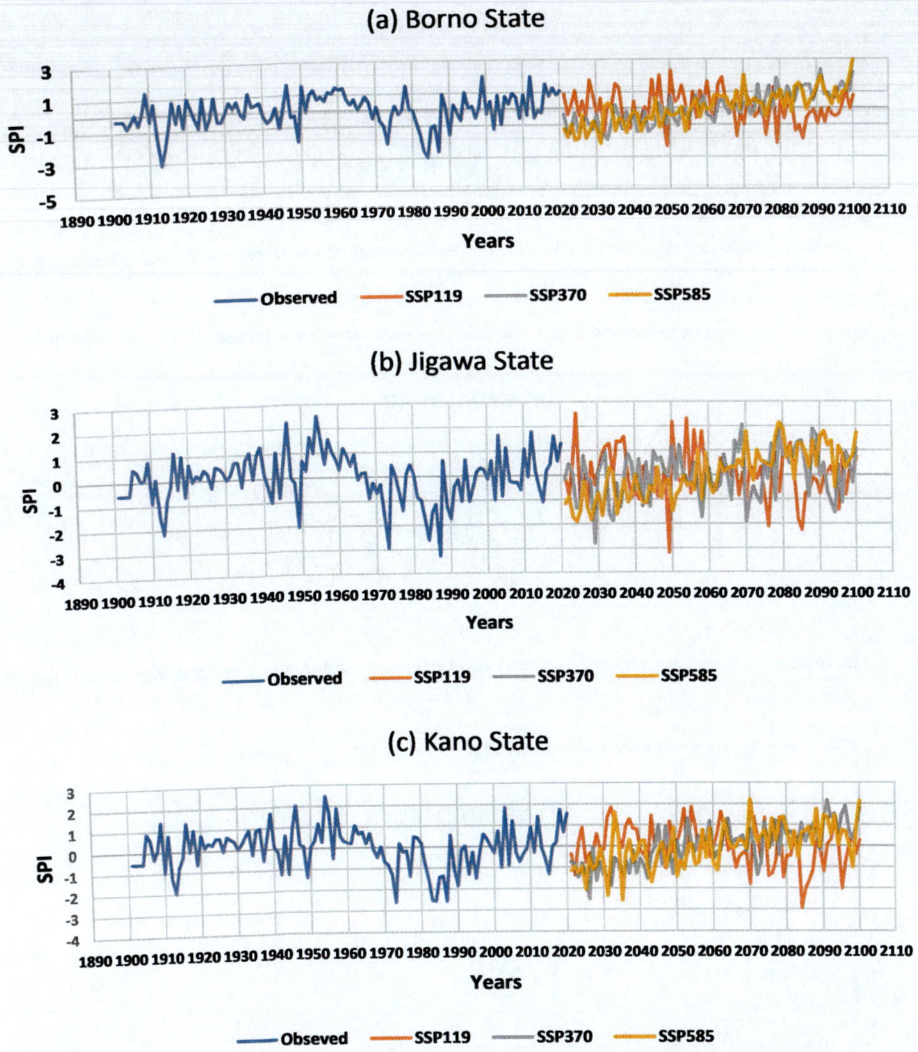

Fig. 19.2 Drought intensity for the observed (1901–2020) and projected (2021–2100) periods

Conley 2012). This has been negatively affecting the people who depend on the wetlands for their livelihoods. In addition, it has also been leading to a decline in the bird species that migrate seasonally across the Sahara from Europe (Zwarts et al. 2009). It has been estimated that the benefits of these wetlands and floodplains (US\$9,600–US\$14,500 m^{-3} of water) far outweigh those of the irrigation projects (US\$26–US\$40 m^{-3} of water) (Wetlands International 2017).

Millions of people in the Lake Chad Basin (especially Borno and Yobe States in the case of Nigeria) lost or recorded a significant reduction in their livelihoods due

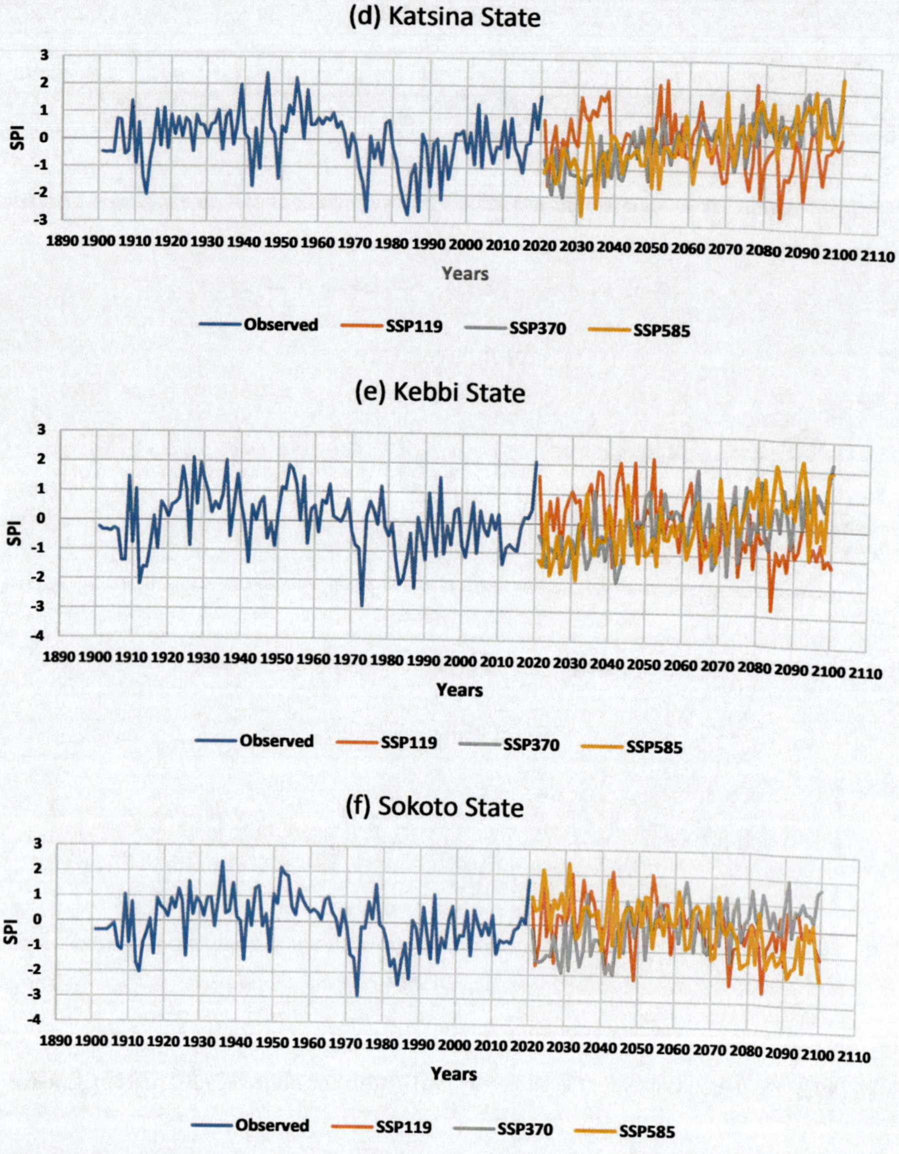

Fig. 19.2 (continued)

to the drought-induced shrinking of Lake Chad (Cole et al. 2017; Vivekananda et al. 2019). In the fishing sector alone, from 1974 to 2012, the shrinking resulted in a decrease in annual fish catch from 220,000 tonnes to about 100,000 tonnes, which adversely affected more than 10 million people, directly and indirectly, dependent on the sector (De Young et al. 2012). Such widespread erosion of livelihood among

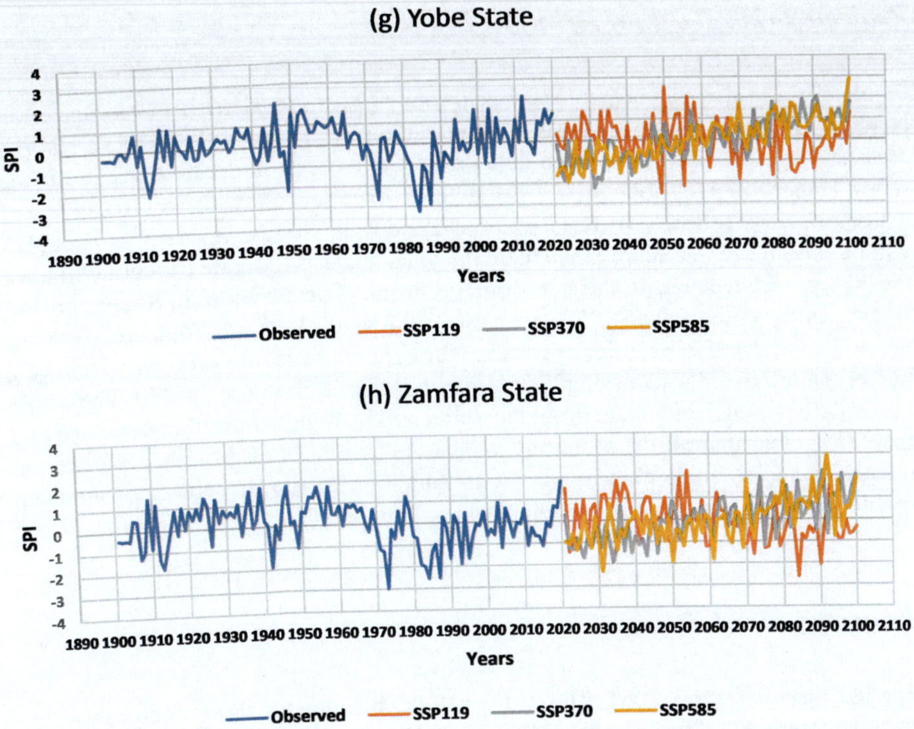

Fig. 19.2 (continued)

the farmers, pastoralists and fishermen in Borno and Yobe States due to shrinking of the Lake Chad is believed to be what provided a suitable breeding ground for Boko Haram ideology in the area (Nett and Ruttinger 2016; David 2017; Skah and Lyammouri 2020). As the Nigerian dry land areas are contiguous to the Sahelian countries like Chad and Niger Republics, drought-induced conflicts and continuous drying of these countries have been leading to an influx of pastoralists into the study area (Werz and Conley 2012). This is believed to be the reason for the proliferation of firearms and people with prior knowledge of handling them in the area, which led to the escalation of armed conflicts including Boko Haram and banditry.

Generally, drought and climate change result in a decrease or total loss of crop yields and pasture, death of livestock as well as a decrease in surface and groundwater due to the decrease in recharge from rainfall in the dry land areas of Nigeria. This has been forcing pastoralists to migrate southward into the central and southern parts of Nigeria, which has been increasing violent conflicts between the pastoralists and farmers.

Projected (2021–2060) Drought Intensity

Figure 19.2a–h also shows that for the near future (2021–2060), SSP119 indicates an increase in extreme drought events in most of the states, especially Yobe and Borno States, severe droughts in Sokoto and Borno States and moderate ones in all the states. Under SSP370, Yobe State is projected to experience the highest frequency of extreme drought followed by Katsina, Jigawa and Kano States. More severe droughts will be experienced in Sokoto than in all the other states. Moderate drought will have the highest frequency in all the states ranging from 13 occurrences in Katsina to 5 in Jigawa. In the case of the SSP585 scenario, only Kano, Katsina, Yobe and Zamfara States are expected to experience extreme drought while severe drought will have a higher spatial coverage involving Yobe, Jigawa, Borno, Katsina and Kebbi. The whole area will experience moderate drought, which will be highly variable over space and time. It is clear that all the scenarios indicate occurrences of droughts of different intensities in the area. This implies a continuation of the present environmental problems and socio-economic predicaments of the people.

Projected (2061–2100)

For the distant future (2061–2100), Fig. 19.2a–h indicates more occurrences of extreme drought in Kano, Jigawa, Sokoto and Zamfara States and fewer occurrences in Borno, Katsina and Kebbi States under the SSP119 scenario. More occurrences of severe drought are projected for Borno, Katsina and Yobe States with less for Kano, Kebbi, Jigawa and Zamfara States and none for Kebbi and Sokoto States. In contrast to SSP119, the SSP370 scenario indicates no occurrence of extreme drought throughout the study area. Jigawa State is the only state to experience severe drought in the distant future. Even the moderate drought is projected to occur only in Jigawa, Kebbi, Kano, Sokoto, Yobe and Zamfara States. Similarly, SSP585 indicates no occurrence of extreme drought in the study area but few occurrences of severe drought in Sokoto and Zamfara States only. Few occurrences of moderate droughts are projected for Sokoto, Kebbi, Kano, Zamfara and Katsina.

Generally, the projected results indicate respite from extreme and severe droughts in comparison with the observed period, especially in the case of SSP370 and SSP 585. This can be good news as it indicates possible improvements in agricultural production, which implies an increase in socio-economic stability in the area. However, the respite from drought could also imply an increase in floods in the area. This is because it has been reported that droughts alternate with floods in this part of the country (Federal Republic of Nigeria 2012).

Drought Frequency

Drought frequency computed using Eq. 19.5 exhibits high spatiotemporal variation in the study area for both the observed and projected periods as shown in Fig. 19.3. For the observed period, the western part of the study area, especially Sokoto, Kebbi, Zamfara and Kano States, has higher frequencies in contrast to the eastern part. For the near future of the projected period (2021–2060), the reverse of the observed period is almost the case. SSP119 indicates high frequencies for Borno, Sokoto and Zamfara States, while SSP370 indicates high frequencies for Katsina and Borno States. In the case of SSP585, it is Jigawa, Borno, Kebbi and Yobe that have the highest frequencies.

The figure indicates a significant reduction in drought frequency during the distant future (2061–2100) in the study area in contrast to the observed period. SSP119 indicates the highest frequencies for Kano and Yobe States. Jigawa and Kebbi have higher frequencies under SSP370 than other parts of the study area. SSP585 indicates the highest drought frequency for Sokoto State only while the other parts have low-to-zero frequencies.

Spatiotemporal variation of drought frequency in the study has implications for both the environment and the people of the area. For the environment, a future

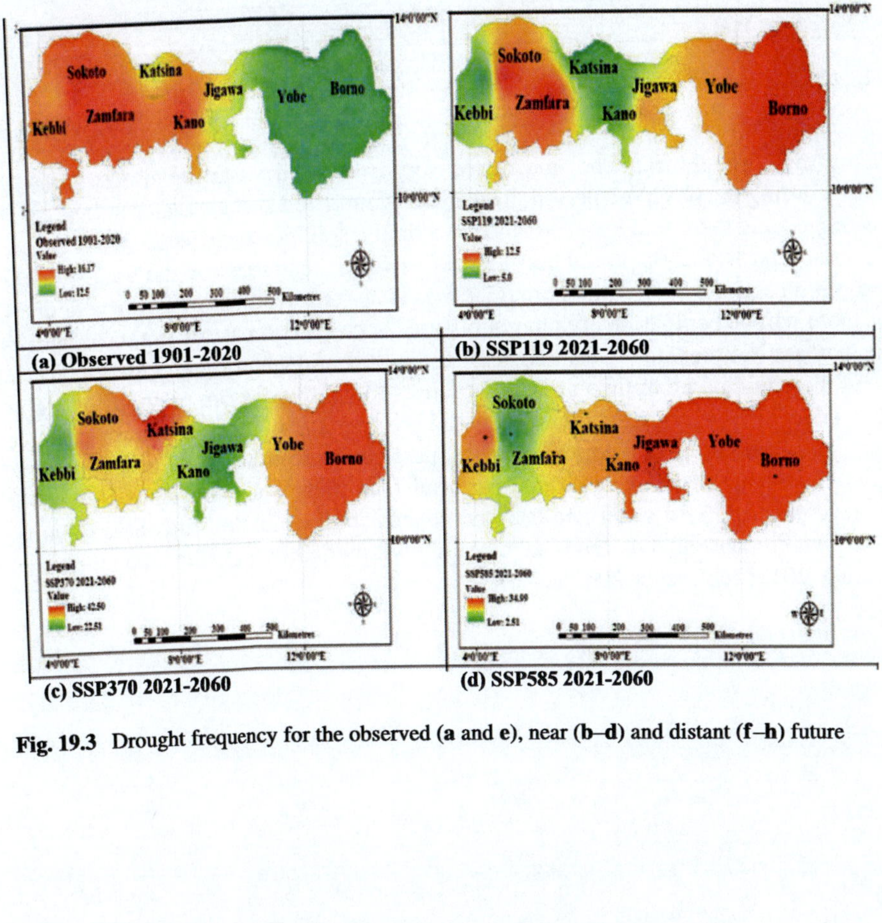

Fig. 19.3 Drought frequency for the observed (**a** and **e**), near (**b–d**) and distant (**f–h**) future

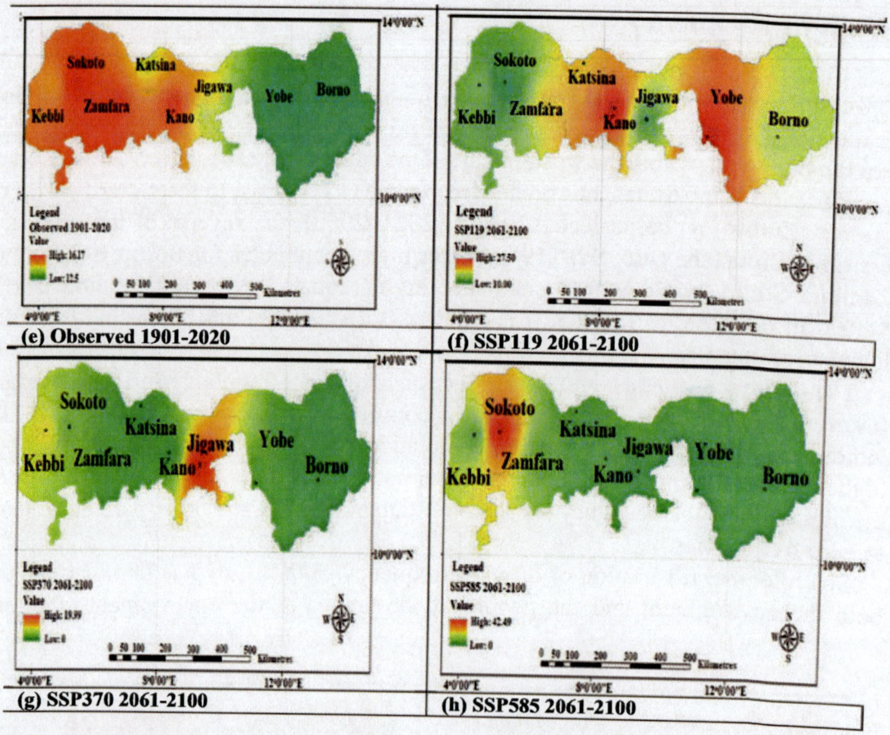

Fig. 19.3 (continued)

decrease in drought frequency means recovery from the devastating effects of desiccation during the observed period. Both underground and surface water will be more recharged and there will be an enhancement in the soil moisture. These will be good for crop and livestock production as well as forest resources in the area. For the people, a respite from frequent droughts means an improvement in their livelihoods, as there will be comparatively more moisture for crops and pasture plants as well as an increase in water supply for irrigation agriculture and watering of animals. Therefore, there will be more time and opportunities for recovery from previous droughts and a decrease in competition for water and grazing lands. This will possibly reduce the rate of migration of people especially pastoralists to other parts of the country. Consequently, a decrease in farmer–pastoralist conflicts is also possible. Because it is the competition for water and land that breeds conflicts in the study area (Nyong, n.d.), which consequently leads to migration within and outside the area (Werz and Conley 2012; McGregor 2017).

Drought Vulnerability

Examining some environmental and socio-economic factors in the study area that determine people's exposure and sensitivity to drought indicates their vulnerability to it. Environmentally speaking, the study area falls within a region characterised as very vulnerable to drought (Federal Republic of Nigeria 2018; United States Agency for International Development [USAID] 2021). The states are also referred to as desertification frontline states, where the rate of desertification is 0.6 kms per year (Federal Ministry of Environment 2001). Soils in the area are generally sandy and have low organic, nitrogen and phosphorus content (Mortimore 2009), which indicate low fertility and low water holding capacity. The area is also the one that receives the lowest amount of rainfall, has the highest rainfall variability and occurrence of drought in the country (Federal Republic of Nigeria 2018). Historically, the area has the highest temperature record in the country. Climate change projections also indicate that the highest increase in temperature in the country will be recorded in the area in future (Federal Republic of Nigeria 2020). These make soils in the area to be very susceptible to excessive loss of moisture through increased evapotranspiration, which exacerbates drought intensity.

Socio-economic factors include the fact that the study area has the largest proportion of the adult population engaged in agricultural production as a primary occupation in the country and the farming system is largely on a smallholding basis. It is also the one with the lowest Subnational Human Development Index, which ranges from 0.339 (Kebbi State) to 0.517 (Borno State) and all are below the national average of 0.540 (Global Data Lab 2021). Furthermore, the states that constitute the study area also have the highest poverty rate (National Bureau of Statistics 2020), highest unemployment rate (National Bureau of Statistics 2021) and highest dependency ratio (National Bureau of Statistics and World Bank 2020) in the country, which ranges from 1.2 (Sokoto State) to 1.4 (Jigawa State); more than the national average of 0.97. Healthwise, the study area is home to the worst malnutrition condition in the country, especially among children and pregnant women (National Bureau of Statistics 2018). This is in terms of Global Acute Malnutrition (GAM), Severe Acute Malnutrition (SAM) and anaemia. Thus, even moderate drought in this area is capable of worsening the situation.

Both the environmental and socio-economic factors indicate how highly vulnerable the people in the study area are to drought. This is because the factors indicate that the people have very few capabilities for adapting to drought, which implies that they are the least able to withstand the rigours of drought or recover from its aftermath without external aid.

Conclusion and Recommendation

Results of this study indicate the significant influence of climate change on the major attributes of drought, frequency and intensity. The historical droughts exhibit high spatiotemporal variations with consequent detrimental effects on the people and the environment. Future scenarios indicate a reduction in drought intensity; however, other factors such as increase in temperature (which significantly reduces soil moisture through increased evapotranspiration) and poor governance may magnify the effects of droughts with lesser intensities. The decrease in the frequency of drought in both the near and distant future may provide the people with the opportunity to rebound from the devastating effects of droughts. However, an increase in people's vulnerability may cancel this benefit. The results also indicate that the people are highly and increasingly becoming vulnerable to drought. When this is coupled with the fact that droughts will continue to occur, the situation becomes a serious cause for concern. Because whenever the people cannot fully recover from the devastating effects of droughts within reasonable times, they are bound to continue being a recruitment pool for Boko Haram and banditry and possibly other dangerous radical Jihadist organisations operating in Nigeria and neighbouring countries. These include Jama'at al.-Ansar al-Muslimeen fi Bilad al-Sudan (Ansaru), Al-qaeda in the Islamic Maghreb (AQIM), Jama'at Nusrat al-Islam wal-Muslimeen (JNIM), Islamic State in the Greater Sahara (ISGS), Al-Mourabitoun and Islamic State-West Africa Province (ISWAP). Another possible consequence is increased migration to Europe by crossing the Sahara. Given the youth population of the area, these could be catastrophic to West Africa.

It is recommended that drought, in particular, and climate change, in general, should be accorded the attention they deserve by both the people and government in the study area. More efforts should be made in the areas of drought monitoring, prediction, early warning, mitigation and adaptation plans as well as adequate and effective implementation of the mitigation and adaptation plans. Sufficient resources should be committed and appropriately used to address the root causes of people's vulnerability to drought, especially poverty, unemployment, illiteracy and general human underdevelopment. As the decrease in drought indicates an increase in rainfall, effective plans for flood monitoring, prediction and early warning should also be produced and implemented.

References

Adejuwon JO, Dada E (2021) Temporal analysis of drought characteristics in the tropical semi-arid zone of Nigeria. Sci African 14:e01016. https://doi.org/10.1016/j.scaf.2021.e01016

AFROSAI (2015) Joint environmental audit on the drying up of Lake Chad. Joint Audit Report. Deutsche Gesellschaft fur Internationale Zusammenarbeit (GIZ) GmbH

Allan RP, Barlow M, Byrne MP et al (2020) Advances in understanding large-scale responses of the water cycle to climate change. Ann N Y Acad Sci 1472:49–75. https://doi.org/10.1111/nyas.14337

Allen RJ, Evan AT, Booth BBB (2015) Interhemispheric aerosol radiative forcing and tropical precipitation shifts during the late twentieth century. J Clim 28:8219–8246. https://doi.org/10.11175/JCLI-D-15-0148.1

Bercos-Hickey E, Nathan TR, Chen S-H (2020) On the relationship between the African Easterly Jet, Saharan mineral dust aerosol, and West African precipitation. J Clim 33:3534–3546. https://doi.org/10.1175/JCLI-D-18-0661.1

Biasutti M, Giannini A (2006) Robust Sahel drying in response to late 20th century forcings. Geophys Res Lett 33:L11706. https://doi.org/10.1029/2006GL026067

Biasutti M, Sobel AH, Camargo SJ (2009) The role of the Sahara low in summertime Sahel rainfall variability and change in CMIP3 models. J Clim 22(21):5755–5771. https://doi.org/10.1175JCLI2969.1

Bonfils CJW, Santer BD, Fyfe JC, Marvel K, Phillips TJ, Zimmerman SRH (2020) Human influence on joint changes in temperature, rainfall and continental aridity. Nat Clim Chang 10:726–731

Byrne MP, Pendergrass AG, Rapp A, Wodzicki KR (2018) Response of the intertropical convergence zone to climate change: Location, width, and strength. Current Clim Change Rep 4:355–370. https://doi.org/10.1007/s40641-081-0110-5

Byrne M, Schneider T (2016) Narrowing of the ITCZ in a warming climate: physical mechanism. Geophys Res Lett 43:11350–11357. https://doi.org/10.1002/2016GL070396

Cole BE, Courtney A, Kaster E, Sheinbaum N (2017) Breaking Boko Haram and ramping up recovery: US engagement in the Lake Chad region 2013 to 2016. United States Institute of Peace

David D (2017) Climate change contributes to Boko Haram uprising. International Institute for Counter-Terrorism. Herzliya, Israel. https://www.ict.org.il/Article/2023/climate-change-contribute-to-boko-haram

De Young C, Sheridan S, Davies S, Hjort A (2012) Climate change implications for fishing communities in the Lake Chad basin: what have we learned and what can we do better? FAO/lake chad basin commission workshop 18–20 November, Ndjamena, Chad

Dey D, Campino AA, Doos K (2021) A complete view of the atmospheric hydrologic cycle. Hydrol Earth Syst Sci. https://doi.org/10.5194/hess-2021-509

Eyring V, Bony S, Meehl GA, Sevior C, Stevens B, Stouffer RJ, Taylor KE (2015) Overview of the coupled intercomparison project phase 6 (CMIP6) experimental design and organisation. Geosci Model Develop Discuss 8:10539–10538. https://doi.org/10.5194/gmdd-8-10593-2015

FAO and ICRISAT (2019) Climate-smart agriculture in Borno state of Nigeria. CSA Country profiles for Africa series. International Centre for Tropical Agriculture (CIAT). International Crops Research Institute for the Semi-Arid Tropics (ICRISAT). Food and Agriculture Organisation of the United Nations

Federal Ministry of Environment (2001) National action programme to combat desertification. Federal Ministry of Environment. www.knowledge.unccd.int/sites/default/files/naps/Nigeria-eng2001.pdf

Federal Republic of Nigeria (2012) National policy on climate change. Federal Ministry of Environment

Federal Republic of Nigeria (2018) National drought plan. Federal Ministry of Environment

Federal Republic of Nigeria (2020) Third national communication (TNC) of the Federal Republic of Nigeria under the United Nations Framework Convention on Climate Change (UNFCCC). Federal Ministry of Environment

Gana AH (2018) Drought and drought mitigation in Yobe State, Nigeria. PhD thesis, University of Wolverhampton. https://wlv.openrepository.com/bitstream/2436/621793/1/HassanGanaPhD.pdf

Giannini A, Saravanan R, Chang P (2003) Oceanic forcing of Sahel rainfall on interannual to interdecadal time scales. Science 302:1027–1030. https://doi.org/10.1126/science.1089357

Gimeno L, Dominguez F, Nieto R, Trigo R, Drumond A, Reason CJC, Taschetto AS, Ramos AM, Kumar R, Marengo J (2016) Major mechanisms of atmospheric moisture transport and their role in extreme precipitation events. Ann Rev Environ Resourc 14:117–141. https://doi.org/10.1146/annurev-environ-110615-08558

Gimeno L, Stohl A, Trigo RM, Dominguez F, Yoshimura K, Yu L, Drumond A, Duran-Quesada AM, Nieto R (2012) Oceanic and terrestrial sources of continental precipitation. Rev Geophys 50:RG4003. https://doi.org/10.1029/2012RG000389

Global Data Lab (2021). Human development indices. https://globaldatalab.org/shdi/shdi/

Harsma RJ, Selton FM, Weber SL, Kliphuis M (2005) Sahel rainfall variability and response to greenhouse warming. Geophys Res Lett 32(1):17702. https://doi.org/10.1029/2005GL023232

Haywood JM, Jones A, Bellouin N, Stephenson D (2013) Asymmetric forcing from stratospheric aerosols impacts Sahelian rainfall. Nat Clim Chang 3:660–665. https://doi.org/10.1038/nclimate1857

IPCC (2018) Global warming of 1.5 °C. An IPCC special report on the impacts of global warming of 1.5 °C above pre-industrial levels and related global greenhouse gas emission pathways, in the context of strengthening the global response to the threat of climate change, sustainable development, and efforts to eradicate poverty. [Masson-Delmotte V, Zhai P, Portner H-O, Roberts D, Skea J, Shukla PR, Pirani A, Moufouma-Okia W, Pean C, Pidock R, Connors S, Matthews JBR, Chen Y, Zhou X, Gomis MI, Lonnoy E, Maycock T, Tignor M, Waterfield T (eds)] In Press

King SE (2016) An examination of Boreal summer Sahel rainfall variability in the context of Tropical Easterly Jet. MSc thesis, Department of Earth, Ocean and Atmospheric Science. Florida State University. https://diginole.lib.edu/islandora/object/fsu%3A405594

Lake Chad Basin Commission (2000) The Lake Chad vision for 2025. Second World Water Forum

Lavaysse C, Flamant C, Janicot S, Knippertz P (2010) Links between African Easterly Waves, midlatitude circulation and interseasonal pulsations of West African heat low. Quart J R Meteorol Soc 135(S1):141–158

Lavaysse C, Flamant C, Evan C, Janicot S, Gaetani M (2016) Recent climatological trend of the Sahara heat low and its impact on the West African climate. Clim Dyn 47:3474–3498. https://doi.org/10.1007/s00382-015-2847-z

Mamalakis A, Randerson JT, Yu J-Y, Pritchard MS, Magnusdottir G, Smyth P, Yu S, Foufoula-Georgiou E (2021) Zonally contrasting shifts of the tropical rain belt in response to climate change. Nat Clim Chang 11:143–151. https://doi.org/10.1038/s41558-020-00963-x

McGregor A (2017) The Fulani crisis: communal violence and radicalization in the Sahel. CTS Sentinel 10(2). www.ctu.usma.edu/the-fulani-crisis-communal-violence-and-radicalization-in-the-sahel

McKee TB, Doesken NJ, Kliest J (1993) The relationship of drought frequency and duration to time scales. In: Proceedings of the 8th conference on applied climatology, 17–22 January, Anaheim, CA, Boston, MA: American Meteorological Society, pp 179–184

Mischell E, Lee J-E (2022) Observed zonal variations of the relationship between ITCZ position and meridional temperature contrast. Climate 10:30. https://doi.org/10.3390/cli10030030

Monerie P-A, Sanchez-Gomez E, Gaetani M, Mohino E, Dong B (2020) Future evolution of the Sahel precipitation zonal contrast in CESM1. Clim Dyn 55:2801–2821. https://doi.org/10.1007/s00382-020-05417-w

Mortimore M (2009) Adapting to drought: farmers, famines and desertification in West Africa, 2nd edn. Cambridge University Press

Mortimore M, Batterbury S (2013) Adapting to drought in the West African Sahel. In: Boulter S, Palutikof J, Karoly K, Guitart D (eds) Natural disasters and adaptation to climate change. Cambridge University Press, pp 235–244

NAERLS and FMARD (2020) Wet season agricultural performance in Nigeria. NAERLS

National Bureau of Statistics (2018) National nutrition and health survey 2018. Report on the nutrition and health situation of Nigeria. National Bureau of Statistics

National Bureau of Statistics (2020) 2019 poverty and inequality in Nigeria. Executive summary. National Bureau of Statistics

National Bureau of Statistics and World Bank (2020). Nigeria living standards survey (2018/2019). National Bureau of Statistics and World Bank

National Bureau of Statistics (2021) Labour force statistics: unemployment and underemployment report. National Bureau of Statistics
Niang C, Mancho AM, Garcia-Garrido VT, Mohino E, Rodriguez-Fonseca B, Curbelo J (2020) Transport pathways across the West African Monsoon as revealed by Lagrangian coherent structures. Sci Rep 10:12543. https://doi.org/10.1038/s/51598-020-69159-9
Nett K, Ruttinger L (2016) Insurgency, terrorism and organised crime in a warming climate: analysing the links between climate change and non-state armed groups. Climate Diplomacy Report. Adelphi
Nyong A (n.d.). Climate-related conflicts in West Africa. Report from Africa: population, health, environment and conflict. ECSP Report (12), 36–43 https://www.wilsoncenter.org/sites/default/files/media/documents/publication/Nyong12.pdf
Olagunju TE (2015) Drought, desertification and the Nigerian environment: a review. J Ecol Natl Environ 7(7):196–209
Pomposi C, Kushnir Y, Giannini A, Biasutti M (2020) Toward understanding the occurrence of both wet and dry Sahel seasons during El Nino: The modulating role of the global oceans. J Clim 33:1193–1207. https://doi.org/10.1175/JCLI-D-19-0219.1
Raj J, Bangalath HK, Stenchikov G (2019) West African monsoon: current state and future projections in high-resolution AGCM. Clim Dyn 52:6441–6461. https://doi.org/10.1007/s00382-081-4522-7
Rigaud KK, de Sherbinin A, Jones B, Abu-Ata N, Adamo S (2021) Groundswell Africa: deep dive into internal climate migration in Nigeria. The World Bank
Sanusi AR, Apampa S, Sotinrin A (2013) Socially inclusive sustainable development in a climate stressed northern Nigeria: a case study of Jigawa State. Heinrich Böll Foundation Nigeria. pp 7
Schneider T, Bischoff T, Hang GH (2014) Migrations and dynamics of the intertropical convergence zone. Nature 513:45–53. https://doi.org/10.1038/nature13636
Skah M, Lyammouri R (2020) The climate change-security nexus: case study of the lake chad basin. Research paper. Policy Centre of the New South
Soldatenko S (2019) Estimated impacts of climate change on Eddy meridional moisture transport in the atmosphere. Appl Sci 9:4492. https://doi.org/10.3390/app9234992
Sori R, Nieto R, Drumond A, Stojanovic M, Gimeno L (2019) On the connection between atmospheric moisture transport and dry conditions in rainfall climatological zones of the Niger River Basin. Water 11:622. https://doi.org/10.3390/w11030622
Thomas N, Nigam S (2018) Twentieth-century climate change over Africa: seasonal hydroclimate trends and Sahara desert expansion. J Clim 31:3349–3370. https://doi.org/10.1175/JCLI-D-1-0187.1
Trenberth KE (2011) Changes in precipitation with climate change. Climate Res 47:123–138. https://doi.org/10.3354/cr00953
United States Agency for International Development (2021) Nigeria water resources profile overview. United States Agency for International Development. https://winrock.org/wp-content/uploads/2021/08/Nigeria_Country_Profile_Final.pdf
UNOPS (2014) United Nations Office for Project Services Demonstration project document – wings over wetlands. www.wetlands.org/portals/I/.../demo_project_document
Vivekananda J, Wall M, Sylvestre F, Nagarajan C (2019) Shoring up stability: addressing climate and fragility risks in the Lake Chad region. Adelphi
Watts MJ (1977) Conference on the aftermath of drought in Nigeria. Savanna 6(1):81–83
Watt-Mayer O, Frierson DMW (2019) ITCZ width controls on Hadley cell extent and Eddy-driven jet position and their response to warming. J Clim 3:1151–1166. https://doi.org/10.1175/JCLI-D-18-0434.1
Weiss H (2003) Migrations during the times of drought and famine in early colonial northern Nigeria. Studia Orientalia 95:1–29
Werz M, Conley L (2012) Climate change, migration, and conflict in northwest Africa. Heinrich Boll Stiftung and Center for American Progress

Wetlands International (2017) Water shocks: wetlands and human migration in the Sahel. Wetlands International. The Netherlands

World Bank (2020) Concept project information document (PID) – Agro-climatic resilience in semi-arid landscapes (ACReSAL) – P175237 (English)

Zhou W, Xie S-P, Yang D (2019) Enhanced equatorial warming causes deep-tropical contraction and subtropical monsoon shift. Nat Clim Chang 9:834–839. https://doi.org/10.1038/s41558-019-0603-9

Zwarts L, Bijlsma RG, van der Kamp J, Wymenga E (2009) Living on the edge: Wetlands and birds in a changing Sahel. KNVV Publishing

Chapter 20
GIS-Based Vulnerability Evaluation of Climate Change Hazards of Flood and Erosion Using an Integrated IVFRN-DEMATEL-ANP Decision Model

E. C. Chukwuma, C. C. Okonkwo, L. C. Orakwe, and Michael O. Dioha

Abstract Climate change is an evident environmental problem today, with the rise in the incidence rate and intensity of hazards such as flooding, heatwaves, drought, tornadoes, erosions, etc., affecting every part of the world. In Anambra State, climate change effects are a threat to food security as flood and erosion hazards destroy arable lands and crops. As such, hazard vulnerability maps of the state are required to assist hazard risk management and to address the issue of climate change across the state. Studies on the weakness of applying GIS as stand-alone technique in evaluating vulnerability by researchers are established. As such, this study applied a geospatial technology based ensemble model of the Interval Value Fuzzy Rough Number (IVFRN), Decision Making Trial and Evaluation Laboratory (DEMATEL), and Analytical Network Process (ANP) methods in vulnerability assessment of climate change hazards of flood and erosion. The IVFRN eliminated uncertainty in the decision making process. The DEMATEL assessed and determined the relative significance of our risk factors. The ANP determined the final weights of the factors based on their significance. The overall climatic vulnerability index map of the study area was produced by overlaying the flood vulnerability map and the erosion vulnerability map. The study results indicate that the state is very vulnerable to climate change effects, with about 71% classified as having medium and very high vulnerability zones. Anambra West was identified as been the Local Government Area (LGA) most vulnerable with about 85% of the entire area categorized as having very high and medium vulnerable zones. It is suggested that adequate

E. C. Chukwuma (✉) · C. C. Okonkwo · L. C. Orakwe
Faculty of Engineering, Department of Agricultural and Bioresources Engineering, Nnamdi Azikiwe University, Awka, Nigeria
e-mail: ec.chukwuma@unizik.edu.ng

L. C. Orakwe
e-mail: lc.orakwe@unizik.edu.ng

E. C. Chukwuma
Research Fellow, Future Africa, University of Pretoria, Pretoria, South Africa

M. O. Dioha
Department of Global Ecology, Carnegie Institution for Science, Stanford, CA 94305, USA
e-mail: mdioha@carnegiescience.edu

J. C. Egbueri et al. (eds.), *Climate Change Impacts on Nigeria*, Springer Climate,
https://doi.org/10.1007/978-3-031-21007-5_20

drainage systems, good soil conservation practices and structural and non-structural approaches should be integrated to mitigate the hazards.

Keywords Vulnerability assessment · Climate change · Flood · Erosion · GIS

Introduction

Climate change is a major environmental issue with resounding consequences facing the world currently, and its effects have grown profoundly over the years. The phenomenon of climate change can be defined as variations in the earth's climate due to human factors or natural processes which are occurring or predicted to occur (Adedeji et al. 2014; Obasi and Uwanekwu 2015). This phenomenon has amplified the frequency and intensity of extreme events and its associated natural hazards, with the impact of these hazards also increasing over the years as a result of urbanization and poor human planning (Frigerio et al. 2016). It is further predicted that the phenomenon will exacerbate multi-risk exposure by altering spatial distribution, magnitude and frequency of hazardous events with a World Bank report stipulating that over 3.8 million km^2 and 790 million people worldwide are exposed to high risk to two or more hazards and over 0.5 million km^2 and 105 million people to at least three hazards (Gallina et al. 2016). The effects of climate change cut across various sectors of human development such as shelter, health, livelihood and life in general, with its negative impacts reflecting on the depletion of natural resources which is a major threat to food security (Ezegwu 2015). Developing countries like Nigeria are at a greater risk to climate change impacts due to their dependency on oil resources and this has several effects, both directly and indirectly (Pandey and Bardsley 2015). For such a country to achieve sustainable development, it is of huge importance to understand and address the impacts of climate change as its negative impacts are quite detrimental to the environment which is a key driver to development.

The adaptation and mitigation of the negative effects of climate change is a major issue due to the complexities and uncertainties associated with climate factors which have necessitated the development of various decision support tools in the fields of environmental management, disaster risk reduction, and impact assessment, to address the issue (Nkoana et al. 2018). Amongst a variety of these decision support tools, vulnerability assessment has emerged as a universal one used to determine probable harm to occur in human and ecological systems of value due to the effects of climate change (Preston et al. 2011). The application of GIS (Geographic Information Systems) for vulnerability assessment has grown over the years in light of the numerous advantages it offers such as data layering, querying, geo-referencing, and visualization (Woodruff et al. 2018). The factors of vulnerability assessment such as climatic, hydrologic, topographic and anthropogenic factors all vary spatially and as such, GIS can be employed to assess these factors and identify areas that are vulnerable. Despite the advantages of applying GIS in vulnerability assessment, these GIS techniques have a major disadvantage. These techniques alone cannot determine with some level of accuracy the different vulnerability factors of hazards relatively (Tella and Balogun 2020). As such, statistical, process-based and ensemble models have been integrated with GIS to overcome this disadvantage and improve vulnerability assessment. This integration is pertinent for more efficient geospatial modeling of climate change hazards toward better decisions and policies in relation to climate change issues. Hence, based on hazards of flood and erosion, this study performed an overall climatic vulnerability assessment of Anambra state, Nigeria using Remote Sensing, GIS and an integrated IVFRN, DEMATEL, and ANP models. The specific objectives were to apply the proposed methodology in the vulnerability assessment of the hazards of flood and erosion; and based on both hazards, develop an overall climatic vulnerability index of the study area using the fuzzy overlay function in the GIS environment. Flood and Erosion are hazards of great consequences in the study area and has been reviewed in subsequent sections. While various studies have assessed both hazards individually in the study area, there is no multi-hazard assessment that addresses uncertainty in decision making, evaluates the significance of risk factors and assigns significance based weights for the development of an overall climatic vulnerability index. As the world continues to deal with the impacts of climate change, a reliable method for determining areas most prone to climate change hazards is of paramount importance. This will serve as an efficient decision making tool for the implementation of climate change adaptation strategies.

Climate Stressors in Nigeria and Anambra State

Climate stressors can be described as extreme weather events which occur as consequences of climate change. An extreme event occurs when the obtained value of climate variable is greater or less than the threshold value (Akande et al. 2017). These extreme weather events are largely responsible for hazards such as flooding, erosions, heatwaves, and wildfires.

Extreme Precipitation

Increased greenhouse gas conditions and climate change have been predicted to impact the hydrological cycle and consequently amplify the frequency and magnitude of heavy precipitation events (Odekunle 2010). These heavy precipitation events will have a significant effect on soil characteristics, crop productivity rate and water resources as well as the magnitude of hazards such as flooding, heatwaves, erosion etc. In various parts of the world, investigations have been done to examine the effects of climate change on hydro-meteorological processes as the detection of rainfall patterns presents important data for comprehending climate, which is very useful in proposing and designing efficient management strategies for regional water resources (Oloruntade et al. 2017). In Nigeria, researchers have documented results regarding rainfall trends in various parts of the country. Mage and Tyubee (2019) observed a zonal pattern of increase in annual number of days for extremely heavy and heavy events from eastern to western parts of the Middle Belt Region. In the Niger River Basin Development Authority Area, Babatolu et al. (2014) observed an increasing trend in annual heavy rainfall amounts and frequency of heavy rainfall events. Daramola et al. (2017) observed an increasing trend in rainfall over all climatic zones of Nigeria with the Guinea zone recording the highest rainfall. Gbode et al. (2019) also observed a significant increase in annual total precipitation in some stations across the Guinea and Sahel zones. Egor et al. (2015) observed an increase in annual rainfall in Calabar, Port Harcourt, Owerri, Enugu, Benin City and Uyo. These studies show that there is increase in the amount and frequency of annual rainfall in various part of the country.

Temperature Variability

Extreme temperature events such as prolonged hot or cold spells can seriously impact our environment and society. These events bring about damage to lives and properties and endanger certain species' existence (Eresanya et al. 2018). Temperature variations can influence other variables of the environment and stimulate health problems due to the relationship between human health and weather parameters of solar radiation, temperature, and wind; where high air temperature in combination with intense radiation may cause heat stroke, amongst other health issues (Onifade and Olaseni 2019). Based on the report by Intergovernmental Panel on Climate Change, there are strong indications of a global temperature increase since 1950, and this is responsible for the increase in heatwaves and intensification of hurricanes and other tropical storms (Mohammed et al. 2015). Temperature variations have also been studied in various parts of Nigeria. Eresanya et al. (2018) observed a significant increase in warm time hours over the year and a decrease in the cold time hours in Ikeja, Osogbo and Maiduguri. Onifade and Olaseni (2019) observed a gradual rising (upward) fluctuation in temperature trends of Warri, Delta State, Nigeria. Abdussalam (2015) observed a statistically significant, spatially coherent trends in temperature indices of North West geopolitical zone of Nigeria with indications of temperature increases in the region. Abatan et al. (2017) observed an increase in the annual and seasonal mean minimum temperatures of Ibadan. The observations made from these studies indicate that there is a gradual rising (upward) fluctuation in temperature trends in various parts of the country.

Flooding Events

A flood can be largely described as a naturally occurring extreme weather event that leads to the overflow of huge quantities of surface water over land that is not usually swamped (Adeoye et al. 2009). Flooding is amongst the most devastating natural hazards in Nigeria, claiming many human lives and destroying properties valued at millions of Naira directly and indirectly in the country yearly (Ishaya et al. 2009). The 2012 flood in the country was reported by the National Emergency Management Agency (NEMA) as a national disaster, with over 2.3 million people affected, over 363 people killed. and 34 out of 36 states of the federation being affected with Anambra state as the worst hit (Efobi and Anierobi 2013). The state is particularly vulnerable to flooding as Obioji and Eze (2019) noted that 2012, 2013, and 2018 floods affected almost all parts of the state, especially Anambra West, Anambra East, Ayamelum, Onitsha North, Onitsha South, Awka North, and Ogbaru, with the 2013 Action Aid research survey establishing that 53.2% of the landmass of Anambra State are subjected to flood disaster on an annual basis. The causes of flood hazards in the state as documented by researchers (Don 2013; Onwuka et al. 2015a, b) include extreme precipitation, inadequate drainage systems, population explosion, climate change and poor city planning with reference to building layouts and other structures. The effects of flooding in the state include damage to infrastructure, reduction of beauty of the environment, displacement from homes, submerging of farmland and destruction of crops, loss of income, overcrowding in shelter camps and associated health problems, increase in poverty level and death (Ebuzoeme 2015; Okonkwo and Onyeizugbe 2017; Onwuka et al. 2015b). Table 20.1 shows the Flood Disaster Statistics in Anambra State.

From the Table, it shows from 2011 to 2018, a total of 190, 849 persons were affected, and about 27,647 persons were displaced. The most devastating year is 2012, with 79,485 affected persons and 12,870 displaced persons.

Table 20.1 Flood disaster statistics in Anambra State (2011–2018) (Obioji and Eze 2019)

Year	No of people affected	No of people displaced
2011	1,014	472
2012	79,485	12,870
2013	34,281	2,112
2014	2,904	544
2015	1,033	376
2016	3,117	723
2017	4,684	969
2018	64,331	9,581
Total	**190, 849**	**27,647**

Erosion

Soil erosion is a major socio-economic and environmental issue that constitutes degradation of land, loss of soil productivity, hampers the sustainable development of rural areas, and the stability and health of the general society (Igwe et al. 2017). The erosion phenomenon is mainly attributed to the action of floods or running water, and it can be described as the detachment of soil particles from the earth's surface and the subsequent transportation and deposition of the detached soil particles by wind and water (Isife 2019). The process of soil water erosion can be distinguished into four and these include splash, sheet, rill, and gully. The process starts off with raindrops falling on the bare or sparsely vegetated soil surface with the force of raindrops detaching soil particles (splash), then the flowing water transports the detached soil in a thin layer (sheet), finally, when the runoff from the sheet process gathers and gain velocity, it creates rills and gullies as it detaches more soil particles (Mezie and Nwajuaku 2020). This process can occur as a slow one that continues relatively unobserved or as a fast one that occurs at a disturbing rate resulting in the serious loss of soil (Aja et al. 2017). Figure 20.1 shows a pictorial view of erosion menace in the study area.

Erosion is an evident environmental issue in the south-eastern part of Nigeria (Egwuonwu et al. 2019) which has destroyed various life forms, properties, arable lands, communication and transportation systems, groundwater and surface sources (Egboka et al. 2019). Anambra state is documented as one of the worst hit erosion states in South-East Nigeria, with over 500 active erosion sites which cut across several rural communities in the state (Nwobodo et al. 2018; Egwuonwu et al. 2019).

Fig. 20.1 Gully Erosion site at Nanka, Anambra state (Iliyasu et al. 2019)

The causes of erosion in the state can be attributed to natural causes such as rainfall intensity and inherent soil properties; and anthropogenic factors, which include road construction without drainage, deforestation and the disposal of solid waste in waterways (Chibuzor et al. 2020; Iliyasu et al. 2019; Mezie and Nwajuaku 2020). The effects of this hazard in the state include the destruction of residential houses, loss of lives and farmlands, creating psychological trauma amongst others (Iliyasu et al. 2019).

Effect of Climate Change on Agriculture in Anambra State

The effect on global agriculture is quite significant with future impacts projected to worsen due to temperature rise and unreliable precipitation (Nnadi et al. 2019). Agriculture is very important to Nigeria's economy, contributing immensely to the rural population's well-being, offering sustenance to about 90% of the rural populace, and accounting for nearly 25% of GDP, about 65% of the workforce of Nigeria and about 5% of total exports (Udemezue 2017). The country's agriculture is particularly susceptible to the impact of climate change due to its high dependency on climate variables such as rainfall, sunlight, temperature, and relative humidity, as they are the major factors that drive crop production and yield (Obasi and Uwanekwu 2015). The impact of climate change on the agricultural sector of the country cuts across various sectors of agricultural production. For example, extreme high precipitation destroys arable lands, impairs cultivated crops, increases the growth and spread of weeds and causes greater post-harvest loss while low precipitation results in drier farmlands, reduction in water level for groundwater exploitation, escalation in irrigation demand, reduction in crop production and yield, etc. (Ogbuabor and Egwuchukwu 2017). Climate change impacts on agriculture also include a decline in the supply capacity and accessibility of protein sources such as egg, meat, milk and other animal produce such as hides and skins, forest produces etc. (Ogbuabor and Egwuchukwu 2017). Anambra state is a significant contributor to the nation's agriculture with neighboring states depending on her for food. The impact of climate change has however affected farmers in the state who contend with climate variations in their quest to achieve optimum yield for increased food supply (Nnadi et al. 2019). Flooding in the state has affected inhabitants who live along riverine areas and depend solely on agriculture, disrupting their agricultural activities and putting them at severe risk of losing their means of livelihood (Nnadi et al. 2019). Rice production which the state is renowned for in the Southeast has been negatively affected as a result of heavy downpours experienced due to climate change and this has culminated in an increase in cost of production and subsequent increase in cost of distribution (Nnadi et al. 2019). Maize production in the state has also been negatively affected by climate change as issues of flooding, erratic rainfall, strange pests, and diseases have led to a decrease in crop yield (Ogbodo et al. 2018).

Vulnerability Assessment of Climate Change

The study of natural hazards gave birth to the vulnerability concept, and it has been applied in the assessment of climate change's influence on systems with applications that describe the degree of potential harm a subsystem or system may experience when subjected to system disturbance (Dong et al. 2015). As defined by the IPCC, vulnerability describes the extent of a system's susceptibility and its inability to withstand the antagonistic impacts as a result of climate change and associated extreme climate events. A system's exposure can be described as the nature and extent of its subjection to major climatic variations. Exposure is a function of physical landscape characteristics such as local climate, water, soils, terrain and ecosystems; remoteness from the hazard; and size of the population (Lioubimtseva 2015). Sensitivity describes the system is affected negatively or positively by climate-related stimuli with effects that may be indirect such as damages associated with increased flood frequency or direct such as variations in crop yield (GIZ 2014). Sensitivity is dependent on several socio-economic and biophysical variables that include occupations, population demographics, land use practices, agricultural crops, food and water availability, etc. (Lioubimtseva 2015). A system's capability to adjust to variations in climate and associated extremes, abate probable damages as well as cope with the repercussions is referred to as its Adaptive Capacity. It depends on socio-economic factors, including economic development, access to amenities such as education, technology, markets, and insurance; socio-cultural and political factors, etc. (Lioubimtseva 2015).

The dependency of farmers on harvests from crop and livestock production for their income renders them vulnerable as vulnerability increases when the impacts of climate variations such as sea level rise, climate extremes and climate-related hazards are adversely affecting the natural resource base (Brzoska and Fröhlich 2015). It is, therefore, necessary to plan adaptation and mitigation measures that minimize the effects of climate change as it poses a significant threat to national development (Dumenu and Obeng 2016). The vulnerability assessment of a valued utility to one or more climate-related hazards for a given population is an integral tool for planning in environmental decision making as it identifies highly vulnerable communities, assigns adaptation resources, provides a better understanding of systemic weaknesses, monitors the effects of adaptation measures, communicates risk, and justifies policy to the public (El-zein and Tonmoy 2015).

Application of GIS for Vulnerability Assessment of Climate Change Hazards

To effectively monitor the environment and human activities, GIS integration with remote sensing is a powerful tool as satellite-borne instruments are known to provide quantitative data at high resolution for spatial and temporal assessments with repetitive coverage, providing data for earth surface condition measurements (Ezezue et al. 2017). The application of GIS for vulnerability mapping has attracted a lot of interest in Nigeria over the years. Anyanwu and Udoh (2016) used GIS techniques to investigate flood vulnerability in Abia State. The researchers considered vulnerability as a function of adaptive capacity, sensitivity, and exposure. The constituents for each of the functions were examined, recognized, weighted, and combined in GIS environment to create three separate layers; the layers were overlaid to produce the final flood vulnerability map. Adebayo (2016) processed demographic, physical, and land use data using GIS to generate a flood risk map of the lower and upper Niger river basin. Results obtained from the study showed that over 1000 settlements and 15 million people were vulnerable to floods. Oyetunji et al. (2019) used GIS to assess the flood vulnerability of areas along Ikpoba River in Benin City. Rivers, buildings, and road layers were digitized to generate a vulnerability map with the creation and analysis of a spatial database to determine buildings at risk. Results obtained showed that 17 buildings were highly vulnerable, 73 buildings moderately vulnerable, 124 buildings lowly vulnerable, and 1885 buildings were at no risk at all. Iko-ojo et al. (2020) determined the susceptibility of the Taraba state to erosion using geomorphological and GIS Analysis. The study integrated GIS and Remote sensing and relied on the conditioning factors of Land use, Soil, Elevation/Aspect, Erosivity Variables; Stream Power index (SPI), Topographic Wetness Index (TWI), plan Curvature, and Length Slope Factor (LSF) to produce its results. Udoh (2015) developed a multi-hazard vulnerability map of Akwa Ibom state to erosion and flooding hazards based on the IPCC framework. GIS was used to generate adaptive capacity, sensitivity, and exposure maps and then integrated them to obtain the multi-hazard vulnerability map. Results obtained in the study shows that parts of the South and North East of the state were highly and moderately vulnerable, whereas North West was lowly vulnerable. The results obtained from these studies show that GIS is a valuable tool for assessing and visualizing the vulnerability of natural hazards.

Application of Multi-criteria Decision Making Models in Vulnerability Assessment

One of the several methods developed for vulnerability assessment is the synthetical index method which involves the selection of various factors that reflects credible indicators; the use of weighted average, expert scoring, principal component analysis, and application of hierarchical analysis methods to ascertain the weight of each indicator; and providing the vulnerability estimate based on corresponding formula (Dong et al. 2015). A common model for this method of vulnerability assessment is the Multi-Criteria Decision Making (MCDM) model. This model employ a set of alternatives, mathematical tools, and decision criteria to create and simplify judgments and forecasting problems (Ali et al. 2020). In Nigeria, various researchers have employed the MCDM models in vulnerability assessment of hazards such as Flood (Azua et al. 2019; Daniel et al. 2016; Eguaroje et al. 2015; Emmanuel et al. 2015; Njoku et al. 2018; Okonufua et al. 2019; Tella and Balogun 2020), and Erosion (Dike et al. 2015; Odunuga et al. 2018; Ogbonnaya et al. 2020). To improve the efficiency of the MCDM models in vulnerability assessment of natural hazards, they have been modified to give birth to hybrid models, one of which is the combination of DEMATEL model with the ANP method. The DEMATEL model or technique is a valuable method for examining the interrelationship between conditioning factors of natural hazards and determining their relative significance (Rahman et al. 2017), while the ANP method helps to determine their relative weights. The main benefit of this hybrid MCDM is that it determines the most significant factor in an intricate multi-criteria decision structure through the use of matrices to transform the interdependency of conditioning factors into causal relationships (Ali et al. 2020). This hybrid MCDM model has a major limitation: inconsistency or bias in the pairwise comparison which will lead to uncertainties and imprecision (Tella and Balogun 2020). To deal with this limitation, various algorithms have been incorporated into various MCDM processes for several purposes and such algorithms include Neutrosophic sets (Liu et al. 2018; Abdel-Basset et al. 2019; Nabeeh 2020), Fuzzy sets (Hategekimana et al. 2018; Kanani-Sadat et al. 2019), Interval Valued Fuzzy Rough Numbers (IVFRN) (Pamučar et al. 2018, 2019; Roy et al. 2019) and Interval Rough Numbers (Pamucar et al. 2017; Hatefi and Tamošaitienė, 2019; Wang et al. 2019). Hence, this study seeks to employ a GIS-based hybrid MCDM model (IVFRN-DEMATEL-ANP) to determine the climatic vulnerability index of Anambra State based on the threats of Flood and Erosion.

Research Methodology

This present study implemented the use of Remote Sensing data, GIS, and the integration of IVFRN, DEMATEL, and ANP methods in vulnerability evaluation of climate change hazards.

The Study Area and Data Acquisition

The model developed was validated using Anambra State of Nigeria. The State is surrounded by various states, for instance, Delta State is located in the western part of the state, in the south by Imo State, in the east by Enugu State and in the north by Kogi (Agbo et al. 2015). The state has a landmass of about 4,415.54 km^2 and lies between latitudes 5° 40′ and 6° 48′ North and longitudes 6° 35′ and 7° 50′ East (Emmanuel et al. 2015). The flowchart in Fig. 20.2 shows the methodological flow used to determine the climatic vulnerability index of the study area. Various datasets from both primary and secondary sources were obtained for this study. Rainfall data for a period of

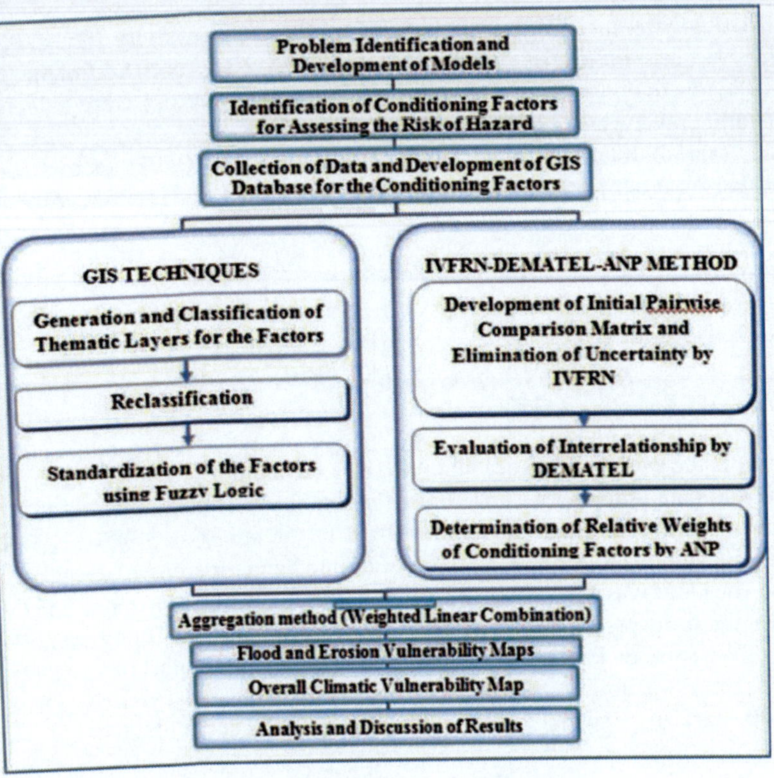

Fig. 20.2 Methodological flowchart of the study

26 years (1991–2016) of the study area was obtained from the World Bank climate database. Soil data of the study area was obtained from the Harmonized World Soil Database (HWSD) found in Food and Agricultural Organization (FAO) website. The Digital Elevation Model (DEM) and Landsat Imagery of the study area were obtained from the United States Geological Survey (USGS). The administrative map of the study area was obtained from the Department of Surveying and Geo-informatics, Nnamdi Azikiwe University, Awka.

Analysis, Processing and Delineation of Conditioning Factors

To carry out the vulnerability assessment of the two hazards, certain conditioning or causative factors were used based on reviewed literature and availability of data. To model Flooding, seven conditioning factors (Criteria) were considered, and these criteria include Elevation (A_1), Slope (A_2), Soil (A_3), Land Use (A_4), Rainfall Intensity (A_5), Drainage Density (A_6), and Drainage Distance (A_7). To model Erosion, nine conditioning factors of Elevation (B_1), Slope (B_2), Soil (B_3), Land Use (B_4),

Normalized Difference Vegetation Index (B_5), Rainfall Erosivity Index (B_6), Topographic Wetness Index (B_7), Stream Power Index (B_8), and Sediment Transport Index (B_9) were considered. The various conditioning factors were selected based on literature review and availability of data.

The Rainfall Intensity and Rainfall Erosivity Index (REI) factors were both delineated from our rainfall data using Eqs. (20.1) and (20.2) below, respectively.

$$\text{Rainfall Intensity} = \frac{\sum_{i=1}^{12} P_i^2}{P} \tag{20.1}$$

$$REI = \sum_{i=1}^{12} 1.735 \times 10^{\left[1.5\log_{10}\left(P_i^2/P\right) - 0.08188\right]} \tag{20.2}$$

where P_i is the average for i-th month and P is the annual average rainfall.

The soil data obtained was used to delineate the Soil factors for both flood and erosion vulnerability assessment. The DEM of the study area was used to delineate the Elevation and Slope factors for both flood and erosion vulnerability assessments. The DEM was also processed to delineate the Drainage Distance and Drainage Density factor maps for flood vulnerability assessment. The Topographic Wetness Index (TWI), Stream Power Index (SPI), and Sediment Transport Index (STI) factors for erosion vulnerability assessment were also de-lineated from the DEM using Eqs. (20.3), (20.4), and (20.5), respectively (Kanani-Sadat et al. 2019).

$$TWI = ln(A_s/\tan\beta) \tag{20.3}$$

$$SPI = A_s \times \tan\beta \tag{20.4}$$

$$STI = (m+1) \times (A_s/22.13)^m \times \sin(\beta/0.0986)^n \tag{20.5}$$

where m is the contributing area exponent set to 0.4,n is the slope exponent set to 1.4, A_s is the upstream contributing area, and β is the slope gradient (in degrees).

Three Landsat 8 imageries covering the study area were obtained, composited and classified to delineate the Land Use factor map for Flood and Erosion vulnerability assessment. The normalized differential vegetation index (NDVI) factor map for Erosion vulnerability assessment was also delineated from the Landsat 8 imagery of the study area using Eq. (20.6) (Kanani-Sadat et al. 2019).

$$NDVI = \frac{IR - R}{IR + R} \tag{20.6}$$

where IR is the infrared portion of the electromagnetic spectrum (Band 5) and R is the red portion of the electromagnetic spectrum (Band 4). All the conditioning factors for the vulnerability assessment of both hazards were delineated in the GIS environment.

After delineation, the factors were standardized to create a basis for their integration by employing the fuzzy membership function. The three fuzzy membership classes of Fuzzy Gaussian, Fuzzy Large, and Fuzzy MS Small were employed in this study for the standardization and fuzzification. The fuzzification which was done based on the influence of each factor in the occurrence of the hazard is shown in Table 20.2 for Flood and Table 20.3 for erosion.

Table 20.2 Fuzzification of the conditioning factors of flood Hazard

Hazard factors	Fuzzy membership	Initial values
Elevation (m)	Fuzzy MS small	313–388 238–313 163–238 88–163 13–88
Slope (degree)	Fuzzy MS small	0–1.3634 1.3635–3.5788 3.5789–6.4760 6.4761–11.2477 11.2478–43.4571
Soil	Fuzzy large	Fluvisol = 1 Nitrosol = 2 Plinthosol = 3 Gleysol = 4 Alluvial = 5
Land use	Fuzzy large	Urban = 1 Agriculture = 2 Barren Land = 3 Forest = 4 Water = 5
Rainfall intensity (mm)	Fuzzy large	2,630.2007–2,720.1345 2,720.1346–2,810.0684 2,810.0685–2,898.2727 2,898.2728–2,981.2886 2,981.2887–3,071.2224
Drainage density (m^{-1})	Fuzzy large	0–0.2921 0.2922–0.6548 0.6549–1.0780 1.0781–1.5818 1.5819–2.5693
Drainage distance (m)	Fuzzy MS small	0–500 500–1000 1000–1500 1500–2000 2000–2500

Table 20.3 Fuzzification of the conditioning factors of Erosion Hazard

Hazard factors	Fuzzy membership	Initial values
Elevation (m)	Fuzzy large	313–388 238–313 163–238 88–163 13–88
Slope (degree)	Fuzzy large	0–1.3634 1.3635–3.5788 3.5789–6.4760 6.4761–11.2477 11.2478–43.4571
Soil	Fuzzy large	Alluvial = 1 Gleysol = 2 Plinthosol = 3 Nitrosol = 4 Fluvisol = 5
Land Use	Fuzzy large	Water = 1 Forest = 2 Urban = 3 Barren Land = 4 Agriculture = 5
NDVI	Fuzzy Gaussian	−0.0915–0.01908 0.01909–0.1051 0.1052–0.1560 0.1561–0.2034 0.2035–0.3561
Rainfall erosivity Index	Fuzzy large	79,919.6953–82,587.5516 82,587.5517–85,255.4078 85,255.4079–87,923.2640 87,923.2641–90,591.1203 90,591.1204–93,258.9765
TWI	Fuzzy large	3.6026–7.3194 7.3195–9.3017 9.3018–11.7796 11.7797–15.0008 15.0009–24.6647
SPI	Fuzzy large	−5.5676– −0.5881 −0.5882– −0.1879 −0.1880–0.0343 0.0344–0.2566 0.2567–5.7696
STI	Fuzzy large	0–41.4713 41.4714–311.0347 311.0348–1,161.1963 1,161.1964–3,131.0829 3,131.0830–5,287.5903

Vulnerability Assessment Model

The proposed assessment model for this study combines the three methods of IVFRN, DEMATEL, and ANP to determine the relative weight of the conditioning factors for both flood and erosion vulnerability assessment. The data that serve as input for the model is a set of pairwise comparison matrix obtained from a questionnaire survey. The survey involved the distribution of questionnaires (see Appendix 1 and 2) to experts with many years of experience in the field of environmental engineering and sciences, and their view about the impact of each criterion on the other in the considered hazard was collated based on a predefined fuzzy scale as shown in Table 20.4.

The first phase of the model involves the application of the IVFRN method. By combining the advantages of both the fuzzy set and the rough number and addressing their disadvantages (Roy et al. 2019), this method eliminates imprecision and uncertainty in decision making process. The method transforms our pairwise comparison matrix into our initial decision matrix I^d. The first step in this method involves the transformation of the obtained triple fuzzy numbers $a_{ij}^k, b_{ij}^k, c_{ij}^k$ into $\left(a_{ij}^k, b_{ij}^k, c_{ij}^k\right)$. Where the IVFRN $\left(a_{ij}^k, b_{ij}^k, c_{ij}^k\right) = \left[\left(a_{ij}^L, a_{ij}^U\right), \left(b_{ij}^L, b_{ij}^U\right), \left(c_{ij}^L, c_{ij}^U\right)\right]$. Here a_{ij}^L, b_{ij}^L and c_{ij}^L symbolizes the lower boundaries, and a_{ij}^U, b_{ij}^U and c_{ij}^U symbolizes the upper limit of the rough sequences of the fuzzy triple $a_{ij}^k, b_{ij}^k, c_{ij}^k$. Next we aggregate the IVFRNs to determine the average, hence we obtain our IVFRN decision matrix $\overline{\overline{I}}$ as shown in Eq. (20.7) below.

$$\overline{\overline{I}} = \begin{bmatrix} \overline{\overline{I}}_{11} & \overline{\overline{I}}_{12} & \cdots & \overline{\overline{I}}_{1n} \\ \overline{\overline{I}}_{21} & \overline{\overline{I}}_{22} & \ldots & \overline{\overline{I}}_{2n} \\ \vdots & \vdots & \ddots & \vdots \\ \overline{\overline{I}}_{n1} & \overline{\overline{I}}_{n2} & \cdots & \overline{\overline{I}}_{nn} \end{bmatrix}_{n\times n} \tag{20.7}$$

where $\overline{\overline{I}}_{ij} = [(\overline{\overline{a}}_{ij}^L, \overline{\overline{a}}_{ij}^U), (\overline{\overline{b}}_{ij}^L, \overline{\overline{b}}_{ij}^U), (\overline{\overline{c}}_{ij}^L, \overline{\overline{c}}_{ij}^U)]$. Here $\overline{\overline{a}}_{ij}^L$, $b_{ij}^{=L}$ and $c_{ij}^{=L}$ symbolizes the lower boundaries, and a_{ij}^U, b_{ij}^U and c_{ij}^U represents the upper boundary of the aggregated

Table 20.4 Predefined fuzzy scale. (*Source* Kanani-Sadat et al. 2019)

Linguistic values	Triple fuzzy numbers
No impact (NI)	(0, 0, 0.25)
Very low impact (VLI)	(0, 0.25, 0.5)
Low impact (LI)	(0.25, 0.5, 0.75)
High impact (HI)	(0.5, 0.75, 1.0)
Very high impact (VHI)	(0.75, 1.0, 1.0)

sequences of our IVFRN, and n symbolizes the number of criteria (conditioning factors).

The final step in the IVFRN method involves the defuzzification of the IVFRN decision matrix $\overline{\overline{I}}$ and this is done to aid the interpretation of data. The matrix $\overline{\overline{I}}$ is defuzzified by employing Eq. (20.8) to obtain the initial decision matrix I^d as shown in Eq. (20.9).

$$I^d = \left(1 + \frac{\overline{\overline{a}}_{ij}^L + \overline{\overline{c}}_{ij}^U}{2}\right) \times \left(\frac{\overline{\overline{a}}_{ij}^L + \overline{\overline{a}}_{ij}^U + 2\overline{\overline{b}}_{ij}^L + 2\overline{\overline{b}}_{ij}^U + \overline{\overline{c}}_{ij}^L + \overline{\overline{c}}_{ij}^U}{8}\right) \quad (20.8)$$

$$I^d = \begin{bmatrix} I_{11}^d & I_{12}^d & \cdots & I_{1n}^d \\ I_{21}^d & I_{22}^d & \cdots & I_{2n}^d \\ \vdots & \vdots & \ddots & \vdots \\ I_{n1}^d & I_{n2}^d & \cdots & I_{nn}^d \end{bmatrix}_{nxn} \quad (20.9)$$

The second phase of our assessment model is the application of the DEMATEL method using the matrix I^d as input data. This method is employed for investigating the interrelationship between conditioning factors which helps to determine their influences (Kadoic et al. 2019). The DEMATEL model transforms our initial decision matrix I^d into our total decision matrix T^d. The first step in this phase involved the normalization of the matrix I^d to obtain the matrix N^d as shown in Eq. (20.10) below.

$$N^d = \begin{bmatrix} N_{11}^d & N_{12}^d & \cdots & N_{1n}^d \\ N_{21}^d & N_{22}^d & \cdots & N_{2n}^d \\ \vdots & \vdots & \ddots & \vdots \\ N_{n1}^d & N_{n2}^d & \cdots & N_{nn}^d \end{bmatrix}_{nxn} ; N_{ij}^d = \frac{I_{ij}^d}{\lambda}; \lambda = 1 \Big/ \max_{1 \le i \le n} \left(\sum_{j=1}^{n} I_{ij}^d\right) \quad (20.10)$$

The next step involves the determination of our total relationship matrix T^r, which describes the direct and indirect relationship between the factors. The matrix T^r is determined using Eq. (20.11), where I is an identity matrix.

$$T^r = \begin{bmatrix} T_{11}^r & T_{12}^r & \cdots & T_{1n}^r \\ T_{21}^r & T_{22}^r & \cdots & T_{2n}^r \\ \vdots & \vdots & \ddots & \vdots \\ T_{n1}^r & T_{n2}^r & \cdots & T_{nn}^r \end{bmatrix}_{nxn} ; T_{ij}^r = N_{ij}^d \times \left(I - N_{ij}^d\right)^{-1} \quad (20.11)$$

To investigate the interrelationship between the conditioning factors, we determine our $\tilde{R}$ factor, which is the summation of the rows of the matrix T^r and our $\tilde{S}$ factor, which is the summation of the columns of the matrix T^r. The equations for $\tilde{R}$ and $\tilde{S}$ are shown in (20.12) and (20.13) below.

$$\tilde{R} = \left[\underset{i}{\tilde{R}}\right]_{n\times 1} = \sum_{j=1}^{n} T_{ij}^{r} \tag{20.12}$$

$$\tilde{S} = \left[\underset{i}{\tilde{S}}\right]_{1\times n} = \sum_{i=1}^{n} T_{ij}^{r} \tag{20.13}$$

Based on Eqs. 20.12 and 20.13, the factors $\tilde{R}+\tilde{S}$ and $\tilde{R}-\tilde{S}$ are further determined. $\tilde{R}-\tilde{S}$ is the most significant of the two as it describes the influence of each factor in the overall system. An obtained positive value of $\tilde{R}-\tilde{S}$ indicate that such a factor will influence other factors and as such is categorized as "causes". In contract, an obtained negative value of $\tilde{R}-\tilde{D}$ indicate that such a factor will be influenced by other factors and as such is categorized as "effects".

The third and final phase of this model is the application of the ANP method to determine the relative weight of the conditioning factors. The ANP method employed involves the application of matrices referred to as super-matrices to determine the dependencies of the conditioning factors. These super-matrices are based on the stochastic column principle, which states that the summation of the elements of each column must be equal to 1 (Pamucar et al. 2017). The ANP method uses the total relationship matrix T^r as input data to determine the relative weights. The first step in the ANP method is the creation of an unweighted super-matrix. The unweighted super-matrix W^α is obtained from the matrix T^r by defining an α-cut threshold to eliminate minor influences from T^r. A general schema of the matrix W^α is shown in Eq. (20.14). Here, if $T_{ij}^{r} < \propto$, then $W_{ij}^{\propto} = 0$, otherwise $W_{ij}^{\propto} = T_{ij}^{r}$

$$W^{\propto} = \begin{bmatrix} W_{11}^{\propto} & \cdots & W_{1j}^{\propto} & \cdots & W_{1n}^{\propto} \\ \vdots & & \vdots & & \vdots \\ W_{i1}^{\propto} & \cdots & W_{ij}^{\propto} & \cdots & W_{in}^{\propto} \\ \vdots & & \vdots & & \vdots \\ W_{n1}^{\propto} & \cdots & W_{nj}^{\propto} & \cdots & W_{nn}^{\propto} \end{bmatrix} \tag{20.14}$$

The second step involves the determination of our weighted super-matrix and this is done by normalizing the unweighted super-matrix $W^{\propto}$ as shown in Eq. (20.15) below.

$$W^{w} = \begin{bmatrix} W_{11}^{w} & W_{12}^{w} & \cdots & W_{1n}^{w} \\ W_{21}^{w} & W_{22}^{w} & \cdots & W_{2n}^{w} \\ \vdots & \vdots & \ddots & \vdots \\ W_{n1}^{w} & W_{n2}^{w} & \cdots & W_{nn}^{w} \end{bmatrix}_{n\times n} ;\; W_{ij}^{w} = W_{ij}^{\propto}/\tilde{d}_i;\; \tilde{d}_i = \sum_{i=1}^{n} W_{ij}^{\propto} \tag{20.15}$$

The final step involves the determination of the limit of the weighted super-matrix, and this is done by multiplying the matrix W^w by itself multiple times. This process is denoted as $\lim_{k\rightarrow\infty} = W^k$, where W denotes the limit super-matrix and k denotes any number. The limit of the weighted super-matrix gives us the relative weights of the conditioning factors. For more information about IVFRN-DEMATE-ANP assessment model, see (Chukwuma et al. 2021; Deveci et al. 2020; Pamučar et al. 2018, 2019).

Final Vulnerability Maps and Validation

After the determination of the relative weights of the conditioning factors, they were aggregated to delineate the vulnerability maps for flood and erosion hazards. In this study, the weighted linear combination method was employed for the aggregation as shown in Eq. (20.16) below.

$$V^I = \sum_{i=1}^{n} W_i C_i \tag{20.16}$$

where parameter V^I is known as the vulnerability index, W_i represents the relative weight of each factor and C_i stands for the appropriate score of each factor. The ultimate vulnerability index map obtained was classified into five discrete categories of "very high", "high", "medium", "low" and "very low". The produced vulnerability maps for both flood and erosion hazards were validated using the Area under the Curve of Receiver Operator Characteristic (AUC-ROC) method.

Overall Climatic Vulnerability Map

Finally, to geospatially model the study area's vulnerability to climate change, GIS model of threats of flooding and erosion was overlaid to determine the overall climatic vulnerability index. The fuzzified vulnerability maps of flood and erosion were overlaid in the GIS environment using the Fuzzy tool (Fuzzy overlay) to delineate the study area's vulnerability. To run the overlay, the "SUM" overlay type was employed to combine the multiple effects of both hazards.

Result and Discussion

The identification and the use of the conditioning factors were based on expert review, literature appraisal, and available data from online sources. As stated earlier, nine conditioning factors were utilized for erosion vulnerability assessment, and they include Rainfall Erosivity, SPI, STI, TWI, Soil, Land Use, NDVI, Slope, and Elevation. While seven conditioning factors were used for flood vulnerability assessment and they include Land Use, Rainfall Intensity, Drainage Distance, Soil, Drainage Density, Slope, and Elevation. Figure 20.3 shows all the developed thematic maps used in the development of both erosion and flood models. Considering the space and scope of this study, details on the description of the functionality of each of the thematic layers utilized in this study could be found elsewhere (Chukwuma et al. 2021). The result obtained from the investigation of the interrelationship between the conditioning factors for both flood and erosion is shown in Table 20.5. The result indicates that Rainfall Intensity has the highest priority for the flood category, with a final weight of 0.2204. This conforms to the major role rainfall plays as a catalyst for the annual flood episodes experienced globally. These flood episodes usually occur during months of intense rainfall both in amount and duration of rainfall. In Aguleri and Umuleria areas of the state, Onwuka et al. (2015a) identified rainfall intensity as the major cause of the 2012 flood episode. For the erosion category, the SPI has the highest priority with a weight of 0.1941. This factor plays a major role in soil erosion as it is the cumulative effect of water flow path, flow accumulation and slope which describes the potential energy surface water has to erode soil particles (Andualem et al. 2020). It largely determines the ability and capacity of flowing water to detach and transport soil particles. For the flood conditioning factor, other weights are drainage Distance (0.1989), Land Use (0.1941), Slope (0.1118), Drainage Density (0.0990), Soil (0.0927) and Elevation (0.0831). While the various weights for the erosion conditioning factors are STI (0.1851), Rainfall Erosivity (0.1245), Slope (0.1057), TWI (0.1018), Land use (0.0996), Elevation (0.0880), Soil (0.0810) and NDVI (0.0203).

Production of Flood and Erosion Vulnerability Maps

The resultant maps for the flood and erosion vulnerability assessment are shown in Fig. 20.3. Further analysis of the flood vulnerability map showed that Anambra West LGA was the most vulnerable area of the study area with 95% of the LGA vulnerable to flood hazards. This is in agreement with the previous works by Emmanuel et al.

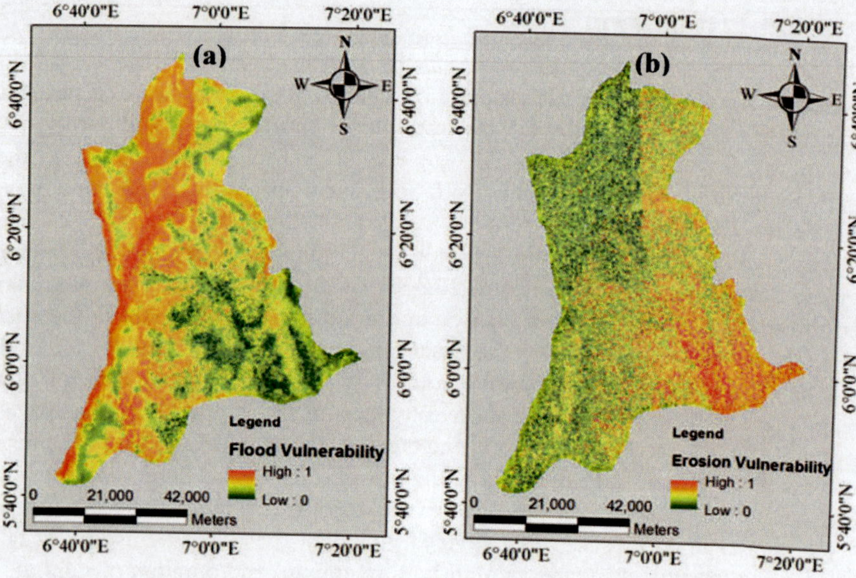

Fig. 20.3 Vulnerability maps of the study area. **a** Flood vulnerability map; **b** Erosion Vulnerability map

Table 20.5 Conditioning factors and final weights for erosion and flood hazards

Conditioning factors (Erosion)	Final weights	Conditioning factors (Flood)	Final weights
Rainfall Erosivity (A_1)	1.245×10^{-1}	Rainfall intensity (B_1)	2.204×10^{-1}
SPI (A_2)	1.941×10^{-1}	Soil (B_2)	9.27×10^{-2}
STI (A_3)	1.851×10^{-1}	Land Use (B_3)	1.941×10^{-1}
TWI (A_4)	1.018×10^{-1}	Drainage distance (B_4)	1.989×10^{-1}
Soil (A_5)	8.10×10^{-2}	Drainage density (B_5)	9.90×10^{-2}
Land use (A_6)	9.96×10^{-2}	Slope (B_6)	1.118×10^{-1}
NDVI (A_7)	2.03×10^{-2}	Elevation (B_7)	8.31×10^{-2}
Slope (A_8)	1.057×10^{-1}		
Elevation (A_9)	8.80×10^{-2}		

(2015) and Nkeki et al. (2013), which identified Anambra West as the most inundated LGA in the state during the 2012 flood episode. The high vulnerability of the LGA to flood hazards can be attributed to its close proximity to the River Niger, which makes the area to be readily inundated by excess water from the river during periods of high rainfall intensity. The erosion vulnerability map was also analyzed further with results showing that Orumba North and South were the most erosion vulnerable LGAs in the state with 81 and 96% vulnerability respectively. This is in agreement with the work by Egbueri et al. (2022), which observed that the Southern part of Anambra State (where Orumba North and South are located) was characterized by high soil loss rate. Similarly, Ajibade et al. (2020), also observed high vulnerability to erosion which extends from the central region to the southern region of the state. The high erosion vulnerability recorded in this part of the state can be attributed to high rainfall events and the abundance of steep slopes in the area. The cumulative effect of both factors as described by the SPI factor is a major trigger for soil erosion. Steep slopes tend to favour runoff by increasing the velocity of flowing water, thereby increasing their ability to erode and transport soil particles.

The produced flood vulnerability map was validated using the AUC-ROC method and an AUC value of 94.6% was the result, which falls within the range of excellent predictive performance. The erosion vulnerability map was equally validated with the same method and an AUC value of 89.7% was the result, which falls within the range of good predictive performance. This shows that the proposed assessment model in this study is an efficient tool for investigating the vulnerability of the study area to both flood and erosion hazards.

Production of Climatic Vulnerability Map

The study's overall climatic vulnerability index map is the outcome of overlaying the flood vulnerability map and the erosion vulnerability map. The obtained result is shown in Fig. 20.4. The obtained map shows that 351.03 km^2, that is 8 % of the study area, is within the zone represented as a very low vulnerable region, 989.21 km^2 representing 22% is classified as a low vulnerable zone, 1247.40 km^2 (27%) is within the medium vulnerable region, 1303.17 km^2 (29%) is within the high vulnerable zones, and 668.06 km^2 (15%) is within the very high vulnerable zone. Considering landmass, the most vulnerable LGA to climate change is Anambra West, with about 622.57 km^2 representing 85% of its total area classified under the vulnerable zones of very high, high, and medium categories. As noted earlier, the LGA's close proximity to a major river and its tributaries, makes it vulnerable to the effects of changes in climate. Ranasinghe (2016) pointed out that climate change influenced variations in environmental factors such as mean sea level, wave conditions, storm surge, and river flows, which shape the world's coastlines, will certainly have a significant effect on these coastal areas. He and Silliman (2019) further noted that the coastal zones of the world has become increasingly threatened by global climate change with stressors such as rising and warming seas, intensifying storms and droughts, and acidifying oceans. A study by Uddin et al. (2019) supports this notion with results that indicate that climate vulnerability in the coastal region of Bangladesh will be enhanced due to geographic settings, dense population, and poverty. Coastal areas around the world have been identified in several studies to be the most vulnerable to the effects of climate change. In Bangladesh, Dasgupta et al. (2015) observed that the coastal area of the country was very vulnerable to climate change induced soil salinization.

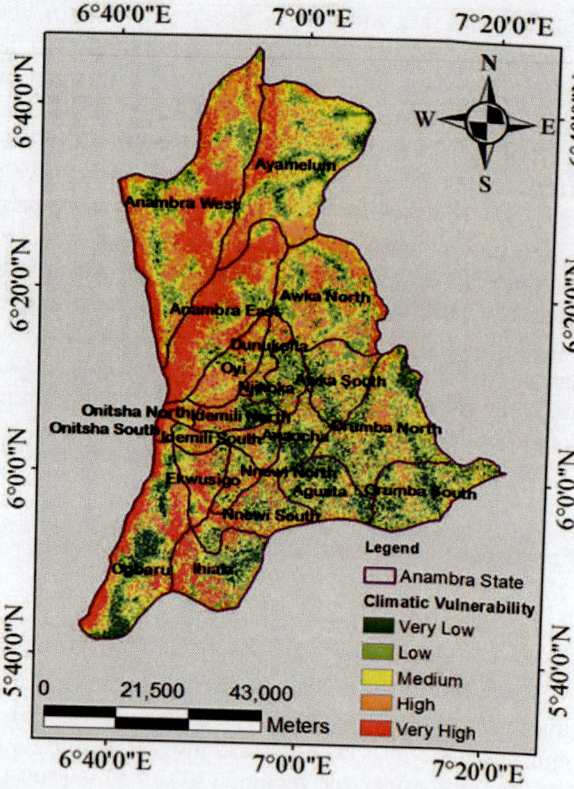

Fig. 20.4 Overall climatic vulnerability map

Hoque et al. (2016) also identified the coastal areas of the mega-deltas in Vietnam and Bangladesh-India to be the most vulnerable in South East Asia to salinisation of drinking water sources due to climate change. Ellison (2015) investigated the vulnerability of mangroves to impacts of climate change impacts and observed that the Tikina Wai area of Fiji had inherent vulnerability due to its location on a subsiding coastline with a low tidal range. In Georgia, USA, Binita et al. (2015) identified the coastal counties of the state to be most vulnerable to climate change. Jiang et al. (2015) also observed that climate change induced increase in risk of salmonellosis was more pronounced in the coastal areas of Maryland, USA.

Another issue worthy of note is that Anambra West area of the study area is predominantly a rural area and as such lacks the coping capacity to adapt to the effects of climate change. This constitutes a major environmental problem in the area, as future projections indicate that extreme events such as flooding will most likely increase in Sub-Saharan Africa (Serdeczny et al. 2017).

Other LGAs more vulnerable to climate change hazards include Ayamelum with about 441.85 km^2 (79%), Anambra East with 337.77 km^2 (89%), Awka North with 290.51 km^2 (76%), and Ogbaru with 249.47 km^2 (64%) exposed to the impact of climate change.

Conclusion

This study investigated the vulnerability assessment of climate change using Remote Sensing, GIS and the integrated IVFRN-DEMATEL ANP model. The proposed assessment model was employed to investigate flood and erosion hazards individually. The resultant flood and erosion vulnerability maps were then overlaid to determine the overall vulnerability to climate change. The investigation of flood hazard was based on the seven conditioning factors of Land Use, Drainage Density, Rainfall Intensity, Drainage Distance, Soil, Slope, and Elevation. The result of the flood hazard assessment shows that Rainfall Intensity factor has the highest priority amongst other factors and Anambra West is the most vulnerable area of the study area. Erosion hazard was similarly investigated based on the nine conditioning factors of Rainfall Erosivity, Sediment Transport Index, Stream Power Index, Topographical Wetness Index, Soil, NDVI, Land use, Slope and Elevation. The result of the assessment of erosion hazard shows that the SPI factor has the highest priority amongst other factors, with Orumba North and South as the most vulnerable areas of the state to erosion hazard. The proposed assessment model proved to be an efficient model for assessing both flood and erosion hazards with AUC values of 94.6% and 89.7% respectively, for the produced flood and erosion vulnerability maps. The overall climatic vulnerability map shows that the study area is vulnerable to climate change and its effects with 70% of the region lying between medium and very high vulnerable zones. The climatic vulnerability map also showed that Anambra West is the most vulnerable LGA to climate change. The result of this study highlights the vulnerability of the study area to climate change and the need for implementation of adequate adaptation strategies in the area. While the result of this study is case specific, the proposed model can be employed to other areas with similar climate dependent hazards. The proposed model in this study can be a reliable decision making tool for determining areas prone to the effects of climate change to aid the implementation of climate change adaptation strategies.

Limitations of Study, Recommendations, and Perspectives for Future Research

While the proposed model in this study was able to determine areas vulnerable to the effects of climate change, the adopted methodology gave rise to certain limitations.

The study relied on only the hazards of flood and erosion to model the overall vulnerability of the study area to climate change. Future studies should consider other hazards such as heat waves, which is also a hazard of importance in the study area. This study did not also consider the resilience of the study area to climate change. This plays a major role in determining the actual vulnerability and such, future studies should consider this. While the proposed model proved to be efficient for assessing erosion hazards, the AUC value of 89.7% obtained in its validation shows there is still room for improvement. Future studies should consider the conditioning factors not considered in this study which might have a significant influence on the occurrence of erosion hazards.

Appendix

Appendix 1: Sample of the questionnaire for flood modelling

QUESTIONNAIRE FOR FLOOD MODELLING IN ANAMBRA STATE

The purpose of this survey is to determine the degree of influence each conditioning factor (criteria) of flooding has on the other. For this survey, we will consider 8 factors of flooding and they include Rainfall Intensity (C_1), Soil (C_2), Land Use (C_3), Drainage Distance (C_4), Drainage Density (C_5), Slope (C_6) and Elevation (C_7). In Table 2, please state the degree of influence the *i*th factor has on the *j*th factor with respect to flood occurrence. The scale of evaluation in Table 1 below will be used for this evaluation.

Table 1

Linguistic values	Corresponding fuzzy triples
No Influence (NO)	(0, 0, 0.25)
Very Low Influence (VL)	(0, 0.25, 0.5)
Low Influence (L)	(0.25, 0.5, 0.75)
High Influence (H)	(0.5, 0.75, 1.0)
Very High Influence (VH)	(0.75, 1.0, 1.0)

1

Table 2

i \ j	C_1	C_2	C_3	C_4	C_5	C_6	C_7
C_1	0, 0, 0.25						
C_2		0, 0, 0.25					
C_3			0, 0, 0.25				
C_4				0, 0, 0.25			
C_5					0, 0, 0.25		
C_6						0, 0, 0.25	
C_7							0, 0, 0.25

Each factor has no influence on itself, hence $C_1 - C_1$ is 0, 0, 0.25 and $C_2 - C_2$ is 0, 0, 0.25 and so forth and so on.

Thanks for your time and effort as you participate in this survey.

NAME OF EXPERT:

FIELD:

YEARS OF EXPERIENCE:

2

Appendix 2: Sample of the questionnaire for erosion modelling

QUESTIONNAIRE FOR EROSION MODELLING IN ANAMBRA STATE

The purpose of this survey is to determine the degree of influence each conditioning factor (criteria) of Erosion has on the other. For this survey, we will consider 9 factors of erosion and they include Rainfall Erosivity (C_1), Stream Power Index (C_2), Sediment Transport Index (C_3), Topographic Wetness Index (C_4), Soil (C_5), Land Use (C_6), Normalized Difference Vegetation Index (C_7), Slope (C_8) and Elevation (C_9). In Table 2, please state the degree of influence the *i*th factor has on the *j*th factor with respect to erosion occurrence. The scale of evaluation in Table 1 below will be used for this evaluation.

Table 1

Linguistic values	Corresponding fuzzy triples
No Influence (NO)	(0, 0, 0.25)
Very Low Influence (VL)	(0, 0.25, 0.5)
Low Influence (L)	(0.25, 0.5, 0.75)
High Influence (H)	(0.5, 0.75, 1.0)
Very High Influence (VH)	(0.75, 1.0, 1.0)

1

Table 2

i \ j	C_1	C_2	C_3	C_4	C_5	C_6	C_7	C_8	C_9
C_1	0, 0, 0.25								
C_2		0, 0, 0.25							
C_3			0, 0, 0.25						
C_4				0, 0, 0.25					
C_5					0, 0, 0.25				
C_6						0, 0, 0.25			
C_7							0, 0, 0.25		
C_8								0, 0, 0.25	
C_9									0,0, 0.25

Each factor has no influence on itself, hence $C_1 - C_1$ is 0, 0, 0.25 and $C_2 - C_2$ is 0, 0, 0.25 and so forth and so on.

Thanks for your time and effort as you participate in this survey.

NAME OF EXPERT:
FIELD:
YEARS OF EXPERIENCE:

2

References

Abatan AA, Osayomi T, Akande SO, Abiodun BJ, Jr WJG (2017) Trends in mean and extreme temperatures over Ibadan, Southwest Nigeria. Theor Appl Climatol 131(3):1261–1272. https://doi.org/10.1007/s00704-017-2049-1

Abdel-Basset M, Atef A, Smarandache F (2019) A hybrid Neutrosophic multiple criteria group decision making approach for project selection. Cogn Syst Res 57:216–227. https://doi.org/10.1016/j.cogsys.2018.10.023

Abdussalam AF (2015) Changes in indices of daily temperature and precipitation extremes in North-West Nigeria. Sci World J 10(2):18–26

Adebayo HO (2016) Flood risk and vulnerability mapping of settlements within upper and lower Niger River Basin, Nigeria. Ethiop J Environ Stud Manag 9(1):815–828. https://doi.org/10.4314/ejesm.v9i1.2S

Adedeji O, Reuben O, Olatoye O (2014) Global climate change. J Geosci Environ Prot 2:114–122. https://doi.org/10.4236/gep.2014.22016

Adeoye NO, Ayanlade A, Babatimehin O (2009) Climate change and menace of floods in Nigerian cities: socio-economic implications. Adv Nat Appl Sci 3(3):369–377

Agbo FU, Arua RN, Okonkwo EA (2015) Effects of climate variability on the choices of livelihood among farm households in Anambra state, Nigeria. Afr J Agric Res 10(44):4134–4141. https://doi.org/10.5897/AJAR2015

Aja OO, Matthews-Njoku EC, Ani AO, Echetama JA, Godson-Ibeji CC, Owolade EO (2017) Socio-economic determinants of the effects of soil erosion as perceived by rural households in Imo state, Nigeria. Agric Biol J N Am 8(5):193–201. https://doi.org/10.5251/abjna.2017.8.5.193.201

Ajibade FO, Nwogwu NA, Adelodun B, Abdulkadir TS, Ajibade F, Lasisi KH, Fadugba OG, Owolabi TA, Olajire OO (2020) Application of RUSLE integrated with GIS and remote sensing techniques to assess soil erosion in Anambra State, South-Eastern Nigeria. J Water Clim Chang 11(1):407–422. https://doi.org/10.2166/wcc.2020.222

Akande A, Costa AC, Mateu J, Henriques R (2017) Geospatial analysis of extreme weather events in Nigeria (1985–2015) using self-organizing maps. Hindawi Adv Meteorol 1–11. https://doi.org/10.1155/2017/8576150

Ali SA, Parvin F, Pham QB, Vojtek M, Vojteková J, Costache R, Linh NTT, Nguyen HQ, Ahmad A, Ghorbani MA (2020) GIS-Based comparative assessment of flood susceptibility mapping using hybrid multi-criteria decision-making approach, Naïve Bayes tree, bivariate statistics and logistic regression: a case of Topl̆a Basin Slovakia. Ecol Ind 117:106620. https://doi.org/10.1016/j.ecolind.2020.106620

Andualem TG, Hagos YG, Kefale A, Zelalem B (2020) Soil erosion prone area identification using multi-criteria decision analysis in Ethiopian highlands. Model Earth Syst Environ 6(3):1407–1418. https://doi.org/10.1007/s40808-020-00757-2

Anyanwu AJ, Udoh JC (2016) A GIS based flood vulnerability assessment of Abia State, Nigeria. Int J Soc Sci 10(2):64–73

Azua S, Youngu TT, Aliyu YA, Shebe MW, Sule JO (2019) Spatial multi-criteria analysis for mapping of flood vulnerable areas in fagge local government area of Kano State, Nigeria. FUTY J Environ 13(1):23–35

Babatolu JS, Akinnubi RT, Folagimi AT (2014) Variability and trends of daily heavy rainfall events over Niger River Basin development authority area in Nigeria. Am J Clim Chang 3:1–7. https://doi.org/10.4236/ajcc.2014.31001

Binita KC, Shepherd JM, Gaither CJ (2015) Climate change vulnerability assessment in Georgia. Appl Geogr 62:62–74. https://doi.org/10.1016/j.apgeog.2015.04.007

Brzoska M, Fröhlich C (2015) Climate change, migration and violent conflict: vulnerabilities, pathways and adaptation strategies. Migr Dev 37–41. https://doi.org/10.1080/21632324.2015.1022973

Chibuzor SN, Okoyeh EI, Egboka BCE, Onwuemesi AG (2020) Gullying processes in the upper Idemili River catchment area of Anambra state Nigeria. Asian J Geol Res 3(4):64–78

Chukwuma EC, Okonkwo CC, Ojediran JO, Anizoba DC, Ubah JI, Nwachukwu CP (2021) A GIS based flood vulnerability modelling of Anambra state using an integrated IVFRN-DEMATEL-ANP model. Heliyon 7:e08048. https://doi.org/10.1016/j.heliyon.2021.e08048

Daniel GI, Ibidun OA, Henry OU (2016) Flood vulnerability assessment in lagos city using multi–criteria analysis. Greener J Environ Manag Public Saf 5(2):24–32. https://doi.org/10.15580/GJEMPS.2016.2.011316005

Daramola M, Eresanya EO, Erhabor S (2017) Analysis of rainfall and temperature over climatic zones in Nigeria analysis. J Geogr Environ Earth Sci Int 11(2):1–14. https://doi.org/10.9734/JGEESI/2017/35304

Dasgupta S, Hossain M, Huq M, Wheeler D (2015) Climate change and soil salinity: the case of coastal Bangladesh. Ambio 44(8):815–826. https://doi.org/10.1007/s13280-015-0681-5

Deveci M, Özcan E, John R, Covrig CF, Pamucar D (2020) A study on offshore wind farm siting criteria using a novel interval-valued fuzzy-rough based Delphi method. J Environ Manag 270:110916. https://doi.org/10.1016/j.jenvman.2020.110916

Dike BU, Agunwamba JC, Alakwem O (2015) GIS-based multi-criteria evaluation (MCE) approach for assessment of Urualla gully erosion watershed. Education 2018

Don OVU (2013) The environmental effects of flood disaster in Anambra State. Adv Appl Sci Res 4(1):499–505

Dong Z, Pan Z, An P, Wang L, Zhang J, He D, Han H, Pan X (2015) A novel method for quantitatively evaluating agricultural vulnerability to climate change. Ecol Ind 48:49–54. https://doi.org/10.1016/j.ecolind.2014.07.032

Dumenu WK, Obeng EA (2016) Climate change and rural communities in Ghana: social vulnerability, impacts, adaptations and policy implications. Environ Sci Policy 55:208–217. https://doi.org/10.1016/j.envsci.2015.10.010

Ebuzoeme OD-F (2015) Evaluating the effects of flooding in six communities in Awka Anambra state of Nigeria. J Environ Earth Sci 5(4):26–39

Efobi K, Anierobi C (2013) Urban flooding and vulnerability of Nigerian cities: A case study of Awka and Onitsha in Anambra state, Nigeria. J Law Policy Glob 19:58–64

Egboka BCE, Orji AE, Nwankwoala HO (2019) Gully erosion and landslides in Southeastern Nigeria: causes, consequences and control measures. Glob J Eng Sci 2(4):1–11. https://doi.org/10.33552/GJES.2019.02.000541

Egbueri JC, Igwe O, Ifediegwu SI (2022) Erosion risk mapping of Anambra state in Southeastern Nigeria: soil loss estimation by RUSLE model and geoinformatics. Bull Eng Geol Env 81(3):1–15. https://doi.org/10.1007/s10064-022-02589-z

Egor AO, Osang JE, Uquetan UI, Emeruwa C, Agbor ME (2015) Inter-annual variability of rainfall in some states of Southern Nigeria. Int J Sci Technol Res 4(10):134–154

Eguaroje OE, Alaga TA, Ogbole JO, Omolere S, Alwadood J, Kolawole IS, Muibi KH, Nnaemeka D, Popoola DS, Samson SA, Adewoyin JE, Jesuleye I, Badru RA, Atijosan A, Ajileye OO (2015) Flood vulnerability assessment of Ibadan city, Oyo State Nigeria. World Environ 5(4):149–159. https://doi.org/10.5923/j.env.20150504.03

Egwuonwu GN, Okoyeh EI, Chikwelu EE (2019) 2D near-surface litho-structural geophysical investigation at the Vicinities of two gully-erosion/landslide sites in South—Eastern Nigeria. IOSR J Appl Geol Geophys 7(5):70–75. https://doi.org/10.9790/0990-0705017075

Ellison JC (2015) Vulnerability assessment of mangroves to climate change and sea-level rise impacts. Wetlands Ecol Manag 23(2):115–137. https://doi.org/10.1007/s11273-014-9397-8

El-Zein A, Tonmoy FN (2015) Assessment of vulnerability to climate change using a multi-criteria outranking approach with application to heat stress in Sydney. Ecol Ind 48:207–217. https://doi.org/10.1016/j.ecolind.2014.08.012

Emmanuel UA, Baywood CN, Gift UA, Ojinnaka OC (2015) Flood hazard analysis and damage assessment of 2012 flood in Anambra state using GIS and remote sensing approach. Am J Geogr Inf Syst 4(1):38–51

Eresanya EO, Ajayi VO, Daramola MT, Balogun R (2018) Temperature extremes over selected stations in Nigeria. Phys Sci Int J 20(1):1–10. https://doi.org/10.9734/PSIJ

Ezegwu CN (2015) Climate change in Nigeria: the impacts and adaptation strategies. SSRN Electron J 1–14. https://doi.org/10.2139/ssrn.2543940

Ezezue AM, Abazuonu LC, Enete IC, Odoanyanwu NM (2017) Assessment of urban heat island effects in Awka metropolis Anambra State. Environ Rev 6(1):95–105

Frigerio I, Ventura S, Strigaro D, Mattavelli M, De Amicis M, Mugnano S, Boffi M (2016) A GIS-based approach to identify the spatial variability of social vulnerability to seismic hazard in Italy. Appl Geogr 74:12–23. https://doi.org/10.1016/j.apgeog.2016.06.014

Gallina V, Torresan S, Critto A, Sperotto A, Glade T, Marcomini A (2016) A review of multi-risk methodologies for natural hazards: consequences and challenges for a climate change impact assessment. J Environ Manag 168:123–132. https://doi.org/10.1016/j.jenvman.2015.11.011

Gbode IE, Adeyeri OE, Menang KP, Intsiful JDK, Ajayi VO, Omotosho JA, Akinsanola AA (2019) Observed changes in climate extremes in Nigeria. Meteorol Appl 26:642–654. https://doi.org/10.1002/met.1791

GIZ (2014) A framework for climate change vulnerability assessments. Clim Chang Adapt Rural Areas-India (CCA-RAI)

Hatefi SM, Tamošaitienė J (2019) An integrated FUZZY-DEMATEL-ANP model for evaluating construction projects by considering interrelationships among risk factors. J Civ Eng Manag 25(2):114–131

Hategekimana Y, Yu L, Nie Y, Zhu J, Liu F, Guo F (2018) Integration of multi-parametric fuzzy analytic hierarchy process and GIS along the UNESCO world heritage: a flood hazard index, Mombasa County Kenya. Nat Hazards 92(2):1137–1153. https://doi.org/10.1007/s11069-018-3244-9

He Q, Silliman BR (2019) Climate change, human impacts, and coastal ecosystems in the anthropocene. Curr Biol 29(19):R1021–R1035. https://doi.org/10.1016/j.cub.2019.08.042

Hoque MA, Scheelbeek PFD, Vineis P, Khan AE, Ahmed KM, Butler AP (2016) Drinking water vulnerability to climate change and alternatives for adaptation in coastal South and South East Asia. Clim Chang 136(2):247–263. https://doi.org/10.1007/s10584-016-1617-1

Igwe PU, Ezeukwu JC, Edoka NE, Ejie OC, Ifi GI (2017) A review of vegetation cover as a natural factor to soil erosion. Int J Rural Dev Environ Health Res 1(4):21–28. https://doi.org/10.22161/ijreh.1.4.4

Iko-ojo IV, Joshua MM, Akoji IF (2020) Geomorphology and GIS analysis for mapping gully erosion susceptibility in Taraba State, Nigeria. Int J Trend Sci Res Dev 4(5):1728–1736

Iliyasu AM, Buba YA, Timnan N, Emeson S, Oyekeshi N (2019) Environmental effects of gully erosion in Nigeria: case study of Nanka community, Orumba North, LGA of Anambra State. J Environ Earth Sci 9(10):68–82. https://doi.org/10.7176/JEES/9-10-08

Ishaya S, Ifatimehin OO, Abaje IB (2009) Mapping flood vulnerable areas in a developing urban centre of Nigeria. J Sustain Dev Afr 11(4):180–194

Isife TC (2019) Environmental challenges and protection practices among women of Anaocha local government area of Anambra State Nigeria. AJSD 7(3):137–149

Jiang C, Shaw KS, Upperman CR, Blythe D, Mitchell C, Murtugudde R, Sapkota AR, Sapkota A (2015) Climate change, extreme events and increased risk of salmonellosis in Maryland, USA: evidence for coastal vulnerability. Environ Int 83:58–62. https://doi.org/10.1016/j.envint.2015.06.006

Kadoic N, Divjak B, Redep NB (2019) Integrating the DEMATEL with the analytic network process for effective decision-making. CEJOR 7(3):653–678. https://doi.org/10.1007/s10100-018-0601-4

Kanani-Sadat Y, Arabsheibani R, Karimipour F, Nasseri M (2019) A new approach to flood susceptibility assessment in data-scarce and ungauged regions based on GIS-based hybrid multi criteria decision-making method. J Hydrol 572:17–31. https://doi.org/10.1016/j.jhydrol.2019.02.034

Lioubimtseva E (2015) A multi-scale assessment of human vulnerability to climate change in the Aral Sea Basin. Environ Earth Sci 73:719–729. https://doi.org/10.1007/s12665-014-3104-1

Liu F, Aiwu G, Lukovac V, Vukić M (2018) A multicriteria model for the selection of the transport service provider: a single valued neutrosophic DEMATEL multicriteria model. Decis Mak: Appl Manag Eng 1(2):121–130. https://doi.org/10.31181/dmame1802128l

Mage JO, Tyubee BT (2019) Spatial variation and pattern of daily rainfall intensity in the middle belt region of Nigeria. Curr J Appl Sci Technol 34(1):1–8. https://doi.org/10.9734/CJAST/2019/v34i130114

Mezie EO, Nwajuaku AI (2020) Investigation into the causes of gully erosion in parts of Anambra State. J Sci Technol Res 2(1):139–149

Mohammed MU, Abdulhamid A, Badamasi M, Ahmed M (2015) Rainfall dynamics and climate change in Kano, Nigeria. J Sci Res Rep 7(5):386–395. https://doi.org/10.9734/JSRR/2015/17098

Nabeeh N (2020) A hybrid neutrosophic approach of DEMATEL with AR-DEA in technology selection. Neutrosophic Sets Syst 31(2):17–30. https://digitalrepository.unm.edu/cgi/viewcontent.cgi?article=1452&context=nss_journal

Njoku CG, Efiong J, Uzoezie AC, Okeniyi FO, Alagbe AO (2018) A GIS multi-criteria evaluation for flood risk–vulnerability mapping of Ikom local government area, cross river state. J Geogr Environ Earth Sci Int 15(2):1–17. https://doi.org/10.9734/jgeesi/2018/40527

Nkeki FN, Henah PJ, Ojeh VN (2013) Geospatial techniques for the assessment and analysis of flood risk along the Niger-Benue basin in Nigeria. J Geogr Inf Syst 5:123–135. https://doi.org/10.4236/jgis.2013.52013

Nkoana EM, Verbruggen A, Hug J (2018) Climate change adaptation tools at the community level: an integrated literature review. Sustainability 10(796):11–18. https://doi.org/10.3390/su10030796

Nnadi OI, Liwenga ET, Lyimo JG, Madukwe MC (2019) Impacts of variability and change in rainfall on gender of farmers in Anambra, Southeast Nigeria. Heliyon 5:1–14

Nwobodo CE, Otunwa S, Ohagwu VA, Enibe DO (2018) Farmers use of erosion control measures in Anambra state, Nigeria. J Agric Ext 22(3):174–184. https://doi.org/10.4314/jae.v22i3.17

Obasi IO, Uwanekwu GA (2015) Effect of climate change on maize production in Nigeria. J Agric Econ Rural Dev 2(1):22–25

Obioji JN, Eze KT (2019) Disaster management in Nigeria: a study of Anambra State, 2011–2019. Int J Innov Dev Policy Stud 7(4):83–103

Odekunle TO (2010) Trends in the extreme daily rainfall amount and the frequency of rainy days in Sokoto, Sudano-Sahelian ecological zone of Nigeria. Ife Res Publ Geogr 9(1):1–14

Odunuga S, Ajijola A, Igwetu N, Adegun O (2018) Land susceptibility to soil erosion in Orashi catchment, Nnewi South, Anambra State, Nigeria. Proc IAHS 376:87–95. https://doi.org/10.5194/piahs-376-87-2018

Ogbodo AJ, Anarah SE, Abubakar SM (2018) GIS-based assessment of smallholder farmers perception of climate change impacts and their adaptation strategies for maize production in Anambra State. In: Corn-production and human health in changing climate. IntechOpen, pp 115–134. https://doi.org/10.5772/intechopen.79009

Ogbonnaya I, Ugwuoke IJ, Onwuka S, Ozioko O (2020) GIS-based gully erosion susceptibility modeling, adapting bivariate statistical method and AHP approach in Gombe town and environs Northeast Nigeria. Geoenvironmental Disasters 7(32):1–16. https://doi.org/10.1186/s40677-020-00166-8

Ogbuabor JE, Egwuchukwu EI (2017) The impact of climate change on the Nigerian economy. Int J Energy Econ Policy 7(2):217–223

Okonkwo AU, Onyeizugbe RU (2017) Disaster vulnerability, severity of flood losses and information dissemination in Ogbaru local government area of Anambra. Int J Adv Agric Environ Eng 4(1):1–5. https://doi.org/10.15242/IJAAEE.C0417030

Okonufua E, Olajire OO, Ojeh VN (2019) Flood vulnerability assessment of Afikpo South local government area, Ebonyi State, Nigeria. Int J Environ Clim Chang 9(6):331–342. https://doi.org/10.9734/IJECC/2019/v9i630118

Oloruntade AJ, Mohammad TA, Aimrun W (2017) Rainfall trends in the Niger-south basin, Nigeria, 1948–2008. Pertanika J Sci Technol 25(2):479–496

Onifade YS, Olaseni VB (2019) Determining temperature extreme in Warri city, Niger-delta region, Nigeria. Int J Environ Clim Chang 9(3):160–166. https://doi.org/10.9734/IJECC/2019/v9i330105

Onwuka SU, Ikekpeazu FO, Onuoha DC (2015a) Assessment of the causes of 2012 floods in Aguleri and Umuleri, Anambra East local government area of Anambra State Nigeria. Br J Environ Sci 3(1):43–57

Onwuka SU, Ikekpeazu FO, Onuoha DC (2015b) Assessment of the environmental effects of 2012 floods in Umuleri, Anambra East local government area of Anambra State, Nigeria. Int Res J Nat Sci 3(1):1–15

Oyetunji AB, Ogunkolu AB, Hamza AI, Njoku MC, Ogbole M (2019) Application of geographical information system for flood vulnerability mapping along Ikpoba River Benin City, Nigeria. Ethiop J Environ Stud Manag 12(5):485–496. https://ejesm.org/doi/v12i5.1

Pamucar D, Mihajlovic M, Obradovic R, Atanaskovic P (2017) Novel approach to group multi-criteria decision making based on interval rough numbers: Hybrid DEMATEL-ANP-MAIRCA model. Expert Syst Appl 88:58–80. https://doi.org/10.1016/j.eswa.2017.06.037

Pamučar D, Petrović I, Ćirović G (2018) Modification of the best-worst and MABAC methods: a novel approach based on interval-valued fuzzy-rough numbers. Expert Syst Appl 91:89–106. https://doi.org/10.1016/j.eswa.2017.08.042

Pamučar D, Ćirović G, Božanić D (2019) Application of interval valued fuzzy-rough numbers in multi-criteria decision making: the IVFRN-MAIRCA model. Yugosl J Oper Res 29(2):221–247. https://doi.org/10.2298/YJOR180415011P

Pandey R, Bardsley DK (2015) Social-ecological vulnerability to climate change in the Nepali. Appl Geogr 46:74–86. https://doi.org/10.1016/j.apgeog.2015.09.008.Social-Ecological

Preston BL, Yuen EJ, Westaway RM (2011) Putting vulnerability to climate change on the map: a review of approaches, benefits, and risks. Sustain Sci 6:177–202. https://doi.org/10.1007/s11625-011-0129-1

Rahman NA, Tarmudi Z, Rossdy M, Muhiddin FA (2017) Flood mitigation measures using intuitionistic Fuzzy Dematel method. Malays J Geosci 1(2):1–5. https://doi.org/10.26480/mjg.02.2017.01.05

Ranasinghe R (2016) Assessing climate change impacts on open sandy coasts: a review. Earth-Sci Rev 160:320–332. https://doi.org/10.1016/j.earscirev.2016.07.011

Roy J, Pamučar D, Kar S (2019) Evaluation and selection of third party logistics provider under sustainability perspectives: an interval valued fuzzy-rough approach. Ann Oper Res:1–46. https://doi.org/10.1007/s10479-019-03501-x

Serdeczny O, Adams S, Baarsch F, Coumou D, Robinson A, Hare W, Schaeffer M, Perrette M, Reinhardt J (2017) Climate change impacts in Sub-Saharan Africa: from physical changes to their social repercussions. Reg Environ Chang 17(6):1585–1600. https://doi.org/10.1007/s10113-015-0910-2

Tella A, Balogun A (2020) Ensemble fuzzy MCDM for spatial assessment of flood susceptibility in Ibadan Nigeria. Nat Hazards 104(3):2277–2306. https://doi.org/10.1007/s11069-020-04272-6

Uddin N, Islam AKMS, Bala SK, Islam GMT, Adhikary S, Saha D, Haque S, Fahad GR, Akter R (2019) Mapping of climate vulnerability of the Coastal region of Bangladesh using principal component analysis. Appl Geogr 102:47–57. https://doi.org/10.1016/j.apgeog.2018.12.011

Udemezue JC (2017) Climate change and adaptation by yam farmers in Anambra West local government area in Anambra State, Nigeria. Br J Res 4(2)

Udoh JC (2015) Multi-hazard vulnerability mapping: an example of Akwa Ibom State Nigeria. Eur Sci J 11(29):293–300

Wang Y, Hong H, Chen W, Li S, Pamucar D, Gigovic L, Drobnjak S, Bui DT, Duan H (2019) A hybrid GIS multi-criteria decision-making method for flood susceptibility mapping at Shangyou, China. Remote Sens 11(1):62. https://doi.org/10.3390/rs11010062

Woodruff et al., 2018Woodruff S, Vitro KA, Bendor TK (2018) GIS and coastal vulnerability to climate change. Compr Geogr Inf Syst:236–257. https://doi.org/10.1016/B978-0-12-409548-9.09655-X

Chapter 21
Spatio-Temporal Characteristics of Agricultural Drought in North Central Nigeria

N. M. Ekpeni, A. O. Ayeni, F. Oni, O. Adeaga, and E. O. Oladipo

Abstract Agricultural drought involves a decrease in soil moisture available below the optimal level as required by crops resulting in impaired growth and reduced yields. This study examines the spatio-temporal characteristics of agricultural droughts from 1968–2018 over North Central (Niger, Benue, Kwara, Plateau, Abuja and Kogi states). Data used include monthly rainfall, maximum and minimum temperature and evapotranspiration (Minna, Makurdi, Jos, FCT and Lokoja) that were used in the study and the data gotten covered a period of 50 years (1968–2018) from the Nigerian Meteorological Agency (NIMET) and Nigerian Bureau of Statistics (NBS). The Reconnaissance Drought Index (RDI) and Standardized Precipitation Index (SPI) were chosen considering the advantages to characterize occurrences of drought into mild, moderate, severe and extreme drought conditions. This also characterizes specific temporal drought events with the spatial interpolation being carried out by the Inverse distance weighting interpolation method. Trend analysis shows that 80% of the states, such as Niger, Benue, Kwara, Plateau, Kogi states and Abuja showed an increasing trend for the period 1968–2018. Major changes from drought analysis in North Central Nigeria during the major wet and dry episodes for the period of study were shown in the analysis. This study shows that the different events of drought in the stations looked into were mostly "Near Normal" droughts (0.00–0.99) and mild droughts (0 to–0.99). The most common years that experienced a deficit in rainfall are 1970, 1971, 1973, 1982, 2015, 2016, 2017 and 2018, while the years with an increase in rainfall are 1996, 1997, 2008, 2009, 2010 and 2012. The Reconnaissance Drought Index (RDI) recorded higher drought events compared to the Standardized Precipitation Index (SPI). More so, the first growing season (APR–JUN) for both SPI and RDI recorded more drought events than the second growing season (JUL–SEPT). Farmers and other stakeholders in the study region need to be proactive, especially during the first growing season (APR–JUN) in drought adaptation and mitigation planning.

Keywords Agriculture · Drought characteristics · Indices · Occurrences · North Central · Nigeria

---

N. M. Ekpeni (✉) · A. O. Ayeni · F. Oni · O. Adeaga · E. O. Oladipo
Department of Geography, University of Lagos, Lagos, Nigeria
e-mail: maurexangel@yahoo.com; nekpeni@unilag.edu.ng

J. C. Egbueri et al. (eds.), *Climate Change Impacts on Nigeria*, Springer Climate,
https://doi.org/10.1007/978-3-031-21007-5_21

Introduction

Drought is when rainfall is abnormally lower than normal leading to dry weather persistence that results in water deficiency. Droughts as a natural hazard can aggravate water-related problems leading to negative impacts on the environment, people's health, the ecosystem and agriculture (Payus et al. 2020; Wilhite 1993). The worst hit by drought is rain-fed agriculture which happens to be the mainstay of the rural economy (Okorie 2003; Gidey et al. 2018). From past studies on drought, a number of definitions of drought have been put together (Oladipo 1985), ranging from drought being seen as a period of hunger, famine, starvation, death and failing water supplies to a prolonged and abnormal soil moisture deficiency which further includes monthly or annual precipitation less than some particular percentage of normal (Oladipo 1985).

Droughts are synonymous with dryness. Drought analysis most times are region, application or type specific, hence difficult to describe (Quiring and Papakryiakou 2003). The different ways in which drought is looked at can be categorized under two headings: conceptual and operational. Conceptually speaking, drought is formulated in general terms thereby helping people to understand its concept and draw up policies. The operational approach helps people to determine drought frequency, intensity and duration of drought for a given historical period (NDMC 2006).

Agricultural drought can be defined as a situation when soil moisture is not sufficient to sustain crop yield and production (Maracchi 2000). An agricultural drought is said to have set in when soil moisture is no longer available to plants which affect crop yield, and hence agricultural productivity and economic profitability. Agricultural drought indicators include precipitation, temperature, soil moisture and crop yield (Mannochi et al. 2004).

The number of livestock that died during 1972–1973 represents about 300,000 of the total livestock population in the north-eastern part of Nigeria. Again, in the year 1987, in many parts of northern Nigeria, drought impact which led to huge crop losses at the end of the growing season was recorded (Eze 2018).

Drought Indices are important measuring tools used for simplicity in characterizing and monitoring drought for ease of communication. These indicators help expatriate and simplify drought information to make it easily comprehensible so as to outline and display the relative or absolute change over the period under study in a region.

Examining drought occurrences with the use of drought indices would enhance drought monitoring and mitigation of drought impacts on the environment, and possibly proffer drought management strategies to help withstand climate change threats.

Some studies ranging from drought impacts on health, water resources, causes and effects on the land, mitigation have been carried out by various researchers (Oladipo 1993; Obaje et al. 2013; Olagunju 2015; Um et al. 2017; Yue et al. 2018). To date, not much appropriate technique has been used to quantitatively examine drought occurrences, especially in the study area (Eze 2018). It is against this background that this research seeks to quantitatively examine drought events using the Standardized

Precipitation Index (SPI) and Reconnaissance Drought Index (RDI) in the study area. These drought indices will help to better prepare and monitor drought in other to reduce its impacts on the environment which will be of more benefit to stakeholders and policy makers.

Materials and Methods

Rainfall data was collected on a monthly basis from the Nigerian Meteorological Agency (NIMET) and Nigerian Bureau of Statistics (NBS), for a period of 50 years which spanned between 1968 and 2018. Drought Indices Calculator (Drinc) is a specialized user-friendly software package created with the purpose of providing a simple interface for the calculation of drought indices. Drinc also includes a standard unit of measurement for the Estimation of Potential Evapotranspiration (PET) through Temperature based methods (Hargreaves, Blaney-Criddle and Thornthwaite), that can be used for the calculation of RDI but for this study, the Hargreaves was used to generate Potential Evapotranspiration (PET). For the Standardized Precipitation drought index (SPI), only rainfall parameter with the latitude of the state was infused into the Drinc software to generate estimated drought condition, while for the Reconnaissance Drought Index (RDI), rainfall and temperature data were loaded into the Drinc software to generate drought indices to examine the changes in drought events in the study area. Descriptive graphs were used to model the trend(s) in the rainfall data for the selected stations of the North Central Region of Nigeria. The Spatial variation of agricultural drought in the region was described using Inverse Distance Weighting (IDW) interpolation method.

Study Area Description

The study was carried out in the North Central part of Nigeria which lies between latitudes 8°00′–9°30′N and longitudes 6°30′–8°00′ E, (see Fig. 21.1). Eight (8) states make up North Central Nigeria of which six (6) states, namely: Kwara, Kogi, Plateau, Benue, Niger and Abuja were sampled due to the unavailability of data. See Fig. 21.1 below. North Central Nigeria enjoys a tropical continental climate characterized by wet and dry seasons. Mean annual rainfall ranges between 1200 mm and 1500 mm. The climate of north central Nigeria is characterized by a rainy season which extends from April to October and a dry season starts in December and lasts till March. The rains climax usually in August. The months of November through January are marked by cold and dry weather conditions (i.e., the harmattan) under the influence of the northeast trade wind (Nweze et al. 2014).

A critical meteorological driver of drought is precipitation but rising temperatures are playing an increasingly strong role in influencing drought severity. Intensification of drought conditions combined with hotter temperatures may lead to a decline in

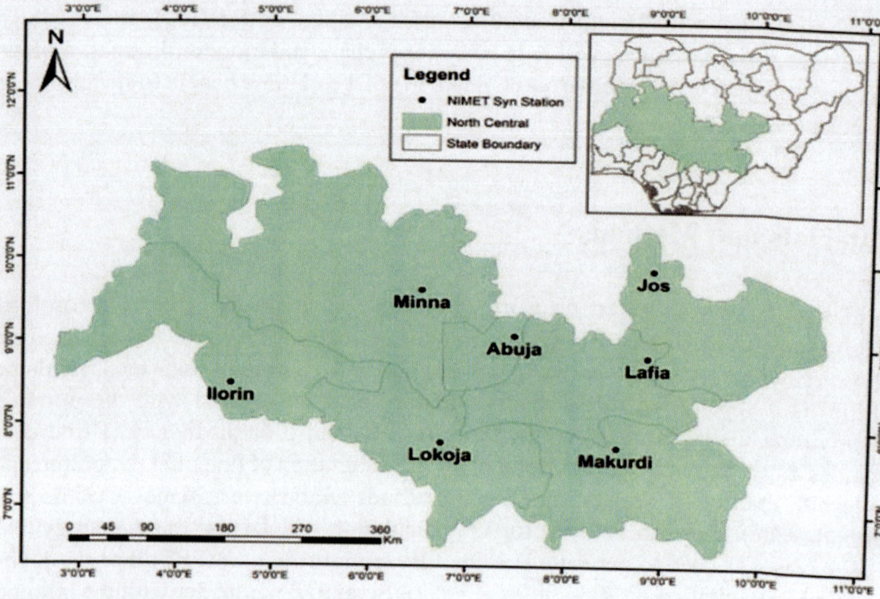

Fig. 21.1 Map of the study area (Author)

primary production and ecosystem services. To predict and manage drought events at regional scales, there is, therefore, the need for assessments and monitoring of the ecosystem responses to drought through the use of drought indices.

The study area comprises of porphyroblastic gneiss, granite gneiss, undifferentiated schists and biotite granite to mention but a few (James and Tundun 2020). Soil resources of the area are either porous, coarse-grained sandy or lateritic usually grey or reddish in colour, the soil is generally low in fertility (Iloeje 2007). The soil supports a wide variety of crop species including cereals such as Maize, Rice, Millet and Sorghum (Olayemi et al. 2014). The family's farm average subsistence food requirements are gotten from the crop species listed above, which help in growing the income in the region.

Data Analysis

In order to assess the spatial and temporal changes and measure the characteristics of agricultural drought in the study area, two agricultural drought indices were used: Standardized Precipitation Index (SPI) and Reconnaissance Drought Index (RDI). Details of the procedures for the computation of the two indices are fully shown with their corresponding formulas below.

The Standardized Precipitation Index (SPI)

McKee et al. proposed that standardized precipitation and its usage has been on the rise for the past two decades due to its solid theoretical development, robustness and versatility in drought analyses (Redmond 2002). SPI can be calculated for different time scales using precipitation as its only parameter. The equations calculating the SPI for a chosen time scale are as follows:

$$SPI = (Xik - Xi)/\sigma i \tag{21.1}$$

where

σ_i = Standardized deviation for the ith station

X_{ik} = Precipitation for the ith station and kth observation

X_i = Mean precipitation for the ith station

Reconnaissance Drought Index (RDI)

The Reconnaissance Drought Index (RDI) was developed utilizing the ratios entrenched on the cumulative precipitation to potential evapotranspiration (P/PET) for different time scales. It is a common drought index used for monitoring drought occurrences due to its sensitivity and resilience. The initial form of the index (α) of RDI is calculated for different time scales (K) months:

$$\alpha'(\mathrm{f})\mathrm{k} = \sum_{j=1}^{\mathrm{k\,pij}} \mathrm{I} = \mathrm{I(I)N\ and\ j = I(I)K} \tag{21.2}$$

$$\sum_{j=1}^{\mathrm{k}} \mathrm{PET} \tag{21.3}$$

in which P_{ij} and PET_{ij} are the precipitation and potential evapotranspiration of the j_{th} month of the i_{th} year and N is the total number of years of the available data (Tigkas 2008). The standardized Reconnaissance Index (RDI_{st}) would be adopted because the results from RDI_{st} are comparable with SPI. The standardized equation for RDI_{st} is given below

$$\mathrm{RDI}_{\mathrm{st}}^{(\mathrm{f})} = \frac{\mathrm{Y(f)} - \overline{\mathrm{Y}}}{\sigma \overline{\mathrm{Y}}} \tag{21.4}$$

where y_i is the in (α'^{i}_{k}); y_k is the arithmetic mean and σ_{yk} is the standard deviation.

Positive values of RDI_{st} indicate wet periods, while negative values indicate dry periods compared with the normal conditions of the area (Tsakiris et al. 2008).

Estimation of Potential Evapotranspiration can be obtained using the simple Hargreaves method. This method was chosen to generate PET for RDI because it is simple to use and requires temperature data and latitudinal coordinates for the location. Thus, becoming a temperature-based model. The resulting equation is

$$ET = 0.0023RA\ TD1/2(Ta + 17.8) \tag{21.5}$$

Ra = Expressed in equivalent evaporation unit can be obtained from tables.
Ta = Mean air temperature.

Interpolation is when the geology of the unsampled area is estimated using existing data observations available, which would be used for spatial changes of drought events (severe, moderate, mild, etc.) in each state at different time scales for the different years. Inverse distance weighting (IDW) is a spatial interpolation method having a set of known scattered points. With IDW, points that are close to each other are more alike than points that are farther apart (Schut 1976).

Weights are proportional to the inverse distance between the location for the interpolation and the sample points raised to the power value p. As a result, as the distance increases, the weights decrease rapidly. Again, as p increases, the weights for distant points decrease rapidly. The surrounding points will affect the prediction if the p value is very high.

Results and Discussion

Temporal Variation in Precipitation in All Stations in North Central Nigeria (1968–2018)

From Fig. 21.2., Rainfall in Abuja for the period of study was a fluctuating one with a peak year in 2016 with rainfall amount of 293.57 mm and with a sharp decrease in 2018 with 135.30 mm which makes it the year with the lowest amount of rainfall in the station.

Figure 21.3 revealed that the Annual rainfall in Ilorin was highly variable, with a sharp peak in 2014 having 1990.5 mm of total rainfall amount and the lowest rainfall amount of 575.2 mm was received in the year 1970.

According to Fig. 21.4, Jos depicted a variable rainfall total. The year 1969 was recorded as the year with a highest rainfall amount of 1591.1 mm, while the lowest rainfall amount of 797.3 mm has gotten in 1994.

Figure 21.5 shows that the Annual rainfall in Lokoja was also very variable with 1517.5 mm being the highest amount of rainfall received in 1999, while the lowest amount of rainfall of 461.4 mm was recorded in 1973.

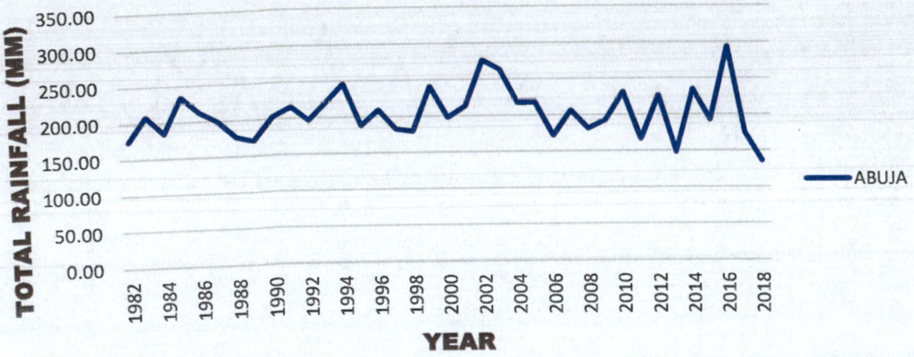

Fig. 21.2 Annual rainfall in Abuja

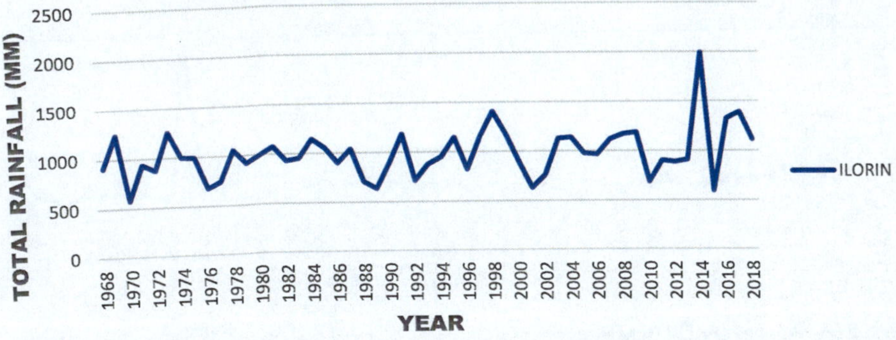

Fig. 21.3 Annul rainfall in Ilorin

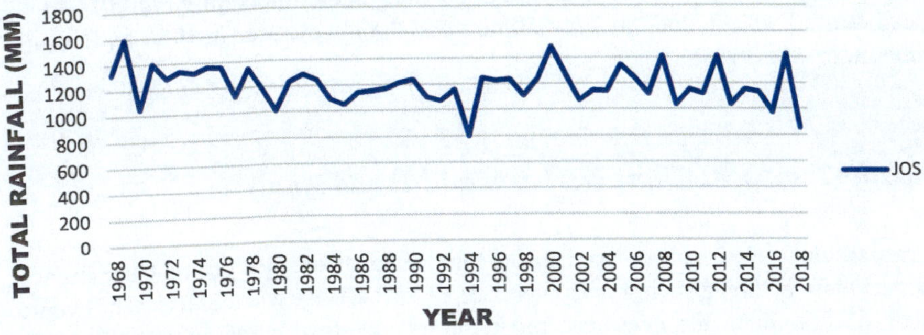

Fig. 21.4 Annual rainfall in Jos

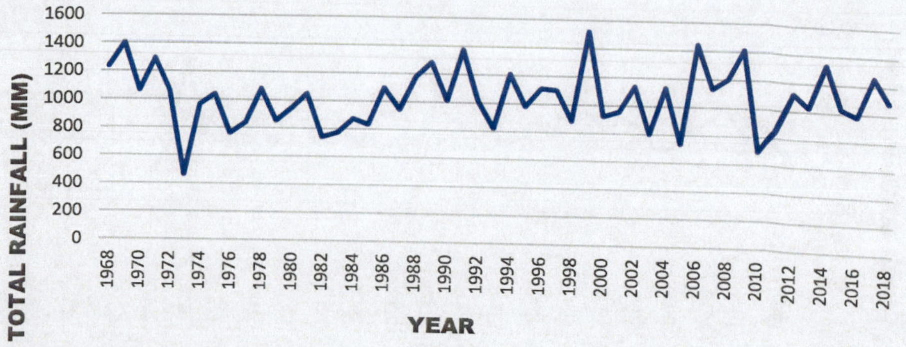

Fig. 21.5 Annual rainfall in Lokoja

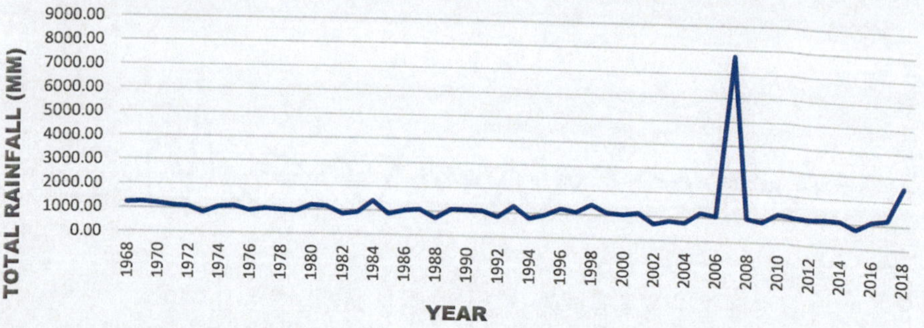

Fig. 21.6 Annual rainfall in Makurdi

From Fig. 21.6., Makurdi also experienced a gradual fluctuating rainfall with the peak rainfall year in 2007 at 7744.60 mm and the lowest year in 1988 with rainfall amount of 728.20 mm.

Spatio-Temporal Characteristics and Trend Analysis

From Table 21.1, it can be seen that a total of forty-six (46) drought occurrences occurred in the first growing season with SPI (APR–JUN), whereas a total of twenty-six (26) drought occurrences occurred in the second growing season with SPI (JUL–SEPT). While with RDI (APR–JUN), the first growing season recorded fifty-six (56) drought events, and in the second growing season (JUL–SEPT), fifty-one (51) drought events were displayed. In total RDI recorded one-hundred and seven (107) total drought events (covering both seasons), higher drought event compared to SPI

Table 21.1 Summary of the drought events in the region using SPI and RDI drought indices for both growing seasons

Category	SPI (APR–JUN)	Sum	SPI (JUL–SEPT)	Sum	RDI (APR–JUN)	Sum	RDI (JUL–SEPT)	Sum
Mild drought	1968, 1969, 1972, 1973, 1974, 1975, 1977, 1978, 1979, 1980, 1981, 1982, 1983, 1984, 1985, 1986, 1987, 1988, 1989, 1990, 1993, 1998, 1999, 2000, 2001, 2002, 2004, 2005, 2006, 2007, 2009, 2010, 2011, 2012, 2013, 2015, 2016	37	1968, 1970, 1972, 1978, 1979, 1980, 1981, 1984, 1986, 1988, 1989, 1990, 1992, 1993, 1996, 1997, 1998, 1999, 2001, 2003, 2007, 2009, 2011, 2012	24	1968, 1969, 1970, 1971, 1972, 1973, 1974, 1975, 1976, 1977, 1978, 1979, 1980, 1981, 1982, 1983, 1984, 1985, 1986, 1988, 1989, 1991, 1992, 1993, 1994, 1995, 1998, 1999, 2000, 2001, 2002, 2004, 2005, 2006, 2007, 2008, 2009, 2010, 2011, 2012, 2013, 2015, 2018	43	1968, 1970, 1971, 1972, 1973, 1975, 1976, 1977, 1978, 1979, 1980, 1981, 1982, 1984, 1986, 1988, 1989, 1990, 1991, 1992, 1993, 1994, 1995, 1996, 1997, 1998, 2000, 2001, 2002, 2003, 2004, 2006, 2007, 2008, 2009, 2011, 2012, 2013, 2016, 2017, 2018	41
Moderate drought	1973, 1987, 1995, 2006, 2011	5	1980, 2010	2	1973, 1974, 1975, 1977, 1987, 1989, 1995, 2006, 2011	9	1997, 1980, 1985, 2009, 2010	5
Severe drought	1970, 2015	2	NILL	0	1971, 2015	2	1994, 2014	2
Extreme drought	1971, 1982	2	NILL	0	1970, 1982	2	1987, 1994, 2002	3
TOTAL		46		26		56		51

which recorded seventy-two (72) total drought events covering both seasons. Furthermore, the first growing season (APR–JUN) for both SPI and rdi recorded more drought events than the second growing season (JUL–SEPT), one of the reasons is that the second growing season records peak rainfall.

Discussion

The temporal evolution patterns of growing seasons 1 and 2 for RDI and SPI are shown in Figs. 21.7 and 21.8. In north central Nigeria, (growing season 1 for RDI and SPI), several moderate drought episodes were detected throughout the study period and these episodes were prolonged in time, with critical and extreme situations. Short and more dry events were found mainly during season 1 scantily throughout the 2000s to 2016. The behaviour and frequency of drought of growing season 1 (APR–JUN) are more comprehensive than that of growing season 2 (JUL–SEPT) for both indices. The most severe drought episodes were recorded at the beginning, the mid 1990s and through the 2000s of the periods analyzed. In this region, as in the above-mentioned ones, practically1/2 of the region suffered mild drought, 1/3 suffered moderate drought and 1/5 suffered extremely dry conditions (see the distribution of growing season 1 and 2 from Figs. 21.7 and 21.8). Hasanean and Almazroui (2015) and Mwangi (2016) and Anarul et al. (2018) observed similar results in their studies in Saudi Arabia, Kenya and Bangladesh, respectively.

In general, RDI showed more drought conditions in both growing seasons; more moderate drought and more severe drought conditions throughout the period of study and in the study area than SPI. Though the SPI has been the popular drought index used over the years because it is based on only the primary factor which controls drought conditions precipitation due to the unavailability of agro-climatic data. The analysis shows that drought conditions have been aggravated over the last five years

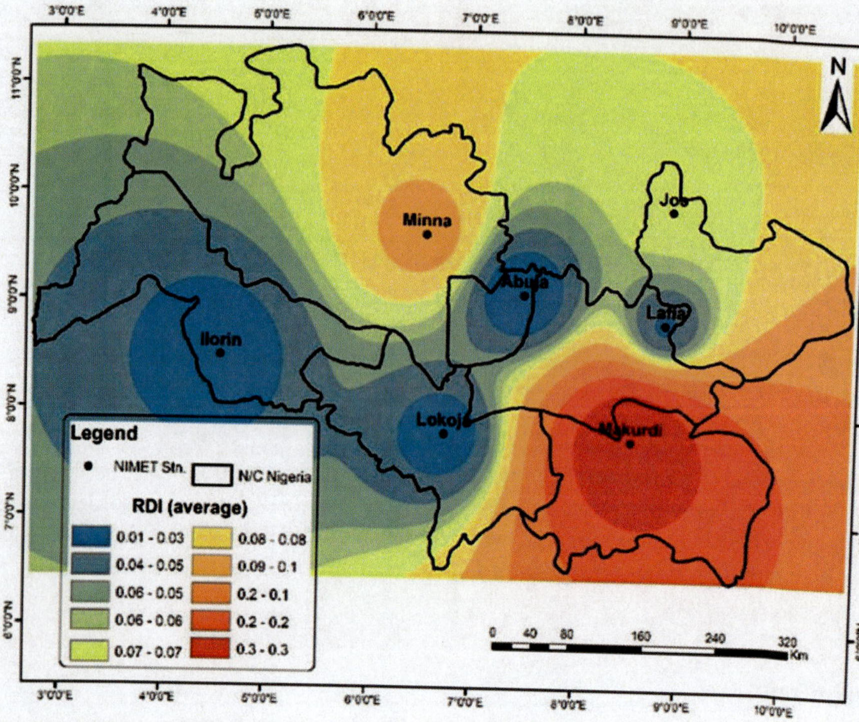

Fig. 21.7 The temporal evolution patterns of growing season 1 and 2 (APR–JUN and JUL–SEPT) using average RDI

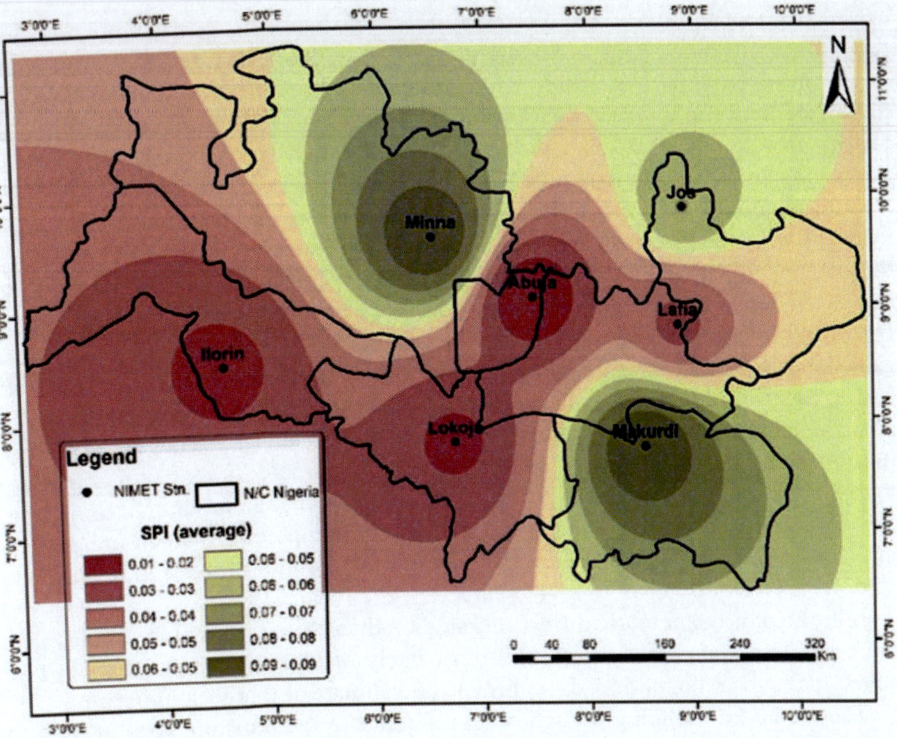

Fig. 21.8 The temporal evolution patterns of growing season 1 and 2 (APR–JUN and JUL–SEPT) using average SPI

in the region. During the period 2013–2018 with regard to drought duration, area affected, agricultural produce, livestock and economic losses. In general, the region was mainly characterized by near normal (1968–2018). Similar results are observed by Zehtabia et al. (2013), Cai et al. (2015), Jang (2018), Mlenga and Jordaan (2019) using SPI and RDI for drought analysis.

Summary and Conclusion

In this research: The Standardized Precipitation Index (SPI) and the Reconnaissance Drought Index (RDI). The evolution of agricultural drought was investigated in North Central Nigeria, for the period 1968–2018. Climate change in north central Nigeria is complex and responds sensitively to global warming. Drought relief resulted from the local climate becoming warmer and wetter in north central, as indicated by the precipitation-based drought indices. SPI-based evaluations of drought severity in the region have lessened in recent years. In contrast, the results of this study indicate that RDI-based drought severity has been aggravated in the last few years as a result of temperature rise. The North Central region has been experiencing increased drought severity since 2015 as a consequence of a significant increase in temperature (higher evaporative demand) with an insignificant increase in precipitation. The major difference between the precipitation-based indices and RDI is the meteorological factors used to calculate the indices, i.e., P and P-PET, respectively. A relatively fluctuating trend with temperature and precipitation has been observed in North Central Nigeria due to global warming. The reduction of precipitation was more evident during the first growing season APR–JUN), hence more drought episodes in this season while precipitation increased in the second growing season (JUL–SEPT) and so a reduction in drought events was seen because rainfall is just beginning in the first growing season while maximum amount of rainfall is received during the second growing season. SPI considers only climatic water supply, whereas RDI considers both climatic water supply and demand. Precipitation is a critical meteorological driver of drought, but rising temperatures are playing an increasingly strong role in influencing drought severity. RDI therefore is considered a more suitable index than SPI for studying drought severity under climate change looking at the figures above since it incorporates both precipitation and temperature which are directly affected by climate change; it recorded most of the drought events and it was more comprehensive in terms of drought characterization than SPI.

More so, it is better to use different drought indices with different thresholds with more than one data input which will most likely display the indices that are best suited to the location, period, drought type and climate of that location/station.

The method of defining drought with reference to the moisture state of the soil has some advantages in monitoring and analyzing agricultural production. Going forward, this should be given priority to better monitor and prepare for natural hazard and how it affects crop yield.

References

Anarul HMA, Al-Mamun IM, Dong-Ho J (2018) Precipitation concentration in Bangladesh over different temporal periods. Adv Meteorol (1849050), 18 p

Cai W, Zhang Y, Chen Q, Yao Y (2015) Spatial patterns and temporal variability of drought in Beijing-Tianjin-Hebei metropolitan areas in China. Adv Meteorol (289471), 14

Eze JN, Aliyu U, Alhaji-Baba A, Alfa M (2018) Analysis of farmers' vulnerability to climate change in Niger state, Nigeria. Int Lett Soc Hum Sci 82:1–9

Gidey E, Dikinya O, Zenebe A (2018) Modelling the Spatio-temporal meteorological drought characteristics using the standardized precipitation index (SPI) in Raya and its Environs, Northern Ethiopia. Corpus ID: 133993046

Hasanean H, Almazroui M (2015) Rainfall: features and variations over Saudi Arabia: A review. Climate 3(3):578–626

Iloeje NP (2007) Characterization of surface layer turbulence across a West African tropical climate belt. Atmos Clim Sci 10(3). A New Geography of Nigeria. 6th Edition, Longman Nigeria Plc, Ikeja

James EE, Tundun K (2020) Regional groundwater studies using high-resolution aeromagnetic data in Abuja and environs. Modibo Adama University, Yola, North-Central Nigeria

Jang D (2018) Assessment of meteorological drought indices in Korea using RCP 8.5 Scenario. Water 10:283

Mannocchi F, Francesca T, Vergni L (2004) Agricultural drought: Indices, definition and analysis. Universita degli studi di Perugia

Mlenga DH, Jordaan AJ (2019) Monitoring droughts in Eswatini: a spatiotemporal variability analysis using the Standard Precipitation Index. Jamba (potchefstroom, South Africa) 11(1):712

Mwangi M (2016) Diverse drought spatiotemporal trends, diverse Etic-Emic perceptions and knowledge: implications for adaptive capacity and resource management for Indigenous Maasai-Pastoralism in the Rangelands of Kenya. Climate 4(2):22

NDMC (2013) U.S. Drought Monitor, National Drought Mitigation Center (NDMC). http://drought.unl.edu/MonitoringTools/USDroughtMonitor.aspx

Nweze N, Tsue PT, Okoye CU (2014) Vulnerability of farming households to environmental degradation in developing countries: evidence from North Central Nigeria. J Econ Sustain Dev

Okorie FC (2003) Studies on Drought in the sub-Saharan Region of Nigeria using remote sensing and precipitation data. University of Hyderabad, India, JNCASR-Costed Fellowship Programme

Oladipo EO (1985) A comparative performance analysis of three meteorological drought indices. J Climatol 5:655–664

Oladipo EO (1993) Some aspects of the spatial characteristics of drought in northern Nigeria. Nat Haz 8:171–188

Olagunju TE (2015) Drought, desertification and the Nigerian environment: A review. J Ecol Nat Environ 7(7):196–209

Olayemi IK, Idris BA, Ejima IAA, Adeniyi K, Azubuike CU (2014) The climate of north-central Nigeria and potential influence on mosquito (Diptera: Culicidae) vectorial capacity, for disease

Quiring SM, Papakryiakou TN (2003) An evaluation of agricultural drought indices for the Canadian prairies. Agric Meteorol 118(1–2):49–62

Redmond KT (2002) The depiction of drought. Bull Amer Meteorolo Soc 83:1143–1147

Tigkas D (2008) Drought Characterisation and monitoring in regions of Greece. European Water 23(24):29–39

Tsakiris G, Nalbantis I, Pangalou D, Tigkas D, Vangelis H (2008) Drought meteorological monitoring network design for the reconnaissance drought index (RDI). In: Franco Lopez A (ed), Proceedings of the 1st international conference "Drought Management: scientific and technological innovations". Option Méditerranéennes, Series A, Zaragoza, Spain, No. 80, 12–14 June 2008, pp 57–62

Um M, Kim Y, Park D, Kim J (2017) Effects of different reference periods on drought index estimations from 1901 to 2014. Hydrol Earth Syst Sci 21:4989–5007

Wilhite DA (1993) The enigma of drought. Drought Assessment, management, and planning. Theory and case Studies. Kluwer Academic Publishers, Boston, MA, pp 3–15

Yue Y, Shen S, Wang Q (2018) Trend and variability in droughts in Northeast China based on the reconnaissance drought index. Water 10(318):1–17

Zehtabia G, Karimi K, Fard SNN, Mirdashtvan M, Khosravi H (2013) Comparability analyses of the SPI and RDI meteorological drought indices in South Khorasan province in Iran. Int J Adv Biol Biomed Res 1(9):981–992

Chapter 22
Influence of Seasonal Changes on the Quality of Water Resources in Southwestern Nigeria: A Review

Daniel A. Ayejoto, Johnbosco C. Egbueri, Johnson C. Agbasi, Michael E. Omeka, Chinanu O. Unigwe, Vincent E. Nwazelibe, Joshua O. Ighalo, and Chaitanya B. Pande

Abstract Water demand has risen dramatically in southwestern Nigeria over the last decades as a result of urbanization, agricultural activities, increasing populations, rapid industrialization, and economic growth. Increasing carbon dioxide and other greenhouse gases in the air are causing weather patterns that have profound consequences for the water resources. These greenhouse gases are projected to modify the atmosphere's thermal balance, producing temperature rises and alterations in a variety of other climatic factors, which would impair water quality. In various climate areas and major rivers across Nigeria, changes in agricultural and land-use patterns, and over-exploitation of water resources are now changing the hydrological cycle. For applicable regional long-term development initiatives, an assessment of available water resources in terms of the quality and predicted impacts of climate change and variation is crucial. This study summarizes current research into the effects of seasonal change on water resources and analyzes the significance of these effects for future water development. It is based on an analysis of estimates of seasonal

D. A. Ayejoto (✉)
Department of Environmental and Sustainability Sciences, Texas Christian University, Fort Worth, TX, United States
e-mail: ayejotodaniel@gmail.com; daniel.ayejoto@tcu.edu

J. C. Egbueri · J. C. Agbasi
Department of Geology, Chukwuemeka Odumegwu Ojukwu University, Uli, Nigeria

M. E. Omeka
Department of Geology, University of Calabar, Calabar, Nigeria

C. O. Unigwe
Boone Pickens School of Geology, Oklahoma State University, Stillwater, OK, United States

V. E. Nwazelibe
Department of Geology, Albert Ludwig University of Freiburg, Freiburg im Breisgau, Germany

J. O. Ighalo
Department of Chemical Engineering, Nnamdi Azikiwe University, Awka, Nigeria

C. B. Pande
Sant Gadge Baba Amravati University, Amravati 444602, MS, India

J. C. Egbueri et al. (eds.), *Climate Change Impacts on Nigeria*, Springer Climate,
https://doi.org/10.1007/978-3-031-21007-5_22

fluctuations and water quality. According to these estimates, yearly runoff in southwestern Nigeria will continue to be on the increase if prompt action is not taken, groundwater recharge will rise, and water quality—as measured by greenhouse gas concentrations—will suffer under plausible climatic variability. The prospects for sustainability of surface and groundwater resources in Nigeria are also discussed, taking into account the limitations placed by climate change as well as future research directions.

Keywords Water resources · Seasonal change · Southwestern Nigeria · Climate · Water Quality

Introduction

Water resources (surface and ground water) are one of the most essential aspects of environmental resources that are threatened by anthropogenic activities on the earth. The ecological environment and public health are both influenced by water quality (Lu et al. 2015; Shi et al. 2017). Water contamination has become a major issue in Nigeria's southwestern region as a result of fast population increase, industrialization, and urbanization. Different physicochemical properties are commonly used to determine the quality of water. These properties fluctuate dramatically as a result of many forms of pollution, seasonal changes, and groundwater use (Ramakrishnaiah et al. 2009). While much has been made of the problems with water quality caused by human activities, it is crucial to remember that other natural phenomena such as seasonal fluctuations can also cause water contamination. data, regularity, and quantity, land cover, geography, terrain, soil composition, and river flow have all been proven to worsen the issues of water quality management.

Water quality varies according to the seasons and temperature (Ioryue et al. 2015; Ling et al. 2017; Dey et al. 2021). Seasonal changes in water quality may have both good and harmful effects. Temperature differences are associated with different seasons, and temperature shifts due to seasonal variation cause water resources to shift in terms of quality. Because of their fragility and limited adaptive skills, developing nations like Nigeria will be disproportionately impacted by climate variation. As a result, emerging countries with larger population densities and growth rates will face far greater climate change threats. These effects may be felt more strongly in metropolitan areas, such as south-west Nigeria, where the demand on societal, commercial, and sustainable development is considerably greater (Pickett et al. 2001; Shahid et al. 2016). Several studies have been published in the literature to analyze the influence of seasonal change forecasts on the hydrology of different regions in other parts of Nigeria (Efe et al. 2005; Agbaire and Obi 2009; Essien-Ibok et al. 2010; Offem et al. 2011; Eni 2015; Adedeji et al. 2019; Shiru et al. 2020a; Raimi et al. 2022).

Despite recent advances in our knowledge of atmospheric mechanics and huge climatic phenomena, the climatic consequences of greenhouse gases remain unclear

(Clifton et al. 2018; Qiuhong 2019; Shen et al. 2020; Vicuña et al. 2021; Raulino et al. 2021). Variations in regional hydrological cycles, and the resulting outcomes on the amount and quality of regional water resources, is one of the greatest significant implications of climate change scenarios (Sýs et al. 2021). Nonetheless, their implications are underappreciated. Current hydrology studies' finding shows that probable seasonal changes resulting by rise in greenhouse gas levels will modify runoff time and volume, alter water levels and accessibility, and affect water quality (Cherkauer et al. 2021; Akinwumi et al. 2021; Dallison et al. 2022). This situation increases the danger of significant environmental and public health disruptions, as well as having far-reaching effects for future water management in the country.

According to research, rising temperatures and decreased rainfall may impair net recharge and have an impact on water resource levels (Lal et al. 2018; Wu et al. 2020; Nielsen-Gammon et al. 2020). However, there has been little research done on the hydrological effects of anticipated seasonal change in Nigeria's southwestern areas. As a result, comprehensive temperature and rainfall rate statistics are required for water resource management and to analyze the influence of seasonal change. Due to the changing climate, the Nigerian water sources will be put under greater strain. Spontaneous run-off to the ocean currently wastes well over 60% of the rainfall annually in the nation (Galadima et al. 2011; Bernardo et al. 2022). It is critical to consider how potential the seasonal change would affect the severity, geographical and temporal distribution of rainfall, evaporation rates, and temperatures in Nigeria's many modern agricultural areas and waterways.

With Lagos being the largest populated state and one of the world's largest quickest emerging metropolises, south west Nigeria remains the most vulnerable region in terms of water pollution. The region gets its water from both the country's surface and groundwater sources. Even though some states in the region, such as Lagos, Oyo, and Ogun, appear to have adequate water resources in terms of quantity, considering the intensity of rain they receive and the numerous waterways that transfer into them as a consequence of climate change, many people in this region nevertheless dearth access to potable water. If present migration trends and Nigeria's birthrates persist, the southwest area of Nigeria will suffer a rapid seasonal change in environment. Water resource quality would be jeopardized due to diminishing water level (Longe 2011) and widespread degradation of freshwater in numerous parts of the nation (Olufemi et al. 2010). Changes in rainfall patterns and rising temperatures may exacerbate this problem (Shiru et al. 2020b).

The tuning of global seasonal models is primarily subjective in affluent countries with a well-distributed data collection network. However, information on seasonal variation is often lacking in impoverished nations, where seasonal variability has a greater impact due to low adaptability and response capability (Ziervogel et al. 2006, 2010). The influence of seasonal change on Nigeria's water resources from scaled down climate projections, particularly in the southwestern region of the country, is scarce or non-existent in the literature. In order to assess the variations and levels of complications on water quality in this location, we compared two situations, one in the dry (temperature) and one in the wet seasons (rainfall). An effort was made here to provide a critical summary of the probable effects of climate change on surface water and groundwater resources in south-west Nigeria. Based on recent literature and available statistics, we identify the main areas of susceptibility and sources of ambiguity. We gave an analysis that progresses from recent trends that have been well-observed to more speculative model simulations and explanations. The first section discusses recent seasonal change findings (temperature and rainfall). The second portion looks at previous and expected future trends in water resources, determines if recent changes are uncommon or within typical climatic variability. The implications of expected climate change and variability, as well as related hydrological occurrences and regional water resource susceptibility to seasonal change, are discussed, along with significant risks and adaption measures that have also been considered and evaluated.

Study Area

The review was conducted in Southwestern Nigeria (Fig. 22.1), which has a tropical humid climate. The wet or rainy period (April–October) and the temperate or dry period (November–March) are two distinct seasons in the region (Adejuwon and Odekunle 2006). In aspects of weather and climate, the study area is different from other areas in the country. Except for Lagos, which has a high level of urbanization as a result of large businesses located in the state, the region comprising of Ondo, Osun, Oyo, Ekiti, and Ogun states is mostly an agro-ecological zone, where farming is the predominant employment. The typical lowest temperature for the hot season is 20 °C to 22 °C, whereas the mean daily temperature is 32.6 °C (Ayanlade et al. 2017; Odekunle 2006; Odekunle et al. 2005). Three major weather flows affect the climate of Southwestern Nigeria: The Tropical Eastwards, Tropical Mediterranean (Tm) rising air, and the Tropical Central (Tc) rising air. The tropical eastwards originate in the eastern and flow through the upper orbit along the climatic tropics, leaving a frigid airflow behind. Tm and Tc frequently collide at crossing discontinuity (CD), a sloped floor. According to findings by Odekunle et al., the existence of CD movements influences rainfall patterns in the Southwestern part of Nigeria (Odekunle et al. 2005).

The heavy industries, vegetation, soil, and farming methods in these areas all contribute to climate change. Rainfall occurs in two phases. As a result, a year is separated into two key seasonal changes for this study: early rainy season (April–Mid August) and post rainy period (late August–October). The dry period occurs frequently in August (Adejuwon and Odekunle 2006).

Observed Recent Climate Changes in Southwestern Nigeria

Temperature rises, unpredictable rainfall, rising sea levels, and more occurring severe weather scenarios have all impacted Nigeria's climate, causing fresh water

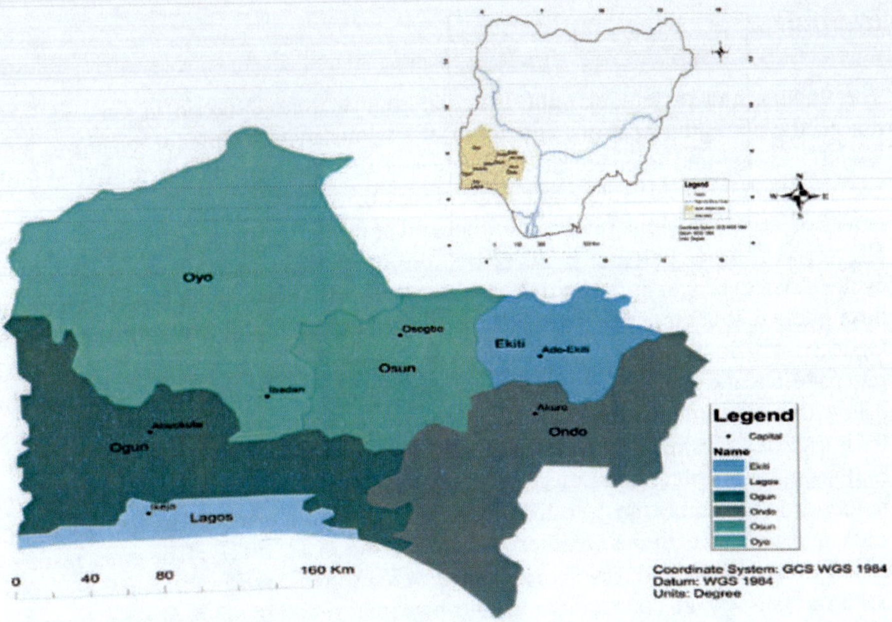

Fig. 22.1 Map of Southwestern Nigeria showing the study areas (*Source* Akinyemi et al. 2020)

problems (Elisha et al. 2017; Ebele and Emodi 2016; Olaniyi et al. 2013). Rainfall frequency and severity have increased, causing substantial urban overflow and flooding in numerous places of Nigeria (Enete 2014). In the southwest, rainfall is projected to increase, and rising sea levels are projected to intensify flooding and water contamination (Akande et al. 2017; Ebele and Emodi 2016).

Over the last few years, the image of global seasonal change, as well as the situation in Nigeria, has become clearer. There seems to be little uncertainty that the country overall, and the southwest area especially, have gathered enough data in coming to ascertain temperature and rainfall trends in the past decades. In recent years, rainfall in the Southwest has also been lower than typical. A decline in the quarterly daily temperature range, which is also the variation quarterly peak and the lowest temperature, has been connected to global warming (Lobell et al. 2007). The elevation from both the distinct seasonal mean lowest temperature over Ibadan, a major region in Southwest Nigeria, for example, revealed that continuous warming is occurring, which has been linked to the man-made greenhouse effect (Egbinola and Amobichukwu 2013; Abatan et al. 2018). Two significant elements determining seasonal changes in Nigeria will be explored in depth in this section: rainfall in the rainy period and temperature in the dry period.

Rainfall

The volume and pattern of rainfall in the southwestern Nigeria is very variable due to the strength, location, and route of various meteorological systems. As a result, seasonal and inter-annual fluctuation define the region's rainfall (Egbinola and Amobichukwu 2013; Ojo and Olurotimi 2014; Layi-Adigun et al. 2020; Shiru et al. 2020a, b). Previous research has looked at rainfall variations in Southwestern Nigeria over time. Balogun et al. (2016) discovered that rainfall variation with a cycle of five to six years can be linked to tropical and extra-tropical forces that occur throughout distinct months, periods, and years in Odeda Local Government Area of Ogun State in South-western Nigeria. Ojo and Olurotimi (2014) investigated Tropical rain pattern features over two sites in southwestern Nigeria using two years of rainfall data with a one-minute time interval. This research was carried out in order to express the intricacies of rainfall intensity in this location, as well as their impact on satellite-Earth communication capabilities. Another recent study used probability distribution functions to determine the most likely maximum flooding from discharge data using statistical tool of extreme weather events (Adeola et al. 2021). The runs testing, correlation analysis, and wavelet resonance were used in this work to better comprehend rainfall-sewage connections in four proximal rivers in southwestern Nigeria. The greatest Severe Annual Rainfall in the record is expected to take place once every 31 years with a 0.03 likelihood, according to the study. Between Yewa—Ikeja and Yewa—Abeokuta, wavelet consistency has a clear meaningful correlation with periods (Adeola et al. 2021). In southwestern Nigeria, Akinyemi and his colleagues compared Multi-satellite rainfall readings with data from ground-based meteorological stations (Akinyemi et al. 2020). Monthly average rainfall for each site is shown in Fig. 22.2 for the years 1998–2016. Rainfall was found to be typically lower in January, February, March, November, and December, with totals under 50 mm. This scenario was discovered through land and surface data analysis. All of the rainfall observations suggest that the rainy season rises in June and September, with a brief dry period in August (Fig. 22.2) (Akinyemi et al. 2020). The findings also indicated that virtually every satellite measurements match land rainfall measurements. These situations are significantly more prevalent during the dry seasons, when all datasets revealed a greater degree of similarity during the dry seasons but a greater degree of variation during the rainy seasons (Akinyemi et al. 2020).

The annual rainfall results gathered from the Nigerian Meteorological Agency (NIMET) are shown in Table 22.1 (Oke et al. 2015). Abeokuta supplied the least mean annual rainfall of 1196.54 mm to the drainage area, with upper and lower bounds of 1604.40 and 712.20 mm, respectively, according to the statistical analysis of yearly rainfall data (Table 22.2). With an average yearly rainfall of 1751.05 mm, Lagos generated the most annual rainfall to the drainage region throughout the study period. The yearly rainfall totals were 2525.90 mm for the maximum and 996.20 mm for the minimum. With a standard error of 410.53 mm, rainfall in Lagos deviates significantly from the average. This is because Lagos State is a coastline region that benefits from the Atlantic Ocean's tropical Maritime Air Mass (Oke et al. 2015).

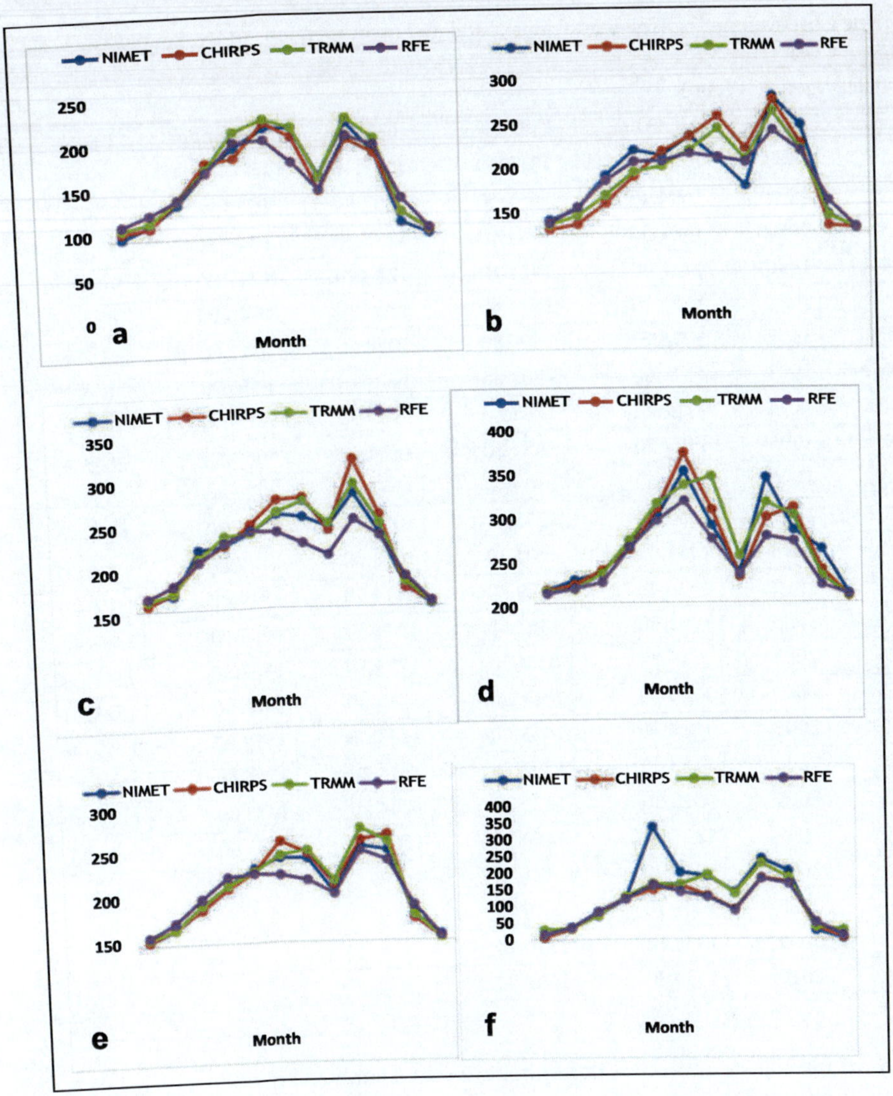

Fig. 22.2 Monthly average rainfall changes in Abeokuta, southwest Nigeria (**a**), Ado-Ekiti (**b**), Akure (**c**), Ikeja (**d**), Osogbo (**e**) and Ibadan (**f**) (1998–2016) (*Source* Akinyemi et al. 2020)

To analyze the pattern of rainfall variability and its consequences on water use in the equatorial metropolis of Ogbomoso, Oyo State, Nigeria, a 37-year rainfall data was obtained from NIMET (Ogunbode and Ifabiyi 2019). To establish the trend, multivariate statistical approaches were used to examine the data. The simple regression analysis and M–K both indicated an excellent pattern in rainfall in the study region across the timeframe, with R^2 of 0.001 and M–K showing clearly sets at the significant value of 0.05 (Fig. 22.3). The findings of the study suggested that the

Table 22.1 Descriptive Statistics of Annual Rainfall in some regions of the Southwestern Nigeria (*Source* Oke et al. 2015)

Rainfall per year (mm)						
	Year	Abeokuta	Oshogbo	Ibadan	Ijebu-ode	Lagos
1	1987	1277.10	1368.60	1285.10	1501.90	2057.00
2	1988	1604.40	1464.50	1397.80	1653.20	2432.10
3	1989	1401.00	1191.10	1223.70	1445.50	1702.60
4	1990	1111.00	1411.10	1283.90	1713.80	2030.20
5	1991	1173.10	1684.40	1369.30	1646.90	1722.90
6	1992	1076.70	1404.80	1088.60	1611.80	1357.60
7	1993	1193.60	1263.80	1478.10	1456.90	1828.20
8	1994	712.20	1207.80	905.90	1556.90	1602.30
9	1995	1177.50	1452.50	1534.70	1643.80	2025.30
10	1996	1471.60	1349.40	1653.70	2032.40	2535.90
11	1997	1354.90	1231.10	1097.10	1705.70	1936.80
12	1998	1118.40	1203.30	1018.00	1172.80	1232.90
13	1999	1530.40	1529.30	1814.90	1819.40	2056.90
14	2000	1201.90	1367.80	1244.50	1655.00	1458.90
15	2001	849.20	1014.70	1289.70	1462.30	1110.00
16	2002	1235.10	1361.00	1515.30	1426.50	1523.20
17	2003	1214.70	1633.00	1143.00	1572.00	1549.00
18	2004	1153.30	1393.70	1314.20	1766.60	2117.40
19	2005	917.50	1130.20	1204.20	1473.30	1745.60
20	2006	1157.20	1469.70	1260.10	2043.60	996.20
21	2007	1615.70	1421.70			
22	2008	1408.70	1610.10			
23	2009	1351.10	1310.20			
24	2010	1450.70	1691.80			

city acquired greater rainfall throughout the time analyzed period. As a result, both surface and underground water sources were replenished in a natural way (Ogunbode and Ifabiyi 2019). Although average rainfall for a period is equally necessary to water consumers, it may not always reflect their needs. The 'rainfall reliability' is a metric which has been established to assist farmers in predicting how much rain they will have over the course of a season (Laughlin et al. 2003; Murphy and Timbal 2008; Ogungbenro and Morakinyo 2014). The reliability is defined as the number of seasons wherein every month of a different climatic or growing period obtains at most a given amount or share of the normal rainfall. It shows the possibility of getting useful rainfall variations throughout the regular season, but it does not mean that dependability will decrease if average rainy season falls.

Table 22.2 Statistical analysis of yearly rainfall data (*Source* Oke et al. 2015)

	Annual Rainfall				
	Abeokuta	Oshogbo	Ibadan	Ijebu-Ode	Lagos
N Valid	20	20	20	20	20
Missing	0	0	0	0	0
Mean	1196.54	1356.59	1306.09	1618.00	1751.05
Median	1185.55	1368.20	1284.50	1627.80	1734.25
Std *Deviation*	216.90	165.41	217.98	204.12	410.53
Variance	47,047.55	27,361.25	47,516.62	41,660.00	168,536.16
Minimum	712.20	1014.70	905.90	1172.80	996.20
Maximum	1604.40	1684.40	1814.90	2043.60	2535.90

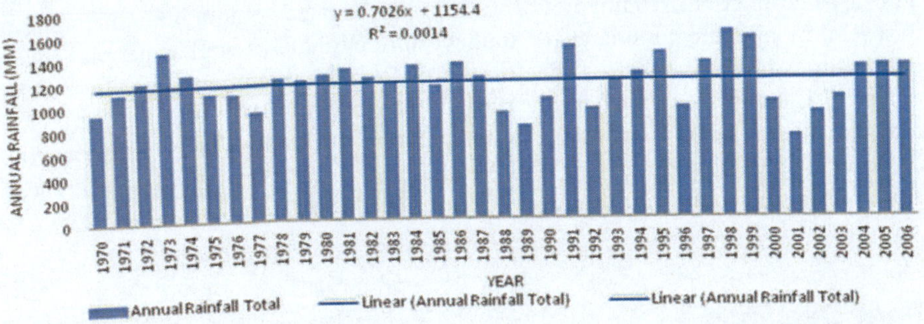

Fig. 22.3 Mean annual rainfall pattern in Ogbomoso from 1970–2006 (*Source* Ogunbode and Ifabiyi 2019)

Temperature

Nigeria's average temperature has been rising for years, with the trend accelerating in recent years. Drought intensity is influenced by climate change because increased temperature results in more evaporation. The average yearly temperature in Nigeria is 26.9 °C, with mean monthly temperatures spanning from 24 °C (December, January) to 30 °C (April) (Efe and Eyefia 2014). Both daily T_{max} and, though little dramatic, daily minimum temperatures have experienced similar patterns and accelerated warming (T_{min}). Several studies in Nigeria have found strong correlations between rainfall and temperatures. From 1971 to 2000, Akinsanola and Ogunjobi (2014) analyzed records of air temperature and rainfall from 25 meteorological sites in Nigeria to analyze temperature and precipitation variations. The data was examined for rapid variations in temperature and rainfall levels across Nigeria, and also seasonal and spatial patterns. Atmospheric temperature abnormalities between −0.2 and −1.6 were common in the first decade of 1971–1980, but only five stations showed positive deviations in the second decade of 1981–1990, while the majority of the country was standard with indications of warming in the third decade of 1991–2000. Eludoyin (2016) also looked at spatial variability in temperature and humidity over Nigeria, using data from 18 climatic zones over a 59-year timespan (1951–2009), and found that the mean, minimum, and maximum temperatures have all risen since 1951, while relative humidity has decreased. Based on variations in rainfall and temperatures between 1970 and 2012, Egbinola and Amobichukwu (2013) examined climatic variance in the Ibadan region. They employed Linear Regression to calculate the value of variable Y (total rainfall or average minimum temperature) for a specific amount of variable X (time), and Correlation Analysis to assess the link between total rainfall and average minimum temperature across the duration of the study. In addition, the study looked at how total rainfall and the average lowest temperature varied across the study area. The findings reported that the overall rainfall has been increasing over the study period. In addition, the average lowest temperature is trending increasing. In fact, total rainfall and average minimum temperature have a positive correlation. Abuloye et al. (2017) investigated temperature fluctuations in several bioclimatic parameters of southwestern Nigeria, as well as people' rising temperatures protective factors. Their research found that temperatures climbed by around 1 °C at most locations during 1961–1990 and 1991–2010, and that the region near the Atlantic coast is experiencing temperature extremes, which has intensified in the 1991–2010 timeframe (Fig. 22.4). Table 22.3 depicts temperature differences at a number of locations in southwest Nigeria (1961–2010).

Water Resources

Surface Water Resources

Nigeria has rich natural water resources, as evidenced by her high annual rainfall, large surface water bodies such as creeks, streams, and reservoirs, as well as extensive subsurface water reserves whose quantity and circulation have yet to be properly analyzed (Fig. 22.5). Surface water availability in the country is 215 cubic kilometers per year (Meludu and Akaninyene 2021; Ngene et al. 2021). The country stretches along the river for the most part, which, including the River Benue, divides the country into 3 distinct localities. A countrywide system of drainage basins is provided by the Cross River, Imo, Sokoto, Ogun, Anambra, and Kaduna rivers, as well as various streams and tunnels, lagoons, and wetlands (Fig. 22.6). Southwestern Nigeria has been considered to rely on surface water because the bulk of the states in this region have more water than the rest of the country. That so, it is worth noting that this region's surface water has been discovered to be severely contaminated as a result of high human activity, such as urbanization, industrialization, agriculture, and a variety of other factors. In the southern part of Nigeria, a series of studies have been conducted to tackle water pollution using various methodologies and models. For example, in measuring the surface water quality in the Ikare metropolis

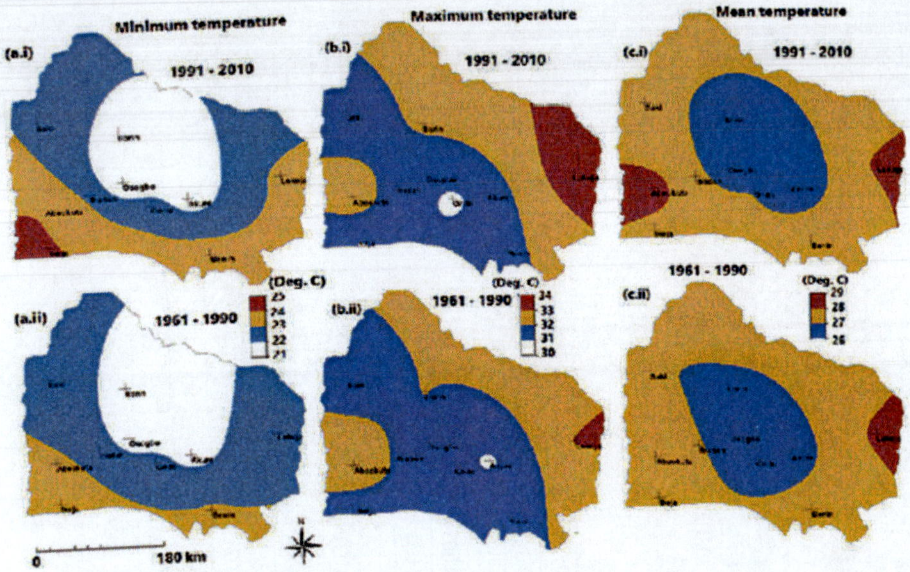

Fig. 22.4 Variation in minimum, mean, and maximum temperatures in southwest Nigeria from 1961 to 1990 and 1991 to 2010. The position of the meteorological station whose data were adjusted is indicated by the cross (*Source* Abuloye et al. 2017)

of southwestern Nigeria, Oladipo et al. (2021) contrasted two widely obtainable statistical techniques: fuzzy logic (FL) prediction and analytical approaches for the water quality index (WQI). They determined that FL rated 8.3% of the waters as "moderate" for drinking, but WQI rated the complete research region's waters as "bad". Ekute and Etim (2021) investigated the physicochemical parameters and heavy metal concentrations in Ota industrial estate effluents to ascertain their quality and influence on the Oruku River's surface water quality in Ota, southwestern Nigeria. Their findings showed that agricultural residues, urban run-offs, and home wastes are the primary sources of pollution of surface waters, albeit to a minor extent. The surface water was found to be suitable for irrigation, recreation, and industrial use, but not for drinking, according to WHO, USEPA, CQC, and FEPA guidelines for drinking water quality, irrigation quality, and industrial usage.

In Ado-Ekiti and its environs, Olowe et al. (2016) sought to determine the occurrence of water—borne diseases and analyze the microbiological suitability of drinking water reserves. The frequency of waterborne s was determined to be 27.7%, according to the findings. The total content of escherichia coli in many water samples collected from diverse freshwater resources in Ado-Ekiti in a large colony formation shows that the city's high rate of waterborne illness is due to contaminated water supply. As a result, they suggested that authorized authorities should make an effort to undertake quality assessments of water sources on a regular basis to facilitate access to secure, high-quality drinking water. Adeola Fashae et al. (2019) analyzed the effects of land use pattern on surface water quality in the southwestern Nigerian

Table 22.3 Temperature variables' mean, range, variation, trend (change in values over time), and temporal fluctuations in southwest Nigeria (1961–2010) (*Source* Abuloye et al. 2017)

	Station	Mean (Min–Max) (°C)	CV (%)	Trend (a ± b(x)) (R2 in parenthesis)	1961–1990 (mean) (°C)	1991–2010 (mean) (°C)	Variation ANOVA	
							P-value	F-value
Minimum temperature	Abeokuta	23.6 (15.7–26.5)	5.8	0.01x + 21.3 (0.1)	23.2	23.9	0.002	609.8
	Akure	21.4 (13.2–23.8)	6.7	−0.001x + 21.8 (0.01)	21.6	21.4	0.001	39.9
	Benin	22.9 (18.4–26.3)	4.3	0.003x + 22.2(0.2)	23.1	23.5	0.002	252
	Ibadan	22.3 (16.4–29.8)	5.2	−0.003x + 21.5 (0.2)	22.6	22.8	0.001	14.7
	Ikeja	23.3 (20–29)	5.7	0.005x + 22 (0.3)	23.5	24.1	0.002	399.1
	Ilorin	21.3 (14.5–25)	7.8	0.002x + 20.7 (0.05)	21.7	21.7	0.535	0.7
	Lokoja	22.8 (14.1–27.7)	9	0.001x + 22.4 (0.01)	22.7	23.3	0.002	129.8
	Ondo	22.2 (18.5–25.1)	4.1	0.002x + 21.8 (0.07)	22.3	22.5	0.001	57.9
	Osogbo	21.1 (13.5–24.3)	7.5	0.003x + 21.1 (0.001)	21.5	21	0.001	122.9
	Saki	22.7 (19.5–25.4)	4.8	0.002x + 21.7 (0.03)	22.6	22.8	0.003	44.8
	Summary	22.4 (13.5–29.8)	6		22.48	22.7	0.06	167.2
Maximum temperature	Abeokuta	32.7 (28.0–38.1)	8	0.003x + 31.7 (0.03)	32.7	32.9	0.002	9.44
	Akure	31.0 (26.5–36.5)	8	−0.01x + 32.9 (0.03)	31.4	30.9	0.002	92.68
	Benin	31.4 (27.0–37.0)	6	0.002x + 31 (0.02)	31.7	31.7	0.749	0.102
	Ibadan	31.3 (23.7–38.0)	8	0.002x + 30.8 (0.02)	31.6	31.9	0.003	18.17
	Ikeja	30.9 (26.9–35.3)	6	0.002x + 30.5 (0.02)	31	31.4	0	66.33
	Ilorin	32.2 (27.5–37.9)	8	−0.001x + 32.6 (0.01)	32.2	32	0.031	4.666
	Lokoja	33.1 (28.0–39.1)	7	−0.001x + 33.1 (0.001)	33.7	33.1	0.002	110.9
	Ondo	30.9 (26.1–34.2)	8	0.03x + 30.0 (0.03)	30.9	31.4	0.003	57.87

(continued)

Table 22.3 (continued)

	Station	Mean (Min–Max) (°C)	CV (%)	Trend (a ± b(x)) (R2 in parenthesis)	1961–1990 (mean) (°C)	1991–2010 (mean) (°C)	Variation ANOVA P-value	F-value
	Osogbo	31.2 (25.0–37.2)	8	0.001x + 30.9 (0.01)	31.3	31.5	0.001	11.26
	Saki	31.6 (25.0–37.2)	7	0.002x + 30.8 (0.004)	31.5	31.7	0.013	6.165
	Summary	31.6 (25–38.1)	7		31.8	31.9	0.08	37.8
Mean temperature	Abeokuta	28.1 (23.9–32.2)	5.9	0.004x + 26.5(0.05)	27.9	28.4	0.002	138.9
	Akure	26.2 (22.4–30.1)	5.3	−0.03x + 27.3 (0.04)	26.5	26.1	0.001	128.6
	Benin	27.2 (24.3–31.3)	4.9	0.002x + 26.5 (0.07)	27.4	27.6	0.003	28.6
	Ibadan	26.8 (23.4–31.4)	5.9	0.03x + 26.1 (0.1)	27.1	27.3	0.003	21.46
	Ikeja	27.2 (21.6–31.3)	5.4	0.003x + 26.2 (0.1)	27.2	27.7	0	205.2
	Ilorin	26.8 (22.3–30.7)	6.1	0.001x + 26.2 (0.01)	26.9	26.8	0.041	4.185
	Lokoja	28.0 (23.9–32.9)	6.2	0.001x + 26.2 (0.003)	28.2	28.2	0.571	0.32
	Ondo	26.5 (23.5–38.9)	5.2	0.002x + 25.9 (0.1)	26.6	26.9	0.002	100.9
	Osogbo	26.2 (23.1–30.2)	5.5	0.001x + 26.0 (0.01)	26.4	26.3	0.119	2.44
	Saki	27.1 (24.8–30.4)	5.2	0.002x + 26.2 (0.01)	27	27.2	0.001	20.42
	Summary	27 (22.3–38.9)	5.5		27.1	27.3	0.07	65.1

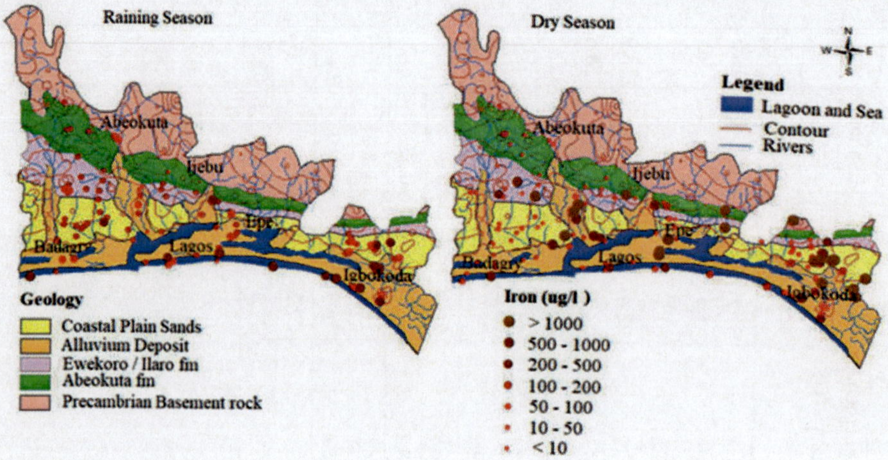

Fig. 22.5 Spatial distribution of Fe^{3+} in $\mu g/L$ (*Source* Aladejana et al. 2020)

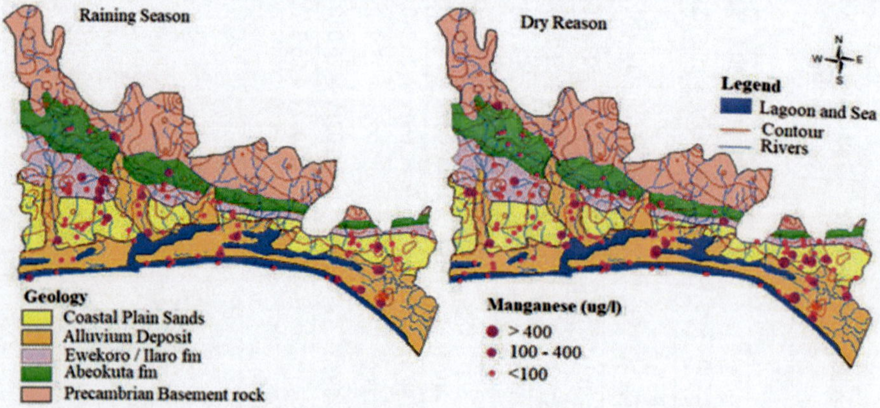

Fig. 22.6 Spatial distribution of Mn^{2+} in $\mu g/L$ (*Source* Aladejana et al. 2020)

city of Osogbo. Their research goals were to define current land use types, investigate water quality variability across land use types, assess surface water quality using the water quality index, and relate water quality variables to World Health Organization (WHO) criteria. According to the findings, the quality of surface water in the research region and its surrounds is extremely bad. As a result, a continuous comprehensive water quality management scheme of surface waters throughout the region and its surrounds should be implemented. The Obafemi Awolowo University, Ile-Ife, dumpsite was investigated geophysically and hydrochemically to determine the possibility of subsurface and surface water pollution. Although the evaluated physiochemical characteristics, excluding turbidity, were within acceptable limits for drinkable water, heavy metal concentrations such as Cd, Fe, and Pb were much higher. As a conclusion, the study's findings suggest that the subsurface and surface water inside and surrounding the trash superfund site are contaminated (Joe-Ukairo and Oni 2018).

Groundwater

In Nigeria, groundwater is a primary component of residential and commercial water. According to Akujieze et al. (2003), Nigeria possesses about 50 million trillion liters of groundwater per year for a demography of 128 million people. In 2001, the domestic water usage requirement was around 6.0 billion liters per year, showing that there is ample water resource capacity (Ezenwaji et al. 2016). Despite the vast amount of water resources in Nigeria. Access to clean drinking water is believed to be constrained, with just 69% of Nigerians having access to fundamental water supply system (Ngene et al. 2021). Despite the country's enormous water sources, which are estimated to be 286.2 km^3/year, these reserves are not evenly distributed across the country and are sensitive to climatic variation (Ngene et al. 2021). The southwestern and northern part of the country is particularly hard hit, with yearly precipitation averaging between 100 and 250 mm and a large population that has culminated in water deficit (Ezeabasili et al. 2014). Additionally, expanding populace and industrialization are exerting additional strain on available groundwater resources, resulting in conflict for water resources throughout many sectors and limited access to water including locations where water is plentiful (Nwankwoala and Ngah 2014; Merem et al. 2017). In accordance with these processes, a radical shift in groundwater management from a standard, segmented strategy to integrated water resources management (IWRM) is required, with the idea that putting its theory into action will result in the appropriate use of groundwater sources (Agyenim and Gupta 2012; Ngene et al. 2021).

Several studies using various integrated models and approaches have been conducted to assess groundwater quality and management in southwestern Nigeria. Using an integrative geological research and hydro-chemical assessment of water samples, Odipe et al. (2018) evaluated the environmental effects on groundwater quality of an open sanitary landfill in Akure, southwestern Nigeria. The integrated results revealed remarkably low total resistance at the subsurface and corroded surfaces, and the hydro-chemical analysis of water samples suggested that the leachate had contaminated the soil and groundwater within the sanitary landfill. In Ajao industrial estate, Nigeria, Ukah et al. (2020) used the entropy-weighted water quality index (EWQI) to investigate the impact of leachate toxic substances on groundwater quality. To effectively determine the level of toxic metal contamination in groundwater, three environmental pollution risk evaluation methodologies were combined. The bulk of the groundwater samples (85.71%) were found to be acceptable drinking water, whereas 14.29% were found to be unfit for use. They proposed that much sensitization campaigns aimed at protecting groundwater in this area be implemented. Certain groundwater samples in Ota, Ogun State, Nigeria, were also tested for quality over the course of three years (2018, 2019 and 2020).

Assessing the samples for human consumption using the water quality index (WQI) revealed that the groundwater is suitable, with WQI < 50, and using the average daily dosage (ADD) and hazard quotient to determine health hazards revealed that the groundwater poses no non-carcinogenic dangers (Enyoh et al. 2021). To represent the proper health of groundwater sources in Ota, regular assessment encompassing more parameters was suggested.

In Abeokuta, Nigeria, Emenike et al. (2018) investigated the degrees of groundwater quality metrics utilized for commercial, household, and consumption uses. The findings indicated that the water quality measures had significant regional differences, with pH and temperature having the lowest variance at 5.9 and 4.5, respectively. The majority of the water samples (71%) were mildly acidic, signifying that complicated rock layers were dissolving. Water standard deviations were in the following order: Na > Mg > Fe > EC, implying connection with sodium-rich rocks. Total dissolved solids concentrations in the water samples were found to be extremely elevated, posing a major potential hazard. The results of the study were considered useful in helping water management agencies comprehend the hydrochemistry of the region's groundwater prospects in handling them efficiently and sustainably. A study used multivariate analyses of Principal Component Analysis (PCA) and Cluster Analysis (CA) to analyze the likely probable contamination contributor to groundwater quality in the foundation rocks of Osun State, South-Western Nigeria. Rusting of source rocks, discharge from sewer systems and landfills, outflow of sediments, rigidity, chemicals from farmlands, and chlorine poisoning have all been cited as probable causes of pollutants to groundwater safety (Awomeso et al. 2020).

Based on land use, Odukoya and Ifarajinmi (2021) evaluated groundwater quality in Lagos, southwest Nigeria, with a focus on health risk evaluation of both crucial and eight potentially toxic elements (PTEs) in groundwater. In 2.94%, 1.5%, 1.5%, 29.4%, 4.4%, and 4.7% of groundwater samples, respectively, the concentration levels of As, Cd, Cr, Pb, Ni, and Fe were significantly greater than guideline values. The pollutant index was used to classify groundwater in the study site, ranging from not contaminated (CI < 1) to very highly polluted (CI > 5). Industrial disposal, untreated sewage, landfills, and sludges were all cited as possible sources of groundwater pollution. The health risk assessment study showed the values less than 1 indicating that majority of the residents were exposed to non-carcinogenic health risk.

Seasonal Change and Water Resources

Despite Nigeria's vast water resources, seasonal change presents a serious risk to the country's water resources, involving variations in rainfall size and influence, aquifer recharge, storms, and drought crises, as well as pollution of surface and ground water resources (Haider 2019; Shiru et al. 2020a, b). Aside from the obvious threat of seasonal change on water availability and fitness for use, assessing the effects of seasonal variation on water resources is much more difficult (Odjugo 2009; Goyal et al. 2018; Hamed et al. 2018). Indirect effects are linked to removal of water from the reservoir and raising flow that might alter hydraulic conductivity, storage space, and water compression rate (Hamed et al. 2018). The extraction of groundwater reserves in southwestern Nigeria is one of the most sensitive issues. As a result, many places in the research area have seen the cumulative influence of arid areas and seasonal change, resulting in significant depletion of groundwater supplies.

The seasonal variation study of groundwater quality was carried out near the Ajakanga garbage dump in Ibadan, Southwestern Nigeria, with the goal of determining its appropriateness for household and irrigation use (Ganiyu et al. 2018). For the physicochemical study, ten groundwater samples were taken in both dry and wet periods. The main ions in all examined groundwater samples were determined to be below the WHO (2007) and NSDWQ (2007) guideline levels. Nevertheless, elevated concentrations of various water quality measures were found in water samples 1 and 10, which could be related to waste diffusion towards the dumpsite's southern portion, proximity to the dumping site, nutrient run, and fertilization (Ganiyu et al. 2018). Badmus et al. 2021 in their recent study employed a combination of topography (ERT) and a groundwater hydrological approach to show the seasonal variance of salinization in Ogun River, which covers seven major villages. The intrusion was found to be more severe during the dry period, owing to a drop in groundwater quantity caused by limited rainfall and probably exacerbated by a high percentage of groundwater extraction. The projected water quality index (WQI) for the rainy and dry periods demonstrates high water quality in the majority of water samples collected in the study region, with the exception of samples collected in the southern part of the study region, which were observed in the poor water quality category (Badmus et al. 2021).

Omopariola and Adeniyi investigated the geographical and seasonal variability in toxic metals (Fe, Zn, Cr, and Ni) contents in groundwater sources in the Ayedaade Area, southwestern Nigeria, using 25 sampling stations. They also evaluated the viability of the researched groundwater sources for domestic use and matched the water quality to the World Health Organization's (WHO) minimum drinking water quality guidelines. The average values of lead were found to be comparatively greater in the rainy period than in the dry period, but the averages of nickel were discovered to be considerably higher in the dry weather than in the rainy season. In summary, the heavy metal concentrations in water were estimated to be greater than the WHO drinking water requirements, indicating a major health hazard as a result of toxic impacts of these metals, and thus the water was unsuitable for the area's residents (Omopariola and Adeniyi 2019). To analyze the groundwater quality of the near waters reservoir of the Eastern Dahomey Basin in southwestern Nigeria, Aladejana et al. 2020 investigated the climatic impact of the changes in biogeochemical activity for the redox-responsive ions and metals Fe^{2+}, Mn^{2+}, SO_4^{2-}, and NO_3. Figures 22.5 and 22.6 show a major distribution map of metals (Fe and Mn) that shows a higher abundance in dry period samples collected compared to wet season water samples. Changes in the strength of hydrochemical processes like as absorption, oxidation, and minerals deposition could be to blame for the greater concentration (Aladejana et al. 2020). Their research found a correlation between climate change and groundwater quality in nearshore basins, as well as the necessity for proactive groundwater management policies and strategy to prevent further degradation.

The presence and seasonal changes of 37 pharmaceuticals relating to 19 medicinal groups were examined in a wastewater treatment facility (Alausa STP) and surface water in Lagos, Southwest Nigeria. During the dry season, the average and median levels of all medicines found in the Odo-Iya Alaro River are much higher than at any season of the year. Concentrations are often higher during low flow conditions, when sewage discharge makes up a larger percentage of river flow. As a result, there were statistical significance variations between the dry and the rainy season's peak (Ogunbanwo 2021). This is perhaps the most comprehensive research of medicines in African rivers till present, and it has revealed their presence at significant levels. Adekunle and Isaac (2021) used descriptive statistics and a covariance matrices approach to analyze the influence of the water quality measures on each other in the Federal University of Agriculture Abeokuta (FUNAAB) setting. The quality of groundwater in the FUNAAB and its environs were found to vary somewhat between the dry and wet periods, according to the findings (Adekunle and Isaac 2021). It was also discovered that 36% of the variables in the dry period covariance dataset had a negative correlation among themselves, while 64% had a positive correlation, contrary to 31% negative and 69% positive in the rainy season dataset. This means that the reliance of the water quality metrics on each other was just 5% different throughout the whole data set of the dry and rainy seasons (Adekunle and Isaac 2021).

The concentrations of trace metals (Pb, As, Ni, Cr, Zn, Cu, and Cd) were evaluated in 72 water samples taken from River Atuwara during the dry and wet seasons of 2018, and a human health hazard evaluation tool, HHRISK code, was used to predict the health hazards related with drinking water from the Atuwara River (Emenike et al. 2020). The levels of trace metals in the river were found to be Cd < Ni < Pb < Cr < Cu < As < Zn in the rainy season and Cd < Pb < Ni < Cu < Cr < As < Zn in the dry period, according to the findings. Contamination of the Atuwara River was attributed to industrial wastes, agricultural practices, and water exploitation. According to the health risk findings, up to 1 in 10 people and 1 in 20 children may develop cancer due to consuming and being exposed to River Atuwara water throughout a lifespan. These findings demonstrated that the loss of sustainable access to safe drinking water in several regions of the country is a genuine and chronic threat that should be addressed (Emenike et al. 2020). Ameloko and Ayolabi used a time-frame study to examine the upward movement of leachate from 17 sampling points surrounding the Olusosun dumpsite in SW Nigeria throughout the rainy and dry periods. The majority of the physicochemical parameters investigated were found to be within the standard limits (Ameloko and Ayolabi 2018). When analyzing the wet and dry period data, the dry period analysis showed a rise in total TDS, EC, temperature, hardness, Ca^{2+}, Mg^{2+}, K^+, Na^+, Cl, NO_3, Zn, Cu, Pb, and Cr, especially in samples near the landfills. In comparison to the rainy season, mean SO_4^{2-}, PO_4^{3-}, Fe, pH, and Ni levels were lower in the dry season analysis. This pattern could be based on the argument that, due to the scarcity of water during the dry season, pollutant buildup in groundwater persisted unabated given the ongoing disintegration of compostable units over time, resulting in a generally heavy proportion of physicochemical properties. The result further revealed that water quality in the research area was far less in the dry season than in the wet season, and that treated water should be prioritized during the dry season prior to ingestion and other household activities to prevent transmitted infections. In summary, the findings of the geological investigation will aid in making judgments about how to better protect groundwater resources in the study region (Ameloko and Ayolabi 2018).

With 45 samples gathered from covered bore holes during dry and rainy seasons, integrated methods of water quality index, multivariate statistical and geospatial techniques were used to investigate the quality of ground water and hydrogeochemical operations in three geographic areas around Lagos, Nigeria, namely Ikorodu, Epe, and Badagry (Akoteyon et al. 2018). During the dry period, 95.6% of the samples were shown to exhibit pH levels underneath the minimal drinking water recommendation. With the exception of one location at Igando, where the rainy season WQI indicates inadequate drinkable water, all of the locations had good water quality during the rainy season. The unsuitability of water in this study area was caused by dumpsite land use. During the rainy season, 22.2% and 24.4% of the samples, respectively, trigger intense and serious degradation. The elevated pH, TDS, TH, and HCO_3^- values in the research area were linked to the serious forms of rusting (Akoteyon et al. 2018). For acceptable groundwater use in the region, a regulation that ensures sufficient security of groundwater contamination from effluent, salt, and human influences is required.

Although a number of research have been performed to mitigate the impacts of seasonal variation in water quality in southwestern Nigeria, as stated above, these studies are still in their early stages, and more data, in terms of field information, is required. This will also help with proper model evaluation for current water quality conditions. However, it is obvious from the aforementioned studies that the seasonal change menace is genuine, and so the repercussions of climate change events are numerous and severe. Future climatic change may have a larger effect on developing countries like Nigeria, whose economy is heavily reliant on agriculture and has been under strain due to rising populace and corresponding needs for energy, freshwater, and foods. Despite the controversy surrounding the exact amount of changing climate and its potential consequences, especially on local levels, steps should be done to foresee, avoid, or reduce climate change's sources and negative consequences.

Conclusion and Recommendation

Nigeria's future water resources are still undetermined as a result of climate change, and water policy makers cannot presume that things will stay the same in the future. Seasonal change must be taken into account in water management. This review is part of a larger effort to comprehend the probable influences of seasonal change and to determine the most pressing research issues. While there is agreement that human activity is increasing, there is little agreement about the anticipated influence on water quality as a result of seasonal variations, particularly changes in intense occurrences. It is, however, not too soon to begin thinking about sustainable adaptability options. Seasonal variation impacts will have to be factored into the design of future freshwater resources, disposal, and treatment processes, and new operational techniques may be necessary. The detection of expanding consequences of changing climate conditions on water quality is one of the research's valuable outcomes. The southwestern zone, particularly Lagos, has faced climate risks defined by stresses on the ecology and increased human density along coastal regions, based on the frequency of severe occurrences. Climate change variables such as excessive rainfall, floods, increased temperature, and greenhouse gases pose a hazard to water quality in these locations. Thus, proper actions must be made to foresee and alleviate the undesirable effects of seasonal variation on water quality.

Seasonal variation effects must be considered alongside air pollution policies, which could contribute to water management via raising awareness among heavy industries in southwestern Nigeria. As amines are used in carbon extraction methods at energy plants, they may increase ammonia discharges, increasing nitrate accumulation and finding their way into water as runoff. As a result, industries must ensure that their industrial activities are well regulated so that they meet standard recommendations while having no negative health consequences later on. Land-use changes and extended growing seasons may boost fertilizer consumption, resulting in leakage into water bodies, rivers, and streams, raising the danger of eutrophication and habitat loss. Most of the risks could be minimized by the use of low-toxicity chemicals and improved management. Regulations including forestry and agricultural control have the potential to lower unhealthy effect such as heavy metals and PCBs, and therefore should be prioritized.

The foregoing studies illustrate that a number of planned and unexpected mitigation initiatives could have serious repercussions for water quality. Consequently, attempts to address detrimental water quality impacts will have to incorporate activities across all sectors of the country that are responsible for managing air, land, and water resources. Seasonal change must also be taken into account in any water management plan aimed at sustainable development principles.

References

Abatan AA, Osayomi T, Akande SO, Abiodun BJ, Gutowski WJ (2018) Trends in mean and extreme temperatures over Ibadan. Southwest Nigeria. Theoret Appl Climatol 131(3):1261–1272

Abuloye AP, Nevo AO, Eludoyin OM, Popoola KS, Awotoye OO (2017) An assessment of effective temperature, relative strain index and dew point temperature over Southwest Nigeria. J Climatol Weather Forecasting 5:192. https://doi.org/10.4172/2332-2594.1000192 Page 2 of 10 J Climatol Weather Forecasting, an open access journal ISSN: 2332-2594 Volume 5 Issue 1 1000192. Figures, 2, 3

Adedeji HA, Idowu TA, Usman MA, Sogbesan OA (2019) Seasonal variations in the physico-chemical parameters of Lake Ribadu, Adamawa State Nigeria. Int J Fish Aquat Stud 7(1):79–82

Adekunle AA, Isaac IU (2021) Multivariate analysis of seasonal variation in groundwater quality within Federal University of agriculture Abeokuta, Southwestern Nigeria

Adeola Fashae O, Abiola Ayorinde H, Oludapo Olusola A, Oluseyi Obateru R (2019) Landuse and surface water quality in an emerging urban city. Appl Water Sci 9(2):1–12

Adejuwon JO, Odekunle TO (2006) Variability and the Severity of the "little Dry Season" in southwestern Nigeria. J Clim 19(3):483–493

Adekunle AA, Isaac IU, Multivariate analysis of seasonal variation in groundwater quality within federal University of Agriculture Abeokuta, Southwestern Nigeria

Adeola OA, Adeyemi O, Onyemaenu V (2021) Rainfall–runoff in conterminous tropical river basins of Southwestern Nigeria. African Geograph Rev, 1–16

Agbaire PO, Obi CG (2009) Seasonal variations of some physico-chemical properties of River Ethiope water in Abraka, Nigeria. J Appl Sci Environ Manag 13(1)

Agyenim JB, Gupta J (2012) IWRM and developing countries: implementation challenges in Ghana. Phys Chem Earth Parts a/b/c 47:46–57

Akande A, Costa AC, Mateu J, Henriques R (2017) Geospatial analysis of extreme weather events in Nigeria (1985–2015) using self-organizing maps. Adv Meteoro. https://doi.org/10.1155/2017/8576150

Akinsanola AA, Ogunjobi KO (2014) Analysis of rainfall and temperature variability over Nigeria. Global J Human-Soc Sci B Geogr Geo-Sci Environ Disaster Manag 14(3):1–17

Akinwumi AM, Adewumi JR, Obiora-Okeke OA (2021) Impact of climate change on the streamflow of ala river, Akure Nigeria. Sustain Water Resour Manag 7(1):1–11

Akinyemi DF, Ayanlade OS, Nwaezeigwe JO, Ayanlade A (2020) A comparison of the accuracy of multi-satellite precipitation estimation and ground meteorological records over Southwestern Nigeria. Remote Sens Earth Syst Sci 3(1):1–12

Akoteyon IS, Balogun II, Soneye ASO (2018) Integrated approaches to groundwater quality assessment and hydrochemical processes in Lagos. Nigeria. Appl Water Sci 8(7):1–19

Akujieze CN, Coker S, Oteze G (2003) Groundwater in Nigeria–a millennium experience–distribution, practice, problems and solutions. Hydrogeol J 11(2):259–274

Aladejana JA, Kalin RM, Sentenac P, Hassan I (2020) Assessing the impact of climate change on groundwater quality of the shallow coastal aquifer of eastern Dahomey basin, southwestern Nigeria. Water 12(1):224

Ameloko AA, Ayolabi EA (2018) Geophysical assessment for vertical leachate migration profile and physicochemical study of groundwater around the Olusosun dumpsite Lagos, south-west Nigeria. Appl Water Sci 8(5):1–15

Awomeso JA, Ahmad SM, Taiwo AM (2020) Multivariate assessment of groundwater quality in the basement rocks of Osun State, Southwest. Nigeria. Environ Earth Sci 79(5):1–9

Ayanlade A, Radeny M, Morton JF (2017) Comparing smallholder farmers' perception of climate change with meteorological data: A case study from southwestern Nigeria. Weather Climate Extremes 15:24–33

Badmus GO, Akinyemi OD, Gbadebo AM, Oyedepo JA, Folarin GM (2021) Assessment of seasonal variation of saltwater intrusion using integrated geophysical and hydrochemical methods in some selected part of Ogun Waterside, Southwest. Nigeria. Environ Earth Sci 80(3):1–14

Balogun II, Sojobi AO, Oyedepo BO (2016) Assessment of rainfall variability, rainwater harvesting potential and storage requirements in Odeda Local Government Area of Ogun State in Southwestern Nigeria. Cogent Environ Sci 2(1):1138597

Bernardo B, Candeias C, Rocha F (2022) Characterization of the dynamics of leachate contamination plumes in the surroundings of the Hulene-B Waste dump in maputo. Mozambique. Environments 9(2):19

Cherkauer KA, Bowling LC, Byun K, Chaubey I, Chin N, Ficklin DL, Williamson TN (2021) Climate change impacts and strategies for adaptation for water resource management in Indiana. Clim Change 165(1):1–20

Clifton CF, Day KT, Luce CH, Grant GE, Safeeq M, Halofsky JE, Staab BP (2018) Effects of climate change on hydrology and water resources in the Blue Mountains, Oregon, USA. Climate Serv 10:9–19

Dallison RJH, Williams AP, Harris IM, Patil SD (2022) Modelling the impact of future climate change on streamflow and water quality in Wales, UK. Hydrological Sciences Journal, (just-accepted)

Dey S, Botta S, Kallam R, Angadala R, Andugala J (2021) Seasonal variation in water quality parameters of Gudlavalleru Engineering College pond. Current Res Green Sustain Chem 4:100058

Ebele NE, Emodi NV (2016) Climate change and its impact in Nigerian economy. J Sci Res Rep 10(6):1–13

Efe SI, Ogban FE, Horsfall MJ, Akporhonor EE (2005) Seasonal variations of physico-chemical characteristics in water resources quality in western Niger Delta region, Nigeria. J Appl Sci Environ Manag 9(1):191–195

Efe SI, Eyefia OA (2014) Urban warming in Benin city, Nigeria. Atmospheric and climate sciences

Egbinola CN, Amobichukwu CA (2013) Climate variation assessment based on rainfall and temperature in Ibadan, South-Western, Nigeria. J Environ Earth Sci 3(11):32–45

Ekute BO, Etim EU (2021) Effect of Ota Industrial Estate effluents on surface water quality of Oruku River, Ota, South Western Nigeria. J Appl Sci Environ Manag 25(9):1671–1675

Elisha I, Sawa BA, Ejeh UL (2017) Evidence of climate change and adaptation strategies among grain farmers in Sokoto State, Nigeria. IOSR J Environ Sci Toxicol Food Technol 11(3):1–7

Eludoyin OM (2016) Air temperature and relative humidity areal distribution over Nigeria. Ife Res Publicat Geogr 10(1):134–145

Emenike PC, Nnaji CC, Tenebe IT (2018) Assessment of geospatial and hydrochemical interactions of groundwater quality, southwestern Nigeria. Environ Monit Assess 190(7):1–17

Emenike PC, Tenebe IT, Neris JB, Omole DO, Afolayan O, Okeke CU, Emenike IK (2020) An integrated assessment of land-use change impact, seasonal variation of pollution indices and human health risk of selected toxic elements in sediments of River Atuwara. Nigeria. Environ Pollut 265:114795

Enete IC (2014) Impacts of climate change on agricultural production in Enugu State, Nigeria. J Earth Sci Climatic Change 5(9):234. https://www.omicsonline.org/open-access/impacts-of-climate-change-on-agricultural-production-in-enugu-state-nigeria-2157-7617.1000234.php?aid=32633

Eni D (2015) Seasonal ARIMA modeling and forecasting of rainfall in Warri Town, Nigeria. J Geosci Environ Protect 3(06):91

Enyoh CE, Enyoh EC, Ohiagu FO (2021) Evaluation of Some Groundwater Sources in Ota, Ogun State, Southwestern Nigeria. World News Nat Sci 36:99–113

Essien-Ibok MA, Akpan AW, Udo MT, Chude LA, Umoh IA, Asuquo IE (2010) Seasonality in the physical and chemical characteristics of Mbo River, Akwa Ibom State, Nigeria. Nigerian J Agric Food Environ 6(1):60–72

Ezeabasili ACC, Okoro BU, Ezeabasili AI (2014) Water resources: management and strategies in Nigeria. AFRREV STECH: Int J Sci Technol 3(1):35–54

Galadima A, Garba ZN, Leke L, Almustapha MN, Adam IK (2011) Domestic water pollution among local communities in Nigeria-causes and consequences. Eur J Sci Res 52(4):592–603

Ezenwaji EE, Eduputa BM, Okoye CO (2016) Investigations into the residential water demand and supply in Enugu metropolitan area. Nigeria. Amer J Water Resour 4(1):22–29
Ganiyu SA, Badmus BS, Olurin OT, Ojekunle ZO (2018) Evaluation of seasonal variation of water quality using multivariate statistical analysis and irrigation parameter indices in Ajakanga area, Ibadan. Nigeria. Appl Water Sci 8(1):1–15
Goyal MK, Surampalli RY (2018) Impact of climate change on water resources in India. J Environ Eng 144(7):04018054
Joe-Ukairo A, Oni AG (2018) Geophysical and hydro-chemical investigations of Oke Asunle dump-site in Ile-Ife, Southwestern Nigeria for subsoil and surface water pollution. J Health Pollut 8(20)
Haider H (2019) Climate change in Nigeria: impacts and responses
Hamed Y, Hadji R, Redhaounia B, Zighmi K, Bâali F, El Gayar A (2018) Climate impact on surface and groundwater in North Africa: a global synthesis of findings and recommendations. Euro-Mediterranean J Environ Integr 3(1):1–15
Ioryue IS, Wuana RA, Augustine AU (2015) Seasonal Variation in water quality parameters of river Mkomon Kwande local government area, Nigeria
Lal M, Sau BL, Patidar J, Patidar A (2018) Climate change and groundwater: impact, adaptation and sustainable. Int J Bio-Resour Stress Manag 9(3)
Laughlin GP, Zuo H, Walcott J, Bugg AL (2003) The rainfall reliability wizard—a new tool to rapidly analyse spatial rainfall reliability with examples. Environ Model Softw 18(1):49–57
Layi-Adigun BO, Ufoegbune GC, Makinde AA, Ahmad SM, Oyelakin JF, Dada VO (2020) Rainfall characteristics in the humid and transition zones of southwest, Nigeria. J Meteorol Climate Sci 18(1):62–69
Ling TY, Gerunsin N, Soo CL, Nyanti L, Sim SF, Grinang J (2017) Seasonal changes and spatial variation in water quality of a large young tropical reservoir and its downstream river. J Chem
Lobell DB, Bonfils C, Duffy PB (2007) Climate change uncertainty for daily minimum and maximum temperatures: a model inter-comparison. Geophys Res Lett 34(5)
Longe EO (2011) Groundwater resource potential in the Coastal Plain Sands Aquifers, Lagos, Nigeria. Res J Environ Earth Sci 3(1):1–7
Lu YL, Song S, Wang RS et al (2015) Impacts of soil and water pollution on food safety and health risks in China. Environ Int 77:5–15
Meludu NT, Akaninyene PO (2021) Analysing the impacts of Covid-19 on water resources and household food security. University-Led Knowledge and Innovation for Sustainable Development, 133
Merem EC, Twumasi Y, Wesley J, Isokpehi P, Shenge M, Fageir S, Nwagboso E (2017) Analyzing water management issues using GIS: the case of Nigeria. Geosciences 7(1):20–46
Murphy BF, Timbal B (2008) A review of recent climate variability and climate change in southeastern Australia. Int J Climatol J Royal Meteorol Soc 28(7):859–879
Ngene BU, Nwafor CO, Bamigboye GO, Ogbiye AS, Ogundare JO, Akpan VE (2021) Assessment of water resources development and exploitation in Nigeria: a review of integrated water resources management approach. Heliyon 7(1):e05955
Nielsen-Gammon JW, Banner JL, Cook BI, Tremaine DM, Wong CI, Mace RE, ... Kloesel K (2020) Unprecedented drought challenges for Texas water resources in a changing climate: what do researchers and stakeholders need to know? Earth's Future 8(8):e2020EF001552
NSDWQ (2007) Nigerian standard for drinking water quality. Nigerian Industrial Standard NIS 554, Standard Organization of Nigeria p 30
Nwankwoala HO, Ngah SA (2014) Groundwater resources of the Niger Delta: Quality implications and management considerations. Int J Water Resour Environ Eng 6(5):155–163
Odekunle TO, Balogun EE, Ogunkoya OO (2005) On the prediction of rainfall onset and retreat dates in Nigeria. Theoret Appl Climatol 81(1):101–112
Odekunle TO (2006) Determining rainy season onset and retreat over Nigeria from precipitation amount and number of rainy days. Theoret Appl Climatol 83(1):193–201

Odipe OE, Ogunleye RA, Sulaiman M, Abubakar SS, Olorunfemi MO (2018) Integrated geophysical and hydro-chemical investigations of impact of the ijemikin waste dump site in Akure, Southwestern Nigeria, on groundwater quality. J Health Pollution 8(18)
Odjugo PAO (2009) The impact of climate change on water resources: global and Nigerian analysis. FUTY J Environ 4(1):59–77
Odukoya AM, Ifarajinmi WT (2021) Assessment of selected major and trace elements in groundwater of Lagos based on land use and implication on human health. Appl Water Sci 11(3):1–11
Offem BO, Ayotunde EO, Ikpi GU, Ochang SN, Ada FB (2011) Influence of seasons on water quality, abundance of fish and plankton species of Ikwori Lake, South-Eastern Nigeria. Fish Aquac J 13:1–18
Ogunbanwo OM (2021) Seasonal variations in the occurrence and distribution of pharmaceuticals in the Odo-Iya Alaro river, Lagos State, Southwest Nigeria. Int J Adv Res Rev 6(2):39–67
Ogunbode TO, Ifabiyi IP (2019) Rainfall trends and its implications on water resources management: a case study of Ogbomoso city in Nigeria. Int J Hydrol 3(3):210–215
Ogungbenro SB, Morakinyo TE (2014) Rainfall distribution and change detection across climatic zones in Nigeria. Weather Climate Extremes 5:1–6
Ojo JS, Olurotimi EO (2014) Tropical rainfall structure characterization over two stations in southwestern Nigeria for radiowave propagation purposes. J Emerg Trends Eng Appl Sci 5(2):116–122
Oke MO, Martins O, Idowu OA, Aiyelokun O (2015) Comparative analysis of groundwater recharge estimation value obtained using empirical methods in Ogun and Oshun River Basins. IFE J Sci 17(1):53–63
Oladipo JO, Akinwumiju AS, Aboyeji OS, Adelodun AA (2021) Comparison between fuzzy logic and water quality index methods: a case of water quality assessment in Ikare community. Southwestern Nigeria. Environ Challenges 3:100038
Olaniyi OA, Ojekunle ZO, Amujo BT (2013) Review of climate change and its effect on Nigeria ecosystem. Int J African Asian Studies-an Open Access Int J 1:57–65
Olowe BM, Oluyege JO, Famurewa O (2016) Prevalence of waterborne diseases and microbial assessment of drinking water quality in ado-ekiti and its environs, Southwestern. Nigeria. Prevalence 10:15
Olufemi AG, Utieyin OO, Adebayo OM (2010) Assessment of groundwater quality and saline intrusions in coastal aquifers of Lagos Metropolis, Nigeria. J Water Resour Prot 2(10):849
Omopariola OA, Adeniyi IF (2019) Statistical investigation of the spatial and seasonal variations of heavy metal contents of groundwater sources in Ayedaade area, Southwestern, Nigeria. J Environ Pollut Hum Health 7(2):100–106
Pickett ST, Cadenasso ML, Grove JM, Nilon CH, Pouyat RV, Zipperer WC, Costanza R (2001) Urban ecological systems: linking terrestrial ecological, physical, and socioeconomic components of metropolitan areas. Annu Rev Ecol Syst 32(1):127–157
Qiuhong TANG (2019) Global change hydrology: Terrestrial water cycle and global change. Sci China Earth Sci 63(3):459–462
Raimi MO, Sawyerr OH, Ezekwe CI, Salako G (2022) Toxicants in water: hydrochemical appraisal of toxic metals concentration and seasonal variation in drinking water quality in oil and gas field area of rivers State, Nigeria
Ramakrishnaiah CR, Sadashivaiah C, Ranganna G (2009) Assessment of water quality index for the groundwater in Tumkur Taluk, Karnataka State. India. E-J Chem 6(2):523–530
Raulino JB, Silveira CS, Lima Neto IE (2021) Assessment of climate change impacts on hydrology and water quality of large semi-arid reservoirs in Brazil. Hydrol Sci J 66(8):1321–1336
Shahid M, Dumat C, Khalid S, Niazi NK, Antunes P (2016) Cadmium bioavailability, uptake, toxicity and detoxification in soil-plant system. Rev Environ Contam Toxicol 241:73–137
Shen M, Huang W, Chen M, Song B, Zeng G, Zhang Y (2020) (Micro) plastic crisis: un-ignorable contribution to global greenhouse gas emissions and climate change. J Clean Prod 254:120138

Shi P, Zhang Y, Li Z, Li P, Xu G (2017) Influence of land use and land cover patterns on seasonal water quality at multi-spatial scales. CATENA 151:182–190

Shiru MS, Shahid S, Dewan A, Chung ES, Alias N, Ahmed K, Hassan QK (2020a) Projection of meteorological droughts in Nigeria during growing seasons under climate change scenarios. Sci Rep 10(1):1–18

Shiru MS, Shahid S, Shiru S, Chung ES, Alias N, Ahmed K, Sediqi MN (2020b) Challenges in water resources of Lagos mega city of Nigeria in the context of climate change. J Water Climate Change 11(4):1067–1083

Sýs V, Fošumpaur P, Kašpar T (2021) The impact of climate change on the reliability of water resources. Climate 9(11):153

Ukah BU, Ameh PD, Egbueri JC, Unigwe CO, Ubido OE (2020) Impact of effluent-derived heavy metals on the groundwater quality in Ajao industrial area, Nigeria: an assessment using entropy water quality index (EWQI). Int J Energy Water Resour 4(3):231–244

Vicuña S, Vargas X, Boisier JP, Mendoza PA, Gómez T, Vásquez N, Cepeda J (2021) Impacts of climate change on water resources in Chile. In: Water resources of Chile. Springer, Cham, pp 347–363

World Health Organization (WHO) (2007) Guidelines for drinking water quality, 3rd edn. WHO, Geneva

Wu WY, Lo MH, Wada Y et al (2020) Divergent effects of climate change on future groundwater availability in key mid-latitude aquifers. Nat Commun 11:3710. https://doi.org/10.1038/s41467-020-17581-y

Ziervogel G, Bharwani S, Downing TE (2006) Adapting to climate variability: pumpkins, people and policy. In Natural resources forum, Vol 30, No 4. Oxford, UK: Blackwell Publishing Ltd, pp 294–305

Ziervogel G, Shale M, Du M (2010) Climate change adaptation in a developing country context: the case of urban water supply in Cape Town. Climate Dev 2(2):94–110

Chapter 23
A Consideration of the Climatic Drivers, Focal Points and Challenges of Soil Erosion, Land Degradation, Landslides and Landscapes in Nigeria

Felicitas U. Iwuchukwu, Ugochukwu Ewuzie, Oluwaseun J. Ajala, Victor E. Ojukwu, Innocent C. Nnorom, Johnbosco C. Egbueri, Chaitanya B. Pande, and Joshua O. Ighalo

Abstract Climate change being the anomaly in the climate system is a major concern in Nigeria. Landslides, landscapes, soil erosion, temperature increment, flooding and drought are proofs of climate change in Nigeria. The climate and environmental impact of its degradation is a popular and significant issue amongst local and international stakeholders. The degradation can be from natural, physical or human factors. Climate change impact differs across various locations in the country such as visible landslides in Niger Delta and erosion gullies in some southern eastern parts of the country like Abia, Anambra and Imo states. These impacts can be grouped under sectoral (agriculture, fisheries, food and water security, forestry, health, economy, energy), demographic (gender, age groups, income groups and occupation), geographic (high precipitation in southwest and southeast regime and low precipitation in the north) and security (communal clashes and conflicts over resources) impacts. When these issues are not addressed, the anomalies can possibly

F. U. Iwuchukwu (✉) · V. E. Ojukwu · J. O. Ighalo
Department of Chemical Engineering, Nnamdi Azikiwe University, P. M. B. 5025, Awka, Nigeria
e-mail: uiwuchukwu@unizik.edu.ng

U. Ewuzie · I. C. Nnorom
Department of Pure and Industrial Chemistry, Abia State University, P. M. B. 2000, Uturu, Nigeria

O. J. Ajala
Department of Industrial Chemistry, Faculty of Physical Sciences, University of Ilorin, P. M. B. 1515, Ilorin, Nigeria

Department of Pure and Applied Chemistry, Ladoke Akintola University of Technology, P. M. B. 4000, Ogbomosho, Nigeria

J. C. Egbueri
Department of Geology, Chukwuemeka Odumegwu Ojukwu University, Uli, Nigeria

C. B. Pande
Center for Climate Smart Agriculture and Water Management, Rahuri and Sant Gadge Baba Amravati University, Rahuri, India

J. O. Ighalo
Department of Chemical Engineering, University of Ilorin, P. M. B. 1515, Ilorin, Nigeria

J. C. Egbueri et al. (eds.), *Climate Change Impacts on Nigeria*, Springer Climate,
https://doi.org/10.1007/978-3-031-21007-5_23

lead to global warming, alternation in rainfall patterns and regimes, devastated farmlands resulting from floods and landscapes which will not stop at creating havoc to Nigerians' livelihood but also affect properties and lives. This research review focuses on the drivers, focal points and challenges of soil erosion, land degradation, landslides and landscapes in Nigeria and how they are affected by the climate change. The challenges encountered in the course of land preservation such as non-implementation of findings, poor funding of research and projects, and poor attitude to environmental degradation issues were also highlighted. Contributions on measures and opportunities for curbing these environmental issues were also reviewed with practical reports of specific occurrences in different regions in Nigeria.

Keywords Climate change · Erosion · Environment · Topography · Nigeria

Introduction

The increasing demand for demographical pressures and treasures which gave rise to disturbing environmental stress and instability in ecology raised concerns for natural resources management. Couple of years ago, land use-land cover changes have taken visible directions. These changes and impacts are more pronounced in countries that are not fully developed as a result of rapid population growth and consequent exploitation of resources. Due to poverty, many people are pushed into overexploiting resources like fisheries, water and forests for survival and afterwards create a huge problem in the long term (Adelodun et al. 2021). Hence, the need for constant auditing with respect to environmental protection and its inhabitants (Abbas and Fasona 2012).

Additionally, there are also the issues associated with domineering effects of climate change. Currently, strategic innovations are needed in the construction of buildings and surroundings that are environmentally friendly. These will help to reduce climate change effects (Akinola et al. 2020). The efforts are not restricted to only buildings but also the total management of the environment, city planning and infusion of environmental resilience strategies (Adelodun et al. 2022) that will subsequently support the environment in overcoming natural disasters.

Soil erosion is a type of surface process that remodels the landscape of the earth and causes environmental problems (Abdulfatai 2014a, b). It is also one of the major contributors of land degradation. Land degradation is the shedding of productive capacity of the soil. It can occur through soil, water and biota. One of the major effects of land degradation is soil organic carbon loss which makes it a big contributor to climate change through emissions increment and carbon sinks reduction. While climate change increases the rate of land degradation via heat stress and variability of rainfall, land degradation reduces carbon sinks. Deforestation and agriculture are major drivers of land degradation through deficient use of agricultural resources, cultivated land expansion and soil loss on lands that are cultivated. The interaction of land degradation and climate change poses a great and urgent challenges for humanity in the ecosystem and its environment with the consequences emphasized on food production, food security and also natural resource conservation.

Landslide is the external and downward movement of the rock, man-made fills and soil which is caused by either natural or human factors (Inyang 2014a, b). Although landslides are naturally occurring, their frequency in some parts is induced by human activities. Some of the natural factors include very heavy storms, volcanic eruptions and earthquakes while physical factors can be inadequate or abandoned drainages and water. Places or regions with high water tables are prone to landslides. Other factors like excessive deforestation in search of woods, cultivation, induced vibration from machinery or traffic and, inadequate road maintenance and repairs also contribute to landslides as they reduce slope stability. All human-induced factors can be regulated with adequate awareness programs. Landscapes are triggered by a lot of causes like abiotic condition variabilities such as climate, topography, past and recent human pattern of settlement and natural disturbance dynamics.

In order to critically evaluate how climate change affects the environment, there is a high need to analyze the causes of vulnerabilities of social, economic, geographical and ecological systems within a country. This will help with the decisions as to the type of interventions and aids a particular country will require such as finance and awareness to successfully adapt to the climatic changes. These vulnerabilities can generally be referred to as soil degradation or land degradation or global environmental problems.

This research review focuses on the drivers, focal points and challenges of soil erosion, land degradation, landslides and landscapes with the focus on Nigeria, specific regions densely affected and how they are affected by climate change, the health hazards and measures to take in preventing them. Relevant literature was properly done and a secondary data was collected. This review provides a comprehensive report on them. The aim is to provide historically an overview of regional landmarks in Nigeria that gave rise to climate change by analyzing how climate change impacts on land issues and vice versa. The paper concludes on measures and opportunities for curbing these environmental issues.

The Status of Land Degradation, Soil Erosion, Landslides and Landscapes in Nigeria

Soil Erosion

Soil erosion is one of the natural occurrences which alters the earth's surface and landscape and constitutes one of the global environmental problems (Abdulfatai et al. 2014; Ighodaro et al. 2013). It involves detachment, breakdown, transport, redistribution and deposition of sediments leading to loss of soil organic carbon, disruption of soil structure and depletion of effective rooting depth in some cases (Lal 2003, 2019). Soil erosion can be classified into two namely; slow geological erosion which gives rise to fertile soil and soil erosion which is due to anthropogenic perturbation leading to degradation of soil fertility and soil structure (Lal 2003; Sartori et al. 2019). It is adjudged the most serious mechanism of land degradation in the tropics (Abdulfatai et al. 2014a, b). Soil erosion has imparted some environmental challenges to the Nigerian state which include but are not limited to loss of farmland, the threat to vegetation, destruction of lives and properties, land infertility and land barrenness due to land detachability, liquefaction and gullying processes (Abdulfatai et al. 2014a, b; Egbueri et al. 2021; Egbueri et al. 2021; Mbah et al. 2016). Igwe (2012) identified factors influencing soil erosion to include topography, climate, geology, soil factors, and vegetation among others.

Land Degradation

Land degradation in Nigeria is one of the most important environmental change processes on land which leads to the loss of valuable land resources (Abbas and Fasona 2012; Thiombiano and Tourino-Soto 2007). This is usually an offshoot of natural factors or different anthropogenic activities that force useful land resources to depreciate to an unfavorable level (Abbas and Fasona 2012; Ajayi 2015; Eni 2012). These natural factors range from water erosion to wind erosion, flood hazard, pollution, land use and tillage methods, and the nature of agro-technologies (Igwe 2012). Another anthropogenic activity that can lead to land degradation may also include the mining of gold and other minerals (Adeoye 2016). Land degradation effects can be very complex due to the strong link with people's health, socio-economic environment pattern and quality (Adewuyi and Baduku 2012). In addition, land use has changed greatly leading to more buildings and fewer and fewer arable lands for cultivation (Akinnagbe and Umukoro 2011; Musa et al.2011; Okorafor et al. 2017b). This has resulted in congestion, food scarcity, erosion, blocked drainage, and deforestation among other negative impacts (Musa et al. 2011).

Landslides and Landscapes

Landslide is the mass movement of debris or earth down a slope (Bamisaiye 2019a, b). The major triggers of this occurrence are both human and natural factors ranging from precipitation, and slope instability to construction, mining, quarrying and excavation (Bamisaiye 2019a, b; Inyang 2014a, b). Among the causes are the slow and steady decrease in rock strength due to fracturing and water escaping into pore spaces and cracks (Bamisaiye 2019a, b). The landscape found in Ukpabi and Nsukka in the South Eastern part of Nigeria is moderate to strongly acidic with the topsoil containing more macronutrients than the subsoil (Ezeaku and Anikwe 2006). Research has shown that landslides normally occur at the beginning of the rainy season characterized by short duration but high-intensity rainfall (Igwe et al. 2014). There have been cases of landslides in Nigeria namely; Umuchiani, Anambra state in 2005, Bayelsa community in 2011 leading to the displacement of over 250 families and the death of two persons, respectively (Inyang 2014a, b).

Focal Points of Climate Change Pressures on Soil Erosion, Land Degradation, Landslides and Landscapes in Nigeria

Land Degradation and Soil Erosion

Land use-land cover changes (LULCC) have become pronounced especially in developing countries because of unprecedented population explosion and subsequent over-exploitation of natural resources good instance of which is the upper Eyiohia river catchment in Nigeria (Abbas and Fasona 2012; Obiahu and Elias 2020b). LULCC plays an important role in detecting emerging spatial patterns and directions of land degradation which will provide useful data for addressing the challenge of land degradation and soil erosion (Abbas and Fasona 2012). Many states in Nigeria are currently under the control of this environmental monster and particularly the south-eastern states (Abdulfatai et al. 2014a, b). And of all the states in southeast Nigeria, Anambra is the worse hit with epicenters in Agulu, Nanka and Oko communities (Abdulfatai et al. 2014a, b).

Mining activities such as gold mining have led to the accumulation of the raw material in the soil which over time will escape the subsurface to the groundwater and so adversely affect plants and animals leading to growth inhibition and liver problems, respectively (Adeoye 2016). The concomitant change in climate, especially in the Sudano Sahel region, has led to variability in rainfall and temperature regimes (Adesina and Odekunle 2011b, a). The Northwest Benin Owena River Basin (NBORB) has undergone serious land cover change as a result of deforestation due to agricultural expansion (Aladejana et al. 2018). This expansion is largely due to poverty and the dependence of locals on natural resources.

Land degradation in peri-urban Kaduna has led to increased cost of farming and the non-biodegradable material is buried in the soil permanently constituting health hazards (Adewuyi and Baduku 2012). Flooding is one of the agents of land degradation which leads to erosion, pollution and removal of vegetation leading to food insecurity in the long or short-run (Adewuyi and Olofin 2014b; Ajayi 2015; Ayinde et al. 2011). The worse hit by this development is states that are located along major rivers like Niger and Benue rivers (Adewuyi and Olofin 2014b). Studies carried out in Iguosa near the Anambra basin revealed that geomorphological characteristics, weakly developed structures, slope steepness, wrong use of land and vegetation cover are responsible for the environmental and climatic changes in the area (Aigbadon et al. 2021a; Ajibade et al. 2020). The unconsolidated nature of the soil, less cohesion, high permeability and low plasticity above and below the clay lithologic units (gliding planes) are responsible for these climatic changes (Aigbadon et al. 2021a). In the southwestern part of Nigeria (Ogun state), land degradation is usually a function of land use patterns hence affecting soil productivity leading to food insecurity and loss of livelihood (Senjobi and Ogunkunle 2011). Land degradation in northern Nigeria is majorly due to the building of dams which has led to the loss of major floodplains leading to a mass exodus of peasant farmers from their homelands (Barbier 2000).

Soil erosion, one of the offshoots of climate change, is reported to be widely spread in the southeastern part of the country, particularly in Anambra state which makes management and conservation techniques difficult and expensive to use (Ajibade et al. 2020; Li and Fang 2016). This is mainly caused by the land clearing and tillage method maximum being observed at sub-watersheds even when graded channel terraces were constructed to minimize the same (Ehigiator and Anyata 2011). This phenomenon is more pronounced in the south-eastern part of Nigeria, particularly Omambala (Anambra) and Imo State with rare cases recorded in other parts of the federation (Ofomata 1985). Gullies which are the most visible type of erosion and have their root cause linked to vegetation, geomorphology, torrential rainfall, which also results from microclimatic changes. is heavily present in the southeast (Igwe 2012). In the southeastern part of Nigeria, the impact of erosion is not the same in the five states, for instance, erosion impact is mild in Ebonyi state than in the likes of Anambra, Abia, Enugu and Imo states (Igwe and Fukuoka 2010). In the southwestern part of Nigeria, the story of the destructive impact of soil erosion is pronounced in the Okitipupa area of Ondo state due to irregular rainfall events (Obasi 2013).

Landslides and Landscapes

Nigeria has a tropical climate with two precipitation regions namely; low precipitation in the North and high precipitation in the Southeast and Southwest (Haider 2019). The advent of climate change has led to extreme weather conditions leading to aridity, drought (Mbah et al. 2016), flooding and erosion (Haider 2019; Igwe 2012; Mbah et al. 2016). Climate change is generally a result of an increase in the concentration of GHGs (Mustapha et al. 2012), HFCs and CFCs (Nzeh and Eboh 2011) in the earth's atmosphere (Idowu et al. 2011). Recently, land degradation has been linked to the loss of biomass due to the combined effects of both natural and human factors (Adenle et al. 2020). It has been shown that the amount of degraded land in Nigeria is as large as the landmass of Ghana (Adenle et al. 2020a, b). One of the places affected by land degradation is the Nigerian Guinea Savannah which has over 52% of it lost to degradation via human activities and natural impacts (Adenle et al. 2020a, b). Also, the Sahelian zone of Nigeria is not spared as there is an overwhelming impact on land use, ecosystem and biodiversity (Mustapha et al. 2012). The natural factors responsible for climate change may include; the lack of eccentricity of the earth's orbit and change in the obliquity of the plain of elliptic and changes in orbital procession (Nzeh and Eboh 2011).

Climate variables such as temperature and rainfall determine the characteristics of the surface and affect fluxes of energy, moisture and particulates that modulate meteorological factors (Adewuyi and Baduku 2012; Audu 2012; Onyenechere 2010). Among the factors responsible for climatic changes in our environment are greenhouse gases emissions from industries and the use of fossil fuels for electricity and automobile (Akinola et al. 2020; Bello et al. 2012). These factors combine and lead to variation in rainfall in some parts of the country with the coastal regions having high amount of rainfalls as opposed to other regions (Akpodiogaga-a and Odjugo 2010; Apantaku et al. 2013; Haider 2019). The health of the persons living close to these areas have been affected by diseases such as malaria, common cold and fever (Apantaku et al. 2013), grossly decreasing the level of agricultural output and undermining food safety efforts (Oyedele et al. 2020) around the world and Nigeria in particular (Apata 2011).

Nigeria has become vulnerable to climate change because of our weather-sensitive or rain-fed (Mbah et al. 2016) agricultural practice (Onwuchekwa Raphael Iheke and Winner Chiagoziem Agodike 2016; Ogbuabor and Egwuchukwu 2017), yet, research has shown that there is a net positive impact of climate change in Nigeria, especially in the southern part of Nigeria (Apata 2011; Ayinde et al. 2011; Bello et al. 2012; Enete and Amusa 2010; Fasona et al. 2013). Although landscapes are negatively affected by natural and anthropogenic factors, yet, simple landscape evaluation and preference studies can be used to shop for those with comparable economic advantage (Chokor and Mene 1992). Struggle for scarce resources has increased as a result of climate variability, and communal and ethnic crises (farmers-herders clash) over the claim for land and mineral resource control (Fasona and Omojola 2005; Sayne 2011). The impact of climate change has affected the livelihood of Nigerians, particularly the Niger Deltans, leading to crop production, livestock production, fisheries, and balance of nature to mention only a few (Abbas and Fasona 2012; Idowu et al. 2011; Nzeh and Eboh 2011).

Landslides are mainly found in the southwestern part of Nigeria (Bamisaiye 2019a, b). The hazard and the risk associated with a landslide in Okemesi, Nigeria, are largely due to lithology, proximity to the fault, presence of abundant foliation planes, slope angle, and slope aspect among others (Bamisaiye 2019a, b). Global and regional climate models show that terrestrial ecosystem transformation is a key cause of climate change (Fasona et al. 2013). The radical interaction of climatic, environmental and socio-economic factors constitutes climatic risks (Fasona et al. 2013). The climate of the Niger Delta region is greatly affected by atmospheric pollution, oil spillage and attendant deforestation (Ituen and Alonge 2009; Jimoh et al. 2011).

Drivers of Land Degradation in Nigeria

Land degradation (that is, landslides, landscapes and soil erosion) in Nigeria is caused by two major factors. These factors are; natural and man-made influences. The climate is the most important natural component, whereas anthropogenic refers to human-made actions. Climate change refers to all long-term changes in climate, whether caused by human actions or natural variations (Gupta et al. 2021). Any considerable shift in climate over time, whether caused by natural variability or human action, is largely regarded as the most serious environmental concern currently facing our planet (Pletterbauer et al. 2018). Unpredictable rainfall; rising temperatures and drought; increased chance of hazards such as floods, landslides, and severe cycloids that can result in hurricanes and typhoons are only a few examples of significant changes.

Nigeria, on the other hand, is divided into two regions: the Northern region and the Southern regions. The map of Nigeria is shown in Fig. 23.1 (Ajala et al. 2020a, b). These two have various climate conditions, resulting in different levels of land degradation, landslides, landscape erosion, and soil erosion from one zone to the other (Prăvălie 2021). Land degradation, landslides, landscape, and soil erosion may be more prevalent in one region than in the other, and vice versa. In addition, a variety of anthropogenic activities are to blame for some of these issues. Anthropogenic is a term used to describe the effects of humans on the environment such as man-made factors that have an impact on climate change (Prăvălie 2021). As a result, both the natural and artificial factors have an influence on each other, causing all of these effects. Different drivers of land degradation, landslides, landscape, and soil erosion would need to be considered across Nigeria's two regions.

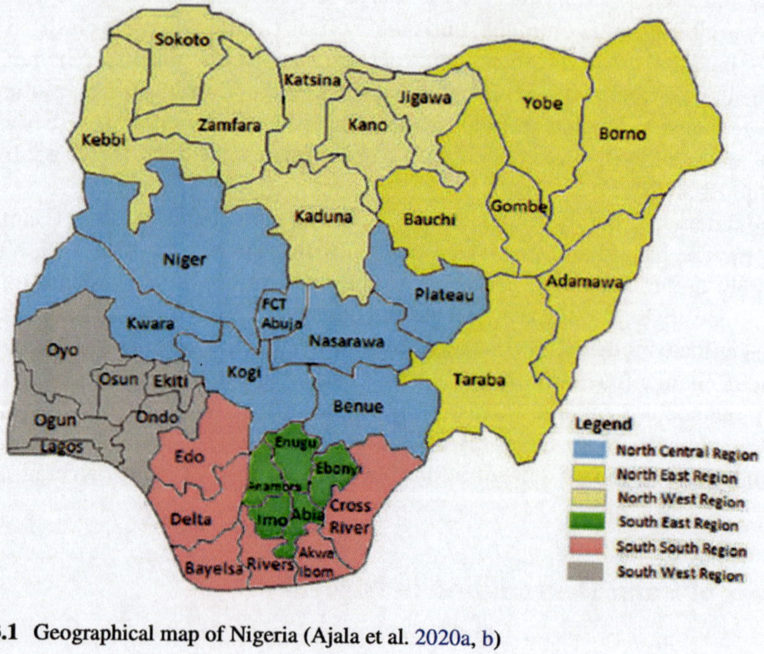

Fig. 23.1 Geographical map of Nigeria (Ajala et al. 2020a, b)

The Northern Region

This region comprises of three geopolitical zones. Namely; North-east, North-west and North-central with six states (Borno, Adamawa, Yobe, Taraba, Bauchi and Gombe), seven states (Jigawa, Kaduna, Kano, Kastina, Kebbi, Sokoto and Zamfara) and six states (Benue, Niger, Nasarawa, Plateau, Kogi and Kwara) respectively (Modibbo et al. 2021). This region has experienced high level of land degradation whereby 80% of their indigenous plant species are either completely disappeared or mitigated toward south regions where climatic conditions are more favorable (Ibrahim et al.2018). This shows that the unfavorable climatic condition has greatly contributed to land degradation. In the equatorial region, the annual precipitation is always on increase while the temperature decreases along the coasts. This can be attributed to the change in tree species diversity and decrease in the tree density which results from land degradation (Ibrahim 2017). The vegetation decrease intensified via the weakening of positive feedbacks between vegetation and precipitation in the midst of evapo-transpiration reduction (Uprety et al. 2019). There are numerous symptoms of land degradation such as; sheet erosion (inter-dunes), wind erosion (dune crests) and sedimentation which have greatly contributed to reduction in productivity of agricultural land and range land (Ibrahim et al. 2018).

The potential evapo-transpiration is 2–4 mm per annum while the peak of the rainfall is 600 mm per annum but in most cases it is as low as 300 mm per annum (Greenwood 2021). The climatic condition in this region is very hot with short rainy seasons between three and five months (June to September) while the dry season lasted for seven to nine months (October to May) (Stojanovic et al. 2020). During the summer months of April and May, for example, the mean monthly maximum temperature in the northeastern city of Maiduguri may approach 100 °F (38 °C), yet frosts may occur at night (Bissolli et al. 2018). The humidity is high in the north, although it drops during the harmattan (the hot, dry northeast trade wind), which blows for more than three months in the north but only two weeks along the Nigerian coast (Ayanlade et al. 2019).

In the other hand, increase in temperature has been reported to positively increase the growth of sugarcane positively, though this has contributed to the need of raise in the demand of water through irrigation and prevalence of disease, weeds and insect pests of sugar cane (Suleiman 2020). The northern area of the forest belt is covered in tropical grassland, which is dotted with baobab, tamarind, and locust bean trees. In the far north, the savanna opens up and is characterized by scattered stunted trees and short grasses (Suleiman 2020). The Lake Chad region has semi-desert conditions, with acacia and doum palm species aplenty (Pfadenhauer and Klötzli 2020). The open savanna in the north is also known for gallery forests (narrow forest zones along rivers). Continuous farming, overgrazing, and bush burning as man-made factors have eliminated vegetation in heavily inhabited portions of the savanna, such as those near the towns of Sokoto, Kano, and Katsina (Pfadenhauer and Klötzli 2020). The nearly complete extinction of plant life in the far north has aided the Sahara's gradual southward advance. Camels, antelopes, hyenas, lions, baboons, and giraffes formerly roamed the plains, while red river pigs, forest elephants, and chimps roamed the jungle (Kurowska et al. 2022). Leopards, golden cats, monkeys, gorillas, and wild pigs were among the animals present in both the forest and the grassland. Today, these species can only be found in protected areas like Bauchi State's Yankari National Park, Taraba State's Gashaka Gumti National Park, Kwara State's Kainji Lake National Park (Chakraborty 2021). The largest family of animals is rodents, which include squirrels, porcupines, and cane rats. Guinea fowl abound on the northern savanna. Quail, vultures, kites, bustards, and gray parrots are among the other prevalent birds. Crocodiles, hippopotamuses, and a wide variety of fish can be found in the rivers (Baker et al. 2020).

The Southern Region

This region comprises of three geopolitical zones. Namely, South-west, South-east and South-south with six states (Ekiti, Lagos, Ogun, Ondo, Osun, Oyo), five states (Imo, Anambra, Abia, Enugu, and Ebonyi) and six states (Akwa-ibom, Bayelsa, Cross-river, Delta, Edo and River), respectively (Ehinomen 2019). The length of the rainy season in the south is extremely high compared to the north which resulted to soil erosion and even flooding which has contributed great hazardous effect on the regions. Several properties worth of billion including lives have been lost as a result of flood. Soil erosion also has caused gullies and rills which makes paddocks cultivation imposable. It has also reduced the soil ability to retain nutrients and water. The rainy season lasts from March to November (Roffe et al. 2020). The south receives more rain, particularly in the southeast, which receives more than 120 inches (3,000 mm) every year, compared to roughly 70 inches (1,800 mm) in the southwest. Rainfall reduces as you get further away from the shore (Ezenekwe 2019). The temperature and humidity are usually stable throughout the year. On the coast, mean monthly maximum temperatures remain consistent throughout the year, hovering around 90 °F (32 °C) in Lagos and 91 °F (33 °C) in Port Harcourt; mean monthly minimum temperatures hover around 72 °F (22 °C) in Lagos and 68 °F (20 °C) in Port Harcourt (Aroh 2021).

Along the coast and in the Niger delta, mangrove and freshwater swamps can be found. The marshes give place to lush tropical rainforests a short distance inland (Elisha and Felix 2021). The oil palm, which is economically significant, grows wild and is typically preserved when the forest is cleared for cultivation. The natural forest vegetation has been replaced by open palm bush in the more heavily inhabited portions of the southeast. Cocoa and rubber plantations have replaced significant sections of the forest in the southwest (Phil-Eze et al. 2000).

Drivers of Land Degradation in Nigeria

Land degradation (that is, landslides, landscapes and soil erosion) in Nigeria is caused by two major factors. These factors are; natural and man-made influences. The climate is the most important natural component, whereas anthropogenic refers to human-made actions. Climate change refers to all long-term changes in climate, whether caused by human actions or natural variations (Gupta et al. 2021). Any considerable shift in climate over time, whether caused by natural variability or human action, is largely regarded as the most serious environmental concern currently facing our planet (Pletterbauer et al. 2018). Unpredictable rainfall; rising temperatures and drought; increased chance of hazards such as floods, landslides, and severe cycloids that can result in hurricanes and typhoons are only a few examples of significant changes.

Nigeria, on the other hand, is divided into two regions: Northern and Southern regions. The map of Nigeria is shown in Fig. 23.2 (Ajala et al. 2020a, b). These two have various climate conditions, resulting in different levels of land degradation, landslides, landscape erosion, and soil erosion from one zone to the other (Prăvălie 2021). Land degradation, landslides, and landscape, soil erosion may be more prevalent in one region than in the other, and vice versa. In addition, a variety of anthropogenic activities are to blame for some of these issues. Anthropogenic is a term used to describe the effects of humans on the environment such as man-made factors have an impact on climate change (Prăvălie 2021). As a result, both the natural and artificial factors have an influence on each other, causing all of these effects. Different drivers of land degradation, landslides, landscape, and soil erosion would need to be considered across Nigeria's two regions.

The Northern Region

This region comprises of three geopolitical zones. Namely; North-east, North-west and North-central with six states (Borno, Adamawa, Yobe, Taraba, Bauchi and Gombe), seven states (Jigawa, Kaduna, Kano, Kastina, Kebbi, Sokoto and Zamfara) and six states (Benue, Niger, Nasarawa, Plateau, Kogi and Kwara) respectively (Modibbo et al. 2021). This region has experienced high level of land degradation whereby 80% of their indigenous plant species are either completely disappeared or mitigated toward south regions where climatic conditions are more favourable (Ibrahim et al. 2018). This shows that the unfavourable climatic condition has greatly contributed to land degradation. In the equatorial region, the annual precipitation is always on increase while the temperature decreases along the coasts. This can be attributed to the change in tress species diversity and decrease in the tree density which results from land degradation (Ibrahim 2017). The vegetable decrease intensified via the weakening of positive feedbacks between vegetation and precipitation in the midst of evapo-transpiration reduction (Uprety et al. 2019). There are

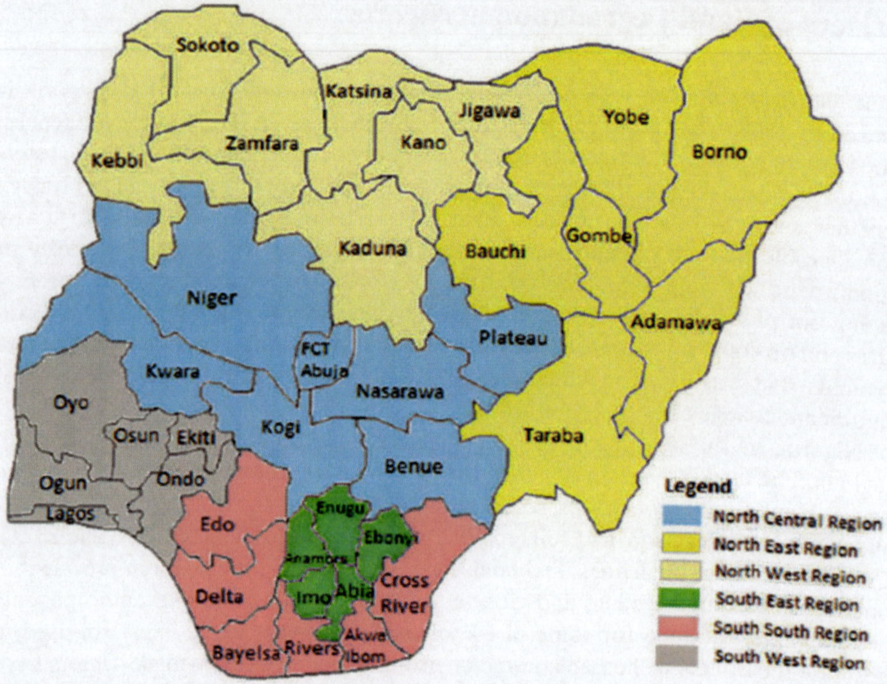

Fig. 23.2 Geographical map of Nigeria (Ajala et al. 2020a, b)

numerous symptoms of land degradation such as; sheet erosion (inter-dunes), wind erosion (dune crests) and sedimentation which has greatly contributed to reduction in productivity of agricultural land and range land (Ibrahim et al. 2018).

The potential evapo-transpiration is 2–4 mm per annum while the peak of the rainfall is 600 mm per annum but in most cases it is as low as 300 mm per annum (Greenwood 2021). The climatic condition in this region is very hot with short rainy season between three to five months (June to September) while the dry season lasted for seven to nine months (October to May) (Stojanovic et al. 2020). During the summer months of April and May, for example, the mean monthly maximum temperature in the northeastern city of Maiduguri may approach 100 °F (38 °C), yet frosts may occur at night (Bissolli et al. 2018). The humidity is high in the north, although it drops during the harmattan (the hot, dry northeast trade wind), which blows for more than three months in the north but only two weeks along the Nigerian coast (Ayanlade et al. 2019).

In the other hand, increase in temperature has been reported to positively increase the growth of sugarcane positively, though this has contributed to the need of raise in the demand of water through irrigation and prevalence of disease, weeds and insect pests of sugar cane (Suleiman 2020). The northern area of the forest belt is covered in tropical grassland, which is dotted with baobab, tamarind, and locust bean trees. In the far north, the savanna opens up and is characterized by scattered stunted trees and short grasses (Suleiman 2020). The Lake Chad region has semi-desert conditions, with acacia and doum palm species aplenty (Pfadenhauer and Klötzli 2020). The open savanna in the north is also known for gallery forests (narrow forest zones along rivers). Continuous farming, overgrazing, and bush burning as man-made factors have eliminated vegetation in heavily inhabited portions of the savanna, such as those near the towns of Sokoto, Kano, and Katsina (Pfadenhauer and Klötzli 2020). The nearly complete extinction of plant life in the far north has aided the Sahara's gradual southward advance. Camels, antelopes, hyenas, lions, baboons, and giraffes formerly roamed the plains, while red river pigs, forest elephants, and chimps roamed the jungle (Kurowska et al. 2022). Leopards, golden cats, monkeys, gorillas, and wild pigs were among the animals present in both the forest and the grassland. Today, these species can only be found in protected areas like Bauchi State's Yankari National Park, Taraba State's Gashaka Gumti National Park, Kwara State's Kainji Lake National Park (Chakraborty 2021). The largest family of animals is rodents, which include squirrels, porcupines, and cane rats. Guinea fowl abound on the northern savanna. Quail, vultures, kites, bustards, and gray parrots are among the other prevalent birds. Crocodiles, hippopotamuses, and a wide variety of fish can be found in the rivers (Baker et al. 2020).

The Southern Region

This region comprise of three geopolitical zones. Namely, South-west, South-east and South-south with six states (Ekiti, Lagos, Ogun, Ondo, Osun, Oyo), five states (Imo, Anambra, Abia, Enugu, and Ebonyi) and six states (Akwa-ibom, Bayelsa, Cross-river, Delta, Edo and River), respectively (Ehinomen 2019). The length of the rainy season in the south is extremely high compare to the north which resulted to soil erosion and even flooding which has contributed great harzards effect on the regions. Several properties worth of billion and including lives has been loss has a result of flood. Soil erosion also has caused gullies and rills which makes paddocks cultivaltion impossble. It has also reduced the soil ability to retain nutrients and water. The rainy season lasts from March to November (Roffe et al. 2020). The south receives more rain, particularly in the southeast, which receives more than 120 inches (3,000 mm) every year, compared to roughly 70 inches (1,800 mm) in the southwest. Rainfall reduces as you get further away from the shore (Ezenekwe 2019). The temperature and humidity are usually stable throughout the year. On the coast, mean monthly maximum temperatures remain consistent throughout the year, hovering around 90 °F (32 °C) in Lagos and 91 °F (33 °C) in Port Harcourt; mean monthly minimum temperatures hover around 72 °F (22 °C) in Lagos and 68 °F (20 °C) in Port Harcourt (Aroh 2021).

Along the coast and in the Niger delta, mangrove and freshwater swamps can be found. The marshes give place to lush tropical rainforests a short distance inland (Elisha and Felix 2021). The oil palm, which is economically significant, grows wild and is typically preserved when forest is cleared for cultivation. The natural forest vegetation has been replaced by open palm bush in the more heavily inhabited portions of the southeast. Cocoa and rubber plantations have replaced significant sections of forest in the southwest (Phil-Eze et al. 2000).

Responses to Land Degradation, Soil Erosion, Landslides, and Landscapes in Nigeria

There is ample evidence from both previous and recent studies on the extent of land degradation in Nigeria (Adenle et al. 2020; Okoye 1998). Evidently, soil erosion is creating several gullies and causing landslides in the country. For instance, in the Southeastern region alone, there are about 2,300 gully erosion sites distributed across Abia, Anambra, Ebonyi, Enugu, and Imo States in approximately 13.0, 30.4, 10.9, 26.1, and 19.6 percentages, respectively (Egboka et al. 2019). Hence, people residing in urban, peri-urban, and rural areas across Nigeria, especially in the Southeast, are conversant with soil erosion, land degradation, and landslides. The case in point is that most residential buildings in these areas are either threatened by erosion or landslides, while the constructed highways are in jeopardy because of the above-mentioned phenomena. Land degradation, soil erosion, and landslides are highly connected and almost always lead to one another. Consequently, any control or eradication measure taken on any of them has a beneficial impact on the others. In addition, managing landscapes promotes appropriate land use and management, ensuring an optimum constellation of land uses and management schemes. This integrated configuration guarantees a steady flow of ecosystem services and offers advantageous prospects to the inhabitants of a given landscape to better their livelihoods.

In Nigeria, empirical information on how land degradation, soil erosion, and landslides are being managed is scarce in the literature. These experiential data are paramount for designing programs needed for environmental management. Much earlier, Okoye (1998) had indicated that lots of current evidence were established for controlling soil erosion, for instance, emanated from developed countries and are not readily practicable, especially by farmers. Regardless, some mitigation techniques to landslides, for example, can be implemented globally even though their mechanisms, influences, and management may be controlled by the area's geomorphology, geohydrology, and geology (Igwe and Una 2019). In addition, various studies on land degradation continue to enumerate several mitigation and adaptation measures in the form of recommendations and proposals instead of instances of what has already been accomplished (Abdulfatai et al. 2014; Adenle et al. 2020a, b). For instance, regarding measures to tackle soil erosion and land degradation, researchers have recommended the planting of vegetation or close-growing grasses such as Bahama (*Cynodon dactylon*) or carpet grasses (*Axonopus compressus*) (Okorafor et al. 2017a); avoiding uncontrolled and unnecessary topsoil removal, continuous cropping, overgrazing, and indiscriminate dumping of refuse, which may cause drainage obstruction

(Abdulfatai et al. 2014a, b; Egboka et al. 2019); creating awareness on the immediate and continuing effects of extreme farming practices that lead to soil erosion (Adewuyi and Olofin 2014a); and enactment of construction laws and implementation of fierce punishment for offenders (Egboka et al. 2019). Other approaches that can be employed, especially for gullies, include the Revetment Panel and Erosion Control bags and Silt Terrace Erosion Prevention inventions (Adesina and Odekunle 2011b, a).

Furthermore, for landslides control and mitigation, it is recommended that landslides be protected with impermeable materials and ensured that surface water is channeled away from them to stem further sliding; application of biotechnical stabilization, a combination of biological and mechanical elements, to prevent and forestall slope failures (Aigbadon et al. 2021b); planting of trees, such as bamboo trees, to stabilize the slopes (Igwe and Una 2019); and government orders to prevent building in landslide-prone areas (Aigbadon et al. 2021b).

Nevertheless, to showcase the strides that researchers, individuals, and government institutions have made in response to these environmental phenomena, we discussed the adopted measures in phases based on the studies that reported the implementation of actual mitigation, eradication, or adaptation approaches to soil erosion, land degradation, and landslides. Figure 23.3 illustrates the different phases, which will assist in shedding light on the successful response approaches, those that need more attention, and the ones that have been effective to combat erosion, land degradation, and landslides.

Fig. 23.3 Conceived actual multifarious response diagram

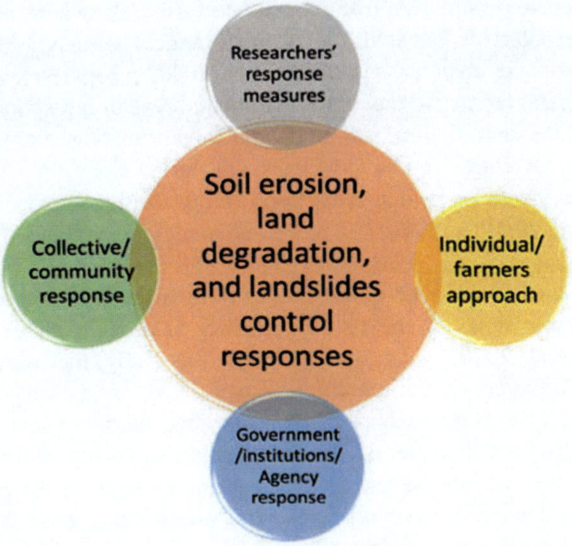

Researchers' Response Measures

This response approach is in the form of gathering information, identifying, and understanding the geotechnical, geomorphological, geological, and hydrogeological features of the areas where soil erosion, landslides, and land degradation are suspected, experienced, or being combated. Research may not be seen as a direct mitigating response to these environmental events, yet it plays a leading role in combating them. Therefore, data from evidence-based research should be among the tools, if not the first, to be included in consolidated mitigation or preventive plan for these environmental occurrences.

Moreover, literature has indicated that soil erosion is the primary cause of land degradation, which results in soil fertility loss (Ajibade et al. 2020; Odunuga et al. 2018). Therefore, in estimating soil loss for determining optimal land protection practices and mitigating land degradation, researchers in Nigeria have used models such as Universal Soil Loss Equation (USLE), Revised Universal Soil Loss Equation (RUSLE), and Soil and Water Assessment, Tool (SWAT), and European Soil Erosion Model (EUROSEM), among others (Egbueri et al. 2022) (Ajibade et al. 2020; Egboka et al. 2019; Obiahu and Elias 2020a). Recently, the combination of Global Information System (GIS) and remote sensing technology with the USLE or RUSLE was brought to the limelight because it offers easy and improved results for soil erosion assessment for various locations (Ajibade et al. 2020; Obiahu and Elias 2020a). Although some artificial intelligence models, such as support vector machines (SVM), artificial neural networks (ANNs), and adaptive neuro-fuzzy inference systems (ANFIS), have been used elsewhere and they depicted encouraging results (Mosavi et al. 2018), their practical use in Nigeria has not been widely reported for soil erosion assessment (Egbueri 2021; Egbueri and Igwe 2021). In the following paragraphs, we have provided actual research responses to combating soil erosion, land degradation, and landslides in various Nigerian locations.

A study conducted in Orashi Basin, Anambra State, Southeastern Nigeria, to evaluate its land susceptibility to erosion indicated that the basin's soil types were mainly loam (27.07%) and sandy clay loam (25.94%) (Odunuga et al. 2018). These soil classes had erodibility factors (K-factor) of 0.70 and 0.48, respectively, which showed a higher susceptibility to erosion. It is noted that as the erodibility factor approaches 1, the land propensity to erosion increases and vice versa (Ajibade et al. 2020). Using the results of this study, the researchers categorized the lands in the communities within this basin as having very low, medium, high, and very high erosion susceptibility after some mathematical calculations (Odunuga et al. 2018). Subsequently, based on land use susceptibility assessment, the light forest areas within the basin were determined to have the greatest erosion susceptibility followed by the built areas. This process has made it easier to create a mitigation commencement plan in the area.

In a related study, Ajibade et al. (2020) utilized the USLE method to estimate soil loss due to erosion in Anambra State. They estimated the rainfall-runoff erosivity (R-factor), slope length component, slope steepness component, K-factor, cover and management factor, and the support practice component. These biophysical factors were instrumental in characterizing the soil properties for integrated soil erosion and degradation control. For example, the R-factor ranging from 2.73 to 3.71 was spatially distributed for clear identification of the areas with low to high erosivity. Similarly, three soil types, Dystric Nitosols, Gleysols, and Calcaric Fluvisols, having varying sand, silt, and clay compositions were determined, which have a direct relationship with the K-factor. Thus, with high sand component and low silt and clay compositions, the K-factor was high and vice versa. As a result, the Central to Southwestern part of Anambra State (with Dystric Nitosols soil type) had the highest K-factor (0.8), Gleysols occurring within the northwestern and a small part of the southwest had 0.06, whereas northeast and southeast areas, bearing the Calcaric Fluvisols, has a K-factor of 0.05 (Ajibade et al. 2020). Again, this study provides very useful information regarding focus areas to consider during actual erosion tackling exercises. Additionally, studying land use land cover changes impact on soil erosion in Afikpo North, Ebonyi State, Obiahu and Elias (2020a) applied the RUSLE parameters combined with satellite remote sensing and GIS to ascertain areas, within the evaluated land use types, that possessed extreme soil erosion losses. As a result, a proper management planning and land prioritization were suggested in advance. These included making ridges across slopes, protecting land-use types with stone bunds, and reforestation of bare lands with suitable trees.

Furthermore, the research response measure for landslides has been carried out in various parts of the country with more visibility in the southeastern and southwestern regions (Aigbadon et al. 2021b; Egboka et al. 2019; Igwe 2015; Igwe and Una 2019). For instance, in Okemesi, Ekiti State, Southwestern Nigeria, a susceptibility map was designed to assist in landslides' early warning and palliative measures (Bamisaiye 2019a, b). This study considered landslide susceptibility factors such as slope aspect, distance to fault, elevation, slope, lineament, and lineament density to configure a landslide susceptibility map for the study area. Hence, following the susceptibility map, high and very high landslide-prone areas were mapped and landscaped. As a result, buildings were advised to be located about 500 m away from such areas, whereas farming activities were cautioned against.

More so, in the same Southwest region, Aigbadon et al. (2021b) studied the landslides characteristics of Iguosa environs, Edo State, to propose suitable mitigation responses to salvage the impacted land areas and forestall future episodes. Therefore, soil samples from the study area were analyzed to determine some geotechnical indices, such as coefficient of permeability, plastic and liquid limits, plasticity index, angle of internal resistance, cohesion, and grain size. Based on the outcome of the geotechnical analysis, the researchers submitted some mitigation measures for landslide-susceptible areas. These include the application of impermeable materials to cover visible landslides, rerouting surface water away from sliding areas, constructing sound drainage systems, locating houses away from the sliding areas, employing bioengineering and biotechnical slope methods, and afforestation.

Additionally, about 120 landslides in the popular Nanka environs, Anambra State, Southeastern Nigeria, were investigated for adequate management options (Igwe and Una 2019). They determined analogous geotechnical indices as done by Aigbadon et al. (2021b), coupled with slope movement, depths, slope angles, elevation, width, and lengths of landslide. Thus, they found a positive association between slope angle and potential slip mass volume, i.e., the areas that possess higher slope angles are likely to have higher potential slip volume, resulting in lower slope stability. The study further discovered that the landslides were majorly shallow with depths ranging between 0.2 and 1.8 m. Hence, the researchers proposed planting of grasses and trees, such as the bamboos, and vegetation turfing to assist in stabilizing the slopes, which is the most vital mitigation measure. Further, there are other similar studies in the Southeast (Igwe 2015; Igwe et al. 2013) detailing geotechnical analysis of soils within landslides areas to establish indices necessary for mitigations responses. More importantly, the preceding instances, therefore, are a clear indication that such analyses are a major mitigation step to take before any physical palliation should be considered. Without this step, wrong responses could be applied, leading to a waste of resources and time, without yielding the desired result.

Individual/Farmers' Approach

Besides recommendations to people regarding erosion control and farmers concerning land restoration, literature exists on actual approaches farmers and individuals have employed to mitigate these environmental events. The approaches utilized by farmers were made known through questionnaires, personal communications, and one-on-one interviews. To illustrate, Iheke and Agodike (2016a) examined farmers' adoption of climate change mitigating options. Therein, soil erosion was denoted as a manifestation of climate change. Thus, the respondents, farmers in Owerri Agricultural Zone, Imo State, confirmed the adoption of various erosion control measures including drainage/flood barrier construction, planting of trees, and zero/minimum soil tillage, which minimized soil erosion, especially during heavy rainfall (Iheke and Agodike 2016b). In another report, the people of Oko and Amucha in Anambra State, Southeastern Nigeria, successfully utilized Bamboo trees to stabilize slopes in their areas (Igwe and Una 2019). Also, in Ogbomoso, Southwest Nigeria, the farmers adopted vetiver grass technology for erosion control because of its affordability, effectiveness, and ease of cultivation (Ewetola et al. 2021).

Further, in a similar study, the farmers in Kogi State, North Central, enlisted soil erosion and poor yield of crops (evidence of land degradation) as consequences of climate change and then enumerated their mitigation/ adaptation responses (Opaluwa et al. 2020). The recruited households adopted the planting of cover crops, changing their planting and harvesting dates, and religious belief and prayers to improve their farm yields. In addition, to mitigate the impact of erosion, they constructed drainage around their homes and farms, used minimum tillage systems, and engaged in reforestation. The majority of the respondents also indicated that they resort to migration, probably when they can no longer control or adapt to the impacts of climate change (Opaluwa et al. 2020).

Additionally, there is a widespread use of carpet grasses (*Axonopus compressus*) and other close-growing grasses in combating slopes and erosion in various parts of Nigeria. Nowadays, individuals purposefully plant these grasses around their homes for both aesthetic and erosion and landslide mitigation reasons. In most rural and peri-urban areas, carpet grasses are seen planted in front of buildings instead of flooring with concrete materials. They also constructed ditches around buildings for easy passage of water to avoid creating gullies that could escalate with time. More importantly, individuals have attested to the efficacy of this mitigation response to slope failures and erosion, particularly against those occurring at small scale around homes.

Collective/Community Response

While individuals can tackle slopes and erosions within their neighborhood, collective efforts yield more satisfying results, particularly, on commonly shared roads, community halls, worship centers, and markets. In most states in Nigeria, environmental sanitation is observed every last Saturday of the month. On such days, people gather and work collectively to clear drainage systems for easy flow of water. They also engage in planting close-growing grasses around worship centers and market squares to combat erosion and slides. Personal experience shows that these collective mitigation responses against land degradation, erosion, and landslides are taken seriously so that no movement of vehicles, opening of shops, or the rendering of any other form of services are allowed during the sanitation hours (usually 8:00–11:00 h.). Moreover, in rural areas, there is a monetary penalty for non-participation in environmental sanitation. All these efforts are actual measures that communities have adopted to mitigate or control any form of these challenges peculiar to them.

Government/Institutions/Agency Response

Most serious and colossal landslides and erosions in Nigeria are being prevented or mitigated by the government in collaboration with institutions and agencies. Specifically, the Federal Government of Nigeria, through the support of the World Bank, has tackled erosion, landslides, and land degradation issues in various states of the federation. These projects have been carried out through the Nigeria Erosion and Watershed Management Project (NEWMAP) (https://newmap.gov.ng/). So far, the response of the government, through NEWMAP, has been encouraging and seems to spread across the six geo-political zones in Nigeria. The success stories of soil

Fig. 23.4 Erosion mitigation measures by the government in **a** Ogbomoso, Oyo state, **b** Idimu, Lagos State, and **c** Ezinihitte Mbaise, Imo State

erosion, landslide, and land degradation prevention and mitigation by NEWMAP is well documented (https://newmap.gov.ng/). More so, Fig. 23.4 shows some of the erosion control measures undertaken or executed by NEWMAP in Oyo, Lagos, and Imo States.

In the management of landslides and erosion, NEWMAP has employed the services of local consultants that have utilized conventional and engineered construction material to reclaim degraded lands by erosion. One of the consultants, Maccaferri Nigeria Limited, provided palliation to gully erosion in Ekehuan Community, Edo State (https://www.maccaferri.com/ng/gully-erosion-in-ekehuan/). They built a concrete drainage system and check dams (at intervals) using Maccaferri gabions made of steel wire and galvanized with an alloy. The gabions reduced the erosion velocity and allowed slow deposition to reclaim the gully site. In addition, they used Reno mattresses and geotextiles to reinforce the gully bed, which in turn aid bioremediation. Figure 23.5 shows the gully site before and after mitigation. Although the government has contributed significantly to controlling these environmental challenges, more is expected. The rapid climate change has induced rapid failures; thus, the pace of government response is expected to match the rate of the land failures.

Fig. 23.5 Ekehuan gully site before mitigation (**a**), the constructed concrete drainage (**b**), the check dam constructed with gabions and engineered materials (**c**), and the revegetated gully

Challenges in Addressing Land Degradation, Soil Erosion, Landslides, and Landscapes in Nigeria

There are many challenges militating against the solutions to these environmental occurrences. Some of the challenges are outside the control of the individuals, community, and government but the bulk lies within the confines of communities/government—attitude issues and not doing enough, especially concerning the mismanagement of funds. The major challenges can be summarized to include the following subsections.

Non-implementation of Research Findings

The preceding section has provided evidence that geotechnical investigations are mandatory for land and landscape restoration. It is also evident that through research, factual reports rather than speculations are vital for cost-effective and timely intervention to soil erosion, land degradation, and landslides. Sadly, most research findings on these all-important issues in Nigeria are, sometimes, not considered or implemented. It appears that most research works/publications in Nigeria on such issues are only for academic purposes such as promotion, recognition and award, and academic achievement. Moreover, most often, the committees set for the implementation of most rural or community projects are less informed and interested in what they would gain; hence, they care less about the nitty–gritty of the project, thereby failing to address the environmental problem. For instance, Igwe and Una (2019) opined that the challenges of landslides in Nanka, Southeast Nigeria, persist due to the inability of the local governments to integrate landslide hazard information into their long-term strategies.

Poor Funding of Research and Projects

Inadequate research funding has been a perennial issue in Nigeria. Ewuzie et al. (2021) decried the fact that researchers in Nigeria are almost always funding their research works, which likely leads to poor research outputs. Such geotechnical studies require huge funds to execute. Funding also includes providing the required equipment to the research institutions like the universities, enabling them to engage in groundbreaking research that would offer panaceas to our environmental challenges. Notwithstanding, non-collaboration among researchers within and between institutions should be discouraged because teamwork encourages sharing of knowledge and research instruments/equipment, forming formidable research partners, and producing quality research outputs (Ewuzie et al. 2021).

On the other hand, executing land restoration, erosion and slides control, and protection or rebuilding of natural landscapes require enormous capital. Most often, these projects are least on the priority list of the sitting government, thereby receiving poor attention and funds. Sometimes, the funds apportioned to these projects are allegedly distributed among the project committee members that only a small fry is left for the project implementation. More so, poor project costing due to inadequate evaluation of the project scale may result in poor budgeting. In all, these unsatisfactory performances tend to jeopardize the brilliantly designed prevention and mitigation plans for environmental challenges.

Employing the Services of Inexperienced Contractors, Resulting in Poor Execution of Projects

The challenging factors in addressing land degradation, soil erosion, landslide, and landscape in Nigeria are interrelated and, evidently, the ripple effect is huge. First hand experience has shown that most drainage systems in some areas in Nigeria are wrecked immediately after construction. Eyewitnesses had even alleged the destruction of gutters and roads while the construction work was still going on. This poor performance can only be attributed to the unsophistication of the contracting firm. Notwithstanding, the process of awarding the contract, in the first instance, is to be faulted. However, experience has equally shown that, sometimes, the panel that should probe such a reprehensible act may constitute or have an alliance with the owners of the contracting firm; thus, meting out less severe punishment or inconsequential penalty.

Further, poor project execution may stem from the non-assessment of the geotechnical, geomorphological, or geological soil components of the area under construction. Probably, the construction is usually based on mere assumptions or experience from past locations. However, it has been noted that an area's geomorphology, geohydrology, and geology largely determine the kind of management to be employed (Igwe and Una 2019). Therefore, awarding of contracts for environmental restoration and management should be based on merit and technical know-how, rather than favoritism.

Poor Attitude to Environmental Degradation Issues

Sustainable agricultural practices by farmers play important roles in the effective restoration or use of farmlands. However, these practices are stimulated by the farmers' perceptions, inclinations, knowledge, and attitudes (Liao et al. 2022; Šūmane et al. 2018). Poor attitudes, therefore, can overshadow farmers' knowledge and proclivity to pay attention to signals revealing land degradation. Sometimes, insufficient land to observe bush fallow contributes to poor land management. In addition, some farmers in the rural areas are not knowledgeable enough about the practice of crop rotation; thus, resort to monocropping and constant tilling of the soil before planting. These attitudes or, probably, inadvertent practices jeopardize the efforts to maintain or restore agricultural farmlands.

Additionally, Nigerians' general attitudes toward environmental maintenance and management are poor. For instance, the rate of deforestation in most cities is a consequence of the notion that the natural forest landscape is evidence of under-development (Ogunbode et al. 2018). Planting of trees, leading to land revitalization, is not the concern of many individuals; thus, destroying the initial well-planned landscapes in many states of Nigeria. Unfortunately, those living in the rural areas and peri-urban centers seem not to be appreciative of their natural landscape; hence, resorting to tree-felling without replanting to protect the environment. Furthermore, residents' involvement in environmental degradation, especially concerning waste disposal into the drainages is a constant scenario. Some residents are in the habit of discharging wastes inside gutters, especially, during rainfall with the hilarious perception that the erosion is swallowing the refuse. This ugly practice clogs drainage systems and forces water out of their channels to further exacerbate land degradation and frustrate degradation abatement.

Finally, the lack of political will and poor government attention to environmental degradation issues is a substantial challenge towards mitigating them. More so, there is always a case of embezzlement of ecological funds, fund diversions, and other corrupt practices that hinder project execution when the contract is awarded. Sometimes, the contractors are owed, resulting in the abandonment of the projects. In addition, there are also allegations of media propaganda for uncompleted projects being commissioned without completion in order to embezzle the remaining project funds. All these factors and more are posing threats to the efforts of maintaining, reclaiming, and rebuilding lands and landscapes.

Opportunities for Addressing Climate Change-Related Environmental Hazards in Nigeria

Interventions on innovative land degradation: Various environmental, social and economic benefits can be obtained from new innovations. Some measures make available options like agroforestry, inclusion of fruit trees plantation beneficial. These plantations enable and support food security during climate change. In addition, biochar application boosts soil water retention capacity and drought resilience (Li 2021).

Concise landscape approach: This approach takes cognizance of varying agro-ecological zones in Nigeria to address the multi-dimensional nature of land degradation.

National plan actions and Climate change policies: Several interventions like intensive landscape approach and early warning systems need to be infused into the national development budget and planning in order to ensure political commitment and resource allocation.

Conclusions

Land use changes and land use caused mainly by degradation of forests, deforestation and agricultural expansions reduce food availability for the overly populated environment, cause increment in green house gas emissions as well as biodiversity loss. Land generally plays a crucial role in climate change as a sink and also as a carbondioxide source. Hence, the need for its proper preservation and protection. Conclusively, it can be stated that enablement of environmental policy, total landscape approach investment, solidification of early warning systems (EWSs), and capacity building will contribute a lot towards addressing the wide multifaceted land degradation issues.

References

Abbas II, Fasona MJ (2012) The human perception of land degradation in a section of Niger delta. Nigeria. Marine Sci 2(5):94–100

Abdulfatai I, Okunlola I, Akande W, Momoh L, Ibrahim K (2014a) Review of gully erosion in Nigeria: causes, impacts and possible solutions. J Geosci Geomatics 2(3):125–129

Abdulfatai IA, Okunlola IA, Akande WG, Momoh LO, Ibrahim KO (2014b) Review of gully erosion in Nigeria causes, impact and possible solutions. J Geosci Geomatics 2(3):125–129. https://doi.org/10.12691/jgg-2-3-8

Adelodun B, Ajibade FO, Ibrahim RG, Ighalo JO, Bakare HO, Kumar P, ... Choi KS (2022) Insights into hazardous solid waste generation during COVID-19 pandemic and sustainable management approaches for developing countries. J Mater Cycle Waste Manag 23(6):2077–2086.https://doi.org/10.1007/s10163-021-01281-w

Adelodun B, Ajibade FO, Ighalo JO, Odey G, Ibrahim RG, Kareem YF, ... Choi KS (2021) Assessment of socioeconomic inequality based on virus-contaminated water usage in developing countries: a review. Environ Res 192:110309. https://doi.org/10.1016/j.envres.2020.110309

Adenle AA, Eckert S, Adedeji OI, Ellison D, Speranza CI (2020a) Human-induced land degradation dominance in the Nigerian Guinea savannah between 2003–2018. Remote Sens Appl Soc Environ 19:100360

Adeoye NO (2016) Land degradation in gold mining communities of Ijesaland, Osun state. Nigeria. Geojournal 81(4):535–554

Adesina FA, Odekunle TO (2011a) Climate change and adaptation in Nigeria: some background to Nigeria's response–III. Paper presented at the international conference on environmental and agriculture engineering

Adesina FA, Odekunle TO (2011b) Climate change and adaptation in Nigeria: Some background to Nigeria's response—III. 2011b international conference on environmental and agriculture engineering, 15, 146–154

Adewuyi TO, Baduku AS (2012) Recent consequences of land degradation on farmland in the peri-urban area of Kaduna Metropolis, Nigeria. J Sustain Dev Africa 14(3):154–193

Adewuyi TO, Olofin EA (2014a) spatio-temporal analysis of flood incidence in nigeria and its implication for land degradation and food security. J Agric Sci 6(2). https://doi.org/10.5539/jas.v6n2p150

Adewuyi TO, Olofin EA (2014b) Spatio-temporal analysis of flood incidence in nigeria and its implication for land degradation and food security. J Agric Sci 6(2):150

Aigbadon GO, Ocheli A, Akudo EO (2021a) Geotechnical evaluation of gully erosion and landslides materials and their impact in Iguosa and its environs, southern Nigeria. Environ Syst Res 10(1):1–17

Aigbadon GO, Ocheli A, Akudo EO (2021b) Geotechnical evaluation of gully erosion and landslides materials and their impact in Iguosa and its environs, Western Anambra Basin. Nigeria. Environ Syst Res 10(36):1–17. https://doi.org/10.1186/s40068-021-00240-6

Ajala OJ, Ighalo JO, Adeniyi AG, Ogunniyi S, Adeyanju CA (2020a) Contamination issues in commercially available water in Nigeria: a mini-review. Sustain Water Resour Manag 6(6):112–122. https://doi.org/10.1007/s40899-020-00478-5

Ajala OJ, Ighalo JO, Adeniyi AG, Ogunniyi S, Adeyanju CA (2020b) Contamination issues in sachet and bottled water in Nigeria: a mini-review. Sustain Water Resour Manag 6(6):1–10

Ajayi A (2015) Land degradation and the sustainability of agricultural production in Nigeria: a review. J Soil Sci Environ Manag 6(9):234–240

Ajibade FO, Nwogwu NA, Adelodun B, Abdulkadir TS, Ajibade TF, Lasisi KH, ... Olajire OO (2020) Application of RUSLE integrated with GIS and remote sensing techniques to assess soil erosion in Anambra State, South-Eastern Nigeria. J Water Climate Change 11(S1):407–422

Akinnagbe OM, Umukoro E (2011) Farmers' perception of the effects of land degradation on agricultural activities in Ethiope East Local Government Area of Delta State. Nigeria. Agriculturae Conspectus Scientificus 76(2):135–141

Akinola AO, Opoko AP, Ibem EO, Okagbue HI, Afolabi AO (2020) Climate change adaptation and mitigation strategies in Lagos, Nigeria: built environment professionals' perspective. Int J Eng Adv Technol 9(3):1273–1282

Akpodiogaga-a P, Odjugo O (2010) General overview of climate change impacts in Nigeria. J Hum Ecol 29(1):47–55

Aladejana OO, Salami AT, Adetoro O-IO (2018) Hydrological responses to land degradation in the Northwest Benin Owena River Basin, Nigeria. J Environ Manage 225:300–312

Apantaku S, Seriki I, Aromolaran A, Apantaku F, Adebanwo A (2013) Climate change and rural households health in Ijebu North East area of Nigeria. Int J Sust Dev World 20(4):302–308

Apata TG (2011) Effects of global climate change on Nigerian agriculture: an empirical analysis. CBN J Appl Stat 2(1):31–50

Aroh J (2021) Soil fertility status and sugarcane growth performance in the mangrove ecosystem of Nigeria sustainable intensification for agroecosystem services and management. Springer, pp 543–613

Audu E (2012) An analytical view of temperature in Lokoja, Kogi State, Nigeria. Int J Sci Technol 2(12):856–859

Ayanlade A, Atai G, Jegede MO (2019) Variability in atmospheric aerosols and effects of humidity, wind and InterTropical discontinuity over different ecological zones in Nigeria. Atmos Environ 201:369–380

Ayinde O, Muchie M, Olatunji G (2011) Effect of climate change on agricultural productivity in Nigeria: a co-integration model approach. J Hum Ecol 35(3):189–194

Baker LR, Che J, Teneke VN, Kadala E, Uba MS, Geoffrey N, Haskainu C (2020) Common hippopotamus in Nigeria: New census data and literature review confirm the conservation importance of sites outside protected areas. Aquat Conserv Mar Freshwat Ecosyst 30(10):1996–2003

Bamisaiye O (2019a) Landslide in parts of southwestern Nigeria. SN Appl Sci 1(7):1–12

Bamisaiye OA (2019b) Landslide in parts of southwestern Nigeria. SN Appl Sci 1(7). https://doi.org/10.1007/s42452-019-0757-0

Barbier EB (2000) The economic linkages between rural poverty and land degradation: some evidence from Africa. Agr Ecosyst Environ 82(1–3):355–370

Bello O, Ganiyu O, Wahab M, Afolabi M, Oluleye F, Ig S, … Abdulmaliq S (2012) Evidence of climate change impacts on agriculture and food security in Nigeria. Int J Agric Forestry 2(2):49–55

Bissolli P, Ganter C, Li T, Mekonnen A, Sanchez-Lugo A (2018) 7. Regional climates. Bullet Amer Meteorol Soc 99(8):S193–S193

Chakraborty SK (2021) Diversity and conservation of wildlife associated with rivers: an eco-ethological analysis riverine ecology, Vol. 2. Springer, pp 287–441

Chokor BA, Mene SA (1992) An assessment of preference for landscapes in the developing world: case study of Warri, Nigeria, and environs. J Environ Manage 34(4):237–256

Egboka BCE, Orji AE, Nwankwoala HO (2019) Gully erosion and landslides in southeastern Nigeria: causes, consequences and control Measures. Global J Eng Sci 2(4). https://doi.org/10.33552/gjes.2019.02.000541

Egbueri JC (2021) Use of joint supervised machine learning algorithms in assessing the geotechnical peculiarities of erodible tropical soils from southeastern Nigeria. Geomechanics Geoeng. https://doi.org/10.1080/17486025.2021.2006803

Egbueri JC, Igwe O (2021) Development of a novel numerical indicator (DLPI) for assessing the detachability and liquefaction potentials of soils in erosion-prone areas. Model Earth Syst Environ 7(4):2407–2429

Egbueri JC, Igwe O, Unigwe CO (2021) Gully slope distribution characteristics and stability analysis for soil erosion risk ranking in parts of southeastern Nigeria: a case study. Environ Earth Sci 80(7):1–19

Egbueri JC, Igwe O (2022) Assessing the role of soil engineering properties in gully development and enlargement in southeast Nigeria using geostatistical and novel indexical techniques. Environ Earth Sci 81:7. https://doi.org/10.1007/s12665-021-101

Egbueri JC, Igwe O, Ifediegwu SI (2022) Erosion risk mapping of Anambra State in southeastern Nigeria: soil loss estimation by RUSLE model and geoinformatics. Bull Eng Geol Environ 81:91. https://doi.org/10.1007/s10064-022-02589-z

Ehigiator O, Anyata B (2011) Effects of land clearing techniques and tillage systems on runoff and soil erosion in a tropical rain forest in Nigeria. J Environ Manage 92(11):2875–2880

Ehinomen C (2019) Creation of states in Nigeria and their financial viability in advancing development. Ife Soc Sci Rev 27(1):60–81

Elisha OD, Felix MJ (2021) Destruction of coastal ecosystems and the vicious cycle of poverty in Niger delta region. J Global Agric Ecol 11(2):7–24

Enete AA, Amusa TA (2010) Challenges of agricultural adaptation to climate change in Nigeria: a synthesis from the literature. Field actions science reports. J Field Actions 4

Eni I (2012) Effects of land degradation on soil fertility: a case study of Calabar South, Nigeria. Environ Land Use Planning, 21–34

Ewetola EA, Fanifosi GE, Ezekiel AA, Adetona AA, Oyewole FM, Onatoye NM, … Akinniku AK (2021) Farmers' awareness of the potential of vetiver grass for soil erosion control on Ogbomoso Agricultural Zone farmlands, south-western Nigeria. Bullet Nat Res Centre 45(1). https://doi.org/10.1186/s42269-020-00483-w

Ewuzie U, Aku NO, Nwankpa SU (2021) An appraisal of data collection, analysis, and reporting adopted for water quality assessment: a case of Nigeria water quality research. Heliyon 7(9):e07950. https://doi.org/10.1016/j.heliyon.2021.e07950

Ezeaku PI, Anikwe MA (2006) A model for description of water and solute movement in soil-water restrictive horizons across two landscapes in south eastern Nigeria. Soil Sci 171(6):492–500

Ezenekwe UR (2019) Nigeria's physical and natural resource endowment

Fasona M, Tadross M, Abiodun B, Omojola A (2013) Some implications of terrestrial ecosystems response to climate change for adaptation in Nigeria's wooded savannah. Environ Dev 5:73–95

Fasona MJ, Omojola A (2005) Climate change, human security and communal clashes in Nigeria. Paper presented at the International Workshop on Human Security and Climate Change, Oslo, Norway

Greenwood N (2021) 4. Weather and climate the Sinai. University of Texas Press, pp 51–69

Gupta P, Singh J, Verma S, Chandel AS, Bhatla R (2021) Impact of climate change and water quality degradation on food security and agriculture. Water Conservation in the era of global climate change. Elsevier, pp 1–22

Haider H (2019. Climate change in Nigeria: impacts and responses

Ibrahim YZ (2017) Vegetation and land cover change in the context of land degradation in sub-Saharan West Africa. University of Leicester

Ibrahim YZ, Balzter H, Kaduk J (2018) Land degradation continues despite greening in the Nigeria-Niger border region. Global Ecol Conser 16:e00505

Idowu AA, Ayoola SO, Opele A, Ikenweiwe N (2011) Impact of climate change in Nigeria. Iranica J Energy Environ 2(2):145–152

Ighodaro ID, Lategan FS, Yusuf SF (2013) The impact of soil erosion on agricultural potential and performance of Sheshegu community farmers in the Eastern Cape of South Africa. J Agric Sci 5(5):140

Igwe C (2012) Gully erosion in southeastern Nigeria: role of soil properties and environmental factors. InTech Open Science, 157–171

Igwe O (2015) The geotechnical characteristics of landslides on the sedimentary and metamorphic terrains of South-East Nigeria, West Africa. Geoenviron Disasters 2(1). https://doi.org/10.1186/s40677-014-0008-z

Igwe O, Fukuoka H (2010) Environmental and socio-economic impact of erosion in Nigeria, West Africa. Int J Erosion Control Eng 3(1):102–109

Igwe O, Mode W, Nnebedum O, Okonkwo I, Oha I (2013) The analysis of rainfall-induced slope failures at Iva Valley area of enugu state nigeria. Environ Earth Sci 71(5):2465–2480. https://doi.org/10.1007/s12665-013-2647-x

Igwe O, Mode W, Nnebedum O, Okonkwo I, Oha I (2014) The analysis of rainfall-induced slope failures at Iva Valley area of Enugu State. Nigeria. Environ Earth Sci 71(5):2465–2480

Igwe O, Una CO (2019) Landslide impacts and management in Nanka area, Southeast Nigeria. Geoenviron Disasters 6(1). https://doi.org/10.1186/s40677-019-0122-z

Iheke OR, Agodike WC (2016a) Analysis of factors influencing the adoption of climate change mitigating measures by smallholder farmers in Imo State Nigeria. Scientific Papers Ser Manag Econ Eng Agric Rural Dev 16(1):213–220

Iheke OR, Agodike WC (2016b) Analysis of factors influencing the adoption of climate change mitigating measures by smallholder farmers in IMO state, Nigeria. Sci. Pap. Ser Manag Econ Eng Agric Rural Dev 16:213–220

Inyang DMP (2014a) The role of landslide education in preventing environmental hazards and promoting health in Niger delta region of Nigeria. Int J Sci Environ Technol. ISSN: 2278–3687 (O)

Inyang MP (2014b) The role of landslide education in preventing environmental hazards and promoting health in Niger Delta region of Nigeria. Int J Sci Environ Technol 3(2):427–433

Ituen EU, Alonge AF (2009) Niger delta region of Nigeria. American Society of Agricultural and Biological Engineers, Climate Change and the way Forward

Jimoh H, Ajewole O, Onotu S, Ro I (2011) Implications of land degradation, reclamation and Utilisations in the oil producing areas of Nigeria: perspectives on environmental sustainability and development. Int J Bus Soc Sci 2(22)

Kurowska EE, Czerniak A, Garba ML (2022) Afforestation of transformed savanna and resulting land cover change: a case study of Zaria (Nigeria). Sustainability 14(3):1160

Lal R (2003) Soil erosion and the global carbon budget. Environ Int 29(4):437–450

Lal R (2019) Accelerated soil erosion as a source of atmospheric CO_2. Soil and Tillage Res 188:35–40

Li Z, Fang H (2016) Impacts of climate change on water erosion: a review. Earth Sci Rev 163:94–117

Liao X, Nguyen TPL, Sasaki N (2022) Use of the knowledge, attitude, and practice (KAP) model to examine sustainable agriculture in Thailand. Regional Sustain 3(1):41–52. https://doi.org/10.1016/j.regsus.2022.03.005

Li L, Zhang Y-J, Novak A, Yang Y, Wang J (2021) Role of Biochar in improving sandy soil water retention and resilience to Drought. Water MDPI

Mbah E, Ezeano C, Saror S (2016) Analysis of climate change effects among rice farmers in Benue State, Nigeria. Curr Res Agric Sci 3:7–15

Modibbo UM, Ali I, Ahmed A (2021) Multi-objective optimization modelling for analysing sustainable development goals of Nigeria: Agenda 2030. Environ Dev Sustain 23(6):9529–9563

Mosavi A, Ozturk P, Chau K-W (2018) Flood prediction using machine learning models: Literature review. Water 10(11). https://doi.org/10.3390/w10111536

Musa S, Adeyemi J, Ifatimehin O (2011) A review of the impacts of changing land use on land degradation in Anyigba. Nigeria. Canadian Soc Sci 7(2):236–243

Mustapha S, Undiandeye U, Gwary M (2012) The role of extension in agricultural adaptation to climate change in the Sahelian zone of Nigeria. J Environ Earth Sci 2(6):48–58

Nzeh EC, Eboh OR (2011). Study of technological and farming systems adaptation to climate change in farming communities of Enugu State, Nigeria

Obasi R (2013) Vulnerability of soil erosion in Okitipupa area of Ondo state, Southwest Nigeria: a climatic problem. Int J Engr Sci 2:226–235

Obiahu OH, Elias E (2020a) Effect of land use land cover changes on the rate of soil erosion in the Upper Eyiohia river catchment of Afikpo North Area. Nigeria. Environ Challenges 1. https://doi.org/10.1016/j.envc.2020.100002

Odunuga S, Ajijola A, Igwetu N, Adegun O (2018) Land susceptibility to soil erosion in Orashi Catchment, Nnewi South, Anambra State, Nigeria. Proc Int Assoc Hydrol Sci 376:87–95. https://doi.org/10.5194/piahs-376-87-2018

Ofomata G (1985) Soil erosion in Nigeria: the views of a geomorphologist. 7 th Inaugural Lecture. University of Nigeria, Nsukka, Nigeria

Ogbuabor JE, Egwuchukwu EI (2017) The impact of climate change on the Nigerian economy. In J Energy Econ Policy 7(2):217–223
Ogunbode CA, Henn L, Tausch N (2018) Context-appropriate environmental attitude measurement in Nigeria using the Campbell paradigm. Environ Dev Sustain 22(3):2141–2158. https://doi.org/10.1007/s10668-018-0281-1
Okorafor OO, Akinbile CO, Adeyemo AJ (2017a) Soil Erosion in South Eastern Nigeria a review. Scientific Res J 5(9):30–37
Okorafor OO, Akinbile CO, Adeyemo AJ (2017b) Soil erosion in South eastern Nigeria: a review. Scientific Res J (scirj) 5(9):30–37
Okoye CU (1998) Comparative analysis of factors in the adoption of traditional and recommended soil erosion practices in Nigeria. Soil Tillage Res 45:251–263
Onyenechere E (2010) Climate change and spatial planning concerns in Nigeria: remedial measures for more effective response. J Hum Ecol 32(3):137–148
Opaluwa HI, Opeyemi G, Eleojo MJ (2020) Determinants of choice of climate change adaptation strategies among rural farming households in lokoja local government Area, Kogi State, Nigeria. J Agribusiness Rural Dev 57(3). https://doi.org/10.17306/j.Jard.2020.01324
Oyedele OA, Akinyemi MO, Kovač T, Eze UA, Ezekiel CN (2020) Food safety in the face of climate change: consequences for consumers. Croatian J Food Sci Technol 12(2):280–286
Pfadenhauer JS, Klötzli FA (2020) Zonal vegetation of the tropical zone with summer rain global vegetation. Springer, pp 179–258
Phil-Eze P, Elekwachi W, Bosco-Abiahu L, Collins HW, Muktar A, Omobolaji OA, Eze IC (2000) Vegetation change detection in the Niger delta region of nigeria using remote sensing and GIS Techniques from 2000 to 2020
Pletterbauer F, Melcher A, Graf W (2018) Climate change impacts in riverine ecosystems. Riverine ecosystem management. Aquatic Ecol Ser 8:203–223
Prăvălie R (2021) Exploring the multiple land degradation pathways across the planet. Earth Sci Rev 220:103689
Roffe SJ, Fitchett JM, Curtis CJ (2020) Determining the utility of a percentile-based wet-season start-and end-date metrics across South Africa. Theoret Appl Climatol 140(3):1331–1347
Sartori M, Philippidis G, Ferrari E, Borrelli P, Lugato E, Montanarella L, Panagos P (2019) A linkage between the biophysical and the economic: assessing the global market impacts of soil erosion. Land Use Policy 86:299–312
Sayne A (2011) Climate change adaptation and conflict in Nigeria: JSTOR
Senjobi B, Ogunkunle A (2011) Effect of different land use types and their implications on land degradation and productivity in Ogun State, Nigeria. J Agric Biotechnol Sustain Dev 3(1):7–18
Stojanovic M, Liberato ML, Sorí R, Vázquez M, Phan-Van T, Duongvan H, ... Gimeno L (2020) Trends and extremes of drought episodes in Vietnam sub-regions during 1980–2017 at different timescales. Water 12(3):813
Suleiman M (2020) The 'War' against corruption in Nigeria: the lesson of Singapore and the neglected strategies. Int J Comparat Law Legal Philosophy (IJOCLLEP) 2(2)
Šūmane S, Kunda I, Knickel K, Strauss A, Tisenkopfs T, Rios IdI, ... Ashkenazy A (2018) Local and farmers' knowledge matters! How integrating informal and formal knowledge enhances sustainable and resilient agriculture. J Rural Studies 59:232–241.https://doi.org/10.1016/j.jrurstud.2017.01.020
Thiombiano L, Tourino-Soto I (2007) Status and trends in land degradation in Africa Climate and land degradation. Springer, pp 39–53
Uprety DC, Reddy V, Mura JD (2019) Climate change and agriculture. Springer

Chapter 24
A Review on the Influence of Rainfall in the Formation and Expansion of Gullies in Southeastern Nigeria

Michael E. Omeka, Johnbosco C. Egbueri, Chinanu O. Unigwe, Johnson C. Agbasi, Daniel A. Ayejoto, and Vincent E. Nwazelibe

Abstract The occurrence and expansion of gullies in southeastern Nigeria has remained an environmental impediment, despite several successive attempts by policymakers and environmental stakeholders in curbing the menace. Climatic variability (variability in temperature and rainfall patterns), soil type and geomorphology are peculiar drivers of gully expansion in southeastern Nigeria. Research findings on gully erosion triggers, quantification, modeling and management have been highlighted. Key triggering mechanisms have been based on the hydrological framework, geomorphology and human factors, while modeling and management approaches have been based on conventional and non-conventional monitoring techniques and crisis management approaches. Based on a review of previous and recent literature, and available statistics, this chapter provides a detailed assessment of climatic variation on the impact of rainfall distribution patterns and intensity on the formation and expansion of gullies in southeastern Nigeria using time series data on rainfall patterns (e.g. trends in annual rainfall, trend of variability and change in seasonal monthly rainfall, variability in the number of rain days, onset and cessation, and trends of the number of rain days and the anomalies) for various periods. A detailed analysis of progress from recently well-observed trends to more hypothetical model simulations and explanations for gully erosion prediction, mitigation, and management with respect to climatic variability are also discussed.

M. E. Omeka (✉)
Department of Geology, University of Calabar, P.M.B. 1115, Calabar, Cross River, Nigeria
e-mail: omeka.ekuru@unical.edu.ng; michaelekuru@gmail.com

J. C. Egbueri · J. C. Agbasi
Department of Geology, Chukwuemeka Odumegwu Ojukwu University, Uli, Anambra, Nigeria

C. O. Unigwe
Department of Geology and Geophysics, Alex Ekwueme Federal University, Ikwo, Ebonyi, Nigeria

D. A. Ayejoto
Department of Chemistry, University of Lagos, Akoka, Yaba, Nigeria

V. E. Nwazelibe
Department of Geology, Albert Ludwig University of Freiburg, Breisgau, Germany

J. C. Egbueri et al. (eds.), *Climate Change Impacts on Nigeria*, Springer Climate,
https://doi.org/10.1007/978-3-031-21007-5_24

Keywords Climate change variability · Southeastern Nigeria · Gully erosion · Sustainable development

Introduction

Rainfall-mediated erosion is widely known as the major trigger of the formation and expansion of rills and gullies globally. Rills occur when runoff water forms small pathways as it moves down a slope. In terms of depth, rills are usually less than 0.3 m deep. When they occur at a depth greater than 0.3 m, they are known as gullies. Gully erosion, mostly caused by rainfall variability occurs as the last stage in the rain splash-sheet-rill-gully erosion processes. Among these stages of erosion, gully erosion is described as the most impactful, as it causes more damage to soil (Abdulfatai et al. 2014). Land degradation by gullies is attributed to a combination of factors such as urbanization, climate change, industrialization, deforestation, population increase, and poor land use and management. Several studies have recorded other rainfall-induced land degradation occurrences, such as landslides, in the sandy terrain of southeastern Nigeria as well as the mountain ranges of northern Nigeria (Igwe 2012, 2014). Among the most extensive and devastating types are those of Northern Nigeria's mountain ranges, characterized by steep metamorphic and igneous slopes. However, in the southeastern parts of Nigeria, low-density landslides, followed by frequent scars of slumps and short-runout movements have been recorded. This has been attributed to the unconsolidated sandy terrain of the region (Igwe 2014). Some other studies have agreed that the response of the area to gully and rills development is chiefly tied to its varying geographical architecture and soil composition (Ogboi and Odeh 2012; Anejionu et al. 2013a, b; Abdulfatai et al. 2014). Lowlands and plains, highlands, and uplands make up the terrain of the area. Due to these variations in landforms and soil type, identifying the chief triggering mechanism for gully and rills expansion has become very dicey and subjective among different researchers. For instance, the highlands, (consisting of cuesta landscapes), tend to be resistant to erosion mainly as a result of the composition of the soil. However, they can serve as channels for the movement or discharge of water that can ruin the architecture of lowlands (Igwe 2014). While much has been made of the causes of gullies and rills as a result of variation in landforms, soil composition, and river flow characteristics, urbanization, industrialization, deforestation, population increase, and poor land use and management, it is important to take into account the possibility that other phenomena such as climatic fluctuations (in rainfall) can also induce the formation and expansion of gullies and rills in southeastern Nigeria. Data regarding rainfall variation, intensity, and quantity have been shown to influence gullies' formation and expansion.

Gully erosion is regarded as a severe environmental issue given its huge impact on the ecosystem and natural environment (Rahman et al. 2009). It is estimated that 1.1 billion soil (in hectares) are known to be lost to gullying globally, accounting for about 56% of the total land mass (Poesen 2011). Data have revealed that gullies and erosion-prone regions in Southeastern Nigeria increased from approximately 1.33% (1,021 km^2) in 1976 to around 3.7% (2,820 km^2) in 2006, making it the country's most afflicted region (The World Bank 2012). There have been many published reports on the impact of gullies as a major trigger of environmental degradation in southeastern Nigeria. These impacts are inexhaustible as there are deleterious to both the environment and the economy of the region. Most literature has reported debris deposition and pollution of aquatic ecosystems caused by incessant runoff from agricultural fields (Anejionu et al. 2013a, b; Abdulfatai et al. 2014). River and stream eutrophication caused by fertilizer runoff from agricultural fields has also been reported (Ogboi and Odeh 2012). Most of the arable lands within the region have been degraded resulting in reduced agricultural sustainability of the area. For instance, the reported incident in Oko village in Anambra State, where extensive gullies have grown into craters, causing danger to over 826 households and taking up most of the arable lands calls for serious concern (Obiadi et al. 2010). These impacts have been mostly traced to the variability in climate change in the last decades (Nnadi et al. 2019). According to empirical data and supporting climate models, there has been variability in the global climate for the past 100 years. According to reports by the Nigeria Meteorological Agency (NIMET 2017), dramatic changes in weather patterns (in terms of flooding and varying rainfall patterns) have been experienced in the past decades. Supporting evidence from data collected from 1941 to 2000 indicated a long-term increase in temperature in some regions. On average, there was an increase in temperature from about 26.2 °C (in 1952) to 27.6 °C over the past years (NIMET 2017).

According to climatic projections, these changes are likely to increase rather rapidly over time (Adeleke et al. 2018; Thakural et al. 2018). Future impacts from climate change are predicted to even get worse due to increasing temperature rise and unpredictability in global rainfall patterns (Ochieng et al. 2016). Because temperature rise and precipitation variability are precursors of climate change variability, it therefore suffices to say that rising temperatures and variability in rainfall patterns are projected to accelerate since global warming is a consequence of increasing anthropogenic activities (land use pattern, bush burning, deforestation urbanization, road construction, etc.) that in turn influences rainfall patterns.

As reported by some researchers, hydrological framework and geomorphology (surface and subsurface water level changes, topography, slope gradients, plant cover), and human activities (road construction, land use practices, and poor surface drainage facilities), can have an impact on gully creation and growth (Igwe 2012; Obiadi et al. 2014; Okoyeh et al. 2014). However, not much research has been carried out on the influence of rainfall patterns in the formation and expansion of rills and gullies in the Nigerian southeastern region. Keeping this in perspective, there is a need for comprehensive statistical data on the rate and pattern of rainfall and temperature changes for adequate environmental management and resource sustainability. Adequate analysis of seasonal changes will serve as the major precursor of this variation within the region. Given the impending climatic variations, it can be projected that arable lands for agriculture and construction in Nigeria, and most especially in the southeastern region of the country will be put under greater strain in the nearest future. According to studies and supporting documents, the average long-term annual rainfall within the Southeastern region from 1986–2016 occurred at 1849 mm per year. The maximum yearly rainfall from 1989 and 1992 was recorded at 1920 mm, while in 1987, the minimum rainfall occurred at 177 mm (Nadi et al. 2019). According to a farmers' perception report on rainfall pattern variability in Ogbaru L.G.A, Anambra State, Southeastern Nigeria, the region has been experiencing downward and upward trends in yearly rainfall patterns (Nadi et al. 2019). These results on the variability in rainfall trend also agree with those of Ifeka and Akinbobola (2015) for a time period between 1991 to 2010 in the area.

It is therefore critical to consider holistically how potential variability in climate would impact the signatures of rainfall patterns in the area and how these changes can in turn influence the expansion of rills and gullies in the region. With Anambra state being one of the most populous states in the region with a rapid increase in population, urbanization, and socio-economic activities, and standing out as one of the states with the highest vulnerability to gullies and floods (Akanwa and Ezeomedo 2018), southeastern Nigeria accounts for the country's most vulnerable area to land degradation activities. Most reports on gully and flood disasters have been mostly recorded in this area. For instance, Agulu-Nanka in Anambra State has recorded some of the most devastating impacts of gullying. About 250 tons per hectare of arable lands have been reported lost to gully erosion (Kalu 2001; Ezeigwe 2015). It is feared that if the Nigerian population continues to persist over time, coupled with increasing socio-economic activities, the southeastern region will likely experience greater seasonal changes, with available arable lands degraded even further, thereby putting more stress on the economy and livelihood of the people. The problem may further be exacerbated by rising temperatures and changes in rainfall patterns.

A relationship can therefore be established between rainfall patterns, climate, and soil type as relevant triggers of gully erosion (Fournier 2000; Morgan 2000) in southeastern Nigeria. The region experiences very high periods of rainfall activities with rainy periods spanning for as long as eight months with only four months of the dry season (Inyang 2000; Ezechi 2000). The intense and extensive rainfall activities together with the reported weak lithological units within this area increase its vulnerability to gullying processes. The lithology of the southeastern region is dominated mainly by unconsolidated sand units. Some studies carried out in this region have agreed that the morphology of gully erosion has a strong relationship with the soil type and the morphology and architecture of gullies follow the trend and intensity of rainfall flow (Akanwa and Ezeomedo 2018).

In developed countries of the world with adequate data distribution networks, data on climate change variability is well documented. However, in developing countries like Nigeria for instance, where the impacts of climate change are greatly felt, there is a lack of data on seasonal variation (Ziervogel et al. 2006). The influence of seasonal variability in Nigeria, stemming from climate change projections, especially in the ravaged southeastern region is grossly inadequate or absent in the literature. Therefore, in this chapter, a holistic assessment of the variation in rainfall patterns and intensity of the expansion of gullies has been provided. Based on a review of recent literature and available statistics, this chapter provides a detailed assessment of climatic variation on the impact of rainfall distribution patterns and intensity on the formation and expansion of rills and gullies in southeastern Nigeria using time series data on rainfall patterns (e.g. trends in annual rainfall, trend of variability and change in seasonal monthly rainfall, variability in the number of rain days, onset and cessation, and trends of the number of rain days and the anomalies) for various periods. An analysis of the climate change perception on rainfall patterns by different population sizes within communities in this region has also been discussed. A detailed analysis of progress from recently well-observed trends to more hypothetical model simulations and explanations are also discussed. The first section discusses a review of the occurrences of gully erosion in Southeastern Nigeria concerning seasonal changes, geology, soil type, and topography. The second part takes into account previous and expected future trends from various empirical evidence of impacts of rainfall patterns on gully development; to know if recent variations are within typical climatic variability. Considerations on the impacts of climate change variability on gullying and rills expansion as well as future research prospects towards environmental protection and sustainability have been discussed as well.

Study Area Description

This review was focused on the southeastern region of Nigeria. The region is made of five (5) major states including Enugu, Anambra, Abia, Ebonyi, and Imo (Fig. 24.1), and is located between 4° 47′ 35" N and 7° 7′ 44" N latitudes, and 7° 54′ 26" E and 8° 27′ 10" E longitudes. The area is approximately 29,095 square kilometers; occupying 3.19% of Nigeria's total area. According to reports from the 2006 National population commission (NPC 2006), the region can be described as densely populated with a total population density of 16.32 million. The area consists of a higher ratio of males to females with the male population estimated at 8.31million (50.92%) while the female population composes of 8.01million (49.08%) (NPC 2006). Relatively high annual temperature ranges are observed in this region (Nwagbara and Ibe 2015). The average daily maximum temperature occurs at about 27 °C annually. Between February and April, the temperature is usually higher. The mean daily maximum temperature usually varies in the region with temperature increases from the southern parts to the north, attributable to the variation of the southern Atlantic Ocean winds (Nwagbara and Ibe 2015; Nnadi et al. 2019). Higher rainfall intensities occur within the southern parts compared to the north, with the southern parts recording rainfall intensities as high as 4000 mm while the north records intensities of 1500 mm. Due to the high rainfall activities the region is usually heavily vegetated; with vegetation

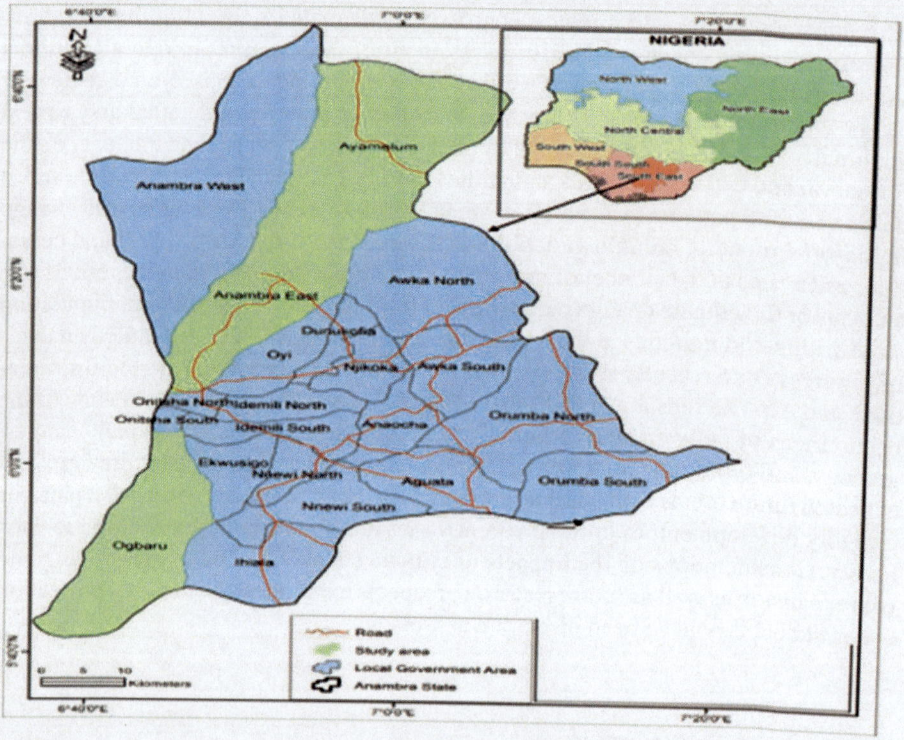

Fig. 24.1 Map of southeastern Nigeria showing the study areas. (Modified after Nnadi et al. 2019)

typical of tropical rainforest. Although within the northern fringes of the area, the vegetation is characteristic of that of the Guinea Savanna (Nwagbara and Ibe 2015). The soil type in the area can be said to be influenced by the climate, as the soils are known to be leached with a conspicuous reddish-brown colour (Chukwu et al. 2009a, b).

The major erosion sites in the study area that are notorious for their dimension and destruction are located in the southern parts of the region dominated by derived savanna. These sites are found mainly in the following towns of southern parts: Agulu, Aguata, Anaocha, Idemili North, Idemili South, Nnewi North, Nnewi South, Ihiala, Udi, Nnewi, Ideani Oraukwu, Oko Nkpor, Ekwulobia/Oko, Enugu, Ukehe, Ezimo, Obollo-Afor, Amucha, Njaba, Isiukwuato, Njikoka, and parts of Orumba South LGAs. These areas record higher annual rainfall with total annual rainfall averaging at over 2062 mm; as a consequence, gullying activities are more pronounced in these areas (Ofomata 1988; Egboka 2004; Igwe 2012; Nwagbara and Ibe 2015; Ifeka and Akinbobola 2015).

Review of Prevalent Factors Influencing the Occurrences and Expansion of Gullies in Southeastern Nigeria

Variation in Rainfall Patterns, Geomorphology, and Soil Type

A study carried out in Akpo community in Aguata LGA of Anambra State revealed that the morphology of gully erosion has a strong relationship with the soil type and the morphology and architecture of gullies follow the trend and intensity of rainfall flow (Akanwa and Ezeomedo 2018). Using information from satellite imageries, a dendritic pattern characterized by the irregular spread of branches of a network in a V-shaped pattern was observed. This observation was verified by field observations. The V-shaped channels were revealed to be composed of high sand contents. The poor cohesive strength of the sands resulted in the inability to support elevated vertical sidewalls resulting in a high tendency of collapse (Egboka and Okoyeh 2019; Akanwa and Ezeomedo 2018). This architectural framework is an indication that the surface soils show higher susceptibility to erosion compared to subsurface soils as a result of surface runoff giving rise to high overland flows that dissect through channels (Kirkby and Bracken 2009). In a recent study carried out on the spatial variability of soil loss using soil properties and different erosion causative indices in Anambra state, rainfall was identified as the highest contributor to soil loss with a total soil mass of 1,1134.7 ton/ha/yr loss compared to other indices such as soil moisture index, land surface, temperature, and normalized difference vegetation index (Ayadiuno et al. 2022). This study further identified areas within the southern parts of the region as major vulnerable areas to gully erosion. Due to the prevalence of higher grounds in the southern region, the impact of rainfall seems to be more pronounced within these parts compared to the northern fringes dominated by plains (Ayadiuno et al. 2022).

There is also a relationship between the lineament density, topography, soil moisture index, and rainfall intensity concerning gully expansion. Ordinarily, gully erosion from rainfall is supposed to be higher in areas such as Aguleri, Umueze, and Anam, which are confined by large watersheds, and streams, with clay materials making up the soil type compared with the areas within the upland such as Agulu, Awka, and Enugu-Ukwu with more porous soil types with higher infiltration capacity. However, due to high elevation within these areas, more aggressive runoffs and gullying are experienced (Ayadiuno et al. 2022). The implication of this is that topography is also an important index in gully development. Findings by Ufot et al. (2016), revealed that soils within this region have reduced organic matter content, low base, and high permeability, making them vulnerable to increased erosion (Fig. 24.2).

It is important to note that a relationship also exists between lineament density and gully expansion as a result of rainfall. Soils within the southeastern region of the country are known to be highly friable with lower cohesion and clay content. Due to the friable nature of the soils in this region, the development of fault lines is a common feature (Okoyeh et al. 2014). The fault lines can serve as pathways for runoff. A study

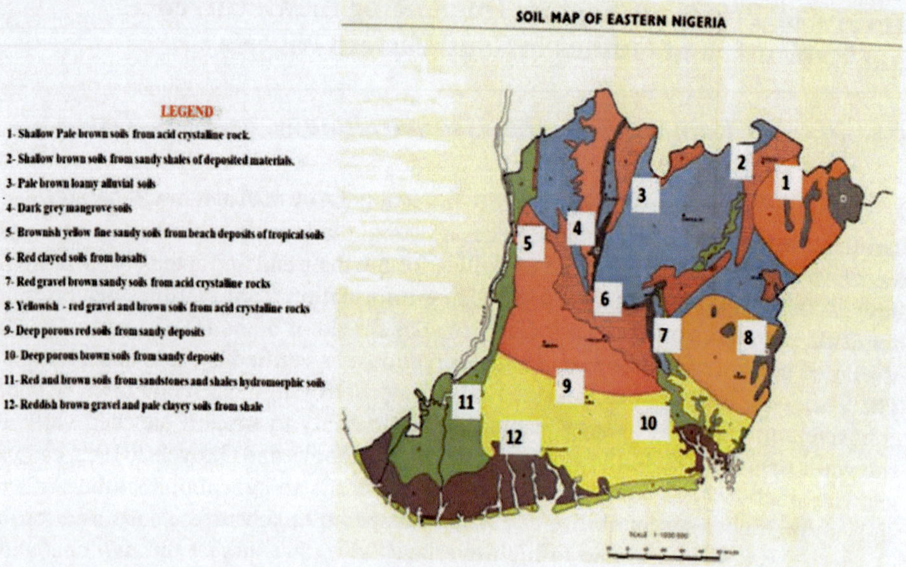

Fig. 24.2 Map showing the distribution of soils in Southeastern Nigeria. (*Source* FDLAR 1990)

by Ayadiuno et al. (2022) revealed that faults are more prevalent in the erosion-devastated areas of southeastern Nigeria. Igwe and Egbueri (2018) and Hong et al. 2006) agreed that soil shear strength has a relationship with their erosion potential. The study by Ayadiuno et al. (2022) showed a strong correlation between erosion-prone areas and fault lines especially in areas around Ozubulu (Ekwusigo LGA), Ihiala, Okija, Nnewi North, Nnewi South, and Ogbaru LGA, as gullying processes seemed to increase towards areas with increasing fault lines. Their study also reported positive linearity between erosivity density, topography, and lineament density. Areas observed to occur within the upland (e.g. Awka, Onitsha North, Anaocha, Ihiala, Nnewi North, Nnewi South, and Aguata LGAs) characterized by high lineament density with friable soils seemed to have high erosivity density with high gully features compared to areas within the plains like Anambra East, Ayamelum, Anambra West, Ogbaru, and pockets of areas within Awka North LGAs.

Geology and Soil Characteristics

The subsurface geology of southeastern Nigeria is composed of diverse components such as beach deposits, basement complex rocks, mangrove swamp deposits, coastal plain sands, shale-sandstones, as well as alluvium deposits (FDLAR 1990). Soils within southeastern Nigeria are characteristically heterogeneous; consisting of loosely compacted red-earth materials associated with clayey-loam, shales, and sandstones (Egede 2013). The shales and sandstones make up the parent materials of the soils (Ezemonye and Emeribe 2012). The shales are known to be highly porous, acidic, and associated with low organic matter. The low organic matter content is attributed to high leaching as a result of the high rainfall intensity that is characteristic of the area. The influence of geology on gully expansion in southeastern Nigeria has been widely documented. Igwe (2012) noted that due to the sandy geological formation of the southeastern region, the worst cases of gully are recorded compared to other sub-Saharan African regions with shale formations. The geology of southeastern Nigeria is characterized by coastal plain sands and false-bedded sandstones especially in areas around Nanka and Bende Ameki. It has been observed that the characteristics of the subsurface geology determines the properties of the surface soils and bears the intrinsic behaviour of the parent material (Nwajide 2013). How the soil behaves during erosion depends on the formative parent material. Igwe (1999) noted that surface soils from igneous, limestone and dolomite origin were more resistant to gully erosion than their clay and sandstone counterparts. Conversely, well sorted soils (comprising fine and large particles) tend to be more resistant to erosion; as greater force will be required to transport the large particles while the cohesive properties of the fines will inhibit further transport of materials.

Anthropogenic Factors

Despite factors such as topography and soil type acting as some of the major natural triggers of gully erosion, land use patterns and improper farming activities can be considered to exacerbate gully expansion in southeastern Nigeria. Bush burning as a result of continuous farming tends to reduce the vegetation cover, thereby exposing the surface soil to runoff. The felling of trees coupled with the high rainfall events results in soil detachment and gullying processes (Egede 2013). In Ukpor community, in the Nnewi South local government area of Anambra state, gully erosion expansion has been attributed to bush burning, tree felling and quarrying activities, which are common practices within the area (Ubuoh et al. 2013).

Climatic Factors

According to FAO (1990), soil erosion is caused by the action of heavy rainfall on surface soil elements, which diminishes or changes the vegetation cover. Rainfall erosivity resulting in huge quantities of runoff (that commence the formation of streams and channels giving rise to gullies and expansion) may be attributed to high rainfall intensity. The impact of climate change variability on rainfall distribution patterns has been assessed in Anambra state using monthly and annual meteorological time series data for a period of 31 years (Nnadi et al. 2019). The significant trends in climate change were also assessed using parametric and non-parametric statistical tests and compared with farmers' respondents' perceptions of rainfall variability for the same period. Results from both statistics and farmers' respondents agree that high variation occurred, as there seemed to be significant differences in rainfall patterns such as rainfall onset, early cessation of yearly rainfall (resulting in the alteration in patterns of growing seasons), as well as regular flooding and drought in recent times than the past thirty years. These findings are in agreement with Onyeneke et al. (2018) who noted that the variation in climatic conditions in the southeastern region during the last decade has been manifested through the irregular rainfall pattern such as early halt in rainfall as well as incessant flooding and droughts compared to three decades ago. These findings can be correlated with the increasing incidences of gully processes in this region as reported in some works of literature (Igwe 2012; Egede 2013; Abdulfatai et al. 2014; Ezeabasili et al. 2014). They have been an observed increase in the expansion of gullies during the wet season due to runoff streams formed during this time. A report by the National Emergency Management Agency (NEMA 2018) agreed that the incessant issue of the ever-increasing floods, storms, and rainfall intensity, in Anambra State in the last decades can be traced to climate change variability. Apparently, these processes are the major divers of gully formation and expansion within the same period. Many studies have agreed that there is strong positive linearity existing between soil loss due to erosion and rainfall intensity, the total amount of ground cover, and runoff, accounting for 86% of erosion variability (Ofomata 2000; Othieno and Laycock 2007).

The duration of rainfall experienced in this region is another important parameter to consider as a triggering force of rills and gully expansion. In this context, there exists linearity between rainfall patterns, soil type, and climate change (Fournier 2000; Morgan 2000). Many works of literature have reported very high rainfall period occurrences in southeastern Nigeria; with rainfall periods persisting as long as eight months (Ezechi 2000; Ofomata 2000; Obiadi et al. 2010; Igwe 2012). The intense and extensive rainfall activities together with the unconsolidated soil and weak lithological units within this area enhance gullying processes. Based on farmers' perception of gully erosion factors in the Akpor community in Aguata LGA of Anambra state, 63.2% of the farmers agreed that rainfall is responsible for gully erosion in the area, while 3.16%, 6.32%, 16.8%, and 10.0% of the farmers attributed gully erosion occurrences to erodibility, agricultural activities, deforestation, and soil type respectively (Akanwa and Ezeomedo 2009).

Observed Climate Variability in Southeastern Nigeria

In the context of time and space, climate variability involves a deviation in the average state of the climate over some time. The period includes measurable continuous variables such as months, seasons, and years, and these variables can be measured either on a short or long-term basis. Variations in climate have had their toll on the rainfall patterns and temperature in Nigeria (Nwaerema and Ajiere 2020). The rising temperature for one has elevated the rate of evapotranspiration and as a consequence, increased the amount of condensation, resulting in high precipitation and flooding events in the country during the past decades. According to the bulletin report from the Nigeria Meteorological Agency on climate change (NIMET 2017), global warming has increased the atmospheric water holding capacity thereby resulting in the increased precipitation events; and these changes seem to be pronounced in the southeastern region of the country. The chief triggers of global warming have been widely attributed to greenhouse gases with carbon dioxide as the major driver. However, these temperature rises seem to vary from one region to another. Evidence of spatial temperature variation in Nigeria have been observed by many studies (Oguntunde et al. 2012; Akinsanola and Ogunjobi 2014; NIMET 2017). These studies have shown that over the last century there has been an upsurge in air temperature, evidenced in varying phases of cooling and warming. Therefore, to have a comprehensive overview of the pattern of the present and future climate projections, it is critical to understand the trend in temperature and rainfall variation in the country especially in the southeastern region where the impacts seem to be more overwhelming.

Relatively, the rate of rainfall in the last decade in the southeast seems to be higher than typical compared to the last three decades. Consequently, this has been simultaneously accompanied by rising temperature in the region attributable to global warming (Onyeneke et al. 2018; Nwaerema and Ajiere 2020). The upsurge of the mean monthly and annual values of temperature and rainfall intensity in Anambra has confirmed that indeed there has been an increase in warming (Onyeneke et al. 2018). Hence in this section, the two most influential variables of seasonal change (temperature and rainfall) have been discussed comprehensively. While establishing a link between these variables and the expansion of rills and gullies, consideration of four states within the southeastern region have been established.

Temperature Variation

In the last decades, Nigeria has experienced an upward variation in average temperature, with rising trends that seemed to be increasing in the last decade. This has been accompanied by incessant drought events in some regions as a consequence of increased evapotranspiration from temperature increases. Nigeria records a mean annual temperature of 26.9 °C, and a monthly average temperature ranging between 24 °C (in the months of December-January) and 30 °C (in April). Several studies have agreed that there is a strong positive linearity between rainfall and temperature variations. For instance, time-series data within a period of 1971–2000, from rainfall and air temperature collected from about 25 meteorological stations have been analyzed to assess the variations in temperature and precipitation (Akinsanola and Ogunjobi 2014). Based on the result of the analysis, an anomaly of −0.2 and −0.6 were observed within the first decade (i.e. from 1971–1980). The second decade (1981–1990) recorded positive variation from five stations. However, most parts of the country showed relative indications of warming in the third decade (1991–2000). In another study carried out on the spatial variability in humidity and temperature across the country using time series data for a period of 59 years, it was discovered that the average, minimum, and maximum temperatures in the country have been on the rise since 1951. However, they were a recorded decrease in relative humidity for the same period (Eludoyin 2016). In southeastern Nigeria, Nwaerema and Ajiere (2020) have assessed the distribution, pattern, trends, and anomalies of rainfall and temperature variations in four states (Abia, Cross River, Enugu, and Imo). Their study involved the collection and partitioning of time series data for three decades (1987–1996; 1997–2006; and 2007–2016). This research employed descriptive statistics (mean, maximum, minimum), and 3rd order polynomial trend analysis (using simple regression analysis) to analyse the decadal and annual temperature and rainfall anomalies in the four states (Table 24.1). Time series data for a period of 30 years (1987–2016) were collected from the Nigeria Meteorological Agency on climate change (NIMET), from four meteorological stations representing each state. Their results showed that the first decade (1987–1996), experienced the highest rainfall intensity in Imo State. In the second decade (1997–2007) Abia experienced high rainfall intensity, while in the third-decade rainfall intensity was higher in Cross River and Enugu States respectively. The mean annual temperature for 30 years (1987–2016) showed a persistent temperature increase in all the states from 2001 to 2002 (Table). Apparent fluctuations in the polynomial trends in temperature and rainfall patterns were observed from the results of decadal distribution trends for the four states. Cross River, Enugu, and Abia States showed positive trends in rainfall patterns, while negative trends were observed for Imo State (Fig. 24.3). This study concluded that climate variability from rainfall and temperature distribution patterns is evident in the southeastern region of Nigeria and the observed variation in rainfall and temperature patterns are in perfect agreement with the IPCC (2007, 2014) climatic projections in Africa. This shows evidence of climate variability and that high extreme temperature variations seem to influence high rainfall intensity and by extension increased gullying occurrences.

Diagi and Weli (2017) analyzed temperature fluctuations with relation to climate change variability in Ebonyi State, Southeastern Nigeria using temperature time series data for a 31 years period (1984–2015). Result from this study found that the area experienced a positive trend in temperature variation over the study period, with an annual temperature increase of 0.0037 °C (Fig. 24.4). Significant increase in temperature was observed within two decades (1984–1993 and 1994–2003), however, between 2004 and 2013, a slight increase was observed in the first decade (i.e. 1984–1993) (Table 24.2).

Rainfall Variation

The rainfall pattern of southeastern Nigeria is defined by seasonal upward and downward inter-annual trends (Ogungbenro and Morakinyo 2014). Generally, bi-modal rainfall distribution patterns are observed for the southeastern region. The first peak

Table 24.1 (a–c) Decadal distribution of minimum, maximum and mean Temperature (°c) across four southeastern States. (*Source* Nwaerema and Ajiere 2020)

Period	Abia	Cross River	Enugu	Imo
(a)				
1987–1996	22.35	23.01	22.49	23.07
1997–2006	22.8	23.14	22.42	23.84
2007–2016	22.77	23.29	22.59	23.69
(b)				
1987–1996	31.03	30.63	32.23	31.94
1997–2006	31.63	30.93	32.34	31.56
2007–2016	31.52	30.96	32.39	31.7
(c)				
1987–1996	26.7	26.8	27.4	27.5
1997–2006	27.2	27.0	27.4	28.1
2007–2016	27.1	27.1	27.5	28.1

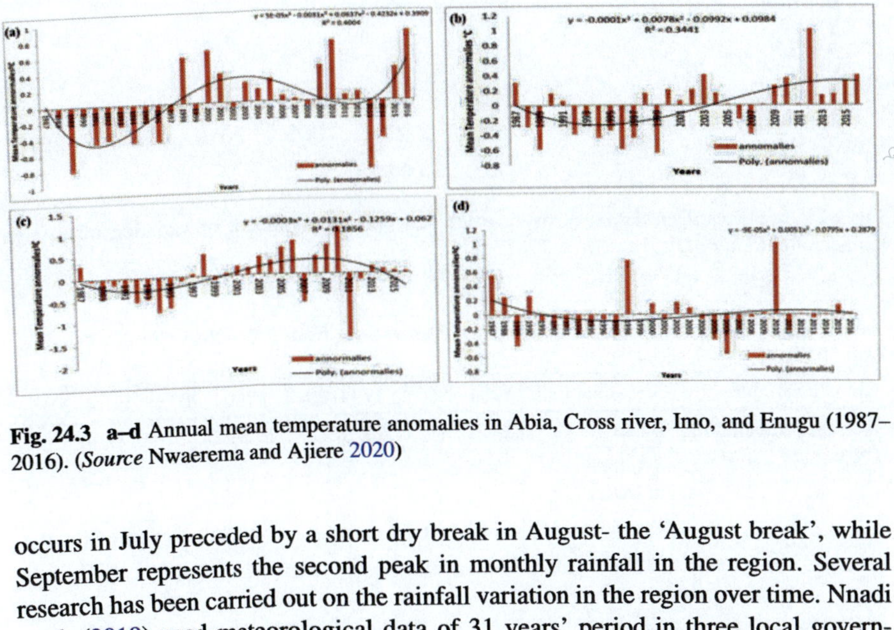

Fig. 24.3 a–d Annual mean temperature anomalies in Abia, Cross river, Imo, and Enugu (1987–2016). (*Source* Nwaerema and Ajiere 2020)

occurs in July preceded by a short dry break in August- the 'August break', while September represents the second peak in monthly rainfall in the region. Several research has been carried out on the rainfall variation in the region over time. Nnadi et al. (2019) used meteorological data of 31 years' period in three local government areas (Ayamelum, Anambra east and Ogbaru) in Anambra state to assess the impacts of climate change variability on rainfall patterns. Through the use of students' t-test, chi-square, annual rainfall anomaly (ARA), annual rainfall trends, and standardized annual rainfall anomaly (SARA), the significant trend in climate change on rainfall distribution patterns in the area over a period of 31 years (1986–2016)

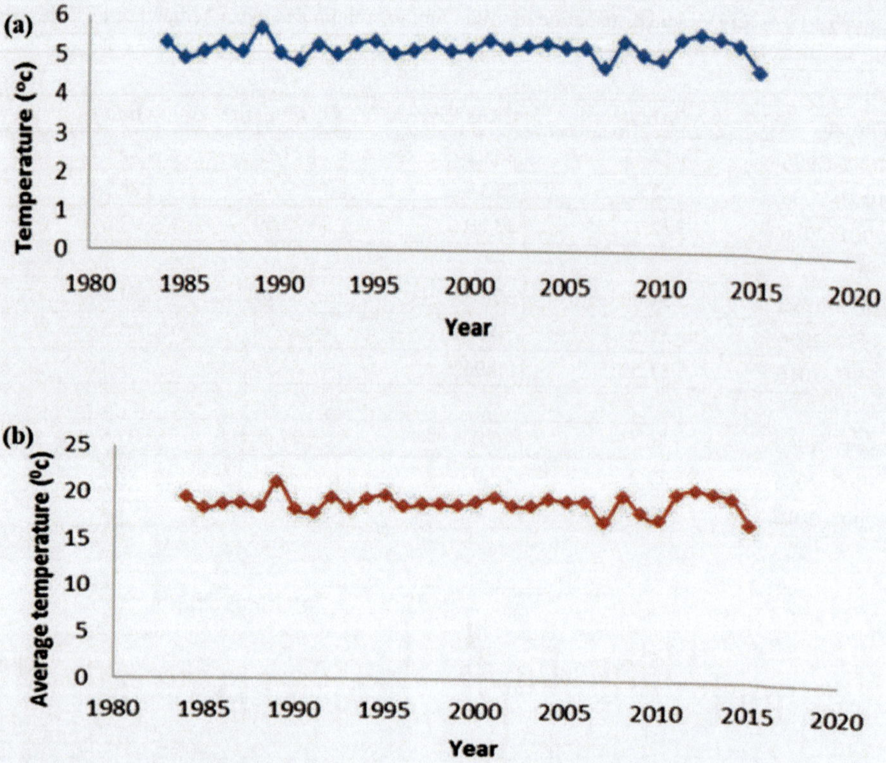

Fig. 24.4 **a** Standard Deviation of average temperature, **b** Coefficient of variation of average temperature. (*Source* Diagi and Weli 2017)

Table 24.2 Variation in temperature pattern for Ebonyi State from 1984–2015. (*Source* Diagi and Weli 2017)

Climate variable	Statistical tool	1984–1993	1994–2003	2004–2013	2014–2015
Average temperature (°C)	Mean	27.32	27. 50	27.40	27.30
	Standard deviation	5.08	5.14	5.16	4.98
	Coefficient of variation (%)	18.61	18.69	18.82	18.25

was assessed. To obtain the standardized annual and seasonal anomalies in rainfall patterns, the results of the mean, standard deviation, and standard errors were used. Results showed that both ARA and SARA were <0.00 for the three locations implying that they were a prevalence of more dry years than rainy years within the observed period (Fig. 24.5). However, the rainfall intensity was higher than typical from 2007–2016. These results correlate with the farmers' response to perception on climate variability with relation to gully formation and expansion within this area, as most of the farmers agreed that a high incidence of flooding and gullies were experienced within this period. A wide inter-annual variability in rainfall patterns was observed for the three locations with their coefficient of determination (R^2) occurring at $R^2 = 0.0003$, $R^2 = 0.0071$ and $R^2 = 0.0014$, for Ogbaru, Anambra East and Ayamelum respectively (Figs. 24.5 and 24.6). Wide range of variations were also observed on the onset and cessation of rainy seasons in the area; with R^2 values for onset and cessation of rainfall occurring at $R^2 = 0.006$ and $R^2 = 0.009$ respectively. These results seem to agree with a study by Oloukoi et al. (2014), that changes in the local climate, such as delay in the onset of rain, decreased total annual rainfall amount and early cessation of rainfall are affecting agricultural production and livelihood in southeastern Nigeria.

Using time series data for 90 years (1910–1999), together with non-parametric statistical tests and probability density function, Ogungbenro and Morakinyo (2014) explored the average monthly and yearly rainfall variation in rainfall distribution pattern for nine locations within the Guinea climatic zone representing southeastern region (Port Harcourt, Ikom, Uyo, Onitsha, Owerri, Ogoja, Enugu, and Calabar). The finding indicated an upward shift from dry to wet season in all locations within this climatic zone. A recent study by Okechukwu and Mbajiorgu (2020) assessed the spatial trends in evapotranspiration and rainfall patterns in five southeastern

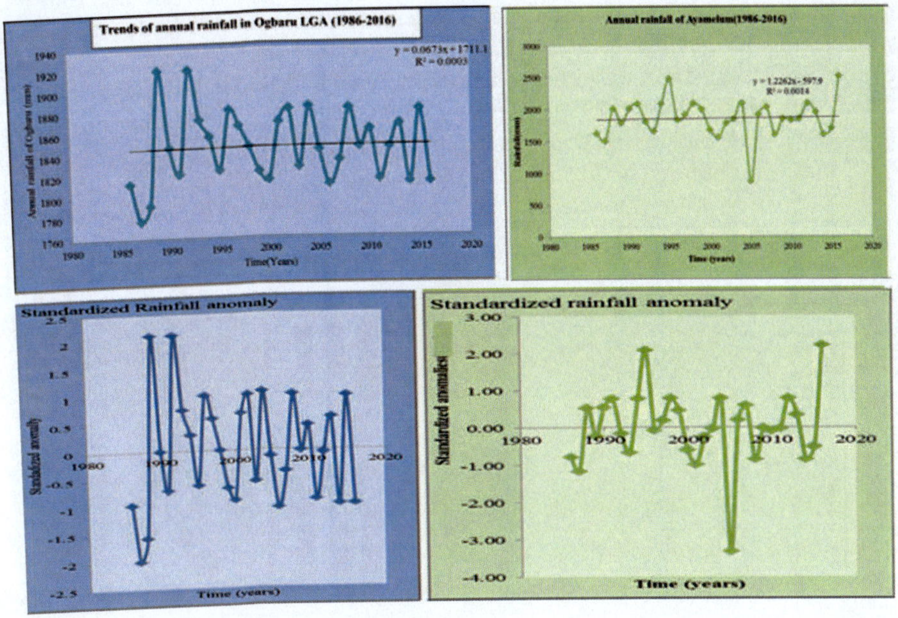

Fig. 24.5 a–b Trends of annual rainfall variability and change in Ogbaru and Ayamelum LGA (1986–2016). *Source* Meteorological Station (2018)

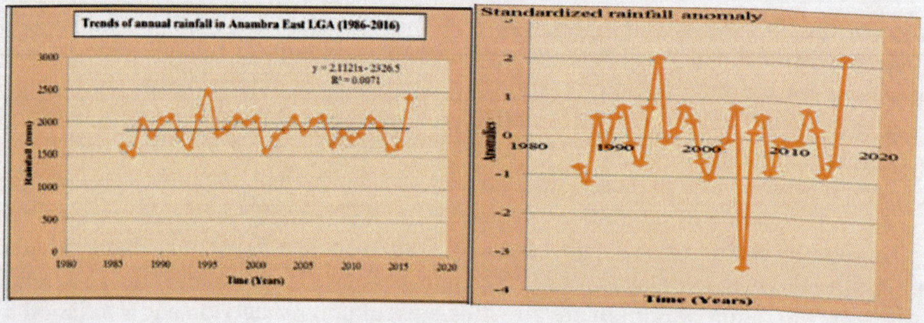

Fig. 24.6 Trends of annual rainfall variability and change in Anambra east LGA. (*Source* Nnadi et al. 2019)

states (Anambra, Ebonyi, Imo, Abia, Enugu) using geostatistical interpolation techniques such as Kriging and Inverse Distance Weighting (IDW) overlay analysis. It was found that they appear to be an upward gradual increase in rainfall from the northern parts of the region to the southern parts. However, evapotranspiration seems to increase from the south to the north. The high evapotranspiration events occurring during the dry seasons accounted for the upward increase in temperature values, with temperature values occurring at 1.1%. The annual rainfall amount from the north to the south occurred in decreasing order as: 2203 mm >2058 mm >1885 mm >1764 mm >1663 mm for Abia, Imo, Ebonyi, Anambra, and Enugu States, respectively (Fig. 24.7b). The long-term average monthly rainfall across the states decreased from 17 mm in January to 291 mm in July while decreasing to 253 mm in August. However, a peak was observed in September at a value of 320 mm. A steady decrease of 13 mm was observed in the month of December (Fig. 24.7a).

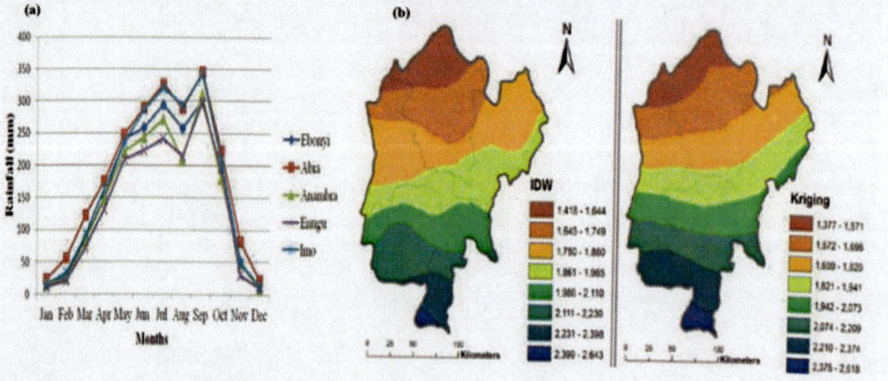

Fig. 24.7 **a** average distribution of monthly rainfall across the study area, **b** Spatial variation of average annual rainfall

Modeling and Quantification of Gullies with Respect to Rainfall: Implications for Future Prospects

Historically, the quantification of gully erosion in Nigeria, spanning from 1930–1985 has been focused on conventional field survey and mapping, using runoff plots, and simulated soil loss techniques. The approach was geared towards understanding the interaction occurring between the soil-water-erosion in the field on a regional scale (Skyes 1940; Grove 1951; Stamp 1938). However, during 1985 to the recent, enormous attempts have been made to integrate geostatistics, numerical indices and geopspatial models in an attempt to broaden the scale of soil erosion research (Ofomata 1985; Nwajide 1992; Ezemonye and Emeribe 2012; Fagbohun et al. 2016; Egbueri and Igwe 2022; Egbueri et al. 2021, 2022). The major focus of these approaches has been on using multidisciplinary and multidimensional approaches for erosion assessment based on a wide scale of perception of local, regional and global scales. Consideration of the global climatic variability as an influence on gully expansion and environmental sustainability have been the major focus. Most of these works have been focused majorly in the gully-prone southeastern region of the country. Table 24.3 shows a summary of the various gully erosion quantification methods used by researchers in southeastern Nigeria. As observed by Chiemelu et al. (2013), on the spatial variability of landforms in the Omagba erosion site in Onitsha, Anambra State, it was demonstrated that Digital Terrain Models (DTMs) can be used to derive 3-dimensional information on the spatial relationships of landforms, and that this model can provide supplementary information for the interpretation of gully distribution patterns and rate. Their study also showed that detailed information on areas impacted by sheet-rill-and gully erosion, and the rate of expansion can be adequately ascertained when this model is combined with aerial photographs. In order to assess and identify the most gully erosion degraded farmlands in southeastern Nigeria, Onweremadu (2006) integrated field sampling techniques and information from morphological landscape changes. The common drawback in the use of conventional mapping technique is the need for cartographic skills and difficulty and time constraints in interpretation. However, with the emergence of more advanced techniques, information on space and time can be obtained and will provide a robust visual representation for erosion quantification. The use of runoff plots has also been efficient in evaluating rainfall induced erosion sites. A common drawback in the application of this technique is in the fact that it alters the natural hydrological framework of the field due to the creation of artificial boundaries, hence the real-time erosion conditions may be distorted (Lal 2001). Additionally, information on the extrapolation beyond the observed area may be incorrect (Li et al. 2017). The erosion plots on natural fallow vegetation of varying ages in southeastern Nigeria have been constructed by Iwara and Ewa (2018). Their studies found that the highest amounts of rainfall and runoff as well as the incidences of gullies were observed between the months of July to September.

Table 24.3 Summarized erosion quantification and management in Southeastern Nigeria. (modified after Okenmuo 2021)

Location	Type of erosion	Method of quantification	Result
Abia State	Gully	GIS and GPS	The major area of gully dispersion is dominated by false-bedded sandstone mostly where the slope appeared to be >15°
Port Harcourt, Calabar, Onitsha, Uyo, Owerri, Enugu, Ogoja and Ikom	Sheet	Rainfall erosivity and USLE	Rainy seasons are accompanied by high erosion causing rains. Owerri, Port Harcourt and Calabar showed very high erosivity index
Anambra	Sheet	RUSLE and GIS	Average value of estimated soil loss was given at 214.82t ha^{-1}; High rainfall erosivity combined with high slope factor and decreasing vegetal cover
Enugu and Anambra States	Sheet, interill and Rill erosion	USLE and SLESMA	Coincidental and significant relationship between the USLE estimated maps and extent of actual gullying on ground
Orlu-Okigwe axis of Imo State	Gully	Field and LANDSAT data	Surface phenomenon of washing away of loose top soils is not the only factors responsible for gullies but also deep-seated near surface structural weakness along gully axis
Anambra Basin	Gully	Remote sensing, field and Geotechnical analysis	Gully development on steep slopes and non-vegetated areas are facilitated by friable and very permeable nature of Nanka and Ajali formations

(continued)

Table 24.3 (continued)

Location	Type of erosion	Method of quantification	Result
Orashi Catchment, Amambra State	Not known	Expert judgement, field and remote and sensing	Vegetation and slope are the major erosion triggers

A quantitative analysis of erosion can be carried out using erosion models. This model can either be conceptualized or empirical (Ganasri and Ramesh 2016) following inductive logic and observations from environmental factors. Over the last decades, empirical models such as the Revised Universal Soil Loss Equation (RUSLE), the Universal Soil Loss Equation (USLE), the Erosion Productivity Impact Calculator (EPIC), and the Unit Stream Power Based Erosion/Deposition model (USPED) have been widely used in gully erosion modeling. The RUSSLE model is a modification of the USLE model to include the erosivity component of rainfall. Fagbohun et al. (2016) successfully employed the RUSLE model to estimate the rate of soil loss by gully and identify critical gully erosion-prone areas in the Anambra sub-basin, southeastern Nigeria. Their results showed that about 39.49% of the total area (1804.39 km^2) experienced an erosion rate of 0–10 t/ha/yr. High rainfall intensity, low vegetation cover and morphological characteristics (moderate to high slope density) were identified as the major triggers of gully erosion in the area. Recently, studies by Obidai et al. (2011), Egbueri and Igwe (2022), and Egbueri et al. (2021, 2022) used a combination of geostatistical, machine learning, and novel numerical indexical models to predict the erodibility potential of soils to gullying, by taking into account several factors such as soil engineering properties, anthropogenic factors, and environmental factors, as potential triggers of soil erosion. To evaluate the role of soil engineering properties in the erosion of 10 main gullies in southern parts of Anambra State, Egbueri and Igwe (2022) used multiple geostatistical modeling techniques (such as correlation analysis, hierarchical clustering, and factor analysis) and new indexical models (e.g. sand-gravel ratio, entropized soil loss rate, erodibility factor, and sand–fines ratio) and artificial neural networks. The eroding materials are weak and have low erosion potential qualities, according to their findings. Therefore, their studies concluded that the soils' engineering attributes are the key triggers of the erodibility potential of soils in this area. In a similar line, Egbueri et al. (2022) used geoinformatics and the Revised Universal Soil Loss Equation (RUSLE) to estimate and predict soil loss rate in Anambra state by identifying the most erosion-prone locations. Soil erodibility (K), Rainfall erosivity (R), plant cover (C), anti-erosion methods (P), and topography (LS), were the five erosion factors included in the RUSLE model. The soil loss data revealed that roughly 5% (242.2 km^2), 25% (1211 km^2), 30% (1453.2 km^2), and 40% (1937.6 km^2) of the overall area had extremely low, low, medium, and severe erosion risks respectively. The soil loss in the western and northern parts of the state was extremely low, low, and moderate. The eastern and southern parts, on the other hand, had a high rate of soil loss. In Anambra, the most critical soil loss variables were identified as, K, LS and R.

Large gully sites can be detected using data from remote sensing and satellite imageries. Real-time and spatial information can be detected using this technique without necessarily disturbing the studied area. Due to the undulating terrain of the southeastern region, this method has been widely used and is considered the most suitable. Igbokwe et al. (2008) have successfully employed the Landsat ETM and Sat-1 image for a comparative assessment of gully development and land use patterns in southeastern Nigeria. Akanwa and Ezeomedo (2018) combined remote Sensing and GIS techniques to assess the relationship between field topography and gully occurrence and expansion in Akpo community in Anambra State. This study found that gully erosion in the area occurred in a dendritic network pattern, with the network pattern attributed to runoff from rainfall. A recent study by Ayadiuno et al. (2022), investigated the spatial variability of soil loss to gully erosion in Anambra State, using ArcGIS. The study identified rainfall as the highest contributor to soil loss with a total soil mass of 1,1134.7 ton/ha/yr loss, compared to other indices such as soil moisture index, land surface, temperature, and normalized difference vegetation index.

In other parts of the world such as Iran, the rainfall and gully erosion potential of a watershed in Kor-chamriz, Farz province was assessed using different climatic scenarios (Roshan and Negahban 2015). The study made use of the Arnoldus and Sepaskhah 's methods of humidity-soil to predict the climatic variability on gully erosion for the period of 2002–2060. The precipitation and temperature data were stimulated using the output of the general circulation climatic models (HadCM3). The data were then downscaled using the Lars-WG software. This study found that all future climate change scenarios showed a decreasing trend of gully erosion. It was therefore concluded that in the coming decades, the ratio of rainfall-induced gully erosion in the basin will decrease compared to the real erosion scenario of the watershed. Although this model has not been used in Nigeria, it can therefore be projected that due to its ability to use a set of parameters to generate randomly selected time series data based on fictitious climatic variables, this model will be critical in gully erosion prediction in southeastern Nigeria, and will serve as a future prospect tool for climate change variability on gully expansion. Apparently, due to the peculiarity of this model, it will immensely aid in proffering future erosion management and mitigation measures. Conversely, the radionuclide tracer method is another advanced erosion quantification technique which has not been used in the southeastern region. This model has been widely used in other parts of the world to quantify the rate of gully erosion at different time scales. This model uses the rate of adsorption and redistribution of soil particles following the events of erosion (Collins and Walling 2004). The model compares observed soil losses to a particular reference value- if the observation shows losses compared to the reference value, erosion is indicated, while observations higher than the reference value indicates deposition.

This model has been used in the southwestern region of Nigeria by Junge et al. as an alternative method to the conventional soil erosion prediction techniques.

Gully Erosion Mitigation and Management: Implication for Future Prospects

Due to the variability in the different gully triggering mechanisms in southeastern Nigeria, adequate measures to reduce the effect of the impact of gully will cut across crop management practices, gully erosion control measures, and landscape management. The amount and type of vegetation cover may influence erosion and runoff (Okenmuo 2021). High vegetation cover will intercept raindrops and inhibit infiltration and runoff, thereby improving the soil's organic matter content. Topography and soil type have been identified as one of the major influencing factors of gully expansion in southeastern Nigeria. Due to the highly friable nature of the sands in Southeastern Nigeria, soil detachment from uplands can occur on a slope as low as 5% (Ofomata 1985). In areas where the lithology appears to be stable with highly resistive materials, the occurrence of highlands and cuestas may enhance pathways for high runoff thereby obscuring the lowland areas around the slope toes (Ofomata 1975).

Over the years, in an attempt to curb the menace of gully erosion in the southeast, the farmers and environmental stakeholders have adopted practices such as construction of ditches, soil stabilizing stones, embankments, agroforestry, rainfall harvesting (using large reservoirs), among others. The government through various intervention programs have put in a lot of effort towards gully management; through sensitization programs and workshops geared towards gully mitigation. Apparently, not much have been achieved in curbing the menace. Perhaps, a look at more promising techniques will serve as better alternatives/future prospects on gully erosion mitigation in southeastern Nigeria. Doing this would involve taking into detailed consideration the major climatic variables of gully erosion: temperature and rainfall patterns. Although a lot of research have been carried out to this effect in southeastern Nigeria, however, some advanced techniques are yet to be tested. Moreover, the data is scarce and often times inaccurate. In other parts of the world, more advancements have been made towards gully erosion prediction using rainfall variability patterns and temperature changes. For instance, in Iran, the Arnoldus and Sepaskhah 's methods of humidity-soil have been used to predict the climatic variability on gully erosion for a period of 2002–2060. This study concluded that future climate change scenarios will give rise to a decreasing trend of gully erosion expansion. A take home from the use of this model will be that knowing the future rainfall pattern in the erosion-ravaged southeastern Nigeria will afford decision makers the ease of planning and mitigating the events of gully erosion.

Conclusion

Gully erosion in southeastern Nigeria has left remarkable fingerprints on the modification of the landscape of the region. It is projected that with the impending climatic variability, the menace of gullying in the region is feared to worsen. The incidences of gully erosion in southeastern Nigeria seem to be evolving due to processes and events such as soil type, topography, geology, and variation in rainfall patterns. To understand gully erosion processes and characteristics, great efforts have been made to quantify and model the event. A lot of research have been carried out to this effect, however, some advanced techniques are yet to be tested. Moreover, the data is scarce and often times inaccurate, which is evident in the prevalence of critically high gully events in the region. Another problem could stem from the non-inclusion of proactive (from cutting-edge climatic prediction models) and preventive measures into gully erosion management plans. A knowledge of this would be integral to researchers and environmental stakeholders in developing an all-inclusive conservation and management approach towards gully prediction and management.

References

Abdulfatai IA, Okunlola IA, Akande WG, Momoh LO, Ibrahim KO (2014) Review of Gully erosion in Nigeria: causes, impacts, and possible solutions. J Geosci Geomat 2(3):125–129

Adeleke ML, Al-Kenawy D, Nasr-Allah AM, Murphy S, El-Naggar GO, Dickson M (2018) Fish farmers' perceptions, impacts and adaptation on/of/to climate change in Africa (the case of Egypt and Nigeria). Theory and Practice of Climate Adaptation. Springer, Cham, pp 269–295

Akanwa A, Ezeomedo IC (2018) Changing climate and the effect of gully erosion on Akpo community farmers in Anambra state. J Ecol Nat Resour 2(6):1–12

Akinsanola AA, Ogunjobi KO (2014) Analysis of rainfall and temperature variability over Nigeria. Global J Human-Soc Sci B Geogr Geo-Sci Environ Disaster Manag 14(3):1–17

Anejionu OCD, Nwilo PC, Ebinne ES (2013a) Long term assessment and mapping of erosion hotspots in southeastern Nigeria. A paper presented at FIG Working Week 2013a Environment for Sustainability Abuja, Nigeria, 6th–10th May 2013

Anejionu OCD, Nwilo PC, Ebinne ES (2013b) Long term Assessment and mapping of erosion hotspots in southeastern Nigeria. A paper presented at FIG Working Week 2013b Environment for Sustainability Abuja, Nigeria, 6th–10th May 2013

Ayadiuno RU, Ndulue DC, Ndichie CC, Mozie AT, Phil-Eze PO, Onyekwelu AC (2022) Geospatial analysis of soil erosion susceptibility and causative factors in Anambra State, South East. Nigeria. Scientific Rev 8(1):5–32

Chiemelu N, Okeke F, Nwosu K, Ibe C, Ndukwu R, Ugwuoti A (2013) The role of surveying and mapping in erosion management and control: case of Omagba erosion site, Onitsha Anambra State, Nigeria. J Environ Earth Sci 3(11):11–18

Chukwu GO, Chude LA, Ekpo IE (2009a) Wetlands of south eastern Nigeria: extent and characteristics. Nigerian J Agric Food Environ 5(2–4):42–46

Chukwu GO, Chude LA, Ekpo IE (2009b) Wetland of southeastern Nigeria: extent and Characterisitics. Nigerian J Agric Food Environ 5(2–4):42–46 Ebonyi State, South-eastern Nigeria, 1984–2015

Diagi B, Weli V (1984–2015) Analysis of trend and variability of temperature in Ebonyi State, South-eastern Nigeria

Egboka BCE (2004) Distress call and plea to the senate committee for urgent Actions Against Floods, Soil/Gully erosion/landslides Disasters in the southeastern Nigeria, paper presented to senate committee on Environmental; roads/Erosion senate delegation
Egbucri JC, Igwe O (2022) Assessing the role of soil engineering properties in gully development and enlargement in southeast Nigeria using geostatistical and novel indexical techniques. Environ Earth Sci 81(1)1–24
Egboka BCE, Okoyeh EI (2019) Review and assessment of environmental impacts of ecological disasters on biodiversity in Anambra state. Nigeria. Biodivers Int J 3(2):53–58
Egbueri JC, Igwe O, Ifediegwu SI (2022) Erosion risk mapping of Anambra State in southeastern Nigeria: soil loss estimation by RUSLE model and geoinformatics. Bull Eng Geol Env 81(3):1–15
Egbueri JC, Igwe O, Unigwe CO (2021) Gully slope distribution characteristics and stability analysis for soil erosion risk ranking in parts of southeastern Nigeria: a case study. Environ Earth Sci 80(7):1–19
Egede EA (2013) Threats and mitigation of soil erosion and land degradation in Southeast Nigeria. J Environ Earth Sci 3(13):95–102
Eludoyin OM (2016) Air temperature and relative humidity areal distribution over Nigeria. Ife Res Public Geogr 10(1):134–145
Ezeabasili ACC, Okoro BU, Emengini EJ (2014) Relative erodibilities of some soils from Anambra basin. Sky J Soil Sci Environ Manag 3(8):83–90
Ezechi JI (2000) The influence of runoff, lithology, and water table on the dimensions and rate of gullying processes in eastern Nigeria. In: International symposium of gully erosion under global change. Gath University of Leuven, Belgium, pp 16–19
Ezeigwe PC (2015) Evaluation of socio-economic impacts of gully erosion in Nkpor and Obosi. Environ Res 7(7):34–38
Ezemonye MN, Emeribe CN (2012) Rainfall erosivity in southeastern Nigeria. Ethiopian J Environ Stud Manag 5(2):112–122
Fagbohun BJ, Anifowose AY, Odeyemi C, Aladejana OO, Aladeboyeje AI (2016) GIS-based estimation of soil erosion rates and identification of critical areas in Anambra sub-basin, Nigeria. Model Earth Sys Environ 2(3):1–10
FAO (1990) The state of food and agriculture 1990. (FAO Agriculture Series, no. 23) ISBN 92-5-102989-X
FDLAR (1990) Federal Department of land Resources: the reconnaissance soil survey of Nigeria. Soil Report 5:377–389
Fournier FT (2000) Land use soil conversion in parts of Onitsha and Owerri Province. Geological survey of Nigeria. Bulletin No. 21
Ganasri BP, Ramesh H (2016) Assessment of soil erosion by RUSLE model using remote sensing and GIS-a case study of Nethravathi Basin. Geosci Front 7(6):953–961
Grove AT (1951) Land use and soil conservation in parts of Onitsha and Owerri Provinces. Geol Surv. Nigeria. Bull No. 21;74
Ifeka A, Akinbobola A (2015) Trend analysis of precipitation in some selected stations in Anambra state. Atmos. Clim Sci 5:1–12
Igbokwe, Akinyede JO, Dang B, Alaga T, Ono MN, Nnodu VC, Anike LO (2008), Mapping and monitoring of the impact of gully erosion in southeastern Nigeria using satellite remote sensing and geographic information system. Int Arch Photogram Remote Sens Spat Inf 38:113–126
Igwe CA (2012) Gully erosion in southeastern Nigeria: role of soil properties and environmental factors. InTech. Research on Soil Erosion, G. Danilo, Ed. doi: https://doi.org/10.5772/51020
Igwe O (2014) The compressibility and shear characteristics of soils associated with landslides in geologically different localities— case examples from Nigeria. Arab J Geosci. https://doi.org/10.1007/s12517-014-1616-3
Igwe O, Egbueri JC (2018) The characteristics and the erodibility potentials of soils from different geologic formations in Anambra State, Southeastern Nigeria. J Geol Soc India 92(4):471–547
Intergovernmental panel on climate change (IPCC). Synthesis Report. Contribution of Working Groups I, II and III to the fifth assessment report of the intergovernmental panel on climate

change [Core Writing Team, R.K. Pachauri and L.A. Meyer (eds.)]. IPCC, Geneva, Switzerland. Climate Change 2014, pp151

Inyang IT (2000) Climate Zonation of the Rain Forest Zone of Eastern Nigeria. Geological Society of Nigeria Annual Report 5127

Kalu AC (2001) Soil Erosion and Landslides: 21st Century Issues and Challenges to Rural Development

Kirkby MJ, Bracken LJ (2009) Gully processes and gully dynamics. Earth surface processes and landforms: the J British Geomorphol Res Group 34(14):1841–1851

Lal R (2001) Soil degradation by erosion. Land Degrad Dev 12:519–539

Li Y, Bai X, Tian Y, Lou G (2017) Review and future research directions about major monitoring method of soil erosion. IOP Conf Ser Earth Environ Sci 63:1–15

Morgan RDC (2000) Soil Erosion. In: Davidson J (ed) Donald D. Topics in Applied Geomorphology London, Longman Group Ltd, pp 91–99

National Emergency Management Agency (NEMA, 2018) Rainfall Pattern, Flooding, Storms an interview with the Director of (NEMA) Walson Ibarakumo on Climate change

NIMET (Nigerian Meteorological Agency) (2017) 2017 Seasonal Rainfall Prediction (SRP). Retrieved from. http://www.nimet.gov.ng/sites/default/files/publications/NiMet%202017%20SRP_2.pdf

Nnadi OI, Liwenga ET, Lyimo JG, Madukwe MC (2019) Impacts of variability and change in rainfall on gender of farmers in Anambra. Southeast Nigeria. Heliyon 5(7):e02085

Nwajide CS (1992) Gullying in the Idemili River catchment, Anambra State, Nigeria: theory and cure. In: Natural hazards in West and Central Africa. Vieweg+ Teubner Verlag, Wiesbaden, pp 149–162

Nwaerema P, Ajiere S (2020) Regional mapping of land surface temperature (LST), land surface emissivity (LSE) and normalized difference vegetation index (NDVI) of south-south coastal settlements of rivers State in Nigeria. World News of Natural Sciences, 28

Nwagbara MO, Ibe OK (2015) Climate change and soil conditions in the tropical rainforest of southeastern Nigeria. J Geogr 3(1):99–106

Nwajide CS (2013) Geology of Nigeria's sedimentary basins. CSS Bookshop Limited

Obiadi II, Nwosu CM, Ajaegwu NE, Anakwuba EK, Onuigbo NE, Akpunonu EO, Ezim OE (2010) Gully Erosion in Anambra State, southeast Nigeria: Issues and solutions. Int J Environ Sci 2(2):796–804

Obiadi II, Nwosu CM, Ajaegwu NE, Anakwuba EK, Onuigbo NE, Akpunonu EO, Ezim OE (2014) Gully Erosion in Anambra State, South East Nigeria: issues and solution. IJES 2(2):796–804

Ochieng J, Kirimi L, Mathenge M (2016) Effects of climate variability and change on agricultural production: the case of small scale farmers in Kenya. NJAS Wagening. J Life Sci 77:71–78

Ofomata GEK (1975) Landform regions. Nigeria in maps, Eastern States. Ethiopia publishing house Benin, pp 33–34

Ofomata GEK (1985) Soil erosion in Southeastern Nigeria: the views of a geomorphologist. Inaugural Lect Ser 7

Ofomata GEK (1988, April) The management of soil erosion problems in south eastern Nigeria. In: Proceedings of International symposium on Erosion in SE Nigeria, pp 3–12

Ofomata GEK (2000) Missing Links in the management of Soil Erosion Problems in Nigeria. In:

Ogboi E, Odeh H (2012) Erosion problems in Ika environs: causes and forms. J Environ Sci Manag 4:53–61

Ogungbenro SB, Morakinyo TE (2014) Rainfall distribution and change detection across climatic zones in Nigeria. Weather Climate Extremes 5:1–6

Oguntunde PG, Abiodun BJ, Gunnar L (2012) Spatial and temporal temperature trends in Nigeria 1901–2000. Meteorol Atmos Phys 2012(118):95–105. https://doi.org/10.1007/s00703-012-0199-3

Okechukwu ME, Mbajiorgu CC (2020) Determination of crop coefficients and spatial distribution of evapotranspiration and net irrigation requirement for three commonly cultivated crops in South-East Nigeria. Irrig Drain 69(4):743–755

Okenmuo FC (2021) Erosion quantification and management: Southeastern Nigeria Case Study. In: Landscape architecture framed from an environmental and ecological perspective. IntechOpen
Okoyeh EI, Akpan AE, Egboka BCE, Okeke HI (2014) An assessment of the influences of surface and subsurface water level dynamics in the development of Gullies in Anambra State, Southeastern Nigeria. Earth Interact 18:1–24
Oloukoi G, Fasona M, Olorunfemi, Fm Adedayo V, Elias P (2014) A gender analysis of perceived climate change trends and ecosystems-based adaptation in the Nigerian wooded savannah. Gend Clim Change 24(3). https://www.tandfonline.com/doi/full/, https://doi.org/10.1080/10130950.2014.949477
Onweremadu EU (2006) Assessment of mined soils in erosion-degraded farmlands in southeastern Nigeria. Estud Biol 28(65):59–67
Onyeneke RU, Igberi CO, Uwadoka CO, Aligbe JO (2018) Status of climate-smart agriculture in southeast Nigeria. GeoJournal 83(2):333–346
Othieno RT, Laycock EP (2007) Monitoring soil organic carbon erosion with isotopic tracers, two case studies on cultivated tropical catchments with steep slopes Laos, Venezuela. Advances in soil science. CRC Press, Boca Raton, Florida, USA
Poesen J (2011) Challenges in gully erosion research. Landform Anal 17:5–9
Rahman EL, Abd MA, Ali RR, Hussain MA, ElSemey MA (2009) Remote Sensing and GIS based physiography and soils mapping of the Idku-Brullus Area. North Delta, Egypt Egypt J Soil Sci 49(3):209–432
Roshan G, Negahban S (2015) Modeling of the effects of climate change on rainy and gully erosion potential of Kor-chamriz watershed in Fars province. Model Earth Syst Environ 1(3):1–12
Skyes RA (1940) History of anti-erosion work in Udi. Farm Forestry 1:3–6
Stamp LD (1938) Land utilization and soil erosion in Nigeria. Geogr Rev 28:32–45
Thakural LN, Kumar S, Jain SK, Ahmad T (2018) The impact of climate change on rainfall variability: a study in central himalayas. In: Singh VP, Yadav S, Yadava RN (eds), Climate change impacts. Springer Singapore, pp 181–192
The World Bank. "Nigeria—Erosion and Watershed Management Project" 2012–04–12
Ubuoh EA, Akhionbare WN, Oweremadu E, Onifade OA (2013) Characterization of soil quality in erosion prone environment of Ukpor, Nnewi-South LGA of Anambra State. Nigeria. Inter J Adv Appl Sci 2(1):1–8
Ufot UO, Iren OB, Chikere Njoku CU (2016) Effects of land use on soil physical and chemical properties in Akokwa area of Imo State, Nigeria. Int J Life Sci Scientific Res 2(3):273–278
Ziervogel G, Nyong A, Osman B, Conde C, Cortés S, Downing T (2006) Climate variability and change: implications for household food security. Assessment of impacts and adaptations to climate change (AIACC): Washington, DC, USA

Chapter 25
The Impact of Seasonal Changes on the Trends of Physicochemical, Heavy Metal and Microbial Loads in Water Resources of Southeastern Nigeria: A Critical Review

Johnson C. Agbasi, Johnbosco C. Egbueri, Daniel A. Ayejoto, Chinanu O. Unigwe, Michael E. Omeka, Vincent E. Nwazelibe, Joshua O. Ighalo, Chaitanya B. Pande, and Abdulamid A. Fakoya

Abstract The quality of water consumed by humans has a direct impact on their health and life span. Globally, seasonal changes have been identified as one of the major factors influencing water quality. Southeastern Nigeria is a rapidly developing region, experiencing an increase in human activities. The current review article analysed works that studied surface and groundwater resources in Southeastern Nigeria during the rainy and dry seasons. An in-depth literature analysis was employed to achieve the objectives of this study. To the best of the author's knowledge, this is the first article to review the impacts of seasonal changes on the quality of water resources in Southeastern Nigeria. The findings of the literature review revealed that

J. C. Agbasi (✉) · J. C. Egbueri
Department of Geology, Chukwuemeka Odumegwu Ojukwu University, Uli, Nigeria
e-mail: johnsonagbasi@gmail.com

D. A. Ayejoto
Department of Environmental and Sustainability Sciences, Texas Christian University, Fort Worth, TX, USA

C. O. Unigwe
Boone Pickens School of Geology, Oklahoma State University, Stillwater, OK 74078, USA

M. E. Omeka
Department of Geology, University of Calabar, Calabar, Nigeria

V. E. Nwazelibe
Department of Geology, Albert Ludwig University of Freiburg, Freiburg im Breisgau, Germany

J. O. Ighalo
Department of Chemical Engineering, Nnamdi Azikiwe University, Awka, Nigeria

C. B. Pande
Sant Gadge Baba Amravati University, Amravati, MS 444602, India

A. A. Fakoya
School of Meteorology, University of Oklahoma, Norman, OK, USA

J. C. Egbueri et al. (eds.), *Climate Change Impacts on Nigeria*, Springer Climate,
https://doi.org/10.1007/978-3-031-21007-5_25

seasonal variations have a significant impact on water resources in the region. Variations in rainfall and temperature influenced the levels of physicochemical parameters, heavy metals, and microbial loads in surface and groundwater resources of the study region. Moreover, seasonal changes also influenced anthropogenic activities (particularly mining and farming) in the region, which in turn altered the water quality. Studies in the region revealed that the wet season is usually the peak of farming activities. Prolonged rainfall and an increase in rainy days during the wet season led to runoff from farms and a decline in mining activities. Seasonal variations in natural processes and anthropogenic activities resulted in different water quality patterns in the region. The findings of this review study is significant to policymakers, water managers and academics interested in public health, water quality, climate change and risk assessment.

Keywords Climate change · Seasonal variation · Southeastern Nigeria · Water quality · Water resources

Introduction

All living things need water to survive. Humans use water in their homes, farms, and industries for multiple purposes. The two major sources of water are surface and groundwater. The quality of water resources is affected by human activities, natural processes, and seasonal changes. Due to the bio-accumulative nature, persistence and toxicity of heavy metals in the environment, they have been identified as one of the primary contaminants of water resources across the world, with negative impacts on the ecosystem and human health (Sharma and Agrawal 2005; Nagajyoti et al. 2010). Nevertheless, studies have shown that heavy metals have some health benefits when they are within their permissible limits in water resources (Agbasi and Egbueri 2022; Ayejoto et al. 2022; Egbueri et al. 2022). Conventional and advanced methods (principal component analysis, factor analysis, and Pearson's correlation analysis) have been used globally to identify the sources of pollution in water resources (Egbueri and Agbasi 2022a, b; Ganiyu et al. 2018; Lkr et al. 2022; Gomes et al. 2019). Sources of heavy metals in water include both natural processes (rock weathering and volcanic eruption) and human-induced processes (use of agrochemicals, emissions from industrial waste and mining) (Nagajyoti et al. 2010). The consumption of poor quality water exposes water users to carcinogenic and non-carcinogenic health risks (Agbasi and Egbueri 2022; Ayejoto et al. 2022; Egbueri et al. 2022). Several parameters (anions, cations, metals and microorganisms) are analysed to evaluate the quality of water resources. Due to the numerous parameters that are considered, water quality indices (pollution index of groundwater, water pollution index and overall index of pollution) were introduced. Water quality indices have proven to be a reliable means for classifying water samples (Onyemesili et al. 2020; Enyigwe et al. 2021; Karmakar et al. 2021; Egbueri et al. 2021). These indices harmonise the results of the parameters analysed and come up with a value. The descriptive significance of the value is used to characterise water samples based on the classification system of the water quality index.

Nigeria is found in Western Africa and is divided into six major geopolitical zones: north-central, north-east, north-west, south-east, south-west, and south–south. However, the south-eastern zone which is the focus of this study comprises of five states which includes Abia, Anambra, Ebonyi, Enugu and Imo. Anambra state is well known for trading. One of the largest markets in West Africa is found in Onitsha, Anambra State. The largest automobile market in Africa is also found in Nnewi, Anambra state (Agbasi and Egbueri 2022). Enugu state, which is referred to as "Coal City," has large deposits of coal. Thus, there are a lot of mining activities going on in the region. In Ebonyi state, there are active mine sites where barites are extracted (Omeka et al. 2022). On the other hand, Imo state has the best recreational establishment in Southeast Nigeria and, as a result, it gets a lot of visits from people in other parts of the country. Agriculture is the main source of livelihood for a good number of people in Abia state. Thus, there are a lot of agricultural activities going on in Abia state. The mentioned anthropogenic activities that exist in southeastern Nigeria contribute greatly to the economic growth of the region. However, these activities have some negative effects on the environment. Studies have shown that waste generated from the markets, mines, industries, recreational centres, and farms in the region has led to degradation in the quality of water resources (Agbasi and Egbueri 2022; Ezenwaji et al. 2022; Omeka et al. 2022; Adumanya et al. 2022; Odo et al. 2022).

Climate is a region's averaged long-term weather (rainfall and temperature) trend over a period of time, usually 30 years (IPCC 2021). In simple terms, climate change can be defined as an irregularity in the usual pattern of rainfall and temperature in an area. Climate change has been identified as a global issue and has attracted the attention of many research studies (Olabi and Abdelkareem 2022; Perera and Nadeau 2022; Sesana et al. 2021; Seddon et al. 2020). Globally, climate change has a significant negative influence on people's health, particularly in underdeveloped nations (Ahmed et al. 2020). Studies have shown that anthropogenic activities (industrialisation, increasing population, deforestation, bush burning, etc.) are the major causes of climate change (Amanullah et al. 2020; Ahmed and Aliyu 2019; Raghuvanshi and Ansari 2016; Odjugo 2013). In many places in the world, climate change has resulted in rising temperatures, significant precipitation, flooding, severe droughts, and heatwaves (Sun and Wang 2022; Ridder et al. 2022; Amanullah et al. 2020; Ahmed et al. 2020). The aforementioned processes lead to accumulation of contaminants (heavy metals, agrochemicals, microorganisms, etc.), in water systems across the world. Because of this, water-borne illnesses like dengue fever and chikungunya are resurfacing (Ahmed et al. 2020). Research has confirmed that climate change is evident in Nigeria (Ayanlade et al. 2022; Zvobgo et al. 2022; Shiru et al. 2021). Olufemi (2018) argues that in 2017 alone, Nigeria witnessed a series of climate-related disasters, ranging from increased health risk, declining agricultural productivity, biodiversity loss, drying lakes, famine, conflicts or social unrest, poverty, worsening food insecurity, heat stress, declining soil capacity for agricultural production, increased natural disaster, extreme weather events, among others (Anabaraonye et al. 2018). The most peculiar effect of climate change in Nigeria is declining agricultural productivity arising from irregular rain and unpredictable farming calendar (Anabaraonye et al. 2018). A crucial first step in identifying whatever action to be made to offset the detrimental effects of climate change on agriculture is understanding how farmers perceive climate threats (Onyeneke et al. 2021). Between 2008 and 2011, a survey was carried out in southeast Nigeria using a multi-stage random sampling method (Ozor et al. 2012). The survey's findings revealed that the majority of participants (84%) were aware of some of the characteristics of climate change, including uncertainty at the start of the growing season, extreme weather events like flooding and droughts, pests, diseases, weed infestations, and land degradation (Ozor et al. 2012). In order to decrease post-harvest losses, farmers have implemented a variety of adaptation measures, such as increasing weeding, altering the timing of farm operations, and processing crops (Ozor et al. 2012). Despite the fact that most of the respondents were aware of government policies intended to safeguard the environment, the majority of them concurred that these regulations were not being successfully enforced (Ozor et al. 2012). Ighalo and Adeniyi (2020) also reported that inadequate implementation of environmental policies has resulted in a decline in the quality of Nigeria's water resources. Studies on climate variations have predicted that negative effects of climate change could worsen in Nigeria and even in other regions of the world (Ayanlade et al. 2022; Anabaraonye et al. 2018; Ozor et al. 2012).

Seasonal change is usually identified with an alteration in the rate of rainfall, temperature, evapotranspiration and river flows. Studies have shown that reduced stream flow might increase pollutant concentrations (Fan et al. 2012). Prathumratana et al. (2008) discovered that various water quality parameters (such as chemical oxygen demand (COD), biochemical oxygen demand (BOD) and ammonium) have positive associations with hydrological factors (such as stream flow, precipitation and mean water level). Pollution concentrations in surface waters are often greater during the dry season than during the wet season, owing to dilution by a considerable quantity of rainfall and stream flow from upstream during the wet season (Fan et al. 2012; Nas and Nas 2009; Zhang et al. 2007). The diluting effects of accumulated pollutants and their transfer to the receiving water are all influenced by rainfall intensity and runoff during the wet season (Tsihrintzis and Hamid 1997; Maniquiz and Lee 2010). Increased runoff and soil erosion caused by higher rainfall intensities, for example, might increase phosphorus transport, particularly phosphorus particles (Quinton et al. 2001). In a rainfall simulation research, Sporre-Money et al. (2004) and Shigaki et al. (2007) discovered that the concentrations of reactive and particulate phosphorus in runoff increased with higher rainfall intensity. Phosphorus runoff from agricultural areas has also been linked to freshwater eutrophication (Shigaki et al. 2007). Additionally, studies have revealed that during the rainy season, there is an increase in sediments and particles transported into open water bodies (Warrier et al. 2022; Oh and Choi 2022; Hachad et al. 2022). This leads to a rise in the total suspended solids (TSS) in the water body. High concentrations of suspended solids reduce transparency and photosynthesis; this also leads to the death of aquatic plants and animals (Malik and Rath 2022; Gutiérrez et al. 2022; Sivakumar et al. 2022). The death of these organisms releases potentially toxic substances into water systems and could likely degrade the water quality (Ding et al. 2022; Waters et al. 2022; Zhan et al. 2022; Yang et al. 2022). Li and Zhang (2010) analyzed seasonal fluctuations of dissolved trace elements and heavy metals in the Upper Han River in China. The findings revealed that the changes in season significantly influenced the concentration of heavy metals, with most variables showing greater concentrations during the wet season. In Southeastern Nigeria, a study in Ekulu River, Enugu state during the wet season agrees with Li and Zhang (2010), as it revealed that all heavy metals (Fe, Mn, As, Cd, Cr, Cu, Ni, Pb and Zn) content in water samples analysed surpassed the acceptable human intake limits (Ifediegwu et al. 2021). Physicochemical analysis of groundwater samples collected from boreholes in Onitsha, Anambra state during a transitional phase between the wet and dry season (August–December) revealed that they were all above the World Health Organization (WHO 2009) standards for drinking water (Chinye-Ikejiunor 2021). A study by Duru et al. (2012) in Otamiri river, Imo State, during the rainy season investigated heavy metal and bioload levels in the Otamiri river. Apart from total iron upstream, the researchers discovered that all other detected heavy metals (Cn, Mn and Zn) in river water were within the maximum permitted values. The study also found high bioload levels per microbe category for total fungal count, total viable bacteria count, total coliform count and total heterotrophic bacteria count. Similar findings were also reported by another study carried out in the dry season to assess the quality of groundwater from different parts of Southeastern Nigeria for potable use (Omaka et al. 2017).

Intake of contaminated water has resulted in an outbreak of water-borne diseases such as cholera, diarrhoea, and other gastrointestinal ailments in Nigeria (Forstinus et al. 2016; Oguntoke et al. 2009). In response, there has been an increase in the number of research carried out yearly to access the quality of water resources in Nigeria (Egbueri 2022; Anyanwu et al. 2022; Unigwe et al. 2022; Egbueri and Agbasi 2022a, b, c). However, there have not been sufficient review studies on research projects that have evaluated the status of water resources in Southeastern Nigeria. A recent review on water quality monitoring and assessment in Nigeria was conducted by Ighalo and Adeniyi (2020). The pollution sources of groundwater, surface water, rainfall, and commercially accessible water in the region were discussed. In the same year, a review on drinking water quality in Nigeria towards attaining the WHO sustainable development goal six was conducted (Nwinyi 2020). The research looked at the pollution extent of drinking water sources in Nigeria. Previous reviews on water resources that cut across Southeastern Nigeria have not paid much attention to the impacts of seasonal changes on water quality. Therefore, the present study focuses on the influence of seasonal changes on the quality of surface and groundwater resources in Southeastern Nigeria. This comprehensive review aims to (1) compare research works that have been conducted with regard to the quality of surface and groundwater resources in Southeastern Nigeria, and (2) depict the signatures and trends impacted by seasonal changes in the water quality of the region. It is hoped that the findings of this research will provide insights to water managers, researchers, and water users to aid effective monitoring and assessment of water resources with regard to seasonal change. It is also anticipated that this research will foster studies on the identification of relationships between seasonal variations and water quality worldwide.

Materials and Methods

Study Area

The research region, Southeastern Nigeria, is generally located between latitudes 4° and 7° north of the Equator and longitudes 5° and 10° east of the Greenwich meridian (Nwagbara 1997), (Fig. 25.1). The Atlantic Ocean, the Republic of Cameroon, and Delta State border it on the south, east, and west, respectively, while the middle belt states of Kogi and Benue border it on the north (Ezemonye and Emeribe 2012). The Anambra River, Imo River, Niger River, Oji River, Njaba River, Nkisi River, Agulu Lake, Oguta Lake, Nike Lake, and other water bodies drain the area (Igbokwe et al. 2008; Egbueri and Igwe 2021). According to Ajiere and Nwagbara (2018), the region has a total size of about 75,488 km^2, with tropical rainforest comprising 60% of that (Njoku 2006). Because tropical rainforest covers most of southern Nigeria, the soils are a combination of ferrasol, vertisols, and hydromorphic soils (Areola 1983). The dry (November–March) and wet (April–October) seasons are experienced throughout the study area (Nganje et al. 2015). According to Koppen's classification, the climate of the region may be classed as tropical (Ajiere and Nwagbara 2018). The tropical maritime and tropical continental air masses regulate the climate there. Air temperatures in Southeastern Nigeria are consistently high throughout the year. Southeastern Nigeria is one of the country's six geopolitical zones. The geopolitical zones are key divides in modern Nigeria that were not totally carved out based on geographical setting, but rather states with comparable cultures, ethnic groupings and shared history (Agbor and Ashabua 2018).

Statistical analysis of climate data for Southeastern Nigeria in 2020 revealed that the highest mean annual rainfall in Southeastern Nigeria was 183.73 mm in Owerri, Imo state and the lowest mean annual rainfall was 144.59 mm in Awka, Anambra state (Ibeje and Ekeleme 2020). The highest mean annual air temperature was 28.063 °C in Enugu and the lowest mean annual air temperature was 26.396 °C in Abakaliki, Ebonyi state (Ibeje and Ekeleme 2020). This is because Enugu is closer to northern Nigeria, which is a semi-arid region with a characteristic high mean annual temperature (Balogun et al. 2012). The greatest annual mean relative humidity was 79.025% in Umuahia, Abia state, while the lowest annual mean relative humidity was 70.289% in Awka, Anambra state (Ibeje and Ekeleme 2020).

The regional hydrogeology of Southeastern Nigeria is separated into three primary groupings. The first (Lower Hydrogeological Group) is found in an area dominated by shaly formations (Adelana et al. 2008). The Asu River Group, Awgu Shale, Ekenkpon Shale, Eze-Aku Shale, Nkporo Shale, Enugu Shale, New Ntetim Marl and Mfamosing Limestone Formation, are members of the Lower Hydrogeological Group. The Ajali, Mamu, and Nsukka Formations (Middle Hydrogeological Group) feature notable sandy horizons, forming the second group (Adelana et al. 2008). The third group (Upper Hydrogeological Group) includes the Imo Shale (Paleocene–Lower Eocene), Bende-Ameki/Nanka Sand (Middle Eocene), Benin Formations and Ogwashi-Asaba (Adelana et al. 2008).

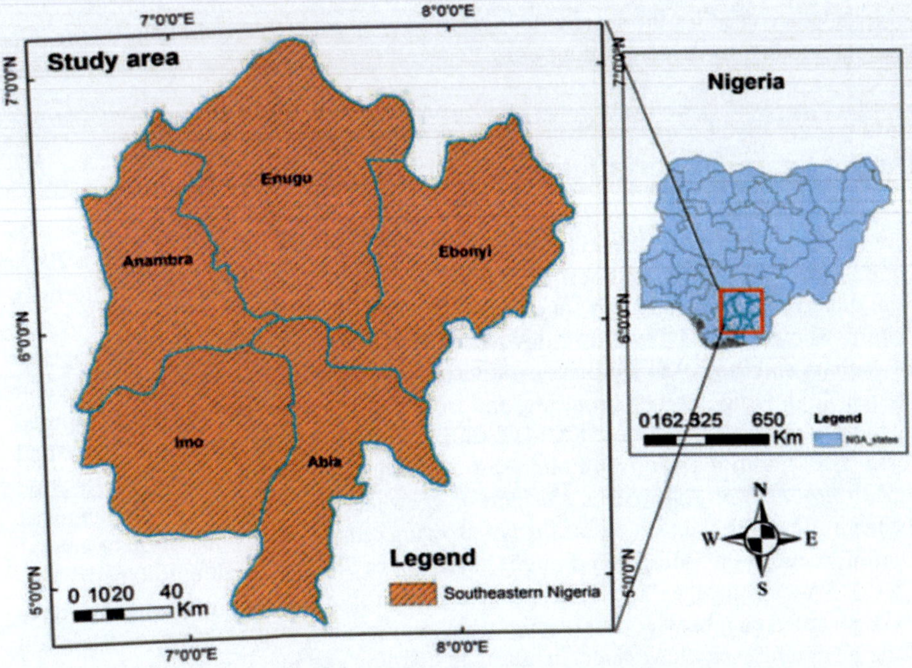

Fig. 25.1 Map of southeastern Nigeria showing the five states (*Source* Okenmuo 2021)

The major source of livelihood for people in Southeastern Nigeria is trading, followed by agriculture and white-collar jobs. The main market in Onitsha, Anambra state, is one of the largest in West Africa (Agbasi and Egbueri 2022). Nnewi, Anambra state, is home to Africa's leading automobile market. There has been an increasing rate of socioeconomic development in Southeastern Nigeria, which has attracted investors. Over the years, a lot of macro- and microindustries, modern markets, banks, hotels, recreational centres, commercial houses and schools have been established in Southeastern Nigeria. These investments have boosted the economy of the region. Presently, in Anambra state, the private/individual contribution to the gross domestic product is at 98%. However, just like every developing region, some cities in Southeastern Nigeria are facing challenges like climate change, overpopulation, environmental degradation, and environmental hazards.

Data Generation

The data utilised for discussing the seasonal variations in Southeastern Nigeria came from a thorough search of literature published on the issue over the years on Google Scholar. The keywords "water quality", "rainfall", "temperature", "seasonal changes", "dry and wet season", "groundwater", "surface water", and "Southeastern

Nigeria" were used for the search. This allowed us to filter down the relevant articles without leaving out any important ones.

Recent Seasonal Variations in Southeastern Nigeria

Rainfall and temperature are important climate change indicators (Nwagbara 2008; Uguru et al. 2011). For long-term water resource management, spatial patterns in rainfall and evapotranspiration are critical (Okechukwu and Mbajiorgu 2020). Their patterns show a broad upward or downward trend in climate variables across time (Nwagbara and Ibe 2015). Farmers in Nigeria experience climate change as irregular rainfall, high temperatures, droughts, and frequent floods. Agriculture is hampered by these circumstances. Enete (2014) found that, with the exception of pepper and cassava, all traditional crops in Enugu saw a significant field drop as rainfall became more irregular. Men and women's responses show statistically significant changes at the beginning of the rainy season; shift of growing seasons; early cessation of annual rainfall; recurrent flooding and drought in Anambra State, according to Nnadi et al. (2019). Water shortages and food instability disproportionately impacted women, and they suffered a heavier cost of migration as a result of climate changes. Rainfall duration and intensity have risen, resulting in flooding and massive runoffs in several regions of Nigeria (Enete 2014). Farmers in Ebonyi state have noted the influence of climate change on rice farming as a result (Diagi and Nwagbara 2018). Farmers noticed that the swamp size had grown, meaning that there was more space accessible for swamp rice farming. The majority of rice farmers ascribed the rise to climate change, owing to increased rainfall in recent years compared to prior years. The majority of farmers in Imo state have noticed an increase in floods in recent years, which implies climate change (Okoro et al. 2016). According to Ozor et al. (2015), the perceived implications of climate change on farming in Imo state include decreased agricultural yields and soil fertility, drought occurrences and increasing heatwaves. According to Nwaiwu et al. (2013), temperature and rainfall have been increasing in Southeast Nigeria, and as a result, farmers in the area have been growing choice cropping systems and cover crops.

Rainfall Variations: A Clear Signature of Climate Change

Trend identification, according to Ifeka and Akinbobola (2015), is a hot topic in both hydrology and climatology. Trend identification facilitates the examination of climate change scenarios and improves climate impact studies. Monthly rainfall data taken from the Nigerian Meteorological Agency (NIMET) Synoptic Station at Amawbia, Awka, Anambra State was used to investigate the rainfall pattern in Awka, Anambra State from 2011 to 2020 (Omoja et al. 2021). The yearly rainfall pattern shows a drop from 2012 to 2014 and a rise from 2016 to 2019. According to the analysis, 2019 had the most yearly rainfall, while 2013 had the lowest.

Variations in rainfall observed in Imo State during a 30-year period (1988–2017) indicates change in climate (Okorie et al. 2019). The state's mean annual rainfall data for each of the 30 years were plotted in a single line graph, and the graph reveals the state's large and dramatic range in mean annual rainfall over the inquiry period (Okorie et al. 2019). The mean annual rainfall in Imo State during the first decade of 1988–1997 shows a lot of variation, with the lowest mean of 181.9 mm in 1993 and the greatest mean of 246.8 mm in 1990 (Okorie et al. 2019). For the second decade (1998–2007), there was continuous and substantial variability in mean annual rainfall. The lowest mean value of 136.7 mm was recorded in 1998, while the greatest mean value of 267.4 mm was recorded in 2006. The lowest and highest mean values recorded in the entire study were found to occur in the second decade. The substantial fluctuation in mean annual rainfall persisted drastically in the third decade (2008–2017), indicating climatic variation in the state with respect to the rest of Southeast Nigeria. The greatest average rainfall was 236.2 mm in the middle of the decade (2012), while the lowest average rainfall was 164.6 mm in 2013. As a result of the investigation, climate change in Imo State has been confirmed. This finding backs up the claims of Nnaji (2009), that climate change has expressed itself in sub-Saharan Africa through rainfall variability. Furthermore, the manifestations of climate change in Imo state as reported by Okorie et al. (2019), are consistent with previous findings by Adewuyi et al. (2015), Duru and Chibo (2014), Ladan (2014), Zakari et al. (2014) and Enaruvbe and Yusuf (2016), all of which concluded that climate change is a reality in Nigeria.

Okechukwu and Mbajiorgu (2020) reported on long-term climatic daily data for 17 weather stations in and around Southeastern Nigeria received from NIMET Lagos and the Centre for Basic Space Science, Nsukka during a 21-year period (1993–2013). The average monthly rainfall in Southeastern Nigeria followed a similar trend, with small differences in values as shown in Fig. 25.2a–e. In decreasing order from the north, the state-wide average annual rainfall for Abia, Imo, Ebonyi, Anambra and Enugu was 2203 mm, 2058 mm, 1885 mm, 1764 mm and 1663 mm, respectively. This is because the northern part of the study region (Enugu) is closer to a semi-arid region with a characteristic high mean annual temperature (Balogun et al. 2012). The rainfall pattern shown in Fig. 25.2a–e clearly distinguished the two seasons in the area: dry (November–March) and rainy (April–October) seasons.

According to a study conducted by Osugiri et al. (2019) on the aggregate amount of rainfall in Southeastern Nigeria, there is substantial inter-annual variability in rainfall volume from 1972 to 2014, resulting in a very low and negligible correlation (0.049) between rainfall volume and time. The amount of rainfall in the area increased somewhat but not significantly (Fig. 25.3). This result is consistent with the findings of Onyeneke and Mmagu (2015) and Babatunde et al. (2011), who claimed that while aggregate rainfall in Nigeria's rainforest and coastal regions will not change, the intensity and pattern will. The trend of rainfall intensity in Southeast Nigeria shows that the number of rainy days has decreased significantly between 1972 and 2014 (Fig. 25.4) and this agrees with the trend of mean annual rainfall in Imo State,

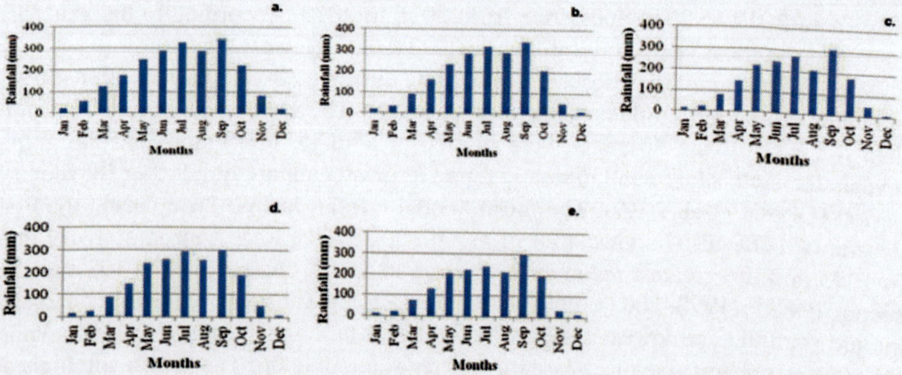

Fig. 25.2 Distribution of average monthly rainfall from 1993 to 2013 in **a** Abia State, **b** Imo State, **c** Anambra State, **d** Ebonyi State and **e** Enugu State (*Source* Okechukwu and Mbajiorgu 2020)

Southeast Nigeria from 1988 to 2017 (Okorie et al. 2019). The findings of Osugiri et al. (2019) also agree with Onyeneke and Mmagu (2015), who found that the number of rainy days in Southeast Nigeria is declining and would continue to decline. Because of the growing volume of rainfall on a decreasing number of wet days, dangers such as erosion and flooding, may become more common in the area (Osugiri et al. 2019). Data on rainfall patterns collected and analysed for 42 years (1972–2014) uncovered that there were no evident large gaps in rainfall distribution over the months within the period, but there was evident variation in the intensity of rainfall within the months in a year, resulting in sudden spells that left farmers unprepared for the onset and/or end in rainfall in a year (Osugiri et al. 2019).

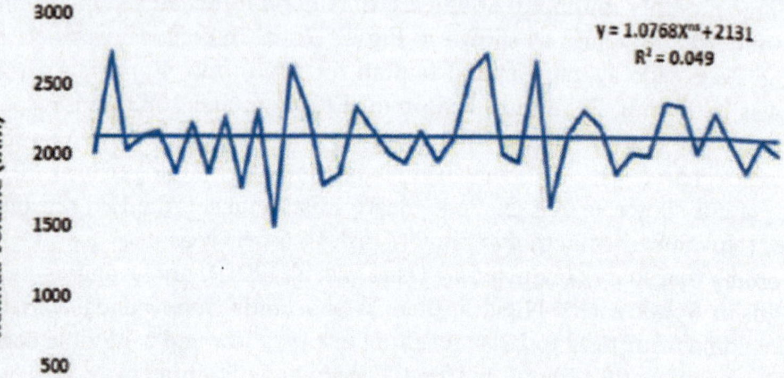

Fig. 25.3 Trend result of rainfall volume in southeast Nigeria from 1972 to 2014 (*Source* Osugiri et al. 2019)

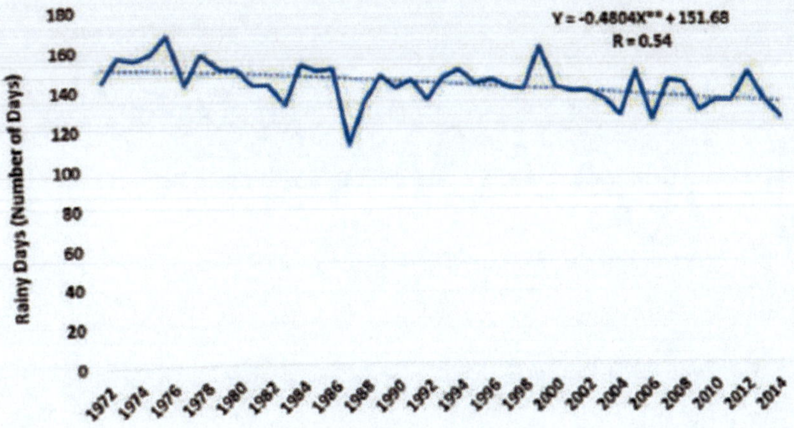

Fig. 25.4 Trend result of rainfall intensity in southeast Nigeria from 1972 to 2014 (*Source* Osugiri et al. 2019)

Temperature Variations: A Clear Signature of Climate Change

The rise and fall in temperature are notable signs of climate change. There have been reports in the United States, United Kingdom, China, India, Canada, Japan, South Africa, and other parts of the world with regards to increasing temperature conditions (Dong et al. 2022; Zeng et al. 2022; Malhi et al. 2021; Conway et al. 2021). Temperature and time have a statistically significant coefficient of correlation of 78.7%, meaning that temperature has a considerable positive association with time (Osugiri et al. 2019). As a result, temperature fluctuations are heavily influenced by time. In Southeastern Nigeria, temperature data from 1972 to 2014 reveals a statistically significant upward trend (Fig. 25.5). Numerous studies also confirm that in Southeastern Nigeria, there is a steady increase in surface temperature (Onyeneke and Mmagu 2015; Chidiebere-Mark et al. 2014; Nwaiwu et al. 2013; Okorie et al. 2012; Babatunde et al. 2011; WACDI 2011; Nwajiuba and Onyeneke 2010). Chikezie et al. (2016) discovered a three-decade pattern of sunlight fluctuation in southern Nigeria (1984–2014). It revealed an inconsistency in the movement of sunlight levels. In 1984, the average daily sunlight length was over 5 h, rising to 6 h in 1991, 2002 and around 7.5 h in 2013. (Fig. 25.6). This might be linked to the growing temperature trend caused by the global warming impact, which has been a prominent feature in the recent decade. This indicates that warming is occurring in Southeast Nigeria. It is anticipated that the temperature of Southeastern Nigeria will continue to rise in the coming years if there are no significant measures taken to combat global warming. Studies have also shown that, on a global scale, a continual rise in temperature is expected (Mathew 2022; Cavanagh et al. 2022; Nashwan and Shahid 2022; Ghazi and Jeihouni 2022).

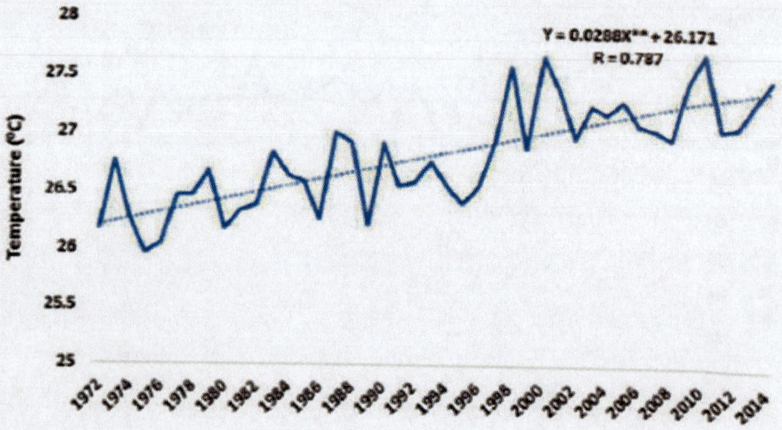

Fig. 25.5 Trend result of temperature in Southeast Nigeria from 1972 to 2014 (*Source* Osugiri et al. 2019)

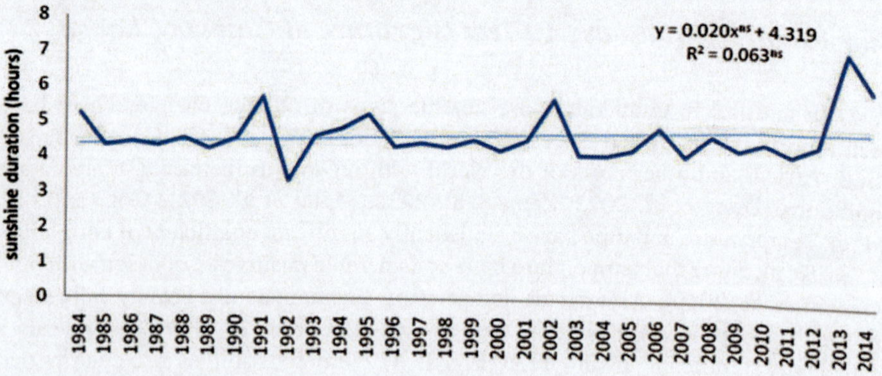

Fig. 25.6 Trend of Sunshine duration variation in South-East, Nigeria from 1984 to 2014 (*Source* Chikezie et al. 2016)

Water Resources

Surface Water Resources

Prior to the discovery of groundwater, the majority of people in the world relied greatly on surface water. In developing countries like Nigeria, millions of people still depend on surface water for survival. In Southeastern Nigeria, there are a lot of large surface water bodies, which include the famous River Niger, Agulu Lake, Oguta Lake, etc. These bodies of water are used orally and dermally for drinking, bathing, cooking, recreational activities, etc. Therefore, the need to ascertain the quality of surface water is important. Interestingly, a good number of research papers have evaluated the quality of surface water resources in the research area.

Odo et al. (2021) looked at the microbiological purity and physicochemical features of the Waterside River in Ogbor-Hill, Abia State, Nigeria. Except for heavy metals (Pb, Mn, Hg and Fe), all physicochemical characteristics measured were within Nigerian Industrial Standards and WHO guidelines. Because of the health risks caused by heavy metals, the Waterside River is deemed unsuitable for consumption. Before using water from the Waterside River in Ogbor-Hill Aba, Nigeria, it must be properly treated. A human health risk evaluation of polycyclic aromatic hydrocarbons in surface water resources was undertaken in Ifite Ogwari, Anambra State, Nigeria (Okafor et al. 2022). According to the study, residents of Ifite Ogwari would suffer from unfavourable health conditions like cancer over time. Thus, researchers and government agencies must employ a holistic strategy to reduce ecological and health risks. Okoye et al. (2022) found that the spring water in Oji River LGA, Enugu State, Nigeria, is unsafe for drinking without sufficient treatment. Water treatment procedures such as liming, boiling, and ozonation were advised. Water samples were also taken from the Ogbajarajara River in Imo State, Nigeria. The findings of the study showed that both adults and children are at the risk of developing cancer (Ihenetu et al. 2022). Many nations rely on the mining of natural resources for their economic survival, but the process has far-reaching environmental and human health repercussions, especially when it is carried out without sufficient sustainability measures (Uwa et al. 2022). Uwa et al. (2022) investigated the consequences of mining activities on the Omege River in Ebonyi State, Nigeria. The findings imply that drinking Omege River water can cause both non-carcinogenic and carcinogenic health issues in both children and adults. Furthermore, the study found that children were more vulnerable to heavy metal exposure than adults while drinking water from the Omege River.

The majority of reports on surface water in Southeastern Nigeria reveal that most of the surface water in the region is unsafe for human consumption and many of the water users are not aware of this fact.

Groundwater Resources

The increasing need for an alternative source of water led to the discovery of groundwater. Globally, there is a high preference for groundwater over surface water. This is because surface water is exposed to multiple sources of contamination when compared to groundwater. A good number of people in Southeastern Nigeria depend on groundwater as their main source of water. Several studies have evaluated the quality and health risks of groundwater resources in the region.

In the Onicha Local Government Area of Ebonyi State, Nigeria, Ojukwu et al. (2022) used a conventional approach to analyse the physicochemical and microbiological characteristics of water samples from streams, boreholes and hand-dug wells. The quality was computed using the weighted arithmetic water quality index approach. The borehole water source was found to have the best water quality in the research region, with a WQI score of 7.58, indicating exceptional water quality. The stream water source came in second with a WQI of 39.43, indicating acceptable water quality, while the hand-dug well water quality was unsuitable for consumption with a WQI of 169.17. As a result, the study suggests that water sources be appropriately treated to an acceptable level before use. Borehole water was analysed bacteriologically in certain villages in Owerri West, Southeastern Nigeria, using conventional microbiological procedures such as total coliform count, total faecal coliform count, and total heterotrophic bacterial count (Ewelike et al. 2022). The overall bacteria count was higher than the WHO's standard for drinking and other home uses. Furthermore, pathogenic organisms discovered from water samples have the potential to cause major health issues in people. As a result, the research suggested that borehole water be thoroughly treated before consumption, as well as other preventative measures, to reduce the potential health concerns connected with its usage. The impact of heavy metal contamination in groundwater resources near an abandoned limestone quarry pit in Enugu State, Nigeria, was successfully assessed using statistical index analysis employing contamination factor, metal pollution index, enrichment factor and ecological risk (Eyankware and Akakuru 2022). The results of the ecological risk assessment indicated that ecological dangers associated with Pb, Zn, Cu, Fe, As, Co, Mn, Ni and Cd were typically modest. The pollution load index indicated that there is no pollution in the area. Regardless, pre-treatment of water resources was strongly advised.

Industrial pollution is a serious environmental concern in Southeastern Nigeria due to the uncontrolled flow of industrial waste into the environment (Ifeanyichukwu et al. 2022). Ifeanyichukwu et al. (2022) conducted a study on harmful components in groundwater resources near factories in Nnewi north L.G.A., Southeastern Nigeria. The findings of the study revealed that groundwater supplies were polluted largely by anthropogenic sources. It is suggested that before consumption, the groundwater should be treated. Ohaegbuchu et al. (2019) evaluated the groundwater quality in Abia State's Isiala-Ngwa North LGA. High nitrite and pH levels were shown to be significant indicators of physical and chemical contamination. Environmental interventions, such as community-based health worker public health education and awareness, and sensitisation campaigns on the sustainable use of inorganic fertiliser are anticipated to improve groundwater integrity in the region (Ohaegbuchu et al. 2019). It was recommended that the government build more water filtration plants to ensure that everyone in the area has safe drinking water; sewer drains should be kept away from water supply drains to prevent wastewater from leaching into ground water; and groundwater quality should be monitored on a regular basis in all areas (Ohaegbuchu et al. 2019).

A good number of studies on groundwater in Southeastern Nigeria revealed that the quality of groundwater in the study region is unsuitable for human consumption and the pollution sources are majorly anthropogenic.

Seasonal Variations and Water Resources

Studies have shown that seasonal variations have an effect on the quality of water resources (Muhammad and Ullah 2022; Odabaşi et al. 2022; Verla et al. 2018; Alum et al. 2021). In order to identify the impact of seasonal variations on various water quality parameters in Southeastern Nigeria, studies that analysed water resources (groundwater and surface water) during the wet and dry seasons across the five states (Abia, Anambra, Ebonyi, Enugu and Imo) in the region were reviewed. The comprehensive results of the analysis of surface water and groundwater in the study region are presented in Tables 25.1, 25.2, 25.3, 25.4, 25.5 and 25.6. The results are discussed in subsequent sections.

Table 25.1 Signatures of seasonal variations in surface and groundwater quality in Abia state (*Sources* Ndubuisi et al. 2021 and Ubechu et al. 2021)

Parameters	Surface water (Ndubuisi et al. 2021)		Groundwater (Ubechu et al. 2021)	
	Dry season	Wet season	Dry season	Wet season
Temperature (°C)	31.88	29.1	–	–
pH	5.52	6.316	–	–
Electrical conductivity (μS/cm)	23.334	40.2	–	–
Dissolved oxygen (mg/kg)	2.954	5.99	–	–
Total dissolved solids (mg/L)	3.642	4.41	–	–
Total suspended solids (mg/L)	4.998	5.334	–	–
Nitrate (mg/L)	0.924	0.846	–	–
Sulphate (mg/L)	24.356	25.092	–	–
Phosphate (mg/L)	1.212	1.5242	–	–
Sodium (mg/L)	1.408	2.034	–	–
Potassium (mg/L)	0.7182	0.726	–	–
Magnesium (mg/L)	2.126	2.386	–	–
Calcium (mg/L)	23.604	29.54	–	–
Arsenic (mg/L)	–	–	0.017	0.142
Cadmium (mg/L)	0.052	0.012	0.119	0.036
Copper (mg/L)	0.346	0.166	0.300	1.256
Chromium (mg/L)	–	–	0.874	0.134
Iron (mg/L)	0.136	1.032	0.119	0.573
Manganese (mg/L)	0.412	0.042	–	–
Nickel (mg/L)	–	–	0.206	0.235
Lead (mg/L)	0.216	1.142	0.448	0.358
Zinc (mg/L)	2.25	2.784	–	–

Table 25.2 Signatures of seasonal variations in surface and groundwater quality in Anambra state (*Sources* Uzoukwu et al. 2004 and Chukwura and Igwe 2021)

Parameters	Surface water (Uzoukwu et al. 2004)		Groundwater (Chukwura and Igwe 2021)	
	Dry season	Wet season	Dry season	Wet season
Temperature (°C)	28.300	27.233	–	–
pH	5.876	6.206	6.292	4.937
Flow rate (cm/s)	13.860	15.700	–	–
Electrical conductivity (μS/cm)	21.100	18.866	287.33	550.610
Dissolved oxygen (mg/kg)	4.166	4.256	38.720	7.952
Biological oxygen demand (mg/L)	0.333	0.436	4.560	0.569
Chemical oxygen demand (mg/L)	–	–	15.16	7.0025
Turbidity (NTU)	10.333	20.66	–	–
Total dissolved solids (mg/L)	11.000	10.333	217.04	202.500
Total hardness	9.333	6.666	–	–
Total alkalinity	13.333	19.333	–	–
Chloride (mg/L)	7.800	10.900	51.654	95.4
Nitrate (mg/L)	1.253	2.140	3.009	2.61075
Sulphate (mg/L)	1.000	2.333	17.112	22.9
Phosphate (mg/L)	0.260	0.156	–	–
Sodium (mg/L)	1.476	1.193	–	–
Potassium (mg/L)	0.390	0.440	–	–
Magnesium (mg/L)	0.195	0.115	–	–
Calcium (mg/L)	1.210	0.486	–	–
Cupper (mg/L)	–	–	0	0.1
Chromium (mg/L)	–	–	0.1333	0.0512
Iron (mg/L)	0.183	0.320	0.1956	0.197
Manganese (mg/L)	0.010	0.003	0.1944	0.095
Nickel (mg/L)	–	–	0.3231	0.2921
Lead (mg/L)	0.016	0.009	0.0063	0.26
Zinc (mg/L)	0.026	0.049	1.3597	0.036

Influence of Seasonal Variations on Surface Water Quality

Physical and Chemical Parameters

In Abia state, Ndubuisi et al. (2021) analysed the Amuzuta Stream, Umuahia, South-east Nigeria, during the dry and rainy seasons. The physical parameters analysed in the study were temperature, pH, TSS, electrical conductivity (EC), dissolved oxygen (DO) and total dissolved solids (TDS). The findings of the study show that

Table 25.3 Signatures of seasonal variations in surface and groundwater quality in Ebonyi state (*Sources* Ezekwe et al. 2013 and Obiekezie et al. 2006)

Parameters	Surface water (Ezekwe et al. 2013)		Groundwater (Obiekezie et al. 2006)	
	Dry season	Wet season	Dry season	Wet season
Temperature (°C)	35.5	32.8	27.52	26.73
pH	7.2	6.93	7.62	7.8
Electrical conductivity (µS/cm)	5270	255	1229.5	413.67
Dissolved oxygen (mg/kg)	–	–	2.13	1.6
Turbidity (NTU)	30	11	9.32	61.97
Total dissolved solids (mg/L)	11,126.2	180	639.0	217.0
Total suspended solids (mg/L)	–	–	14.54	5.56
Total hardness	29,920	105.6	–	–
Total alkalinity	124	32	349.83	234.25
Salinity (%)	3	0	–	–
Chloride (mg/L)	1155	1	50.85	43.82
Nitrate (mg/L)	0.17	0.63	–	–
Sulphate (mg/L)	27.2	21.2	57.99	76.17
Phosphate (mg/L)	0.13	0.04	–	–
Potassium (mg/L)	–	–	0.29	0.11
Magnesium (mg/L)	700	18.7	–	–
Sodium (mg/L)	–	–	84.17	66.67
Calcium (mg/L)	9234	1115	11.42	17.1
Cadmium (mg/L)	BDL	BDL	0.004	0.006
Copper (mg/L)	–	–	0.006	0.006
Iron (mg/L)	2.43	0.24	0.02	0.03
Manganese (mg/L)	0.78	0.49	–	–
Lead (mg/L)	0.24	BDL	0.01	0.01
Zinc (mg/L)	1.07	0.06	0.001	0.001

the concentrations of all the analysed physical parameters except temperature were higher during the rainy/wet season. The decline in temperature in the region might have been a result of greater rainfall and reduced daylight hours, both of which occur during the rainy season. Rivers in Southeastern Nigeria have seen similar temperature changes (Verla et al. 2018; Duru and Kenneth 2014; Akubugwo and Duru 2011). The decrease in pH seen during the dry season might be the result of the death and decomposition of some aquatic living forms, which release proteins in the form of ammonia. Dissolved ammonia causes a significant decline in pH of water resources (Ndubuisi et al. 2021; Akubugwo and Duru 2011).

The physicochemical properties of the upstream (UU), midstream (MU), and downstream (DU) courses of the Ubu River in Anambra State were analysed by

Table 25.4 Signatures of seasonal variations in surface and groundwater quality in Enugu state (*Sources* Okoye et al. 2022 and Nnaemeka-Okeke et al. 2019)

Parameters	Surface water (Okoye et al. 2022)		Groundwater (Nnaemeka-Okeke et al. 2019)	
	Dry season	Wet season	Dry season	Wet season
Temperature (°C)	26.777	25.252	30.4	30.32
pH	6.847	7.555	6.13	5.40
Colour (TCU)	8.088	12.266	–	–
Electrical conductivity (μS/cm)	416.565	174.983	133.9	189.6
Biological oxygen demand (mg/L)	2.975	1.933	5.155	4.288
Chemical oxygen demand (mg/L)	–	–	7.541	5.207
Turbidity (NTU)	–	–	9.743	10.80
Total dissolved solids (mg/L)	339.05	207.15	62.50	127.4
Total suspended solids (mg/L)	169.64	349.040	164.2	186.7
Total solids (mg/L)	–	–	226.7	311.8
Total hardness	119.388	173.777	111.7	118.9
Total alkalinity	14.653	30.988	36.48	29.07
Total acidity	5.765	12.396	0.347	2.739
Ammonium (mg/L)	0.620	0.220	–	–
Nitrate (mg/L)	0.916	1.910	–	–
Sulphate (mg/L)	–	–	21.78	31.27
Phosphate (mg/L)	0.19	0.786	–	–
Sodium (mg/L)	–	–	11.81	11.97
Potassium (mg/L)	–	–	6.408	10.17
Cadmium (mg/L)	0.001	0.005	–	–
Copper (mg/L)	0.012	0.572	–	–
Iron (mg/L)	0.207	0.090	1.244	1.499
Mercury (mg/L)	0.000	0.003	–	–
Lead (mg/L)	0.017	0.048	0.000	0.000
Zinc (mg/L)	0.055	0.018	1.317	1.437

Uzoukwu et al. (2004). The mean values of the UU, MU, and DU parameters were used in the evaluation of the study. The physical and chemical parameters studied were temperature, pH, EC, TDS, DO, BOD, flow rate, turbidity, total hardness and alkalinity. The result of the study revealed that the level of temperature, EC, TDS and total hardness in the Ubu River was on a decline during the wet season, whereas there was an increase in the level of pH, flow rate, DO, BOD, turbidity and alkalinity during the wet season. The variation in DO was possibly as a result of runoff from agricultural soils containing phosphate and nitrogen compounds, which produces a drop in dissolved oxygen, commonly known as eutrophication (Uzoukwu

Table 25.5 Signatures of seasonal variations in surface and groundwater quality in Imo state (*Sources* Ihenetu et al. 2022 and Braide et al. 2016)

Parameters	Surface water (Ihenetu et al. 2022)		Groundwater (Braide et al. 2016)	
	Dry season	Wet season	Dry season	Wet season
Temperature (°C)	28.40	25.29	30.7	29.7
Colour	1.30	11.80	–	–
pH	5.37	5.57	5.4	5.3
Electrical conductivity (µS/cm)	100.68	194.74	30	20
Dissolved oxygen (mg/kg)	5.42	8.87	–	–
Total dissolved solids (mg/L)	63.80	122.17	20	0
Chloride (mg/L)	–	–	0.18	0.15
Nitrate (mg/L)	21.39	21.90	12.3	7.2
Sulphate (mg/L)	0.43	0.54	0	0
Phosphate (mg/L)	60.27	60.74	1.74	0.20
Potassium (mg/L)	5.19	5.63	–	–
Sodium (mg/L)	6.08	7.19	–	–
Calcium (mg/L)	3.49	4.00	160	100
Cadmium (mg/L)	0.00	0.00	–	–
Chromium (mg/L)	–	–	4	7
Cupper (mg/L)	0.08	0.07	–	–
Iron (mg/L)	6.28	8.42	0.0	0.0
Manganese (mg/L)	0.07	0.06	–	–
Nickel (mg/L)	0.00	0.00	–	–
Lead (mg/L)	0.40	0.69	–	–
Zinc (mg/L)	1.15	2.29	0.12	0.01

et al. 2004). Increased rainfall during the rainy season may have diluted the phosphate and nitrogen molecules in the river, resulting in a higher quantity of dissolved oxygen. Similarly, rising turbidity levels during the rainy season might be ascribed to increased rainfall, which resulted in erosion and the movement of solid material into the river by flooding.

Surface water proximal to the Mayo and Amikol Pb–Zn Mines at Ihetutu, Ebonyi State, was sampled by Ezekwe et al. (2013). The physical and chemical parameters analysed include pH, temperature, turbidity, EC, TDS, salinity, alkalinity, and hardness. The concentration of all parameters increased during the dry season. The decline in the number of rainy days possibly impacted the mining activities in the region as the workers could now fully work on the site. Increased mining activities may have possibly led to an increase in the concentration of the physical and chemical parameters.

Table 25.6 Signatures of seasonal variations of microbial load in surface and groundwater, southeastern Nigeria (*Sources* Ndubuisi et al. 2021 and Braide et al. 2016)

	Surface water (Ndubuisi et al. 2021)		Groundwater (Braide et al. 2016)	
Parameters	Dry season	Rainy season	Dry season	Rainy season
Total heterotrophic Bacterial Count (THBC)	16.084	9.466	–	–
Total Fungi Count (TFC)	0.268	0.058	1300.000	0.000
Total Feacal Coliform Count (TFCC)	0.252	0.030	–	–
Total Coliform Count (TCC)	0.254	0.252	0.000	0.000
Total aerobic count (TAC)	–	–	8,000,000.000	1600.000
Total hydrocarbon count (THC)	–	–	10,000.000	0.000
Total salmonella-shigella count (TSSC)	–	–	0.000	0.000

In the Oji River Local Government Area, Enugu State, physicochemical characteristics of spring water samples were conducted by Okoye et al. (2022). The physical and chemical parameters studied include pH, colour, temperature, EC, TSS, TDS, BOD, total hardness, total acidity, and total alkalinity. The results showed that the levels of pH, colour, TDS, total hardness, total acidity, and total alkalinity in the spring increased during the rainy season, whereas temperature, EC, TSS, and BOD levels increased during the dry season.

A study conducted in Ogbajarajara River, Imo state by Ihenetu et al. (2022), analysed six physical parameters, which included temperature, DO, EC, pH, TDS, and colour. The findings of the research revealed that all physical parameters except temperature increased during the wet season. Ndubuisi et al. (2021) also reported similar seasonal variations in the Amuzuta Stream, Umuahia. Greater TDS levels in the Ogbajarajara River, indicate increased pollution from runoff of soils in the region as a result of higher precipitation rates during the rainy season (Ihenetu et al. 2022).

Anions

The levels of nitrate, sulphate, and phosphate were analysed in the Amuzuta Stream, Umuahia, Abia State by Ndubuisi et al. (2021). The study reported that there was a rise in sulphate and phosphate levels during the wet season and a rise in nitrate levels during the dry season. Higher rainfall rates, which resulted in runoffs from agricultural fields, livestock and human wastes, may have aided seasonal variance in nitrate (Ndubuisi et al. 2021).

The anions studied by Uzoukwu et al. (2004) in the Ubu River, Anambra State, are sulphate, nitrate, phosphate and chloride. When the results of the anions analysed in the rainy and dry seasons were compared, some changes were noticed. There was a rise in sulphate, nitrate and chloride during the rainy season, while phosphate concentrations increased in the dry season. The rise in the level of sulphate in the Ubu River during the rainy season agrees with the finding of Ndubuisi et al. (2021) in the Amuzuta Stream.

In the surface water close to Mayo and Amikol Pb–Zn Mines at Ihetutu, Ebonyi State, sulphate, nitrate, phosphate and chloride were analysed during the dry and wet seasons by Ezekwe et al. (2013). The results indicated that sulphate, phosphate and chloride levels were higher during the dry season, while nitrate levels increased in the wet season. Similar to the seasonal variations observed in the physical and chemical parameters analysed in the region, activities in the mines seemed to have also influenced the anions. Seasonal changes in the level of nitrate agree with the findings of Ndubuisi et al. (2021), which revealed that an increase in rainfall rates brought about runoff through fertilised farmlands, industrial discharges, domestic effluents, sewage sludge, leachates from refuse dumps, leguminous soil, decayed vegetables and animal matter. All of which could have contributed to the rise in nitrate levels during the wet season.

Nitrate, phosphate and ammonium were amongst the anions analysed by Okoye et al. (2022) during a study on spring water samples located in Enugu State. The result of the research showed that during the rainy season, the concentration of nitrate and phosphate increased in the water. This could be due to the use of phosphate, nitrate and potassium fertilisers in the growth of bitter leaf and ugu leaf near the locations (Okoye et al. 2022). The fertiliser and other pesticide residues could have infiltrated the soil particles and rock strata, poisoning the spring water aquifers. Nonetheless, during the dry season, the level of ammonium increased in the spring.

In the Ogbajarajara River, Imo state, the anions analysed were nitrate, phosphate and sulphate (Ihenetu et al. 2022). The results indicated the level of nitrate, phosphate and sulphate in the Ogbajarajara River increased during the wet season. Seasonal changes observed in the levels of nitrate and phosphate in the study agree with the findings of Okoye et al. (2022) in spring water samples located in Enugu State.

Cations

The cations studied by Ndubuisi et al. (2021) in the Amuzuta Stream, Abia State, were Ca, Mg, K and Na. The level of cations in the Amuzuta Stream all increased during the wet season.

In the Ubu River, Anambra state, the study by Uzoukwu et al. (2004) showed that Na and Mg concentrations increased during the dry season. A reduction in precipitation during the dry season could have possibly limited dilution by water runoff input into the Ubu River, which resulted in an increase in Na and Mg levels (Uzoukwu et al. 2004). Notwithstanding, the concentration of potassium and calcium increased during the wet season.

Calcium and magnesium were analysed in surface water close to Mayo and Amikol Pb–Zn mines at Ihetutu, Ebonyi State (Ezekwe et al. 2013). The findings of the research showed that Ca and Mg concentrations in surface water were higher in the dry season. According to Ezekwe et al. (2013), the Mayo mines are only active during the dry season. During the dry season, natural activities in the mining region, such as mobilisation of NaCl from brines and evaporating deposits, may have led to high Ca and Mg concentrations.

The concentrations of Ca, Mg and K were analysed in the Ogbajarajara River, Imo state, in a study by Ihenetu et al. (2022). The outcome of the research indicates that the level of all cations in the Ogbajarajara River increased during the wet season. The findings of the study agree with the state of the Amuzuta Stream, which was analysed by Ndubuisi et al. (2021).

Heavy Metals

Ndubuisi et al. (2021) studied seasonal changes in the concentration of copper (Cu), cadmium (Cd), manganese (Mn), zinc (Zn), iron (Fe) and lead (Pb) in the Amuzuta Stream, Abia State. The findings of the study revealed that Cu, Cd and Mn levels rose during the dry season. While Zn, Fe and Pb levels increased during the wet season. The elevated amounts of iron and lead found in the Amuzuta Stream during the rainy season might be due to mineral deposits from runoff on the surrounding soil and earth crust.

In Anambra state, Uzoukwu et al. (2004) studied heavy metals (Mn, Fe, Zn and Pb) in the Ubu River. The findings of the study showed that in the dry season, Mn and Pb concentrations increased, while during the wet season, Fe and Zn levels decreased. The rise in Fe and Zn content during the rainy season was possibly due to the increased contributions from leachates, erosion and water runoff during the wet season (Uzoukwu et al. 2004).

Heavy metals (Fe, Mn, Zn, Pb and Cd) were studied in surface water close to Mayo and Amikol Pb–Zn Mines at Ihetutu, Ebonyi State (Ezekwe et al. 2013). The result of the analysis showed that the concentrations of Fe, Mn, Zn and Pb increased in surface water during the dry season. The likely reason for the observed seasonal variation is that during the dry season, dilution of the surface water from the intake of water runoff does not occur. The effect of seasonal changes on Cd was not captured in the research because the concentration of Cd in both seasons was below the detectable limit.

In Enugu State, Okoye et al. (2022) did a study on six heavy metals (zinc, iron, lead, cadmium, mercury (Hg) and copper) in spring water. The research showed that the concentration of Fe and Zn in the spring increased during the dry season, whereas levels of Pb, Cd, Hg and Cu were on the rise during the wet season.

Research by Ihenetu et al. (2022) in Imo state observed the behaviour of seven heavy metals (Nickel (Ni), Cu, Mn, Zn, Fe, Pb and Mg) in the Ogbajarajara River. The findings of the study show that levels of Cu and Mn in the Ogbajarajara River rose during the dry season, whereas the concentration of Fe, Zn and Pb increased during the wet season. Moreover, the seasonal variation in the concentrations of Ni and Mg was not captured in the research because of the negligible concentration of both elements in the Ogbajarajara River. Mineral residues on the soil with a high quantity of iron might have influenced the seasonal fluctuations observed in iron.

Microorganisms

The total and faecal coliform bacteria tests are the most important indicators of drinking water safety (Ndubuisi et al. 2021). The tests described detect the total coliform bacteria concentration, which is linked to the presence of disease-causing organisms. In a study by Ndubuisi et al. (2021), the biological characteristics (which include total faecal coliform count, total faecal count, total heterotrophic bacteria count and total coliform counts) of the Amuzuuta stream, Southeastern Nigeria, were investigated. Seasonal variations in the results indicated higher mean microbial load values in the dry season compared to the values in the rainy season. This may be related to temperature variances (Ndubuisi et al. 2021). Microorganisms are more active at specific temperatures (20–40 °C) (Eszter 2011). This temperature range is predominant during the dry season, and it supports the conclusion of Ndubuisi et al. (2021).

Influence of Seasonal Variations on Groundwater Quality

Physical and Chemical Parameters

In Anambra state, Chukwura and Igwe (2021) studied groundwater proximal to automobile junk waste in Obosi and Onitsha environs, Southeast Nigeria. The physical and chemical parameters considered in the study include pH, EC, DO, TDS, COD and BOD. The result of the study identified that the concentration of all physical and chemical parameters in groundwater, except EC, increased during the rainy season.

In Ishiagu, Ebonyi State, Obiekezie et al. (2006) studied the seasonal variations in physiochemical qualities of groundwater proximal to a rock mining site. pH, TSS, TDS, DO, temperature, turbidity, conductivity and alkalinity were among the parameters of interest. The result of the study showed that the levels of pH and turbidity increased during the wet season, whereas TSS, TDS, DO, temperature, conductivity and alkalinity levels rose during the dry season. The fact that temperature and DO in groundwater have a similar seasonal pattern is thought to be owing to their inverse relationship (Obiekezie et al. 2006). When compared to the dry season, the rainy season has lower mean levels of conductivity, TSS, TDS and DO. This might be due to the rain's diluting impact and the fact that mining activities are decreased during the rainy season. (Obiekezie et al. 2006).

Another study on groundwater was conducted in a coal-enriched environment, Enugu State, by Nnaemeka-Okeke et al. (2019). The physical and chemical parameters analysed were temperature, EC, pH, TDS, TSS, total solid (TS), total acid, total alkaline, BOD, COD, turbidity and total hardness. The seasonal variations observed in the physical and chemical parameters studied include an increase in the concentration of EC, TDS, TSS, TS, total acid, turbidity and total hardness during the wet season; and a rise in the levels of temperature, pH, total alkaline, BOD and COD during the dry season. The minor increase in temperature during the dry season might be attributed to solar energy thermal storage (Nnaemeka-Okeke et al. 2019). The increased intrusion of aquifers with water-containing chemical fertilisers and municipal and industrial wastes may have led to the pH fall during the rainy season (Nnaemeka-Okeke et al. 2019). An increase in the content of ionisable minerals delivered into groundwater during the rainy season may be the cause of an increase in EC (Nnaemeka-Okeke et al. 2019). The rise in TSS and turbidity during the rainy season might be attributed to more silt and clay being transferred into groundwater by rainfall runoff on the terrain (Nnaemeka-Okeke et al. 2019). The flushing and percolation of water carrying dissolved acidic radicals into the groundwater during the rainy season might explain the rise in acidity loading (Nnaemeka-Okeke et al. 2019). Percolation of biodegradable organic matter and leaching of inorganic iron and/or manganese into groundwater aquifers might explain the increased BOD mean value during the dry season (Nnaemeka-Okeke et al. 2019). The longer build-up of seeping organic leachate into the aquifers during the dry season may explain the greater mean COD concentration (Nnaemeka-Okeke et al. 2019).

In a study by Braide et al. (2016) at Egbema, Imo State, Nigeria, 20 groundwater samples were collected from commercial boreholes. The physical parameters analysed in the study were temperature, pH, EC and TDS. The seasonal variation observed in the study showed that the mean value of all physical parameters was higher during the dry season. The dry season's high EC values can be ascribed to higher temperatures, which increase evaporation and result in a greater concentration of components in the water (Braide et al. 2016).

Anions

The study carried out in Obosi and Onitsha environs, Southeast Nigeria, by Chukwura and Igwe (2021), assessed the concentrations of chloride (Cl), sulphate (SO_4) and nitrate (NO_3) in groundwater. Seasonal variations observed in their research revealed that the level of NO_3 increased during the dry season, whereas the levels of Cl and SO_4 rose during the rainy season.

In Ishiagu, Ebonyi state, chloride and sulphate were the anions considered in the study by Obiekezie et al. (2006). The results of the study showed that chloride levels increased in groundwater during the dry season, while sulphate levels rose in the rainy season.

The study by Nnaemeka-Okeke et al. (2019) in Enugu city analysed the concentration of sulphate in groundwater. The result of the research showed that the level of sulphate increased during the wet season and this agrees with the findings of Obiekezie et al. (2006).

In Egbema, Imo state, four anions (Cl, NO_3, phosphate and sulphate) were examined in twenty groundwater samples (Braide et al. 2016). The mean values of the anions in the dry and wet seasons show that the levels of chlorine, nitrate and phosphate increased in groundwater during the dry season. Seasonal variations in sulphate could not be ascertained because changes in the levels of sulphate in both seasons were negligible.

Cations

The study carried out in Ishiagu, Ebonyi state, by Obiekezie et al. (2006) analysed potassium, calcium and sodium. The result of the analysis with regard to seasonal variation showed that the levels of potassium and sodium in groundwater increased in the rainy season, while the levels of calcium increased in the dry season.

In Enugu city, Nnaemeka-Okeke et al. (2019) studied sodium and potassium in groundwater. The seasonal variation observed in the study showed that potassium and sodium levels increased during the wet season. Observed seasonal variations in potassium may be due to an increase in runoff and agricultural activities in the wet season. Percolation of dissolved potassium from fertilisers into groundwater during the rainy season, in particular (Nnaemeka-Okeke et al. 2019).

In Egbema, Imo state, the concentration of calcium in groundwater was analysed during the wet and dry seasons (Braide et al. 2016). The result of the analysis indicated a clear seasonal variation. The level of calcium was higher during the dry season compared to the dry season.

Heavy Metals

Ubechu et al. (2021) studied the seasonal variations of nine heavy metals (cadmium, lead, chromium, mercury, aluminium, copper, nickel, arsenic and iron) in groundwater close to Enyimba Dumpsite, Aba, Southeastern Nigeria. The result of the analysis showed that the levels of cadmium, lead, chromium, mercury and aluminium rose during the dry season, while the levels of copper, nickel, arsenic and iron increased during the wet season. Phosphate fertilisers and sewage sludge are the two main sources of cadmium in the environment (WHO 2011). Increased runoff and use of phosphate fertilisers in the wet season possibly led to the seasonal variation of cadmium in their study. Increased runoff and leaching of decomposed industrial waste (lead-based paints, pipes, lead batteries, etc.) into the aquifer during the rainy season could have led to the observed changes in lead. Copper has poor mobility and does not react well with water because when exposed to water, it forms a thin protective metallic layer, which inhibits its reactivity with water (Obasi and Akudinobi 2020). Copper's poor reaction with water may have contributed to the drop in copper levels during the rainy season.

In Anambra state, Chukwura and Igwe (2021) analysed heavy metals in groundwater. The metals analysed include Ni, Mn, Pb, Cu, Zn, Fe and Cr. The result of the analysis revealed that Pb, Cu and Fe concentrations increased in groundwater during the rainy season, while Ni, Mn, Zn and Cr concentrations increased in the dry season. Some heavy metals have been shown to scavenge manganese in the aqueous environment, according to studies (Arinze et al. 2015; Wenfeng et al. 2005). This could be responsible for the drop in Mn during the rainy season.

The study conducted in Ebonyi state by Obiekezie et al. (2006) analysed the concentration of iron, lead, cadmium, zinc and copper in the dry and wet seasons. The findings of the study revealed that iron and cadmium levels rose in the wet season. Seasonal variations in the concentration of lead, zinc and copper in groundwater could not be identified due to negligible changes in their levels during both seasons.

Nnaemeka-Okeke et al. (2019) analysed for iron, zinc and lead in 12 groundwater samples in Enugu urban. The result of the analysis revealed that iron and zinc levels in groundwater increased during the rainy season. The spike in zinc levels during the rainy season might be due to the leaching of some zinc-based fertilisers into groundwater (Nnaemeka-Okeke et al. 2019). Seasonal changes in lead could not be judged in their study due to negligible differences in lead concentration in the wet and dry seasons.

In Imo state, the concentration of Fe, Zn and Cr was analysed in groundwater by Braide et al. (2016). The findings of the study showed that Zn levels increased during the dry season, while Cr levels heightened during the wet season. Seasonal changes in Fe could not be identified in their study due to negligible changes in the level of Fe in dry and wet seasons.

Microorganisms

Standard microbiological techniques were applied during the microbiological analysis of groundwater samples collected in Imo state, Southeastern Nigeria (Braide et al. 2016). The parameters analysed were total aerobic count (TAC), total coliform count (TCC), total hydrocarbon count (THC), total fungal count (TFC) and total salmonella-shigella count (TSSC). Seasonal variation in the levels of TCC and TSSC in their study was negligible. Nevertheless, seasonal variation was evident in the cases of TAC, THC and TFC. The microbial load reveals that there was a rise in the levels of TAC, THC and TFC during the dry season. Similar findings were also reported by Ndubuisi et al. (2021) in the Amuzuta stream, Southeastern Nigeria. Rise in microbial load during the dry season could be a result of temperature variation.

Conclusion

There are clear signatures of seasonal variations in the quality of water resources across Southeastern Nigeria. Variations in temperature and rainfall are the two major drivers of seasonal change, which in turn influence the levels of physicochemical parameters and microbial loads in water resources. The majority of the physicochemical parameters and heavy metals analysed in water did not follow a regular pattern with regard to seasonal change in parts of the study region. It was observed that, besides seasonal changes, other factors, especially anthropogenic activities (particularly mining and farming), influenced the quality of the water in the region. This possibly explains the irregular patterns of the physicochemical parameters and heavy metals in the water resources. Nevertheless, there were some clear patterns recognised. It was observed that in surface water resources, the majority of the studies reported an increase in the levels of microbial loads, temperature and manganese during the dry season and a rise in pH, nitrate, sulphate, phosphate, calcium, potassium and iron during the wet season. On the other hand, a review of groundwater quality in the study region showed that the majority of studies reported an increment in microbial load and pH during the dry season and a rise in iron level during the rainy season. Conclusively, it suffices to say that, after a thorough literature review, an increase in microbial load and pH during the dry season and a rise in iron level during the rainy season are the most obvious signatures of seasonal changes observed in analysed water resources in Southeastern Nigeria. It is also pertinent to note that some heavy metals (e.g. cadmium, lead, chromium, mercury, aluminium, copper, nickel, arsenic, etc.), which are mostly influenced by anthropogenic activities, did not show regular patterns in both seasons.

Challenges, Recommendations and Future Perspectives

A lot of hurdles were faced in the course of the present study. Firstly, a good number of studies that reported on the state of water resources in Southeastern Nigeria did not identify the sampling season. This limited the number of studies that were relevant to discovering the seasonal variations in the region. Therefore, we urge researchers to ensure that the season of water sample collection is clearly stated in the methodology section to aid a better understanding of the influence of seasonal variation on the quality of water resources globally. Secondly, some studies that analysed surface and groundwater in the region mixed up the results of both water sources. This made some quality research articles not to be considered in the present study. Hence, we encourage future research on water resources to endeavour to present comprehensive results of laboratory analysis for a better understanding of water resources quality. Thirdly, a few studies analysed surface and groundwater during both seasons. We understand that funding was possibly the major factor. Thus, we encourage the government, research institutes, schools, and other research establishments to allocate more funding to research on water resources in Nigeria to aid in understanding the outcomes of seasonal changes on the quality of water resources in the region.

Many of the local inhabitants who make use of the water resources in Southeastern Nigeria for survival are unaware of the health risk exposure. The government, water managers and parastatals need to organise seminars to enlighten the people on the risks involved in consuming unsafe water. Additionally, in Southeastern Nigeria and other parts of the world, there is a need for people to be educated on the need and different ways to treat water before use. In the present study, it was revealed that the quality of groundwater is better than surface water. However, most inhabitants of rural communities in Southeastern Nigeria rely primarily on surface water. Therefore, we also call on the government to provide an alternative source of water like boreholes in rural communities, as some surface waters are severely polluted. Anthropogenic activities were also found to be amongst the factors influencing the quality of water resources, alongside variations in rainfall and temperature. The agencies responsible for regulating human activities, protecting and sustaining natural resources, and monitoring and accessing water resources need to collectively ensure that water resources in the region are protected from natural and anthropogenic source pollution. In addition, we also beckon on non-governmental and governmental agencies responsible for the implementation of climate change policies in Nigeria, and other parts of the world to do the needful for the sustainability of human resources and preservation of natural resources for the present and future generation. We also encourage the adoption of more environmental practices, especially with the use of energy. Renewable and environmentally friendly energy sources like solar energy, wind energy and hydro energy should be embraced.

This review article indicates that seasonal and climate change aggravate existing water quality problems in Southeastern Nigeria. Thus, there is a need to incorporate seasonal and climate change in water resources management and planning. It is hoped that the findings of this study will promote more research towards understanding the impacts of seasonal and climate change on the quality of water resources in different regions.

References

Adelana SMA, Olasehinde PI, Bale RB, Vrbka P, Edet AG, Goni IB, (2008) An overview of the geology a hydrogeology of Nigeria. In: Adelana SMA, McDonald AM (eds), Applied groundwater studies in Africa, Leiden, Netherlands: CRC Press/Balkema, vol 13, pp 171–198

Adewuyi SA, Folorunso B, Okojie LO, Akerele D (2015) Effect of climate change on food crop production and vulnerability assessment in Oyo State. J Econ Int Financ 7:18–24

Adumanya OCU, Osuji CN, Nwinee SA, Ofurum OC (2022) Physicochemical and bacteriological properties of packaged water sold in Imo State, Nigeria: a case study of Owerri municipal council. J Pollut Monit Evaluat Studies Control 1(1):4–8

Agbasi JC, Egbueri JC (2022) Assessment of PTEs in water resources by integrating HHRISK code, water quality indices, multivariate statistics, and ANNs. Geocarto Int. https://doi.org/10.1080/10106049.2022.2034990

Agbor CA, Ashabua DA (2018) Basic education curriculum in South-South Nigeria: challenges and opportunities of quality contents in French language learning. Global J Educ Res 17:97–101

Ahmed T, Zounemat-Kermani M, Scholz M (2020) Climate change, water quality and water-related challenges: a review with focus on Pakistan. Int J Environ Res Public Health 17(22):8518

Ahmed YA, Aliyu I (2019) Climate Change Induced challenges on deforestation: the needs to educe mitigation measures in Nigeria. Analele Universităţii Din Oradea, Seria Geografie 29(2):64–76

Ajiere SI, Nwagbara MO (2018) Effect of climate change on maize (Zea-mays) and cassava (Manihot-esculenta) yields in selected states of South Eastern Nigeria. J Agric Crops 4(2):8–15

Akubugwo EI, Duru MK (2011) Human activities and water quality: a case study of Otamiri river, Owerri, Imo State, Nigeria. Glob Res J Sci 1:48–53

Alum OL, Okoye CO, Abugu HO (2021) Quality assessment of groundwater in an agricultural belt in eastern Nigeria using a Water Quality Index. Afr J Aquat Sci 46(3):304–318

Amanullah, Khalid S, Khan HA, Arif M, Altawaha AR, Adnan M, Fahad S, Shah A, Parmar B (2020). Effects of climate change on irrigation water quality. In: Environment, climate, plant and vegetation growth. Springer, Cham, pp 123–132

Anabaraonye B, Okafor JC, Hope J (2018) Educating farmers in rural areas on climate change adaptation for sustainability in Nigeria. Int J Sci Eng Res 10(4):1391–1398

Anyanwu ED, Jonah UE, Adetunji OG, Nwoke OB (2022) An appraisal of the physicochemical parameters of Ikwu River, Umuahia, Abia State in South-eastern, Nigeria for multiple uses. Int J Energy Water Resour, 1–8

Areola O (1983) Soil and vegetal resources. In: Oguntoyinbo JS, Areola OO, Filani M (eds) A geography of Nigeria development. Ibadan, Heinemann Education Books, pp 105–126.

Arinze IE, Igwe O, Una CO (2015) Analysis of heavy metals' contamination in soils and water at automobile junk markets in Obosi and Nnewi, Anambra State, Southeastern Nigeria. Arab J Geosci. https://doi.org/10.1007/s12517-015-2001-6

Ayanlade A, Oladimeji AA, Okegbola OM, Eludoyin AO, Eslamian S, Ayinde AF, Perkins PE (2022) Effect of climate change on water availability and quality: an assessment of socio-resilience in Nigeria. In: Disaster risk reduction for resilience. Springer, Cham, pp 245–262

Ayejoto DA, Agbasi JC, Egbueri JC, Echefu K (2022) Assessment of oral and dermal health risk exposures associated with contaminated water resources: an update in Ojoto area, southeast Nigeria. Int J Environ Analyt Chem. https://doi.org/10.1080/03067319.2021.2023515

Babatunde JA, Salami AT, Tadross M (2011) Developing climate change scenarios, biophysical impacts and adaptation strategies in Nigeria. A final report submitted to Nigerian environmental study/action team (NEST) as part of the building nigeria response to climate change (BNRCC) project

Balogun I, Adegun O, Adeaga, O (2012) An assessment of rainfall Erosivity in parts of Eastern Nigeria: a case study of Owerri and Enugu. Hydrology for disaster management

Braide W, Nwachukwu J, Adeleye SA, Egbadon EO (2016) Effects of gas flaring on the physico-chemical and microbiological quality of water sources in Egbema, Imo State, Nigeria. Int Lett Nat Sci, 56

Cavanagh AP, South PF, Bernacchi CJ, Ort DR (2022) Alternative pathway to photorespiration protects growth and productivity at elevated temperatures in a model crop. Plant Biotechnol J 20(4):711–721

Chidiebere-Mark NM, Eze CC, Nwankwo OO, Onyeneke RU, Ukwandu JC, Ejike RD, Enyia CO (2014) Perception and adaptation of food crop farmers to climate change in Imo State, Nigeria. Paper presented at the 14th annual national conference of the nigerian association of agricultural economists, federal university of technology, Akure, 24th–27th February, 2014

Chikezie C, Ibekwe UC, Ohajianya DO, Orebiyi JS, Henri-Ukoha A, Ukoha II, Osuji MN, Anthony G (2016) Climate change and perceived climate hazards: a trend analysis in Southeast Nigeria. Int J Weather Climate Change Conserv Res 2(1):1–10

Chinye-Ikejiunor N, Iloegbunam GO, Chukwuka A, Ogbeide O (2021) Groundwater contamination and health risk assessment across an urban gradient: case study of Onitcha metropolis, south-eastern Nigeria. Groundw Sustain Dev 14:100642. https://doi.org/10.1016/j.gsd.2021.100642

Chukwura UO, Igwe O (2021) Seasonal evaluation of the hydrogeochemical interpretations of groundwater quality: a case study from automobile junk waste in Obosi and Onitsha environs. Southeast Nigeria. Arabian J Geosci 14(4):1–24

Conway G, Joshi A, Leach F, García A, Senecal PK (2021) A review of current and future powertrain technologies and trends in 2020. Transp Eng 5:100080

Diagi BE, Nwagbara MO (2018) Perceived impact of climate change on swamp rice cultivation by farmers in Ebonyi State, Southern Nigeria. Archi Current Res Int 14(2):1–10

Ding S, Dan SF, Liu Y, He J, Zhu D, Jiao L (2022) Importance of ammonia nitrogen potentially released from sediments to the development of eutrophication in a plateau lake. Environ Pollut 305:119275

Dong Z, He Y, Ren Y, Wang G, Chu D (2022) Seasonal and year-round distributions of Bactrocera dorsalis (Hendel) and its risk to temperate fruits under climate change. Insects 13(6):550

Duru Pat N (2014) Flooding in Imo State Nigeria: The socio-economic implication for sustainable development. Human Soc Sci Lett 2(3):129–140

Duru KC, Nwanekwu KE, Adindu EA, Odika PC (2012) Heavy metal and bioload levels of Otamiri river, Owerri, Imo State, Nigeria. Arch Appl Sci Res 4(2):1002–1006

Egbueri JC, Agbasi JC (2022a) Data-driven soft computing modeling of groundwater quality parameters in southeast Nigeria: comparing the performances of different algorithms. Environ Sci Pollut Res 29(25):38346–38373. https://doi.org/10.1007/s11356-022-18520-8

Egbueri JC, Agbasi JC (2022b) Combining data-intelligent algorithms for the assessment and predictive modeling of groundwater resources quality in parts of southeastern Nigeria. Environ Sci Pollut Res. https://doi.org/10.1007/s11356-022-19818-3

Egbueri JC, Ameh PD, Enyigwe MT, Unigwe CO (2021) Entropy-based analysis of the impact of environmentally sensitive elements on groundwater quality of the Ameka region of southeast Nigeria: medical geology implications. Anal Lett 54(7):1193–1223. https://doi.org/10.1080/00032719.2020.1797076

Egbueri JC, Ayejoto DA, Agbasi JC (2022) Pollution assessment and estimation of the percentages of toxic elements to be removed to make polluted drinking water safe: a case from Nigeria. Toxin Rev. https://doi.org/10.1080/15569543.2021.2025401

Egbueri JC, Igwe O (2021) The impact of hydrogeomorphological characteristics on gullying processes in erosion-prone geological units in parts of southeast Nigeria. Geology Ecology Landscapes 5(3):227–240. https://doi.org/10.1080/24749508.2020.1711637

Egbueri JC (2022) Predicting and analysing the quality of water resources for industrial purposes using integrated data-intelligent algorithms. Groundw Sustain Dev. https://doi.org/10.1016/j.gsd.2022.100794

Egbueri JC, Agbasi JC (2022c) Performances of MLR, RBF-NN, and MLP-NN in the evaluation and prediction of water resources quality for irrigation purposes under two modeling scenarios. Geocarto Int. https://doi.org/10.1080/10106049.2022.2087758

Enaruvbe GO, Yesuf GU (2016) Spatial analysis of flood disaster in Delta State. Nigeria. Ife Res Publ Geogr 11(1):52–58

Enete IC (2014) Impacts of climate change on agricultural production in Enugu State, Nigeria. J Earth Sci Climate Change 5:234

Enyigwe MT, Onwuka OS, Egbueri JC (2021) Geochemical distribution, statistical and health risk assessment of toxic elements in groundwater from a typical mining district in Nigeria. Environ Forensics. https://doi.org/10.1080/15275922.2021.1907822

Eszter T (2011) Soil CO_2 Emission measurements in different soil use systems PhD thesis at Cornviscus University of Budapest, pp 1–2

Ewelike NC, Okoli CA, Echendu MN, Enekwa JU (2022) Bacteriological assessment of borehole water in some communities in Owerri West, Southeastern Nigeria. GSC Biol Pharmaceut Sci 18(3):177–181

Eyankware MO, Akakuru OC (2022) Appraisal of groundwater to risk contamination near an abandoned limestone quarry pit in Nkalagu, Nigeria, using enrichment factor and statistical approaches. Int J Energ Water Res https://doi.org/10.1007/s42108-022-00186-0

Ezekwe IC, Ezekwe AS, Chima GN (2013) Metal loadings and alkaline mine drainage from active and abandoned mines in the Ivo River basin area of southeastern Nigeria. Mine Water Environ 32(2):97–107

Ezemonye M, Emeribe C (2012) Rainfall Erosivity in Southeastern Nigeria. Ethiopian J Environ Studies Manag 5(2). https://doi.org/10.4314/ejesm.v5i2.1

Ezenwaji NE, Andong FA, Nnachi IA, Aniede D, Okwuonu ES, Ndefo JC (2022) Factors that influence the toxicity levels of metals in water and soils: a case study of Lake Nike, Enugu. Nigeria. Aquatic Sci 84(2):1–11

Fan X, Cui B, Zhang K, Zhang Z, Shao H (2012) Water quality management based on division of dry and wet seasons in Pearl River Delta, China. CLEAN—Soil, Air, Water 40(4):381–393. https://doi.org/10.1002/clen.201100012
Forstinus NO, Ikechukwu NE, Emenike MP, Christiana AO (2016) Water and waterborne diseases. Rev Int J Tropical Diseases Health 12:1e14
Ganiyu SA, Badmus BS, Olurin OT, Ojekunle ZO (2018) Evaluation of seasonal variation of water quality using multivariate statistical analysis and irrigation parameter indices in Ajakanga area, Ibadan. Nigeria. Appl Water Sci 8:35. https://doi.org/10.1007/s13201-018-0677-y
Ghazi B, Jeihouni E (2022) Projection of temperature and precipitation under climate change in Tabriz. Iran. Arabian J Geosci 15(7):1–11
Gomes C, Sperandio D, Borges N, Barbosa A, Viçozzi A, Dias GP (2019) Water quality assessment through PCA analysis. Int J Adv Eng Res Sci 6(3):202–210
Gutiérrez JE, Gutiérrez-Hoyos N, Gutierrez Benedetti JS, Vives MJ, Sivasubramanian V (2022) Clarification of cyanotoxins in El Guajaro Reservoir, Colombia using a microalgae-based consortium MPMC. J Chem Technol Biotechnol
Hachad M, Yarahmadi H, Duy SV, Sauvé S, Prévost M, Dorner S (2022) Occurrence and partitioning behavior of E. coli and wastewater micropollutants following rainfall events. Resour Environ Sustain, 100067
Ibeje AO, Ekeleme A (2020) Statistical analysis of climate data of Southeast Nigeria. Umudike J Eng Technol 6(2):82–87. https://doi.org/10.33922/j.ujet_v6i2_10
Ifeanyichukwu KA, Okolo CM, Odoh A (2022) evaluation of toxic elements in groundwater in the industries area of Nnewi North South-Eastern Nigeria. Environ Contaminants Rev (ECR) 5(1):31–34. https://doi.org/10.26480/ecr.01.2022.31.34
Ifediegwu SI, Ozoko DC, Aganigbo IC (2021) Multivariate statistical and GIS methods for the assessment of heavy metal toxicity in Ekulu River, Southeastern, Nigeria. Int J Energ Wat Resour https://doi.org/10.1007/s42108-021-00133-5
Ifeka A, Akinbobola A (2015) Trend analysis of precipitation in some selected stations in Anambra State. Atmosph Climate Sci 05(01):1–12. https://doi.org/10.4236/ACS.2015.51001
Igbokwe JI, Akinyede JO, Dang B, Alaga T, Ono MN, Nnodu VC, Anike LO (2008) Mapping and monitoring of the impact of gully erosion in Southeastern Nigeria with satellite remote sensing and Geographic Information System. Int Arch Photogram Remote Sens Spat Inf Sci 37(part B8):865–872
Ighalo JO, Adeniyi AG (2020) A comprehensive review of water quality monitoring and assessment in Nigeria. Chemosphere 260:127569
Ihenetu SC, Njoku VO, Ibe FC, Li G, Chinweuba A, Enyoh CCE (2022) Determination of pollutions in the surface of water samples from Ogbajarajara river, Nigeria by spectrophotometer and atomic absorption spectrometry before evaluation of health risk assessment. Analyt Methods Environ Chem J 5(01):5–21
IPCC (2021) Annex VII: Glossary [Matthews JBR, Möller V, van Diemen R, Fuglestvedt JS, Masson-Delmotte V, Méndez C, Semenov S, Reisinger A (eds)]. In Climate change 2021: the physical science basis. contribution of working group I to the sixth assessment report of the intergovernmental panel on climate change [Masson-Delmotte V, Zhai P, Pirani P, Connors SL, Péan C, Berger S, Caud N, Chen Y, Goldfarb L, Gomis MI, Huang M, Leitzell K, Lonnoy E, Matthews JBR, Maycock TK, Waterfield T, Yelekçi O, Yu R, Zhou B (eds)]. Cambridge University Press, Cambridge, United Kingdom and New York, NY, USA, pp 2215–2256. https://doi.org/10.1017/9781009157896.022
Karmakar B, Singh MK, Choudhary BK, Singh SK, Egbueri JC, Gautam S, Rawat SK (2021) Investigation of the hydrogeochemistry, groundwater quality and associated health risks in industrialized regions of Tripura, northeast India. Environ Forensics. https://doi.org/10.1080/15275922.2021.2006363
Ladan SI (2014) An appraisal of climate change and agriculture in Nigeria. J Geogr Regional Plann 7(9):176–184

Li S, Zhang Q (2010) Risk assessment and seasonal variations of dissolved trace elements and heavy metals in the Upper Han River, China. J Hazard Mater 181(1–3):1051–1058. https://doi.org/10.1016/j.jhazmat.2010.05.120

Lkr A, Singh MR, Puro N (2022) Spatio-temporal influence on river water chemistry of Doyang river, Nagaland, India, using multivariate techniques. Int J Environ Sci Technol (2022).https://doi.org/10.1007/s13762-021-03897-9

Maniquiz MC, Lee SY, Kim LH (2010) Multiple linear regression models of urban runoff pollutant load and event mean concentration considering rainfall variables. J Environ Sci Chin 22(6):946–952

Malhi GS, Kaur M, Kaushik P (2021) Impact of climate change on agriculture and its mitigation strategies: a review. Sustainability 13(3):1318

Malik DS, Rathi P (2022) The influence of Physico-chemical parameters on habitat ecology and assemblage structure of freshwater phytoplankton in Tehri Reservoir Garhwal (Uttarakhand) India. Proceedings of the national academy of sciences, India Section B: Biological Sciences, 1–11

Mathew MD (2022) Nuclear energy: a pathway towards mitigation of global warming. Prog Nucl Energy 143:104080

Muhammad S, Ullah I (2022) Spatial and seasonal variation of water quality indices in Gomal Zam Dam and its tributaries of south Waziristan District, Pakistan. Environ Sci Pollut Res 29:29141–29151

Nagajyoti PC, Lee KD, Sreekanth T (2010) Heavy metals, occurrence and toxicity for plants: a review. Environ Chem Lett 8:199–216

Nas SS, Nas E (2009) Water quality modeling and dissolved oxygen balance in streams: a point source Streeter-Phelps application in the case of the Harsit Stream, Clean—Soil Air Water 37 (1):67–74

Nashwan MS, Shahid S (2022) Future precipitation changes in Egypt under the 1.5 and 2.0 °C global warming goals using CMIP6 multimodel ensemble. Atmosph Res 265:105908

Ndubuisi CC, Nwakanma C, Onyeizu UR (2021) Assessment of water quality status and benthic fauna diversity of Amuzuta stream, Umuahia. Southeast Nigeria. Nigeria Agric J 52(3):209–221

Njoku JD (2006) Analysis of the effect of global warming on forest of southeastern Nigeria using remotely sensed data. A PhD Thesis: Imo State University, Owerri, Nigeria

Nganje TN, Hursthouse AS, Edet A, Stirling D, Adamu CI (2015) Adamu Assessment of the health risk, aesthetic and agricultural quality of rainwater, surface water and groundwater in the shale bedrock areas, Southeastern Nigeria Water Qual. Expo Health 7 (2):153–178

Nnadi OI, Liwenga ET, Lyimo JG, Madukwe MC (2019) Impacts of variability and change in rainfall on gender of farmers in Anambra. Southeast Nigeria. Heliyon 5(7):e02085. https://doi.org/10.1016/j.heliyon.2019.e02085

Nnaemeka-Okeke RC, Eze-Steven PE, Ugwu CC (2019) Evaluating the impact of climate change on the quality of ground water—case study of a coal enriched environment in Enugu Urban. In: Laryea S, Essah E (eds) Procs west Africa built environment research (WABER) conference, 5–7 August 2019, Accra, Ghana, 536–550

Nwagbara MO (1997) Temperature characteristics and variability over cities in southeastern Nigeria (1965-1994). An MSc Thesis: Department of Geography, University of Port-Harcourt, Port-Harcourt

Nwagbara MO (2008) Landcover change in relation to climate change in northern Nigeria using GIS techniques. A PhD Thesis: Department of Geography and Planning, Abia State University, Uturu, Nigeria.

Nwagbara MO, Ibe OK (2015) Climate change and soil conditions in the tropical rainforest of southeastern Nigeria. J Geogr Earth Sci 3(1):99–106

Nwaiwu IOU, Ohajianya DO, Orebiyi JS, Ibekwe UC, Lemchi LJ, Onyeagocha SUO, Odoemena B, Utazi CO, Osuagwu CO, Tasie CM (2013) Climate change trend and appropriate mitigation and adaptation strategies in Southeast Nigeria. GJBAHS 3(1):120–125

Nwajiuba CU, Onyeneke R (2010) Effects of climate change on the agriculture of Sub-Saharan Africa: lessons from southeast rainforest zone of Nigeria. Paper presented at the 10th global conference on Business and Economics, St. Hugh's Colleges University Press

Nwinyi O, Uyi O, Awosanya E, Oyeyemi I, Ugbenyen A, Muhammad A, Alabi O, Ekwunife O, Adetunji C, Omoruyi I (2020) Review of drinking water quality in Nigeria: towards attaining the sustainable development goal six. Ann Sci Technol 5(2):58–77. https://doi.org/10.2478/ast-2020-0014

Obasi PN, Akudinobi BB (2020) Potential health risk and levels of heavy metals in water resources of lead–zinc mining communities of Abakaliki, southeast Nigeria. Appl Water Sci 10(7):1–23

Obiekezie SO, Okereke JN, Anyalogbu E, Okorondu SI, Ezejiofor TIN (2006) Underground water quality of rock mining in Ishiagu, Ebonyi State, Nigeria. Estudos de Biologia 28(63)

Odabaşi SÜ, Ceylan Z, Şentürk İ, Akbal F, Bakan G, Büyükgüngör H (2022) Investigation of spatial and seasonal variation of water quality along the mid-Black Sea coast (from Sinop to Ordu) of Turkey, by multivariate statistical techniques. Regional Studies in Marine Science, 102169

Odjugo PAO (2013) Analysis of climate change awareness in Nigeria. Scientific Res Essays 8(26):1203–1211

Odo SE, Emmanuel CP, Nwosu NH, Chinomso FG (2021) Evaluation of the bacterial quality and physicochemical properties of waterside River Ogbor-Hill Aba. Nigeria. Path Sci 7(12):1019–1025

Odo SN, Odo MK, Anyanwu ED, Ikechukwu PI (2022) Spatial and temporal variations of physical-chemical parameters of the Akor river in Ikwuano local government of Abia state, Nigeria. Limnology and Freshwater Biology, 1217–1228

Oguntoke O, Aboderin O, Bankole A (2009) Association of water-borne diseases morbidity pattern and water quality in parts of Ibadan City, Nigeria Tanzania. J Health Res, 11

Oh H, Choi JH (2022) Changes in the dissolved organic matter characteristics released from sediment according to precipitation in the Namhan River with Weirs: a laboratory experiment. Int J Environ Res Public Health 19(9):4958

Ohaegbuchu HE, Ijeh BI, Aigba PI, Ozurumba OG (2019) Assessment of groundwater quality in Isiala-Ngwa North LGA of Abia State, Nigeria. Int J Agric Earth Sci 5(1)

Ojukwu CK, Chukwu-Okeah GO, Ezekwe CI, Mmom PC (2022) Assessment of major water sources in Onicha local government area of Ebonyi State Nigeria using water quality index. Amer J Eng Res 11(1):60–66

Okafor VN, Omokpariola DO, Igbokwe EC, Theodore CM, Chukwu NG (2022) Determination and human health risk assessment of polycyclic aromatic hydrocarbons (PAHs) in surface and ground waters from Ifite Ogwari, Anambra State, Nigeria. Int J Environ Analyt Chem, 1–23

Okechukwu ME, Mbajiorgu CC (2020) Spatial distribution of rainfall and reference evapotranspiration in southeast Nigeria. Agric Eng Int 22(1)

Okenmuo FC (2021) Erosion quantification and management: Southeastern Nigeria case study. In: Ergen M, Ergen YB (eds), Landscape architecture framed from an environmental and ecological perspective. Intech Open. https://doi.org/10.5772/intechopen.99551

Okorie FC, Njoku JD, Onweremadu EU, Iwuji MC (2019) Climate change and food crop production in Imo State Nigeria: an analysis of 30-year period. African J Soc Behav Sci 9:(2)

Okorie FC, Okeke I, Nnaji A, Chibo C, Pat-Mbano E (2012). Evidence of climate variability in imo state of Southeastern Nigeria. J Earth Sci Eng 2:544–553

Okoro J, Olaolu MO, Igbokwe EM (2016) Climate change mitigation and adaptation strategies used by farmers in Imo state. Nigeria. J Agric Ext 20(2):130–142

Okoye ECS, Onuorah SC, Okoye LC, Nwadiogbu JO (2022) Effect of seasonal variations on the physicochemical characteristics of spring water in Oji River LGA, Enugu State Nigeria. Arch Agric Environ Sci 7(1):86–92

Olabi AG, Abdelkareem MA (2022) Renewable energy and climate change. Renew Sustain Energy Rev 158:112111

Olufemi M (2018) Climate change and effects on present-day Nigeria. Nigerian Tribune. www.tribuneonline.ng. Accessed 29th June 2022

Omaka ON, Aghamelu OP, Ike-Amadi CA, Ofoezie RC (2017) Assessment of the quality of groundwater from different parts of southeastern Nigeria for potable use. Environ Earth Sci 76(9). https://doi.org/10.1007/s12665-017-6680-z

Omeka ME, Igwe O, Unigwe CO (2022) An integrated approach to the bioavailability, ecological, and health risk assessment of potentially toxic elements in soils within a barite mining area. SE Nigeria. Environ Monit Assess 194(3):1–30

Omoja UC, Okpalaku BN, Uchechukwu UN, Obiekezie TN (2021) Variability of Rainfall in Awka, Anambra State Nigeria. IOSR J Appl Phys 13(4):50–56

Onyemesili OC, Egbueri JC, Ezugwu CK (2020) Assessing the pollution status, ecological and health risks of surf-ace waters in Nnewi urban, Nigeria: implications of poor waste disposal. Environ Forensics. https://doi.org/10.1080/15275922.2020.1850564

Onyeneke RU, Mmagu CJ (2015) Crop Farmers' understanding of climate change, and adaptation practices to climate change in southeast Nigeria. Proceedings of the Conference in honour of Professor Chinedum Uzoma Nwajiuba held at the Oak of Mamre Library, Imo State, Nigeria, July 19 and 20

Onyeneke RU, Amadi MU, Njoku CL, Osuji EE (2021) Climate change perception and uptake of climate-smart agriculture in rice production in Ebonyi State. Nigeria. Atmosphere 12(11):1503

Osugiri II, Ben-Chendo GN, Okwara MO, Anyanwu GU, Utazi CO (2019) Analysis of the trends of climatic variables in Southeast Nigeria. Int J Current Res 11:3910–3913. https://doi.org/10.24941/ijcr.31970.05.2019

Ozor N, Madukwe MC, Enete AA, Amaechina EC, Onokala P, Eboh EC, Ujah O, Garforth C (2012) A framework for agricultural adaptation to climate change in Southern Nigeria. Int J Agric Sci 4(5):243–251

Ozor N, Umunakwe PC, Ani AO, Nnadi FN (2015) Perceived impacts of climate change among rural farmers in Imo State. Nigeria. African J Agric Res 10(14):1756–1764

Perera F, Nadeau K (2022) Climate change, fossil-fuel pollution, and children's health. N Engl J Med 386(24):2303–2314

Prathumratana L, Sthiannopkao S, Kim KW (2008) Relationship of climatic and hydrological parameters to surface water quality in the lower Mekong River. Environ Int 34:860–866

Quinton JN, Catt JA, Hess TM (2001) The selective removal of phosphorus from soil: is event size important?. J Environ Qual 30(2):538–545

Raghuvanshi R, Ansari MA (2016) Farmer's awareness about climate change and adaptation practices: a review. Res Rev J Agric Sci Technol 5(3):41–51

Ridder NN, Ukkola AM, Pitman AJ, Perkins-Kirkpatrick SE (2022) Increased occurrence of high impact compound events under climate change. npj Climate Atmosph Sci 5(1):1–8

Seddon N, Chausson A, Berry P, Girardin CA, Smith A, Turner B (2020) Understanding the value and limits of nature-based solutions to climate change and other global challenges. Philos Trans R Soc B 375(1794):20190120

Sesana E, Gagnon AS, Ciantelli C, Cassar J, Hughes JJ (2021) Climate change impacts on cultural heritage: a literature review. Wiley Interdisciplinary Rev Climate Change 12(4):e710

Sharma RK, Agrawal M (2005) Biological effects of heavy metals: an overview. J Environ Biol 26:301–313

Shigaki F, Sharpley A, Prochnow LI (2007) Rainfall intensity and phosphorus source effects on phosphorus transport in surface runoff from soil trays. Sci Total Environ 373(1):334–343. https://doi.org/10.1016/j.scitotenv.2006.10.048

Shiru MS, Shahid S, Park I (2021) Projection of Water availability and sustainability in Nigeria due to climate change. Sustainability 13(11):6284

Sivakumar R, Prasanth BR, Ramaraj M (2022) An empirical approach for deriving specific inland water quality parameters from high spatio-spectral resolution image. Wetlands Ecol Manage 30(2):405–422

Sporre-Money JL, Lanyon LE, Sharpley AN (2004) Low-intensity sprinkler for evaluating phosphorus transport from different landscape positions. Appl Eng Agric 20(5):599–604

Sun X, Wang G (2022) Causes for the negative scaling of extreme precipitation at high temperatures. J Climate, 1–37

Tsihrintzis VA, Hamid R (1997) Modeling and management of urban runoff quality: a review. Water Resour Manage 11:137–164

Ubechu BO, Israel HO, Amadi CC, Ofor IJ (2021) Seasonal variations of cyanide and heavy metals in groundwater around Enyimba Dumpsite, Aba Southeastern Nigeria

Uguru MI, Baiyeri KP, Aba SC (2011) Indicators of climate change in the derived savannah niche of Nsukka, south-eastern Nigeria

Unigwe CO, Egbueri JC, Omeka ME (2022) Geospatial and statistical approaches to nitrate health risk and groundwater quality assessment of an alluvial aquifer in SE Nigeria for drinking and irrigation purposes. J Indian Chem Soc 99(6):100479. https://doi.org/10.1016/j.jics.2022.100479

Uwa CU, Nwachukwu AN, Olagunju TE, Babalola SO, Sowechi EE (2022) Environmental pollution assessment of surface water in omege community using chemical data and risk-based evaluation. Polish J Nat Sci 36(4):419–431

Uzoukwu BA, Ngoka C, Nneji N (2004) Monitoring of seasonal variation in the water quality of Ubu river in Ekwusigo and Nnewi local government areas of Anambra State. Nigeria. Environ Manag 33(6):886–898

Verla AW, Verla EN, Amaobi CE, Enyoh CE (2018) Water Pollution scenario at river uramurukwa flowing through owerri metropolis, Imo State, Nigeria. Int J Appl Scientific Res (IJASR) 3(3):40–46

WACDI (Women and Children Development Initiative) (2011) Gender dimensions and indigenous knowledge systems for adaptation to climate change in the Southeast Agro-Ecological Zone of Nigeria. Final Project Report submitted to the Nigerian Environmental Study/Action Team (NEST) as part of Building Nigeria's Response to Climate Change Project

Warrier AK, Kulkarni B, Amrutha K, Jayaram D, Valsan G, Agarwal P (2022) Seasonal variations in the abundance and distribution of microplastic particles in the surface waters of a Southern Indian Lake. Chemosphere 300:134556

Waters S, Hamilton D, Pan G, Michener S, Ogilvie S (2022) Oxygen nanobubbles for lake restoration—where are we at? A review of a new-generation approach to managing lake Eutrophication. Water 14(13):1989

Wenfeng T, Fan L, Xionghan F, Qiaoyun H, Xueyuan L (2005) Adsorption and redox reactions of heavy metals on Fe–Mn nodules from Chinese soils. J Colloid Interface Sci 284:600–605

WHO (2011) Edition, F. Guidelines for drinking-water quality. WHO chronicle 38(4):104–108

Yang C, Li J, Yin H (2022) Phosphorus internal loading and sediment diagenesis in a large eutrophic lake (Lake Chaohu, China). Environ Pollut 292:118471

Zakari DM, Mohammad AB, Medugu NI, Sandra I (2014) Impact of climate change on yam production in Abuja, Nigeria. Int J Sci Environ Technol 3:458–472

Zeng L, Keshavarz A, Xie Q, Iglauer S (2022) Hydrogen storage in Majiagou carbonate reservoir in China: Geochemical modelling on carbonate dissolution and hydrogen loss. Int J Hydrogen Energy

Zhan Q, Teurlincx S, van Herpen F, Raman NV, Lürling M, Waajen G, de Senerpont Domis LN (2022) Towards climate-robust water quality management: testing the efficacy of different eutrophication control measures during a heatwave in an urban canal. Sci Total Environ 828:154421

Zhang Q, Shi X, Huang B, Yu D, Öborn I, Blombäck K, Wang H, Pagella TF, Sinclair FL (2007) Surface water quality of factory-based and vegetable-based peri-urban areas in the Yangtze River Delta region. China. Catena 69(1):57–64

Zvobgo L, Johnston P, Williams PA, Trisos CH, Simpson NP (2022) The role of indigenous knowledge and local knowledge in water sector adaptation to climate change in Africa: a structured assessment. Sustain Sci, 1–16

Chapter 26
Impact of Climate Change on Soil Salinity Along Irrigated Farmlands of Jakara River Downstream Minjibir Local Government Area, Kano State, Nigeria

Karimatu Umar Adamu and Binta Lawan Yusuf

Abstract Soil salinization has always been an issue in many places around the world, especially in arid regions where evapotranspiration surpasses rainfall. This study analysed the soil salinity level and variation both laterally and vertically along the irrigated farmlands of the Jakara River downstream. Soil samples were collected at a depth of 0–15 cm along 10 traverse lines from both sides of the river. A total of 40 soil samples were collected from both sides of the stream at an interval of 100, 200, 300 and 400 merers away. Conventional laboratory techniques for determining soil salinity were adopted for electrical conductivity (EC), sodium absorption ratio (SAR), and exchangeable bases (Ca^{2+}, Mg^{2+}, Na^{+} and K^{+}) soil pH and particle size distribution (PSD). Semivariogram analysis was used to identify the spatial variability of soil salinity causes by either stochastic or structural factors, while an ordinary kriging interpolation technique was used to map the spatial distribution of soil salinity. To avoid detrimental effect of salinity on crops, soil parameters that are major indicators should be assessed periodically for better management.

Keywords Climate change · Semivariogram · Kriging interpolation · Soil salinity · Irrigation · Downstream

Introduction

Climate change has the potential to accelerate soil salinity development. Excessive usage of groundwater and synthetic fertilizers leads to soil salt stress (United Nations 2011). Many of the country's most fertile agricultural lands are under threat from aridity processes. Increasing air temperatures, heavy rains, weather patterns and

---

K. U. Adamu (✉)
Geography Department, Sa'adatu Rimi College of Education, Kano, Kumbotso, Nigeria
e-mail: karimatuumaradamu@gmail.com

B. L. Yusuf
Geography Department, Kano State College of Education and Preliminary Studies, Kano, Nigeria

J. C. Egbueri et al. (eds.), *Climate Change Impacts on Nigeria*, Springer Climate,
https://doi.org/10.1007/978-3-031-21007-5_26

insect infestations are all contributors. Increasing air temperatures, heavy rain events, weather patterns and new pest infestations are all important factors contributing to low agricultural growth (Corwin 2020). Many irrigated fields result in a rise in salt levels and salinity buildup. Saline soil is a non-sodic soil encompassing an adequate amount of soluble salt. It commonly occurs in irrigated soil due to the accumulation of salts introduced from the continuous use of irrigation water. In arid and semi-arid regions, soil salinity is considered to be one of the major and widely spread environmental problems, limiting crop production and declining soil productivity (Adamu et al. 2014; Yusuf et al. 2020). Soil salinity is caused by a mismatch between rainfall, temperature, evapotranspiration and water inputs (Wichelns Qadir 2015; Minhas et al. 2020; Soni et al. 2021), Review of Soil Salinity and Sodicity Challenges to Crop Production in the Lowland Irrigated Areas of Ethiopia and Its Management Strategies (Daba and Quraishi 2021), found out that locations with extensive irrigation, salt has a severe impact on crops, causing leaf burn, reduced root development, water stress, stunted growth and crop mortality, resulting in low agricultural yields. However, according to Sarkar et al. (2020), Cuevas et al. (2019), Lal (2015), and Yang et al. (2006), the main causes include intense soil tillage, residue clearance, diminishing soil organic carbon and deteriorating soil health (SOC) and mono-cropping systems. According to Dasgupta et al. (2015), the salinity of soil and groundwater in arid regions of the world is being exacerbated by rising sea levels and climate change. Soil salinity is dynamic globally, affecting more than 100 nations. Only approximately 2.5% of the world's total land surface is arable, and roughly 20–25% of cultivated fields are salty (FAO 2020). Salinization is expected to increase as a result of future climate change scenarios. Crop production is facing a serious challenge with soil salinity in irrigated regions. Worldwide, saline soil conditions have affected the value and output of agricultural land (Asfaw 2016). Excessive soil salinity impairs soil microorganisms' biological activity. Adequate information is essential for enhanced planning and implementation of effective soil reclamation projects (Quan et al. 2013). One of the main issues in Nigeria's large- and medium-scale irrigation projects influencing agricultural yield in irrigated areas is soil salinity. Therefore, several studies on soil salinity have been carried out by various researchers in different parts of the country using several methodological approaches to acquire information related to soil salinity with the aim of finding solutions. In Nigeria, for example, A Review of Soil Salinity and Sodicity Challenges to Crop Production in Ethiopia's Lowland Irrigated Areas and Its Management Strategies (Daba and Quresh 2021). and Salinization in Agricultural Lands: Causes, Effects, and Mitigation (Dikeogu et al. 2021), Soil salinity and its implications for sustainable agriculture in the southern and north-central states of Nigeria (Aizebeokhai et al. 2020), An Assessment of Salinity Levels of Soils and Water in Irrigated Vegetable Farms around Lake-Alau Dam of Maiduguri, Borno State, Nigeria, (Magaji et al. 2020), Soil salinity and its implications on sustainable agriculture in Southern and North-central States of Nigeria (Kayode, Aizebeokhai and Odukoya 2020). Influence of Salinity on Soil Chemical Assessing the Effect of Salinity on Irrigated Land (Aderoju and Festus 2013), Assessing the Effect of Salinity on Irrigated Land (Ankidawa and Auhari 2010). Similar studies on soil salinity status were also conducted in Kano, such as; (Mohammed et al. 2021),

Mapping Spatial Variability of Salinity in Kano River Irrigation Project (KRIP), Nigeria Appraisal and Mapping of Soil Salinity and Sodicity Problems in Sector One of Watari Irrigation Scheme, Kano State (Bashir et al. 2020), Magnitude of Soil Salinity Hazard in Dausa Fadama, Digol Village, Danbatta Local in Kano, (Muzammil 2018), Soil salinity assessment using geostatistical models in some parts of Kano River Irrigation Project Phase I (Yusuf et al. 2020).

The aim of this research is to assess the level of salinity status and to identify the distribution of the soil salinity status in the area.

Materials and Methods

Study Area

Location and Extent

The study area is the Jakara River downstream which is located between latitude 12°06′81″E–12°20′23″E and longitude 8°58′8.27″N–8° 57′7.28″N. The study area is 5 km from the Jakara Dam along the regulated channel of the Jakara River (Fig. 26.1). The Jakara River is located in the north-eastern part of Kano Metropolis in which the wastewater from Kano City is being discharged into it. Jakara dam is the fifth largest of the 22 dams in Kano state, having a total storage capacity of 651,900,000 m^3 and a surface area of 16,590,000 m^2 (Mustapha and Aris 2012).

Geology and Relief

The area falls within the basement complex. These rocks are of Precambrian origin and consist of metamorphic and igneous rocks. Common among the rocks are granites of different descriptions, such as granulated sandstone, magmatites, gneisses and phyllites. The rocks have been subjected to chemical and other forms of weathering to produce clay-rich regolith. The regolith was subjected to widespread lateralization during the fluvial period of the past. There are also extensive sedimentary rocks of cretaceous age, forming sand plains (Nabegu 2010).

Landforms

The landforms are relatively plain, with little granite and few lateritic outcrops; these plains are thought to be originally pediplains, which are now almost completely covered by wind drift material and ancient alluvial on the terraces, and the current savannah incision phase is following an arid, infilling one (Olofin 1987; Nabegu 2010). The area is covered by alluvial channel complexes, a sub-landform type of the

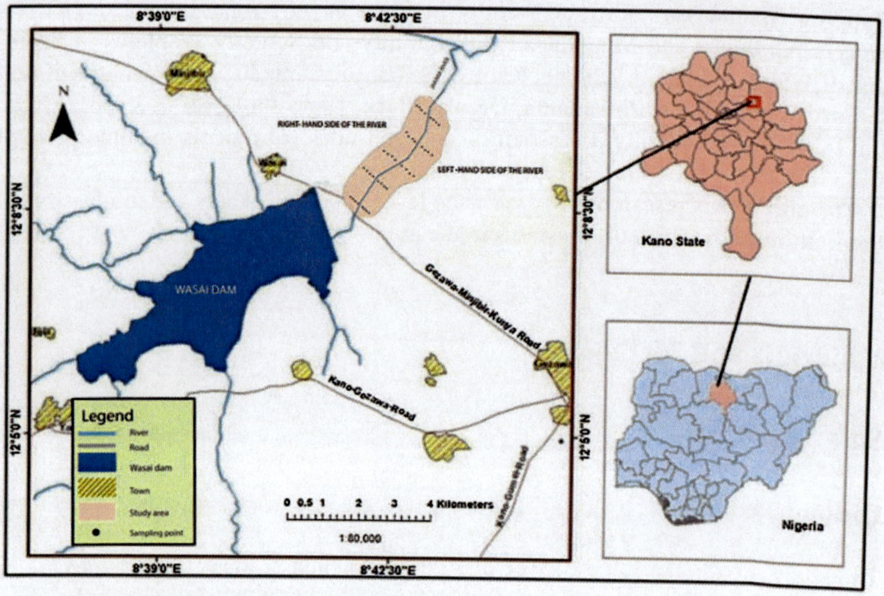

Fig. 26.1 Jakara river downstream

region in which many river systems create alluvial channel complexes consisting of both old and current flats. The old flats are muddy, abandoned flood plains originating from the migration of the rivers over very extensive storm channels (Olofin 1987).

Hydrology and Drainage

The main water source of the Jakara River is the Urban Kano and it has two major sources. The first source of the Jakara River is traced into the city (i.e. the Jakara/Rafin Mallam) and the second source is the Bompai Industrial Estate (i.e. the Getsi). The Jakara River receives many inputs, both anthropogenic and natural, that cause deterioration of the river's water quality; therefore, it is not suitable for many uses. The river runs through Kano city for a length of about 13.5 km. Approximately 27.2 km downstream of the Jakara River, the river was dammed at Wasai in Minjibir local government for the last three decades without proper utilization. Before the dam, this flow disappeared into Yadai and Yakana sand dunes (in the north-eastern part of the region) and which still forms the only longitudinal dune fields in the present Kano State. Other rivers that joined the Jakara at Chedi in Minjibir Local Government are the Agalawa, Gwagwata and Fagalawa Rivers from the north, The Agalawa floodplain has been one of the most important Fadama that was put into intensive cultivation in the region (before the dam) (Abdulhamid 2014). The more humid the micro-climatic zone, the higher the water content. Climate determines the amount of water available on the surface and subsurface. Human activities also have an influence on hydrology and drainage; the more humid the micro-climatic zone, the higher the water content (Olofin 1987).

Climate

The climate of Kano and the research area is tropical wet and dry, designated as AW on the Koppen climate classification (Adamu and Aliyu 2011). The climate is influenced by two primary air masses that come into the nation at various periods of the year. These air masses, the tropical maritime air mass and the tropical continental air mass, collide at a moisture gradient known as the inter-tropical discontinuity (Olofin 1987).

Rainfall

The annual average rainfall of the area is about 884–1200 mm; it is characterized by one peak period (mono-modal), usually in August. The southern parts of the state are wetter, with an average annual rainfall of about 1000 mm. This decreases northwards to about 800 mm around Kano and about 600 mm in the north and north-east of the state (Liman et al. 2014).

Temperature

The average yearly temperature is around 260 °C, but the mean monthly values vary very little seasonally. The temperatures range from 210 °C in the coldest months (December and January) to 310 °C in the warmest months (April and May) (Olofin 1987).

Evaporation and Transpiration

The combination of evaporation and transpiration (evapotranspiration) is very high in the area. Potential evaporation is approximately 1772 mm per year, and mean evapotranspiration by the sunken pan method is approximately 2538 mm per year. However, the actual evapotranspiration, estimated at about 60% of the annual rainfall, is not as high as potential (Olofin 1989).

Soils

The basement complex and the Chad Formation are the two major geologic formations from which the study area's soils are produced. The soils are sandy, at least at the top horizon (0–30 cm). Excluding hydromorphic soils found in river valleys, where the quantity of sand is significantly lower, the sand fraction frequently surpasses 70% on most surfaces. This soil is physically unstable due to the high sand content, which results in poor aggregation. Generally, the mature soils occupy the plain and the immature ones are found on the hill slopes, foot slopes, valleys and valley bottoms. The combined influence of topography and wind-drift material from the desert is what shapes the aggregates of the ferruginous type (Olofin 1987).

Vegetation

The natural vegetation of the area is of the Sudan Savannah type. It is composed of a variety of trees scattered over an expanse of grassland. There are various types of acacias (Albida, nolotica and siyal, among others). Most tree species are adapted to drought through long tap roots, leathery leaves and tiny leaves. Grasses rarely grow taller than 1.5 m at maturity, except in favoured spots. There are also medium-sized trees in the drier interfluves and areas of rocky or dissected terrain (Olofin 1987).

Population and Economic Activity

The population of Minjibir is 213,744 according to the 2016 census. The main types of land use are residential, farming and recreational. Farming occupies most of the land area and 75% of the area is under cultivation of crops such as millet, tomatoes, groundnuts, vegetables, etc. The major economic activity of the people is farming and the cultivation of perishable crops such as onions, tomatoes and pepper.

Materials and Methods

Materials

A global positioning system (GPS) was used to record the coordinates of the sampling points; a soil auger and spade were used for collecting the soil samples, respectively; polythene bags to store soil samples; masking tape and markers for labelling the samples were stored for further analysis.

Methods

Laboratory Tcchnique

Each of the parameters underwent a laboratory analysis. Using a porcelain pestle and mortar, the materials were softly crushed in the lab before being run through a 2 mm filter to remove large fragments. The following parameters were assessed in the fine earth samples (2 mm sieve portion): pH, soil electrical conductivity (EC), exchangeable bases (Ca^{2+}, Mg^{2+}, Na^{+} and K^{+}), exchangeable acidity and soil texture.

Soil pH

The pH was determined in the soil by 0.01 M of $CaCl_2$ using a 1:2:5 soil-to-water ratio (IITA 1979). The soil pH was read with a glass electrode on a pH meter. (JENWEY 3520 MODEL).

Electrical Conductivity

The electrical conductivity (EC) of the soil was determined from the soil ratio of 1:2.5. Bower and Wilcox (1965) described measuring the EC with an electrical conductivity meter (DDS-307 MODEL).

Exchangeable Bases

The exchangeable bases (Ca, Mg, K and Na) were extracted with 1 M ammonium acetate solution (1 M NH_4OAC) solution, buffered at pH 7.0 as described by Anderson and Ingram (1998). The sum of Ca, Mg, K and Na gave the total exchangeable basis. Calcium and magnesium were read on an atomic absorption spectrometer (AAS, 210 VGP MODEL), while K and Na were read using a Flame Photometer (JENWEY PFP7 MODEL).

Ratio of Sodium Absorption

The sodium absorption ratio was calculated as the ratio of Na^{+} to Ca^{+} and $Mg2^{+}$ as shown in the following relationship:

$$SAR = Na^{+}\left(Ca^{2+} + Mg^{2+}\right)/2$$

Particle Size Distribution

Particle size distribution was determined using the Bouycous Hydrometer method (Gee and Or 2002). The textural class was determined using the USDA textural triangle.

Geostatistical Technique

To create distribution maps, a geostatistical technique (Kriging) was used. The dataset was compiled, merged and loaded, and spatial interpolation was performed using the Geostatistic extension, which uses the technique of semi-variogram to measure the spatial variability of a regionalized variable and provides the input parameters for spatial interpolation of Kriging (Webster and Oliver 2007).

Sampling Frame and Techniques

Sampling Frame

The study area is located approximately 5 kms downstream of the Jakara Dam along the regulated channel of the Jakara River. The samples were collected along ten traverse lines from both sides of the river, and the interval between one traverse line and the other was 1 km (see Fig. 26.1). Four-point samples at a depth of 15 cm were collected from both sides of the river, the first at 100 m away from the river, then at 200 m, 300 m and 400 m away from the river, respectively.

Geostatistical Models and Analysis of Soil Salinity

Spatial interpolation was conducted using the Arc GIS 10.1 software's Geostatistical extension. The geographic distribution of soil salinity was created using the semi-variogram of ordinary kriging, one of the interpolation methods used in earth sciences

$$\gamma(h) = \frac{1}{2N(h)} \sum_{i=1}^{N(h)} [z(xi + h)]^2 \tag{26.1}$$

where $\gamma(h)$ is the semivariance for interval class h, N (h) is the number of pairs separated by lag distance h (separation distance between sample positions), Z (x_i) is a measured variable at spatial location i and Z ($x_i + h$) is a measured variable at spatial location i + h. Samples separated by distances closer than the range are related spatially and those separated by distances greater than the range are not spatially related.

Selection of the best fitting model was based on the biggest (maximum) coefficient of determination (R2) will give the best fit model into variogram data. The generation of semivariogram parameters was then carried out for each theoretical model (Stable, Circular, Spherical, Tetraspherical, Pentaspherical, Exponential, Gaussian, Rational Quadratic, Hole Effect, K-Bessel and J-Bessel). The selection is based on the biggest (maximum) coefficient of determination (R2), which provides an indication of how well the model fits into variogram data (should be between 0.5 and 0.99). Important model parameters are the nugget variance (Co), which is the y-intercept of the model that can never be greater than the sill; the sill (Co + C), which is the model asymptote that can never be less than the nugget; the range (A), which is the separation distance over which spatial dependence is apparent; and the nugget/sill ratio or spatial dependency (Co/(Co + C) defines the spatial property (Ahmed 2015).

Geostatistical Technique

Variation of Soil Salinity

The composition of the soil plays an important role in salinity control because it provides the physical barrier between rainfall and groundwater recharge Muzammil (2018). The results of particle size distribution revealed the mean values of 66.89% for sand, 9.36% for silt and 23.74% clay and for the textural class is collectively sandy clay loam. The particle size analysis had shown that the sand is the dominant particle size in the whole study area (vertical and spatial). This might be due to the parent materials of which the soils were developed from. Clay was next to sand in dominance and then followed by silt. The results are similar to those reported by Iyaka and Kakulu (2009) in their study of soils along the irrigated farmlands in Minna, and are similar to the findings of Dawaki et al. (2014) in his study in soils along the Jakara River in which he found the soil to be sandy loam to loamy sand he revealed that sandy texture soils are common in the basins of many rivers, as materials brought about by transporting water are segregated Dawaki et al. (2014). However, according to Mortimore (1989), the predominance of sand particles in arid and semi-arid climates is not uncommon because many of them were for aeolian deposits blown from across several thousands of kilometres.

Using geostatistical analysis, we may determine the soil properties. The spatial attribute is defined by this. When Co (Co + C) has a value, there is a significant geographical dependency. A slight spatial dependency is present when this occurs. The selection of the best-fitting model was based on the biggest (maximum) coefficient of determination (R^2), defined using the models; Stable, circular, rational quadratic, Gaussian, spherical, pentaspherical and tetraspherical K-Bessel and J-Bessel as shown on Table 26.1. The variable is considered as a strong spatial dependence when the value of $Co/(Co + C)$ is less than 0.25, a moderate spatial dependence when this value is between 0.25 and 0.75, and a weak spatial dependence when the value is more than 0.75.

Results of the Geostatistical Analysis

The laboratory data were loaded into Microsoft Excel along with their relevant coordinates (Latitude and Longitude) and then transformed into a GIS environment to identify the spatial distribution of the chosen parameters of the area. Different thematic maps (layers), according to Ahmed et al. (2015) were generated for each of the parameters using kriging interpolation in the Arc GIS 10.1 software. Geostatistical technique (Kriging) was used to create distribution maps for interpolation geographical information (GIS) techniques (Ha, Olson, Brian and Rogerson 2014).

The Kriged maps thus produced were then reclassified according to USDA (Landon 1991; ESU 1991) to obtain the concentration level of each of the measured parameters in the study site. The nugget/sill ratio or spatial dependency Co (Co + C) defines the spatial property. When the value of Co/(Co + C) is less than 0.25, the variable is considered to have a strong spatial dependence; moderate spatial dependence when the value is between 0.25 and 0.75; and weak spatial dependence when the value is greater than 0.75 (Cambardella et al. 1994).

The results displayed in Table 26.1 showed that soil pH followed the circular model display good spatial distribution across the study area with a nugget value of 0 and R2 of 0.8503.

Distribution of pH

The distribution map is shown in Fig. 26.2; it shows about six classes of pH from ultra-acidic to very strongly alkaline according to London (1991) and ESU (1991), a result of slightly Alkaline was found by Dawaki (2015) in soils along the Jakara River, but this is contrary to the findings of Muhammad (2015), which is neutral according to (Landon and ESU 1991).

Table 26.1 Variogram models and parameters of soil salinity properties

Par	Model	Lag	Partial sill(C)	Range	Nugget Co	Sill Co + C	Ratio Co/(Co + C)	Lag Size	RMSE	R2
pH	Circular	12	0.9219	0.0034	0	0.9219	0	0.0004	0.6332	0.8503
EC	Exponential	12	0.461	0.0035	0	0.461	0	0.0003	0.0596	0.9438
SAR	Halo effect	12	0.4264	0.0072	0.3096	0.736	0.7261	0.3096	0.2766	0.9551
ESP	Halo effect	12	22.7528	0.049	2.5485	25.3013	0.112	0.004	0.3693	0.6043
Ca	Rational quad	12	0.6393	0.0241	1.675	2.3143	2.6201	0.0023	1.3814	0.9699
Mg	Tetra spherical	12	8.4674	0.0044	0.001	8.4684	0.0001	0.0005	0.0339	0.9707
Na	Circular	12	0.0493	0.0044	0.0083	0.0576	0.1684	0.005	0.1842	0.9874
K	K-Bessel	12	0.0199	0.0042	0.0125	0.0324	0.6281	0.0005	1.0383	0.1603

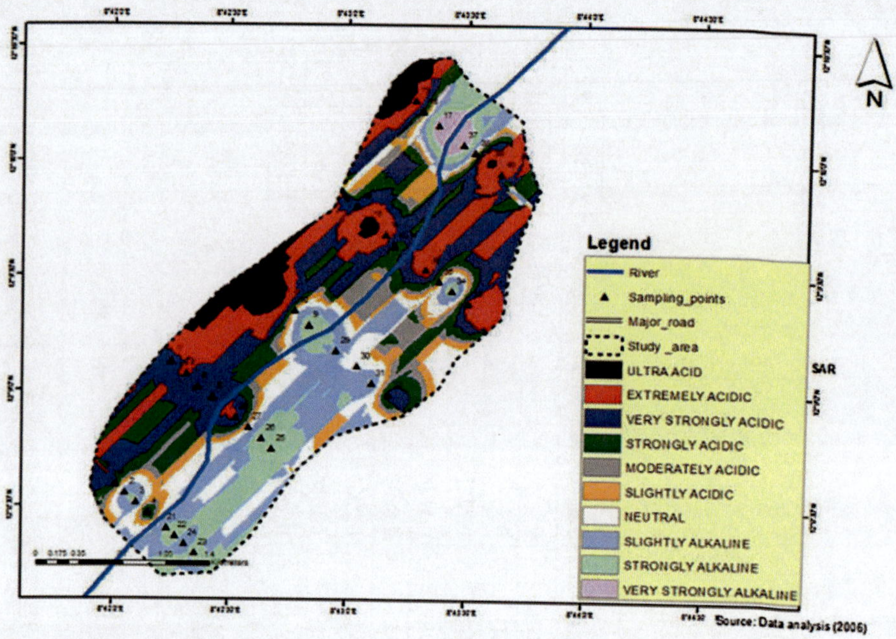

Fig. 26.2 Distribution map of pH

Distribution of EC

The distribution of EC is shown in Fig. 26.3; it followed the exponential model, and has the nugget of 0 and R^2 value of 0.9438. A result of low EC was found in the study area, which is contrary to the findings of Dawaki et al. In his research on soils along the Jakara River, he discovered a range of EC values of 3.79–4.43ds/m, which is slightly moderately saline according to London and ESU 1991. As noted by Binns et al. (2003), a slightly high pH value and an increase in EC value over time observed in some points in this study indicate that there may be a potent salinity hazard over time.

Distribution of SAR

The ratio of sodium to calcium and magnesium in soil solution is measured using the widely used and recognized Sodium Adsorption Ratio (SAR) indicator. The soil is referred to as sodic when the SAR value is more than 13. The distribution of SAR is revealed in Fig. 26.4, which shows low in the whole study area.

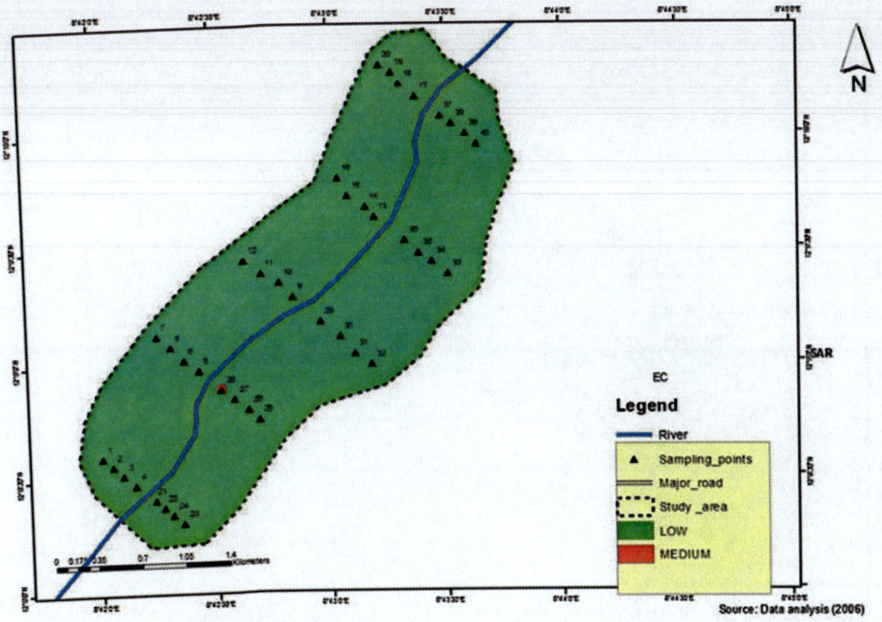

Fig. 26.3 Distribution map of EC

Distribution of ESP

The proportion of sodium absorbed onto clay mineral surfaces to total cation exchange capacity is known as the exchangeable sodium percentage. It indicates the potential for soil sodicity and affects either the structural stability of soil aggregates or the osmotic potential of the soil moisture on biological activities (Patterson and Chapman 1998). The distribution of ESP for the study is revealed in Fig. 26.5, which shows that more than 95% of the soil in the study area is highly saline. This is contrary to the findings of Dawaki, Dikko and Aliyu who obtain the mean of 3.79–4.43 ms/m which is slightly saline according to the London and ESU, 1991 rating.

Distribution of Ca^{2+}

Calcium is a very important divalent cation being natural suppressor of the effect of high levels of sodium and potassium. The distribution of Ca^{2+} is revealed in Fig. 26.6, showing two classes of high and medium which come in close agreement with the findings of Adamu and Dawaki (2008) for the Jakara valley, as well as those reported by Adamu and Dawaki (2008) for some sites along the Basin of the Challawa River,

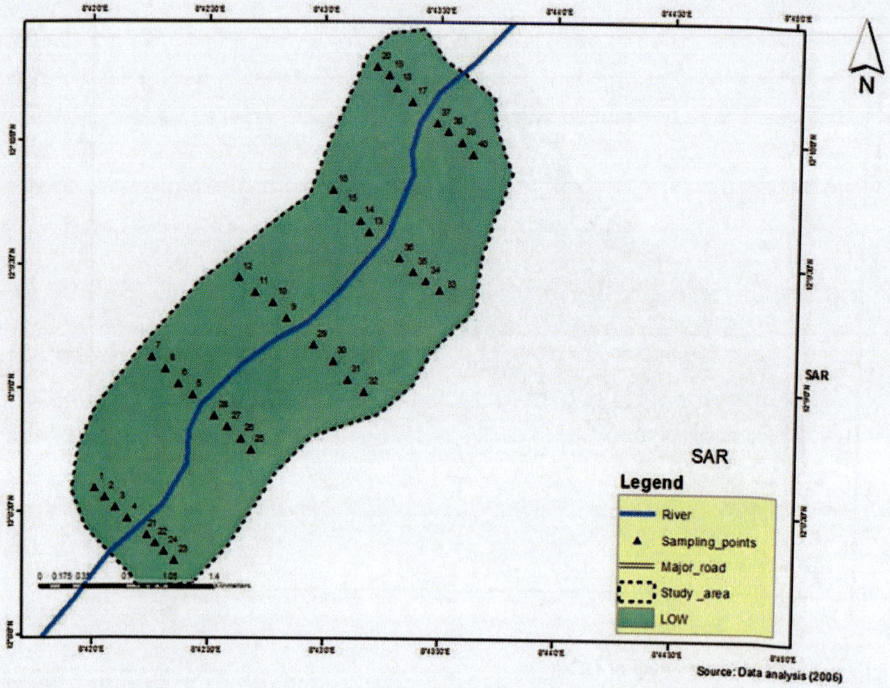

Fig. 26.4 Distribution map of SAR

and is in contrast to the findings of Dawaki et al. (2015) in his study on soil along the river Jakara, in which he found the mean Ca^{2+} value of 2.82 cmol/kg and 3.27 cmol/kg at Akija and Airport road, which can be described as the source of the Jakara.

Distribution Map Mg^{2+}

Magnesium (Mg) is also a divalent cation just like Ca is a natural suppressor of the effect of high level of sodium and potassium.

The distribution of Mg^{2+} in the whole study area is low (London and ESU 1991) the value is contrary to the findings of Adamu and Dawaki (2008) for Jakara; and Adamu and Dawaki (2008) for the Challawa River (Fig. 26.7).

Distribution of Na^{2+}

According to the study, the north-western part of the river has a larger concentration of Na+ than the north-eastern part of the study area. Despite the fact that Landon

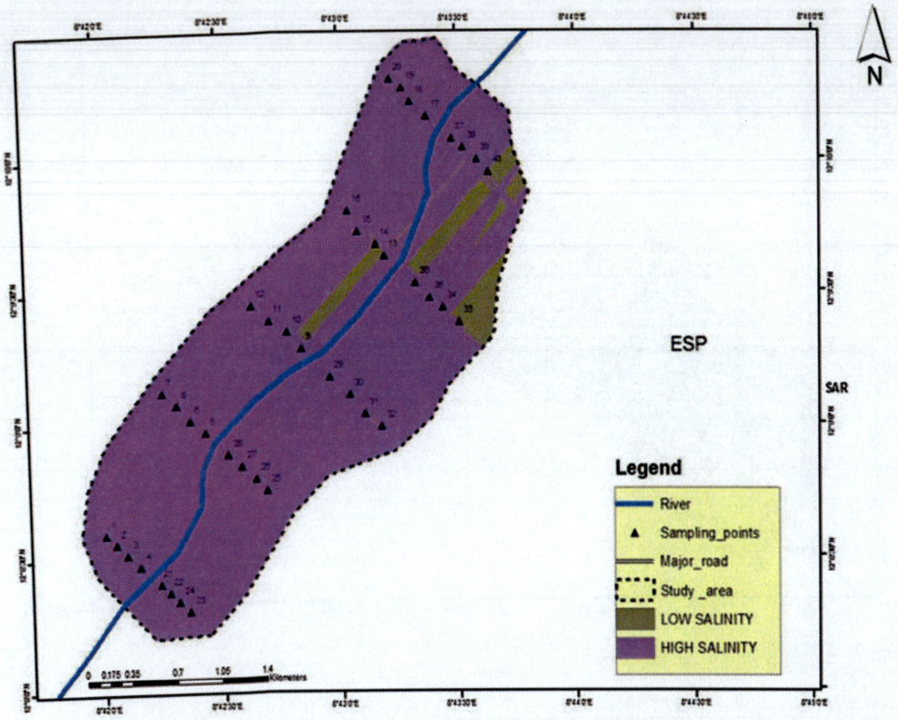

Fig. 26.5 Distribution map of ESP

and ESU (1991) rank sodium as medium throughout the study area, significant levels of sodium are present in the irrigated soils on a regular basis, and this makes it one of the criteria for determining the salt effect in these soils. Adamu and Dawaki (2008) found comparable salt levels in Jakara that range from medium to high. According to Binns et al. (2003), domestic water flowing from the city may include a high amount of sodium, which might account for the sodium content.

The Distribution of K+

The distribution of K+ is revealed in Fig. 26.8, showing three classes' low, medium, high and very high classes. The medium value is found at the edge of the northern and southern parts spreading towards the east and the central parts, there is also a high concentration value inside the medium concentration towards the central part of the study site and the rest of the area is covered by low.

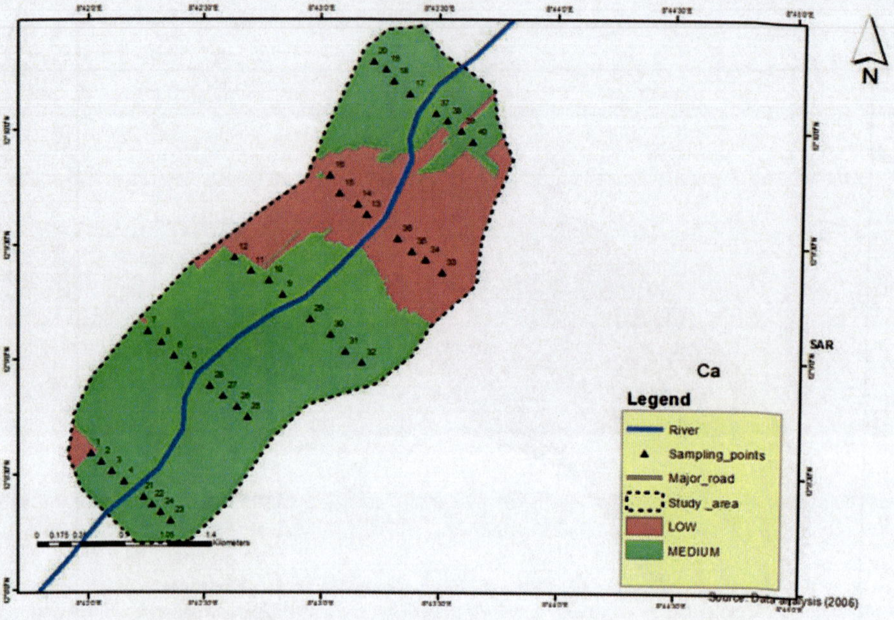

Fig. 26.6 Distribution map of Ca^{2+}

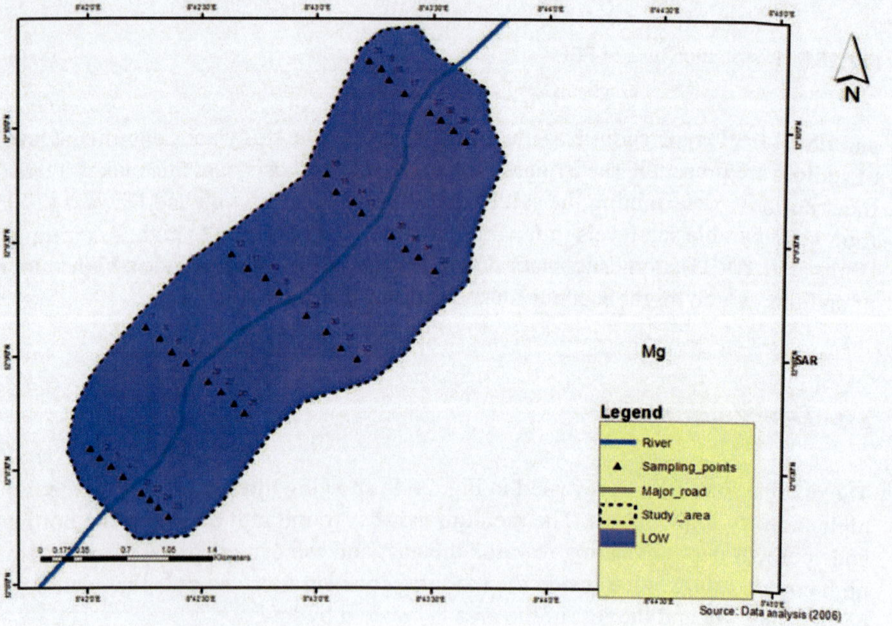

Fig. 26.7 Distribution map of Mg^{2+}

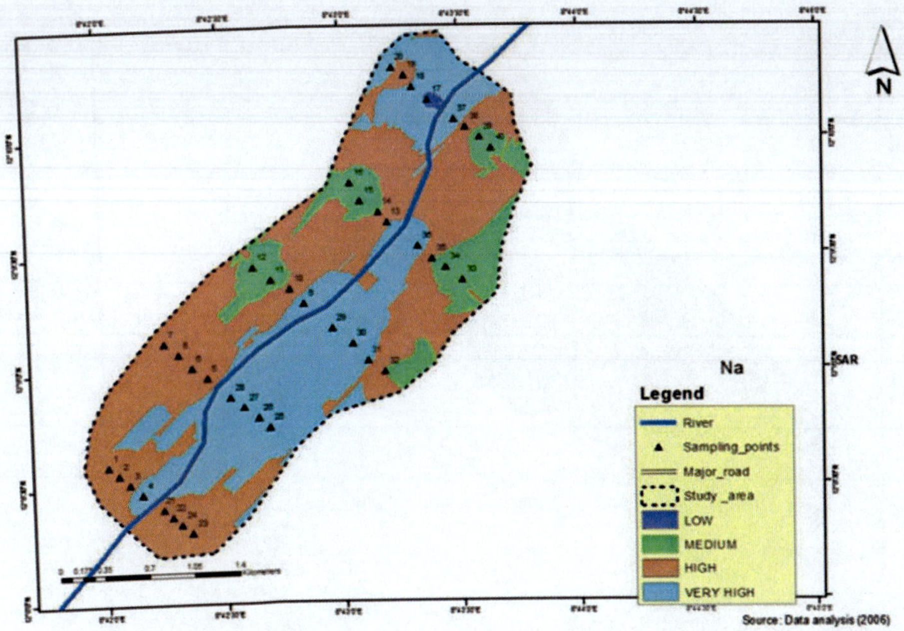

Fig. 26.8 Distribution map of Na^{2+}

The Distribution Map of K+

Limitations, Recommendations and Perspectives for Future Research

The study examines the distribution of salinity in soils along the irrigated farmlands of the Jakara River downstream. Forty soil samples were collected along the irrigated farmlands and were analysed for salinity (electrical conductivity, sodium absorption ratio and exchangeable basis) and other soil parameters (pH and particle size distribution). The mean values of the soil salinity are compared with the ESU 1991 standard rating and it was found that the mean concentration of EC is low, ESP is high, sodium absorption ratio (SAR) is low, exchangeable calcium and sodium are medium in concentration, exchangeable magnesium is low and there is a high concentration of potassium and sodium found at the south-eastern part of the study site. The distribution of K shows a medium to high concentration value inside the medium (Fig. 26.9).

Limitation

The study is limited to only 5 km due to a lack of funds and time.

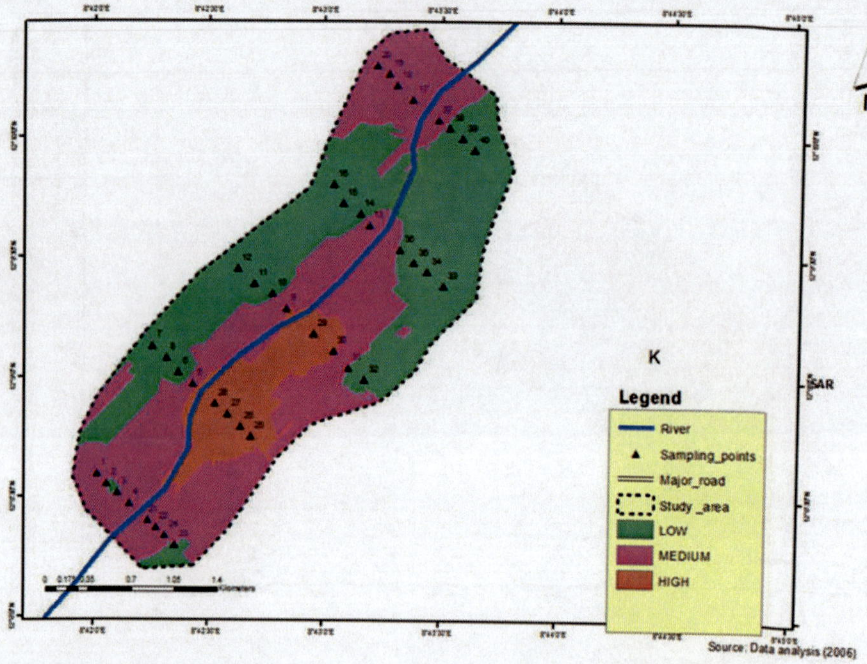

Fig. 26.9 Distribution map of K

Conclusion

Based on the findings of this research it can be seen that the soils along the Jakara River downstream are found to be sodic since it has high ESP and pH values.

Recommendations

1. There is a need for monitoring the soil for excess sodium to avoid salinity and sodicity which can lead to flocculating soils condition which leads to decline in aeration and poor drainage resulting in water logging which affect the normal plant development process.
2. To avoid detrimental effect of salinity on crops, soil parameters that are major indicators should be assessed periodically for better management.
3. There is a need for the farmers to have a good management practice such as good drainage and aeration in order to improve the chemical and physical properties of the soil and selection of tolerant crop.

Further Studies

There is a need for further studies to carry out suitability in the study area and other methods can also be used to assess the soil and water salinity.

References

Abdulhamid A (2014) Drainage hydrology and water resources. In: Tanko AI, Momale SB (eds) Kano environment society and development Adonis and Abbey Publishers Ltd ISBN: 9781909112391

Adamu GK, Aliyu AK, Jabbi AM (2014) Fertility assessment of soils under rice cultivation in Kadawa Garun Mallam Llocal Government Kano State. Acad Res Int 5(1):92–99. ISSN-L: 2223–9944 ISSN-L2223–9944

Adamu M, Aris AZ (2011) Spatial aspect of surface water quality using chemometric analysis. J Appl Sci Environ Sanitation 6(4):411–426

Aderoju DO, Festus AG (2013) Influence of salinity on soil chemical properties and surrounding vegetation of Awe salt mining site Nasarawa State Nigeria. African J Environ Sci Technol 7(12):1070–1075. https://doi.org/10.5897/AJEST20131600. ISSN: 1996–0786 © 2013 Academic Journals http://www.academicjournalsorg/AJEST

Anderson JS, Ingram JIS (1998) Tropical soil biology and fertility a hand book of methods (second edition) information press UK, p 221

Ankindawa BA, Awhari DP (2010) Assessing the effect of salinity on an irrigated land Leonardo. J Sci, 1–8. ISSN 1583-0233

Asfaw E, Suryabhagavan KV, Mekuria A (2016) Soil salinity modelling and mapping using remote sensing and GIS: the case of Wonji sugar cane irrigation farm Ethiopia. J Saudi Soc Agric Sci, 1–9

Bashir AY, Samndi AM, Adam MA, Moji KT, Auwal M (2020) Appraisal and mapping of soil salinity and sodicity problems. In: Sector one of Watari Irrigation Scheme Kano State Nigerian. J Sci. https://doi.org/10.36265/njss2020290207. ISSN-1597–4488 cpublishingrealtime

Binns JA, Maconachie RA, Tanko AI (2003) Water land and health in urban and peri urban Food production: the case of Kano Nigeria land degrade develop14:413–444

Corwin DL (2020) Climate change impacts on soil salinity in agricultural areas. Eur J Soil Sci. https://doi.org/10.1111/ejss13010

Cuevas J Daliakopoulos IN Del Moral F Hueso JJ Tsanis IK (2019) A review of soil-improving cropping systems for soil salinization. Agronomy 9:1–22. https://doi.org/10.3390/agronomy9060295

Daba AW, Qureshi AS (2021) Review of soil Salinity and Sodicity challenges to crop production in the lowland irrigated areas of Ethiopia and its management strategies. https://doi.org/10.3390/land10121377

Dasgupta S, Hossain MM, Huq M, Wheeler D (2015) Climate change and soil salinity: the case of coastal. Bangladesh Ambio. https://doi.org/10.1007/s13280-015-0681-5

Dawaki MU Dikko A U Noma SS and Aliyu U (2015) Effects of Wastewater Irrigation on Quality of Urban Soils in Kano Nigeria International Journal of plant and soil science

Dikeogu TC, Okeke OC, Nwachukwu HGO, Agbo CC (2021) Salinization and Waterlogging in agricultural lands: causes effects and mitigation. Int J Innov Environ Studies Res 9(1):45–57 http://www.seahipajorg/. ISSN: 2354–2918

FAO (2020) Sustainable Food and Agriculture (Online). http://www.faoorg/sustainablity/news/detail/en/c/1274219/

Gee GW, Or D (2002) Particle size analysis, p 2255–293–411. In: Dan JH, Topp GC (eds) Method of soil analysis Part 4 physical and mineralogical methods American society of agronomy

IITA (1979) International institute of tropical Agriculture selected Method of soil AnalysisManual Series No1 Revised Edition IITA IbadanNigeria p70 Istanbul+5 reviewing and Appraising Progress Five Years After Habitat II

Iyaka YA, Kakulu SE (2009) Copper and Zinc contents in urban agricultural soils of Niger State Nigeria. Int Multi-Discip J Ethiopia 3(3)23

Kayode OT, Aizebeokhai AP, Odukoya AM (2020) Soil salinity and its implications on sustainable agriculture in Southern and North-central States of Nigeria. J Earth Environ Sci 655:012077. 4th international conference on science and sustainable development IOP conference series. https://doi.org/10.1088/1755-1315/655/1/012077

Lal R (2015) Restoring soil quality to mitigate soil degradation Sustain 7 5875–5895. https://doi.org/10.3390/su7055875

Liman M, Idris HA, Mohammed UK (2014) Weather and climate. In: Tanko AI, Momale SB (eds) Kano environment society and development Adonis and Abbey Publishers Ltd. ISBN: 9781909112391

Magaji MJ, Samndi AM, Pantami S A, Abdulrahman B L, Buji I B, Abubakar MA (2020) An assessment of salinity levels of soils and water in irrigated vegetable farms around Lake-Alau Dam of Maiduguri Borno State Nigeria. J Agric Econ Environ Soc Sci Interdisciplinary J. http://www.jaeesscomng. ISSN: 2476-8423

Minhas PS, Ramos TB, Ben-Gal A, Pereira LS (2020) Coping with salinity in irrigated agriculture: crop evapotranspiration and water management issues Agric Water Manag 227:105832. https://doi.org/10.1016/jagwat2019105832

Mohammed D, Maina MM, Audu I, Tudun Wada IY, Nasir NK (2021) Remote sensing techniques in mapping spatial variability of Salinity in Kano River Irrigation Project (KRIP) Nigeria Nigerian Journal of Technology (NIJOTECH) 40(4):732–739 wwwnijotechcom Technology No xx xxx xxx, pp 1–4 wwwnijotechcom Print ISSN: 0331–8443 Electronic ISSN: 2467–8821. https://doi.org/10.4314/njtv40ixxx

Mustapha A, Aris AZ (2012) Spatial aspects of surface water quality in the Jakara Basin Nigeria using chemometric analysis. J Environ Sci Health Part A Toxic/Hazard Substances Environ. Eng 47(10):1455–1465. https://doi.org/10.1080/109345292012673305

Muzammil AK (2018) Magnitude of soil Salinity Hazard in Dausa Fadama Digol Village Danbatta Local Government Area Of Kano State FUDMA. J Sci (FJS) 2(4):194–199. ISSN online: 2616–1370 ISSN print: 2645–2944

Nabegu AB (2010) The impact of urbanization on the morphology of Jakara Stream Channel Unpublished PhD Thesis Department of Geography Bayero University Kano

Olofin EA (1987) Some aspects of physical geography of kano region: departmental geography lecture series BUK: Dabis standard printers Zaria road Kano

Quan Q, Shen B, Xie J, Luo W, Wang W (2013) Assessing soil salinity in the fields of Western China using spatial modeling and remote sensing. Acta Agriculturae Scandinavica Section B-Soil Plant Sci 63:4 289–296

Sarkar S, Skalicky M, Hossain A, Brestic M, Saha S, Garai S, Ray K, Brahmachari K (2020) Management of crop residues for improving input use efficiency and agricultural sustainability Sustain 12 1–24. https://doi.org/10.3390/su12239808. Soil Science Glossary Terms Committee (2008) Glossary of soil science terms Soil Science Society of America ASA-CSSA-SSSA

Soni PG, Basak N, Rai AK, Sundha P, Narjary B, Kumar P, Yadav G, Kumar S, Yadav RK (2021) Deficit saline water irrigation under reduced tillage and residue mulch improves soil health in sorghum-wheat cropping system in semi-arid region. Sci Rep 11:1–13. https://doi.org/10.1038/s41598-020-80364-4

United Nations (2011) World population prospects: the 2010 revision In: Comprehensive Tables Department of Economic and Social Affairs Population Division vol I United Nations New York
Webster R, Oliver MA (2007) Geostatistics for environmental scientists, 2nd edn. Wiley
Wichelns D, Qadir M (2015) Achieving sustainable irrigation requires effective management of salts soil salinity and shallow groundwater. Agric Water Manag 157:31–38. https://doi.org/10.1016/jagwat201408016
Yang YM, Liu XJ, Li WQ, Li CZ (2006) Effect of different mulch materials on winter wheat production in desalinized soil in Heilonggang region of North China. J Zhejiang Univ Sci B 7:858–867. https://doi.org/10.1631/jzus2006B0858
Yusuf BL, Mustapha Yusuf AM, Ahmed M (2020) Soil salinity assessment using geostatistical models in some parts of Kano River Irrigation Project Phase I (KRPI). Model Earth Syst Environ. https://doi.org/10.1007/s40808-020-00841-7

Chapter 27
Relationship Between Agricultural Production, Energy Consumption, and Climate Change in Nigeria

Zhou Jun, Udokwu Adaobi Franca, and Kingsley O. Iwuozor

Abstract Energy consumption and agricultural production are major emitters of greenhouse gases that cause climate change. This climate change in turn affects energy sustainability and food security for both present and future generations. In this chapter, the Auto-Regressive Distributive Lag model (ARDL) and the Granger Causality test were used to examine the connection between climate change, energy consumption, and agricultural output in Nigeria for a duration of 28 years (1990–2017), using carbon dioxide gas as a proxy for climate change. According to the short-run ARDL calculation, increasing energy consumption increases carbon dioxide gas emissions, whereas increasing agricultural output decreases carbon dioxide gas emissions. The long-run estimate, on the other hand, demonstrates that increasing energy usage reduces carbon dioxide emissions while increasing agricultural production does not catalyze any significant change in carbon dioxide emissions. The Granger Causality test results revealed that there is a bidirectional relationship between energy consumption and carbon dioxide emissions, but there is a unidirectional relationship between agricultural output and energy consumption. This implies that in order to reduce greenhouse gases, agricultural production and energy consumption should be increased and decreased, respectively. The results of this study show that Nigeria consumes large amounts of energy, which is mainly dominated by non-renewable energy sources, which consequently results in high emissions of greenhouse gases as a result of emissions from the energy sector. In accordance with the research results, this paper puts forward a series of policy recommendations to help promote sustainable development.

Z. Jun · U. A. Franca
School of Management Science and Engineering, Central University of Finance and Economics, Beijing 100081, China

K. O. Iwuozor (✉)
Nigeria Sugar Institute, Kwara State, Ilorin, Nigeria
e-mail: kingsleyiwuozor5@gmail.com; ko.iwuozor@nsinigeria.org

Department of Pure and Industrial Chemistry, Nnamdi Azikiwe University, P.M.B. 5025, Awka, Nigeria

J. C. Egbueri et al. (eds.), *Climate Change Impacts on Nigeria*, Springer Climate,
https://doi.org/10.1007/978-3-031-21007-5_27

Keywords Agriculture · Auto-Regressive Distributive Lag Model · Carbon emission · Granger causality test · Greenhouse gas

Introduction

Global economic and technological development has raised concern about climate change and its effects (Iwuozor 2018, 2019; Kingsley and Ibraheem 2018). Climate change occurs due to the accumulation of emitted greenhouse gases (GHG) from both natural events and human activities (Braide et al. 2020). The greenhouse gases emitted from human activities are believed to have an impact on the natural processes occurring in the atmosphere and are the main contributing factors to the overwhelming warming being experienced presently (Mohmmed et al. 2019). Fluorine, nitrous oxide, methane, and carbon dioxide are the most prevalent greenhouse gases (Slama 2021). Climate change caused by these gases poses a great danger to the sustainability of global requirements in terms of environmental quality, food security, and health (Zheng et al. 2019). These effects are already evident in the increase in average yearly temperatures. Globally, energy consumption is the highest source of greenhouse gas emissions, contributing to almost 73% of the world's total greenhouse gases, while agriculture, land use change, forestry, industrial processes, and wastes emit 12%, 6.5%, 5.6%, and 3.2% of the global greenhouse gases, respectively (Mengpin et al. 2020).

The growing impact of climate change on agriculture has motivated scientists to determine the correlation between agricultural output, energy consumption, and climate change. Waheed et al. (2018) investigated the influence of renewable energy usage, agricultural output, and forests on CO_2 emissions in Pakistan from 1990 to 2014 using the Auto-Regressive Distributive Lag Model (ARDL) (Waheed et al. 2018). Jebli and Youssef (2017) examined the link between per capita renewable energy usage, agricultural productivity increase, CO_2 emissions, and real gross domestic product (GDP) for a group of five North African countries from 1980 to 2011 using panel cointegration techniques and Granger causality tests (Jebli and Youssef 2017). Using data from 1980 to 2013, Qureshi et al. (2016) investigated how energy demand, air pollution, and GHG emissions in Pakistan (Qureshi et al. 2016). The relationships between per CO_2 emissions, real GDP, clean energy and non-renewable energy usage, trade openness ratio, and agricultural value addition in Tunisia from 1980 to 2011 were explored by Ben Jebli and Ben Youssef (2017) using the Vector Error Correction Model (VECM) and Granger causality tests. In the long term, economic growth induces an increase in carbon dioxide emissions, according to research done by Ridzuan et al. (2020) using the ARDL model to examine the relationship between carbon dioxide emissions, economic growth, clean energy, urban growth, and agricultural subsectors from 1978 to 2016 in Malaysia (Ridzuan et al. 2020). From the review of the literature for various countries, most nations' agricultural output has an effect on climate, and vice versa. However, the results of Jebli and Youssef (2017) proved otherwise. These studies show that the connection between agricultural production, energy usage, and climate change is one that cuts across different countries of the globe. Having looked at previous studies in the research area, it can be observed that the relationship between the trios in Nigeria has not been evaluated.

Nigeria is used as a case study in this study. This is motivated by: Firstly, Nigeria is rich in oil and is one of the world's oil-producing countries. She has become over-dependent on fossil fuel energy, with its increasing population also being an important factor (Olanipekun and Adelakun 2020). According to the data from the United State Energy Information Administration, in 2018, oil consumption and natural gas consumption represented 55% and 41% of Nigeria's total energy consumption, respectively, while non-renewable energy consumption represented only 4% of total energy consumption (EIA 2020). This overdependence on fossil fuel energy consumption, however, impacts negatively on the environment as fossil fuel is not a renewable energy source as both its production and consumption processes emit CO_2, which is a major greenhouse gas (Chanchangi et al. 2022). Secondly, The most common means of agricultural waste disposal in Nigeria is by landfill/dumpsites or open controlled burning (Elum et al. 2017). This practice, which conflicts with international standards for sustainable agriculture, is seen to be environmentally damaging. In addition, the presence of insufficient legislative leadership, substandard trash collection tools, fraud, and a bad work ethic on the part of sanitation workers in the country equally limits the development of clean energy development (Adewole 2009). The lack of effective application of sustainable management rules and regulations in Nigeria has been connected to these actions (Adeyemo 2003).

Thirdly, in the rural areas of Nigeria, most of the dwellers engage in various agricultural practices, including the raising of livestock, with most of its participants utilizing basic traditional technologies (Abay et al. 2021; Odoh et al. 2019). This engagement in agriculture also involves practices that emit greenhouse gases into the atmosphere. Fourthly, increasing temperatures, variable rainfall, flooding, drought, resource depletion, and habitat destruction are all signs that Nigeria's climate is changing (Ebele and Emodi 2016; Elisha et al. 2017). These have contributed to poor agricultural output in several parts of the country. Fifthly, after natural gas, hydroelectric power is the most common source of electricity in Nigeria, and it is influenced by climate change (Fakehinde et al. 2019). It is also susceptible to the level, frequency, and geographic distribution of rainfall in addition to temperature. Due to high temperatures and little rainfall, hydroelectric power stations nationwide have reduced transmission capacities. High evapotranspiration brought on by prolonged drought has a negative impact on water volume, lowers hydroelectric power generating capacity, and reduces energy distribution for all stakeholders (Shu et al. 2018).

Energy usage, agricultural production, and the environment must be well-coordinated in order to ensure energy and food security, and a sustainable environment. Future climate change projections include temperature increases of 1.1–2.5 °C by 2060, with a higher increase in the country's northern region, and an increase in the number of extreme heat days to 260 days (USAID 2019). Agricultural output, fisheries, forestry, food production, water availability, wellness, power generation, and the economy are subsequently impacted by these climate changes. In order to give policymakers a foundation and support them in adopting measures to mitigate the emission of greenhouse gases so as to achieve sustainability, improved environmental standards, and food security for the needs of the present and future generations, this chapter examines the interaction between agricultural production, energy usage, and climate change in Nigeria.

Methodology

Description of the Study Area

Nigeria, the most populated country on the African continent, lies between longitudes 4°E–14°E and latitudes 4°N–14°N (Akande et al. 2017). The country, which is situated in West Africa, has an area of approximately 923,770 km^2 (Babama'aji et al. 2022). It lies a short distance from both the Sahara Desert in the north and the Atlantic Ocean in the south. The interaction of air masses from the Atlantic Ocean and the Sahara Desert had a significant impact on Nigeria's atmospheric conditions (Ayanlade et al. 2019). The two seasons experienced in the country are the rainy and dry seasons (Umar et al. 2021). From the southern part to the far northern region, the rainy season gets shorter and shorter. About 1200–3000 mm and 500–750 mm of rainfall in the south and far north each year, respectively (Akpata et al. 2009; Naintoh et al. 2018). The significant degree of annual variation in rainfall in northern regions causes flooding and droughts. Nigeria's average monthly temperature falls between 24 °C (December and January) and 30 °C (April) on an annual basis, with a mean of 26.9 °C (Bissolli et al. 2018; Dada et al. 2021). Nigeria experiences rain all throughout the year, with the heaviest rain falling from April to October and the least from November to March (Ogunrinde et al. 2019; Olujumoke et al. 2016).

Agriculture is central to the livelihood of Nigeria. The country's agricultural sector currently accounts for around 30% of Nigeria's GDP and more than 90% of its jobs (Nwankpa 2017). Small-scale farmers with less than two hectares of farmland, on the other hand, account for nearly 90% of the country's agricultural production (Chinedu et al. 2010). It is on record that in the year 2017, Nigeria produced 5,952,847,305,000 BTU (5.95 quadrillion BTU) of energy, while it consumed 1,540,259,766,000 BTU of energy. The consumed amount accounts for 0.26 percent of total global energy consumption (Worldometer 2018). Energy sources in Nigeria include fuel wood, coal, electricity, oil, and gas. More than 50 million tons of fuel wood are burned in Nigerian communities each year since more than 70% of the population relies on it for energy (Matar 2015).

Data Source

The Auto-Regressive Distributive Lag (ARDL) Model and Granger Causality are used in the empirical study to examine the link between energy usage, agricultural productivity, and greenhouse gas emissions in Nigeria utilizing data from 1990 to 2017, which translates to 28 years of observation. The data was obtained from the United State Energy Information Administration (EIA 2020), Agricultural production and Gross Domestic Product Data from World bank's data and Greenhouse gas emission data from the World Resource Institute (WRI) database (Mengpin et al. 2020). The data was obtained from these sources on a yearly basis. The parameters used in the analysis are: total GHG emissions in million tons of CO_2 equivalent (Mt CO_2e), agricultural value added (value added per worker), and energy usage (kg of oil equivalent per $1000 GDP). The study span (1990–2017) is limited by data availability, which is not unrelated to the poor collection, handling, and management of data in Nigeria (Afolabi et al. 2021).

Model Estimation

One of the important research contents of sustainability theory and low carbon economy theory is the connection between the environment, energy usage, and agricultural production. Based on the theories of agricultural sustainability, low carbon economy, and 3E theory (CO_2 emission, energy consumption, and economic growth), although energy consumption and agricultural activities are important factors of economic growth, they are also factors of environmental problems through the emission of greenhouse gases (Tong et al. 2020). This model can be represented in Eq. (27.1) as

$$GHG_t = f(ENC_t, AP) \tag{27.1}$$

where GHG is the greenhouse gas emission at time t, ENC is the energy consumption with respect to a particular duration (t), and AP is the agricultural value added with respect to a particular duration (t).

The logarithm linear representation of the model can be represented in Eq. (27.2) as

$$\ln GHG_t = C_1 + \beta_1 \ln ENC_t + \beta_2 \ln AP_t + \mu_t \tag{27.2}$$

where ln is a natural logarithm, C_1 is the intercept, β_1 and β_2 are the coefficients of eEnergy consumption and agricultural value added, respectively, and μ_t is the error term (Chandio et al. 2020).

Model of Auto-Regressive Distributive Lag

In 1995, Pesaran and Shin first introduced the ARDL bound test approach to cointegration, which was later modified (Menegaki 2019). The ARDL model is an estimation model that includes the present and delayed values of the predictor variables as well as the delayed outcome of the dependent variable. Other cointegration test techniques, which include the Engle and Granger cointegration technique and the Johansen and Juselius technique, also exist, but according to recent studies such as Waheed et al. (2018) and Akorede and Afroz (2020), the ARDL bound cointegration test is superior to other cointegration techniques for the following reasons. Firstly, in the event that data are not available, it can be applied to a series with a limited sample size. Secondly, it can be used for series integrated in order one I(1) or a combination of order zero I(0) and order one I(1). Thirdly, it can be used for series with different lag lengths. Fourthly, it provides an unbiased evaluation of the long-run model while simultaneously evaluating the short-run and long-run constituents (Chandio et al. 2020). This study employed the ARDL model to test for cointegration between GHG emissions, energy usage, and agricultural production based on the benefits of the ARDL bound test for cointegration. The ARDL model established for this research is given in Eq. 27.3 (Khan and Yahong 2021).

$$\begin{aligned}\Delta \ln GHG_t = \theta_0 + \sum_{i=1}^{p} \alpha_{1i} \Delta \ln GHG_{t-i} + \sum_{i=0}^{q} \alpha_{2i} \Delta \ln ENC_{t-i} \\ + \sum_{i=0}^{r} \alpha_{3i} \Delta \ln AP_{t-i} + \beta_1 \ln GHG_{t-1} + \beta_2 \ln ENC_{t-1} + \beta_3 \ln AP_{t-1} + \mu_t\end{aligned} \tag{27.3}$$

where θ_0 is the intercept, α_1, α_2, and α_3 are the dynamic coefficients of the short run, β_1, β_2, and β_3 are the equilibrium coefficients of the long run, Δ represents the difference operator, and μ_t is the error term.

The joint F-statistics (Wald statistics) for the cointegration technique serves as the foundation for the bound test. According to Menegaki (2019), the steps involved in checking for cointegration based on the ARDL bound test includes firstly determining the optimal lag for the model using either of the available lag information criteria. The most commonly used lag information criteria include the Final Prediction Error (FPE), Hannan–Quinn *Criterion (HQC),* Schwarz *Information Criterion (SIC),* and Akaike *Information Criterion (AIC)*. The following alternative and null hypotheses are determined in the second stage using the Wald test (F-test) (Dritsaki 2017; Stamatiou and Dritsakis 2014):

H0: $\beta_1 = \beta_2 = \beta_3 = 0$ (Absence of long-run relationship).

H1: $\beta_1 \neq \beta_2 \neq \beta_3 \neq 0$ (Presence of long-run relationship).

In the end, the upper and lower limit values are compared to the obtained F-value in the bound test result. The null hypothesis of no cointegration cannot be adopted if the upper bound values are lower than the estimated F-value, indicating the existence of a long-term correlation and vice versa. If the F-values are between the upper bound values and the lower bound values, the inference is inconclusive.

The estimation of both the short- and long-run ARDL models is necessary if cointegration among the variables is confirmed (Dritsaki 2017). With respect to this study, the following estimates are made using the long-run ARDL model.

$$\ln GHG_t = \theta_0 + \sum_{i=1}^{p} \beta_1 \ln GHG_{t-1} + \sum_{i=1}^{q} \beta_2 \ln ENC_{t-1} + \beta_3 \ln AP_{t-1} + \mu_t \tag{27.4}$$

After the long-run model is estimated, the error correction method was used to assess the short-run dynamic coefficients (Stamatiou and Dritsakis 2014). This study's short-run model is estimated as follows:

$$\Delta \ln GHG_t = \theta_0 + \sum_{i=0}^{p} \alpha_{1i} \Delta \ln GHG_{t-i} + \sum_{i=1}^{q} \alpha_{2i} \Delta \ln ENC_{t-i} + \sum_{i=1}^{r} \alpha_{3i} \Delta \ln AP_{t-i} + \delta ECT_{t-i} + \mu_t \tag{27.5}$$

where ECM is the error correction term; α_1, α_2, and α_3 are the model's short-run dynamic coefficients of consolidation to stability, and δ is the rate of adjustment parameter. The ECT indicates the extent of disequilibrium adjustment at each period and it should be statistically significant with a negative coefficient (Dritsaki 2017).

Granger Causality Test

This is a probabilistic method for determining causality across two parameters in a time -series data by examining empirical data sets for trends of correlation (Chandio et al. 2020). The tests involve estimating the model, stating the null hypothesis, choosing the optimal lag, and, finally, testing for the F-value. In order to test for granger causality, the bivariate equations are estimated as follows:

$$\ln GHG_t = c_2 + \sum_{i=1}^{k} \alpha_i \ln GHG_{t-i} + \sum_{j=1}^{k} \beta_j \ln ENC_{t-j} + v_{2t} \tag{27.6}$$

$$\ln ENC_t = c_2 + \sum_{i=1}^{k} \alpha_i \ln ENC_{t-i} + \sum_{j=1}^{k} \beta_j \ln GHG_{t-j} + v_{2t} \tag{27.7}$$

$$\ln GHG_t = c_2 + \sum_{i=1}^{k} \alpha_i \ln GHG_{t-i} + \sum_{j=1}^{k} \beta_j \ln AP_{t-j} + v_{2t} \tag{27.8}$$

$$\ln AP_t = c_2 + \sum_{i=1}^{k} \alpha_i \ln AVP_{t-i} + \sum_{j=1}^{k} \beta_j \ln GHG_{t-j} + v_{2t} \tag{27.9}$$

$$\ln ENC_t = c_2 + \sum_{i=1}^{k} \alpha_i \ln ENC_{t-i} + \sum_{j=1}^{k} \beta_j \ln AP_{t-j} + v_{2t} \tag{27.10}$$

$$\ln AP_t = c_2 + \sum_{i=1}^{k} \alpha_i \ln AP_{t-i} + \sum_{j=1}^{k} \beta_j \ln ENC_{t-j} + v_{2t} \tag{27.11}$$

Null hypothesis:

$$\mathrm{H}_0 = \beta_1 = \beta_2 = \ldots \beta_k = 0$$

where Y = independent variable and X = dependent variable (Alqahtani et al. 2020; Kónya 2006; Tekin 2012).

Results and Discussion

Descriptive Statistics

Table 27.1 displays descriptive statistics results of the various data sets. The outcome of standard deviation for each and every variable shows that they are distributed closely to the mean value. Greenhouse gas and agricultural production have positive skewness. This is consistent with the findings of Si et al. (2021). This indicates that the parameters are skewed to the right, whereas the negative skewness of energy consumption indicates that the parameters are skewed to the left. Since the null hypothesis that the parameters are regularly dispersed could only be accepted, the value of the Jarque–Bera probability indicates that all the variables are normally dispersed.

Unit Root Result

All of the variables have unit roots at levels I(0) and I(1), proving that they are mobile according to the outcome of the unit root test using the Augmented Dickey–Fuller (ADF) test in Table 27.2. However, as can be seen from the probability values, the presence of a unit root in the variable was rejected in the first difference, indicating that the variables are stationary in the first difference. The Phillips–Perron (PP) technique was additionally employed to check for the existence of unit root in the variables in order to validate the outcomes from the ADF technique. As can be seen in Table 27.1, the results of the PP test and the ADF test were identical. Therefore, the variables are all integrated in first order. That is, they are all I(1) variables and meet the requirement for using the ARDL estimation model. In this regard, the ARDL bound test to cointegration will be employed to establish the presence or absence of cointegration between the variable.

Table 27.1 Descriptive statistics of the data set

Statistics	GHG	ENC	AP
Mean	405.4429	193.4518	3202.021
Median	401.2550	200.3450	3031.852
Maximum	481.1700	238.9500	6091.058
Minimum	349.6200	136.8600	1484.249
Std. Dev	36.22449	37.70712	1626.129
Skewness	0.611416	−0.203057	0.423433
Kurtosis	2.507819	1.418238	1.811650
Jarque–Bera	2.027156	3.111383	2.484251
Probability	0.362918	0.211043	0.288770
Observations	28	28	28

Table 27.2 Unit root test result of variables for the ADF and PP techniques

Variables	ADF		PP	
	Constant	Constant and trend	Constant	Constant and trend
LGHG	0.1749 (0.1749)	−0.7309 (0.9601)	−0.0481 (0.9456)	−0.8645 (0.9459)
LENC	0.7158 (0.9902)	−2.7130 (0.2395)	0.5963 (0.9869)	−2.7130 (0.2395)
LAP	0.0764 (0.9577)	−2.4640 (0.3416)	0.0764 (0.9577)	−2.4640 (0.3416)
ΔLGHG	−3.8402 (0.0074)***	−4.1851 (0.0146)**	−3.8402 (0.0074)***	−4.1439 (0.0159)**
ΔLENC	−4.8163 (0.0007)***	−5.0986 (0.0019)***	−4.8163 (0.0007)***	−5.0981 (0.0019)***
ΔLAP	−4.7288 (0.0009)***	−4.6440 (0.0053)***	−4.7259 (0.0009)***	−4.6414 (0.0053)***

Note values above and below are the T statistics and probability values, respectively. ***, **, and * represent 1%, 5%, and 10% degrees of significance, respectively

Table 27.3 Cointegration test result of the ARDL model

Model specification	Suitable Lag length	F-statistics	Presence of cointegration
$F_{GHG}(\ln GHG_t \vert \ln ENC_t, \ln AP_t)$	ARDL (2,4,1)	5.885**	YES
Critical values			
	10%	5%	1%
Lower bound I (0)	3.17	3.79	5.15
Upper bound I (1)	4.14	4.85	6.36

Note ***, **, and * represent 1%, 5%, and 10% degrees of significance, respectively

Cointegration Test

After determining whether the variables had a unit root, as was mentioned above, the ARDL bound analysis was performed to determine whether the variables had cointegration. The process of this test involves first determining the optimal lag for the variables. The optimal lag lengths are selected with regards to AIC. Table 27.3's bound test results show that the calculated F-statistic, which is 5.885 and higher than both the lower bound and higher bound values of the 5% degree of significance, indicates that the null hypothesis of the presence of cointegration among the variables has been accepted in place of the null hypothesis of no cointegration among the variables. The existence of cointegration suggests a long-term connection between agricultural output, energy use, and greenhouse gas emissions. This finding is consistent with the findings of Chandio et al. (2020) in China. The short-run and long-run models will be further assessed in this regard.

ARDL Long-Run and Short-Run Estimation

This study assessed the long-run and short-run coefficients in accordance with the model provided in Eqs. 27.4 and 27.5, respectively, after determining the existence of a long-run relationship between GHG emission, energy usage, and agricultural production. Using the Akaike Information Criteria (AIC), the ideal lag length was determined. Table 27.4 displays the predicted values for both the short-run and long-run coefficients.

Diagnostic Test

Diagnostic tests were run to check for serial correlation, heteroscedasticity, functional form, and normality in order to validate the model's accuracy and the validity of the model assumptions utilized in this work (Saboori et al. 2012). The outcome of the diagnostic tests is depicted in Table 27.5.

The result of the diagnostic tests shows the presence of no serial correlation, functional form correctly specified, and no heteroscedasticity between the residuals because from the probability values, the null hypothesis of no serial correlation, functional forms correctly specified and no heteroscedasticity cannot be rejected. This result shows that the model assumptions are all valid. As can be seen from the Jacque–Bera probability value, the normality test result in Fig. 27.1 indicates that the null hypothesis that the residuals are normally dispersed can be accepted. The residuals are therefore regularly dispersed.

Table 27.4 Results on the long- and short-run estimation of the variables

ARDL (2, 4, 1) Model dependent variable is LGHG				
Variable	Coefficient	Std. Error	T-Statistics	Probability
Short-run coefficients				
C	6.150278	1.368479	4.494245	0.0005
$\Delta \ln GHG_{-1}$	0.308150	0.162524	1.896026	0.0788
$\Delta \ln ENC$	0.425328	0.152979	2.780303	0.0147**
$\Delta \ln ENC_{-1}$	0.670505	0.207193	3.236135	0.0060**
$\Delta \ln ENC_{-2}$	0.481041	0.206251	2.332308	0.0351**
$\Delta \ln ENC_{-3}$	0.402423	0.180284	2.232164	0.0425**
$\Delta \ln AP$	−0.191117	0.051826	−3.687666	0.0024**
ECT_{-1}	−0.792582	0.176444	−4.491970	0.0005**
Long-run coefficients				
$\ln ENC$	−0.389195	0.150146	−2.592106	0.0213**
$\ln AP$	0.043123	0.067531	0.638573	0.5334
R-squared	0.726505	Mean dependent variable		0.006873
Adjusted R-squared	0.606851	S.D. dependent variable		0.032311
S.E. of regression	0.020260	Akaike info criterion		−4.699167
Sum squared resid	0.006567	Schwarz criterion		−4.306482
Log likelihood	64.39000	Hannan–Quinn criteria		−4.594987
F-statistic	6.071719	Durbin-Watson stat		2.009496
Pro b(F-statistic)	0.001389			

Note ***, **, and * represents 1%, 5%, and 10% degrees of significance, respectively

Table 27.5 Result of the diagnostic test of the various models

Diagnostic test	F-statistics	P-value
Serial correlation	0.4605	0.6416
Functional form (Ramsey RESET)	0.0159	0.9875
Heteroscedasticity	1.2219	0.3554

The tests of cumulative sum (CUSUM) and cumulative sum of squares (CUSUM of Squares) were also performed in order to determine whether the model was stable. The CUSUM and CUSUM of Squares plots are both within the 5% critical lines, indicating that the hypothesis that there is no parameter instability in the model is not rejected based on the results of the tests displayed in Fig. 27.2. Generally, the results of the residual diagnostic tests and stability test show that the model's assumptions are valid and the model is stable.

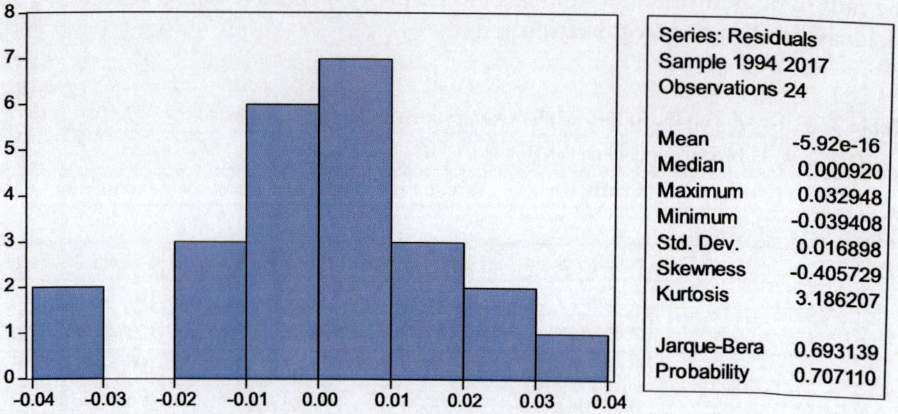

Fig. 27.1 Normality test result of the Jacque–Bera probability

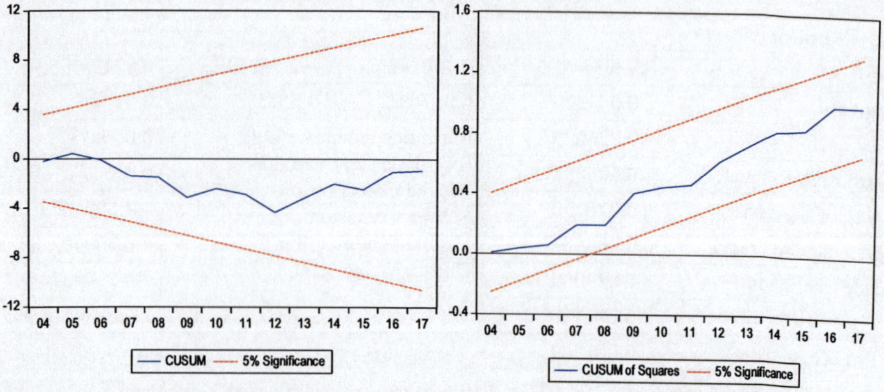

Fig. 27.2 Plot of CUSUM and CUSUM of squares stability test of the ARDL model

Table 27.6 Result of the Granger Causality test

Null hypothesis	F-statistics	Prob. value	Decision
lnENC does not Granger Cause lnGHG	2.74628	0.0477**	Reject hypothesis
lnGHG does not Granger Cause lnENC	3.17220	0.0447**	Reject hypothesis
lnAP does not Granger Cause lnGHG	1.68847	0.2048	Accept hypothesis
lnGHG does not Granger Cause lnAP	0.53566	0.7117	Accept hypothesis
lnAP does not Granger Cause lnENC	6.40129	0.0033**	Reject hypothesis
lnENC does not Granger Cause lnAP	0.22859	0.9181	Accept hypothesis

Note ***, **, and * represents 1%, 5%, and 10% degrees of significance, respectively

Granger Causality Test

The results from the Granger causality test can have four possible outcomes which include X does not granger cause Y but Y granger cause X; Y does not Granger cause X but X granger cause Y (the causality in the two results above are said to be Unidirectional); Y is the granger cause of X and X is the Granger cause of Y (in this cause, there is a feedback system and the causality is said to be bidirectional); X does not Granger cause Y and Y does not Granger cause X (that is there is no causality).

The Granger causality test was applied based on Eqs. 27.9 and 27.10 to ascertain the direction of causality between the variables. Table 27.6 displays the findings of the Granger Causality test.

Table 27.6's Granger causality test results indicate a bidirectional causal relationship connecting energy usage and GHG emissions, a lack of a connection between agricultural production and GHG emissions, and a unidirectional causality between agricultural output and energy usage. From the result shown in Table 27.4, it can be seen that both energy usage and agricultural output have a significant effect on greenhouse gas emission. Chandio et al. (2020) obtained a similar result for China. Energy consumption is statistically significant with a positive coefficient of 0.4253, which means a positive correlation with greenhouse gas emission. This means that a 1% change in energy usage in any direction will result in a 0.43% increase or decrease in greenhouse gas emissions. The results of the lags of energy consumption were also found to be statistically significant and with positive coefficients, which translates that the energy consumed in previous years also has a significant effect on the greenhouse gas emitted in the present year. However, the short-run result of agricultural consumption shows that agricultural consumption has a statistically significant negative correlation with greenhouse gas emission. This implies that a 1% rise in agricultural output would decrease greenhouse gas emission by 0.19% and vice versa. From the result of the Error Correction Term (ECT_{-1}), it can be seen that it meets all the requirements of being significant and with a negative coefficient. The ECT coefficient indicates that disequilibrium in the short run returns to equilibrium in the next year at a rate of 79%.

Only energy consumption is statistically significant at a 5% degree of significance, but with a negative coefficient, according to the long-run estimation results. This suggests that, over time, energy usage and greenhouse gas emissions have a negative relationship. Accordingly, increasing energy use will result in a 39 percent drop in greenhouse gas emissions, and vice versa. Akorede and Afroz (2020) obtained a similar result on the relationship between GHG emission and energy usage. Within the period of study, the average proportion of non-renewable and renewable energy usage was 93.67 and 6.31%, respectively. This over-dependence on non-renewable energy is evident in electricity generation where only 18% of energy generation source is from hydropower and the remaining 82% is generated from nonrenewable energy sources. The impact of energy utilization on CO_2 emission in Nigeria was demonstrated in a study conducted by Akpan and Akpan (2012). Another evidence of high non-renewable consumption in Nigeria is in transportation system which is dominated by road transport in which the engines are mostly powered by diesel combustion and gasses emitted from exhaust pipes.

The result of agricultural production in the short run shows that as agricultural production increases, greenhouse gas emission reduces. This result of a negative impact of agriculture on greenhouse gas in Nigeria might be due to the fact the agricultural industry is mainly populated by local farmers whose activities do not consume a great deal of energy, which minimizes the GHG emissions from the industry. Secondly, according to the World Bank data, the percentage of agricultural land area in total land area as of 2016 was 77.7%. Although it is known that some agricultural practices release greenhouse gases into the atmosphere, plants on these agricultural lands can also act as sinks to remove atmospheric CO_2 through the process of photosynthesis. This result obtained is in line with that obtained by Jebli and Youssef (2017), but did not align with that obtained by Waheed et al. (2018), which observed that an increase in agricultural production led to an increase in CO_2 emission in Pakistan.

The granger causality test results in Table 27.6 indicating that at the 5% significance level, the hypothesis that energy usage is not the granger cause of greenhouse gas emissions is dismissed, implying that energy usage is the granger cause of greenhouse gas emissions. It can also be observed that the null hypothesis that greenhouse gases are not the primary cause of energy usage is dismissed, while the alternative hypothesis that greenhouse gases are the primary cause of energy usage is adopted. In other words, it can be said that a bidirectional causality exists between energy usage and greenhouse gas emission. The result of the pairwise granger causality between agricultural production and greenhouse gas emission shows that no causality exists between agricultural production and greenhouse gas emission as the null hypothesis cannot be dismissed. Finally, the result of the pairwise granger causality between agricultural production and energy usage shows that agricultural production is the granger cause of energy usage because the hypothesis of no granger causality is rejected, while energy usage is not the granger cause of agricultural production as the null hypothesis cannot be rejected.

Conclusions

Various conclusions were derived from this study. The results from the ARDL showed that energy usage has a predictive impact on climate change in the same direction, which implies that a rise or fall in energy consumption will also result in a rise or fall in greenhouse gas emissions. This shows that energy usage/consumption leads to greenhouse gas emissions in Nigeria, especially in a situation where the energy consumed is not clean renewable energy. The Granger causality result, which shows a bi-directional causality between energy usage and greenhouse gas emissions, also confirms the ARDL estimation result. While the ARDL estimation shows that agricultural production has a significant effect on climate change in Nigeria, but in the opposite direction, which translates to a rise in agricultural production, which will lead to a decrease in climate change and vice versa. This implies that in order to reduce greenhouse gases, agricultural production can be increased.

The result of Granger Causality shows that climate change caused by energy consumption also has a consequent effect on energy sources like hydro-energy and agricultural production. Although the result of the Ganger causality shows no causality between agriculture and greenhouse within the period of study but shows a unidirectional causality between agricultural output and energy usage, this implies that agricultural output has a predictive impact on energy usage in Nigeria. This can be seen as agricultural production having an indirect impact on climate change as energy consumption emits greenhouse gases.

The results of the descriptive analysis show that Nigeria consumes large amounts of energy which is mainly dominated by non-renewable energy sources, which consequently results in high emissions of greenhouse gases as emissions from the energy sector contribute approximately 40% of the total greenhouse gases, thereby making it the second highest emitter of greenhouse gases after land use change and forestry. The graph from the descriptive analysis of the total emissions of greenhouse gases in Nigeria shows that Nigeria emits a large amount of greenhouse gases, which contributes to almost 1% of the global greenhouse gases. This excessive emission of greenhouse gases, if not mitigated, will result in extreme climate change, which, according to the United State Agency for International Development, by 2050, will have a large effect on agriculture through an increase in livestock and crop losses, reduced water quality and availability, increased food insecurity, an increase in periods of heat days and also hydropower energy (Cai et al. 2013; Smith et al. 2013).

Disclosure statements

Ethics approval and consent to Participate: Not applicable.

Conflict of Interest: The authors declare that there are no conflicts of interest.

Funding: There was no external funding for the study.

Compliance with Ethical Standards: This article does not contain any studies involving human or animal subjects.

Availability of Data and Materials: Not applicable.

References

Abay KA, Asnake W, Ayalew H, Chamberlin J, Sumberg J (2021) Landscapes of opportunity: patterns of young people's engagement with the rural economy in sub-Saharan Africa. J Dev Stud 57:594–613

Adewole AT (2009) Waste management towards sustainable development in Nigeria: a case study of Lagos state. Int NGO J 4:173–179

Adeyemo OK (2003) Consequences of pollution and degradation of Nigerian aquatic environment on fisheries resources. Environmentalist 23:297–306

Afolabi OO, Leonard SA, Osei EN, Blay KB (2021) Country-level assessment of agrifood waste and enabling environment for sustainable utilisation for bioenergy in Nigeria. J Environ Manag 294:112929

Akande A, Costa AC, Mateu J, Henriques R (2017) Geospatial analysis of extreme weather events in Nigeria (1985–2015) using self-organizing maps. Adv Meteorol 2017

Akorede YF, Afroz R (2020) The relationship between urbanization, CO2 emissions economic growth and energy consumption in Nigeria. Int J Energy Econ Policy 10:491

Akpan GE, Akpan UF (2012) Electricity consumption, carbon emissions and economic growth in Nigeria. Int J Energy Econ Policy 2:292

Akpata ES, Danfillo I, Otoh E, Mafeni J (2009) Geographical mapping of fluoride levels in drinking water sources in Nigeria. Afr Health Sci 9

Alqahtani A, Lahiani A, Salem A (2020) Crude oil and GCC stock markets: new evidence from GARCH co-integration and Granger causality approaches. Int J Energy Sector Manag

Ayanlade A, Atai G, Jegede MO (2019) Variability in atmospheric aerosols and effects of humidity, wind and InterTropical discontinuity over different ecological zones in Nigeria. Atmos Environ 201:369–380

Babama'aji RA, Shaba HA, Adepoju MO, Yusuf MJ, Jantiku J, Eshiet RC, Amanabo UH, Akinola O, Haruna M, Ibrahim H (2022) Mapping and comparison of maize products value chains in Nigeria and Rwanda European. J Agric Food Sci 4:57–65

Ben Jebli M, Ben Youssef S (2017) Renewable energy consumption and agriculture: evidence for cointegration and Granger causality for Tunisian economy. Int J Sustain Dev World Ecol 24:149–158

Bissolli P, Ganter C, Li T, Mekonnen A, Sanchez-Lugo A (2018) 7. Regional climates. Bull Am Meteorol Soc 99:S193–S193

Braide W, Justice-Alucho C, Ohabughiro N, Adeleye S (2020) Global climate change and changes in disease distribution: a review in retrospect. Int J Adv Res Biol Sci 7:32–46

Cai X, Molden D, Mainuddin M, Sharma B, Ahmad M-u-D, Karimi P (2013) Producing more food with less water in a changing world: assessment of water productivity in 10 major River Basins. In: Water, food and poverty in River Basins. Routledge, pp 290–310

Chanchangi Y, Adu F, Ghosh A, Sundaram S, Mallick T (2022) Nigeria's energy review: focusing on solar energy potential and penetration. Environ Dev Sustain 1–42

Chandio AA, Akram W, Ahmad F, Ahmad M (2020) Dynamic relationship among agriculture-energy-forestry and carbon dioxide (CO2) emissions: empirical evidence from China. Environ Sci Pollut Res 27:34078–34089

Chinedu EA, Titus OC, Thaddeus EO (2010) Achieving vision 2020 in Nigeria: a review of the economic and market-oriented business reforms. J Sustain Dev Afr 12:58–71

Dada B, Okogbue E, Ajayi V, Agele S, Yamusa A (2021) Inter-annual variation of weather pattern in the semi arid region of Nigeria and its implication on sorghum production. Int Res J Innov Eng Technol 5:51

Dritsaki C (2017) Toda-Yamamoto causality test between inflation and nominal interest rates: evidence from three countries of Europe. Int J Econ Financ Issues 7:120

Ebele NE, Emodi NV (2016) Climate change and its impact in Nigerian economy. J Sci Res Rep 10:1–13

EIA (2020) Nigeria energy statistics database, in independent statistics & analysis. https://www.eia.gov/international/analysis/country/NGA. Accessed 06 May 2021

Elisha I, Sawa B, Ejeh U (2017) Evidence of climate change and adaptation strategies among grain farmers in Sokoto State, Nigeria. IOSR J Environ Sci Toxicol Food Technol 11:1–7

Elum Z, Modise D, Nhamo G (2017) Climate change mitigation: the potential of agriculture as a renewable energy source in Nigeria. Environ Sci Pollut Res 24:3260–3273

Fakehinde OB, Fayomi OS, Efemwenkieki UK, Babaremu KO, Kolawole DO, Oyedepo SO (2019) Viability of hydroelectricity in Nigeria and the future prospect. Energy Proc 157:871–878

Iwuozor KO (2018) Removal of heavy metals from their solution using polystyrene adsorbent (Foil Take-Away Disposable Plates). Int J Environ Chem 2:10s

Iwuozor KO (2019) Combustion of wood by pyrolysis: a review. Int J Atmos Ocean Sci 3:7

Jebli MB, Youssef SB (2017) The role of renewable energy and agriculture in reducing CO2 emissions: evidence for North Africa countries. Ecol Indic 74:295–301

Khan S, Yahong W (2021) Symmetric and asymmetric impact of poverty, income inequality, and population on carbon emission in Pakistan: new evidence from ARDL and NARDL co-integration Front Environ Sci 9:666362

Kingsley OI, Ibraheem AI (2018) Production of biodiesel and its physiochemical properties produced from Ricinus communis Seeds by trans-esterification process. J Biomater 2:24–30

Kónya L (2006) Exports and growth: Granger causality analysis on OECD countries with a panel data approach. Econ Modell 23:978–992

Matar A (2015) Financial development, economic growth, and energy consumption nexus: a survey literature. J Adv Soc Res 5:01–18

Menegaki AN (2019) The ARDL method in the energy-growth nexus field; best implementation strategies. Economies 7:105

Mengpin G, Friedrich J, Vigna L (2020) 4 charts explain greenhouse gas emissions by countries and sectors. https://www.wri.org/insights/4-charts-explain-greenhouse-gas-emissions-countries-and-sectors. Accessed 06 May 2021

Mohmmed A, Li Z, Arowolo AO, Su H, Deng X, Najmuddin O, Zhang Y (2019) Driving factors of CO2 emissions and nexus with economic growth, development and human health in the Top Ten emitting countries resources. Conserv Recycl 148:157–169

Naintoh MB, Wantim MN, Ndonwi AS (2018) Assessing the impact of climate and change and variability on Irish Potato (Solanum Tuberosum L.) production from 1995 to 2015 in Tubah Sub Division, North West Region, Cameroon. J Cameroon Acad Sci 14:105–119

Nwankpa N (2017) Sustainable agricultural development in Nigeria: a way out of hunger and poverty. Eur J Sustain Dev 6:175–175

Odoh NE, Nwibo SU, Eze AV, Igberi CO (2019) Farm and non-farm income diversification activities among rural households in southeast, nigeria. J Agric Exten 23:113–121

Ogunrinde A, Oguntunde P, Akinwumiju A, Fasinmirin J (2019) Analysis of recent changes in rainfall and drought indices in Nigeria, 1981–2015. Hydrol Sci J 64:1755–1768

Olanipekun BA, Adelakun NO (2020) Assessment of renewable energy in Nigeria: challenges and benefits. Int J Eng Trends Technol (IJETT) 68

Olujumoke AO, Folorunso MA, Valerie G, Christian B (2016) Descriptive analysis of rainfall and temperature trends over Akure, Nigeria. J Geogr Reg Plann 9:195–202

Qureshi MI, Awan U, Arshad Z, Rasli AM, Zaman K, Khan F (2016) Dynamic linkages among energy consumption, air pollution, greenhouse gas emissions and agricultural production in Pakistan: sustainable agriculture key to policy success. Nat Hazards 84:367–381

Ridzuan NHAM, Marwan NF, Khalid N, Ali MH, Tseng M-L (2020) Effects of agriculture, renewable energy, and economic growth on carbon dioxide emissions: evidence of the environmental Kuznets curve resources. Conserv Recycl 160:104879

Saboori B, Sulaiman J, Mohd S (2012) Economic growth and CO2 emissions in Malaysia: a cointegration analysis of the environmental Kuznets curve. Energy Policy 51:184–191

Shu J, Qu J, Motha R, Xu J, Dong D (2018) Impacts of climate change on hydropower development and sustainability: a review. In: IOP conference series: earth and environmental science, vol 1. IOP Publishing, p 012126

Si R, Aziz N, Raza A (2021) Short and long-run causal effects of agriculture, forestry, and other land use on greenhouse gas emissions: evidence from China using VECM approach. Environ Sci Pollut Res 28:64419–64430

Slama RB (2021) Dual anthropogenic origin of global warming through GHGs and IR radiation emissions from artificialized soils. J Atmos Sci Res 4

Smith P, Haberl H, Popp A, Erb K, Lauk C, Harper R, Tubiello FN, de Siqueira PA, Jafari M, Sohi S (2013) How much land-based greenhouse gas mitigation can be achieved without compromising food security and environmental goals? Glob Change Biol 19:2285–2302

Stamatiou P, Dritsakis N (2014) The impact of foreign direct investment on the unemployment rate and economic growth in Greece: a time series analysis. In: International work-conference on time series analysis (ITISE), vol 1, pp 97–108

Tekin RB (2012) Economic growth, exports and foreign direct investment in Least Developed Countries: a panel Granger causality analysis. Econ Model 29:868–878

Tong T, Ortiz J, Xu C, Li F (2020) Economic growth, energy consumption, and carbon dioxide emissions in the E7 countries: a bootstrap ARDL bound test Energy. Sustain Soc 10:1–17

Umar NH, Bora B, Banerjee C, Gupta P, Anjum N (2021) Performance and economic viability of the PV system in different climatic zones of Nigeria. Sustain Energy Technol Assess 43:100987

USAID (2019) Greenhouse gas emissions in the West Africa region. https://www.climatelinks.org/sites/default/files/asset/document/2019_USAID_West%20Africa%20Regional%20GHG%20Emissions%20Factsheet.pdf. Accessed 06 May 2021

Waheed R, Chang D, Sarwar S, Chen W (2018) Forest, agriculture, renewable energy, and CO2 emission. J Clean Prod 172:4231–4238

Worldometer (2018) Energy consumption in Nigeria. Worldometer. https://www.worldometers.info/energy/nigeria-energy/#:~:text=Energy%20Consumption%20in%20Nigeria%20Nigeria%20consumed%201%2C540%2C259%2C766%2C000%20BTU,2017.%20This%20represents%200.26%25%20of%20global%20energy%20consumption. Accessed 25 July 2022

Zheng X, Streimikiene D, Balezentis T, Mardani A, Cavallaro F, Liao H (2019) A review of greenhouse gas emission profiles, dynamics, and climate change mitigation efforts across the key climate change players. J Clean Prod 234:1113–1133

Index

J. C. Egbueri et al. (eds.), *Climate Change Impacts on Nigeria*, Springer Climate, https://doi.org/10.1007/978-3-031-21007-5